Residues and Effluents

Processing and Environmental Considerations

Residues and Effluents

Processing and Environmental Considerations

Proceedings of an International Symposium sponsored by Extraction and Process Division of TMS and The Iron and Steel Society at the 1992 TMS Annual Meeting, San Diego, California, USA, March 1–5, 1992.

Edited by

Ramana G. Reddy
University of Nevada, Reno
Reno, NV, USA

William P. Imrie
Bechtel Group, Inc.
San Francisco, CA, USA

Paul B. Queneau
Hazen Research, Inc.
Golden, CO, USA

A Publication of

TMS
Minerals • Metals • Materials

A Publication of The Minerals, Metals & Materials Society
420 Commonwealth Drive
Warrendale, Pennsylvania 15086
(412) 776-9024

Printed in the United States of America
Library of Congress Catalog Number 91-51084
ISBN Number 0-87339-144-6

Preface

Environmental compliance requirements have had a very significant impact on the technological developments of residues and effluents processing systems for metallurgical operations. The implication of environmental control technology affects the performance of extraction and refining process units and their economics. The compliance pressure will continue to increase on metal industries and will require major innovations in control and processing technology. Of particular interest lately, has been the processing of dusts and heavy metals containing wastes.

This proceedings contains 61 papers, and about half of the papers come from a total of sixteen foreign countries, which exhibits a true international flavor of the symposium. The papers presented are from Australia, Belgium, Canada, China, Finland, France, Germany, India, Japan, Norway, Portugal, Sweden, The Netherlands, United Kingdom, United States of America, and Yugoslavia, attesting to be very significant interest in the subject of metallurgical residues and effluents and their environmentally safe processing technologies.

This international symposium presents environmentally safe technologies and processing methods available and that are considered necessary for further advances in processing of metallurgical residues and effluents. The main topics included are processing of residues and effluents environmental and strategy considerations, characterization, fundamentals, industrial practice and developments, utilization, future trends and innovations. These are organized into four main residues and effluents topical chapters.

1. Environmental and Strategy
2. Characterization
3. Processing
4. Utilization

The editors would like to thank the authors for providing the invaluable scientific information contained in this volume and the session chairpersons for their keen interest and support in making this symposium a success. Finally, special thanks are extended to the TMS Headquarters, particularly Marlene Karl, and Judi Heiles who make it possible to organize and publish all the papers in the book "Residues and Effluents"

Ramana G. Reddy
University of Nevada, Reno
Reno, Nevada

William P. Imrie
Bechtel Group, Inc.
San Francisco, CA

Paul B. Queneau
Hazen Research, Inc.
Golden, CO

Table of Contents

3. PROCESSING

3.1 EAF DUSTS

3.2 NON-FERROUS FLUE DUST

3.3 SLAGS, DROSSES AND SLUDGES

3.4 SCRAP AND OTHER WASTES

4. WASTE UTILIZATION

1.
ENVIRONMENTAL AND STRATEGY

GOVERNMENT REGULATIONS (SARA TITLE III) AND
THE SMALL BUSINESSES

by

*Rajesh K. Mishra & **G. Ram Ramadorai

*Chief Metallurgist, Pease & Curren Inc.,
75 Pennsylvania Ave., Warwick, RI 02888

**Sr. Process Engineer, Mountain States Research
and Development, Vail, AZ 85641

Abstract

The person in charge of Environmental Affairs in a small business has many higher priority projects (such as making money to meet the employee payroll etc.) than filling out and filing the regulatory forms. It is but natural to think that these affairs can be managed by playing ignorant until audited by the regulatory state and federal agencies. Truth is to the contrary; generally, it is too late by then. The fines and legal costs even for failing to file the regulatory forms could deliver a serious blow to the financial well-being of a small business. Forms and reports under SARA Title III are not hard to complete, but they are time consuming. In this paper we will discuss the information and data compilation procedures, simplification of some legal terminology into simple technical language, and tips that help in filling out the forms. Some actual case histories of EPA fines will also be reviewed.

Residues and Effluents - Processing and Environmental Considerations
Edited by Ramana G. Reddy, William P. Imrie and Paul B. Queneau
The Minerals, Metals & Materials Society, 1991

Introduction

On December 4, 1984, a cloud of extremely toxic gas (methyl isocyanate) escaped from a Union Carbide pesticide plant in Bhopal, India. More that 2500 people died and tens of thousands more were injured, some incurring permanent injury. Americans question, "Could it happen here?". A chemical release in West Virginia shortly after the Bhopal tragedy, though not nearly as serious, made the question even more urgent. These events, coupled with other incidents, accelerated the pace of mandatory law promulgation to create a national program to address chemical accidents.

The United States Congress enacted the Emergency Planning and Community Right-To-Know Act, commonly known as the Superfund Amendments and Re-authorization Act of 1986, Title III (SARA Title III). There are six provisions under this title. Four are related to emergency planning/hazardous chemical reporting and the remaining two address trade secrets/penalties and citizen suits.

In a small business, the most logical choice for responsibility for these forms will be a technical person, and more than likely an engineer. Engineers are trained to solve simple to complex industrial problems by the systematic and logical application of science and mathematics. When first exposed to the lawyers' language and strange vocabulary of acronyms such as FIFRA, EPCRA, SARA, SERC etc., one's reaction is simply outrage and disgust; the first question that comes to mind is "Why me?". But our experience shows, if one accepts it as a challenge, it is not as difficult as it appeared at first glance. In the end the effort is worth it and in the succeeding years, it is repetitive and takes a lot less effort.

The Acronyms commonly used in the environmental acts are listed in Appendix 1. A glossary of commonly used terms is indexed in Appendix 2. It is a good idea to familiarize oneself with these terms. These acronyms and terms are used in a fashion similar to mathematical formulas in solving equations.

SARA Title III or CERCLA

The Emergency Planning and Community Right-To-Know Act has two main purposes: to encourage and support emergency planning for responding to chemical accidents; and to provide local governments and the public with information about possible chemical hazards in their communities. Businesses are required by the laws to:

*Write **emergency plans** to protect the public from chemical accidents.

*Set up **procedures to warn and, if necessary, evacuate** the public in case of emergency.

*Provide citizens and local governments with **information about hazardous chemicals and accidental releases of chemicals** in their communities.

*Prepare public **reports on annual releases of toxic chemicals** into the air, water and soil.

In sum, the Act is a guide for dealing with toxic and hazardous materials. Detailed information on the various provisions of this law can be found in the Federal Register (FR) which is available at public or university libraries.

Highlights of the Laws

Emergency Planning (Sections 301-303)
(40 CFR 300 and 355)

*Governor appoints State Emergency Response Commissions (SERC).
*SERC establishes Local Emergency Planning Committee (LEPC).
*LEPC develops local emergency response plans and review them at least annually.
*Facilities must notify SERC and LEPC if they have Extremely Hazardous Substances (EHS) present above Threshold Planning Quantities (TPQ) and participate in emergency planning Form - Tier II.

Emergency Release Notification (Section 304)
(40 CFR 300 and 355)

*Facilities must notify the SERC and LEPC immediately for **accidental releases** of hazardous substances in excess of Reportable Quantity (RQ) and provide written reports on actions taken and medical efforts.
*SERC and LEPC make information available to public.

Hazardous Chemical Reporting (Sections 311-312)
(40 CFR 370)

*Facilities must submit Material Safety Data Sheets (MSDS) and/or lists of hazardous chemicals (HC) on-site above TPQ to SERC and LEPC and local fire departments.
*Facilities submit Emergency and Hazardous chemicals inventory forms **(amounts and locations of chemicals)** to SERC, LEPC, and local fire departments.
*SERC and LEPC make hazardous chemical information available to the public.

Toxic Chemical Release Reporting (Section 313)
(40 CFR 372)

*Covered facilities must submit annual reports on yearly toxic chemical releases to states and EPA.

*EPA establishes a national toxic chemical release inventory based on facility report - Form F.
*States and EPA make information available to the public.

Trade Secrets (Section 322)

*Facilities may claim the need for chemical information trade secrecy, but must substantiate the claim.
*Trade secret information may be disclosed to health professionals.
*Citizens may challenge trade secret claims by petitioning the EPA.

Penalties and Citizen Suits (Sections 325-326)
(To be published)

*The government may asses civil and administrative penalties of **$10,000 to $75,000 per day** against facilities that fail to comply with other provisions.
*Anybody who knowingly and willfully fails to provide emergency release notification is subject to a **$50,000 fine or 5 years in prison.**
*The **SERC, LEPC, State or Federal government** may initiate actions against facility owners or operators for failure to comply with the law. Under section 313 a company can be subject to penalties of up to **$25,000 per day per chemical they should be reporting as of July 1, 1987.**
*States may sue EPA for failure to provide trade secret information.

<u>Where to Start?</u>

If a facility uses or handles chemicals, the first step is to compile a list of all chemicals in excess of 10 lbs. or 2 gallons. Get in touch with the SERC and LEPC and check to see if the chemicals you have tabulated fall under the Hazardous Substances (HS) or Extremely Hazardous Substances(EHS) list. It is a very good idea to talk to the local Fire Marshall since they are always involved in the LEPC.

The substances and their Threshold Planning Quantity (TPQ) (under a different section), are listed in the Title III, List of Lists (4) which can be obtained from EPA publications free of cost. Get all the publications listed in General References; they will prove useful. Chances are, if chemicals used are in excess of 100 lbs., some filing responsibility will be triggered. For example (Table 1) if a facility stores 500 lbs. of Formaldehyde, it has to fulfill all the requirements under Section 302. Also, all the requirements of Sections 311 and 312 must be satisfied. If there is an accidental release of 100 lbs., Section 304 is triggered and if more than 10,000 lbs. are used per year Form R has to be filed under Section 313. Each chemical has to be dealt with separately under Section 313. The EPA publications explain the procedures step by step and are very simple to understand. Once a facility determines that it has to fulfil the requirements, some preliminary work must be done before trying to fill in the different forms.

Basic Data Compilation

1) Make a plot plan of the facility and divide it into a grid system to locate chemicals on site. Find out the Longitude and Latitude of the plant. A local land surveyor might be the best source (Figure 1).

2) Compile a list of all chemicals at the facility 10 lbs./2 gallons or more with CAS#, quantity of hazardous chemical in absolute lbs (% in the mixture is given in the MSDS supplied by the supplier of the chemical) and TPQ/RQ under different sections of the law (Table 2).

3) Re-list all the chemicals (Table 3) with amount of chemical, location (with the plot plan co-ordinates), and suppliers name, address and phone numbers. This list along with MSDS's is submitted to LEPC. From this list HS, EHS are marked and used to fill in Tier 2 form under Section 302.

4) The plot plan, the Table 3-type list, and MSDS are sent to various government agencies in this form or as a part of contingency plans.

5) For filing Form R under Section 313, one has to compile the amount of toxic chemicals "manufactured" (even as a by product of a reaction), "processed" or "otherwise used" by the facility annually. As a rule of thumb, if over 10,000 lbs. of the chemical was used or manufactured (even as a reaction product) annually, check the list of toxic chemicals in Reference (4) and determine from the booklet if Form R must be filed. To fill in Form R, a detailed material balance of toxic chemicals will be required.

The Forms:
Section 302

Tier Two - Emergency and Hazardous Chemical Inventory Form

Figure 2 shows a Tier Two form. The instructions are simple for filling in the form. Use Figure 1, the Plot Plan & vendors'

MSDS's to fill in the form. Determine the average daily amount by using the following formula:

$$\text{Average Daily Balance} = \frac{(n+1)}{2n} \times A \text{ gallons or lbs.}$$

n=days to empty the tank or consume the chemical
A=volume or weight of chemicals (gallons or lbs.)

Section 313
Form R - Toxic Chemical Release Inventory Reporting (Figure 3)

The aim of EPA reporting under this section seems to be to gain a complete mass balance of toxic chemicals. They want to know the distribution of toxic substances consumed or manufactured at the facility, with special emphasis placed on the toxic components discharged into the environment. The following criteria determine the mandatory filing of Form R:

*25,000 lbs. or more of the listed chemical is manufactured or processed over the course of the calendar year.

*10,000 lbs. or more of the listed chemical is otherwise used.

*Facility has 10 or more full-time employees (Part-time employees/subcontractors are also counted as fractions of full- time employees).

The definitions of "manufactured", "processed" and "otherwise used" chemicals are not the same as those commonly used in metallurgical and chemical industries. EPA definitions are:

Manufactured - To produce, prepare, compound or import a listed toxic chemical. The term also includes coincidental production of toxic chemicals, e.g. as a **by product or impurity.**

Process - The preparation of a listed toxic chemical, after its manufacture, for distribution in commerce. Processing is usually the **intentional incorporation** of a toxic chemical into a product. Processing includes preparation of the chemical in the **same physical state or chemical form** as that received by a facility, or preparation that produces a change in physical state or chemical form. The term also applies to the processing of a mixture and other trade name product that contains a listed toxic chemical as one component.

Otherwise Used - Any use of a listed chemical at a facility that does not fall under definitions of a "manufactured" or "process". A chemical that is otherwise used by a facility is **not intentionally incorporated** into a product distributed in commerce.

As is obvious from the definition, most chemicals used in the metallurgical and chemical industries will fall under the "otherwise used" category which has the 10,000 lbs. consumption limit per year. The regulations mandated by the government are written to encompass a large number of industries and therefore language is more like in legal documents. Fortunately, the bulletins produced by agencies are quite simple and help is needed only once in a while from lawyers or environmental specialists to clarify the points. If a facility handles any listed toxic chemical, the SERC will assign it an EPA Identification number. Contact the EPA by telephone (1-800-535-0202) to obtain a TRI facility ID number. Compile total chemical usage from the previous year's purchasing record. A personal computer will prove very helpful for completing this task. Computers are fairly inexpensive now and will save a lot of agony in the future by serving to keep the facility's toxic chemical inventory up to date. We have found that programming the computer with Lotus 1-2-3 has been adequate for our work. As the purchasing department pays the bills, a secretary enters the data into the inventory program at the same time. At the end of the year the remaining inventory is taken, and the computer program calculates all the data that is required for Form R. An example using Sulfuric Acid, Nitric Acid and Hydrochloric Acid inventory is shown in Figure 4. At the end of the year, a measurement of acid in the tank is taken with a dip stick (in inches) and other relevant data is calculated by the computer program. We are in the process of developing a similar program for a complete material balance of toxic substances.

Reports

Sections 301-303 Contingency Plan

One of the important reports to be submitted to various government agencies is the Hazardous Waste Contingency Plan. The plan is designed to meet the responsibilities as a hazardous (Toxic) substance or waste generator or handler under the CFR 40 "Protection of Environment". Basically, the document includes all emergency planning in case of a fire, accidental spill or discharge, explosion, or any other unplanned event at the facility. The same report can also be submitted under the provisions of the OSHA Hazardous Waste Response Operations Plan, 29 CFR 1910.1209. The report must provide response planning for the type of emergency that can be anticipated under a worst case scenario. The document explains how the various agencies such as SERC, LEPC, Hospital, Police should co-ordinate the response with the Facility Emergency Co-ordinator. Important information required before writing this report includes the location of the plant with longitude and latitude (Figure 4) marked clearly on a local detailed map, and a list of Location of Sensitive Receptors (such as Hospitals, Schools, Nursing Homes, etc.) within one and two mile radii (Table 4). It is advisable to ask the LEPC to provide a sample of a Contingency Plan already submitted to them. It may be worthwhile to engage a consultant to prepare this document.

Conclusion

The filing of these forms and reports under SARA is extremely important. One small company was fined $15,000 for not reporting Hydrochloric, Nitric and Sulfuric acid under Section 313 for the year 1987 alone. (They had filed the reports after an EPA inspection in 1988.) They appealed, which was even a bigger mistake. EPA inspectors came back in full force to the facility and fined the firm $75,000 under Sections 302, 304, 311, 312 (and even the LDR-Act). None of the fines was for polluting the environment or illegal dumping. All of the fines were for not filing the paperwork properly. Oh yes - the reader might be interested to know what happened to the $15,000 appeal. SARA Title III cases can be tried only in Federal Court with a presiding judge who works for the EPA. After the hearing, the fine was reduced to $9,000 and of course the attorney fees exceeded the $15,000 fine.

Acknowledgements

The authors acknowledge Pease & Curren Inc., Warwick, RI for permission to publish the information contained in this paper. Special thanks are extended to Dr. Ramana Reddy for encouraging us to write this paper and Karen Hazard for typing the manuscript in time.

TABLE 1

EPA Title III List Of Lists

SARA TITLE III
CONSOLIDATED CHEMICAL LIST
02/13/91

			Section 304			
CAS Number	Chemical Name	Sec. 302(EHS) TPQ	EHS RQ	CERCLA RQ	Sec 313	RCRA Code
50-00-0	Formaldehyde	500		100	313	U122
50-07-7	Mitomycin C	500/10,000		10		U010
50-14-6	Ergocalciferol	1,000/10,000	1			
50-18-0	Cyclophosphamide			10		U058
50-29-3	DDT			1		U061
50-32-8	Benzo[a]pyrene			1		U022
50-55-5	Reserpine			5,000		U200
51-21-8	Fluorouracil	500/10,000	1			
51-28-5	2,4-Dinitrophenol			10	313	P048
51-43-4	Epinephrine			1,000		P042
51-75-2	Nitrogen mustard	10	1		313	
51-75-2	Mechlorethamine	10	1		X	
51-79-6	Urethane			100	313	U238
51-79-6	Carbamic acid, ethyl ester			100	X	U238
51-79-6	Ethyl carbamate			100	X	U238
51-83-2	Carbachol chloride	500/10,000	1			
52-68-6	Trichlorfon			100	313	
52-85-7	Famphur			1,000		P097
53-70-3	Dibenz[a,h]anthracene			1		U063
53-96-3	2-Acetylaminofluorene			1	313	U005
54-11-5	Nicotine	100		100		P075
54-11-5	Pyridine, 3-(1-methyl-2-pyrrolidinyl)-,(S)	100		100		P075
54-11-5	Nicotine and salts			100		P075
54-62-6	Aminopterin	500/10,000	1			
55-18-5	N-Nitrosodiethylamine			1	313	U174
55-21-0	Benzamide				313	
55-63-0	Nitroglycerin			10	313	P081
55-91-4	Isofluorphate	100		100		P043
55-91-4	Diisopropylfluorophosphate	100		100		P043
56-04-2	Methylthiouracil			10		U164
56-23-5	Carbon tetrachloride			10	313	U211
56-25-7	Cantharidin	100/10,000	1			
56-38-2	Parathion	100		10	313	P089
56-49-5	3-Methylcholanthrene			10		U157
56-53-1	Diethylstilbestrol			1		U089
56-55-3	Benz[a]anthracene			10		U018
56-72-4	Coumaphos	100/10,000		10		
57-12-5	Cyanides (soluble salts and complexes)			10		P030
57-14-7	1,1-Dimethyl hydrazine	1,000		10	313	U098
57-14-7	Dimethylhydrazine	1,000		10	X	U098

TABLE 2

REPORTING REQUIREMENTS

CHEMICAL NAME	CAS Number	Location (Map)	CAS# Mixtures (?)	% H.S. in Mix (%)	Maximum Quan of Chemical on Site (lb)	(%) Ingred. of HS (%)	Max. Quant of Absol. HS on Site (lb)	SARA Title III-Sec·302 (EHS) TPQ (lb)	SERC RQ (lb)	LEPC RQ (lb)	SARA Title III-Sec·313 RQ (lb)	Sec·304 EHS (lb)	Sec·304 CERCLA (lb)
Acetylene (Gas)	74-86-2		NO	--	88	NH	88	--	--	10 lbs or 2 Gallons	--	--	--
Alcohol (Ethyl denat.)	64-17-5		NO	--	140	NH	140	--	--	" "	--	--	--
Ammonium Chloride	12125-02-9		NO	--	500	100.0	500	--	--	" "	--	--	5,000
Ammonium Hydroxide	1336-21-6		NO	--	1,200	44.0	528	--	--	" "	--	--	1,000
Alken V-7	138-93-2 137-41-7		YES	1 to 10 1 to 10	600	NH NH	60 60	--	--	" "	--	--	--
Barium Chloride	10326-27-9		NO	--	5	100.0	5	--	--	" "	--	--	--
Copper Sulfate (Solution)	7758-99-5		NO	--	45,000	2.5	1,125	--	--	" "	--	--	10
Calcium Fluoride (Florspar)	7789-75-5		NO	--	3,600	100.0	3,600	--	--	" "	--	--	--
(Bleach) Calcium Hypochlorite	7778-54-3		NO	--	100	100.0	100	--	--	" "	--	--	10
						Reporting Date		Mar.1,Yearly	Mar.1,Yearly	Mar.1,Yearly	July1,Yearly	For Spills Reporting Only	
						Form		Tier Two	Tier Two	Tier Two	EPA - R		

DEM R.I. STATE ← Apr1, Bienial State Form EHC Only | 1-Time Only Tier Two → Fire Dept

TPQ - Threshold Planning Quantity
EHS - Extremely Hazardous Substance
CERCLA - Comprehensive Environmental Response Commission and Liablity Act
SERC - State Emergency Response Commission
LEPC - Local Emergency Planning Commision
NH - Non Hazardous
HS - Hazardous Substance

RQ - Reportable Quantity
EHC - Extremely Hazardous Chemicals

<u>SARA TITLE III - Sec. 302</u>
Tier Two Forms to:
(a) SERC (b) LEPC

<u>SARA TITLE III - Sec. 313</u>
Form R to:
(a) EPA (b) RI-DEM

<u>Biennial Report</u>
(a) Emergency Response Commission State of RI

6/11/90

TABLE 3

LIST OF ALL CHEMICALS

(10 lbs / 2 Gallons or more)

Name of Chemical	CAS Number	Maximum Amount (lbs) Absolute	Location (See attached drawing)	Suppliers Name/Address/Phone Number
A				
Acetylene	74-86-2	88	1-D Receiving 11-C Converter Area	Corp Brothers 401/331-8020 1 Brook Street Providence, RI 02903
Alcohol (Ethyl denatured)	64-86-2	140	9-B Converter Area	CMS 1-800-225-1198 17 Jonspin Rd. Wilmington, MA 01887
Ammonium Chloride	12125-02-9	500	9-C Converter	Kraus Chemical 401/434-4960 156 Valley Street East Providence, RI 02914
Ammonium Hydroxide	1336-21-6	1200	9-C Converter 9-D Converter	Kraus Chemical 401/434-4960 156 Valley Street East Providence, RI 02914
Alken V-7	138-93-2 and 137-41-7	600	11-B Converter	Alken Murray Corp. 212/431-4020 417 Canal Street New York, NY 10013
Airex 77 (Bathroom Cleaner)	None	10	3-B Shipping & Receiving	John R. Casey Co./Richmond 401/331-8020 300 Station Street Cranston, RI 02910
B				
Barium Chloride	10326-27-9	5	9-B Converter	Fisher Scientific 1-800-225-3508 461 Riverside Avenue Medford, MA 02155
C				
Copper (Metal)	7440-50-8	10534	9-E Hallway 6-F Smelter	Kraus Chemical 401/434-4960 156 Valley Street East Providence, RI 02914
Casey Furniture Polish	None	10	3-B Receiving	John R. Casey Co./Richmond 401/331-8020 300 Station Street Cranston, RI 02910

FIGURE 1
FACILITY PLOT PLAN

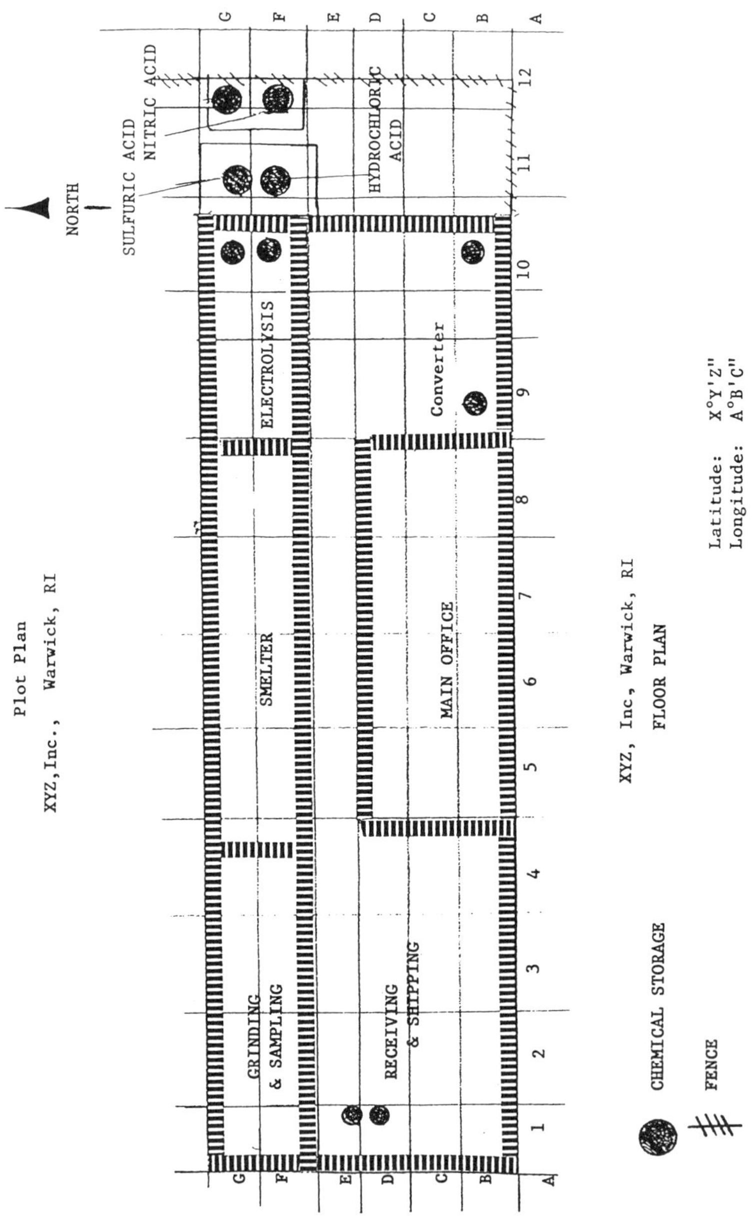

FIGURE 2
TIER TWO FORM

Tier Two
EMERGENCY AND HAZARDOUS CHEMICAL INVENTORY
Specific Information by Chemical

Facility Identification
Name XYZ Inc.
Street 36 Jefferson Blvd.
City Warwick County Kent State RI Zip 02888
SIC Code 3341 Dun & Brad Number 00-110-0001

FOR OFFICIAL USE ONLY — ID# ______ Date Received ______

Owner/Operator Name
Name Z.Z. Smith Phone (401) 222-22
Mail Address 36 Jefferson Blvd., Warwick, RI 02888

Emergency Contact
Name Z.Z. Smith Title V. President
Phone (401) 222-2222 24 Hr. Phone (401) 333-3333
Name Y.Y. Thompson Title C. Chemist
Phone (401) 444-4444 24 Hr. Phone (401) 555-5555

Important: Read all instructions before completing form

Reporting Period From January 1 to December 31, 1990

☐ Check if information below is identical to the information submitted last year.

Chemical Description	Physical and Health Hazards (check all that apply)	Inventory	Container Type / Temperature / Pressure	Storage Codes and Locations (Non-Confidential) *Storage Locations*	Optional
CAS 7664-93-9; Trade Secret ☐; Chem. Name Sulfuric Acid; Check all that apply: Pure ☐ Mix ☐ Solid ☐ Liquid [X] Gas ☐ EHS [X]; EHS Name Sulfuric Acid	Fire ☐; Sudden Release of Pressure ☐; Reactivity [X]; Immediate (acute) [X]; Delayed (chronic) ☐	Max. Daily Amount (code) 03; Avg. Daily Amount (code) 03; No. of Days On-site (days) 365	A 1 4; B 1 5	G-11 Storage in containment - outside the building.; G-10 Below ground in Electrolytic Dept.	☐
CAS 7697-37-2; Trade Secret ☐; Chem. Name Nitric Acid; Check all that apply: Pure ☐ Mix ☐ Solid ☐ Liquid [X] Gas ☐ EHS [X]; EHS Name Nitric Acid	Fire ☐; Sudden Release of Pressure ☐; Reactivity [X]; Immediate (acute) [X]; Delayed (chronic) ☐	Max. Daily Amount (code) 03; Avg. Daily Amount (code) 02; No. of Days On-site (days) 365	A 1 4; R 1 4	F-11 Storage in Containment - outside the building.; G-10 SS Drums in Electrolytic Area	☐
CAS 7647-01-0; Trade Secret ☐; Chem. Name Hydrochloric Acid; Check all that apply: Pure ☐ Mix ☐ Solid ☐ Liquid [X] Gas ☐ EHS [X]; EHS Name Hydrochloric Acid	Fire ☐; Sudden Release of Pressure ☐; Reactivity ☐; Immediate (acute) ☐; Delayed (chronic) ☐	Max. Daily Amount (code) 03; Avg. Daily Amount (code) 02; No. of Days On-site (days) 365	A 1 4; C 1 4	G-11 Storage outside the building in containment; D-1 Shipping Area	☐

Certification *(Read and sign after completing all sections)*
I certify under penalty of law that I have personally examined and am familiar with the information submitted in pages one through 10, and that based on my inquiry of those individuals responsible for obtaining the information, I believe that the submitted information is true, accurate, and complete.

Z.Z. Smith, Vice President
Name and official title of owner/operator OR owner/operator's authorized representative

Signature ______

2/8/91
Date signed

Optional Attachments
[X] I have attached a site plan
☐ I have attached a list of site coordinate abbreviations
☐ I have attached a description of dikes and other safeguard measures

FIGURE 3
EPA Form-R

Form Approved OMB No.: 2070-0093
Approval Expires: 01/94

(Important: Type or print; read instructions before completing form.)

Page 1 of 5

EPA U.S. Environmental Protection Agency

TOXIC CHEMICAL RELEASE INVENTORY REPORTING FORM

Section 313 of the Emergency Planning and Community Right-to-Know Act of 1986, also known as Title III of the Superfund Amendments and Reauthorization Act

Public reporting burden for this collection of information is estimated to vary from 30 to 34 hours per response, with an average of 32 hours per response, including time for reviewing instructions, searching existing data sources, gathering and maintaining the data needed, and completing and reviewing the collection of information. Send comments regarding this burden estimate or any other aspect of this collection of information, including suggestions for reducing this burden, to Chief, Information Policy Branch (PM-223), US EPA, 401 M St., SW, Washington, D.C. 20460 Attn: TRI Burden and to the Office of Information and Regulatory Affairs, Office of Management and Budget Paperwork Reduction Project (2070-0093), Washington, D.C. 20603.

EPA FORM **R**	PART I. FACILITY IDENTIFICATION INFORMATION	(This space for your optional use.)

1.	1.1 Are you claiming the chemical identity on page 3 trade secret? [ ] Yes (Answer question 1.2; Attach substantiation forms.) [ ] No (Do not answer 1.2; Go to question 1.3.)	1.2 If "Yes" in 1.1, is this copy: [ ] Sanitized [ ] Unsanitized	1.3 Reporting Year 19____

2. CERTIFICATION (Read and sign after completing all sections.)

I hereby certify that I have reviewed the attached documents and that, to the best of my knowledge and belief, the submitted information is true and complete and that the amounts and values in this report are accurate based on reasonable estimates using data available to the preparers of this report.

Name and official title of owner/operator or senior management official	
Signature	Date signed

3. FACILITY IDENTIFICATION

3.1	Facility or Establishment Name	
	Street Address	
	City	County
	State	Zip Code
	TRI Facility Identification Number	

WHERE TO SEND COMPLETED FORMS:

1. EPCRA REPORTING CENTER
   P.O. BOX 23779
   WASHINGTON, DC 20026-3779
   ATTN: TOXIC CHEMICAL RELEASE INVENTORY
2. APPROPRIATE STATE OFFICE (See instructions in Appendix G)

3.2	This report contains information for (Check only one): a. [ ] An entire facility b. [ ] Part of a facility.					
3.3	Technical Contact			Telephone Number (include area code)		
3.4	Public Contact			Telephone Number (include area code)		
3.5	SIC Code (4 digit) a.	b.	c.	d.	e.	f.
3.6	Latitude: Degrees	Minutes	Seconds	Longitude: Degrees	Minutes	Seconds
3.7	Dun & Bradstreet Number(s) a.			b.		
3.8	EPA Identification Number(s) (RCRA I.D. No.) a.			b.		
3.9	NPDES Permit Number(s) a.			b.		
3.10	Receiving Streams or Water Bodies (enter one name per box) a.			b.		
	c.			d.		
	e.			f.		
3.11	Underground Injection Well Code (UIC) Identification Number(s) a.			b.		

4. PARENT COMPANY INFORMATION

4.1	Name of Parent Company	4.2	Parent Company's Dun & Bradstreet Number

EPA Form 9350-1 (Rev.1-91) - Previous editions are obsolete.

FIGURE 4
YEARLY CHEMICAL INVENTORY
(On Lotus 1-2-3)

CHEMICAL:	NITRIC ACID				
ID NO:	UN2031				
BAUME:	42.00	PERCENT:	67.18	GAL PER INCH:	38.20
SPECIFIC				YEAR END INCH:	23.00
GRAVITY:	1.41	LB/GALLON:	11.73	YEAR END INV:	590.24 (ABS. HNO3)

RECEIVED MM DD YY	PO NUMBER	SUPPLIER NAME	POUNDS RECEIVED	MONTLY TOTAL	YTD TOTAL	ABSOLUTE CONTENT	CONSUMPTION TO DATE (LBS)
1 1 89	BEGINNING BALANCE		18667.00	18667.00	18667.00	12540	11950
4 1 89		KRAUS CHEMICAL	1875.00	1875.00	20542.00	13800	13210
5 1 89		BORDEN & REMINGTON	23300.00	23300.00	43842.00	29453	28863
10 11 89	8851	KRAUS CHEMICAL	22680.00	22680.00	66522.00	44689	44099
12 1 89		KRAUS CHEMICAL	23150.00	23150.00	89672.00	60242	59651

DATE:19-Sep-91

CHEMICAL:	SULPHURIC ACID				
ID NO:	UN1830				
BAUME:	66	PERCENT:	93.19	GAL PER INCH:	31.2
SPECIFIC				YEAR END INCH:	97.00
GRAVITY:	1.8354	LB/GALLON:	15.29	YEAR END INV:	2820.30 (ABS. H2SO4)

RECEIVED MM DD YY	PO NUMBER	SUPPLIER NAME	POUNDS RECEIVED	MONTLY TOTAL	YTD TOTAL	ABSOLUTE CONTENT	CONSUMPTION TO DATE (LBS)
1 1 89	BEGINNING BALANCE		7486.00	7486.00	7486.00	6976	4156
5 1 89		GEORGE MANN	46260.00	46260.00	53746.00	50086	47266

DATE:19-Sep-91

CHEMICAL:	HYDROCHLORIC ACID				
ID NO:	UN1789				
BAUME:	20	PERCENT:	31.45	GAL PER INCH:	33.10
SPECIFIC				YEAR END INCH:	75.00
GRAVITY:	1.16	LB/GALLON:	9.66	YEAR END INV:	780.75 (ABS. HCL)

RECEIVED MM DD YY	PO NUMBER	SUPPLIER NAME	POUNDS RECEIVED	MONTLY TOTAL	YTD TOTAL	ABSOLUTE CONTENT	CONSUMPTION TO DATE (LBS)
1 1 89	BEGINNING BALANCE		27720.00	27720.00	27720.00	8718	7937
2 1 89		KING	39220.00	39220.00	66940.00	21053	20272
6 1 89		GEORGE MANN	41460.00	41460.00	108400.00	34092	33311

FIGURE 5
XYZ, Inc., Warwick Location Map

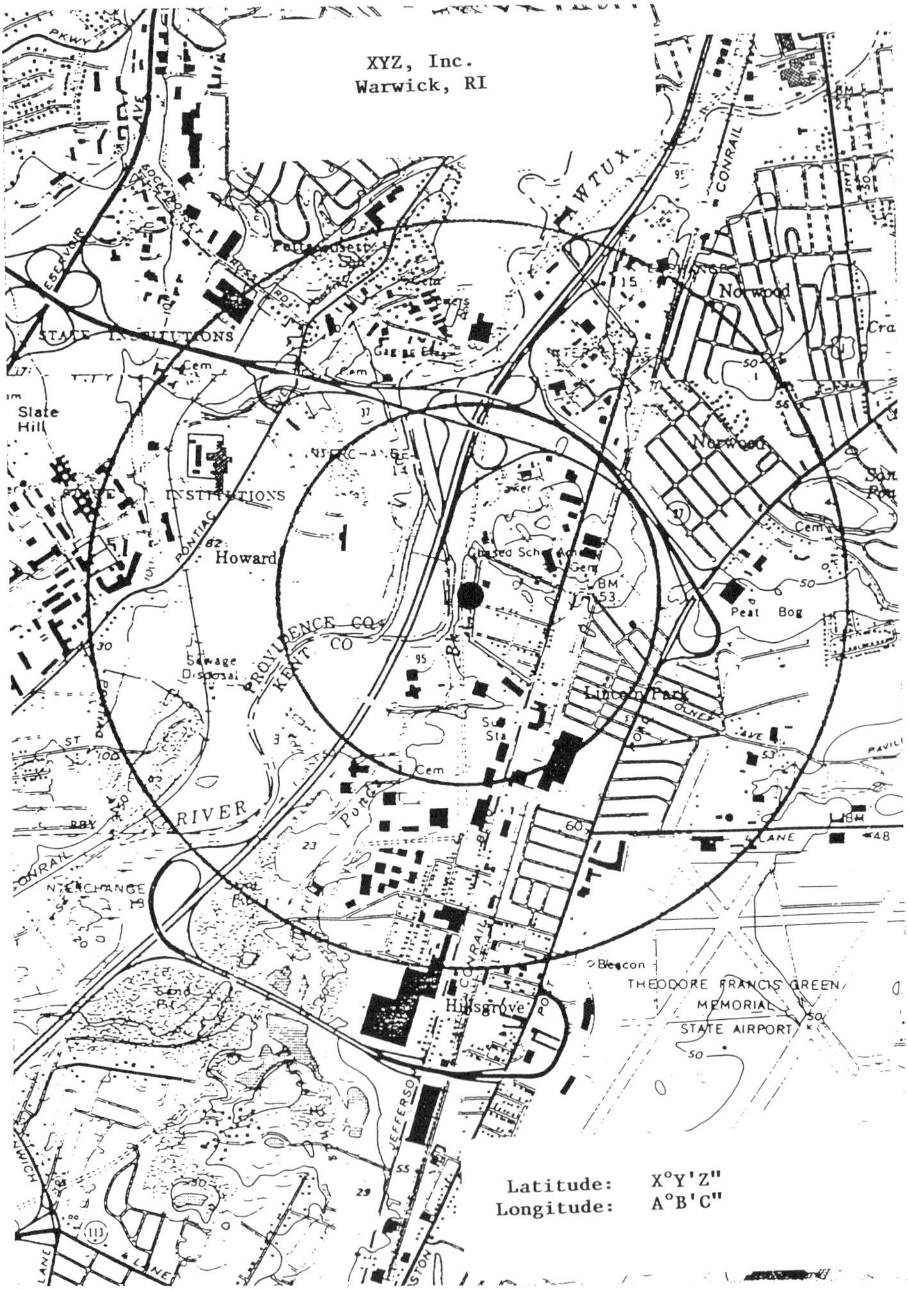

TABLE 4

LOCATION OF SENSITIVE RECEPTORS
WITHIN 1 MILE OF FACILITY

NAME AND ADDRESS	DISTANCE Miles	DIRECTION
RI Dept. Mental Health & Hospitals 600 New London Avenue Cranston, RI	0.92	W
Boy's & Girl's Club of Warwick 42 Frederick Street Warwick, RI	0.94	ENE
Norwood School Norwood Aveenue Warwick, RI	1.03	ENE
Eastern Star Homes of Rhode Island 85 Ethan Street Warwick, RI	0.82	E
Creative Preschool 35 Kilvert Street Warwick, RI	0.86	S
St. Francis School 596 Jefferson Boulevard Warwick, RI	0.77	S
Wonderland Early Learning Center 1081 Pontiac Avenue Cranston, RI	0.94	N
St. William's School 200 Pettaconsett Avenue Warwick, RIWarwick, RI	0.89	ENE

APPENDIX 1

ACRONYMS

ACGIH - American Conference of Governmtl Ind. Hygienists
ANSI - American National Standard Institute
ASTM - American Society for Testing Materials
BACT - Best Available Control Technology
BAT - Best Available Technology
BCT - Best Conventional Technology
BOD - Biochemical Oxygen Demand
BPT - Best Practical Technology
BTU - British Thermal Unit
CAA - Clean Air Act
CAMEO II - Computer Aided Mngmt of Emergency Operations System
CAS - Chemical Abstract Service (Registry Number)
CCW - Constituent Concentration in Waste
CCWE - Constituent Concentrations for Waste Extracts
CEPP - Chemical Emergency Preparedness Program
CERCLA - Comprehensive Env. Response Comp. & Liability Act
CFR - Code of Federal Regulations
CHEMTREC - Chemical Transportation Emergency Center
CMA - Chemical Manufacturers Association
CWA - Clean Water Act
D&B - Dun & Bradstreet
DOT - Department of Transportation
EC - Emergency Co-ordinator
EHS - Extremely Hazardous Substance
EP - Abbr. & prefix for EPA, eg EP-TCLP
EPA ID# - 12 digit number assigned to facilities
EPA - Environmental Protection Agency
EPA HW# - EPA Hazardous Waste Number
EPCRA - Emergency Planning and Community Right-to-Know Act
ERG - Emergency Response Guide Book
FDA - Food and Drug Administration
FEMA - Federal Emergency Management Agency
FIFRA - Fed. Insecticides, Fungicide and Rodenticide Act
FR - Federal Register
GC/MS - Gas Chromatograph/Mass Spectrometer
HC - Hazardous Chemical
HCS - Hazard Communication Standard
HM - Hazardous Material
HMIS - Hazardous Materials Identification System
HMTA - Hazardous Materials Transportation Act
HOC'S - Halogenated Organic Compounds
HS - Hazardous Substance
HSWA - Hazardous and Solid Waste Amendments
HW - Hazardous Waste
IARC - Intl. Agency for Research on Cancer
IATA - International Air Transportation Association
ICAO - International Civil Aviation Organization
IDLH - Immediately Dangerous to Life & Health
IMO - International Maritime Organization
LC50 - Lethal Concentration to 50% of Certain Population
LD - Land Disposal
LD50 - Lethal Dose to 50% of Certain Population
LDF - Land Disposal Facility
LDR - Land Disposal Restrictions (A portion of HSWA)
LEL - Lower Explosive Limit
LEPC - Local Emergency Planning Committee
M/TNP - Mixture/Trade Name Products
MSDS - Material Safety Data Sheet

ACRONYMS CONT'D

MSHA	-	Mine Safety and Health Administration
MTR	-	Minimum Technological Requirements
NOI	-	Not Otherwise Indexed
NOIBN	-	Not Otherwise Indexed By Name
NOS	-	Not Otherwise Specified
NCP	-	National Contingency Plan
NFPA	-	National Fire Protection Association
NIOSH	-	Natl. Inst. of Occupational Safety & Health
NOAA	-	National Oceanic and Atmospheric Administration
NPDES	-	National Pollutant Discharge Elimination System
NPL	-	National Priorities List
NRC	-	Non Reusable Container
NRC	-	National Response Center
NRT	-	National Response Team
NTP	-	National Toxicology Program
ORM	-	Other Regulated Material (for DOT)
OSHA	-	Occupational Safety and Health Administration
PCB's	-	Polychlorinated Biphenyls
PEL	-	Permissible Exposure Limit under OSHA
PH	-	Measurement of Acidity or Alkalinity
POTW	-	Publicly Owned Treatment Works
PPA	-	The Pollution Protection Act
PPB	-	Parts Per Billion
PPM	-	Parts Per Million
PRP	-	Potentially Responsible Party
RA	-	Remedial Action
RACT	-	Reasonably Available Control Technology
RCRA	-	Resource Conservation And Recovery Act
REL	-	Recommended Exposure Limit
RQ	-	Reportable Quantity
RSPA	-	Research & Special Programs Administration
SARA	-	Superfund Amendments and Reauthorization Act of 1986
SCBA	-	Self Contained Breathing Apparatus
SERC	-	State Emergency Response Commission
SIC	-	Standard Industrial Classification Code
SPCC	-	Spill Prevention Control & Countermeasures
SQG	-	Small Quantity Generator
STC	-	Single Trip Container
STEL	-	Short Term Exposure Limit
TC	-	Toxic Chemical
TC	-	Toxicity Characteristics (Waste Codes)
TCLP	-	Toxicity Characteristics Leaching Procedure
Title 3	-	The third part of SARA, or EPCRA
TLC	-	Thin Layer Chromatography
TLV	-	Threshold Limit Value
TOC	-	Total Organic Carbon
TPQ	-	Threshold Planning Quantity
TRI	-	Toxic Release Inventory
TSCA	-	Toxic Substance Control Act
TSD	-	Same as TSDF
TSDF	-	Treatment, Storage and Disposal Facility
TWA	-	Time Weighed Average
UEL	-	Upper Explosive Limit
UIC	-	Underground Injection Well Code ID Number
USDOT	-	U.S. Dept. of Transportation
USEPA	-	U.S. Environmental Protection Agency
UST	-	Underground Storage Tank
VOC	-	Volatile Organic Compound

APPENDIX 2

GLOSSARY OF COMMONLY USED TERMS

Acute Effect	A rapidly developing adverse effect.
Addendum	Supplemental information inserted at the end of a section, similar to an appendix.
Carcinogen	A cancer-causing agent.
Ceiling Limit	The maximum TLV which cannot be exceeded, even momentarily.
CERCLA	(Comprehensive Environmental Response, Compensation, and Liability Act of 1980). The federal statute that authorized "Superfund." Superfund, which is administered by EPA provides funding for cleanups and emergency response actions for hazardous substances at the worst hazardous waste sites in the U.S. CERCLA is also significant because it set the first criteria for notification of emergencies involving hazardous substances.
Chronic Effect	A slowly developing adverse effect.
Closure	Deactivating hazardous waste management portions of a facility.
Combustible Liquid	According to DOT regulations, a combustible liquid has a flash point $\geq$ 100 F but < 200 F.
Containment	Receptacle for holding materials. Primary containment usually refers to tanks, drums or other containers; secondary containment refers to a dike or berm which surrounds the primary containment in order to prevent spills or leaks of material from reaching the environment.
Contingency Plan	A detailed, specifically written plan which describes the precautions and procedures which will be used to minimize the dangers associated with hazardous wastes generated and stored at a facility.
Corrosivity	One of the four EPA characteristics of a hazardous waste. Generally a material is corrosive if it has a pH $\leq$ 2 or $\geq$ 12.5. Also, according to DEM, a solid is corrosive if, when mixed in a 1:1 ratio with distilled water, it results in a pH in either of the above ranges.
Dioxin	A family of chemicals found to be carcinogenic, teratogenic and mutagenic in certain species.
EHS	Extremely Hazardous Substance: Any one of 366 hazardous chemicals on a list compiled by EPA to provide a focus for state and local emergency planning activities.

Environmental Audit — Inspection of a facility to determine its compliance with hazardous waste regulations, pretreatment regulations, air pollution regulations, hazard communication regulations, etc.

EP Toxicity — One of the four EPA characteristics of hazardous wastes. It is designed to simulate conditions in a sanitary landfill and determine if certain constituents leach out of the waste.

Extremely Hazardous Substances — Substances listed in 40 CFR Part 355 Appendices A and B.

Federal Register — A daily U.S. Government publication which contains final & proposed changes to regulations of U.S. Government Agencies, including all EPA environmental regulations.

Flammable — According to hazardous waste regulations, a liquid having a flash point less than 200 F or a solid with a flash point less than 100 F is flammable.

Flammable Liquid — According to DOT regulation, a flammable liquid has a flash point < 100 F.

Groundwater — Water contained in pore spaces of soil, as opposed to surface water bodies such as ponds, lakes, streams, rivers.

Halogen — One of the electronegative elements of Group VII A of the Periodic Table (fluorine, chlorine, bromine, iodine, and astatine).

Halogenated Compound — A chemical compound which has been incorporated with one of the halogen elements, usually chlorine or bromine. many halogenated compounds are listed hazardous wastes, particularly compounds containing chlorine and fluorine, such as trichloroethylene, methylene chloride and trichlorofluoromethane.

Hazardous Chemical (HC) — A material which exhibits any physical or any health hazard. Hazardous chemicals are regulated under 29 CFR 1910.1200 - Hazard Communication and 40 CFR Part 370 - Hazardous Chemical Reporting.

Hazardous Matl. (HM) — Any material listed in 49 CFR 172.101. All EPA hazardous wastes are hazardous materials.

Hazardous Substance (HS) — Substances listed in the Appendix to 49 CFR 172.101 (Hazardous Material Table) when in a quantity in one container which exceeds the listed RQ, and substances listed in 40 CFR 302.4.

Hazardous Waste	EPA hazardous wastes are defined in 40 CFR Part 261. They include listed (F,K,U,P numbers) and characteristic (D numbers) hazardous wastes.
Ignitability	One of the four EPA characteristics of a hazardous waste. A material is ignitable if it is a liquid With a flash point less than 140°F or a spontaneously combustible solid.
Incompatible Wastes	Wastes which, if combined under uncontrolled conditions, may produce heat or pressure, fire or explosion, violent reaction, toxic dusts, mists, fumes or gases, or flammable fumes or gases.
Land Disposal	Placement/disposal of wastes in a landfill, surface impoundment, waste pile, injection well, land treatment facility, salt dome, or salt bed formation, underground mine or cave, or concrete or bunker.
Land Disposal Facility	Facility which is licensed to dispose of hazardous materials on the land, such as chemical landfills, land treatments facilities, surface impoundments etc.
Management Plans	Written plans which describe how a company will manage hazardous waste on site. For a generator, the required management plans are a Contingency Plan and a Training Plan. TSDFs must have more extensive management plans.
Manifest	Shipping paper used in documenting all shipments of hazardous wastes.
Memorandum of Agreement	Agreement between EPA and the state. It establishes a "contract" between two groups.
MSDS	Material Safety Data Sheet. A worksheet required by the U.S. Occupational Safety and Health Administration (OSHA) containing information about hazardous chemicals in the workplace; MSDSs are used to fulfill part of the hazardous chemical inventory reporting requirements under the Emergency Planning and Community Right-to-Know Act.
Mutagen	An agent that causes biological mutations.
National Response Center	A branch of the Coast Guard. Spills of hazardous substances in excess of the listed RQ must be reported to the National Response Center.

Neurotoxin	Chemical which produces its primary toxic effect on the nervous system. Often characterized by behavioral changes and a decrease in motor functions.
NRT	National Response Team. The national team composed of representatives from 14 federal agencies, with emergency planning and response capabilities, including EPA & FEMA.
pH	Measurement of the acidity or alkalinity of a material, represented on a scale of 0-14 with 7 as neutral, 0 as extremely acidic, and 14 as extremely alkaline.
Placard	A diamond shaped sign used on transport vehicles to express the hazards of the vehicle contents.
Reactivity	One of the four EPA characteristics of a hazardous waste. Generally a material is reactive if it contains cyanides or sulfides.
RIDEM	Rhode Island Department of Environmental Management.
Right-to Know	OSHA law which gives employees who work with hazardous materials the right to request information regarding the chemicals they may be exposed to in the workplace.
RIPDES	Rhode Island Pollutant Discharge Elimination System. RIPDES permits must be obtained by companies who discharge wastewater to surface water bodies.
RQ	Reportable Quantity. An amount of a Superfund "hazardous substance" or "extremely hazardous substance" that, if released, must be reported under the emergency release reporting requirements of the Emergency Planning and Community Right-to-Know Act or under those of CERCLA.
Sanitary Landfill	Facility licensed to dispose of non-hazardous materials on land, such as household garbage.
SARA	Superfund Amendments and Reauthorization Act of 1986.
Secondary Containment	Structure used to prevent the migration of spilled materials (e.g., berms, walls, dikes, etc.).
Self-Audit	Suggested procedure which may be followed to establish compliance with federal and state hazardous waste regulations.
SERC	State Emergency Response Commission.

Soft Hammer Waste	Restricted wastes for which EPA did not promulgate treatment standards by scheduled deadlines. These wastes are automatically banned from land disposal after the applicable effective date except under very strict & limited conditions.
Source Reduction	Source reduction usually refers to action taken by a generator to reduce the toxicity or volume of waste generated by modifying processes and/or substituting chemicals.
Specific Gravity	The ratio of the density of a substance to the density of a reference substance. For liquids the reference substance is water (1 g/cc or 8.34 lbs./gal). Liquids with specific gravities less than 1 are lighter than water (e.g., oil, xylene). Liquids with specific gravities greater than 1 are heavier than water (e.g., chlorinated solvents, PCBs).
Teratogenic	Causing fetal malformations or monstrosities.
Title III	The third part of SARA, also known as the Emergency Planning and Community Right-to-Know Act of 1986.
Toxic Chemical	A chemical listed in 40 CFR Part 372 Subpart D.
TPQ	Threshold Planning Quantity. The amount of an extremely hazardous substance present at a facility above which the facility's owner/operator must give emergency planning notification to the SERC and LEPC.
TSCA	Toxic Substances Control Act. Federal statute which gave EPA authority to regulate chemical substances.
TRI	Toxic Release Inventory. A national inventory of annual toxic chemical releases from manufacturing facilities.
Uniform Manifest	National form for use in documenting all shipments of hazardous waste.
Volatile Organic Compound	An organic compound (compound of carbon) with a tendency to pass from the solid or liquid state into the vapor state.
Waste Minimization	Any action taken by a waste generator to reduce the volume or toxicity of hazardous waste generated. It includes reuse, recycling, reclamation, chemical substitution, or process changes which result in a reduction in the quantity or toxicity of waste disposed of on site or off site. See also source reduction.

General References

EPA Publictions: (Can be obtained from EPA - Free of Cost)

(1) Community Right-to-know and Small Business.
(Understanding Sections 311 and 312 or EPCRA of 1986)
Office of Solid Waste and Emergency Response
Washington, D.C., Sept. 1988.

(2) Toxic Chemical Release Inventory Reporting Package
(Most recent one)
(Toxic Chemical Release Inventory Reporting Package Instructions)
Office of the Toxic Substances TS-793, Washington, D.C. 20460, Jan 91, EPA 560/4-91-001.

(3) Reporting Requirements for Continuous Release of Hazardous Substances (A Guide For Facilities and Vessels on Compliance). OSWER Directive: 9360-7-01, Oct. 1990.

(4) Title III, List of Lists
(Consolidated List of Chemicals)
OPTS (OS-120), OSWER (TS-779)
EPA 560/4-90-011, Jan. 1990

(5) Common Synonymns
(For Chemicals Listed under Section 313 of the Emergency Planning and Community Right-To-Know Act.
Pesticides and Toxic Substances TS-779
EPA 560/14-90-005, Nov. 1989.

A Comparative View of Control and Regulating Technologies for Some Primary Smelting Operations

Husamuddin Ahmadzai

Statens Naturvårdsverk, Swedish Environmental Protection Agency,
S-171 85, Solna, Sweden.

Abstract

A survey of regulations and state of the art control technologies is presented relating to the smelting of primary aluminium, copper and recycling of residues. Some costs associated with the alternatives and trends for the coming decade in this sector are identified.

Residues and Effluents - Processing and Environmental Considerations
Edited by Ramana G. Reddy, William P. Imrie and Paul B. Queneau
The Minerals, Metals & Materials Society, 1991

A Comparative View of Control and Regulating Technologies for Some Primary Smelting Operations

Husamuddin Ahmadzai

Statens Naturvårdsverk, Swedish Environmental Protection Agency,
S-171 85, Solna, Sweden.

Introduction

Environmental problems associated with primary non-ferrous smelters are relatable to emissions of acid generating gases (e.g. sulfur dioxide, nitrogen dioxide, hydrogen fluoride), greenhouse producing gases (e.g direct and indirect use of energy by a process), gases leading to atmospheric ozone creation or depletion of the stratospheric ozone layer (e.g. emissions of volatile organic compounds and use of chloro-flurocarbons and halons) and toxic dust and vapour emissions. Emissions of concern originate mainly during the feed preparation, roasting, smelting, converting, electrolyses and co-product processing operations. Emissions can be termed in general as primary process gas or secondary fugitives emissions. Fugitive emissions are releases from uncontained or open sources, such as bulk storage piles, mixing and drying units, transfer facilities, badly joined equipment and accidental spills or uncontrolled testing.

In many smelters, scrap is also charged. In case of chloride/bromide charges with the scarp, emission of toxic organic bromine and chlorine compounds can be expected. Secondary charges eg ashes, dusts etc can lead to additional emissions of mercury (Hg) and toxic emissions.

Only some aspects of concern are addressed within the framework available here.

The basic metallurgical sector in Sweden comprises of two primary non-ferrous smelters, three ore based integrated ferrous works and about thirty specialty steel and secondary smelting facilities. Additionally there is a primary smelter capable of recycling furnace dusts rich in chrome, nickel, lead and zinc contents.

The copper smelter has a franchise to produce approximately 350,000 tonnes per annum primary copper, lead, zinc, and arsenic products. Additional co-products (approximately 10,000 tonnes per year) are precious metals (Au, Ag, Pt-group), compounds of tellurium, selenium and specialty products related to copper, zinc, nickel, arsenic etc. The sulfuric acid and liquid sulphur dioxide capacities for the smelter are 250,000 and 200,000 tonnes/year respectively.

The aluminium smelter operator holds a franchise to produce approximately 100,000 tonnes of aluminium per year. The recycling facility has a franchise to process 70,000 tonnes of dust per year.

Aluminium Smelter

The aluminium industry in general has been particularly successful in their efforts to combine process efficiencies with good environmental control. Hall-Héroult technology is the predominant aluminium smelting process and is expected to dominate well into the next century. In the Hall-Héroult process aluminium is produced by the electrolysis of alumina dissolved in a fused salt cryolite (Na_3AlF_6) with minor addition of other salts, normally fluorides. Two major types of cell are in use at present, i.e. the Söderberg continuous anode where a paste of electrode pitch and coke is baked in situ by the heat of the cell and the prebaked anode. In prebaked cells, the electrodes are prefabricated and added to the cell. There are three variations of the prebaked cell. Modern prebaked cells use automatic breaker/feeders mounted on the center line to mechanize alumina feeding. One type is the centre-work (CWPB), the other a side-work (SWPB) and the third is the point-feeder (PFPB). With point breaker-feeders, it is possible to operate with a nearly constant concentration of alumina in the electrolyte.

In general, emissions from the cells consist mainly of HF, CF_4, SiF_4, CO, CO_2, SO_x, NO_x, particulates (e.g. metal oxides and fluorides), and tar. It is estimated that up to 25 % of the tar content may consist of **P**olycyclic **A**romatic **H**ydrocarbons, **PAH**. The range of PAHs emitted from cells, particularly Söderberg electrode based, is extensive. A non-exhaustive list of PAH is depicted in Table I.

TABLE I - RANGE OF POLYCYCLIC AROMATIC HYDROCARBON SUBSTANCES THAT CAN BE FOUND IN TAR EMISSIONS FROM PRIMARY ALUMINIUM SMELTERS (1).

Naphthalene	Benzo(a)fluorene
2-Methylnapthalene	Benzo(b)fluorene
1-Methylnapthalene	1-Methylpyrene
Biphenyl	Benzo(c)phenanthrene
Acenapthylene	Benzo(a)anthracene
Dibenzofuran	Chrysene/Triphenylene
Fluorene	Benzo(b)fluoranthene
Dibenzothiophene	Benzo(k)fluoranthene
Phenanthrene	Benzo(e)pyrene
Anthracene	Benzo(a)pyrene
Carbazole	Perylene
Methylphenanthrenes	o-Phenylenepyrene
methylanthracenes	
Fluoranthene	Dibenzo(a,h)anthracene
Pyrene	Anthanthrene

The domestic Swedish aluminium smelter utilizes both the point fed pre-baked electrode (PFPB) and the vertical Söderberg electrode (VSE) technologies. The atmospheric emission guidelines relate as depicted in Table II.

TABLE II- EMISSION GUIDELINES RELATED TO ALUMINIUM PRODUCTION

	Guideline		Limiting Value	
	VSE	PFBP	VSE	PFPB
	(kg/tonne$_{Al}$)		(tonne/y)	
Particulate	8.0	1.5	-	60
F_{Tot}	1.0	0.5	-	20
SO_2	-	<6	-	150
Tar	0.2	0*	18	- *

* included in emissions from the Söderberg based technology.

Typical emissions to the ambient air and water from the two technologies are presented in Tables III and IV.

TABLE III- TYPICAL SPECIFIC EMISSIONS FROM SÖDERBERG AND PREBAKE TECHNOLOGIES

	VSE	PFPB
	(kg/tonne$_{Al}$)	
Particulate	3.3	1.0
F_{Tot}	1.0	0.3
SO_2	3.8	0.9
NO_x	0.04	0.3
Tar	0.2	0*
Benzo(a) pyrene	0.01	0

* included in emissions from the Söderberg based technology.

TABLE IV - TYPICAL EMISSION TO THE AQUATIC RECIPIENT FROM AN ALUMINIUM SMELTER

	VSE	PFPB
	(kg/tonne$_{Al}$)	
Suspended Solids	3.7	0.3
F_{tot}	3.6	1.3
Tar	0.2	0.1

Point-fed prebaked (PFPB) operations are currently considered to represent best available technology. Modern operations allow good current efficiencies (93-95 percent), and anode consumption less than 400-430 kg$_C$/tonne$_{Al}$ concurrent with an electrical consumption around 12.5 kWh/tonne$_{Al}$. Good operational practice and labor training have led to fifty percent reduction of labor intensity from 4-6 to 2-3 man-h/-tonne$_{Al}$. New materials and stable operational routines have also enabled extension of average pot-life from 24 - 85 months up to 120 months. Prebaked technology in conjunction with dry-scrubbing of fluorides by aluminium oxide, wet media or semi-dry scrubbing of SO_x, sedimentation/flotation and filtration of waste water, recycling/controlled deposition of spent potlines are expected to lead to considerable reduction of emissions to the environment. Costs associated with modern state of the art technology vary depending on retrofit or establishment of green smelters but are estimated to be in the range of 1000 - 2000 US$/tonne$_{Al}$. An application of best available technology (BAT) can enable a reduction of emissions to the environment as reflected by Table V.

TABLE Va - EMISSIONS TO AIR FROM A 100,000 TPY PRIMARY ALUMINIUM SMELTER

Parameter	Historic	Current[1] (tonnes/year)	BAT
Dust	460	270	< 100
Total fluorides	120	80	< 40
Tar	70	20	0
SO_x	520	300	< 100
NO_x	30	30	< 5

TABLE Vb - EFFLUENT DISCHARGE FROM A 100,000 TPY PRIMARY ALUMINIUM SMELTER WITH BAT

Parameter	Historic	Current[1] (tonnes/year)	BAT
Susp.solids	900	300	< 30
Fluorides	550	300	< 40
Tar	20	15	< 1

[1] Using a combination of VSE and PFPB technologies

The major solid waste of environmental concern pertaining to the primary smelting of aluminium is Spent Pot-lining (SPL) material. Pots or electrolytic cells in aluminium smelters on the average have a life time of 24-85 months whereafter they are dismantled for relining. The annual global production of SPL is estimated to be about 850,000 tonnes. This is equivalent to 43 kgSPL/tonne Al. Good smelter practice can reduce this figure to about 35 kg SPL/ton Al.

SPL material comprises of graphite cathode, refractory materials, penetrated metal and bath components including minor amounts of aluminium nitride, carbides and cyanides. Water may react exothermally with SPL. Carbides react to generate CH_4. Aluminium and sodium metal form H_2. Aluminium nitride forms NH_3. Furthermore, compounds such as carbonates, fluorides, and cyanides dissolve. Excess water can also lead to disintegration of SPL blocks. Since the gaseous reaction products are considered hazardous, storage of SPL requires qualified ventilation.

The most frequent disposal method for SPLs, hitherto, has been by landfilling. Leachates from such land fills may vary considerably depending upon in situ conditions. Table VI shows a range of data from some sites for weathered SPLs.

TABEL VI - COMPOSITION OF LEACHATES FROM WEATHERED SPL DISPOSAL SITES (2).

Parameter		Concentration Range, mg/litre[1]
pH		10 - 12
Total dissolved Solids (TDS)		20,000 - 70,000
Total Suspended Solids (TSS)		10 - 3,000
Total Organic Carbon (TOC)		300 - 500
Carbonates		6,000 - 15,000
Sulphates		800 - 7,000
Fluorid		1,200 - 8,500
Cyanide[2]	-T	150 - 1,250
	-A	0 - 150
	-W	0 - 20
Total Al		20 - 50
Total Fe		50 - 250
Total Na		6,500 - 15,000
Total S		100 - 2,000

1) Except for pH
2) T - Total cyanide - strong acid dissociable
A - Cyanide amenable to chlorination
W - Weak dissociable cyanide

A major portion of the cyanide may be complexed with iron and difficult to oxidize. The primary treatment options for aged leachates include
- iron salt precipitation
- ultraviolet enhanced oxidation
- high temperature chlorination
- polysulphide reaction, and
- biological treatment.

Leaching of fresh SPL may be carried out to produce uncomplexed cyanides. This pretreatment may prevent leaching of CN and F from landfills. Pretreatment can be carried out with $Ca(OH)_2$ to precipitate CaF_2. Cyanide may be oxidized with sodium hypochlorite (NaOCl). Another method uses steam to decompose CN and $CaSO_4$ to generate CaF_2 (2). Some aluminium smelters deposit SPL in the sea. Marine disposal for numerous reasons is considered environmentally objectionable.

A number of technologies have been evaluated for recovering energy and material from SPL. Work is still in progress in this area and not all processes are well documented. Table

VII lists a number of technologies which have been identified.

TABEL VII - TECHNOLOGIES FOR RECOVERY OF MATERIALS AND ENERGY FROM SPENT POT LININGS (2).

Combustion Processes
- Additive & energy source in cement kilns
- Fluidized bed combustion
- Fuel additive in coal fired power plants.

Metallurgical Processes
- Additive in steelmaking processes
- Additive in iron foundry cupola furnaces
- Aluminium cell anode additive, carbon fraction only.

Chemical Processes
- Cryolite recovery by leaching
- Aluminium fluoride recovery by leaching
- Pyrohydrolysis, i.e. combustion in presence of steam to produce aluminium fluoride
- Pyrosulpholysis, i.e. pyrohydrolysis with SO_2 addition.

Other
- Mineral wool, product additive and energy source.

Primary Copper Smelter and Residue Processing Facilities

The main regulations governing emissions from the Swedish primary smelter are summarized below. Table VIIIa depicts the limiting values relevant to the smelter operations. The specific values in Table VIIIb are based on the statutory guidelines (valid as of 1992) correlated to approximately 350,000 tonnes/year production of primary products. The operator is, in the meanwhile, under statutory direction to examine a further reduction of its overall emission of arsenic, cadmium, and mercury.

TABLE VIIIa - OPERATATIONAL GUIDELINES FOR A COPPER SMELTER

	Guidelines 1989-91	1992-93	Limiting Value 1994-
	(tonnes/y (unless otherwise stated))		
SO_2	8000	5000	5000
Dust	400	250	250
Cu	25	10	10
Pb	75	30	30
Zn	75	40	40
Cd	1.6	0.8	0.8
As	15	8	8
Hg	0.6	0.35	0.35

1992-	
Hg (from scrap)**	25 kg/year
Dioxins (")**	0.5 g/year

** during scrap smelting

TABLE VIIIb - SPECIFIC VALUES OF EMISSIONS TO THE ATMOSPHERE FROM THE COPPER SMELTER

No.	Description	Specific Value (g/tonne)	ANNUAL LIMIT (tonne)
1	Particulates (tot)	700	250
2	Cu	30	10
3	Pb	85	30
4	Zn	110	40
5	Cd	2	0.8
6	As	20	8
7	Hg	1	0.35
8	SO_x (SO_2)	14300	5000

Emissions of metals/metallics to ambient water are regulated according to limiting values depicted in Table IX. Regulations apply to cooling water, wash water, rain/precipitation, and process water. The discharge of process and cooling water is at a rate of approximately 3500 m^3/h.

TABLE IX - GUIDELINE AND LIMITING VALUES FOR EMISSIONS OF METALS/METALLICS TO THE AMBIENT WATER

No.	Description	Guideline Value 1989-91 (tonne/y)	(mg/l)	Limiting Value 1992- (tonne/y)
1	Cu	4	0.1	2
2	Pb	3	0.1	2
3	Zn	10	2.0	8
4	Cd	0.5	0.05	0.4
5	As	40	10.0	20
6	Hg	0.2	-	0.1

Emissions to the public sewerage system are regulated according to values shown in Table X.

TABLE X - REGULATORY GUIDELINE VALUES FOR EMISSIONS OF METALS/METALLICS TO THE PUBLIC SEWERAGE SYSTEM.

No.	Description	Guideline Value (mg/l)
1	Cu	1
2	Pb	1
3	As	1
4	Zn	2
5	Cd	0.01
6	Hg	0.002

Noise from industrial activity is regulated such that noise-level at the nearest lodging may not exceed:
- 55 dB(A) day-time (0700-1800 hrs)
- 50 dB(A) evening-time (1800-2200 hrs)
- 45 dB(A) night-time (2200-0700 hrs)
- Max momentary pulse should not exceed 60 dB (A).

Currently, some additional mandated investigations are underway. These studies pertain to:
- further reduction of Cd, As, Hg emissions to the ambient
- treatment of hazardous wastes and suggestions for final repository of waste
- follow-up study pertaining to lung cancer in the area
- follow-up epidemiological studies on reproduction in the area
- exposition of humans to Cd and Hg.

The smelter also generates, annually, approximately 150,000 tonnes slag and 12,000 to 15,000 tonnes waste from the various abatement systems. There are 90 sources of emission to the atmosphere of which 40 are measured regularly. SO_2 is measured continuously at 7 locations, and manually at 10 stations. Dust is measured manually at 40 stations. Emissions of pollutants to the atmosphere and the aquifer during 1990 are listed in Table XI.

TABLE XI- MEASURED EMISSIONS TO THE AMBIENT, 1990

ITEM	AIR (tonnes)	WATER (tonnes)
Dust	250	
SO_2	4 900	
Cu	18	2.3
Pb	52	1.4
Zn	33	5.4
Cd	1.3	0.07
As	4.7	8.1
Hg	0.25	0.03

Ground level sulphur dioxide concentrations are regularly registered in two localities. Concentrations of SO_2 have indicated levels around 3-40 $\mu g/m^3$ (average 24-hr values). There have been inversions, however, and levels in the range 80-180 $\mu g/m^3$ have been measured. The Swedish Environmental Protection Agency's recommended values are 100 $\mu g/m^3$ and 50$\mu g/m^3$ during the winter half-year.

Guidelines for a good environment are application of the precautionary principal and utilization of the best available technology that enables a return of the ambient environment loadings to background levels. Accordingly negotiations are underway with the industry for further reduction and control of emissions of pollutants like sulphur dioxide and toxic matter.

Reduction in the concentration of pollutants can be achieved by dissipation (i.e. dilution) reflected by tall stacks and closed loop control systems. This measure does not reduce emissions of pollutants, only disperses them over a wider area. Effective emission controls are achieved, firstly by an appropriate choice of process technology and thereafter by incorporation of relevant mitigation technology and routines.

New generation process technologies offer high SO_2 concentrations and low dust emissions (Table XII). These advantages are concurrent with low volumetric gas flowrates which are are relatively easier and less expensive to treat.

TABLE XII - A COMPARISON OF SOME ALTERNATIVE TECHNOLOGIES FOR PRIMARY COPPER SMELTING (3,4,5,6).

PARAMETER	PROCESS											
	Flash Harja-valta	Flash Toyo	Flash INCO	MITSU-BISHI	KIV-CET	VANU-KOV (PV)	NOR-ANDA	BAI-YIN	REV-VERB-ATORY	ELECTRIC	ISASMELT	CONTOP
Smelting rate ($t/m^2/d$)	11	9	11	19	8	55-80	20.	8	4	4	100	124
Moisture in charge (%)	0.3	0.3	0.3	0.3	0.3	6-8	5	8	10	10	9.5	0.3
SO_2 in offgas (%)	33	25	80	13.5	75	25-60	10	10	1.5	2-4	3-8	31
Dust in offgas (%)	9	9	-	4.5	12	1-1.5	4	3	11	1	1	1
Furnace campaign (days)	330	360	350	340	330	570	350	240	700	700	630	-

State of the art technology also offers better energy utilization and superior process economics. Table XIII reflects cost savings achieved by application of available modern technology versus conventional technology.

TABLE XIII - CAPITAL COST ESTIMATE (1988 US$) AND ENERGY UTILIZATION FOR A SMALL CAPACITY ANODE COPPER SMELTER (7).

ITEM	REVERBERATORY	ELECTRIC	FLASH	CONTOP
Investment/ tonne Anode Cu, ($/tonne)	1600	1600	1300	1200
Annual Operating Costs(x1000 $)	6000	7000	5500	5000
Energy consumption (GJ/tonne anode)	41	42	32	23

For the control of sulphur dioxide, throw away or regenerative systems are preferred. Such systems are typified by scrubbing e.g lime/limestone (70-99% recoveries depending upon the system), ammonia double alkali process (>90% recovery), sodium sulphite/bisulfite process (90-95% recovery), elemental sulphur processes (up to 50-60 percent recovery), sulphur dioxide or sulfuric acid processes with downstream products (up to 99% recovery).

An estimate of costs associated with some such controls are depicted in Table XIV.

TABLE XIV - SOME COSTS ESTIMATES (1985) FOR A 80 PERCENT REDUCTION OF SO_2 IN SMELTER OFF-GASES (8).

Gas flow-rate $x10^{-3}$ (nm^3/h)	Original amount untreated SO_2 (tonne/y)	Investment cost (M US$)	Total specific cost (US$/tonne$SO_2$)
20	1100	2	600
150	1300	7	2600
160	1600-2400	6-8	1500-3000
250	2000	7-8	1700-2000

Fugitive emissions can be reduced by initiating an inventory of hazardous chemicals and establishing routines for their monitoring, use and their path. For example, leakages from machinery, fuel and chemicals storage, or other sources can be controlled by installing trapping and recovery systems. A dike system is advisable for the oil unloading and storage. Drastic reductions can be achieved by installation of dust suppression systems for transportation; asphalting surfaces; indoors storage of all raw materials and intermediate products; encapsulating the equipment or bath during tapping, charging, transferring, and casting etc. For dust emissions, fabric filters may preferably used for the secondary gases during tapping and charging operations. Filter reverts should be carefully managed. Collection can be carried out in special containers specially constructed for the purpose.

Mercury emissions can be reduced by several established techniques. For example, selenium and carbon based filters are in operations with the ferrous and non-ferrous smelters treating gases ranging from 10,000 up to 120,000 nm^3/h. Mercury elimination is by 90 to 99.9 percent. Outgoing Hg concentration are down to 0.01-0.09 mg/nm^3. Costs estimates vary between US$ 10,000 to US$ 15,000/kg_{Hg}-year.

From an overall point off view, an appropriate choice of technology at a smelter processing upto 350,000 tonnes of primary products can enable reduction of emission levels to those reflected in Table XV.

TABLE XV - REDUCTIONS IN EMISSIONS TO AIR ACHIEVABLE BY EXISTING BEST AVAILABLE AND PROVEN TECHNOLOGIES.

Parameter	Historical	Current (tonnes/year)	BAT
SO_2	11500	5000	2000
Dust	1000	250*	50**
Cu	4	10	2
Pb	100	30	6
Zn	145	40	8
Cd	3	0.8	0.2
As	40	8	2
Hg	0.6	0.35	0.1

* 0.7 kg/tonne product
**0.1 kg/tonne product

The smelter processing dusts and similar residues utilizes a plasma reduction technology. The guidelines for emissions to air have yet to be finalized but the operators proposals are depicted in Table XVI.

TABLE XVI - OPERATOR PROPOSED GUIDELINES FOR EMISSIONS TO THE AIR FROM A FINE-RESIDUE PROCESSING FACILITY UTILIZING PLASMA TECHNOLOGY.

	Guideline	Limiting value
Dust (including fugitives emissions)	0.1 $kg/tonne_{Furnace\ dust,dry}$	7 tonne/y
Hg	0.4 $g/tonne_{Furnace\ dust,dry}$	0.01 tonne/y
HCN	10 mg/nm^3	
NO_x	500 mg/MJ	

The actual emissions from the recycling smelter have been in reality around 0.03-0.08 kg/tonne of furnace dust.

The emission to the aquifier from the recycling smelter is depicted in Table XVII.

TABLE XVII - EMISSION GUIDELINES FOR WATERS (MAX 70,000 M^3/Y) FROM THE PLASMA-BASED DUST PROCESSING SMELTER.

	Description	Guideline Value (mg/l)	Limiting Value (tonne/year)
1	CN	<0.5	0.035
2	Cr^{6+}	<0.1	-
3	Cr_{tot}	0.3	0.025
4	Ni	0.1	0.020
5	Cu	0.2	0.040
6	Pb	-	0.010
7	Zn	0.5	0.100
8	Cd	-	0.001
9	As	-	0.001
10	Hg	-	0.001
11	N_{tot}	-	1.0
12	Suspended Solids	5.0	-
13	F_{tot}	35	1.4

Based on the application of best practicable control technology currently available, it is generally possible to achieve zero discharge of process waste waters to surface or ground waters. Effluents may be impounded, with or without pretreatment, and recycled in most cases. For example, slag mill waste waters can be settled and the supernatant combined with the contact cooling water for treatment by chemical precipitation. The supernatant can receive further filtration, stored and recycled, thus resulting in zero discharge. It may also be preferable in some cases with dry cooling and eventual crushing. This would reduce water requirement, but precautions are required to prevent fuming. Total water containment design is a function of local climatic conditions, nevertheless, surface/rain-storm water handling systems can be designed to virtually contain all precipitations and circulate the water subsequent to impoundment. Table XVIII list reclamation percentages for a typical integrated smelter operating in a dry area.

TABLE XVIII - TOTAL RECLAIMABLE WATER FOR A TYPICAL INTEGRATED SMELTER.

Water consumer	Percentage of reclaimed water %
Mine	0
Concentrator	62
Smelter	91
Power Plant	90

Table XIX depicts historic, current and possible achievable levels of emissions to the water from the domestic smelter. These figures are based on technologies currently being applied and generally available.

TABLE XIX - EFFLUENT DISCHARGE AND LEVELS ACHIEVABLE WITH APPLICATION OF BAT.

	Historic	Current (tonnes/y)	BAT
Total Toxic Metals	50	33	5
Cu	7.2	2	0.2
Pb	3.1	2	0.2
Zn	7.6	8	0.8
Cd	0.4	0.4	0.04
As	31	20	0.6
Hg	0.04	0.1	0.01

With minimal discharge of waste-water, problems with accumulation of salt concentration requires attention. Neutralizations are generally carried out by regulating pH and redox potentials, and eventually flash evaporations. Toxicants like mercury, cadmium, and lead can be precipitated as sulphide for recycling or qualified deposition. For arsenic two widely practiced methods are precipitation of As (V) and removal from solution as either "calcium arsenate" or "ferric arsenate".

Solubility of calcium arsenate has been reported to reach a minimum of arsenic concentration of about 100 mg/l at pH = 10 (9). Addedly atmospheric or in situ CO_2 causes decomposition to $CaCO_3$ in an alkaline solution and renders calcium arsenate residue not stable enough for dumping. Precipitation of basic ferric arsenate with Fe/As molar ratio of 4 (i.e. weight ratio of 3) or more results in 100 to 1000 times lower arsenic solubility over the pH range 3 to 7 than simple ferric arsenate or scorodite ($FeAsO_4.2H_2O$). Disposal of arsenical solids have been reported Fe/As ratios of 8 - 10:1 and 27:1 respectively (9). Irrespective, however, until more information is made available it will remain necessary to continually monitor disposal sites. While adsorption of toxicants by soils and aquatic sediments is an important buffering and fixing process for any leakage, waters can still be polluted. Control of dumps should start with planning of location and configuration and the design must address the objectives that the wastes are sealed to minimize penetration of air and water, and to contain/treat drainage indefinitely. Control of waste stability and dump behavior must be planned on the assumption that leakage will arise at some time in future.

In 1985, one particular case in North America reported the cost of disposal of arsenic bearing residues at a licensed site to be US$ 220 - 230 per wet tonne (9). In Sweden the cost may vary between US$ 155 - 3075 depending upon the stability of the waste. Considerations should therefore be given, whenever possible, to the possibility of generating a product rather than a waste.

Slag has traditionally been vended as a substrate for either landfilling or for sand-blasting purposes. This practice is acceptable provided the granulated slag is innocuous. Disposal of slags must take into account that the possibility of leaching of toxic compounds into soil, groundwater or surfacewater, cannot be overruled if acidity levels increase with a particular application. If slags have a high levels of susceptible toxic metals they should be landfilled. The landfill should be properly designed and isolated to avoid accidental leakages and a management plan for monitoring the landfill should be incorporated.

Finally, disused industrial site should be restored to its original state. In case the infrastructure is to be used for another purpose, the contaminated equipment and infrastructure would require remedial actions. Contaminated residues in that case should be co-deposited with other hazardous waste and stabilized.

Conclusions

Environmental regulation and control is a complex undertaking and requires many inputs. An awareness on behalf of decision makers; definition of the natural environment, its loadings and tolerances; available and developing technology; the interaction between micro and macro economic factors; and monitoring and control are some of the vital issues for sustained and environmentally sound development of the industry.

REFERENCES

1. A. Björseth and L. Wikström, "Kartläggning av PAH-utsläpp från Gränges Aluminium i Sundsvall (Work material SNV-Bm, 1979).

2. Norway, "Primary Aluminium Industry" (2nd Meeting of WGIC, Madrid, January 1990, PARCOM-INDSEC (2/4/1)-E).

3. V.V. Mechev, Pyrometallurgy '87, (London U.K.: The Institution of Mining and Metallurgy, 1987) 783-798.

4. J. Pritchard, private communication with author, MIM Holding, 4 October 1991.

5. A. Fritsch, private communication with author, KHD Humboldt Wedag, 22 August 1991.

6. China National Non-ferrous Metals Industry Corporation, "Baiyin Copper Smelting Process" (Report BCSP Baiyin Gansu, 1989).

7. H. Ahmadzai, "An Economic Assessment of Appropriate Copper Smelting Technology for the Saindak Integrated Mineral Project" (Report to Dept. of Industrial Economics and Management, Royal Institute of Technology, 1987).

8. Boliden Metall AB, "Olika metoder för rening av smältverksgaser från svaveldioxid" (Report Tekniska utredningar, kapital 3, inlaga till koncessionsnämnden, juni 1985).

9. R.G. Reddy, J.L. Hendrix and P.B. Queneau, eds., Arsenic Metallurgy Fundamentals and Applications, (Warrendale PA: The Metallurgical Society, 1988), 3-507.

DEVELOPMENT OF PROCESSES FOR MINIMIZING THE WASTES

Heikki Tuovinen, Maija-Leena Metsärinta, and Launo Lilja

Outokumpu Research Oy
P.O. Box 60, SF-28101 Pori, Finland

Abstract

The total utilization of raw materials and energy is the most important criterion when planning industrial processes. The main goal is clearly to use the raw materials and energy as effectively as possible in order to make the desired products directly in the primary process itself, or to recycle unused materials and energy either as such, or after cleaning back into the process. However, few industrial processes are so straightforward. Most processes generate by-streams of materials and energy, which cannot be recycled directly and it is namely these by-streams which tend to remain as wastes. These by-streams are slags, dusts, vaporized metals, liquids, slurries, sludges, gases etc. Most base metal processes produce all these by-products. Waste water and gas emission controls and regulations are tightening all the time, so new processes must be developed to keep pace with these requirements. These processes generate more solid wastes, which must be utilized or at least processed into a harmless form landfilling. Municipal and hazardous wastes also contain valuable materials and energy, which must be used to minimize landfilling. The development of processes for these purposes places even greater demands on research and development. The same principles and methods developed for mining, ore dressing and metal refining processes can also be adapted for research and development work on processes for waste minimization. However, the problems encountered are much more complicated and less familiar, demanding more versatile and accurate tools and equipment. This paper presents principles, examples and some references of R & D work on processes for waste minimization.

Residues and Effluents - Processing and Environmental Considerations
Edited by Ramana G. Reddy, William P. Imrie and Paul B. Queneau
The Minerals, Metals & Materials Society, 1991

Introduction

The processing of metals from the ore to the final product is quite a complicated operation. Metals exist in nature commonly as mixed sulfides, oxides, carbonates and silicates combined with a sidestone, which can be a mixture of silicates, serpentinites, carbonates etc. Often the metal minerals are partially oxidized and weathered. In many cases they also contain variable amounts of other metals, chlorides, fluorides etc. as impurities.

To extract the valuable minerals from the rock the mined ore is crushed and ground to a suitable grain size. The metal minerals are then concentrated from the ground ore by heavy method, magnetic and electrostatic means or flotation.

Valuable metals are separated from the concentrate either directly or after roasting by leaching and precipitating electrolytically or by hydrogen reduction in the autoclave after purification of the metal containing solution.

Some minerals must be processed or are more easily processed at a high temperature. Sulfidic concentrates are normally processed by autogenous smelting by burning sulphur to sulphur dioxide and by oxidizing to faylite or also partially to magnetite and tapping the metal as matte or as blister. Oxidic ores and concentrates require most often electric smelting (smelting in an electric furnace) and before smelting, preparation like pelletizing, briquetting, sintering, preheating etc.

In smelting the metal, metal alloy or matte and slag are separated and easily vaporized metals and metal compounds are transported with the reaction gases to gas cleaning, where they are condensed and taken out by dust collectors. Energy from the reaction gases is recovered in the waste heat boilers and sulphur dioxide is converted to sulphuric acid or to elemental sulphur when sulfidic concentrates are processed. Heat values of the off-gases from the electric furnace can be utilised for sintering and preheating the furnace feed. From an environmental point of view, the process technologies and problems in the chemical industry, pulp and paper industry, in energy production as well as in the processing of municipal and hazardous wastes are basically similar to the production technologies in the base metal industries. Thus the same methods and equipment can also be adapted in the research and development of processes for these different applications.

Goal Setting

Whether it is a matter of planning new production, improving an old process or making a process more effective by minimizing wastes or developing a new process for handling wastes, the goal setting in all these cases is similar. However, the problems are more complicated and less familiar. As far as research and development for a non-waste process or handling of wastes is concerned, the equipment and devices must also be much more versatile and accurate.

To avoid unnecessary sidetracking and frequent returns to the starting position the goal setting must be based on the most profound and quantitative analysis of the process materials, chemistry and conditions (Fig 1).

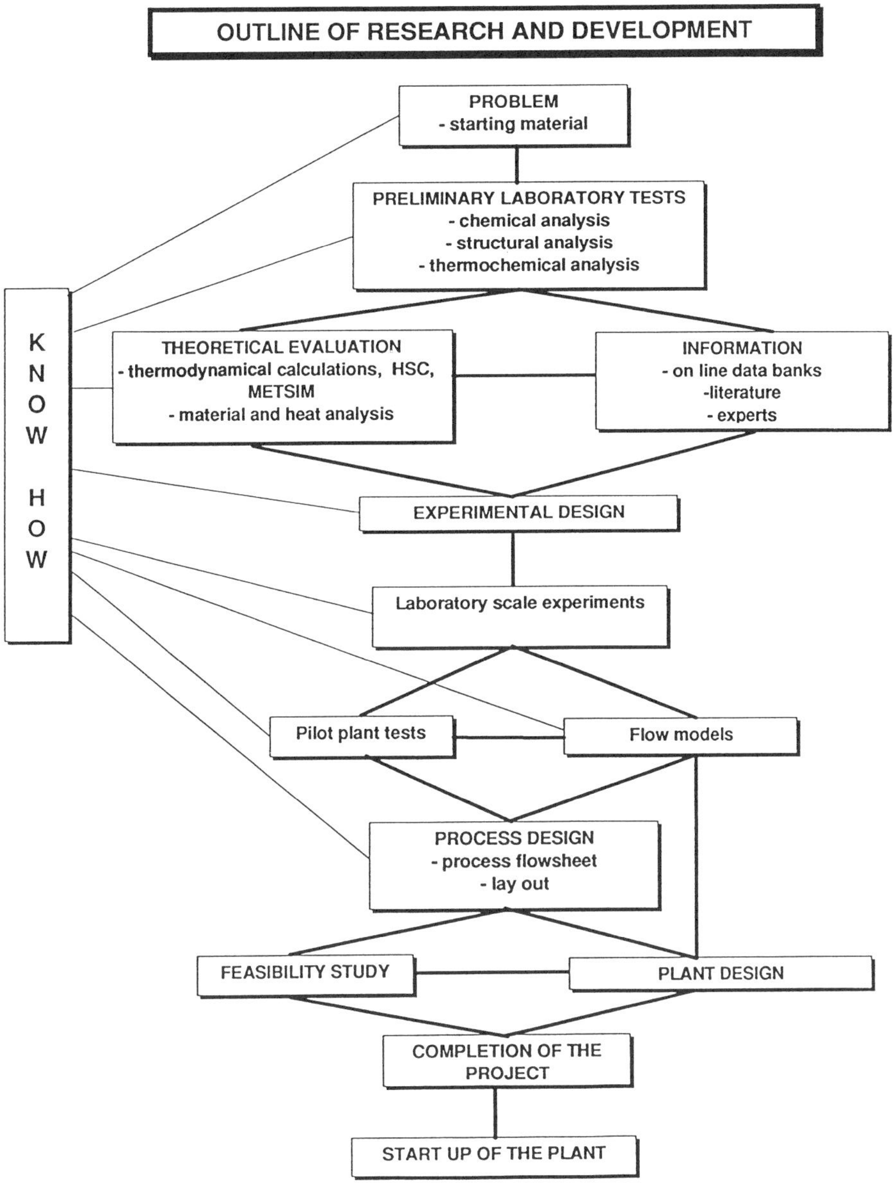

Figure 1 - Development of Processes to Minimize Waste

Analysis of Process Materials, Theoretical Calculations and Laboratory Tests

For the analysis of process materials a great deal of high quality analysers (Fig 2) for identification and specification of basic chemical elements and compounds (Table I and Fig. 3) are needed in addition to conventional wet chemical methods. On the basis of quantitative analyses of the process materials the theoretical thermodynamic balances can be computed (Figs 4 and 5) to specify the process conditions, reaction materials and additives.

The conditions and equipment for the laboratory tests can be selected by comparing the results of the calculation with the phase compositions. The aim of the laboratory tests is to verify the results of the calculations and to specify the reaction kinetics of the process.

Analysis of the laboratory test samples and conditions, calculation of the results and comparison of the test results to the original plan of the goal setting require the same kind of round of evaluation, as for the goal setting and for the laboratory test programme. The results from the laboratory tests and theoretical calculations must also certainly be compared to all the available know-how and to the experience of industrial operations.

Pilot Tests

If the critical examinations on the basis of laboratory tests still support the original goal setting and the project plan, the development of the process can be continued with pilot trials to confirm the reliability of the process operation and to specify the design parameters of the process (Fig 6).

On the basis of the previous analyses, calculations and estimates the process and process equipment can be dimensioned and designed for the pilot trials.

The scale and extent of the pilot operations depend on the character and the requirements of the process under development. Some processes can be developed phase by phase with equipment of some ten liters or kilos, others, for example high temperature processes, must be tested continuously with all the process steps at the same time.

A process can also be tested in the by-stream connected to an operating industrial plant (Figs 7 - 9, Tables II - III and Figs 10 - 16, Table IV).

The duration of a test campaign can vary from one week to a couple of months depending how much knowledge and experience there is available from similar process types.

If it is necessary to clarify by the pilot runs the recycling of liquids, solids or gases, the distribution of trace elements and characterization of saturation limits, test trials will normally take longer.

OUTOKUMPU RESEARCH HAS A BROAD RANGE
OF ANALYTICAL TECHNIQUES IN USE

* CONVENTIONAL WET CHEMICAL ANALYSIS
* ATOMIC ABSORPTION SPECTROSCOPY (AAS)
* X-RAY FLUORESCENCE SPECTROSCOPY (XRF)
* OPTICAL EMISSION SPECTROSCOPY (OES)
* ION AND GAS CHROMATOGRAPHY
* FOURIER TRANSFORM INFRARED SPECTROSCOPY (FTIR)
* OPTICAL MICROSCOPY
* ELECTRON PROBE MICROANALYSIS (EPA)
* X-RAY DIFFRACTION (XRF)
* SCANNING ELECTRON MICROSCOPY (SEM)
* UV- AND IR-SPECTROSCOPY
* THERMOGRAVIMETRY (TG)
* DIFFERENTIAL THERMAL ANALYSIS (DTA)
* THERMODYNAMIC CALCULATING MODELS

Figure 2 - Apparatus and Techniques Needed for Research and Development of New Processes

Table I An Example of Chemical Analysis

Chemical Analyses of Metal containing Wastes

		Fe %	Zn %	Pb %	Cd %	Ni %	Mn %	Mo %	Cu %	Mg %	Al %	Na %	K %
1. (EAF + AOD)-dust	25.6.90	27,7	9,4	0,15	0,029	2,3	2,5	0,34	0,27	2,0	0,29	0,72	1,3
2. Neutralization sludge	9.8.90	8,5	0,007	<0,05	<1 ppm	1,2	0,28	0,16	0,05	0,46	0,17	0,96	0,03
3. Regeneration sludge	9.8.90	6,2	0,003	<0,05	<1 ppm	0,49	0,01	0,04	0,02	0,32	0,39	0,20	0,20
4. Waste water precipitate	9.8.90	9,9	0,65	<0,05	3 ppm	0,09	0,20	0,0006	0,006	10,0	0,56	0,18	0,21

		Ca %	Ti %	V %	Cr %	Cr- %	Si %	F %	Cl %	S %	C %	Moisture %
1. (EAF + AOD)-dust	25.6.90	9,4	0,055	0,018	11,8	0,07	2,6	0,10	0,21	0,30	0,38	0,20
2. Neutralization sludge	9.8.90	21,2	0,01	0,02	5,4	54 ppm	1,3	2,3	28 ppm	6,8	0,65	7,4
3. Regeneration sludge	9.8.90	18,9	0,01	0,005	0,96	34 ppm	1,6	1,7	66 ppm	17,1	0,15	4,8
4. Waste water precipitate	9.8.90	11,0	0,09	0,04	14,4	4 ppm	8,2	0,03	120 ppm	0,33	11,8	1,0

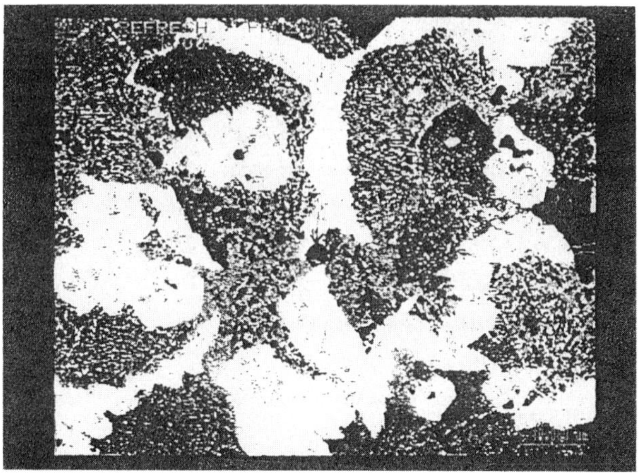

SEM-scanning picture phase distribution

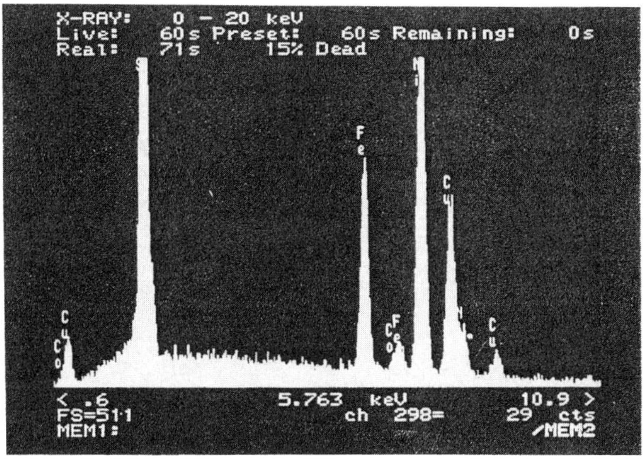

ED-spectrum pentlandite

20.00 kv TILT= .00 ELEV=40.00 AZIM= .00 COSINE=1.000

ED-spectrum pentlandite Ni FSF matte

All elmts analysed. NORMALISED

ELMT	ZAF	%ELMT	ATOM.%
FeK : 1	.987	31.323	25.604
NiK : 1	.910	33.162	25.786
CoK : 1	.915	1.801	1.395
CuK : 1	.874	1.126	.809
S K : 1	.919	32.589	46.406
TOTAL		100.000	100.000

Figure 3 - Quantative Analysis Calculated Using Spectrum

Problem:

At which temperature should $BaSO_4$ be reduced to get BaS as a product, and what is the composition of following species in the gas and solid phases as a function of the temperature:

Ba(g), BaO(g), CO(g), CO_2(g), COS(g), CS(g), CS_2(g), O_2(g), S(g), S_2(g), S_3(g), S_4(g), S_5(g), S_6(g), S_7(g), S_8(g), SO_2(g), SO_3(g), H_2(g), H_2O(g), H_2S(g), BaC_2, $BaCO_3$, BaO, BaO_2, Ba_2O, Ba, $BaSO_4$, C

SOLUTION 1 - MANUAL CALCULATIONS

- **There are 29 species in the system and therefore as many independent reaction** equations and mass balances must be derived
- Search data for all these species from the thermodynamical data books
- Solve the equations using matrix operations.

This method takes several days for the professional to find the compositions as a function of temperature.

SOLUTION 2 - LABORATORY EXPERIMENTS

- Order raw materials
- Construct the laboratory apparatus
- Carry out the experiments at several temperatures
- Analyse the products.

This method takes several weeks.

SOLUTION 3 - HSC CALCULATIONS

- Give elements of the system to HSC program:
  Ba, S, O, C, H
- Give amounts of raw materials to HSC program:

$BaSO_4$	1	mol
C	1	mol
H_2O	1	mol

- Give temperatures for calculations:

Min	0	C
Step	50	C
Max	1200	C

- Run the program to get the results.

For the normal user, these steps can be carried out in 15 minutes

Figure 4 - Simple Example of the HSC Calculations

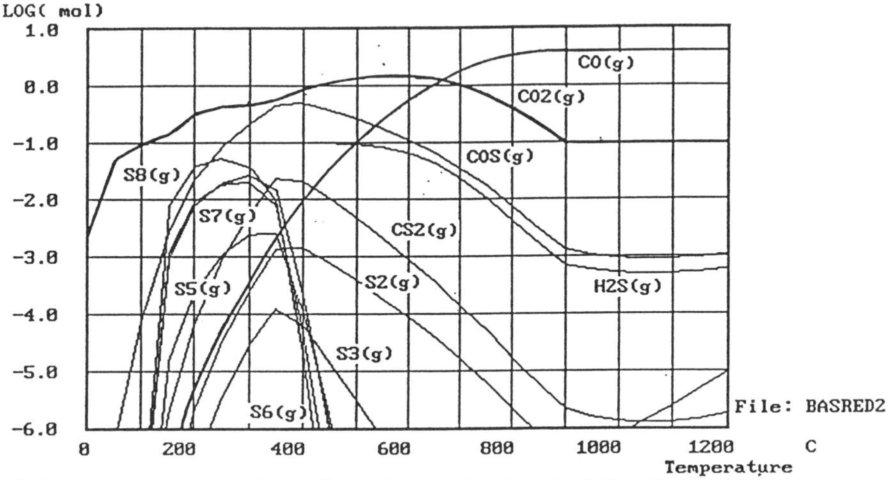

BaSO4 reduction with C (BaSO4 = 1 mol; C = 4 mol; H2O = 0.1 mol)

COMPOSITION OF THE GAS PHASE

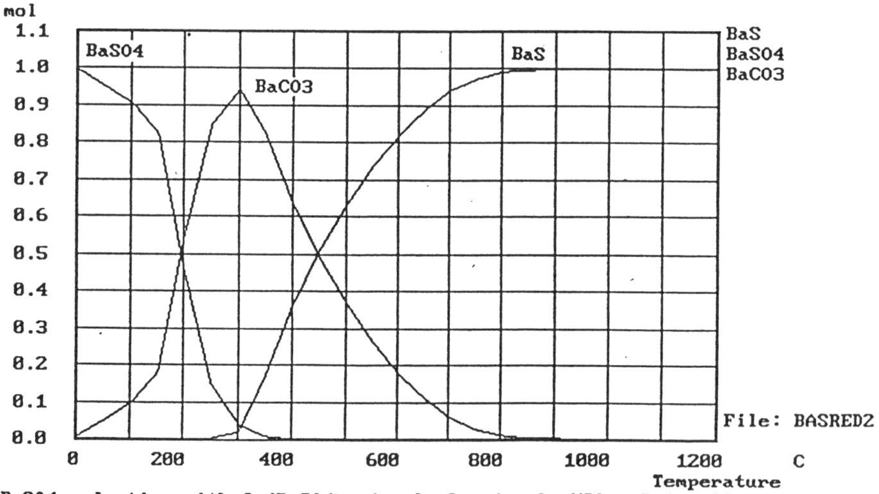

BaSO4 reduction with C (BaSO4 = 1 mol; C = 4 mol; H2O = 0.1 mol)

COMPOSITION OF THE SOLID PHASE

Figure 5 - $BaSO_4$ Reduction with Coal

Figure 6 - Pilot Rotary Kiln for NiCd-Process

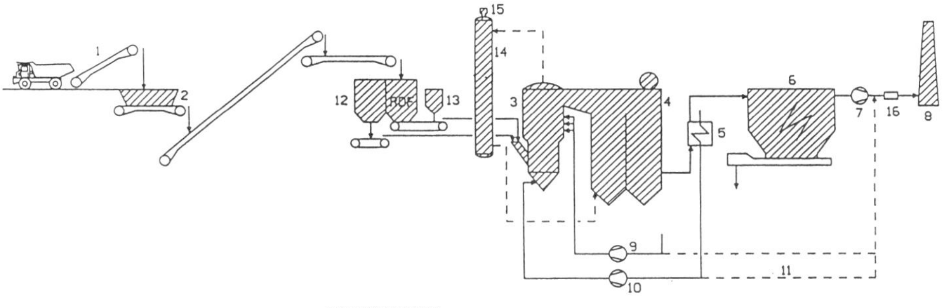

1. FEED CONVEYOR
3. FLUIDIZED BED FURNACE
4. STEAM BOILER
5. AIR PREHEATER
6. ELECTRIC PRECIPITATOR
7. WASTE GAS FAN
8. STACK
9. SECONDARY AIR FAN
10. PRIMARY AIR FAN
12. COAL BIN
13. FEED SYSTEMS FOR LIMESTONE
14. INCINERATOR FOR BYPASS
15. OIL BURNER

Figure 7 - Flowsheet of Waste Burning Tests

COMPOSITION OF THE MUNICIPAL WASTE

Carbon	%	46.2 -49.1
Hydrogen	%	6.5 - 6.8
Nitrogen	%	0.6 - 1.3
Oxygen	%	30.4 - 33.8
Zinc	g/t	100 - 300
Lead	g/t	50 - 100
Cadmium	g/t	0.5 - 9.9
Mercury	g/t	0.2 - 0.82
Chlor	%	0.27 - 1.0
Sulphur	%	0.1 - 0.5

WASTE AFTER PRETREATMENT (RDF-fraction)

16 w-% plastic
13 w-% pasteboard
31 w-% paper
40 w-% others

Figure 8 - Composition of Wastes

Table II Comparison of Ashes Produced in By-stream Combustion Test and Industrial Operations

	Ash in Kokkola	Ash in Turku	Ash in Linköping	Ash in Avesta
Cu %	0.25	no analysis		
Ni %	0.012	no analysis		
Co %	0.008	no analysis		
Zn %	0.33	2.1-2.3	2.4	2.3
Cd %	0.003	0.034-0.035	0.04	0.04
Cr %	0.048	no analysis		
Pb %	0.16	0.7	0.65	1.3
As %	0.010	no analysis		
Mn %	0.13	no analysis		
Ba %	0.34	no analysis		
Hg %	0.0004	0.0020-0.0022	0.0006	0.0007
Cl^- %	0.7	6.1-8.8		
F^- %	0.07	no analysis		

Table III Reduction of the Organic Compounds in the Postcombustion Chamber

		Before postcombustion	After postcombustion
Σ PAH	μg/Nm^3	2.97	0.74
Chlorobenzenes	μg/Nm^3	26.97	5.47
PCB	μg/Nm^3	0.23	0
Chlorophenols	μg/Nm^3	10.96	8.39
Σ PCDD & PCDF (EADON)	ng/Nm^3	34.74	4.35

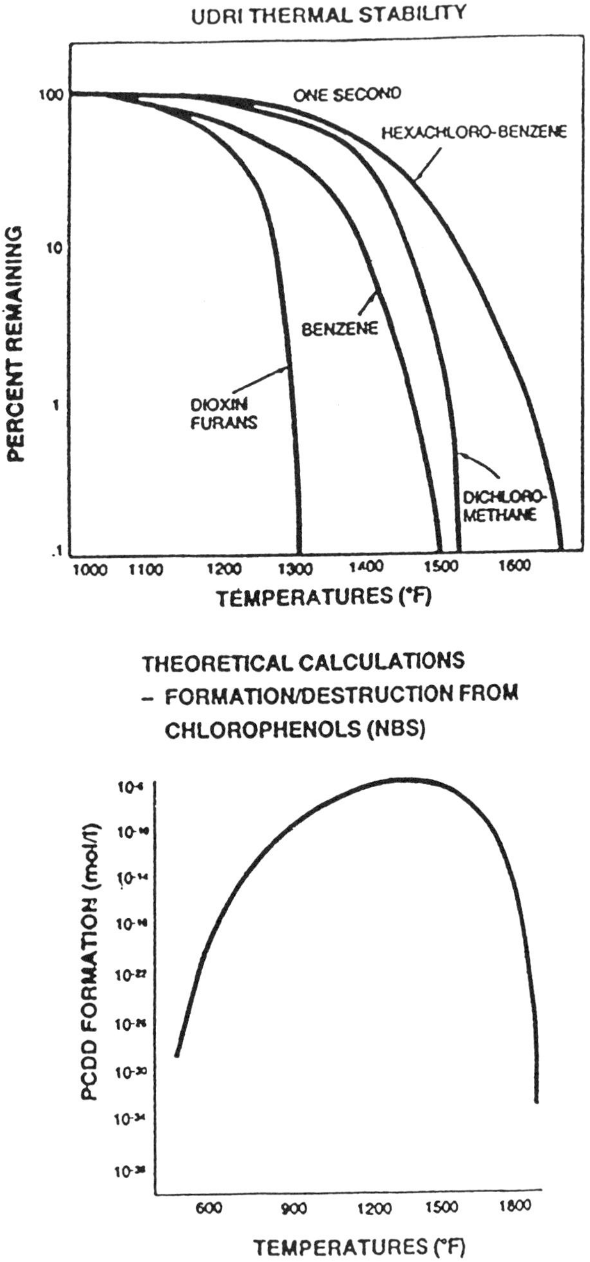

Figure 9 - Thermal Decomposition Characteristics of Selected Hazardous Hydrocarbons

In the tests carried out in the pilot plant, washing was done with an OK scrubber using an alkaline solution and synthetic gas as an additive (500 Nm^3/h). The gas analysis fluctuated as follows:

HCl	381	-	2181 mg/Nm^3
SO_2	576	-	1499 mg/Nm^3
Dust	291	-	1508 mg/Nm^3
Hg	0.130	-	0.205 mg/Nm^3
Pb	2.796	-	11.239 mg/Nm^3
Cd	0.247	-	2.033 mg/Nm^3
Zn	26	-	155 mg/Nm^3

The contents of the emission gases fluctuated as follows:

HCl	3	-	27 mg/Nm^3
SO_2	11	-	96 mg/Nm^3
Dust	42	-	62 mg/Nm^3

The best separating grades for heavy metals were as follows:

Cd	> 90 %
Pb	> 85 %
Zn	> 90 %
Hg	> 60 %

Precipitation tests were carried out on the exhaust sludge using mixed precipitation, in which the Cd content of the solution decreased to below 0.2 mg/l, Pb below 1 mg/l, Zn below 0.3 mg/l and Hg below 0.1 mg/l.

An example of municipal waste combustion plant gas cleaning is shown in the flowsheet.

In other tests, lime milk has been used for neutralizing, which can be used as long as it is taken into consideration when choosing the washing pH range and pumps.

Figure 10 -Gas Cleaning of Municipal Waste Combustion Plant

Table IV Gas Before Cleaning

CO2	**8.4 %**
O2	**10.0 %**
N2	**69.1 %**
H2O	**12.5 %**
HCl	**376 mg/Nm3**
SO2	**297 mg/Nm3**
Hg	**0.157 mg/Nm3**
Dust	**122 mg/Nm3**
- Cd	**0.08 mg/Nm3**
- Pb	**1.4 mg/Nm3**
- Zn	**18.33 mg/Nm3**

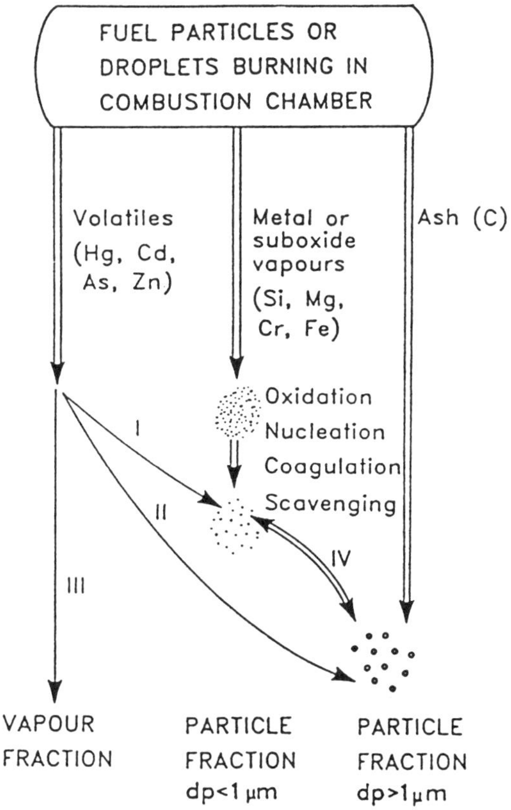

Figure 11 - Shematic Diagram of the Formation of Heavy Metal Aerosol after Combustion of Fossil Fuel or Waste

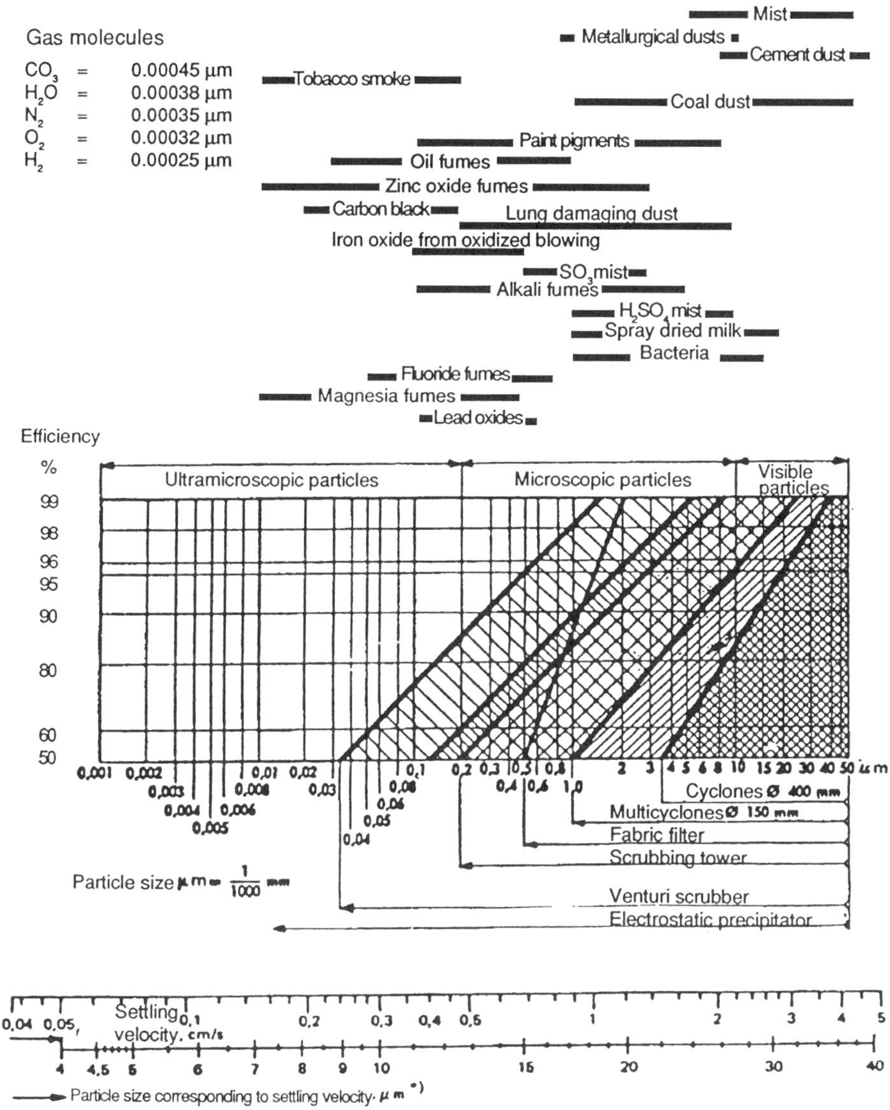

Figure 12 -Types of Collectors for Various Constituents

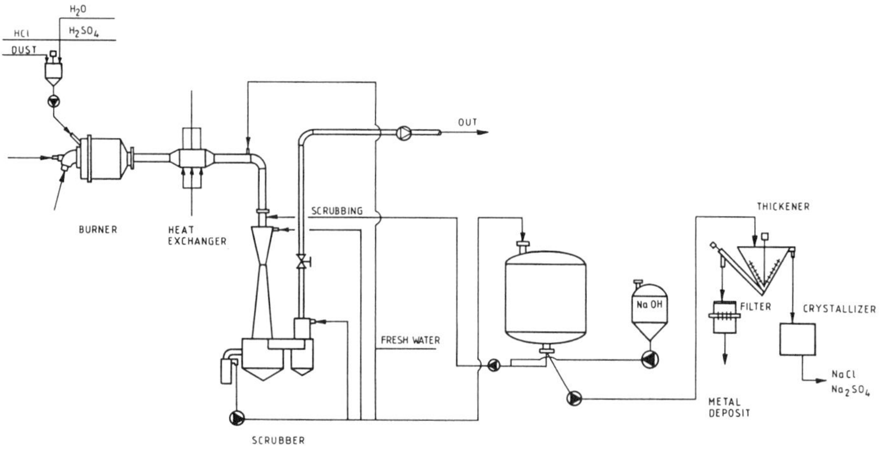

Figure 13 -Gas Scrubbing of the Waste Incinerator

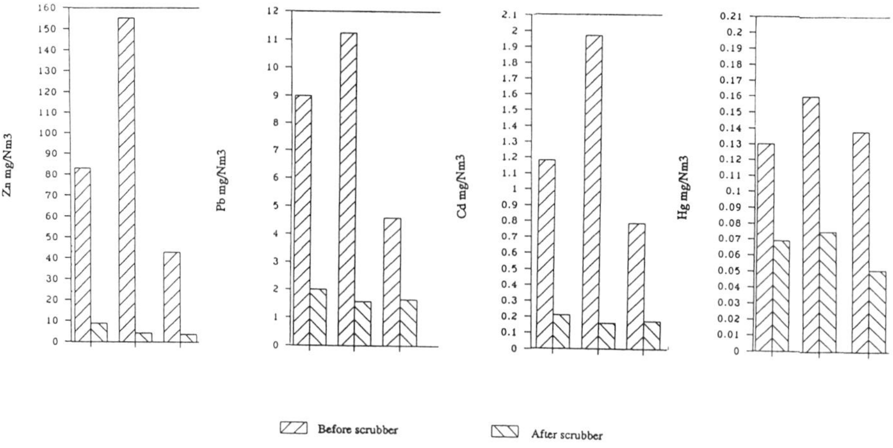

Figure 14 -Typical Recoveries of Heavy Metals Using OK-scrubber

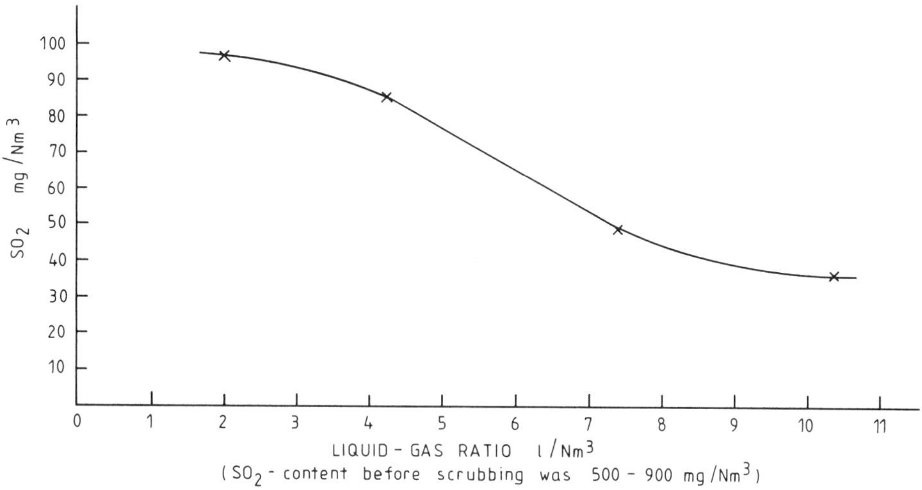

Figure 15 -SO_2 after Scrubber as Function of the Liquid - Gas Ratio

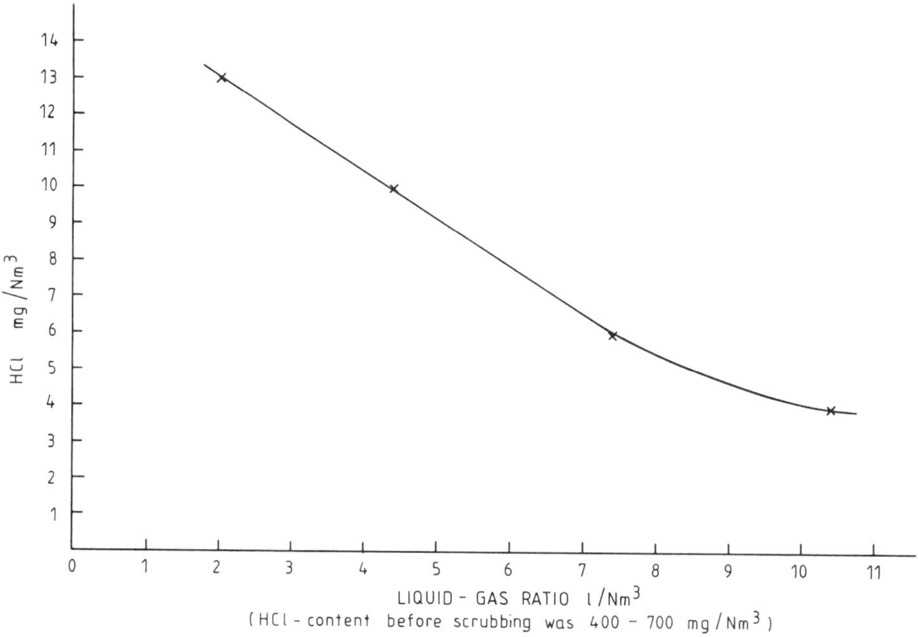

Figure 16 -HCl after Scrubber as Function of the Liquid - Gas Ratio

Modelling of Test Results for Process Design

The basic parameters for process design are obtained from the laboratory and pilot tests. Adapting the test results for production scale requires not only wide experience but also the help of process modelling.

Modelling work can be done either by mathematical or experimental models.

The purpose of the model tests is to analyze a physical phenomenon, most often the flow of a liquid, gas or solid material stream by simulating the process itself. The phenomenon concerned is created by substituting flow media, materials and test equipment, which represent the conditions of the process most closely. These substitute media, model materials and often also the scale of the model are estimated using calculation formulas based on model theory.

Simulation with a mathematical model works principally in a similar way. Difference between physical and mathematical modelling is that in the mathematical models, the coefficients of the interacting equations are specified by theoretical examination based on process results and the empirical backround. However, this kind of knowledge is seldom available adequately, especially when development of a more complicated process is concerned. On the other hand, mathematical modelling can be used in real process conditions.

In physical modelling the model itself represents the computer and the action of the phenomenon solve the definition of the coefficients. The physical model is most often made of transparent plastic. The scale and the flow conditions are specified on the basis of model theory. In brief, this theory is the adjusting of the essential correlations and getting these correlations in a formula of non-dimensional figures and then arranging the conditions so that these figures are kept constant both in the model and the process.

The models are most frequently two-dimensional water models (Fig 17) or three-dimensional water or gas models (Fig 18). Selection of the type of model depends on the character and on the scope of the experiments. To implement the model theory requires wide background and experience.

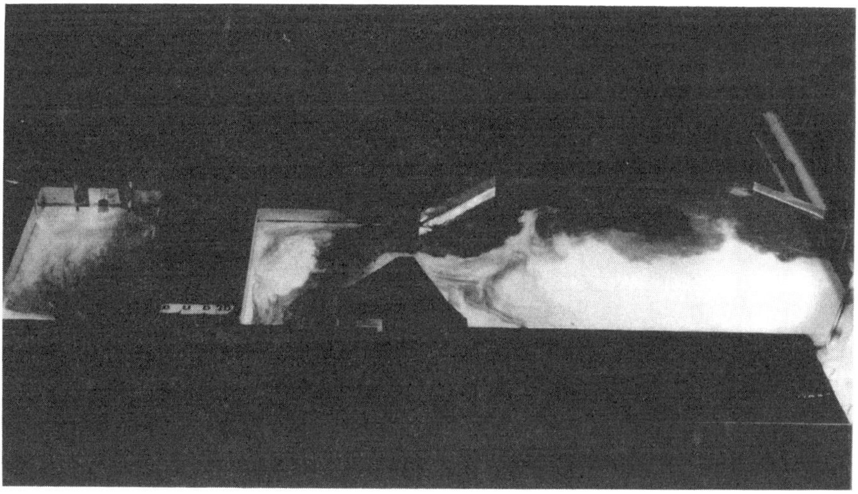

Figure 17 - Two-Dimensional Model

Figure 18 -Three-Dimensional Model

Processes for Waste Minimization

When talking about processes with low emissions one cannot avoid taking into account the blast furnace process, which has become almost wasteless through long tradition and hard competition.

Another similar process is autogenous flash smelting developed to beneficiate sulfidic ores. This process mainly uses its own energy for smelting and also utilizes the sulphur dioxide contents of the off-gases for sulphuric acid or elemental sulphur. The energy of the off-gases is generated as steam. Slag from the smelting process can be utilized as blasting grit and as a raw material for house and roadbuilding. Dusts are normally recycled back into the process.

Flash smelting has been developed and also used for the production of sulphur and sulphuric acid from pyrites. As a by-product energy is beneficiated for power generation and iron oxide is used for steel production.

Sulfred

Based on the knowledge obtained from these processes the Sulfred process has also been developed (Fig 19). Sulfred is a desulfurization process. In this process the sulphur dioxide containing waste gas is treated with a solution containing sodium sulphide. The chemical reactions which take place in the scrubber are completed in the autoclave producing elemental sulphur in molten form and sodium sulphate solution. This solution is regenerated and recirculated into the scrubber. The process is regenerable, so that instead of waste it produces elemental sulphur.

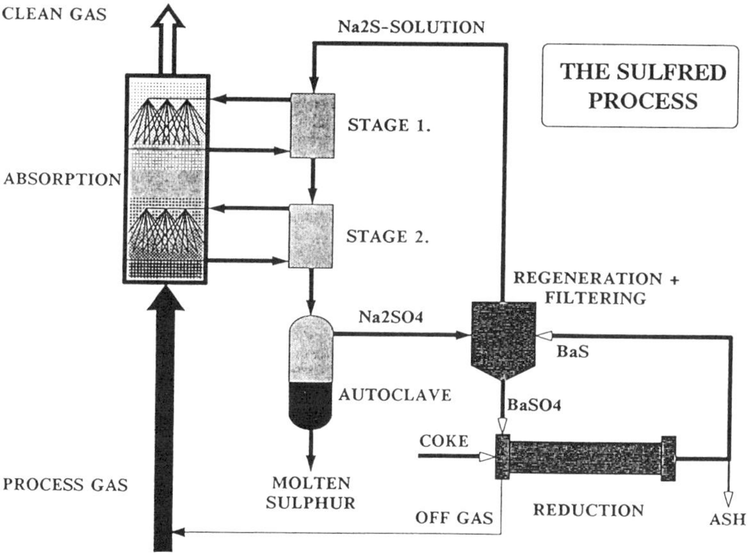

Figure 19 -Flowsheet of the Outokumpu SULFRED Process

Treatment of the Jarosite

Flash smelting can also be adapted for the treatment of sulfate containing residues like jarosite, gypsum, ferrosulphate, (Fig 20). For instance, the dried jarosite and coal powder, gas or are fed into the reaction shaft of the flash smelting furnace and smelted. Smelting of jarosite is highly endothermic. Energy has to be consumed in order to reduce sulphur back to SO_2 and iron to FeO and Fe_3O_4. Slag containing these oxides is insoluble in water and does not pollute the environment. More than 90 % of heavy metals will be turned into dust in this process and the dust will be recycled back to the original metallurgical process. Products from the process are sulphuric acid or elemental sulphur, slag for dams and roadbuilding and heavy metals for further refining.

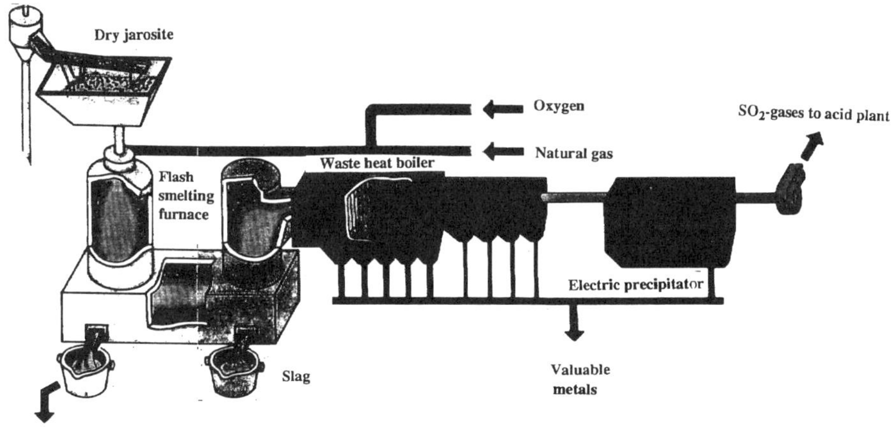

Figure 20 -Flash Smelting Process of Jarosite

Production of the Ferrochromium

The production of ferrochromium in a closed electric furnace is also one typical nonwaste process. In this process, the raw materials and the energy are utilized totally. The carbon monoxide generated in the electric smelting is used for charge preparation in sintering and preheating of the furnace charge and the rest of it is used for other heating purposes in the plant. Metal droplets are separated from the slag and the slag is used for the production of heatresistant castables and for house and roadbuilding materials (Fig 21). The utilization of the furnace gas has required development for an effective gas cleaning process and equipment, the OK-scrubber.

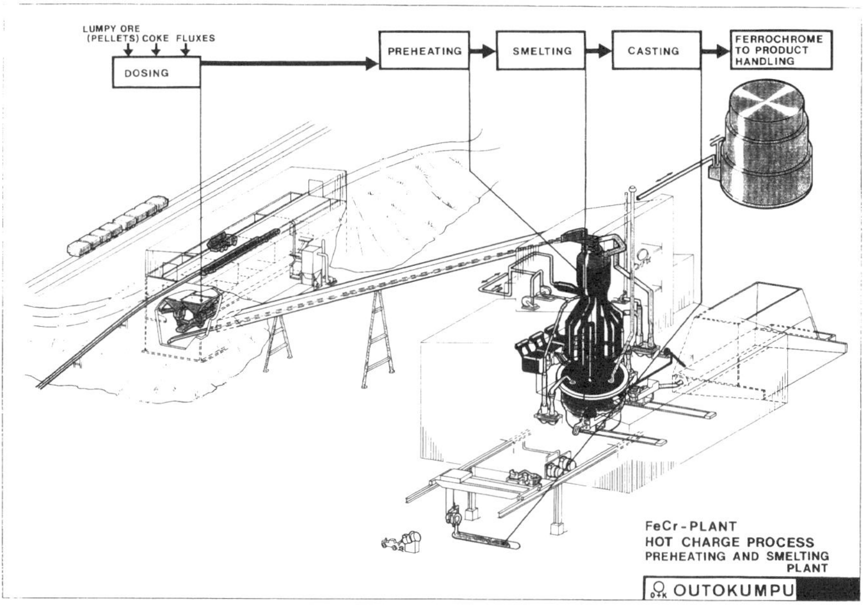

Figure 21.-FeCr Plant Hot Charge Process Preheating and Smelting Plant

Steel Industry Wastes

A process has also been developed for dusts, slurries and sludges from the steel industry. These wastes contain both valuable metals as nickel, chromium and iron and harmful metals as zinc, lead and hexavalent chromium. These wastes are smelted in the electric furnace after briquetting and drying (Fig. 22). This process produces metal containing nickel, molybdenium, chromium and iron and dusts containing lead and zinc can be recycled to the metallurgical industry.

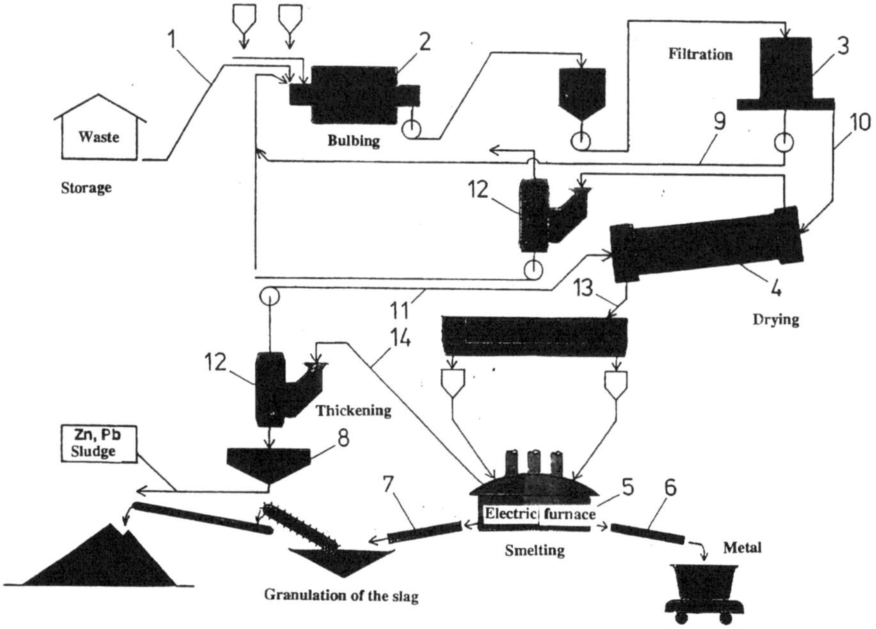

Figure 22 -Treatment of Metal Containing Dusts and Wastes

Batteries, scraps and sediments

The rotary kiln process has been developed for the treatment of hazardous batteries as raw material for both stainless steel and cadmium chemicals (Fig 23). Dried NiCd- sediment and crushed scrap are fed into the rotary kiln and reduced using hydrogen. The process produces ferronickel and evaporated cadmium is condensed and other heavy metals are removed using acid solution scrubbing.

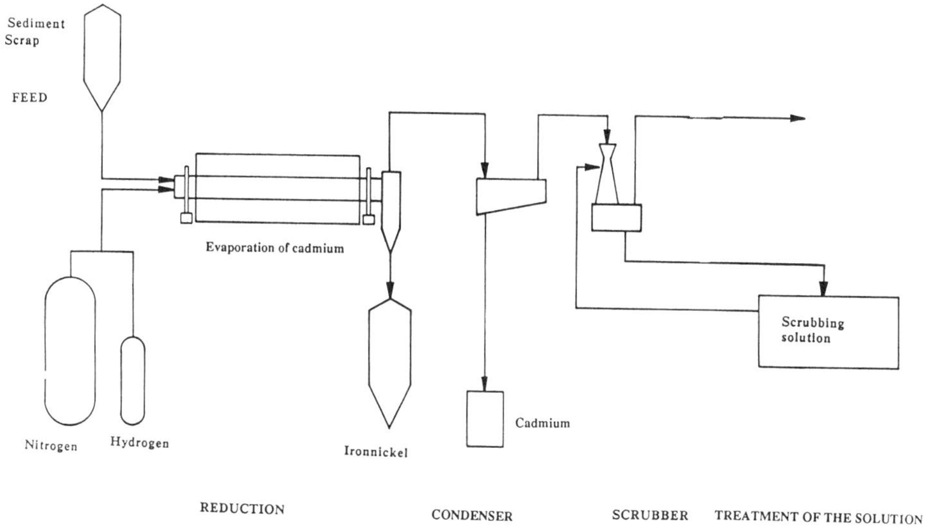

Figure 23 -Process for Treatment of the NiCd Sediment and -Scrap

Municipal wastes

Waste management is a growing urban problem and which is why the biothermal treatment process for municipal wastes has been developed. Waste is pretreated so that organic substances (compost fraction) and combustible subtances (RDF) are separated. Iron scrap is also removed at this stage. The compost fraction is biologically decomposed (the Wabio process , Fig 24), and organic soil and bio gases are gained as final products. Remaining combustibles are fed to a fluidized-bed combustion unit which generates the useful heat for steam production and leaves only inert ash for disposal. Waste gases are cleaned using a high-velocity OK-scrubber, with sodium hydroxide solution and with some additives. Heavy metals are precipitated from the scrubbing solution. The heavy metal precipitate can be recycled into the metallurgical process.

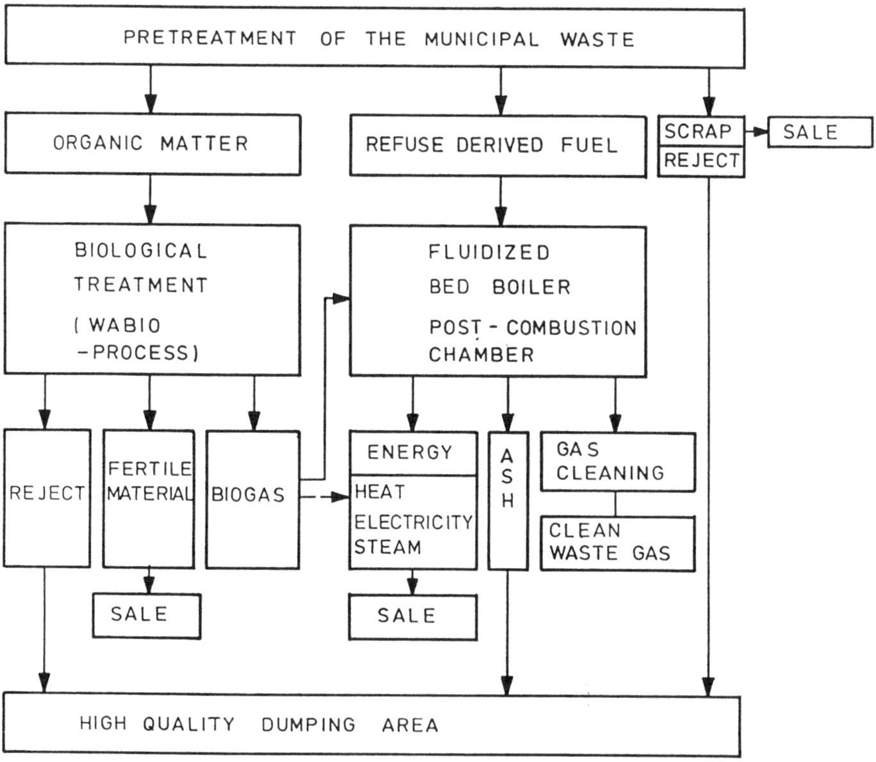

Figure 24 -Biothermal Waste Treatment Process

Conclusion

The technologies used in mining, ore dressing and base metal production, flash smelting, fluidized bed roasting, electric smelting etc. can be modified for different kinds of waste handling such as power generation in fluidized bed boilers, for hazardous and municipal waste incineration and for the beneficiation of slags, dusts, slurries, sludges and various kinds of residues from industry and municipal activities. However, extended and exact research and development technology is needed for this work.

REFERENCES

1. W. R. Seeker et al., "Municipal waste combustion study. Combustion control of organic emissions" (Energy and Enviromental Research Corporation for USEPA, June 1987)

2. K. Larjava, J. Reith and D. Klockow, "Development and laboratory investigations of a denuder sampling system for the determination of heavy metal species in flue gases at elevated temperatures", Intern. J. Environ. Anal. Chem. Vol.38 1989

3. Arthur C. Stern, "Air Pollutants, Their Transformation and Transport", Air Pollution, third edition, vol. 1 1976

4. T. Hyödynmaa, "Esikäsitellyn yhdyskuntajätteen polttokokeilu Outokumpu Oy:n Kokkolan tehtaiden leijupetiuunissa" (Interim report, 1990)

5. M-L Metsärinta et al., "Yhdyskuntajätteenpolttolaitoksen kaasunpesu" (Interim report, 1990)

6. A. Roine, T. Talonen and M-L. Metsärinta, "SO_2 removal from concentrated process gases using Sulfred process" (COPPER 91 Canada 1991)

7. H. Tuovinen, M. Honkaniemi et al., "Recent and future developments in ferroalloy production technology" (Paper for the ILAFA Ferroalloy Congress, Rio Janeiro, 1982)

8. P. Hokkanen et al., "The production of charge chrome using pelletized feed" (Paper for the ILAFA Ferroalloy Congress, Acapulco, 1978)

9. M. Jormanainen, "Refuse Derived Fuels, Fuel Preparation and Compustion in Fluidised Boilers" (Technical Research Centre of Finland, Electrification and power generation, 1989)

10. Talonen et al., "Method for Removing the Sulphur Content of a Weak Gas Containing Sulfurdioxide" US patent 4.861.577

11. Talonen et al., "Method for Removing Sulphur in Elemental Form from Gases Containing SO_2 or SO_2 and H_2S" US Patent 4.937.057

12. Tuovinen et al., "Menetelmä metallurgisessa teollisuudessa syntyvien metallipitoisten pölyjen ja jäteaineiden käsittelemiseksi" SF Patent 58349

13. Tuovinen et al., "Method and apparatus for the purification of gases which contain solid and gaseous impurities" US Patent 4.469.493

14. S. Fugleberg,"Menetelmä metallurgisten jätesakkojen tai pölyjen metallisisällön talteenottamiseksi suspensiosulatusmenetelmällä" FI patent application 910692

15. J. Asteljoki, J. Laulumaa, and L. Lilja, "Tapa ja laitteisto pulverimaisen kiintoaineen kuumentamiseksi ja sulattamiseksi sekä siinä olevien haihtuvien aineosasten haihduttamiseksi suspensiosulatusuunissa" FI patent application 910690

ECONOMIC IMPACT OF TREATMENT OF RESIDUES AND EFFLUENTS ON INVESTMENT DECISIONS

By J.C. Agarwal, F.E. Katrak, M.J. Loreth

Charles River Associates Incorporated

Most metals are produced from minerals containing one or more of the desired metals, some contaminating metals, silicate, oxides, carbonates, and sulfides. Therefore, in the production of metals, there is generally a gaseous effluent in pyrometallurgical processes and a liquid effluent in hydrometallurgical processes. There are also solid residues such as waste ore, tailings, dust, and slag, created by the metal recovery processes. There are two important distinctions between the effluents and the residues:

1. Effluents — gases and liquids — affect the neighboring environment for miles from the point of emission, while residues affect only the soil on which they are stored or dumped.

2. The environmental and economic effects of effluents are known and felt essentially immediately. By contrast, the environmental effects of the residues may not be known for a long time.

For these reasons, the economic impact and the potential corrective actions required are very different for effluents and residues.

RESIDUES

Residues, which we discuss in turn, can be classified as follows:

1. Waste (low-grade ores) and overburden. These are generated during mining.
2. Tailings and red mud from bauxite. These are generated during concentration or purification.
3. Slags, dusts, and pot liners. These are generated during smelting operations.

Residues and Effluents - Processing and Environmental Considerations
Edited by Ramana G. Reddy, William P. Imrie and Paul B. Queneau
The Minerals, Metals & Materials Society, 1991

Waste and Overburden

Mined minerals contain ore, waste (low-grade ores), and overburden. Part of the overburden (mostly topsoil) is saved and used to restore the mined-out operation to an environmentally acceptable state and replanted with the vegetation compatible with its location. Most of these costs are incurred on an ongoing basis.

Large quantities of ores, wastes, and overburden are generated in the minerals industry of the Western world (Table 1). Of nearly 13 billion tonnes of mined materials, roughly one-half is waste and overburden. If only metal-bearing minerals are considered, approximately 5 billion tonnes of ore, waste, and overburden is moved per year; one-half of that is also waste and overburden.

As the waste and overburden are moved away to provide access to the ore, there is a disturbance of the environment. Exploitation of minerals without any disturbance of the earth is essentially impossible. Therefore, the primary requirement is to manage the movement of solids from one area to another in an environmentally sound manner.

The lower-grade ores, which are found to be uneconomical to process, are considered waste. Throughout the history of mining, the definition of low-grade waste ore has changed. New technology has made possible the recovery of valuable mineral entities in waste ores. A good example is the copper industry, where the cutoff grade has steadily dropped. Today, many mines (especially open-pit mines) process ore that has less than 0.3 percent copper; the same ore would have been considered worthless not many years ago. Increasingly, even these wastes are piled into a valley and leached to extract additional amounts of copper by using solution-mining technologies that were unknown even 30 years ago.

The economic impact of recovering metal values in dump leaching operations is generally positive. In spite of the large amount of liquid handled because of the low copper concentration, the operating cost for recovery of copper from dump leaching is about one-half that of a conventional smelter operation.

In the group of metals that occur in nature in low concentration, but are dispersed in a large quantity of gangue, the future may be *in-situ* solution mining. This technology avoids the movement of large quantities of waste ore and overburden and is already proven for the extraction of uranium. Copper is the next best candidate metal for the use of this technology, because a large fraction of the total cost of producing copper is incurred in the handling of solids. With careful application of *in-situ* solution mining, many of the current environmental costs related to tailings disposal and gaseous effluents such as sulfur dioxide could be avoided.

Table 1. Ore and Waste Production by Metal/Mineral for the Western World (Million Tonnes)

Metal/Mineral	Ore	Waste and Overburden	Ore, Waste, and Overburden
Bauxite	80	25	105
Chromite	9	6	15
Coal	2,300	4,200	6,500
Copper	877	1,381	2,258
Gold	286	388	624
Gypsum	100	30	130
Iron Ore	896	75	200
Lead/Zinc	125	75	200
Lime	225	120	345
Manganese	30	15	45
Molybdenum	28	15	43
Nickel	32	16	48
Phosphate	254	365	619
Platinum	30	5	35
Potash	50	10	60
Salt	140	14	154
Silver	25	37	62
Uranium	25	75	100

SOURCE: Charles River Associates, 1990.

Tailings and Red Mud

The three metals with the largest tonnage and revenue — steel, aluminum, and copper — serve as illustrations of three different kinds of tailings. The tailings from steel are generated in the process of concentrating taconites. Tailings amount to roughly 2 to 3 tons per ton of steel produced in North America. In many parts of the world where iron ore quality is high, there are essentially no iron ore tailings generated by the steel industry. There are tailings generated with production of metallurgical coke used in the blast furnace, but the amount is relatively small — 0.10 ton per ton of steel.

In the aluminum industry, the production of red mud is a significant problem. Approximately 2.5 tons of red mud are generated per ton of aluminum. Despite major R&D efforts, there has been no success in finding an outlet for the red mud. The reasons are:

- The difficulty in dewatering red mud; and

- Alkali contamination of the red mud, rendering it unsuitable for most pyrometallurgical treatment.

Red mud can be considered highly contaminated iron ore and a byproduct of the preconcentration step in the production of aluminum. There have been many attempts to use bauxite ore directly in an attempt to avoid the production of red mud, but all of these attempts have proven uneconomical. Therefore, the current process technology for the production of aluminum will remain essentially the same Bayer-Hall process with some improvements.

In the production of copper, approximately 200 tons of tailings is generated per ton of copper, requiring large capital expenditures and electrical energy to grind the ore. Furthermore, a large land area is required to store tailings essentially indefinitely. Except for the possibility of using *in-situ* solution mining technology, the current process technology to beneficiate the ore to a high-grade concentrate will continue to be the dominant technology. The copper or other nonferrous metallic concentrate can then be shipped around the world to a smelter.

Slags, Dusts, or Pot Liners

Slags

Slags are essential to the production of metals for two major purposes:

1. To remove unreduced and *insoluble* impurities such as silica or iron oxide.
2. To partition and remove as much as possible of the contaminating impurities *dissolved* in the produced metal.

Most of the slag produced in the steel industry readily finds a use in the production of cement or as a filler in road beds. However, it is still economical to minimize the concentration of insoluble impurities that increase the quantity of slags produced. For example, the amount of coke used in the blast furnace has steadily declined by nearly 50 percent due to improvements in blast furnace technology. Also, it is no longer necessary to have a large slag volume to decrease the sulfur concentration in the blast furnace iron because of the almost universal adoption of external desulfurization of iron. The quantity of slag produced in the basic oxygen furnace has also declined because of the lower silicon and sulfur contents of the hot metal charged, thereby increasing the metallic yield.

The copper industry has successfully adopted oxygen smelting to cut costs. However, the costs of recovering copper values from the smelter slags (2 to 3 tons of slag per ton of copper) is still high, and requires improvements primarily because of the large investment required in either molten slag processing or grinding and flotation of granulated slag. The aluminum industry produces very little slag as such, because its feed to the smelting operation is essentially pure alumina produced in the Bayer process.

Dusts

There are three major reasons for the production of dusts in metallurgical processing:

- High gas velocity,
- Ultra-fines in the feed material, and
- Volatization of some constituents due to high smelting temperatures.

Generally it is quite economical to minimize the dust formation caused by excessive gas velocity or ultra fines in the feed material. The reason is the high cost of collecting the dust and then agglomerating it and recycling it. The costs can be decreased by making minor design changes in the smelter. However, dust generated by volatization and then condensation of some of the constituents, such as zinc fumes in electric arc steelmaking, would require a separate and expensive processing step to remove the offending volatile constituents, thereby enabling recycling of the captured dust.

Copper smelter dusts also face the same problem of contaminated smelter dust containing bismuth, indium, and arsenic. Even a large copper smelter does not generate enough quantities of the smelter dust to justify an on-site processing plant to recover various constituent metals or to render them environmentally safe. Currently the only economically viable solution appears to be a centralized processing facility with attendant problems of legal responsibilities.

Pot Liners

The aluminum industry faces the problem of the disposal of spent carbon pot liners in which the major culprit has been the cyanides. The presence of fluorides complicates the problem of disposal by simple combustion of the carbon pot liners because fluorides are highly volatile and must be captured from the combustion offgases. Transportation of spent pot liners to a centralized incineration facility could also be hazardous. The optimum solution would be to design a process that combusts the carbon in the spent pot liner at a high temperature, resulting in the formation of a basic slag to capture the fluorides.

EFFLUENTS

Gaseous Effluents

There are two major gaseous effluents of overall importance in the metals industry: sulfur dioxide and nitrogen oxides. Until recently, most of the effort has concentrated on sulfur dioxide abatement. With the passage of acid rain legislation, the control and abatement of NO_x will also become increasingly important. Of all the base metals of economic importance, copper dominates in tonnages of metal and sulfur.

It is instructive to examine the relative prices of three major base metals and the sulfur content per pound of metal in typical concentrates occurring in North America (Table 2).

Table 2. Economic Impact of Sulfur Removal

	Relative Price $/lb of Metal	Relative Sulfur Content, lb of Metal per lb of Sulfur	Revenue Affected, $/lb of Sulfur
Copper	1.00	1.00	1.00
Zinc	0.50	2.00	1.00
Lead	0.30	4.00	1.20

SOURCE: Charles River Associates, 1991.

The implication of Table 2 is that the economic impact of each pound of sulfur removal is relatively constant regardless of the base metal considered. However, the size of the copper industry suggests that it would continue to take the lead in the development of new technologies for decreasing the capital and operating costs of sulfur removal.

The single most important parameter for both capital and operating costs is the volume of hot and dirty gases handled in oxidizing the sulfur contained in the concentrate. For every mole of sulfur, one mole of oxygen will be needed. If no oxygen enrichment of air is used and no infiltration of air is considered, the theoretical amounts of offgases are as follows:

% Oxygen in Oxidizing Stream	Volume of Offgases per Mole of Sulfur Oxidized
21	5
33	3
50	2
66.6	1.5
100	1

Therefore, not only will the volume of offgases decrease substantially, but the usable thermal energy released per mole of sulfur will also increase, thereby increasing the smelting rate and decreasing the investment required.

Another potential advantage of using oxygen in smelting base metal concentrates is a marked increase in the concentration of sulfur dioxide in the offgases. Currently the sulfur dioxide concentration is limited to approximately 10 percent because most base metal smelters convert the sulfur dioxide to sulfuric acid. The 10 percent SO_2 concentration limit is imposed by the exothermic reaction for SO_2 to SO_3 and the temperature tolerance limit of the catalyst bed for the SO_2 conversion reactor.

Making sulfuric acid is not always economical. For example, in the western United States, there is a limited market for sulfuric acid, and the cost of transporting it to the available market for the production of fertilizer is very high.

With the possibility of producing smelter offgas with a higher concentration of SO_2 due to oxygen enrichment smelting, the liquefaction of SO_2 or the production of elemental sulfur could become economical. That process improvement would free North American producers of base metal from dependance on the local sulfuric acid market and permit them to become more competitive against foreign producers who have strong local markets.

There is ample activity in this direction, as evidenced by many new processes under development by companies such as Outokumpu, Noranda, Mt. Isa, and others. The steel industry has already adopted oxygen-based smelting in the changeover from open hearth to basic oxygen furnaces for steelmaking. The new trend in that industry is to develop a continuous direct steelmaking process, eliminating batch processing and thereby decreasing the offgas volume by preventing infiltration of ambient air.

The formation of NO_x in the offgases is directly dependent on the temperature, the concentration of nitrogen and oxygen, and the residence time at high temperatures. Current technology to reduce the oxides of nitrogen is overly expensive. The possible solution lies in substantially decreasing the residence time, or the gas temperature, or both simultaneously, thereby minimizing the formation of NO_x.

Liquid Effluents

Process liquid effluents are not a major *economic* problem in the production of metals. The control of contamination by the runoff water is more of a problem in the mining industry. Neutralization and other chemical treatment to precipitate out the dissolved metals is now practiced almost universally. The use of biological plants which are at home in the prevailing environment to take up the dissolved elements shows promise.

ECONOMIC IMPACTS ON INVESTMENT DECISIONS

The economic impact of controlling the gaseous effluents in the metals industry has been positive in general, although it was first thought to be negative. By introducing oxygen smelting, the industry achieved significant improvements in productivity and operating costs, while controlling the gaseous effluents.

Gaseous pollutants are highly visible and act on the environment at relatively fast rates. The liquid effluents affect the environment at a slower rate and can be controlled if the plant is designed and operated to minimize the discharge of liquids.

For the solids — tailings, dust, and slag — the economic impact is in the future. Moreover, the future changes in regulation might render the currently proposed solution obsolete or unacceptable. For old operations, the size of the solids stockpile is generally unknown. Therefore, it is difficult to design a process to make the solids environmentally safe. However, careful segregation of solids that are nonhazardous from those that might be hazardous would certainly minimize future costs.

If one of the options in considering the fate of an old operation is to close it, then the net present value of the environmental liability is very high, because of the immediate cash outlay to clean up the site. However, if the business is inherently sound and can be operated at modest outflow of cash (including the net present value of increased pension and other liabilities), continued operations might make economic sense. For example, if contaminated solids on a site resulted in environmental closing costs of $100 million, continued operation for seven more years at a break-even cost and at a real discount rate of 10 percent would decrease net present value of those costs to $50 million while saving the plant and its staff for a future turnaround in the business. Success hinges on the outlook for the business, which must be sound enough.

COMPREHENSIVE WATER MANAGEMENT PROGRAM

FOR A PRIMARY COPPER SMELTER

Christine P. Viecelli
Environmental and Safety Supervisor
Phelps Dodge Corporation, Hidalgo Smelter
Playas, New Mexico 88009

Abstract

As Phelps Dodge Corporation addresses the environmental challenges of the 1990's, water management has become a key element in an overall waste minimization strategy. The Phelps Dodge Hidalgo copper smelter has implemented a comprehensive water management program that includes management of both fresh and process water utilized by its pyrometallurgical operation. Two major technological strategies are being implemented. They are the design, construction and operation of a raw water pre-treatment facility to allow increased evaporation within cooling towers thereby minimizing blow down, and the use of a brine concentrator to treat and recycle significant quantities of process water. The end result of this program is a significant reduction in raw water pumped from the underlying aquifer into the facility as well as a decrease in process water which must be impounded or otherwise treated.

Residues and Effluents - Processing and Environmental Considerations
Edited by Ramana G. Reddy, William P. Imrie and Paul B. Queneau
The Minerals, Metals & Materials Society, 1991

Introduction

Within pyrometallurgical operations, water serves as a carrier, a coolant and a solvent. Water is a truly universal resource. It is utilized as seal water in crushers and pumps, for dust suppression and gas scrubbing, in low grade heat extraction from furnace cooling elements and acid plant coolers. It is also used in the process to form sulfuric acid and to cool anode copper, and in energy production as it absorbs process heat within high pressure boilers to form steam which drives turbines to generate electricity. More conventional uses include potable water and sewage disposal. Each of these operations introduce different impurities into the water which must be controlled to prevent equipment fouling, product contamination or environmental concerns. This paper provides an overview of the strategies surveyed and the selection of a unique combination of processes to enhance the Hidalgo smelter waste minimization program as illustrated in Figure 1.

Background

Within pyrometallurgical operations, the story of water is often the story of heat transfer and gas cleaning. Within the closed loop of medium and high pressure boiler operations, make-up water requirements are small and can be economically provided by conventional softening, reverse osmosis and/or condensation. However, low grade heat extracted from furnace cooling elements and acid coolers is dissipated into the environment by evaporation through cooling towers. By nature of the evaporation process, a very pure water enters the environment as the heat dissipates from the system. However, remaining in the cooling water system are the bulk of the impurities: i.e. principally metals, carbonates and sulfates. This remaining solution must be removed or "blown down" from the system before any of the contaminant solubility limits are exceeded to prevent cooling tower or equipment fouling. This realization led to the Hidalgo smelter adopting in the 1980's, a multiple use water strategy which culminated in the following sequence: low grade heat rejection, process use and reclamation, electric furnace off-gas scrubbing, acid plant gas scrubbing and finally water impoundment. Under this scenario, however, the blowdown from the cooling towers remained in excess of 500 gpm.

The next challenge was to reduce the 500 gpm blowdown. A number of technologies were examined. These included precipitation, co-precipitation, chelation, electrolysis, ion exchange, reverse osmosis, crystallization and evaporation. Economics dictated that the appropriate treatment points would include pre-treatment to optimize cooling tower performance and a final concentration and evaporation step to handle low volume but high total dissolved solids (TDS) material.

This has now been effectively accomplished by softening raw water from the underground aquifer before it enters the smelter thereby allowing up to 12 cycles of concentration within the cooling tower . This compares favorably from a previous level of 4 cycles of concentration. The end result is impoundment of approximately 200 gpm or a 60% reduction in waste water at this stage. This water can then be used in the process as required or as an alternative, treated within a brine concentrator which returns 90-95% of the treated water to the softened raw water system or boiler feedwater . The remaining 5-10% constitutes a brine which is impounded in a state-of-the-art double lined pond approved by the state. The end result is the impoundment of a few gpm as opposed to hundreds of gpm.

The double lined pond is designed to exceed existing state environmental standards, primarily to meet any future requirements for groundwater

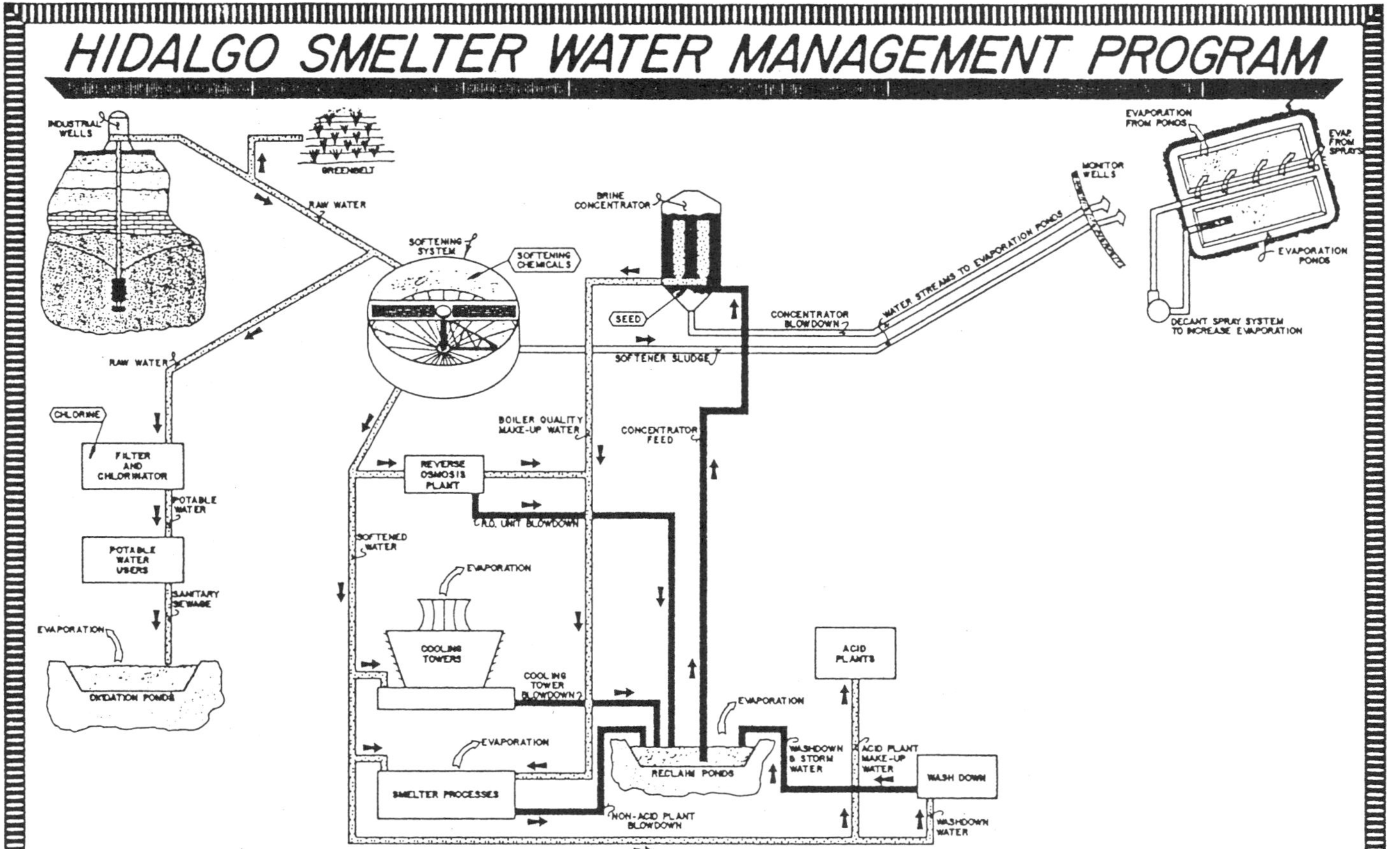
HIDALGO SMELTER WATER MANAGEMENT PROGRAM
INDUSTRIAL WELLS
GREENBELT
RAW WATER
RAW WATER
SOFTENING SYSTEM
SOFTENING CHEMICALS
BRINE CONCENTRATOR
SEED
CONCENTRATOR BLOWDOWN
SOFTENER SLUDGE
WATER STREAMS TO EVAPORATION PONDS
MONITOR WELLS
EVAPORATION FROM PONDS
EVAP FROM SPRAYS
EVAPORATION PONDS
DECANT SPRAY SYSTEM TO INCREASE EVAPORATION
CHLORINE
FILTER AND CHLORINATOR
POTABLE WATER
POTABLE WATER USERS
SANITARY SEWAGE
EVAPORATION
OXIDATION PONDS
BOILER QUALITY MAKE-UP WATER
CONCENTRATOR FEED
REVERSE OSMOSIS PLANT
R.O. UNIT BLOWDOWN
SOFTENED WATER
EVAPORATION
COOLING TOWERS
COOLING TOWER BLOWDOWN
EVAPORATION
SMELTER PROCESSES
NON-ACID PLANT BLOWDOWN
EVAPORATION
RECLAIM PONDS
WASHDOWN & STORM WATER
ACID PLANTS
ACID PLANT MAKE-UP WATER
WASH DOWN
WASHDOWN WATER

FIGURE 1:

protection. The pond lining consists of a four layer system using a 10 ounce Geotextile layer directly on the soil followed by a 60 mil liner, a geonet collection net, and an 80 mil top liner. All materials are high density polyethylene (HDPE). Between the two liners is a leachate collection system. Any drainage of pond water penetrating the top liner will be directed via the collection net to a center ditch and collected in a leak detection sump. Leachate can be then be pumped from the sump and returned to the pond. The pond is also equipped with a spray water system which pumps decant water from the pond into sprays to enhanced evaporation.

Operation

Raw water is pumped from the underlying aquifer through a system of eight industrial wells. The main impurity is silica which can be as high as 45 ppm. Because this impurity has a solubility limit of 176 ppm in the cooling towers, the cycles of concentration can be limited to as few as 3.9.

The treatment process begins with this raw water. First, a side stream is withdrawn for potable water use. It is filtered, chlorinated and pumped into various office and shop areas. A separate system collects the sanitary sewage and pumps it to a four cell oxidation pond for evaporation.

As required, a second side stream sends water to an agricultural green belt.

Pre-treatment

All remaining water pumped from the aquifer is treated to remove the silica. Water enters a 75 foot diameter Eimco reactor clarifier. Magnesium ions in the form of sulfate or chloride are added along with caustic and lime. The silica co-precipitates with the magnesium hydroxide; some calcium carbonate also precipitates. Settling is enhanced by the addition of a polymer. The amount of solids (sludge) in the reactor is monitored and controlled to ensure there is sufficient surface area for the precipitation reactions. The reactor clarifier overflow reports to a clearwell where the pH is adjusted with acid to prevent carbonate scaling in the cooling towers and process equipment. A million gallon tank stores the water until required by one of five process cooling towers. Turbidity, alkalinity and silica are monitored to ensure water quality.

Reactor clarifier underflow is pumped to a 35 foot diameter Eimco thickener with an operating capacity of 120,000 gallons. It is equipped with a lift motor, a torque alarm, and an automatic high torque cut-off. The thickener receives approximately 80 gpm of 0.5-3% solids underflow from the reactor clarifier and discharges 12 gpm underflow containing 7% solids to a double lined pond. The remaining 68 gpm of treated water combines with the reactor clarifier overflow entering the clearwell.

Boiler Make-up Water

A small quantity of treated water is processed through a reverse osmosis unit prior to entering the high pressure boiler feed water system. The two pass reverse osmosis system treats 125 gpm water at an electrical conductivity of 525 micromho/cm and produces 75 gpm deionized boiler feedwater with a conductivity < 1 micromho/cm. The 50 gpm blowdown is returned to the reclamation pond.

Cooling Tower System

The major water users within the smelter are the process cooling towers. Three large towers service the power plant, flash furnace, electric slag cleaning furnace and two acid plants. Two smaller towers support an oxygen plant and the anode casting operation.

Evaporation from this system is 1300 gpm which produces 370 gpm blowdown at 10-12 cycles of concentration. The towers operate on a molybdate corrosion treatment system with blowdown determined by a combination of silica concentration and carbonate/sulfate solubility.

Smelter Processes

The cooling tower blowdown is impounded in an intermediate lined "reclamation" pond which provides "surge" capacity for the various smelter operations. The primary user of this water is the electric slag cleaning furnace scrubber system. Off-gas (1600 ^{o}F) from the electric furnace is cooled by sprays in a packed bed quencher, cleaned in a venturi scrubber, and dewatered in a cyclone chamber with demister pads. The resulting solution is pumped directly to the acid plants where it is used again in a gas cooling and wet scrubber application. This constitutes a significant water bleed because the 500 ^{o}F gas that is cooled and is now water saturated enters the sulfuric acid plant drying tower where it becomes part of sulfuric acid production. The remaining solution from the scrubbing section is collected and sent to a treatment plant for further processing.

The last use for the cooling tower discharge is in the anode cooling operation where molten copper is poured into molds. A light water spray cools the copper below its melting point within 90 seconds so mechanical arms can remove the anode bars and place them into a cooling water tank. Once the bars are cool, they are extracted from the tank by crane and loaded on rail cars for shipment to the refinery. A closed loop recycle is maintained between the anode cooling system and a small cooling tower which blows down to the reclamation pond. Make-up water is supplied from the pre-treated raw water or the reclamation pond.

Brine Concentrator

Because of the highly variable flow to the reclamation pond coupled with variable water quality, a unit operation capable of adjusting to these extremes was required. An additional requirement was that it must produce a quality water which could be returned as a minimum to the cooling towers as make-up to minimize discharge from the smelter. The unit chosen to meet these extremes was a Resources Conservation Company brine concentrator system (Figure 2).

The brine concentrator is capable of producing a high purity condensate that can be used as boiler makeup. It can handle variable flows and contains advanced scale control technology involving seeded recirculating brine which prevents heat transfer surface scaling. With the use of a vapor compressor, the energy consumption is 29-37 BTU per pound of feed solution. Approximately 90-95% of the water entering the unit returns to the smelter. The remaining 5-10% constitutes a concentrated brine solution that is evaporated in the double lined pond.

The brine concentration process begins with the reclamation pond water being pumped to a feed tank where various reagents are added according to the water quality to minimize corrosion and fouling of the system. Decarbonation is

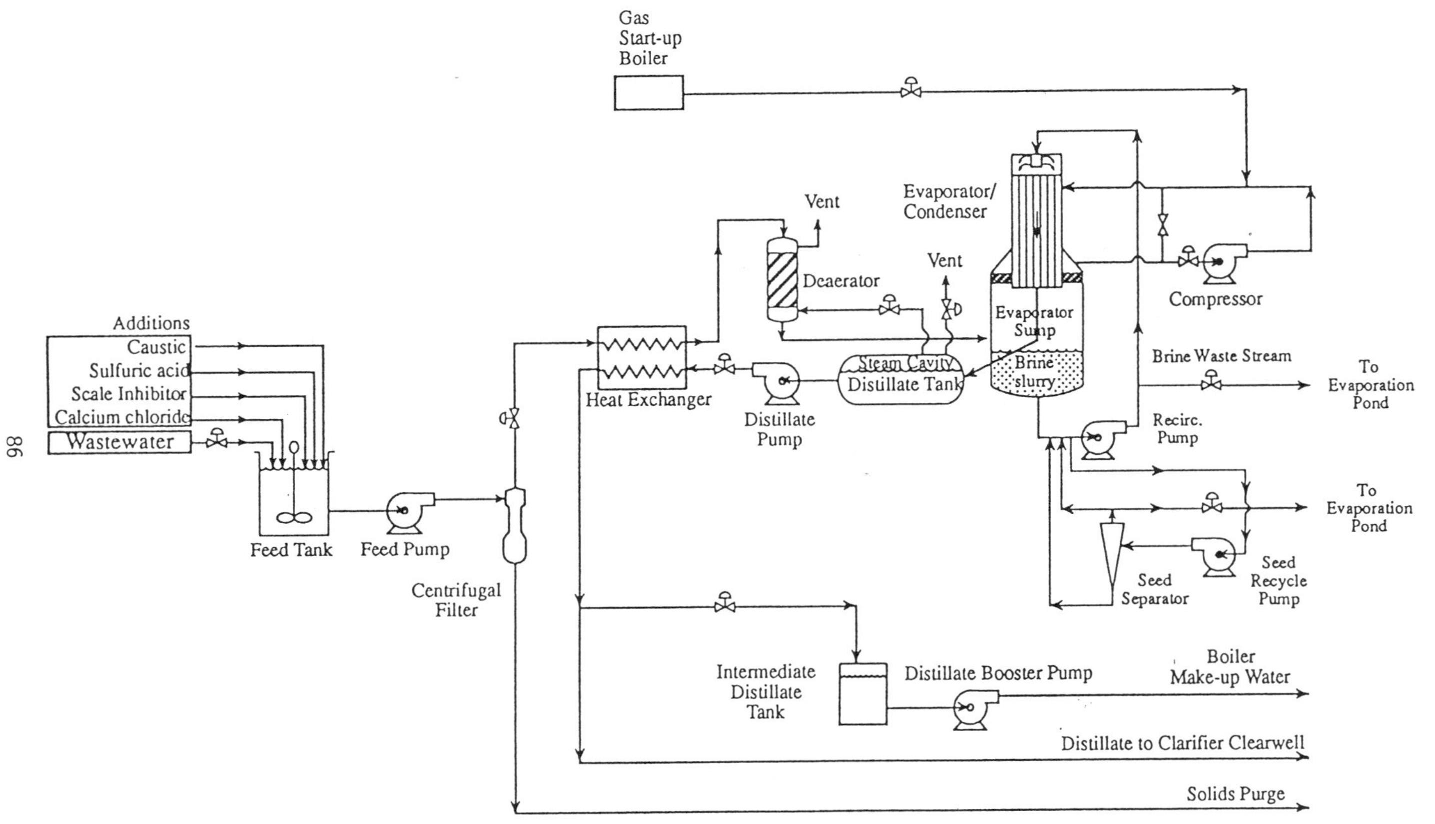

FIGURE 2: HIDALGO SMELTER BRINE CONCENTRATOR

important in this step. Next is suspended solids removal through a centrifugal filter followed by heating and deaeration which removes non-condensible gases.

The solution is now ready for the evaporator. It enters the evaporator sump where it mixes with a concentrated, seeded brine slurry. In a continuous operation, the slurry is pumped to the top of the evaporator into a flood box where it is evenly distributed as a thin film on the inside wall of 1800 titanium tubes. Heat is supplied to the outside of the tubes by a vapor compressor. This allows a portion of the water in the falling thin film on the inside of the tubes to evaporate where it is collected as a vapor in a steam cavity. It then passes through a mist eliminator and into the vapor compressor where its temperature is raised above the boiling point of the recirculating brine. The compressed vapor advances to the outside of the tubes which function as a condenser. As the compressed vapor releases its heat across the hundreds of tubes to the falling thin film on the inside, it condenses as a high quality boiler water. It collects in a distillate tank and is pumped through a heat exchanger where it transfers additional heat to the incoming solution thereby completing the cycle as it returns to the process as "good" water.

The portion of the falling thin film that does not evaporate is collected once again in the seeded brine slurry sump and returned to the top of the evaporator for another cycle. To prevent scaling of the tube walls, a small quantity of solution from both the recirculating slurry brine and the seed separator, which is a liquid cyclone, discharges into the state approved, double lined pond. This permits independent control of total dissolved solids (TDS) and total suspended solids (TSS).

Conclusion

Comprehensive water management is a key element in an overall waste minimization strategy for the Hidalgo smelter. A comprehensive program was implemented in the 1980's to maximize water recycle capability and to minimize raw water input to the facility. In order to further reduce the quantity of waste water, two technological components were designed, constructed and commenced operation in 1991. The first component is pre-treatment of supply well water for cooling towers which has dramatically increased cooling tower cycles of concentration, therefore decreasing the volume of cooling tower blowdown. In addition, cooling tower blowdown is recycled in the plant water circuit for reuse.

All excess water generated from the facility is routed to and contained within a reclamation pond . Several processes within the smelter are designed to reuse water from this pond. The balance is treated in the second technological component of the program: a brine concentrator that produces a high quality water which is returned to the circuit. The amount of residual effluent or concentrated brine from the brine concentrator will represent a significant reduction in the volume of waste water from the facility.

The result of the water management and waste minimization program is a reduction in the amount of fresh water used by the operation as well as a significant reduction in waste water characterized by maximized initial water usage, recycling of water when appropriate, and minimized discharges.

Reference

1. Resource Conservation Company Brochure, Industrial Wastewater Recycling, 1989.

STRATEGIES AND PRACTICES FOR HANDLING MINE WASTES AT THE SUDBURY OPERATIONS OF INCO LIMITED

C.E. Hunt

INCO Limited - Ontario Division
Copper Cliff, Ontario
Canada P0M 1N0

ABSTRACT

Since the turn of the century, copper-nickel sulphide ores have been processed at the Sudbury, Ontario operations of INCO Limited, currently the largest integrated mining, milling, smelting and refining complex in the free world.

Through the years operational, ecomomic, social and legislative changes have dictated the need for research and development directed at mitigating the impact of existing operations on the surrounding natural environment and also the reclamation of areas affected by earlier mining activities.

Approximately one hundred years of mining history and the current annual processing of some 12 million tons of ore presents significant challenges with respect to waste disposal, liquid effluent treatment and land reclamation, particularly in light of rapidly evolving environmental regulations and controls within the Province of Ontario.

Residues and Effluents - Processing and Environmental Considerations
Edited by Ramana G. Reddy, William P. Imrie and Paul B. Queneau
The Minerals, Metals & Materials Society, 1991

INTRODUCTION

Recent and developing legislation and concerns within the Province of Ontario have necessitated increased scrutiny of how and where we process, handle and dispose of mining and metallurgical by-products. Considering its lengthy history in the Sudbury area, INCO Limited must not only do its best to make current operations "environmentally friendly", but deal also with the legacy of several decades of mining activity.

The common denominator with respect to wastes from sulphide ore mining operations is acid mine drainage (AMD). When sulphide bearing materials are exposed to the atmosphere, bacteria oxidize the sulphides to eventually form sulphuric acid. The acid dissolves other waste constituents and results in an effluent of low pH, high metal content and increased suspended particulate. AMD is potentially harmful to the environment.

Historically, waste materials have been utilized for numerous construction applications including railway and roadbeds, backfill and dam construction. Significant amounts lie in abandoned piles at various minesites and as a result there are numerous potential sources of AMD within the Sudbury District. In 1990, a major initiative was introduced to inventory potentially acid-generating materials on INCO Limited property. The program is aimed at identifying and evaluating waste materials for their potential impacts on the surrounding natural environment, their conformity with existing and anticipated environmental legislation, and for finding the most economic alternatives for mitigation of environmental impacts.

EFFLUENT TREATMENT AND MISA REGULATIONS

From February 1990 to January 1991 the Ontario Mineral Industry was subject to perform a stringent monitoring program (MISA Monitoring Regulation) whose purpose was to document plant processes, raw materials and products, and to characterize the resulting process effluents. The monitoring program included analyses for some 200 organic and inorganic parameters as well as acute toxicological testing using rainbow trout and Daphnia magna (water fleas).

This initiative forms part of the Municipal Industrial Strategy for Abatement (MISA) which is Provincial Legislation aimed at the "Virtual Elimination (VE) of persistent toxic contaminants from municipal and industrial discharges into Ontario waterways". The goal of VE will be achieved through the employment of the "Best Available Technology Economically Achievable" (BATEA) for treating various target parameters. Legally enforceable effluent limits (MISA Limits Regulation) which will replace current effluent guidelines are expected to be implemented sometime in 1992. At that time, effluent dischargers will be required to meet or exceed the treatment afforded by the technology chosen by a joint government/industry committee. After a certain amount of time, perhaps five years, technologies will be re-evaluated for their potential to advance the goal of VE. If more suitable technologies have been developed during that time period they will then become the new standard for limit setting. This process will be repeated until the goal of VE is met.

The initial Monitoring Regulation cost INCO Limited an estimated one million dollars. Compliance with the upcoming Limits Regulation will, no doubt, be **significantly** higher. For the time being, the MISA Regulations encompass only minewater, process water from milling operations, smelter/refinery effluents, and stormwater discharges from inactive tailings areas. At some future date, it will be expanded to include **all** waters leaving industrial properties.

Treatment now consists of automatic or batch lime addition to increase the pH of wastewaters and precipitate metals. Removal of suspended particulates is achieved through polymer addition in high-flow treatment plants or by gravity in large settling ponds. Technologies likely to see implementation in the coming years are pH adjustment and ammonia removal as they have a direct bearing on effluent toxicity.

PROGRESSIVE DECOMMISSIONING REGULATIONS

In June 1991 the Provincial Mining Act, administered through the Ontario Ministry of Northern Development and Mines, was amended to include stringent requirements for mine development and closure. Under the Act each mining operation within the Province, regardless of operating status, must be registered with the Ministry and, in a schedule as yet to be determined, must submit a detailed closure plan which complies with specified rehabilitation standards for safety and environmental protection. Guaranteed financial assurance must be provided for site rehabilitation. INCO Limited has registered 38 mining facilities which are operating or temporarily suspended. An additional 32 sites have been registered as abandoned. The legislation does not contain a grandfather clause, so significant demands will be made on time and money in order to bring all sites up to date with respect to the new legislation.

An important aspect of the Mining Act is the concept of "progressive decommissioning". This means that during the life of any project, rehabilitation will be done continually and sequentially throughout development and within reasonable timeframes. Planning for decommissioning will occur in the predevelopment stage for new facilities, thus eliminating future backtracking.

A small pilot project was initiated as a means of establishing identification and evaluation procedures for dealing with wastes located at various sites throughout INCO Limited Ontario Division. It has become readily apparent that much work lies ahead in determining best methods for the collection of representative samples from large piles and the subsequent development of "quick and dirty" testing for the determination of their acid generating potential.

Once AMD status has been established, additional considerations will be weighed. These are the proximity of wastes and runoff streams with respect to existing water treatment facilities, size of the waste pile, and valuable metal content. All the above information will be utilized in making the decision to recycle, relocate, or revegetate to fulfill the decommissioning criteria.

Numerous research initiatives are underway to investigate physical, chemical, and biological means of controlling AMD from existing problem sites. Some of these methods include capping, ion exchange, and engineered wetlands. Difficulties lie in the fact that these methods are currently only feasible or cost effective when small quantities of AMD are to be treated.

WASTE ROCK

The ultimate strategy for the decommissioning of any site, is to handle the material so as to prevent AMD from being generated, thus eliminating the need for treatment. Barring that possibility, two options are available. Runoff from the material can be incorporated into an existing water treatment facility or a treatment system can be tailored specifically for a particular site. The latter alternatives are the least desirable as they will most likely involve long term treatment and monitoring costs.

Methods currently being investigated for the use or disposal of waste rock are recyling as mill feed (for material of sufficient grade) or mine backfill, underwater storage to prevent sulphide oxidation, various capping procedures to prevent weathering, or waste mixing to combine neutralizing agents with those that generate acid. Among the many pertinent considerations are the availability of sufficient space and materials, permanence of the solution, cost effectiveness, and conformity to existing and anticipated environmental legislation.

The recent preliminary evaluation of a remote open pit mining operation for closeout strategies yielded the following options and results. By the end of mining there will be approximately 6 million tons of waste rock stockpiled on-site adjacent to the pit. It is simply not feasible to transport the waste material from this fairly remote area to another site for treatment and disposal. At a cost of approximately 2 dollars per ton the material may be hauled back into the pit upon closeout and then the whole area, for $50,000 to $150,000 per hectare could be capped and revegetated. The estimated total cost of this endeavour could approach $11 million dollars!
Waste mixing would entail an additional cost factor, but the use of neutralizing waste products from other industries, say kiln wastes from the pulp and paper industry, would reduce costs considerably.

There is no guarantee that these actions will eliminate AMD from the waste rock. The site currently possesses a water treatment facility to treat AMD from pit dewatering. The facility consists of automatic lime addition and a series of gravity settling ponds. It was constructed at a cost of approximately one million dollars. It is highly likely that the existing effluent treatment system will be required on an on-going basis.

The above example is an extreme. There are numerous other situations where wastes, closer in proximity to the centre of operations, may be incorporated into existing treatment systems through their use as pit-fill or by diverting wastewater flows.

If the amounts of acid wastewaters can be sufficiently reduced by capping or similar means, then passive treatment systems being developed could be an alternative worthy of consideration, but the pressure is on to act now and utilize whatever methods currently available to solve the AMD dilemma.

SLAG

Smelter slag is generated at a rate of approximately 1.3 million tons per year at INCO Limited's Sudbury operations. Currently, commercial demands for slag products meet or exceed production however this has not always been the case, so millions of tons remain stockpiled.

Leachate test results, when compared with Ontario Regulation 309 (Waste Management) leachate criteria, put slag in the category of non-hazardous waste. This means that slag can be moved and utilized without constraint.

Existing Regulation 309 criteria do not include target concentrations for nickel in leachate. The inclusion of nickel in the Regulation has been discussed and is anticipated in the next few years. If this occurs, then slag might become a "registerable waste" and hence become restricted in its usage. Consequently, research and development may be aimed at altering the slag at the processing level to render it acceptable for varied uses once again or it may be feasible to dispose of slag as mine backfill or simply be stockpiled and revegetated.

During the course of evaluating slag for conformity with anticipated environmental legislation, including decommissioning and effluent regulations, the question was raised as to the quality of runoff water leaving the slag piles. Static tests indicate that slag is not likely to generate acidic run-off, however, limited in-situ testing results have shown elevated metals concentrations in slag-pile drainage. These results are unsubstantiated due to proximity of the test area to the smelter stack and associated atmospheric fallout. Nevertheless, the issue is still under investigation. If slag is proven to be acid generating, then aside from the chemical alteration of slag in-process to render it inert, the management options will be similar to those of AMD generating waste rock.

TAILINGS

Tailings reclamation and water management has been in the forefront of INCO Limited's environmental initiative through the years. The 2,024 hectare Central Tailings Area is one of the largest sulphide tailings areas in the world. Considerable volumes of acid waters are generated in the form of tailings decant, seepage and surface run-off. Catchments have been designed to control seepage as well as contaminated surface run-off from the perimeter dams. Tailings decant water is recycled for use in the milling process. Excess is treated and released into the natural environment.

The tailings area has been the site of numerous on-going AMD research and reclamation studies. Much of the knowledge gained throughout the years will be utilized in developing closeout strategies. Procedures have been developed for tailings coverage using various vegetative covers including grasses and trees. Biological treatment of AMD using engineered wetlands is a promising area of research currently being investigated. Engineering advancements include designs for increased dam integrity and better control of seepage, as well as separation and containment of the high sulphide pyrrhotite portion of the tailings material underwater in the centre of the impoundment area.

The massive central tailings area and associated water management system will form the focal point of strategic planning for decommissioning in the Sudbury area. Where possible, adjacent operations will be incorporated into the water management system and the two main wastewater treatment facilities. Projected needs for wastewater management and treatment are being investigated through an on-going engineering study of the entire tailings area watershed.

CONCLUSION

It is clear that the future brings some tough decisions with regards to strategic planning and the mitigation of acid mine drainage. Ultimately, the mining industry would prefer to find walkaway solutions to waste disposal. Unfortunately, current research has not matured to the point that the accumulated large quantities of wastes can be effectively dealt with, aside from isolation and treatment of wastewaters, far into the future. For now, the strategies will include finding alternatives for smaller existing waste deposits in order to reduce the total number of sites that will require perpetual care and monitoring, and to find ways of altering the wastes to render them inert. It is of paramount importance to encourage and keep abreast of developments in the treatment and control of acid mine drainage.

TREATMENT OF RESIDUES AND EFFLUENTS
IN REFRACTORY METAL INDUSTRY

W. W. Albrecht, B. Gries, H. G. Nadler, W. Rockenbauer

Hermann C. Starck Berlin
P.O.B. 2540
D-3380 Goslar
Germany
-
NRC Inc.
45 Industrial Place
Newton, Massachusetts 02164
USA

I. Introduction

Refractory metals like tungsten, molybdenum, rhenium, tantalum and niobium and their compounds are essential ingredients of our modern technology. Especially in purified form, they are used in special alloys, electronics, optics and ceramics. The processing route from raw material to high purity refractory metals and their compounds includes, as a prime option, the following steps:

- dissolution of raw materials.
- liquid liquid extraction.
- precipitation of high purity intermediate compounds.
- calcination/reduction to high purity oxides/metals.

Especially in the first three steps, major amounts of chemicals like acids or sodium hydroxide are used and, as by-products, substantial quantities of residues and effluents are generated.

The impact of environmental regulations on these by-products and the attempt to reduce or minimize the amount of residues and effluents produced will be discussed in this presentation using Ta/Nb chemistry as the example. The basis of our considerations will be the way these by-products are handled at HCST, a major, and in some cases, the biggest producer of refractory metals (tungsten, molybdenum, rhenium, tantalum, and niobium).

Residues and Effluents - Processing and Environmental Considerations
Edited by Ramana G. Reddy, William P. Imrie and Paul B. Queneau
The Minerals, Metals & Materials Society, 1991

II. Refractory metals and their application.

Let us first review some facts about refractory metals and their application.

	Estimated World Totals	Major Producers in	Major Applications
W	100 mio lbs.	China US Europe Japan	Alloying additive Hard metals Superalloys Lighting/heating elements Heavy metals Electronics
Mo	200 mio lbs.	US Europe China	Alloying additive Catalysts Superalloys Lighting/heating elements Electronics
Re	35,000 lbs.	Chile Europe	Catalysts Superalloys
Ta	2.7 mio lbs. (Ta_2O_5)	Germany US Japan	Electronics Hard metal Alloys Shielding
Nb*	1.9 mio lbs. (Nb_2O_5)	Germany US Japan	Alloys Optics Ceramics

* related to Ta ores

Tungsten

Tungsten is the highest melting metal. It is used for lighting purposes, heavy metal and electric contacts. Tungsten Carbide is the most common hard material for cutting applications in the form of hard metal. Tungsten compounds are important for catalysts and coatings.

Molybdenum

It is estimated that some 80% of the world molybdenum consumption (200 mio lbs. in 1990) was for the steel industry (alloying additive for high grade construction and tool steels) with the balance going into chemicals for catalyst application and Mo metal-products such as superalloys, lighting and heating elements.

Rhenium

The major use of rhenium today is in bi- or multi-metallic catalysts for the manufacture of unleaded petrol. It is also used as an alloying element in nickel base superalloys where Re improves the toughness of these materials at high temperatures.

Tantalum

About 45% of the world tantalum production is used as Ta metal in electronics; 30 - 35% as TaC in hard metals; and the balance as additive in superalloys or various mill products e.g., for the chemical industry.

Niobium

High purity Niobium metal and oxide is mainly used for superalloys, superconductivity, optics and ceramics. Less that 10% of the world's total of niobium is related to the processing of Ta containing ores which is discussed in this presentation.

III. Detrimental Elements/Compounds in the Refractory Metal Chemistry

Before focussing on residues and effluents (again using Ta/Nb as the example), let us consider the detrimental elements and compounds which are connected to the manufacturing processes. The following tables (I, II, III) show major materials which are known to be detrimental to the environment and to human beings and their limits for waste water effluents and hazardous waste.

Detrimental Materials	Detrimental Effects
NH_3	a) Poisons Fishlife b) Potential Negative Effect on Biochemistry
SO_4	Corrosion of Concrete in Water Treatment Systems
F	F-Compounds Decompose Chlorophyll i.e. Reduce Photosynthesis and Lead to Sclerosis of the Bones.
Radioactivity	Genetic Mutation
MiBK	Increases Chemical Oxygen Demand (COD)

Detrimental Materials
in the Ta/Nb Chemical Process

	Western Australia	Lower Saxony	Federal	Pennsylvania	Massachusetts
Hazardous Waste Leachate	< 0,1 Bq/l	see toxicity list Ta-waste	–	same as Massachusetts	see toxicity list Federal 40 CFR 26124 *
Radioactivity Handling	no restriction	< 500 Bq/g for solid material of natural origin	–	no state permit required	no permit required for < 0,5 % U +Th
Radioactivity Disposal	U 246 Bq/g Th 205 Bq/g	U max. 500 Bq/g Th max. 5 Bq/g	natural: 370 Bq/g other : 74 Bq/g	same as Massachusetts	< 10 pCi/g federal guidance

Solid Waste Standards

Table II

	Western Australia	Lower Saxony	Federal	Pennsylvania	Massachusetts
Hazardous Waste Leachate	< 0,1 Bq/l	see toxicity list Ta-waste	–	same as Massachusetts	see toxicity list Federal 40 CFR 26124 *
Radioactivity Handling	no restriction	< 500 Bq/g for solid material of natural origin	–	no state permit required	no permit required for < 0,5 % U +Th
Radioactivity Disposal	U 246 Bq/g Th 205 Bq/g	U max. 500 Bq/g Th max. 5 Bq/g	natural: 370 Bq/g other : 74 Bq/g	same as Massachusetts	< 10 pCi/g federal guidance

Standards of Aqueous Discharges

Table III

In general, the hazardous by-products of the chemical processes shown in Table 1 are subject to strict regulations. This is especially true with respect to radioactivity, fluoride containing emissions, and those materials affecting the chemical oxygen demand in waste water effluents.

Table II and III compare available limits for waste water effluents and waste. These tables have been established especially for the Ta/Nb chemistry which is mainly located in Germany (Lower Saxony), USA (Pennsylvania) and Japan (Fukuoka Prefecture).

Additionally, space limitations at expensive waste disposal sites impose the need to avoid the generation of non-hazardous sludge where ever possible. This last factor is especially important in areas of high population density.

IV. Tantalum and Niobium

Picture 1 describes the Ta/Nb-chemistry in a simplified flow-chart. Natural ores like tantalite and columbite and synthetic concentrates are dissolved to form complex heptafluorides H_2TaF_7 and H_2NbF_7. These fluoroacids are soluble in methylisobutylketone (MIBK), i.e. both elements (Ta,Nb) will be extracted in the organic phase, while most of the dissolved impurities remain at the raffinate.

To separate impurities like W and P, which are extracted together with Ta and Nb, the organic phase is washed with H_2SO_4. This acid is used also to strip Nb from the organic phase which leads to a separation of Ta and Nb. By adding NH_3, Ta/Nb hydroxides are precipitated; and by calcination, Ta and Nb-oxides are formed. K_2TaF_7 is precipitated by adding potassium ions.

Let us look at the by-products: residues originating from the HF/H_2SO_4 treatment still contain these acids and must be carefully neutralized (e.g. by treating with lime) before disposal.

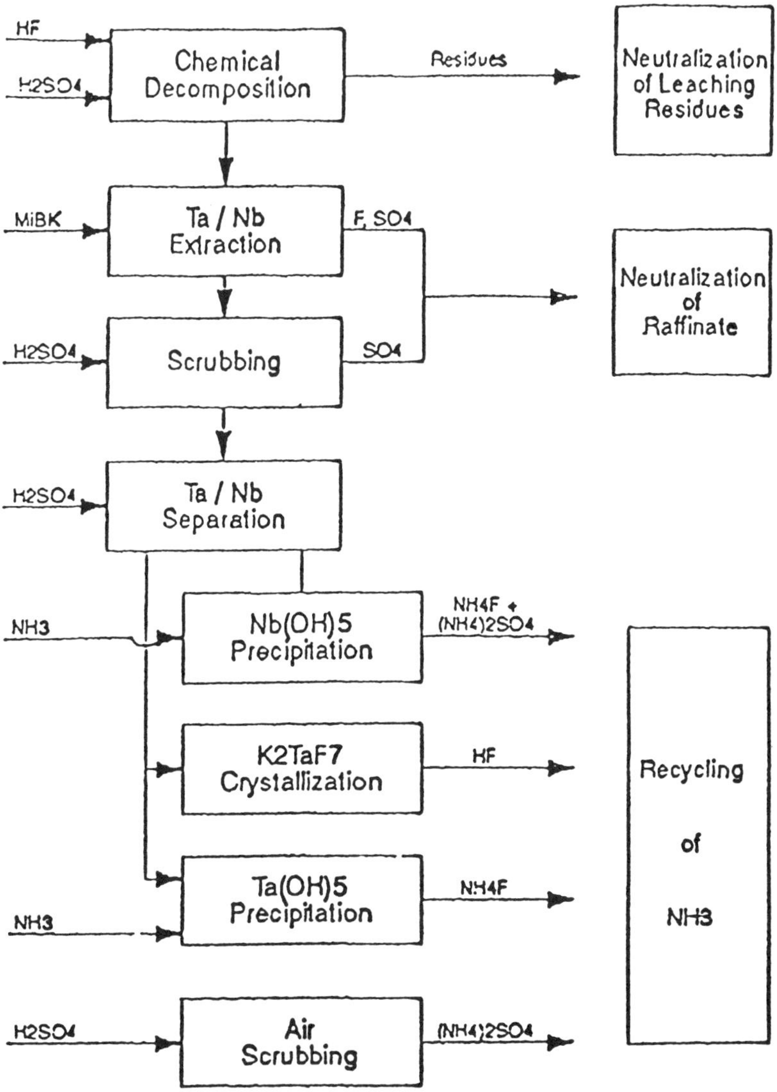

Tantalum/Niobium Chemical Processing

Picture 1

The raffinate, containing HF and H_2SO_4, and the H_2SO_4 containing washing liquid of the liquid liquid extraction process also must be neutralized with lime, and this generates a major amount of CaF_2 and $CaSO_4$ sludge. During the precipitation process, NH_4F and $(NH_4)_2SO_4$-containing filtrates are created, and they also are treated with lime to recycle the ammonia. Once again, a major amount of $CaSO_4/CaF_2$ sludge is created. K_2TaF_7 precipitation effluent leads to CaF_2 after treatment with lime. Air emission will be avoided by keeping HF and NH_3 air streams separate and using NH_3 and H_2SO_4 scrubber separately.

Picture 2 demonstrates these by-products.

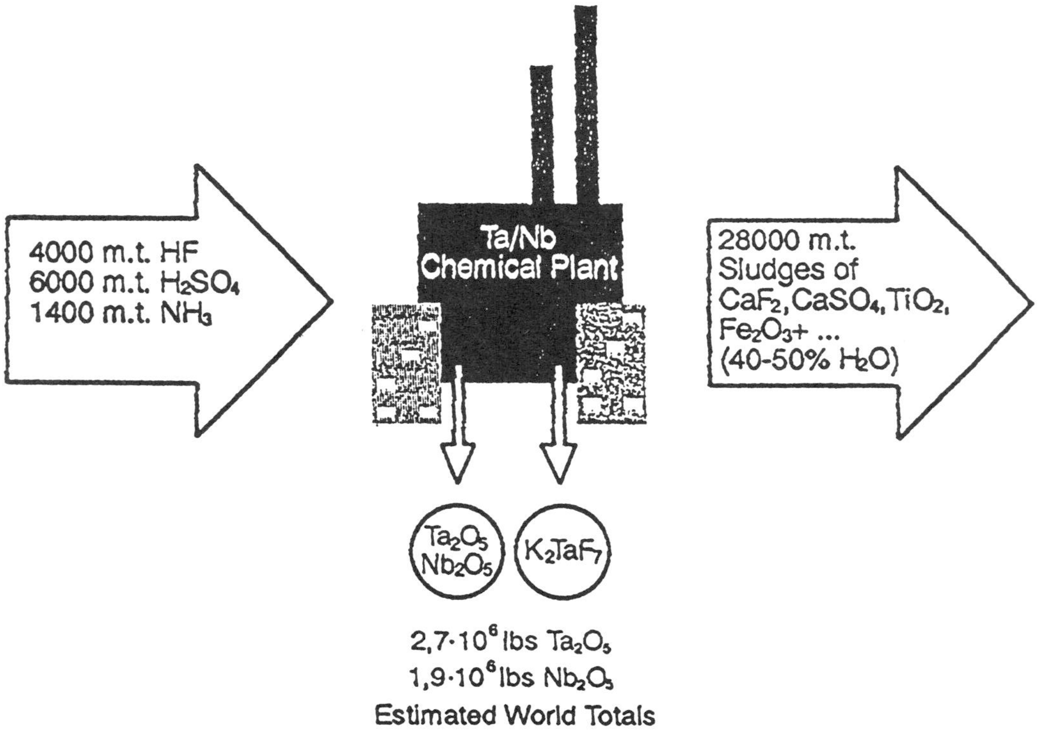

Input of HF, H_2SO_4, NH_3 and Sludge Generation
(Ta/Nb Processing)
Picture 2

The calculation is based on the theoretical assumption that the world production of Ta is combined in one huge plant. If we assume that the world's demand for tantalum in the form of ores and concentrates is about 2.7 million lbs Ta_2O_5 and the amount of niobium contained or added is about 1.9 million lbs Nb_2O_5, then about 4,000 tons of HF (100%), and 6,000 tons of H_2SO_4 (100%) are used to dissolve the ores and concentrates on a world wide basis. In addition, 1,400 tons of NH_3 (100%) are used in this process for precipitation which results in about 28,000 tons of wet (40%) CaF_2 and $CaSO_4$ sludge.

V. Impact of Pollution Control on Ta/Nb Chemistry

Radioactivity must be kept under tight control either by the selection of raw materials to arrange that specified levels in handling or in waste will not be exceeded, or by a costly treatment of raw materials, like that according to a process described in US 4,446,116, US 4,451,438 and DE3,113,335. Special attention must be given to the fact that during ore digestion, U and Th and their nuclides will not be dissolved, which effectively leads to a concentration of these elements.

Regarding the other detrimental compounds, special consideration is given to the reduction of the amount of waste which is generated, and to not exceed specified limits in waste, waste water and emission.

As demonstrated in picture 3, residues of ore digestions are normally neutralized with lime and then disposed.

More than 60% of the total amount of sludges are generated by neutralizing the raffinate of the extraction. These sludges consist of $CaSO_4$, CaF_2 and hydroxides of Fe, Ti, etc. In addition 8,700 tons of CaF_2, $CaSO_4$ are generated by the precipitation or crystallization of Ta_2O_5, Nb_2O_5 and K_2.

One step which reduces the large quantities of sludges is the recovery of CaF_2 and $CaSO_4$ as potential feed stocks for other chemical operations, thus diminishing the sludges worldwide by 30% from 29,300 tons to 20,300 tons (picture 4).

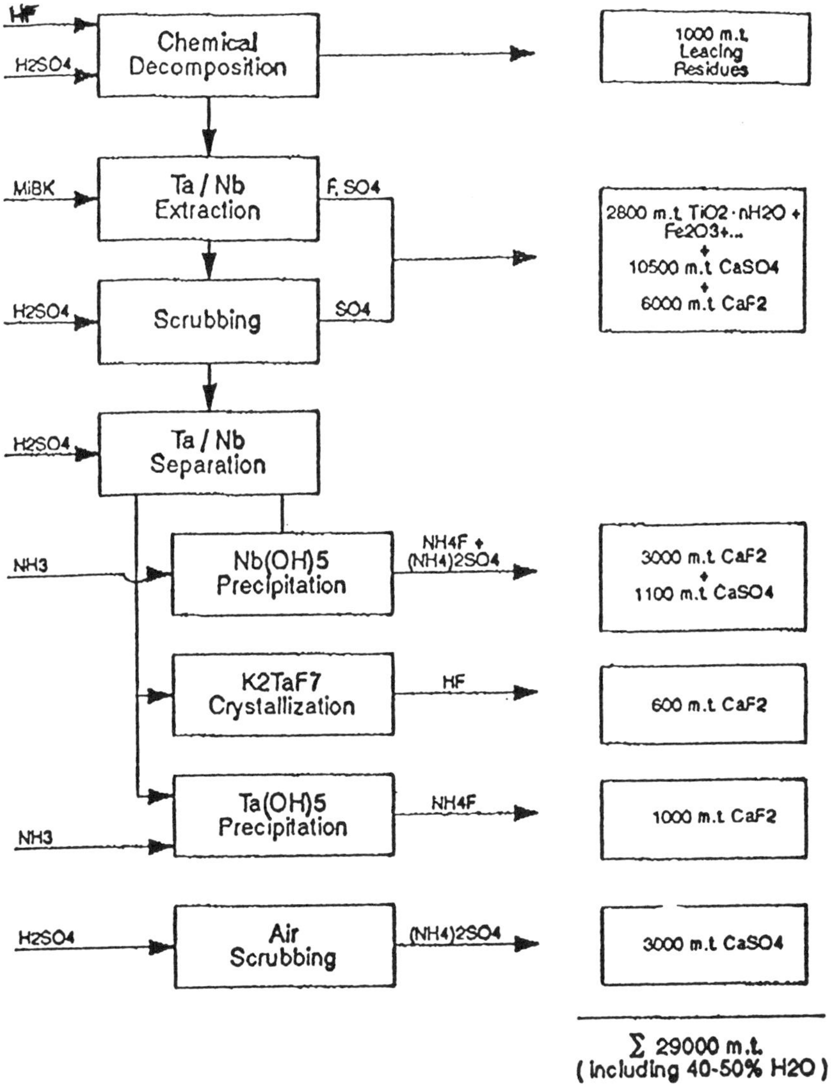

Estimated World Totals
Waste Products in Ta/Nb Processing
(Current Technology)

Picture 3

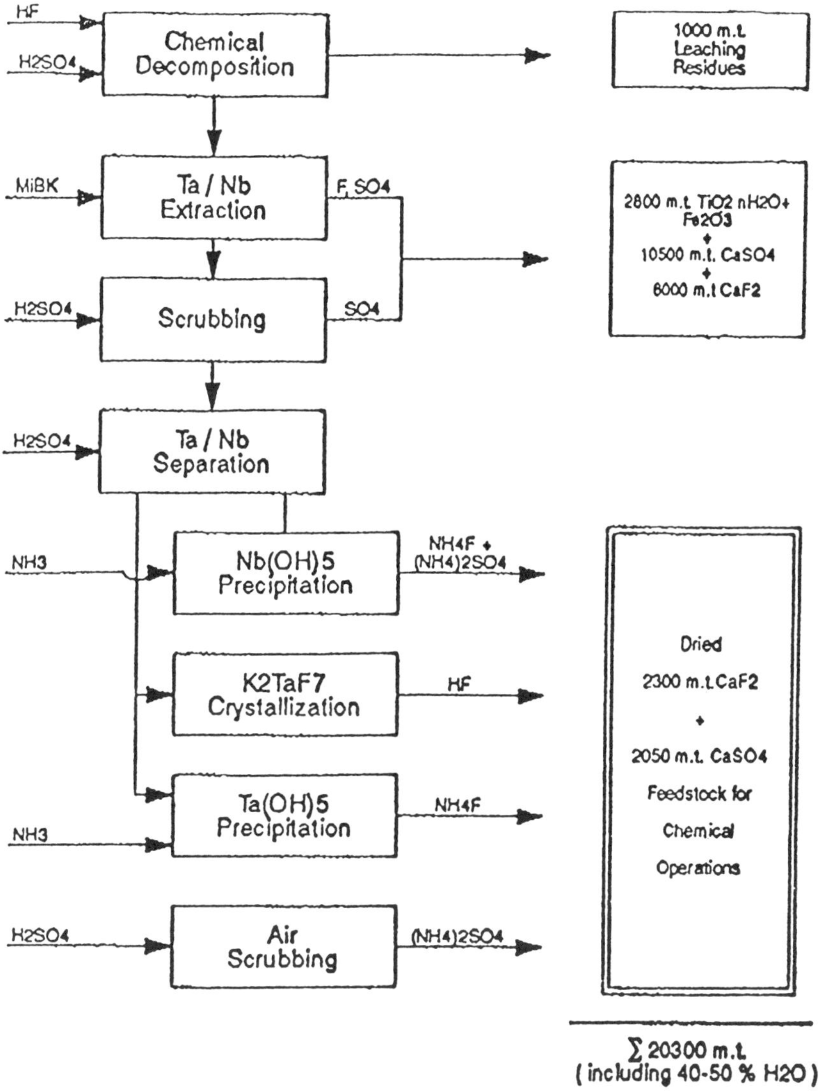

Estimated World Totals
Potential for Waste Reduction
(Short Term)

Picture 4

Let us proceed to picture 5 and consider the recovery of HF and H_2SO_4 by processing the raffinate.

As shown, this raffinate is decomposed by a thermal treatment into the insoluble metallic oxides, HF and H_2SO_4. In a first scrubbing step, H_2SO_4 is recovered, whereas in a second stage, HF is absorbed.

With this potential process, the generation of CaF_2, $CaSO_4$, and hydroxide sludges of about 20,300 tons could be reduced to about 2,400 tons, and in addition, the return to the process of 1,700 tons of HF, and 4,600 tons of H_2SO_4, can be achieved (picture 6).

Further on, the 1000 tons of leaching residues can be recycled in a slag smelting process and as a by-product of this process, used e.g. for road construction.

It should be understood that this HF/H_2SO_4 recycling process does not include changes in the Tantalum/Niobium chemistry.

The items discussed refer concern to the reduction of waste sludges. Attention must be focused on the tight regulation of effluents and emission. A reduction of the NH_3 content to <300 mg/l is achieved by conventional techniques.

NH_4F, $(NH_4)_2SO_4$-containing solutions are treated with lime (picture 7). The filtrates are continuously fed in a column and stripped with live steam and upgraded to about 28% NH_3 content. The recovery is greater than 99%.

With improved equipment and carefully controlled operation NH_3 contents of <50 ppm in the effluent could be guaranteed.

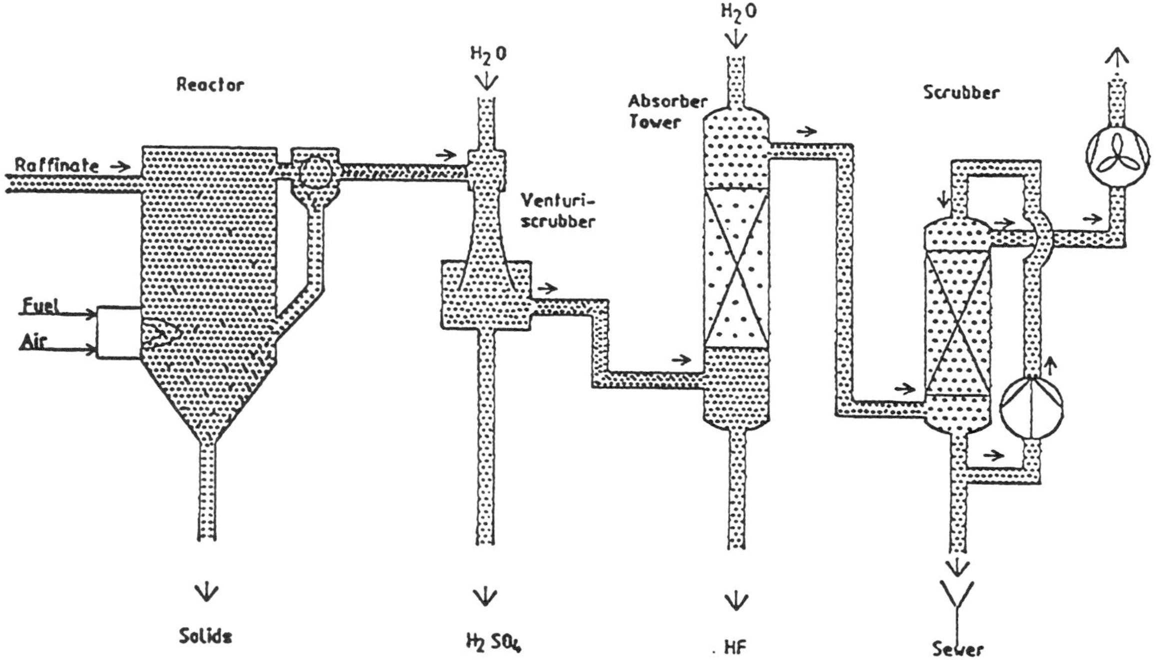

H_2SO_4 + HF Recovery

Picture 5

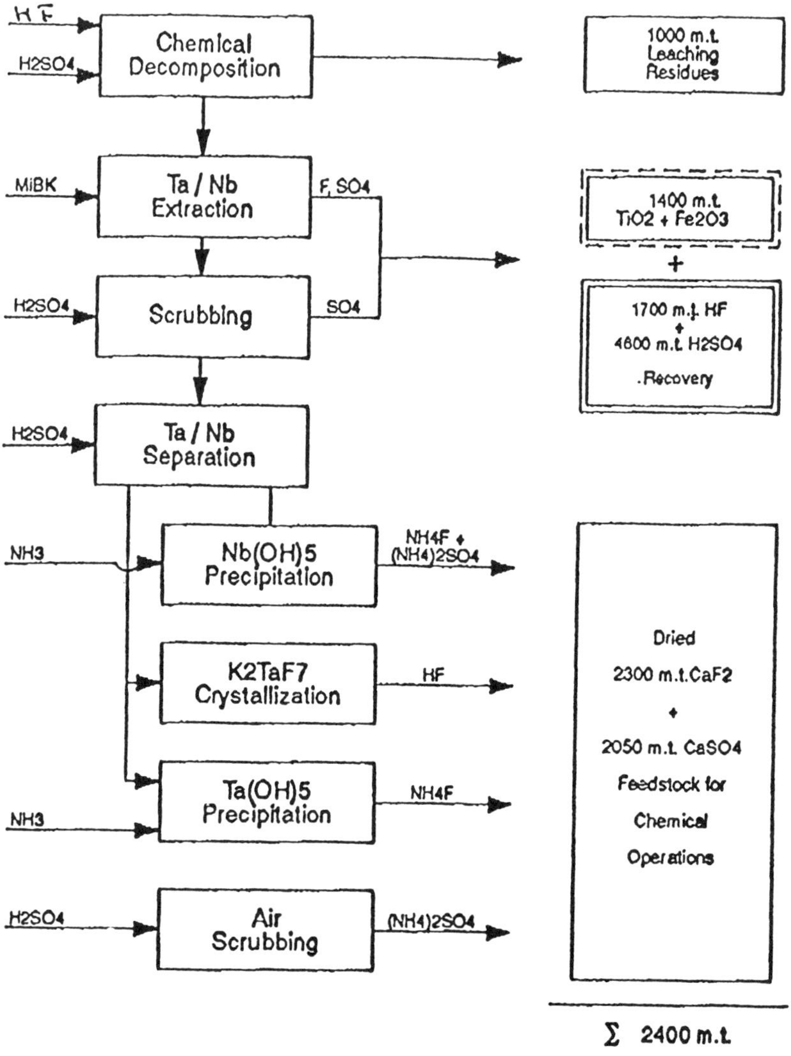

Estimated World Totals
Potential for Waste Reduction
(Long Term)

Picture 6

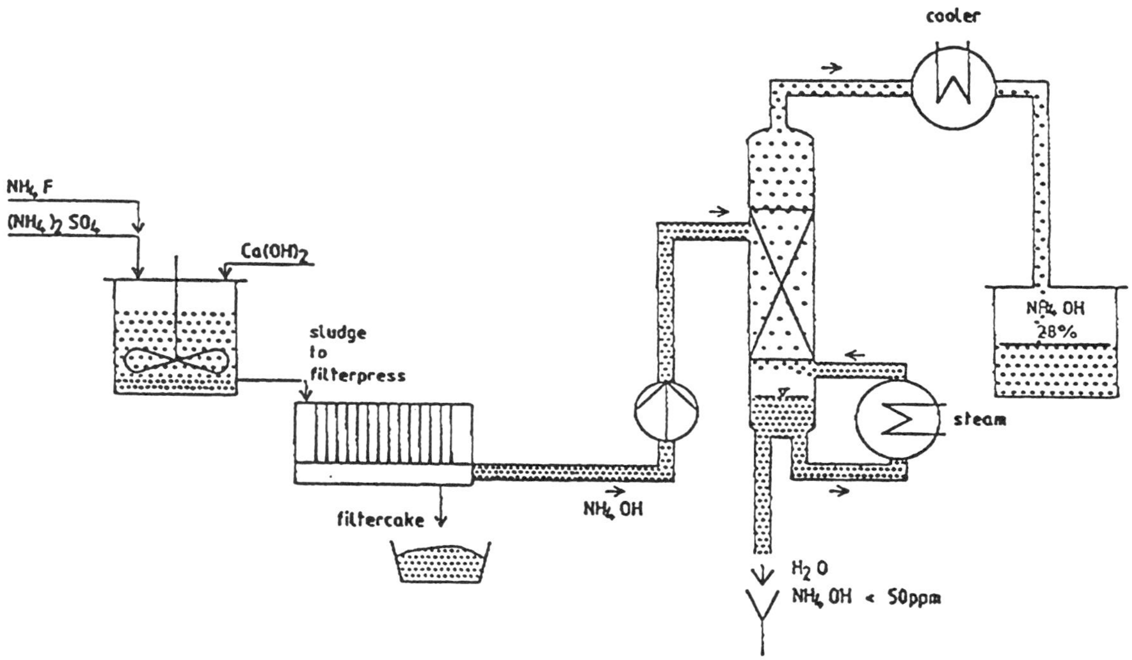

NH_3 Recovery

Picture 7

VI. Conclusion

Environmental protection and waste disposal are topics of ever-growing importance. We can expect that these dynamic subject areas will become - if they are not already - driving forces for manufacturing industries world wide. The refractory metals industry is seeking ways to comply with the increasingly stringent laws passed for environmental protection and the concomitant decrease in the availability of waste disposal areas.

The tantalum and niobium manufacturing process described in this paper comply with current regulations. We can say that, for the most part, we have even reduced the amount of waste and effluent without major change to all the processes described. The situation is the same for our tungsten, molybdenum and rhenium processes also.

It is clear though that the historical imperatives used to design most of these processes were to achieve a high purity product and to minimize manufacturing costs. The driving forces of environmental protection and lowered availability of waste disposal sites can be expected to force reconsideration of both the raw materials used in the respective manufacturing processes and of the manufacturing process itself.

Literature

Clean Air Around the World.
England: International Union of Air Pollution Prevention Associations, 1988.

Albrecht, W. W. and Rockenbauer, W. "Modern Extraction of Tantalum and Niobium," TIC Proceedings November 7-9, 1988. Orlando, Florida, USA.

Beecher, N. and Rappaport A. "Hazardous Waste Management Policies Oversees," Chemical Engineering Progress, May 1990, pp. 30-39.

2.
CHARACTERIZATION

REFLECTED LIGHT AND SCANNING ELECTRON MICROSCOPIC STUDY OF ELECTRIC ARC FURNACE (EAF) DUSTS

Ann M. Hagni and Richard D. Hagni

University of Missouri-Rolla
Department of Geology and Geophysics
Rolla, Missouri 65401

Abstract

Reflected light microscopy (RLM) and scanning electron microscopic-energy dispersive spectroscopic analysis (SEM-EDS) study of electric arc furnace (EAF) dusts have shown that their mineralogy is complex and varied. The particles are largely spherical, and can be seen in RLM sections to be composed of franklinite, and zincite, but they show significant solid solution with each other and may contain small amounts of chromium. These crystalline phases occur as crystals and skeletal growths within Ca-Fe-Si glass. Because chlorine is a deleterious constituent in recovered zinc products from EAF dusts, special attention has been given to chlorine-bearing phases, and they include hydrophylite ($CaCl_2$), lime (CaO), and organic phases. Pyrohydrolysis-treated EAF dusts with silica (SiO_2), tungstite (WO_3), alumina (Al_2O_3), and bunsenite (NiO) additives contain franklinite, scheelite ($CaWO_4$), willemite (Zn_2SiO_4), and hematite (Fe_2O_3).

Residues and Effluents - Processing and Environmental Considerations
Edited by Ramana G. Reddy, William P. Imrie and Paul B. Queneau
The Minerals, Metals & Materials Society, 1991

Introduction

A great deal of interest has been generated in the last twenty years concerning the collection, recovery, disposal and processing of electric arc furnace (EAF) dusts. EAF dust is generated as fumes in the processing of scrap metal, such as automobile scrap, into steel at secondary steel refineries throughout the world. In the United States, these fumes were allowed to dissipate into the air until the Clean Air Act of 1970 required the collection and disposal of such dusts. Worldwide, 5,000,000 tons are collected annually.[1] In the U.S., 500,000 tons are collected each year.[2] For a decade EAF dusts were disposed of in designated landfills; however, in 1980 the Environmental Protection Agency (EPA) classified EAF dust as hazardous due to their lead, cadmium, and chromium contents. No longer could EAF dust be landfilled without additional processing to reduce the lead content to 5 ppm, cadmium to 1 ppm, and chromium to 5 ppm. Because EAF dust often contains 15-20% zinc, processing techniques were sought which both recovered the valuable zinc and reduced or eliminated the hazardous components.[3]

The interest in characterization of minerals and textures in these steel-making dusts has increased because of two reasons: 1) the EPA has tightened their regulations through the Resource Conservation Recovery Act (RCRA) which went into effect August, 1991 for EAF dust, which are now designated hazardous waste KO61, and 2) small amounts (<0.5%) of chlorine and/or fluorine in the dust which will adhere to the cathodes during processing and render them unreusable, creating substantially increased time and costs in processing. RCRA requires ignitability, corrosiveness, reactivity, and toxicity tests to be passed prior to landfill disposal.

A pre-processing technique, pyrohydrolysis, is a de-halogenation process in which the chlorine content is reduced to less than 0.03% and the fluorine is reduced to less than 0.30%. Three microscopy techniques have been utilized to examine the minerals and characterize the pyrohydrolysis-treated EAF dust and the unprocessed EAF dusts. They include reflected light microscopy (RLM), scanning electron microscopy-energy dispersive spectroscopy (SEM-EDS), and cathodoluminescence microscopy (CLM). Reflected light microscopy provides for microscopic analyses of cross-sections of the dust particles, which reveal complex microstructures with varied mineralogical compositions within individual dust particles. SEM-EDS has been utilized to determine the chemical comositions of phases revealed in the cross-sections. Cathodoluminescence microscopy, a non-thermal luminescence technique, has aided in the study of pyrohydrolysis-treated EAF dusts where trace elements have created luminescence in certain mineralogical phases.

Unprocessed EAF Dust

Mineralogy--General

The source of EAF dust, the fumes from electric arc furnaces that process scrap automobile metal, would be expected to yield different chemical and mineralogical compositions from plant to plant and from batch to batch. Eight EAF dust samples have been collected from various industry and research institutions around the country. These samples, although varied in composition, indicate generalities that can be useful.

The most predominant mineralogical phases are the iron spinels, magnetite ($FeFe_2O_4$), franklinite ($ZnFe_2O_4$), and jacobsite ($MnFe_2O_4$), solid solutions of those three spinels, zincite (ZnO), and fayalite (Fe_2SiO_4). In six of the eight samples, two phases are predominant. These two phases are either franklinite and zincite, or jacobsite and fayalite. The remaining two samples have one predominant phase: magnetite or franklinite.

In dust samples containing a high concentration of zinc, franklinite and zincite are the most abundant phases with nearly 50% of each in three different samples. Two additional samples, with lower zinc concentrations, contain predominantly franklinite with 10% zincite and less than 5% zincite, respectively. Dusts containing high manganese form jacobsite and fayalite crystals. A ratio of 6:1 jacobsite to fayalite was observed in those two samples. Fayalite was not found as a predominant phase (>5%) in samples that did not contain jacobsite.

Mineralogy--Specific

EAF dust samples have been implanted in cold-mounting epoxy, ground, polished, and analyzed with reflected light and SEM-EDS. The particle shapes are spheres, occasional broken spheres, rare anhedral agglomerated particles, and coke particles of rectangular shape. The spheres can be composed of a single phase, such as metallic iron or an iron-calcium-silicate glass (Figure 1A.), but are more often composed of numerous crystals of varying texture and composition in iron-calcium-silicate glass matrices (Figures 1D. and 1E.). Magnetite that is partially oxidized to hematite along the octahedral planes is a common occurrence (Figure 1B.). Although most particles are less than 10 μm in diameter, many particles are 40-50 μm and a few are as large as 80-100 μm (Figure 1C.). Agglomerated particles average 500 μm, but spheres of this size do not exist.

The crystalline textures vary from fine, feathery crystals to rhombohedral crystals to cruciform skeletal crystals. Figure 1F. is a 180 μm sphere exhibiting three different crystalline textures enclosed in a calcium-iron-silicon glass matrix. The three texture forms consist of iron spinel solid solutions with different chemical compositions. The internal 50 μm sphere of short spindles forming internal triangles is an iron-manganese spinel: $(Fe_{0.86}\,Mn_{0.14})Fe_2O_4$. The cruciform skeletal crystals toward the edges of the sphere are chromium-bearing iron-manganese-zinc spinels: $(Fe_{0.47}\,Mn_{0.33}\,Zn_{0.20})Fe_2O_4$, that contain 2-3 weight percent chromium. The thin skeletal crystals in the center of the sphere are chromium-bearing magnetite-jacobsite solid solution, $(Fe_{0.61}\,Mn_{0.39})Fe_2O_4$, with 1 weight percent chromium. The rectangular crystals along the right edge of the sphere is chromium-bearing (2 weight percent) iron-manganese-zinc spinel with the formula: $(Fe_{0.50}\,Mn_{0.35}Zn_{0.15})Fe_2O_4$.

A manganese-bearing (3 wt. %) magnetite is shown is Figures 2A. and 2B. These SEM back-scattered photomicrographs show the crystalline structure at 750X and 5000X. Figure 2C. shows franklinite-magnetite solid solution crystals of $(Zn_{0.48}Fe_{0.45}Mn_{0.06})Fe_2O_4$ enclosed in a Ca-Si-Fe glass. Rhombohedral franklinite-magnetite ($(Zn_{0.60}\,Fe_{0.40})Fe_2O_4$) crystals are magnified 5000X in Figure 2D. Although fine, feathery crystals shown in Figure 2E. are difficult to micro-analyze accurately, SEM-EDS analysis indicates they are predominantly jacobsite $(Mn_{0.65}\,Fe_{0.18}\,Zn_{0.18})Fe_2O_4$. The smaller spheres (Figure 2F. is an 8 μm sphere) may also have crystalline structure developed within the sphere, such as this skeletal crystal of chromium-bearing franklinite-jacobsite solid solution.

Of particular interest are the chlorine and fluorine phases. Chlorine concentration is less than 2.0%, and fluorine concentration is less than 1.5%, in the untreated EAF dust. Automated scanning electron microscopy-energy dispersive spectroscopy was used to scan and detect 118 particles of varying sizes for chlorine, iron, calcium,

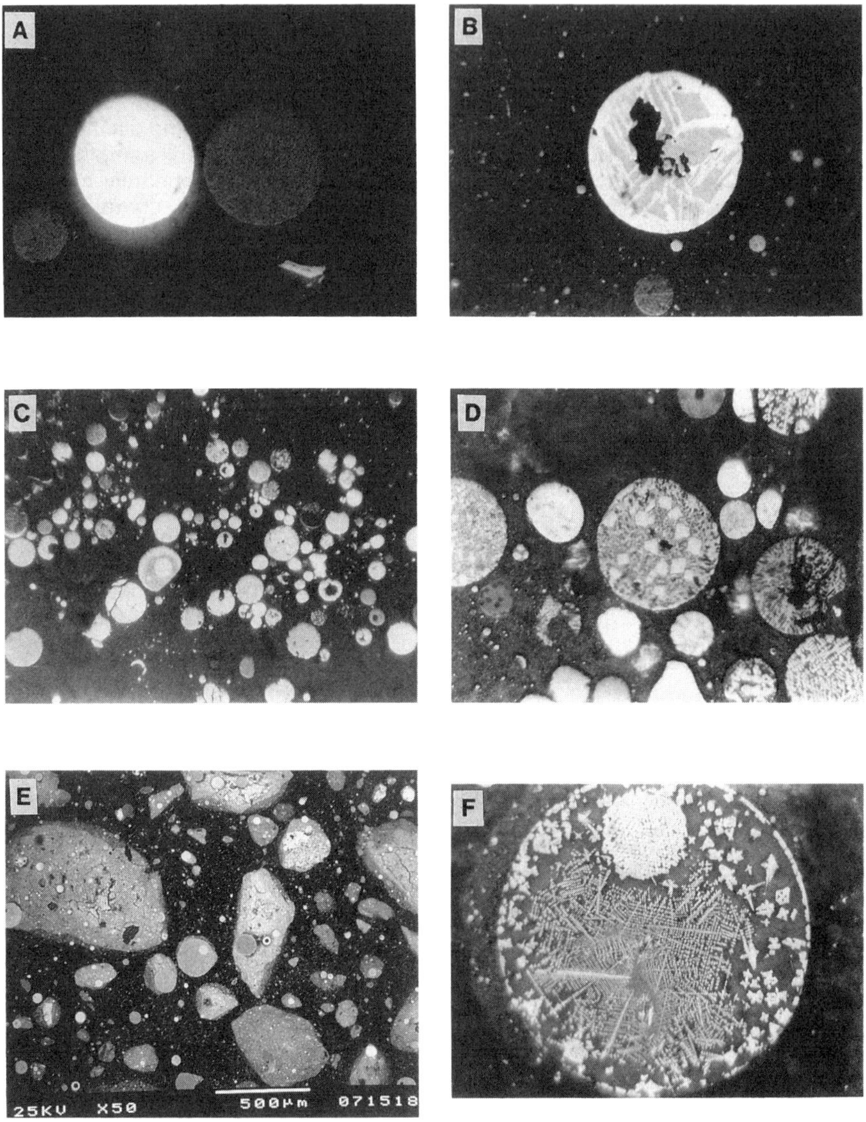

Figure 1. Photomicrographs of untreated EAF dust showing: A. Metallic iron and Ca-Fe-Si glass (50 µm spheres). Reflected light (RL), 640X. B. Magnetite oxidizing to hematite along octahedral planes (56 µm sphere). RL, 640X. C. Photomicrograph of dust particles of varying size and composition. The large center left particle with the internal sphere is 80 µm across. RL, 200X. D. Higher magnification of C. (512X, RL) showing complexity of internal structures. E. SEM back scattered image showing varying sizes and compositions. F. Dust sphere of 180 µm (400X, RL). White internal sphere is (Fe, Mn)Fe_2O_4, thin skeletal crystals in center are chromium-bearing (Fe, Mn)Fe_2O_4, the cruciform skeletal crystals toward the edges of the sphere are chromium-bearing (Fe, Mn, Zn)Fe_2O_4, and the matrix is a Ca-Fe-Si glass.

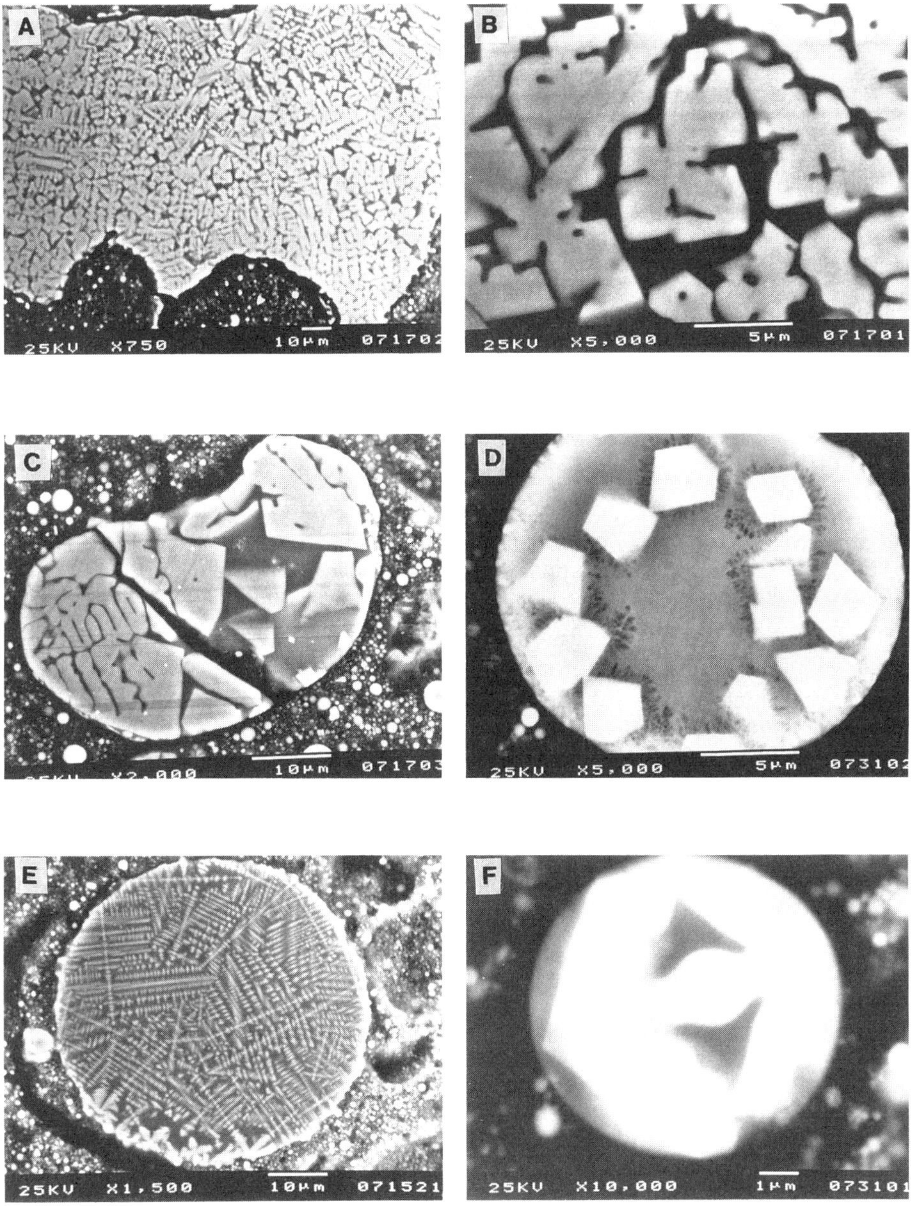

Figure 2. SEM-BSI photomicrographs of untreated EAF dust showing: A. Broken sphere of $(Fe_{0.9}\,Mn_{0.1})Fe_2O_4$ solid solution in a Ca-Si-Fe glass. B. Crystalline structure of A. magnified to 5000X. C. Crystals of $(Zn_{0.5}Fe_{0.4}\,Mn_{0.1})Fe_2O_4$ solid solution in a glass matrix. D. Rhombahedral crystals of $(Zn_{0.6}\,Mn_{0.4})Fe_2O_4$ solid solution in a Ca-Fe-Si-Al glass. The black areas surrounding the rhombs are holes. E. Feathery crystals of iron and zinc-bearing jacobsite crystals in a Ca-Fe-Si glass matrix. F. Skeletal crystal of $(Zn_{0.6}Mn_{0.4})Fe_2O_4$ solid solution with 1% chromium.

manganese, zinc, silicon, chromium, potassium, and aluminum (Figure 3A. and 3B.). Fluorine is too light an element to detect by EDS on this system. Only one of the 118 particles contained substantial chlorine (Figure 3C.). This particle is calcium oxide (CaO) that contains 7% chlorine. Chlorine had been found to be present in CaO and as hydrophylite ($CaCl_2$) in previous examinations of EAF dust.[4]

Chlorine also has been found in small amounts (less than 0.1% to 1.09%) in organic particles and in periclase, MgO (0.6% chlorine). The amounts of chlorine were determined by electron microprobe. The organic material occurs in relatively large particles that range from 75 µm to 350 µm, and that exhibit a rather porous in texture. These particles, due to their high carbon content, are very dark in SEM back scattered image, as can be seen in Figures 3D, E, and F.

Fluorine poses detection difficulties in that less than 1.5% fluorine is present in the unprocessed EAF dust, it is a light element (atomic number 9) and therefore difficult to analyze by SEM-EDS, and fluorine's K_∂ EDS peak is interfered by iron's L_∂ peak.

Pyrohydrolysis-treated EAF Dust

Pyrohydrolysis is a de-halegonation process applied to EAF dusts, in which silica and other oxides are added to the dust in a rotary kiln, the temperature is increased to 850° C or greater, the kiln is rotated, and steam diluted with air is added to carry out the gaseous molecules. The additives act as catalysts in freeing the chlorine and fluorine, which then combine with hydrogen forming HCl and HF. The chlorine is then transported out of the process as a gas.[5] The oxide additives to EAF dusts have included silica (SiO_2), tungstite (WO_3), alumina (Al_2O_3), and bunsenite (NiO). These processed dusts were observed under reflected light, SEM-EDS, and cathodoluminescence to determine the mineralogy and relationships between the various phases. Silica and tungstite added to EAF dust and pyrohydrolysis-treated dust resulted in the formation of franklinite-magnetite solid solution ((Zn, Fe)Fe_2O_4), scheelite ($CaWO_4$), hematite (Fe_2O_3), and willemite (Zn_2SiO_4). The silica and alumina additives produced predominantly franklinite and less than 10% hematite. The silica and bunsenite produced magnetite and a small amount of hematite. This process was successful in reducing the chlorine to 0.03% or less, and fluorine to 0.30% or less.

The tungstite/silica dusts were mineralogically examined for tungsten and silicon to determine how these elements from the additives were combining with other elements. Tungsten was found to have combined with calcium and oxygen to form scheelite ($CaWO_4$). The scheelite particles are locally free from other phases, and would be relatively easily recovered for the valuable tungsten. They may be 1 to 6 µm in size (Figure 4A.), but more commonly form rings 8 µm to 20 µm thick around franklinite-magnetite spheres (Figure 4B.) that are 145 µm to 175 µm in diameter. This scheelite would be difficult to recover without further processing by grinding or chemically freeing the tungsten. The silicon is found to be combining with zinc and oxygen to form willemite (Zn_2SiO_4).

Cathodoluminescence (CL) results from a non-thermal energy source in which certain trace elements give off energy in the form of photons, creating a luminescence. In the tungstite/silica processed dusts, scheelite luminesces a deep purple blue and willemite luminesces a bright green. This luminescence clearly shows that scheelite forming around the iron spinel particles and as large individual particles (Figure 4C.). Although willemite locally forms rings around the scheelite rings or around the iron spinel spheres, willemite is predominantly present as fine-

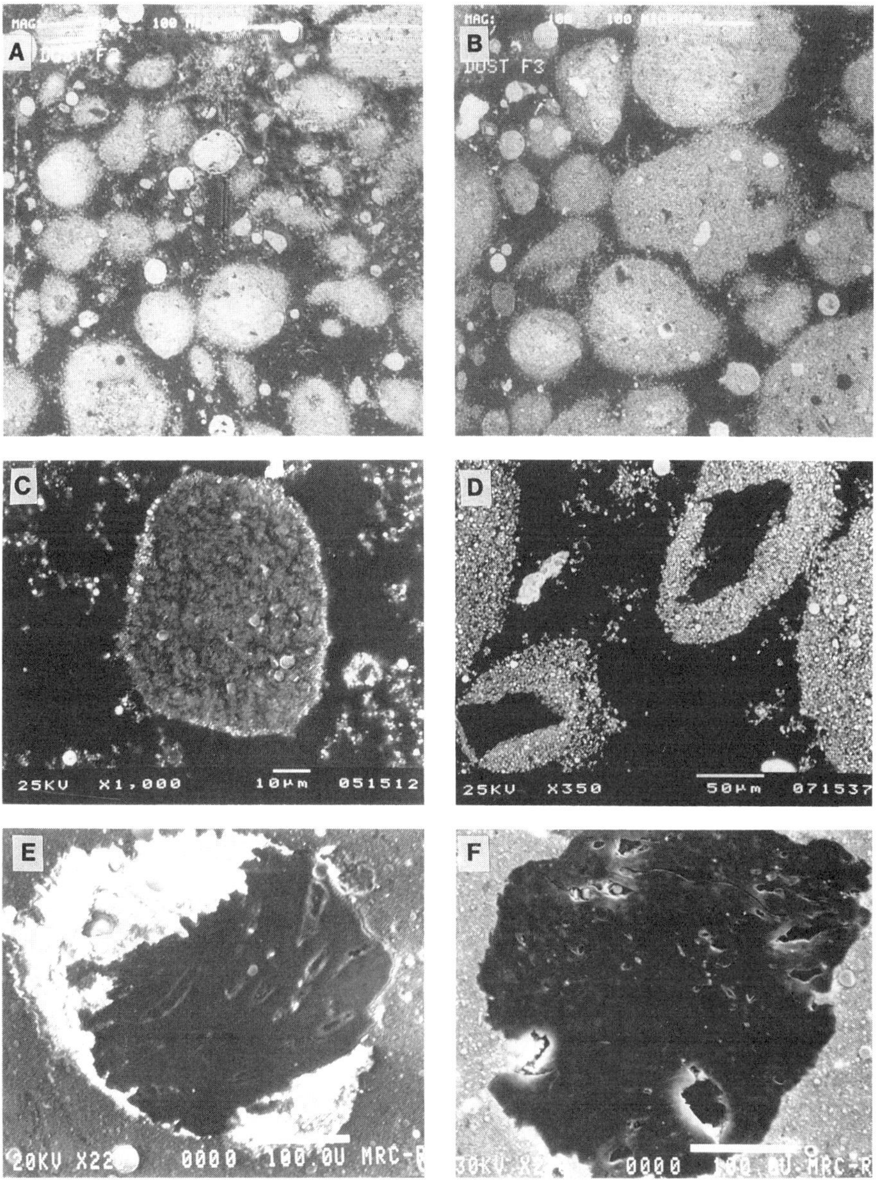

Figure 3. SEM microphotographs of untreated EAF dust showing: A. Scanning area in which 60 dust particles were automatically probed for chlorine content. B. Scanning area in which 58 dust particles were automatically probed for chlorine contents. C. Lime (CaO) particle with 7% chlorine found in B. None of the other particles were chlorine-bearing. D. Two dark organic chlorine-bearing particles; the upper right particle contains 0.3% Cl and the lower left contains less than 0.08% Cl (BSI). E. Porous organic particle containing 0.14% Cl (BSI). F. Organic particle with Cl concentration varying from 0.13 to 1.16% (BSI).

grained loose particles (Figure 4D.). Some of the silicon is contributing to an iron-calcium-silicon phase which luminesces bright red. This phase is not nearly as abundant as either scheelite or willemite. It is found as small clusters of particles or as small particles forming a narrow ring around the larger scheelite particles.

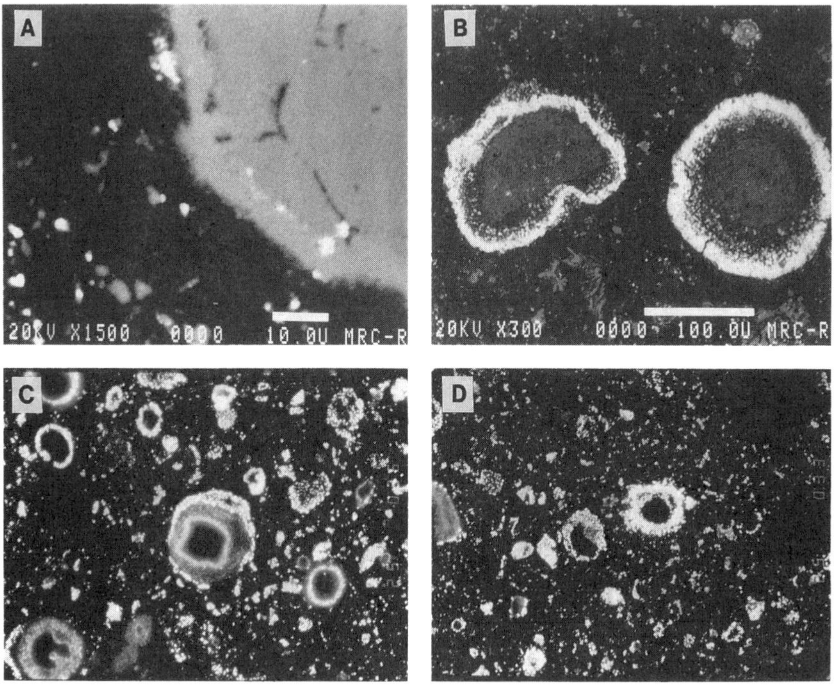

Figure 4. Photomicrographs of EAF dust that has been treated by pyrohydrolysis with silica/tungstite additive that show: A. Small, 1-6 μm particles of scheelite (bright white in SEM-BSI) that are both loose and in the larger franklinite sphere found in the silica/tunstite additive processed dusts. B. Scheelite that forms thick (8-20 μm) rims around 150 μm franklinite spheres. C. Cathodoluminescence photomicrograph showing scheelite (blue in CL, off-white to light gray inner ring in center in black and white) and willemite (green in CL, bright white outer ring in center in black and white). 25X. D. CL of willemite (bright white) and scheelite (light gray particle on left). 25X.

The alumina/silica and the bunsenite/silica pyrohydrolysis-treated EAF dusts exhibit very little luminescence. These processed dusts are characterized primarily by franklinite, magnetite, and hematite.

Conclusions

The techniques employed by this characterization and mineralization study of EAF dusts and their pyrohydrolysis-treated products, reflected light microscopy, SEM-EDS, electron microprobe microscopy, X-ray diffraction and cathodoluminescence, have been extremely beneficial in identifying mineralogical phases, phase relationships, and microtextures of these minute particles. The complete characterization of these phases, their relationships, and the varied textures would not have been possible by other means.

Acknowledgements

This research has been supported by the Department of the Interior's Mineral Institutes program administered by the Bureau of Mines through the Generic Mineral Technology Center for Pyrometallurgy under allotment grant number G1125129. The writers thank John P. Hager and Jerome Downey for providing the samples upon which this process mineralogy study has been based.

References

1. N. A. Barcza and L. R. Nelson, "Technology for the Treatment of Steel-Plant Dusts", World Steel Review, 1 (1) (1991), Mintek Paper No. 8104, 27 p.

2. "Electric Arc Furnace Dust Treatment Technologies", Tech Commentary, Center for Metals Production, Mellon Institute, Pittsburgh, PA, (1988), 5 p.

3. Lehigh University, Bethlehem, Pennsylvania, Characterization, Recovery and Recycling of Electric Arc Furnace Dust, U.S. Department of Commerce (1982), 165 p.

4. A. M. Hagni, R. D. Hagni, and C. Demars, "Mineralogical Characteristics of Electric Arc Furnace Dusts", Journal of Metals, 43 (4) (1991), 28-30.

5. J. P. Downey, "Pyrohydolytic De-halogenation of Electric Arc Furnace Dusts", (Ph. D. thesis, Colorado School of Mines, 1991), 244 p.

REFLECTED LIGHT AND SCANNING ELECTRON MICROSCOPIC CHARACTERIZATION OF DUST FROM LEAD AND COPPER SMELTERS

Musa Karakus and Richard D. Hagni

Department of Geology and Geophysics
University of Missouri, Rolla, MO 65401-0249

Abstract

Reflected light microscopic (RLM) examination of dust particles from lead and copper smelters can effectively characterize their mineralogy, texture, and origin. Scanning electron microscopic-energy dispersive spectroscopic (SEM-EDS) analysis provides confirmation of the optical identifications and chemical data on the phases. Studied lead smelter dust samples were from the: 1) blast furnace, 2) sinter plant scrubber, 3) sinter plant crusher, 4) dross furnace, and 5) refinery. Phases identified included galena fragments, melted PbS, metallic Pb and Cu, Cu-Ni, Cu-Zn, Zn-Ag, Zn-Cu-Ag, Ni-Cu-Sb, and Ni-Cu-Zn-Ag metal alloys, Ni-sulfide, Cu-sulfide, and Cu-Pb-sulfide dross phases, slag and sinter particles, coke fragments, unreacted original feed phases: pyrite, chalcopyrite, sphalerite, siegenite, bravoite, and oxidized particles: iss, bornite, chalcocite, covellite, magnetite, hematite, and goethite. Dust from a secondary copper refinery were found by RLM and SEM to consist of tenorite, cuprite, cassiterite, two silicate phases. Magnetite, hematite, Fe-Al-Cr-Mg spinel, periclase, forsterite, and wollastonite were derived from the furnace linings during initial cleaning.

Residues and Effluents - Processing and Environmental Considerations
Edited by Ramana G. Reddy, William P. Imrie and Paul B. Queneau
The Minerals, Metals & Materials Society, 1991

Introduction

Increased interest in environmental problems connected with the mining industry has focused increasing interest in the characterization of dust produced at various industrial smelters. Smelter dusts typically are very fine-grained and usually have been analyzed only for their chemical constituents. The mineralogy of these dusts typically have been inferred from their chemical compositions. In recent mineralogical studies, we have found that the particles that constitute typical smelter dusts can be well characterized by process mineralogy techniques. Dusts from the sinter and blast furnace plants at the Herculaneum Lead Smelter, Missouri have been studied(1) . EAF dust from secondary steel plants and additional study of dusts from various smelters have also been discussed(2, 3) . The present study reports results of process mineralogical studies of dusts collected from additional locations at the Herculaneum Lead Smelter in Missouri and the initiation of a study of dust from a secondary domestic copper smelter.

These studies have been aimed at a through study of mineral identifications, mineral compositions, and textures shown by those minerals. Reflected light microscopy has been the principal process mineralogy technique used in the studies. Scanning electron microscopy-energy dispersive spectroscopic analysis has been used to further study the dust particles and to determine the chemical compositions of the phases present. Cathodoluminescence microscopy was used to aid in the identification of particles that were derived from the refractory furnace liners.

The purpose of the paper is to illustrate the value of process mineralogy techniques in the study of smelter dusts, demonstrate that the particle sizes are sufficiently large to be adequately studied by reflected light microscopy, exhibit typical textural intergrowths shown by the dust phases, give chemical compositions of the constituent minerals, and provide information regarding the non-hazardous character of those dusts as deduced from their mineralogy.

Herculaneum Lead Smelter Dusts

Dust samples were collected from five locations in the Herculaneum Lead Smelter at Herculaneum, Missouri. The five locations were: 1) blast furnace, 2) sinter plant scrubber, 3) sinter plant crusher, 4) dross furnace, and 5) refinery. The first three types of dust consist mostly of the same phases but the amounts of those phases vary. The last two types of dust differ markedly from the first three and from each other.

Blast Furnace Dust

Blast furnace dust consists largely of sinter particles ($>45\%$), galena fragments (20%), and coke fragments (25%). Other phases include metallic Pb and Cu, dross phases, unreacted original feed particles, and oxidized particles (Figure 1).

Sinter particles have a mineralogy that is similar to that found in the sinter samples from Herculaneum and ASARCO(4) . They are composed of euhedral wollastonite and franklinite crystals enclosed in a Pb-silicate matrix. Minor amounts of PbO (lithargite and massicottite), magnetite, and Pb-Zn-Fe-Ca-Mg silicate phases also have been identified. The particle sizes vary from several tens of micrometers to over 1 mm (Figure 1A and 1B). Galena particles have a rather uniform size distribution, about 3 to 15 micrometers.

Unreacted original feed material, pyrite, sphalerite, chalcopyrite, bravoite, and siegenite are present in minor amounts in the blast furnace dust (about 10 %) (Figure 2A). They have angular shapes and occur in varying sizes. Oxidation of these particles has resulted in the formation of bornite, intermediate solid solution (iss), magnetite, hematite, and goethite. Chalcopyrite commonly is rimmed by iss and bornite due to oxidation. Covellite rimming of bornite and magnetite, and goethite rimming of pyrite, also were commonly observed.

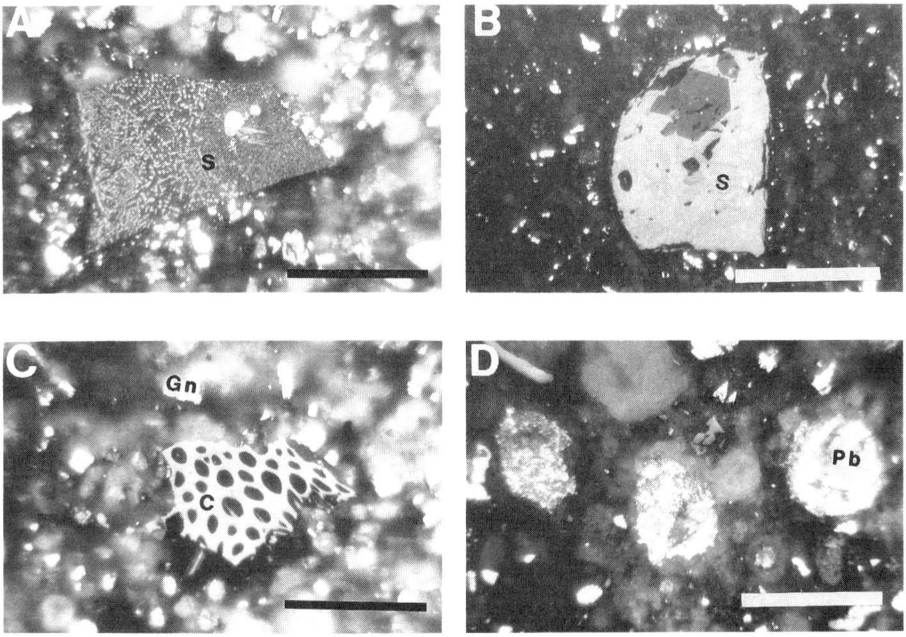

Figure 1. Photomicrographs showing dust particles found in blast furnace dust from Herculaneum, Missouri. A.) Large sinter particle that includes silicate and spinel phases, B.) Large sinter particle that contains euhedral Pb-Zn bearing Ca-Fe-Mg silicate crystals, C.) Cellular structured coke fragment, and D.) three metallic lead spheres. (Reflected light, oil imm.) Scale: 1 mm. S: sinter particle, Gn: galena, C: coke particle, Pb: metallic lead.

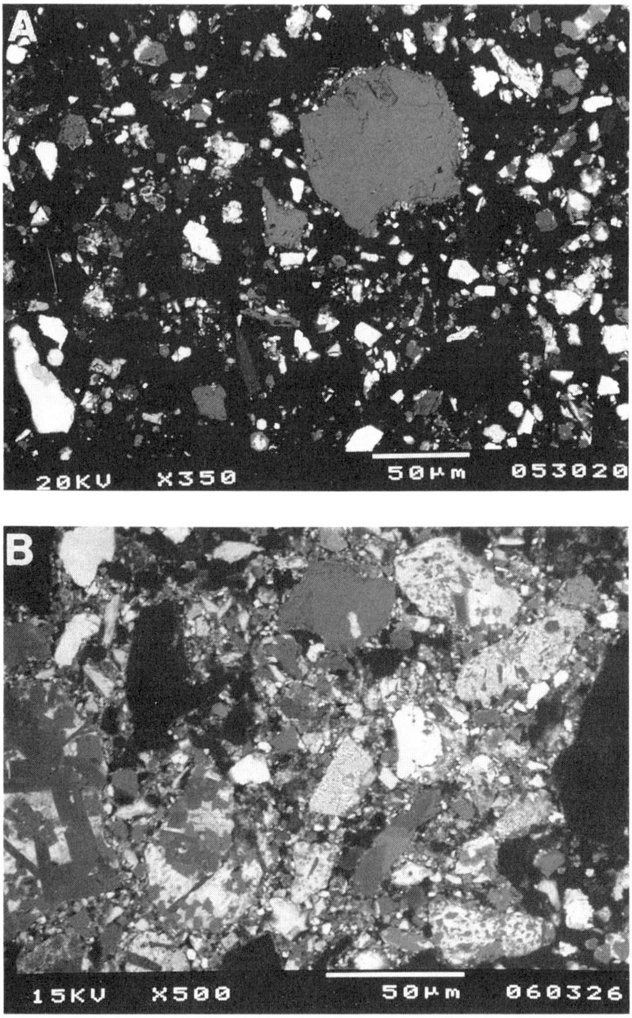

Figure 2. SEM-BSE micrograph of blast furnace dust particles showing: A.) particle size distribution and phase abundance, B.) sinter particle that contain wollastonite and franklinite in Pb silicate matrix and a dross particle showing PbS-Cu sulfide-Pb-Cu sulfide euthectic texture.

Dross phases consist of Cu sulfide, Pb sulfide, Cu-Pb sulfide[5], metallic Pb (Figure 1D), and Ni sulfide, and alloys of Cu, Ni, Zn, and Ag. Dross phases form droplets of melted sulfides and metals typical for dross texture. Eutectic texture between galena and Cu_2S is common (Figure 2B).

Sinter Plant Scrubber Dust

Sinter plant scrubber dust samples contain the same mineral constituents as the blast furnace dusts, but the amounts of those phases are different. Sinter plant scrubber dusts contain much higher galena contents (up to 60 %) and lower amounts of sinter particles (about 15 %) than blast furnace dusts. Original unreacted sulfide particles make up about 10 % of scrubber dust samples. The scrubber dusts also contain about 5 % dross phases, 1 % metallic phases, 3 % oxide particles, and 3 % coke fragments. The size distribution and texture of the dust constituents in the sinter plant scrubber are similar to those found in blast furnace dusts (Figure 3). The average sizes of the sinter and coke particles are smaller than those found in blast furnace dusts.

Figure 3. SEM-BSE micrograph of sinter plant scrubber dust particles showing particle size distribution and phase abundances. Similar size distribution and textures of dust particals are observed.

Sinter Plant Crusher Dust

The sinter plant crusher dusts have the highest galena content, more than 80 % (Figure 4). They also contain less sinter particles (about 7 %). The size distribution and textures of the particles are similar to those found in the sinter plant scrubber dust samples. Very small amounts of coke particles (1-3 %) are present in the sinter plant crusher dusts. Other dust constituents are unreacted and partially reacted sulfides (3-5 %) and oxide particles, including magnetite, hematite, or goethite, (3-5 %).

Figure 4. SEM-BSE micrograph of sinter plant crusher dust particles showing particle size distribution and phase abundances. The galena content is highest in these dusts.

Dross Furnace Dust

Dross furnace dusts differ markedly from the three types described above. Dross furnace dust consists dominantly of fine coke particles (> 50%), dross phases (30 %), sinter particles (3-5 %), and metal and metal alloys (5 %). Unreacted original sulfides and their oxidized products constitute about 10 % of the dross furnace dusts.

Dross phases formed by the melting of the original sulfides, include Pb sulfides, Cu sulfides, Cu-Pb sulfides, and Ni sulfides (Figure 5 and 6). The melted sulfides have formed agglomerated dust particles that have a high sphericity and roundness. Their mineralogical assemblages include bornite, iss, digenite, chalcocite, covellite, galena, and vaesite (Figure 5A and 5B). No Co-sulfide dross phase was identified. SEM-EDS analysis indicates that almost all Cu-sulfide dross phases contain Ni and Zn impurities but they are devoid of Co. Melted sulfides, together with gas reactions in the system, resulted in the formation of iss, bornite, and chalcocite cenospheres that contain magnetite and cuprospinel crystals (Figure 5C).

Unreacted original sulfides include dominantly galena, chalcopyrite, and pyrite together with minor sphalerite, bravoite, and siegenite. They are easily distinguished from their melted and oxidized products by their fragmental and angular shapes (Figure 7A). Their microchemistry differs slightly from the dross phases. Unreacted or partially reacted sulfide particles contain little or no impurities. Reaction between gas and the original solid sulfides results in the formation of iss, bornite, digenite, and oxide minerals. Some chalcopyrite particles are rimmed by iss and bornite (Figure 5D). Exsolution texture between iss and bornite commonly is present. Bravoite and Co-bearing bravoite are present in very minor amounts and form zoned crystals (Figure 7B).

Typical metallic phases and alloys in the dross furnace dust include metallic Pb and Cu, Cu-Zn (43 % Cu, 57 % Zn) alloy, Cu-Ni (85-95 % Cu, 5-15 % Ni) alloy, Zn-Ag (85 % Zn, 15 % Ag) alloy, Cu-Zn-Ag (10-20 % Cu, 60-64 % Zn, 18-26 % Ag) alloy, Ni-Cu-Sb (48-58 % Ni, 16-18 % Cu, 33-35 % Sb) alloy, and Ni-Cu-Zn-Ag (5% Ni, 20 % Cu, 60 % Zn, 15 % Ag) alloy. They occur as large free particles and as smaller particles enclosed in sulfide dross phases (Figure 8A and 8B). No Co alloys were identified.

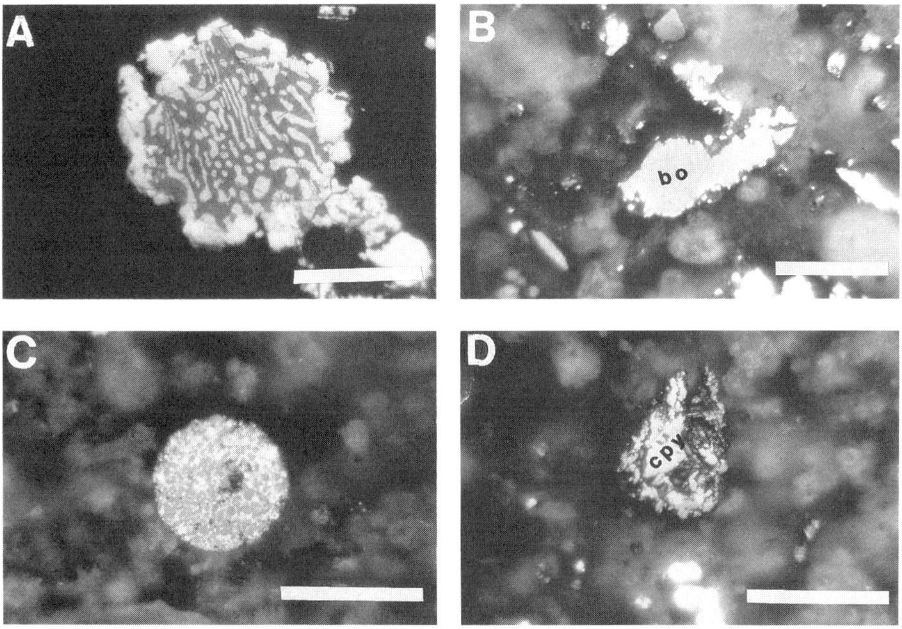

Figure 5. Reflected light photomicrographs of dross furnace dusts of Herculaneum, Missouri showing: A.) Eutectic texture of Cu_2S and Cu-Ni alloy, B.) bornite particle rimmed by vaesite, C.) a chalcocite and magnetite sphere, and D.) original chalcopyrite oxidized to bornite and digenite at the edge. Scale: 100 micrometers.

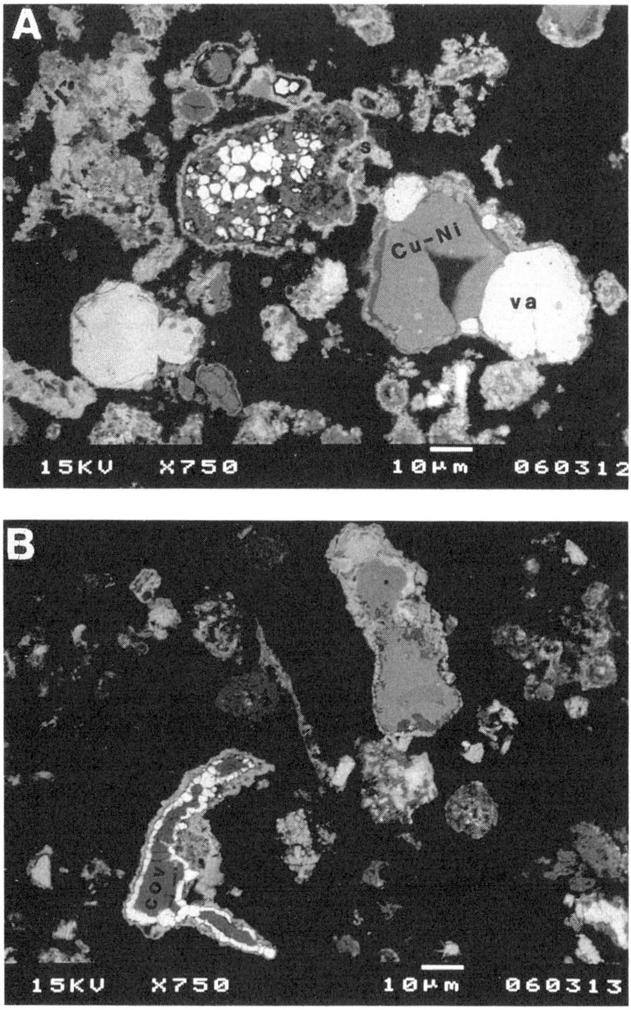

Figure 6. SEM-BSE micrographs of dross phases in dross furnace dusts, showing agglomerated texture of dust particles. A.) agglomerated vaesite, Cu-Ni alloy, and chalcocite coated by metallic lead and slag, and B.) Covellite rimmed by vaesite and a metallic alloy particle. Vaesite: va, covellite: cov, chalcocite: cc, Cu-Ni alloy: Cu-Ni, slag: s.

Figure 7. SEM-BSE micrographs showing unreacted original sulfide phases in dross furnace dusts. A.) A locked particle of chalcopyrite (gray) and galena (white) showing the original ore texture, B.) a zoned bravoite and Co-bearing bravoite unreacted ore particle.

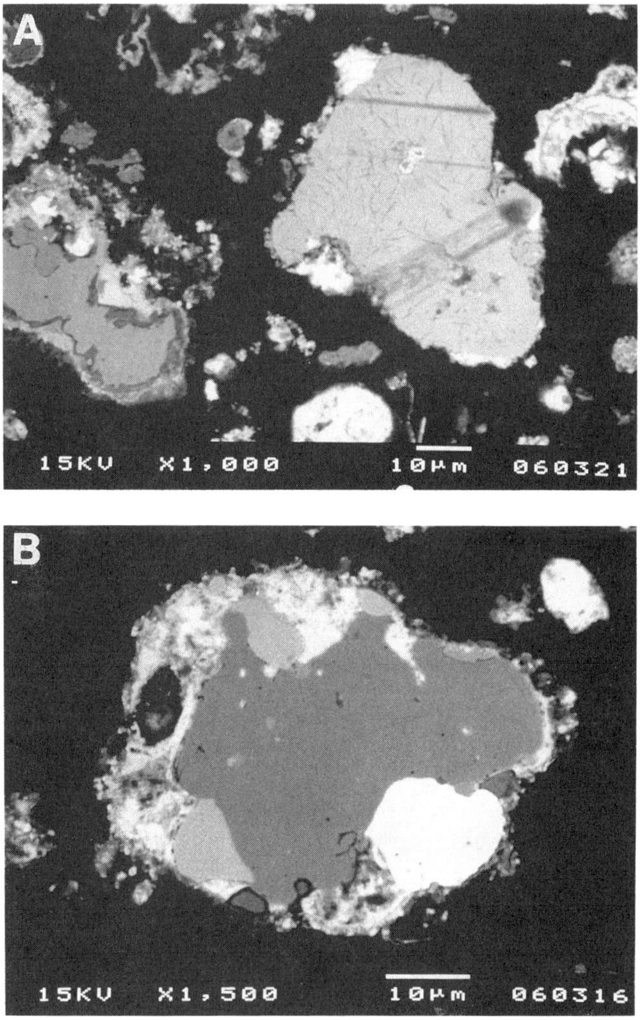

Figure 8. SEM-BSE micrographs showing metallic alloys in dross furnace dusts.
A.) A relatively large free Zn-Ag alloy particle (center), B.) a combined particle of Cu-Zn-Ag (gray), Ni-Cu-Sb (light gray) alloys, and vaesite (white).

Refinery Dust

The main phases in the refinery dust samples are metallic Pb, Zn-Ag, Cu-Zn alloys, and sphalerite (Figure 9A). Minor phases include metallic copper, Cu-Zn sulfide, and Fe-Zn sulfide phases (Figure 9B). Metallic phases occur in very large particles. The compositions of the alloy phases are similar to those described for those particles in the dross furnace dusts.

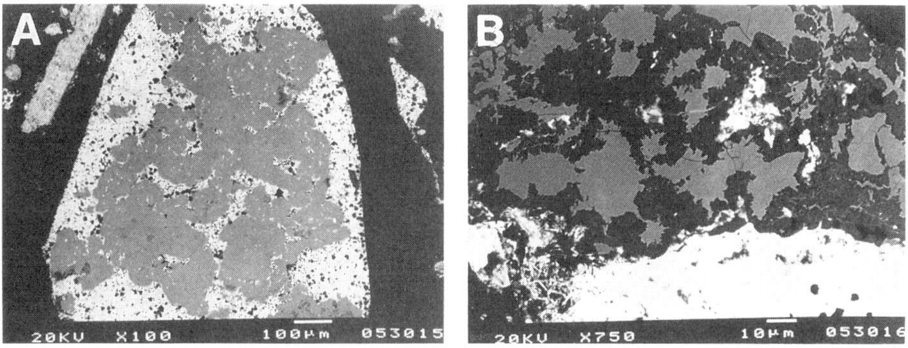

Figure 9. SEM-BSE micrograph of refinery dust samples of Herculaneum, Missouri. A.) Metallic Pb (white) and Zn-Ag alloy (gray). B.) A large dust particle containing metallic Pb (white), Zn-Cu alloy (grey), and Cu-Zn-sulfide phase (dark grey).

Secondary Domestic Copper Smelter Dust

Dust samples recently obtained from a domestic copper smelter were collected from a large horizontal settling chamber in the furnace where dust had collected from the off-take gases. A large accumulation of fume dust was present because the settling chamber had been closed for a long period of time. Although slags from this smelter had been studied[6], no mineralogical study of dust from the plant had previously been undertaken. Samples were collected from the top and bottom of the dust pile.

Dusts from Bottom of Settling Chamber

The bottom dusts consist of large particles of magnetite, hematite, periclase, Fe-Al-Cr-Mg spinel, forsterite, wollastonite, and small amounts of cuprite and tenorite (Figure 10A and 10B). The size and sintered texture of periclase and spinel (Figure 11A and 11B) indicate that these two types of particles were derived from the furnace linings. Magnetite particles are commonly oxidized to hematite and represent a type of rusting that occurred in the furnace. Ca-silicate phases or wollastonite probably are derived from the slag.

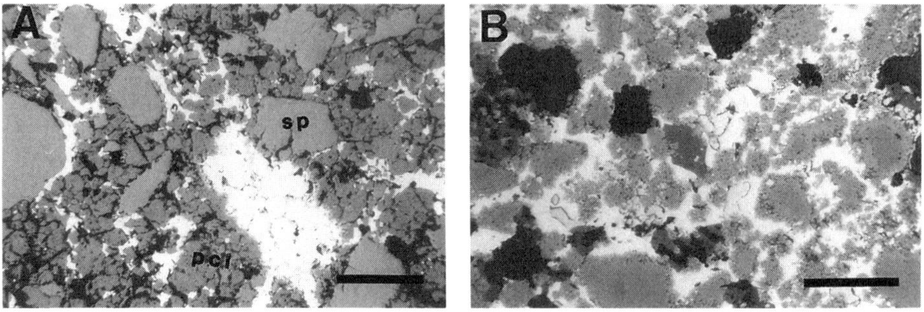

Figure 10. Reflected light micrographs of dusts from a secondary copper smelter, A.) periclase and Fe-Cr-Al-Mg spinel particles enclosed in cuprite, B.) spinel has reacted to form olivine enclosed in a matrix of cuprite. Scale: 1 mm.

Mg-silicate phases are clearly formed as a reaction product between slag and spinel or periclase refractory particles (Figure 8B). These particles were derived from the refractory walls of the furnace during the initial cleaning.

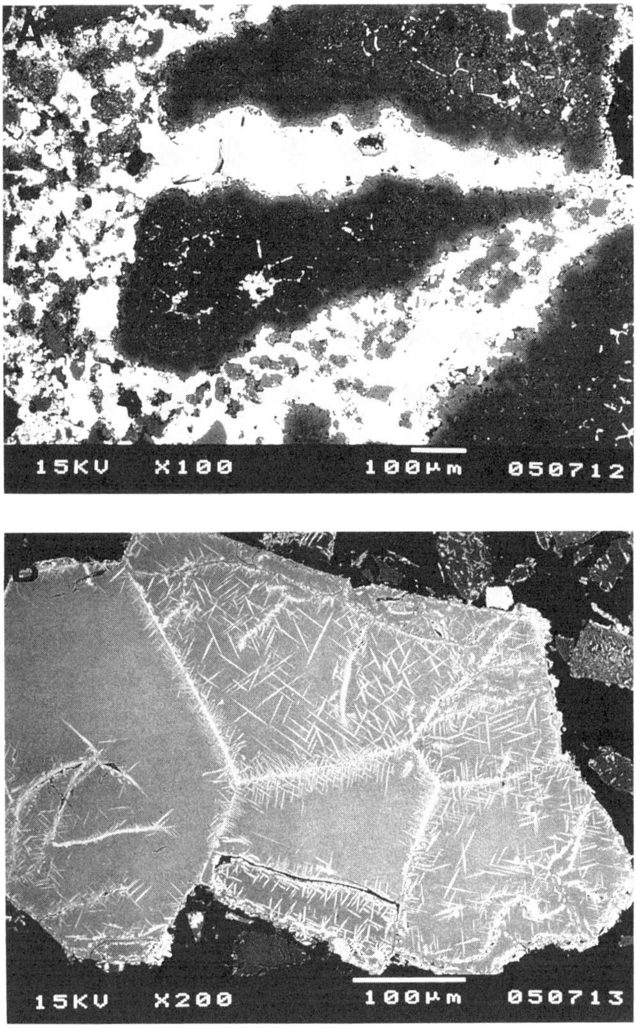

Figure 11. SEM-BSE migrographs of dust particles that are derived from refractory walls: A.) sintered periclase grains in a matrix of cuprite, and B.) a Fe-Cr-Al-Mg refractory spinel particle showing exsolution texture.

Dusts from Top of Settling Chamber

Dusts from the top of the settling chamber contain more than 80 % tenorite. Other dust phases include cuprite (10%), cassiterite, and silicate phases. Copper oxide (tenorite) dust piled up on refractory particles and became agglomerated (Figure 12A and 12B). Aglomerated spheres consist of very small, 3 to 10 micrometers size, tenorite grains (Figure 13A). Small amounts of glassy silicate phase may be present as a bonding material in the agglomerated particles. Cassiterite grains, 3 to 15 micrometers in size, are bound with a glassy phase and agglomerated (Figure 12 D). Cassiterite also occurs as hexagonal plates (Figure 13B). The glass phase is a K-Ca-Al silicate and it contains no deleterious impurities. The only deleterious phase identified is a minor Cu-Zn-rich phase that contains Ca, Fe, Ti, and 3 to 10 % PbO impurities. Tenorite, cuprite, and cassiterite contain no deleterious impurities.

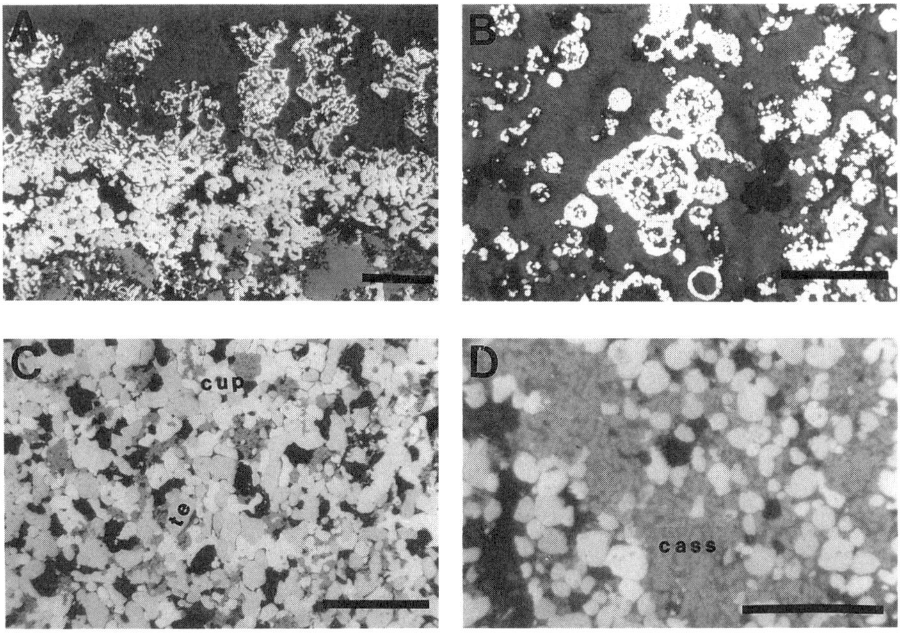

Figure 12. Reflected light micrographs of secondary copper smelter dust taken from top of the settling chamber. A.) Agglomerated tenorite deposited on refractory particles, B.) highly agglomerated tenorite spheres and cenospheres, C.) coarser particles comprised of tenorite, cuprite and cassiterite, and D.) fine cassiterite and tenorite grains bound by glassy phase. Scale: 100 micrometers. Tenorite: te, cuprite: cup, cassiterite: cass.

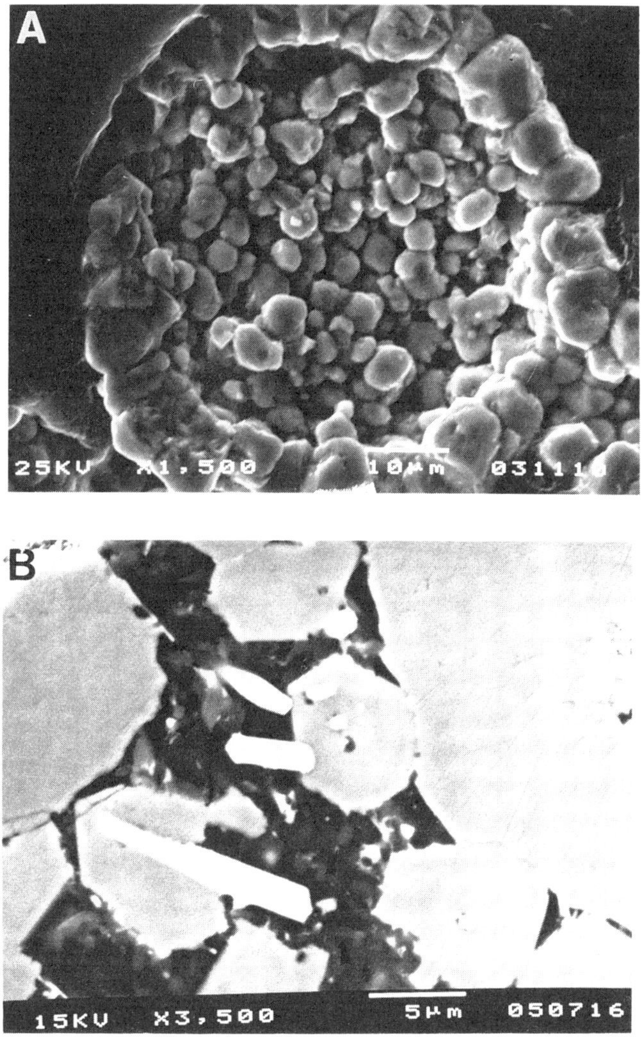

Figure 13. SEM-BSE micrographs showing: A.) a tenorite sphere consisting of individual tenorite grains that are about 3-15 micrometers across, B.) well formed hexagonal needles of cassiterite (white) and tenorite (gray) crystals with very small amounts of a Pb-bearing phase in the matrix (arrowed).

Summary and Conclusions

The employment of process mineralogy techniques in the study of dusts collected from lead and copper smelters has shown that their mineralogy can be effectively characterized by such techniques. The constituent mineral phases present in the dusts can readily be identified by reflected light microscopy. Their compositions can be determined by SEM-EDS. The abundance of each phase can be visually estimated under reflected light, and their modal abundances can be further assessed by automatic image analysis. The grain sizes that are characteristic for individual phases can be measured, and the typical textural intergrowths can be observed by reflected light microscopy. Such information can be important to obtaining a better understanding of the character of the off-gases, and can provide useful information for environmental purposes.

Acknowledgements

This research has been supported by the department of the Interior's Mineral Institutes program administered by the Breau of Mines through the Generic Mineral Technology Center for Pyrometallurgy under allotment grant number G1125129.

References

1. R. D. Hagni and A. M. Hagni, "Determination of the Mineralogy of Dusts near a Missouri Lead Sinter Plant and Blast Furnace by Reflected Light and Scanning Electron Microscopy" EPD Congress 91, Eds. D. R. Gaskell, (Warrendale, PA: TMS, 1991), pp. 821-827.

2. A. M. Hagni, C. Demars, and R. D. Hagni, "Mineralogy of Electric Arc Furnace (EAF) Dusts", EPD Congress 91, Eds. D. R. Gaskell, (Warrendale, PA: TMS, 1991), pp. 801-809.

3. R. D. Hagni and A. M. Hagni, "Mineralogy of Dusts from Lead Sinter Plant, Lead Blast Furnace, and Electric Arc Furnace as Determined by Reflected Light and Scanning Electron Microscopy" ICAM 91, (South Africa: ICAM, 1991), Paper 23, 13 pp.

4. J. M. Gaines and R. D. Hagni, "Mineralogical and Textural Characterization of Good and Bad Lead Sinter by Reflected Light and Electron Microprobe Analysis", ICAM 91, (South Africa: ICAM, 1991), Paper 24, 12 pp.

5. M. Karakus, R. D. Hagni, S. Kang, and A. E. Morris, "Observation and Identification of Dross Phases Precipitated From Lead Bullion", Process Mineralogy IX, Eds. W. Petruk, R. D. Hagni, S. Pignolet-Brandom, and D. M. Hausen (Warrendale, PA: TMS, 1989), pp. 415-428.

6. S. Beutner, R. D. Hagni, and D. G. C. Robertson, "The Mineralogy of Slags from a Secondary Copper Refinary", Process Mineralogy IX, Eds. W. Petruk, R. D. Hagni, S. Pignolet-Brandom, and D. M. Hausen (Warrendale, PA: TMS, 1989), pp. 379-397.

COMPARISION OF AUTOMATED WITH MANUAL TECHNIQUES OF DUST CHARACTERIZATION

Tim B. Vanderwood

Millette, Vander Wood
& Assoc., Inc.
5500 Oakbrook Parkway #200
Norcross, Georgia 30093
404-662-8509

Richard D. Hagni

University of Missouri-Rolla
Department of Geology and Geophysics
Rolla, Missouri 65401
314-341-4616

Abstract

The mineralogy of dusts from several locations in a lead smelting plant, remotely collected dusts and soils near the lead smelter, and dusts from a secondary copper smelter were examined by reflected light microscopy (RLM) to determine their mineralogy, texture, and origins. Individual grains were further analyzed by microbeam techniques. Detailed petrographic analysis of the dust has been supplemented by automated scanning electron microscopy combined with energy dispersive x-ray spectroscopy (ASEM-EDS). This technique allows the rapid sizing and chemical analysis of thousands of individual particles in a dust sample, producing particle size distributions as well as statistically meaningful breakdown of a sample into chemical types (e.g., proportion of galena vs chalcopyrite).

Residues and Effluents - Processing and Environmental Considerations
Edited by Ramana G. Reddy, William P. Imrie and Paul B. Queneau
The Minerals, Metals & Materials Society, 1991

REUSE EVALUATION OF PROCESS DUSTS AND SLAGS BASED ON MINERALOGICAL ANALYSIS

Wolfgang Baum

Pittsburgh Mineral & Environmenalt Technology, Inc.
700 Fifth Avenue
New Brighton, PA 15066

Abstract

Remediation work on thermal process dusts and slags has been primarily based on chemical analyses. These data offer little information for the assessment of treatability and reuse potential. In addition, the lack of mineralogical residue characterization has frequently caused time-consuming and inefficient waste management. This paper will outline the cost-effective and sucessful use of process-oriented mineralogical analyses for treatment and beneficial reuse of various thermal residues. These include a patented process for detoxification of Pb-contaminated bottom ash, the pyrometallurgical treatment of K 061, the reprocessing of lead smelter slag as well as various power plant residues.

Residues and Effluents - Processing and Environmental Considerations
Edited by Ramana G. Reddy, William P. Imrie and Paul B. Queneau
The Minerals, Metals & Materials Society, 1991

SEM CHARACTERIZATION OF LIGNITE FLY ASH

Robert J. Stevenson

McCrone Research Institute
2820 S. Michigan Avenue
Chicago, Illinois 60616

Abstract

The chemical variation of central North American lignite fly ash can be compared using the major and minor oxide analyses and the distribution pattern of the inter-grain chemistries of the individual fly ash have been used to illustrate the variations in bulk sample and individual grain chemistry. This information documents the chemical variability of the lignite fly ashes by geographic region. There is a direct relationship between grains rich in SiO_2 and Al_2O_3 and in the Na_2O content of the grains in lignite derived fly ashes. There is also a direct relationship between grains rich in CaO and the MgO and SO_3 content of the grains in lignite derived fly ashes. These variations are shown using CaO-SiO_2-Al_2O_3 and CaO-Na_2O-MgO ternary diagrams.

Residues and Effluents - Processing and Environmental Considerations
Edited by Ramana G. Reddy, William P. Imrie and Paul B. Queneau
The Minerals, Metals & Materials Society, 1991

A METHOD FOR RAPID CHARACTERISATION OF FLY ASH

C.D.Mishra, H.D.Pandey, B.Prasad & S.R.Mediratta

Research & Development Centre for Iron & Steel,
Steel Authority of India Ltd.,
Ranchi-834002, India

Abstract

The recycling potential for any gainful utilisation of the waste products is dictated by their quality, especially the chemical characterisation. The methods commonly used for the analysis of fly ash e.g. wet chemical, AAS, ICP-AES etc are time consuming. The paper describes a rapid XRF-based method for routine analysis of fly ash which involves preparation of local standards, optimisation of sample preparation technique and use of a simple PC-based software to account for interelement effects. The matrix corrected values are found to agree well with the wet chemical data.

Residues and Effluents - Processing and Environmental Considerations
Edited by Ramana G. Reddy, William P. Imrie and Paul B. Queneau
The Minerals, Metals & Materials Society, 1991

Introduction

The enormous quantity of fly ash generated by coal-based thermal power plants has increasingly been posing an environmentally complex problem. The problem is manifold higher in India where the coal used for power generation is not only rich in ash content (average being 30%) but also varies widely in composition. However, with emphasis, in recent years, on recycling of waste products, the attitude towards fly ash has changed considerably due to the realisation of its potential as a valuable raw material in many applications such as cement-making, brick-making etc.

The major constituents of fly ash are identical to earthy materials such as soil and shales. While the oxides of iron, calcium, aluminium, silicon, magnesium constitute about 98-99% of the composition of fly ash, other elements such as potassium, sodium, titanium, chromium, cobalt etc occur only in traces. Sometimes the improper or incomplete combustion results in the presence of significant amount of carbon in fly ash matrix. However, the composition of fly ash has been seen to vary widely owing to the variation in the coal type and its origin. Since any gainful utilisation of fly ash is dictated by its quality, the determination of variability of chemical composition assumes special significance.

The techniques most commonly used for analysis of fly ash are AAS, ICP-AES or wet chemical as defined in ASTM (311)[1]. All these methods are time consuming since the sample preparation is both tedious and critical to the analysis. Attempts have also been made to analyse samples of fly ash by X-ray fluorescence (XRF)[2,3]. Although XRF instrumentation has higher capital cost, it is cheaper and easier to operate than other instrumental techniques. However, the complex and heterogeneous nature of fly ash requires extra precautions to be taken at the sample preparation stage.

In X-ray fluorescence analysis, grain size, matrix and mineralogical effects impair the analysis results. In order to reduce these problems, samples are crushed to very fine size to obtain pellets or dissolved in a flux to form glossy beads. The pelletisation method, due to various limitations is liable to suffer from inherent inhomogeneities. In addition, in fly ash sample pellets ageing effects[3] have been noticed due to variable ionic character of different constituents resulting in inconsistent mobility of ions. The fusion bead technique of sample preparation overcomes most of these shortcomings of the pelletisation method[4].

The paper outlines an XRF-based method for rapid analysis of fly ash samples. The work involves development of local standards, optimisation of sample preparation procedure and correction for interelement effects using a PC-based software. It is shown that the bead technique of sample preparation is superior and ideally suited for fly ash analysis.

Experimental Details

In the present work, fly ash samples were collected from different hoppers of power plants at Rourkela and Bokaro Steel Plants for preparing local standards. These samples were analysed repeatedly in a laboratory at R&D Centre for Iron & Steel (RDCIS) by wet chemical method. The same samples were also analysed at the National Metallurgical Laboratory (NML), Jamshedpur and National Test House, Calcutta. The average results are given in Table I.

Table I Average Percentage Composition of Fly Ash Samples for Use as Local Standard Reference Material (SRM)

Sl. No.	Sample No.	SiO_2	Fe(T)	Al_2O_3	CaO	MgO	S
1.	663	51.93	8.13	23.29	0.95	0.45	0.110
2.	664	51.36	6.38	23.42	1.04	0.52	0.098
3.	665	51.68	6.72	26.33	1.17	0.61	0.096
4.	666	57.02	5.07	24.25	1.16	0.38	0.032
5.	1240	56.65	5.42	24.66	1.34	0.33	0.053
6.	1241	54.11	4.47	30.89	1.51	0.39	0.043
7.	1242	54.10	9.62	24.88	1.38	0.40	0.220
8.	1243	57.46	4.02	30.27	1.49	0.36	0.046
9.	1244	58.75	7.27	27.11	1.68	0.35	0.027
10.	1245	57.20	4.42	30.98	1.53	0.38	0.027

*Participants: RDCIS,Ranchi; NML,Jamshedpur & National Test House,Calcutta

A simple pressed pellet technique was adopted to prepare pellets for each fly ash sample. One gram of fly ash sample crushed to -300 mesh size was thoroughly mixed with four grams of cellulose binder to ensure homogeneity. The mixed powder was then pressed at 80 Kg/cm^2 pressure in aluminium holder for five minutes in the pellet making machine (HERZOG).

Fusion beads were obtained using the following procedure. One gram of each fly ash sample was thoroughly mixed with ten grams of sodium tetraborate (Borax) and the mixture was heated in a gold-platinum crucible (5% gold and 95% platinum) in a fusion furnace at 1000°C for 10-15 minutes. The melt was stirred briefly to ensure proper dissolution and release of air bubbles. This also prevents the bead from cracking during solidification[5,6].

The pellets and beads were then subjected to XRF analysis using the wave-length dispersive Philips PW1450/20 XRF Spectrometer. Each specimen was analysed for Fe(T), SiO_2, Al_2O_3, CaO, MgO and S with instrumental parameters shown in Table II.

Table II Instrumental Parameters for Analysis of Fly Ash

Phases	kV	mA	Line	Angle 2θ	Crystal	Detector
Fe(T)	50	45	K_α	57.50	LIF 200	FL
SiO_2	50	45	K_α	109.56	PET	FL
Al_2O_3	50	45	K_α	145.42	PET	FL
CaO	50	10	K_α	45.40	PET	FL
MgO	50	45	K_α	137.00	ADP	FL
S	50	45	K_α	76.20	PET	FL

The intensities thus measured by XRF for individual element were plotted against their respective concentrations. It was found that this simple intensity versus concentration plot which can be expressed mathematically as -

$$C_i = A_o + A_i\ I_i\ (i = Fe,\ Ca,\ Al\ etc) \quad (1)$$

did not yield satisfactory correlation in case of pellet specimens. However, samples prepared by the bead technique showed some improvement in correlation.

The wide scatter of points in the plots implies significant absorption and/or enhancement of intensities due to the presence of other elements or due to grain size effect as in the case of pellets. In order to account for the interfering elements, an equation correlating the concentration of the individual element with its own intensity as well as a linear function of other interfering intensities was formulated. It can be expressed as -

$$C_i = A_{io} + A_{ii}\ I_i + F(I_j) \quad (2)$$

$$\text{Where } F(I_j) = \sum_j A_{ij}\ I_j, \qquad i \# j \quad (3)$$

and A_{io}, A_{ii}, A_{ij} are constant and co-efficients of the equation and i, j represent the analyte and the interfering elements respectively.

A personal computer was used to compute A_{io}, A_{ii}, A_{ij} and concentration of individual component in each specimen.

The concentrations of various components in fly ash samples obtained both from pellets and beads are then compared with the wet chemical results both before and after matrix corrections. Typical comparisons are shown in Figures 1-4 for four different constituents viz. MgO, SiO_2, S, Fe(T). The standard error of estimate, R^2-values and compilation of interfering elements for various constituents of fly ash are given in Tables III & IV for pellets and beads respectively.

Table III Standard Error of Estimate and Correlation Co-efficient for Various Fly Ash Specimen (Pellets)

Elements	Range of Analysis (%)	Improvement in R^2-values		Standard error of estimate		Matrix correction required for the elements
		From	To	From	To	
SiO_2	51.36-58.75	0.244	0.967	2.516	0.860	Fe(T),Al_2O_3,CaO,MgO,S
Fe(T)	4.02-9.62	0.900	0.986	0.606	0.359	SiO_2,Al_2O_3,CaO,MgO,S
Al_2O_3	23.29-30.93	0.461	0.910	2.556	0.299	Fe(T),SiO_2,CaO,MgO & S
CaO	0.95-1.68	0.639	0.883	0.169	0.163	Fe(T),SiO_2,Al_2O_3,MgO & S
MgO	0.35-0.61	0.447	0.977	0.075	0.026	Fe(T),SiO_2,Al_2O_3,CaO & S
S	0.027-0.22	0.617	0.948	0.039	0.023	Fe(T),SiO_2,Al_2O_3,CaO & MgO

Table IV Standard Error of Estimate and Correlation Co-efficient for Various Fly Ash Specimen (Bead))

Elements	Range of Analysis	Improvement in R^2-values		Standard error of estimate		Matrix correction required for the elements
		From	To	From	To	
SiO_2	51.36-58.75	0.799	0.992	0.886	0.587	Fe(T),Al_2O_3,CaO,MgO,S
Fe(T)	4.02-9.62	0.984	0.998	0.389	0.223	SiO_2,Al_2O_3,CaO,MgO,S
Al_2O_3	23.29-30.93	0.520	0.875	2.550	0.359	Fe(T),SiO_2,CaO,MgO & S
CaO	0.95-1.68	0.607	0.910	0.161	0.154	Fe(T),SiO_2,Al_2O_3,MgO & S
MgO	0.35-0.61	0.158	0.998	0.101	0.023	Fe(T),SiO_2,Al_2O_3,CaO & S
S	0.027-0.22	0.909	0.998	0.019	0.006	Fe(T),SiO_2,Al_2O_3,CaO & MgO

R^2 - Values of correlation co-efficients.

Results & Discussion

An examination of Tables III & IV reveals that the values of correlation co-efficients (R^2 - values) in each case improves considerably after matrix corrections. This holds good equally in both pellet as well as bead samples. The effect of matrix is particularly more pronounced in the case of MgO analysis in bead samples and SiO_2 analysis in pellet specimens wherein the correlations are observed to be practically non-existent. Application of matrix correction in both cases improves the R^2-values to well within the acceptable range (viz. 0.998 and 0.967 respectively). These trends are reflected in the corresponding figures as well.

A comparison of results obtained from both pellet and bead methods of sample preparation as shown in Figures 1-4 and Tables III & IV clearly demonstrates the superiority of the bead technique over the pelletisation method. While R^2-values both before and after matrix correction are generally higher in bead specimens, a reversal of this trend is observed in the case of MgO analysis before matrix correction. This anomaly, may be attributed to the presence of extra Na atoms introduced in the bead matrix through the use of borax as the flux.

Considerable absorption and enhancement of intensities are observed in the case of analysis of light elements such as Al_2O_3 and MgO. These effects are due to the presence of other heavier and lighter elements. While corrections as given in Eqn.2 seem to account for most of these effects in the case of MgO, it appears to fail marginally in the case of Al_2O_3. The alkali metals in the matrix have been found to affect Al_2O_3 intensity significantly[7]. The results of Al_2O_3 analysis could, therefore, be improved by introducing the effect of alkali metals in the matrix calculation.

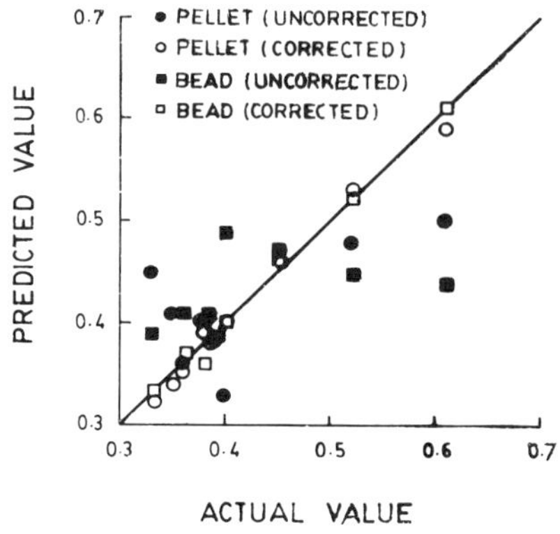

Figure 1 - Chemical versus XRF results for MgO in fly ash

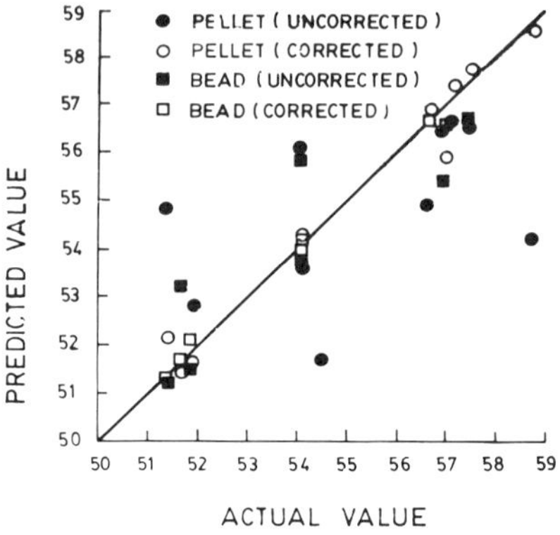

Figure 2 - Chemical versus XRF results for SiO_2 in fly ash

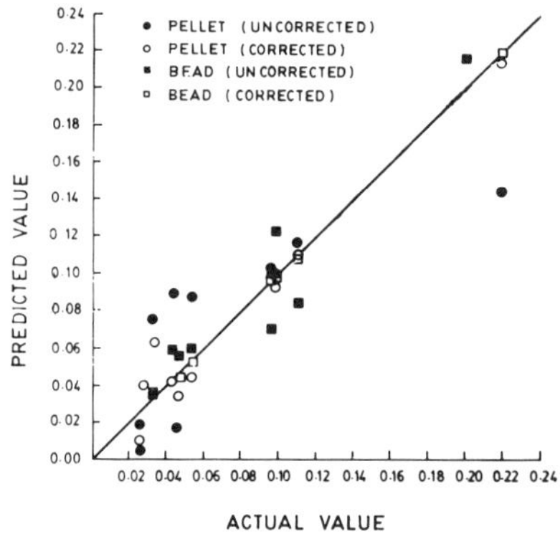

Figure 3 - Chemical versus XRF results for S in fly ash

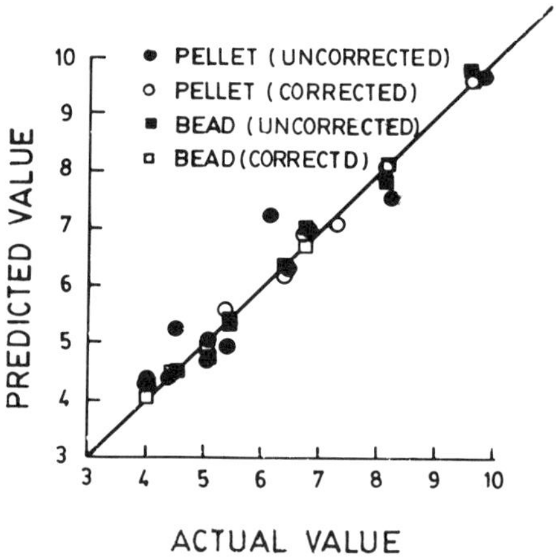

Figure 4 - Chemical versus XRF results for Fe(T) in fly ash

Conclusion

1. X-ray fluorescence is a powerful tool for rapid and accurate estimation of various major constituents of fly ash.

2. The bead technique of sample preparation, yields much better correlation than the pellet method though reasonably acceptable results can be obtained by the latter also.

3. In both cases of sample prepation the interelement effects play important roles in determining the accuracy of results. A simple linear equation (Eqn.2) proposed in the paper is found to be adequate in correcting for interelement effects.

Acknowledgement

The authors are grateful to Mr.S.K.Roy, Director and Dr.V.Ramaswamy, Executive Director for encouragement and support during course of the work and for permission to publish this work. We are also thankful to Shri S.Jha, System Analyst of MTI for his help in developing the software. The assistance rendered by S/Shri R.B. Sinha, Manoj Kumar, B. Khalkho & Nayendra Ram at various stages of work is also gratefully acknowledged.

References

1. American Society for Testing and Materials, Annual Book of ASTM Standards, Part 14, Philadelphia: ASTM (1981).

2. Scoot Schlorholtz and Mustafa Boybay, "Quantitative X-ray Fluorescence Analysis for Fly Ash Samples", Advances in X-ray Analysis, 27 (1984), 497-504.

3. M. Bettielli and P. Taina, "Rapid Analysis of Coal Fly Ash by X-ray Fluorescence Spectrometry", X-ray Spectrometry, 19 (1990), 227-232.

4. H.D.Pandey, C.D.Mishra & B.Prasad, "A Fusion Bead Technique for Analysis of Slags by X-ray Fluorescence", Steel India, 10 (1987), 64-70.

5. C.D.Mishra, H.D.Pandey, B.Prasad & S.R.Mediratta, "Accurate Rapid Analysis of Indian Sinter by XRF", Steel India, 14 (1990), 41-47.

6. C.D.Mishra, H.D.Pandey & B.Prasad, Report No.R&D:62:02:1489:90 (1990).

7. Eugene P.Bertin "Principles and Practices of X-ray Spectrometric Analysis, (New York, Plenum Press, 1975).

3.
PROCESSING

3.1
EAF DUSTS

ELKEM'S MUTI-PURPOSE* FURNACE TEST FACILITY AT MEFOS AND THE INITIAL OPERATING PHASE OF AN ELKEM HTMR SYSTEM FOR EAF DUST

Dr. Goren Lindkvist and Inge Hystad

Elkem Technology a/s
P.O. BOX 4376 Torshov-N 0402 Oslo 4, Norway

Abstract

Elkem has a small industrial scale Muti-Purpose* furnace at the MEFOS Metallurgical Research Facility in Lulea, Sweden. The furnace is 2 meters in diameter with installed power of 4.8 MVA, equipped with oil colled freeze lining, gas tight electrode seals and feed chutes and a slag flow control for continuous tapping.
During tests in June, 1991, the furnace was continuously operated on a high silica slag at loads from 1.5 to 4.0 MW. The slag was then charges to a low silica, high iron and lime composition. Steel plant dust was added, smelted and reduced. The slag was made more ceramic by adding alumobrick and alumina. Data from the start-up and initial operating phase of the Elkem EAF dust plant at Laclede Steel will also reviewed.

Residues and Effluents - Processing and Environmental Considerations
Edited by Ramana G. Reddy, William P. Imrie and Paul B. Queneau
The Minerals, Metals & Materials Society, 1991

AMMONIUM CARBONATE LEACHING OF REDUCED ELECTRIC ARC FURNACE (EAF) DUST

R.L. Nyirenda, E.M.L. Peek, G. Van Weert

Delft University of Technology, Department of Raw Materials Technology
Faculty of Mining & Petroleum Engineering, Mijnbouwstraat 120
2628 RX Delft, The Netherlands

Abstract

One of the problems with ammoniacal-ammonium carbonate leaching of EAF carbon steelmaking flue dust is that zinc present in spinel phases is not leached. This paper presents results of test work in which prior to leaching, EAF dust was reduced with CO containing gas to decompose the spinels into zinc oxide and iron bearing phases, mainly metallic iron and wustite. From such reduction products, the amount of zinc leached increases but the iron co-leached becomes extensive. The effect of various leaching parameters such as liquor concentration and redox potential on zinc and iron dissolution has been investigated. Supplementary experiments utilizing pure zinc oxide and pure iron phases were also carried out. Results show that the co-dissolution of iron from severely reduced EAF dust cannot be prevented. This is because the wustite phase contains zinc which can only be leached when iron also dissolves. Further, the zinc bearing wustite phase is not totally soluble so that the residue cannot be made zinc free. In the absence of air, dissolution of metallic iron hardly occurred, though in leaching experiments of a long duration its transformation to a spinel structure was observed.

Residues and Effluents - Processing and Environmental Considerations
Edited by Ramana G. Reddy, William P. Imrie and Paul B. Queneau
The Minerals, Metals & Materials Society, 1991

Introduction

During the production of carbon and low alloy steel from scrap in electric arc furnaces, 15-20 kg of dust is recovered in gas cleaning units per tonne of steel produced. The presence of leachable Pb, Cd, and hexavalent Cr in the dust has in recent years caused the material to be widely regarded as a toxic waste which is unsuitable for disposal in landfills. In the US, the material even faces a ban from disposal on hazardous waste sites [1].

The classification of EAF carbon steelmaking dust as a hazardous waste has caused a sharp increase in its disposal cost. This has led to the development of processes, several of which have been commercialized, aimed at de-toxifying the material prior to disposal. A crucial aspect to the economics of most of these processes is the recovery of valuable constituents in the dust, mainly zinc and lead. The proposed and commercially available processes have been reviewed in many papers some of which are cited in the references [2-6].

Of the various concepts for processing EAF dust, those based on pyrometallurgy have been more successful than those utilizing hydrometallurgical techniques. A problem which is common with most of the concepts in the latter category is that of low zinc recovery. This also applies to an attempt to utilize ammoniacal ammonium carbonate (AAC) for leaching of EAF carbon steelmaking dust. Whereas zinc present in the dust as an oxide is readily soluble in AAC, that portion present as franklinite and franklinite-magnetite solid solutions, typically 20-50% of total zinc, is not leached. A way of increasing the zinc recovery would be to precede leaching with a reduction step aimed at segregating zinc oxide from these spinel phases. Another major problem with AAC leaching is that lead present in the dust is hardly leached [5]. However, if most of the zinc in the dust were to be recovered then the residue could be leached with another liquor like acetic acid to recover the lead [7].

This paper presents results of bench scale tests in which prior to leaching with AAC solution, EAF dust was reduced with carbon monoxide containing gas and the spinels decomposed into zinc oxide and iron bearing phases, mainly metallic iron and wustite. Previous work [8] had shown that EAF dust reduced at 600 °C for up to 13 hours did not show a change in the XRD pattern when the iron-bearing zinc ferrite decomposition product was magnetite. This observation is similar to results obtained in preliminary work to this study which showed that reduction of the dust to a product containing magnetite only changed the zinc leachability from 59% for the unreduced dust, to not more than 65% for the reduced dust.

Leaching tests for this preliminary work were conducted at a solution temperature of 50 °C, using leach liquor which had 134 gpl total NH_3 and 105 gpl CO_2. These leaching conditions were selected on the basis of information available in the literature on the use of AAC leaching for the production of zinc oxide from low grade zinc ores [9] and car scrap recycling residues [10]. It was unclear however whether these leaching conditions were also optimum with respect to the dissolution of zinc from reduced EAF dust.

For the EAF dust sample utilized in the preliminary work and in this study, Figure 1 shows the amounts of zinc and iron which were leached from the reduced and unreduced dust. The reduced product had wustite and metallic iron as the main iron bearing phases. Details on the EAF dust sample utilized and of the reduction and leaching procedures are described in the next section. Figure 1 shows that for the reduced dust, the amount of zinc leached increased to 88% from 59% but that the amount of iron co-leached rose from 4% to 60%. Since the recovery of zinc from a solution with such a high content of iron is not practically feasible it would be advantageous if iron dissolution from reduced EAF dust could be suppressed.

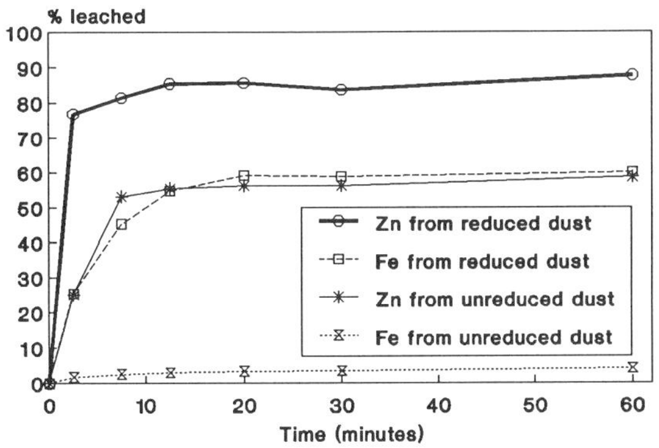

Figure 1 - Zinc and iron leached from 10 grams of reduced and unreduced EAF dust. ($[NH_3]$=134 gpl, $[CO_2]$=105 gpl, liquor volume=600 cm^3, T=50 °C, stirring rate=380 rpm)

Scope of the Present Work

This study was aimed at exploring the possibility of securing selective leaching of zinc from reduced EAF dust. The main factors investigated were the leachant concentration and the solution redox potential. The effect of solution temperature, solid/liquid ratio, and stirring rate on the dissolution of iron and zinc were also considered. In a bid to better understand the dissolution characteristics of zinc and iron from the dust, supplementary tests with pure zinc oxide and iron phases were also carried out. Filtration tests did not however constitute part of this investigation.

Experimental

EAF Dust utilized

The dust sample used had the following chemical analysis as determined by XRF: 18.8% Zn, 28.8% Fe, 2.2% Pb, 0.1% Cd, 0.2% Cr, 8.5% Ca, 5.2% Mn, and 1.6% Mg. The main patterns in the X-ray diffractograms of the material were of the spinel phases and zincite (Figure 2). Physical properties of the dust were typical for this class of material. For instance, the BET surface area was 3-4 m^2/g while the particle size was less than 1 micron.

Reduction

Each reduction experiment was carried out with a pre-dried sample of 22 grams held in an alumina boat. Reductions were done in a tube furnace at a constant temperature of 650 °C using gas with a CO_2/CO ratio of 0.316 and for a duration of 90 minutes. Thereafter, cooling to room temperature in a nitrogen atmosphere followed. The reduction temperature used was a compromise between zinc loss by volatilization at high temperature [11,12] and poor kinetics at low temperature [12].

Nature of the Reduced Dust

Clearly identifiable phases in the reduced product were zincite, wustite, and metallic iron (Figure 2), while XRF analysis yielded the following elemental analyses: 19.5% Zn, 30.1% Fe, 2.5% Pb, < 100 ppm Cd, 0.2% Cr, 9.4% Ca, 5.2% Mn, and 1.9% Mg.

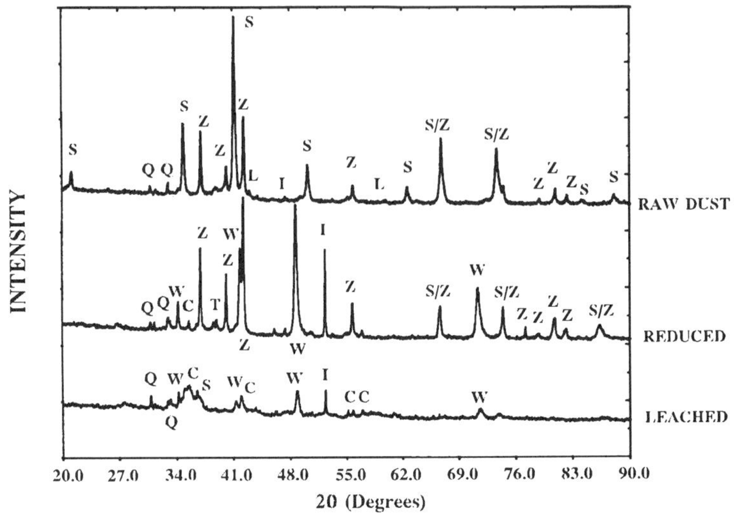

Figure 2 - XRD patterns of (a) initial EAF dust, (b) reduced dust and (c) reduced and then leached dust. (Leaching conditions: T=25 °C, $[NH_3]$=36 gpl, $[CO_2]$=30 gpl)

Legend: C = Calcite; I = Metallic iron; L = Lime; Q = Quartz; S = Spinel; S/Z = Spinel and/or zincite; T = Sphalerite; W = Wustite; Z = Zincite

Leaching

This was done in an oil-jacketed stainless steel cylindrical vessel with an internal diameter of 110 cm. In order to maintain the leach liquor at the required temperature, oil was pumped from a heated bath through the outer shell of the vessel. Leaching was carried out with 600 cm^3 of liquor prepared from pre-analyzed ammonium carbonate and ammonium hydroxide, while samples to be leached weighed 10 grams except when the effect of pulp density had to be investigated. The leaching vessel was fitted with a stirrer, thermometer, and electrodes for the measurement of redox potential. The stirrer had a pitch blade impeller with a diameter of 40 mm.

To ensure that air was kept out, a nitrogen flow of about 0.75 l/min was maintained through the leaching vessel. Ammonia and carbon dioxide loss due to the nitrogen flow was low, however. For instance in a test with 1.5 l/min nitrogen, a loss of 2 grams/hr ammonia was estimated at 55 °C from a liquor with an initial composition of 36 gpl NH_3 and 30 gpl CO_2. This loss was determined by a titration method after bubbling off-gas through HCl solution.

Solution Analysis

Sampling of the solution in order to analyze for zinc and iron was facilitated by means of two narrow tubes equipped with valves. One tube was connected to a nitrogen cylinder, and was used to gently pressurize the vessel freeboard thus causing about 10 cm^3 of sample to spout from the second tube which had one end dipping into the solution. Care was taken to prevent liquid stagnation in the sampling tube by periodically flushing it with nitrogen.

Zinc and iron in solution were determined by atomic absorption spectrophotometry. Using these analyses, the cumulative zinc and iron leached from the material as a percentage of the amount in the initial sample were calculated at any sampling time. The calculations had a volume correction to take account of the diminished liquid in the vessel due to earlier sampling. A similar correction for solids was not applied due to the small amount of solids in sample solutions. As determined from duplicate experiments, the maximum discrepancy in the reported values of iron and zinc leached was 9.6% and 3.6%, respectively.

Results and Discussion

Effect of Selected Variables on Zn and Fe Dissolution from Reduced EAF Dust

Solid/Liquid Ratio. The solid/liquid ratio during leaching was considered at two levels. The leaching conditions utilized for this set of tests and the results obtained are shown in Figure 3. Solid/liquid ratio had negligible influence on the selectivity of zinc and iron dissolution. A ratio of 1/60 was selected for use in further experiments in view of the somewhat higher zinc leachability.

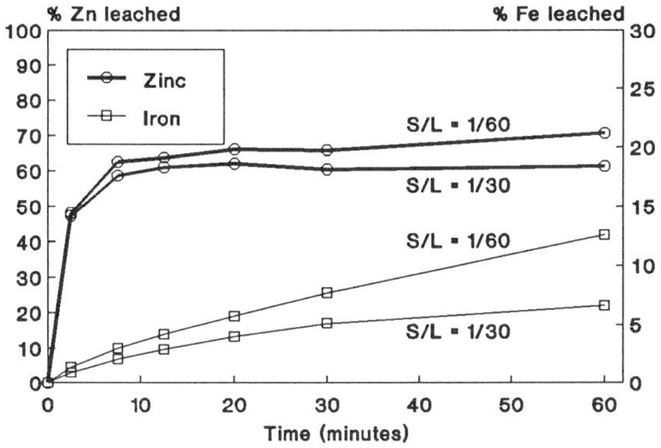

Figure 3 - Effect of solid/liquid (S/L) ratio on the dissolution of zinc and iron from reduced EAF dust. ($[NH_3]$=36 gpl, $[CO_2]$=30 gpl, T=25 °C, stirring rate=380 rpm)

Stirring Rate. The influence of agitation was investigated in a series of tests between 150 and 900 rpm. At 150 rpm, far less solids were present in the sampled liquid than at higher speeds investigated. This observation suggested that solids were not totally suspended at 150 rpm. At higher speeds, the dissolution rates of both zinc and iron were not significantly different. For all stirring rates, the dissolution profiles were similar to those shown in Figure 3. Figure 4 compares for the different agitation rates the solubilized amounts after 60 minutes of leaching. Since agitation speed had no influence on dissolution, as long as the solids were in suspension, it is most likely that the rate limiting step for both zinc and iron is governed by chemical control. A stirring rate of 380 rpm was used in all subsequent experiments.

Ammonia and Carbon Dioxide Concentration. The influence of ammonia and carbon dioxide concentration on the amount of iron in solution is shown in Figures 5 and 6, respectively. The lowest liquor concentrations utilized had about 1.5 times the stoichiometric amount of CO_2 and 8 times the stoichiometric amount of NH_3 required for total dissolution of all iron and zinc in the samples. The results show that under these conditions where more than a stoichiometric amount of the other reagent is present, increasing the concentration of ammonia or carbon dioxide leads to an increase in the amount of iron in solution.

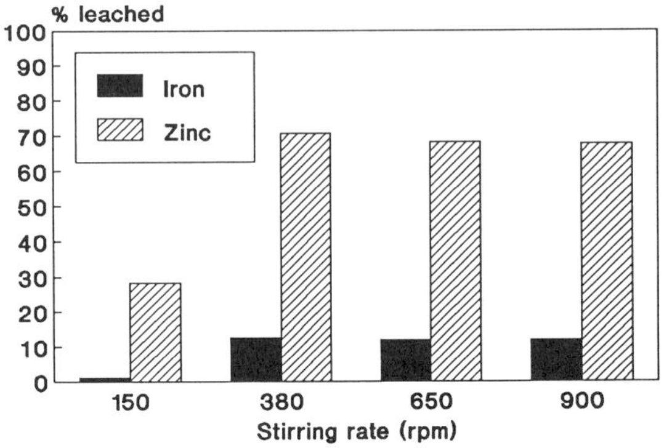

Figure 4 - Effect of stirring rate on leachability. ($[NH_3]$=36 gpl, $[CO_2]$=30 gpl, T=25 °C, time=60 min)

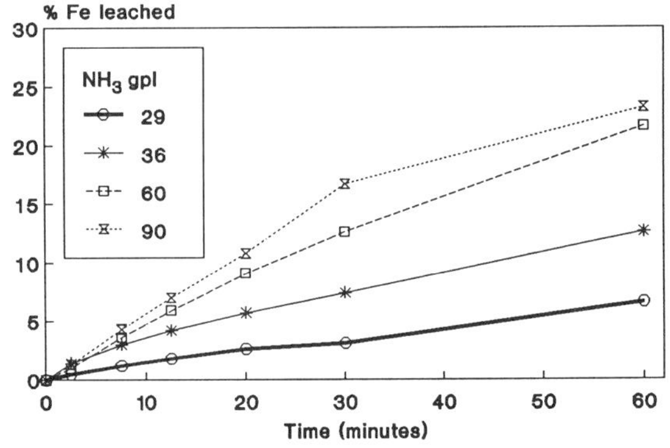

Figure 5 - Influence of NH_3 concentration on the dissolution of iron. ($[CO_2]$=30 gpl, T=25 °C, n=380 rpm)

Except for the leaching test with 29 gpl NH_3 (Figure 7) and possibly the test with 8.1 gpl CO_2 (Figure 8), the dissolution of zinc did not have a dependence on NH_3 or CO_2 concentration as observed for iron. In all cases, zinc dissolution was rapid and after about 15 minutes very little change occurred in the amount of zinc in solution. As reflected in the formula of the zinc complex in solution, $Zn(NH_3)_nCO_3$ with $1 \leq n \leq 6$ [9], dissolution of zinc oxide in AAC solutions occurs in an equimolar ratio with CO_2 so that the lower amount of zinc leached with the 8.1 gpl CO_2 solution is an indication of the poorer kinetics of having only slightly more than the stoichiometric requirement of CO_2 in the system.

The low amount of zinc in the solution with 29 gpl NH_3 could be related to the rather low NH_3/CO_2 ratio of this leach liquor as compared to the other experiments. Whereas the leach liquor with 29 gpl NH_3 had an NH_3/CO_2 molar ratio of 2.5, the other liquors with higher amounts of zinc in solution had molar ratios of 3.1 or greater. The composition of AAC solution for optimum dissolution of non-carbonate zinc minerals like zinc oxide, is known to have an NH_3/CO_2 molar ratio in the range 3 to 3.5 [9].

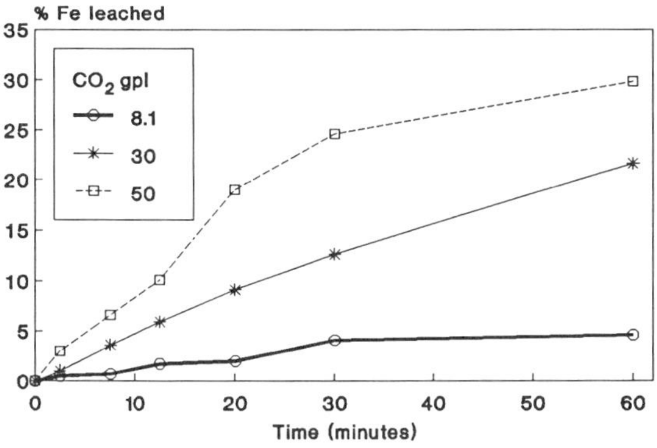

Figure 6 - Influence of CO_2 concentration on the dissolution of iron. ($[NH_3]$=60 gpl, T=25 °C, n=380 rpm)

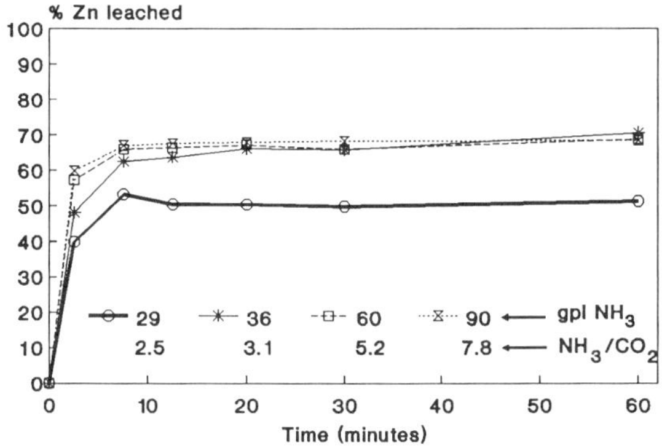

Figure 7 - Effect of NH_3 concentration on the dissolution of zinc. NH_3/CO_2 refers to the molar ratio of the two reagents in the liquor. ($[CO_2]$=30 gpl, T=25 °C, n=380 rpm)

In terms of simultaneously obtaining high zinc and low iron dissolution, it did not appear worthwhile to exceed an NH_3/CO_2 molar ratio of 3.1, and in most subsequent experiments a liquor with 36 gpl NH_3 and 30 gpl CO_2 was utilized.

During the leaching step, zinc present as zinc oxide was completely and rapidly solubilized. The residue had wustite, calcite, and metallic iron as the most abundant phases and a small amount of spinel and other minor phases. As discussed in a later section, most of the zinc present in the leach residue was most likely as a solid solution in wustite. Figure 2 includes the XRD pattern of the residue of leaching reduced EAF dust at 25 °C with a liquor containing 36 gpl NH_3 and 30 gpl CO_2. The chemical analysis of the major elements present in this residue reported: 8.0% Zn, 36.7% Fe, 11.0% Ca, 6.6% Mn, 3.5% Pb.

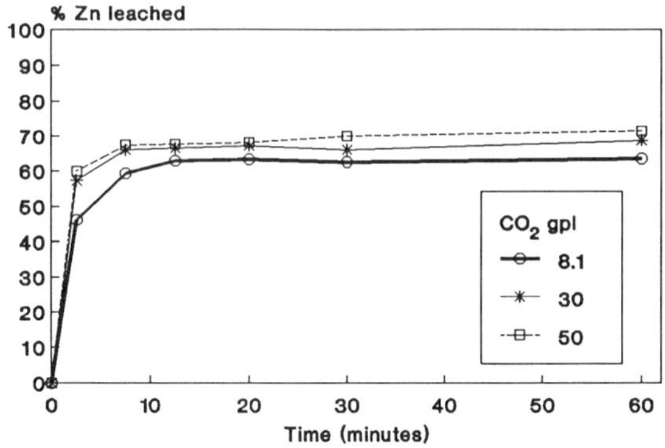

Figure 8 - Effect of CO_2 concentration on the dissolution of zinc. ($[NH_3]$=60 gpl, T=25 °C, n=380 rpm)

Solution Temperature. Using leach liquor with an initial concentration of 36 gpl total NH_3 and 30 gpl CO_2, dissolution experiments were conducted at 25, 40, 55 and 70 °C. At the latter three temperatures, the amount of iron in solution increased for a short period of time to a maximum whose magnitude was dependent on temperature and thereafter decreased. This behaviour is portrayed in Figure 9 which shows the results of the dissolution tests at the four temperatures.

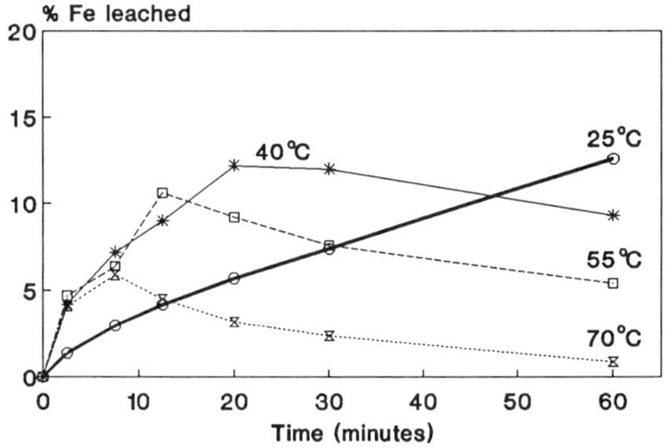

Figure 9 - Effect of leaching temperature on the amount of iron in solution at constant initial concentrations of 36 gpl NH_3 and 30 gpl CO_2. (n=380 rpm)

The variation with time of the amount of iron in solution in the high temperature tests was explained after considering the phases which were present in the leach residues. Whereas at 25 °C the residues had wustite as the main iron bearing phase and only traces of siderite ($FeCO_3$), siderite predominated over wustite in the residues of the 55 °C experiment. The d-values of siderite in the leach residue X-ray diffractograms were in several cases shifted towards higher values thus indicating the possibility of incorporation of elements like zinc and manganese in the crystal structure.

Dissolved iron in ammoniacal ammonium carbonate solution is said to occur as a complex of the type $FeCO_3.(NH_4)_2CO_3$ [13]. Decomposition of this complex in the high temperature tests leading to the precipitation of siderite would explain the decrease of iron in solution after a short leaching time. Due to the relatively low value of reagent loss even at high temperature, it is unlikely that ammonia loss could have significantly contributed to the decomposition of the ferrous ammonium carbonate complex.

In contrast with iron, the tests above 25 °C yielded a marginal decrease in the amount of zinc solubilized (Figure 10). This is a reflection of the thermal stability of the zinc complex, $Zn(NH_3)_nCO_3$, in solution and the larger amounts of NH_3 and CO_2 removal required before basic zinc carbonate precipitation can occur according to the simplified equation:

$$5Zn(NH_3)_4CO_3 + 3H_2O \rightarrow 2ZnCO_3.3Zn(OH)_2 + 3CO_2\uparrow + 20NH_3\uparrow \qquad (1)$$

In the high temperature tests, dissolution of iron was not impeded but merely overtaken by precipitation which occurred simultaneously. Leaching at a high temperature could not therefore be regarded as a way of accomplishing selective leaching of zinc from reduced steelmaking dust. Most further tests were thus restricted to the use of a temperature of 25 °C. Leaching at a high temperature was nevertheless noted as a potential means of obtaining a solution with a relatively low amount of iron.

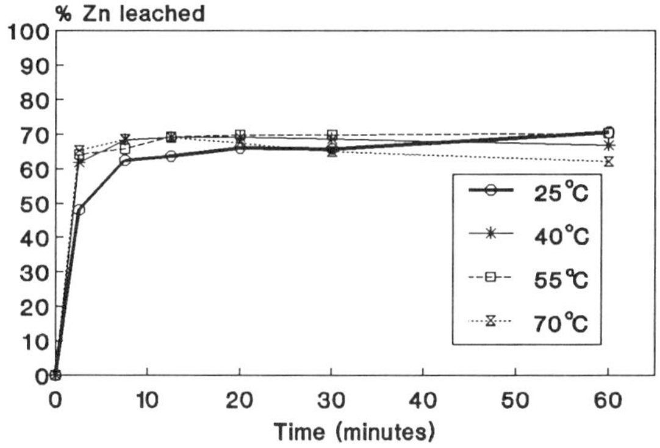

Figure 10 - Effect of leaching temperature on the amount of zinc in solution at constant initial concentrations of 36 gpl NH_3 and 30 gpl CO_2. (n=380 rpm)

Redox Potential. Adjustment of solution potential was accomplished by the use of air and hydrogen peroxide as oxidants. The latter oxidant was added as a 3.5% solution in eight 5 cm^3 portions at selected time intervals. To oxidize the solution with air, the lid was removed from the leaching vessel. The measured solution potential for the set of experiments done is shown in Figure 11.

Oxidation had a negligible influence on the zinc leached but caused a decrease in the amount of iron in solution. Of the two oxidants, hydrogen peroxide gave a lower iron content. The influence of oxidation on iron leached is shown in Figure 12. The residue had the characteristic brownish colour of ferric hydroxide thus indicating that precipitation caused the lower amount of iron determined in solution under oxidizing conditions.

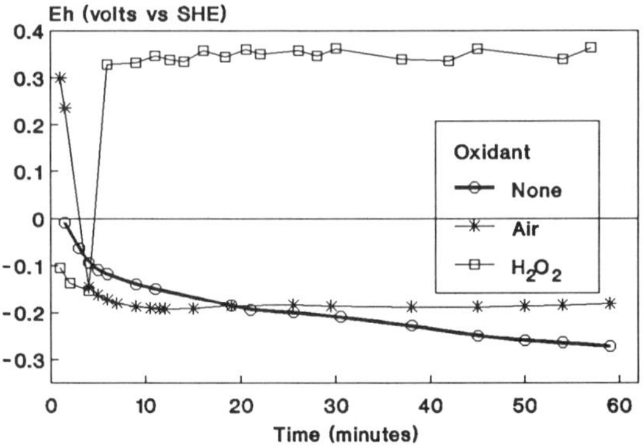

Figure 11 - Redox potential in the course of leaching, with and without the addition of oxidants. ($[NH_3]$=36 gpl, $[CO_2]$=30 gpl, T=25 °C, n=380 rpm)

Concentration, Temperature and Redox Potential. In view of the low amount of iron obtained by the use of a high leaching temperature and by the addition of hydrogen peroxide, these parameters were subsequently used with a concentrated leach liquor which in preliminary tests had solubilized 88% of the zinc and 60% of the iron. Results of this experiment are shown in Figure 13 from which it can be seen that the reduced amount of iron in solution, at a maximum of 36%, was accompanied by a reduced percentage of zinc leached.

Thus from the results of this test, it appeared that only when a high amount of iron is co-leached would it be possible to have a high amount of zinc solubilized. Supplementary experiments were therefore done with pure zinc oxide and pure iron phases in order to clarify this relationship.

Dissolution of Zn and Fe from Pure Zinc Oxide and Pure Iron Phases

Unreduced Mixtures. Intimate powdered mixtures of zinc oxide with each of the iron phases; hematite, magnetite, wustite, and metallic iron were leached under the same conditions. The sample weights were the same at 10 grams, and the Fe/Zn weight ratio nearly equalled that of reduced EAF dust.

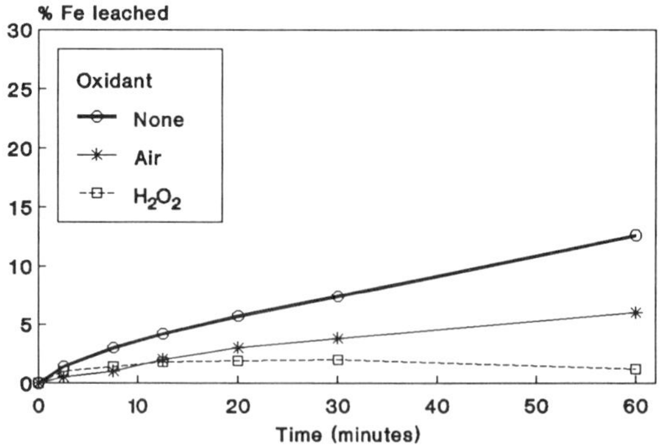

Figure 12 - Effect of oxidants on the amount of iron in solution. ($[NH_3]$=36 gpl, $[CO_2]$=30 gpl, T=25 °C, n=380 rpm)

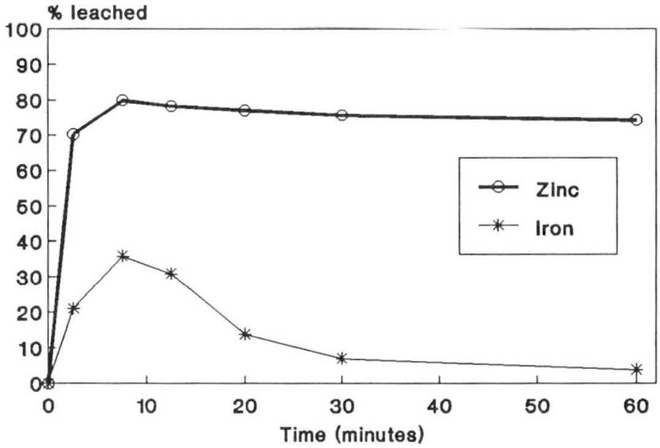

Figure 13 - Zinc and iron in solution while leaching reduced EAF dust with concentrated AAC solution at 50 °C and with the addition of H_2O_2. ($[NH_3]$=134 gpl, $[CO_2]$=105 gpl, n=380 rpm)

In all these experiments, about 90% of the zinc was solubilized after 1 hour while less than 0.1% iron was leached. Because of their similarity, detailed results of these experiments are not presented. Leaching was done at 25 °C with a liquor containing 36 gpl NH_3 and 30 gpl CO_2. The mixture of zinc oxide and metallic iron, though finally giving the same amount of zinc in solution after one hour, was observed to have a lower initial dissolution rate. This was probably as a result of some physical phenomena since upon separate addition of zinc oxide and metallic iron to the leach liquor, the lower initial dissolution rate of zinc did not occur.

The experiments with unreduced mixtures demonstrated that pure hematite, magnetite, wustite, and metallic iron could not have been the cause of high iron dissolution from reduced EAF dust.

Reduced Mixtures. Reduction was carried out under the same conditions as used for EAF dust. The phases present in the reduced samples are shown in Table I, while Table II gives the amounts of zinc and iron solubilized in the leaching step. Leaching conditions were the same as given for reduced EAF dust. At the end of each of these leaching tests, the pulp in the vessel was diluted with distilled water by a factor of 5 and the residues allowed to settle. The clear liquid was subsequently decanted and the residues dried and analyzed by X-ray diffraction. Phases which were identified in the residues are shown in Table III.

The most notable conclusion of data presented in Tables I and II is that the dissolution of iron is high from reduced samples containing wustite. Since the unreduced mixture of ZnO and FeO had less than 0.1% iron leached, it can be postulated that the reduction step causes zinc to enter and destabilize the wustite lattice. Wustite modified in such a manner is then fairly soluble in ammoniacal ammonium carbonate solution. The existence of wustite-ZnO solid solutions has been reported in the literature [14,15], with the wustite lattice containing up to about 17% Zn at 700 °C. The incorporation of zinc causes expansion of the wustite crystal lattice [15] and in the present work X-ray diffraction patterns occasionally indicated expansion of the wustite lattice.

TABLE I: Phases present in reduced mixtures of pure zinc oxide and pure iron phases

Initial mixture	Phases present in reduced mixture
$ZnO/FeO/Fe_2O_3$*	ZnO, wustite, metallic iron
ZnO/Fe_3O_4	ZnO, wustite, metallic iron
ZnO/Fe	ZnO, metallic iron

* Fe_2O_3 present in trace amounts

TABLE II: Dissolution of iron and zinc from reduced mixtures of pure zinc oxide and various iron phases

Leaching time	INITIAL MIXTURE					
	$ZnO/FeO/Fe_2O_3$*		ZnO/Fe_3O_4		ZnO/Fe	
(minutes)	% Zn**	% Fe**	% Zn**	% Fe**	% Zn**	% Fe**
2.5	62	2	56	2	58	0
7.5	80	9	77	5	69	1
12.5	81	13	78	8	75	1
20	82	21	80	15	79	2
30	80	24	76	20	79	3
60	82	25	79	25	84	3

* Fe_2O_3 present in trace amounts
** Weight percent zinc and iron leached from reduced mixtures

TABLE III: Phases present in reduced-and-leached mixtures of pure zinc oxide and pure iron phases

Initial mixture	Phases present in reduced-and-leached material
$ZnO/FeO/Fe_2O_3$*	Magnetite, wustite, metallic iron, hydrozincite [$2ZnCO_3.3Zn(OH)_2$]**
ZnO/Fe_3O_4	Magnetite, wustite, metallic iron, hydrozincite**
ZnO/Fe	Magnetite, metallic iron, hydrozincite**, smithsonite ($ZnCO_3$)**, siderite**

* Fe_2O_3 present in trace amounts
** Precipitation products due to dilution and loss of ammonia

By comparing Tables I and III it can be seen that during the course of leaching, formation of a spinel phase (later identified as magnetite containing zinc) occurred. Furthermore it can be inferred that metallic iron participated in the reaction leading to spinel formation. This phenomenon was investigated further in experiments reported in the next section.

Partial Conversion of Metallic Iron to a Spinel Phase in AAC Solution Containing Zinc. Pure zinc oxide equivalent to the amount present in EAF dust samples considered in earlier experiments was dissolved in leach liquor containing 36 gpl NH_3 and 30 gpl CO_2. This solution which had a pH of 9.52 was then used to leach metallic iron. Residues of the leaching experiment were thoroughly washed with HCl acidified water (pH 3.3) to remove any adsorbed zinc and then dried before examination by XRD and XRF analysis. The residue had the XRD pattern of metallic iron and the pattern common to the spinel phases magnetite and franklinite. The presence of 0.5% Zn in the residue strongly suggested that the spinel phase was magnetite with part of the iron replaced by zinc.

Figure 14 shows the change in solution redox potential during the course of this experiment. When the experiment was repeated using liquor which had been de-oxygenated by bubbling nitrogen gas through, the fall in redox potential did not occur and magnetite was absent in the leach residues. Dissolved oxygen in the leach liquor therefore played a part in the partial transformation of metallic iron to zinc substituted magnetite by oxidizing part of the ferrous ions to the ferric state.

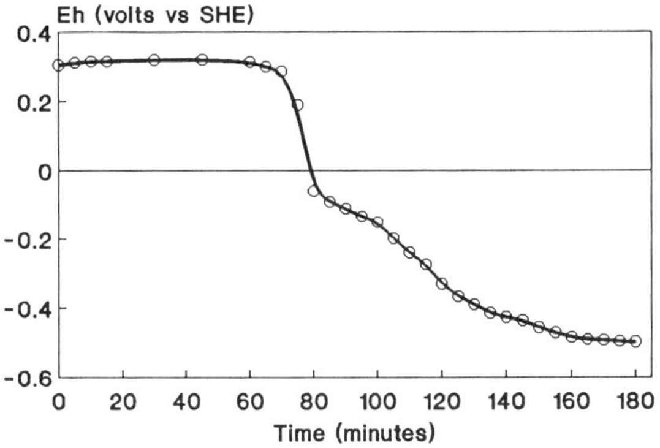

Figure 14 - Change in solution potential while leaching metallic iron with AAC solution containing zinc. ($[NH_3]$=36 gpl, $[CO_2]$=30 gpl, T=25 °C, n=380 rpm)

Mixtures of ferrous and ferric salts in solutions of a pH of about seven and higher are known to lead to the formation of magnetite [16-18] and this technique has been used for the removal of heavy metals like Zn, Cr, and Mn from waste waters by incorporating them in the magnetite formed [18].

Conclusions and Comments

Dissolution of zinc oxide from reduced EAF dust occurs rapidly even in fairly dilute ammonium carbonate solution at room temperature. To leach additional zinc requires the use of a more concentrated solution, however, which is able to leach zinc present in wustite. Such dissolution of wustite results in high amounts of iron in solution which can be removed as siderite or ferric hydroxide. Even then, some zinc would remain in the leach residue as unleached zinc or reprecipitated zinc compounds. The incremental zinc does not therefore appear to warrant the costs of reducing the dust.

Since metallic iron was hardly leached in the absence of air, one way of obtaining a high zinc and low iron leachability would be to severely reduce EAF dust so that wustite is decomposed to metallic iron. This also is a question of economics.

A third approach would be to use ammonium carbonate leaching of as produced EAF dust, to obtain a residue that would be acceptable for landfill disposal. Test work on this approach is being carried out at Delft University of Technology by the principal author.

References

1. A.M. Hagni, R.D. Hagni, and C. Demars, "Mineralogical Characteristics of Electric Arc Furnace Dusts," JOM, 43(4)(1991), 28-30.

2. T.P. McAloon, "So Little Time, So Many Choices," Iron and Steelmaker, 14(8)(1987), 14-20.

3. M. Hanafusa, chairman, "The Management of Steel Industry By-Products and Waste," (Report of the Committee on Environmental Affairs, IISI, Brussels, 1987).

4. E.R. Krishnan, "Recovery of Heavy Metals from Steelmaking Dust," Environmental Progress, 2(3)(1983), 184-187.

5. D.B. Dreisinger, E. Peters, and G. Morgan, "Hydrometallurgical Treatment Options for Carbon Steel Electric Arc Furnace Dusts" (Paper presented at the 2nd International Conference on Separation Science and Technology, Hamilton, Ontario, 1-4 October 1989), 578-585.

6. J.C. Taylor and A.D. Zunkel, "Environmental Challenges for the Lead-Zinc Industry," Journal of Metals, 40(8)(1988), 27-30.

7. D.B. Dreisinger, E. Peters, and G. Morgan, "The Hydrometallurgical Treatment of Carbon Steel Electric Arc Furnace Dusts by the UBC-Chaparral Process," Hydrometallurgy, 25 (1990), 137-152.

8. J.I. Goldstein et al., "Characterization, Recovery and Recycling of Electric Arc Furnace Dust (Section VI.D)" (Project No. 99-26-09886-10 under U.S. Department of Commerce, Lehigh University, 1982).

9. W.J. Wendt, "Ammonia; Ammonium Carbonate Leaching of Low Grade Zinc Ores," Engineering and Mining Journal, 154 (1953), 84-90.

10. F.G. Prado, J.P. Dempsey, and B.W. Wiegers, "High Purity Zinc Oxide Production from Residues in Automobile Scrap Recycling," (Paper presented at the International Symposium on Recycle and Secondary Recovery of Metals, Fort Lauderdale, Florida, 1-4 December 1985), 183-193.

11. R. Pazdej and J.M. Steiler, "New Treatment Possibilities of BF/BOF Zinc (and Lead) Bearing Dusts," (Paper presented at the 1st Technology Conference, Washington D.C., March 1980), 85-103.

12. R.L. Nyirenda, "A Study of the Reduction Step of the Caron Zinc Process," Delft Progress Report, 13 (1989), 49-59.

13. A. Illis, G.C. Nowlan, and H.J. Koehler, "Production of Nickel Oxide from Ammoniacal Process Streams," CIM Bulletin, March 1970, 352-361.

14. J.M. Claude et al., "Preparation and Study of $Zn_xFe_{0.85-x}O$ ($0.085 \leq x \leq 0.170$)," Journal of Solid State Chemistry, 24 (1978), 395-400.

15. G.P. Popov and S.F. Strokatova, "Preparation and Thermodynamic Properties of Solid Solutions of Iron (II) and Zinc Oxides," Russian Journal of Physical Chemistry, 51 (1977), 962-964.

16. J.W. Mellor, A Comprehensive Treatise on Inorganic and Theoretical Chemistry, vol. XIII (London: Longmans, Green and Co., 1934), 739-740.

17. W. Kunda and R. Hitesman, "Recovery of Iron as Magnetite from Aqueous Ferrous Sulphate Solution," Transactions of the Metallurgical Society of CIM, Annual Volume (1977), 118-125.

18. N. Nojiri et al., "Electrolytic Ferrite Formation System for Heavy Metal Removal," Journal WPCF, 52 (1980), 1898-1906.

SOME ALTERNATIVE APPROACHES FOR THE TREATMENT OF ELECTRIC FURNACE STEELMAKING DUSTS

Fernando Castro

University of Minho
Laboratory of Metallurgy
Azurém - 4800 Guimarães - Portugal
FAX - (053) 514400

Abstract

Electric furnace steelmaking dust is a polluting residue that needs to be treated in order to recover metals contained and to produce an environmentally acceptable residue. Some hydrometallurgical approaches for its treatment have been evaluated, namely: caustic soda leaching followed by electrolytic recovery of zinc, ammonia leaching followed by zinc compounds precipitation and acetic acid leaching followed by lead, zinc hydroxide and gypsum precipitation.

Residues and Effluents - Processing and Environmental Considerations
Edited by Ramana G. Reddy, William P. Imrie and Paul B. Queneau
The Minerals, Metals & Materials Society, 1991

Introduction

Electric furnace steelmaking dusts represents, at present time, an ecological problem, especially due to their relatively high zinc and lead contents.
The treatment of these dusts for metal recovery is, very often, an uneconomical task, due to the high processing costs involved and to the nowadays low prices of the metals contained (1)(2)(3).
However, social and local pressures in order to reduce pollution restricts the dumping of this hazardous product and imposes its treatment prior to disposal.
In Portugal, like in the other countries of the EEC, legislation restricts more and more severely the dumping of dangerous, polluting, residues and substances. Electric furnace steelmaking dusts lie in this classification and so its treatment has to be envisaged.
In this work, several ways for the treatment of this kind of residue has been studied, in vue of metal recovery and production of an environmentally innocuous residue.

Characterization of the dust

The dust specimens employed in the experiments have the following chemical composition (in weight %):

Table I - Chemical composition of the dust employed in the experiments

Element	Mean content	range
Zn	17,7	15 - 20
Pb	7,9	6 - 10
Fe	14,0	12 - 16
Ca	31,9	30 - 36
Mn	2,4	2 - 3
K	1,2	0,7 - 1,4
Cl	2,0	1 - 3
Si	2,1	1,5 - 2,5

Experimental work

The experimental work was based on several hydrometallurgical routes. The leaching agents employed were: water, caustic soda, ammonia and acetic acid. The potential ways for metal recovery are presented for

each case.
All the experiments were executed with 100 ml of leaching solution and variable amounts of dust. The temperature has been controlled within ±5ºC and the solid-liquid mixture electromagnetically strirred (1200 rpm). The chemical analysis of the solutions were performed by atomic absorption spectroscopy. In all the experiments, pro-analysis reagents and distilled water have been employed.
The results presented in this text are expressed in terms of the % of each element that has been dissolved. The reference (100%) was the mean content of each element in the dust, shown in Table I. The temperature, time and the solid/liquid ratio are also presented for each test.

Water leaching

The results of the water leaching tests are given in Table II.

Table II - Results of the water leaching experiments

solid/liquid in g/ml x 100	T ºC	time in min.	Zn	Pb	Ca	pH
5	25	300	0	0	17	12.6
25	25	300	0	1	7	13.2
10	95	300	0	32	62	13.8
25	95	60	0	2	34	13.3

These results indicate that water leaching is a good way to decalcify the dust, what could be interesting for further operations.
If this way is to be considered, gypsum precipitation could be an economically interesting solution. However, the presence of lead dissolved may be a problem due to the contamination of gypsum with lead sulphate.

Caustic leaching

Several experiments have been realized utilizing caustic soda as the leaching reagent. In all the cases, the settling of the dust was very slow, what is a consequency of the high viscosity of the caustic solutions.
Table III presents the influence of time and temperature of tests on the dissolution results.

Table III - Influence of temperature and time on the dissolution results

T ºC	time in min.	Zn	Pb	Ca
25	30	9	6	1
25	60	14	7	1
25	180	11	7	2
25	300	15	10	9
40	30	21	8	1
40	60	24	9	4
40	180	54	12	6
40	300	57	17	9
95	30	24	18	1
95	60	42	13	2
95	180	51	13	3
95	300	53	19	3

NaOH conc. = 5 M   solid/liquid ratio = 5%

These results show that an apreciable amount of zinc and lead are dissolved by caustic soda solutions. A certain leaching of calcium was detected in all the experiments, but the results are very scattered. The dissolution of the iron present in the dust was always very weak (less than 0,1% of the total iron), what indicates the high selectivity of these solutions.

Table IV shows the effect of the NaOH concentration on the dissolution rates.

Table IV - Effect of NaOH concentration on the dissolution kinetics

NaOH conc. mol/l	Zn	Pb
2,5	16	17
5	40	11
10	33	8
20	28	4

T = 95 ºC   time = 60 min.   solid/liquid ratio = 5%

These results show that there must be an optimal NaOH concentration in vue of both zinc and lead dissolution kinetics. That concentration should be placed around 5 M (200 g/l).
In Table V the effect of the solid/liquid ratio on the leaching kinetics is shown.

Table V - Effect of the solid/liquid ratio on the dissolution kinetics

solid/liquid in g/ml x 100	Zn	Pb
5	42	13
10	54	10
15	48	9
20	40	11

NaOH conc. = 5 M T = 95 ºC time = 60 min.

As we can see, there is no definite effect of the solid/liquid ratio on the dissolution kinetics.
The results presented in Tables III, IV and V show that it is possible to extract more than 50% of the zinc contained in the dust, by means of caustic soda leaching. This is probably the fraction of the zinc that is present in the form of zincite (ZnO) which is leachable by NaOH solutions (4)(5). The other part of the zinc must be in the form of zinc ferrite which is harder to leach by means of caustic soda.
The solutions produced by these means may attain 40 g Zn/l and 5 g Pb/l and, after filtering, they can be submited to electrolysis for the production of zinc powder (6)(7). Some experiments showed the feasibility of this procedure. Further studies should however be conducted in order to optimise the electrowinning operation parameters and to understand the influence of the impurities present in the electrolyte (Pb and Ca) on those parameters and on the properties of the powder produced.
By caustic soda leaching one can produce a residue containing about:

Zn - 12 %
Pb - 10 %
Fe - 23 %
Ca - 47 %

which, as zinc is in the form of zinc ferrite, must be a non-toxic and dumpable product.

Ammonia leaching

A solution of 150 g/l of ammonium carbonate and 100 g/l of ammonium hydroxide has been employed to leach the dust with a solid/liquid ratio of 5%. The results are shown in Table VI.

Table VI - Results of the ammonia leaching tests

time in min.	Zn
30	31
60	35
180	38

As we can see, the leaching occurs very rapidly and the amount of zinc extracted reaches a relatively high value.
The research on the leaching with ammonia/ammonium carbonate solutions is going to be continued. The solid/liquid separation seems to be very difficult, like in the case of the caustic soda leaching. The leach solution may be treated in order to produce zinc or zinc compounds. Possible routes are (8):

- precipitation of $xZnCO_3 \cdot yZn(OH)_2$ by ammonia evaporation;
- solvent extraction of zinc with stripping for a sulfuric acid solution and electrowinning or ZnS precipitation.

Acetic acid leaching

The acetic acid dissolution experiments were conducted over the residue resulting from the leaching of the dust in hot water (95 ºC) during 5 hours. Its composition was depleted in calcium:

Zn - 25,2 %
Pb - 6,2 %
Ca - 23,0 %
Fe - 13,5 %
Mn - 2,5 %
Si - 2,0 %
Cl - 2,0 %

The acetic acid concentration was, in all the tests, 200 g/l.
The results of the experiments are shown in Table VII.

Table VII - Results of the acetic acid leaching tests

solid/liquid in g/ml x 100	T ºC	time in min.	Zn	Pb	Ca
10	25	60	76	88	97
10	95	30	71	76	98
10	95	60	81	100	100

The residues produced by the acetic leaching have the compositions shown in Table VIII.

Table VIII - Chemical composition of the residues produced by the acetic acid leaching experiments (in wt%)

T ºC	Zn	Pb	Ca	Fe	Mn	Si	Cl
25	15.8	2.0	1.9	40.9	5.6	4.9	<0.5
95	14.8	<0.5	1.4	35.6	5.4	10.0	<0.5

solid/liquid = 10% time = 60 min.

These results show that the acetic acid leaching allows an important elimination of the zinc, lead and calcium present in the dust, producing a residue with very little lead and calcium, but yet a reasonable amount of zinc.
The acetic acid solution could then be treated, for instance, by the following means (2):

- cementation of Pb with zinc powder;
- gypsum precipitation with sulfuric acid, regenerating the acetic acid and enabling its later reutilization;

The zinc rich acetic acid solution can be reutilized for leaching of further dusts, because the reaction:

$$CaO + ZnAc_2 \dashrightarrow CaAc_2 + Zn(OH)_2$$

occurs, allowing the elimination of zinc by precipitation of its hydroxide. This product may be directly sold or dissolved in sulfuric acid for the production of metallic zinc or of other zinc compounds.

The residue of the acetic acid leaching operation may, probably, be classified as a non-toxic one, because all the zinc must be in the stable form of zinc ferrite.

Conclusions

The treatment of electric furnace dusts is possible by means of various hydrometallurgical processes, leading to the production of a non-toxic residue and some valuable materials. Even if it is nowadays, in most of the cases, uneconomical, it is necessary by pollution control regulations. Some routes, namely those based on caustic soda and acetic acid leaching, seems feasible. Further work must be done on the ammonia/ammonium carbonate leaching and on the utilization of other carboxylic acids (like versatic acid, for instance) for the treatment of this kind of dusts. Some combinations of the alternative routes presented may lead to improvements on the final results or reduce the whole processing costs.

References

(1) C.O. Bounds and J.F. Pusateri, "EAF dust processing in the gas-fired flame-reactor process", Lead-Zinc'90, The Met.. Society., 1990, 511-528.

(2) D.B. Dreisinger, E. Peters and G. Morgan, "The hydrometallurgical treatment of carbon-steel electric furnace dusts by the UBC-Chaparral process", Hydrometallurgy, 25(1990), 137-152.

(3) B.I. Wheatley and F.D. Pooley, "Production of zinc powder from arc and smelter flue dusts", Recycling of Metalliferous Materials, The Institution of Mining and Metallurgy, 1990, 291-299.

(4) J. Frenay, "Leaching of oxidized zinc ores in various media", Hydrometallurgy, 15(1985), 243-253.

(5) J.N. Frenay et al., "Recovery of lead and zinc from electric steelmaking dusts by the Cabedeau Process (S.E.R.H.)", Recycle and Secondary Recovery of Metals, The Met. Society. 1985, 195-201.

(6) A.P. Brown, J.H. Melsenheider and Y. Neng-Ping, "The alkaline electrolytic process for zinc production: a critical evaluation", Ind.Eng.Chem.Prod.Res.Dev., 1983, 22, 263-272.

(7) K.D. Sharma and B.R.L. Row, "An electrolytic process for recovery of zinc dust from melting furnace slag", Hydrometallurgy, 13(1985), 377-383.

(8) N.L. Piret and J.F. Castle, "Review of secondary zinc treatment process options", Recycling of Metalliferous Materials, The Institution of Mining and Metallurgy, 1990, 181-219.

Removal of Halogens from EAF Dust by Pyrohydrolysis

Jerome P. Downey　　John P. Hager

Department of Metallurgical and Materials Engineering
Colorado School of Mines
Golden, Colorado 80401

Abstract

In the thermal treatment of electric arc furnace (EAF) dusts, zinc is reduced, volatilized, and ultimately recovered either as liquid metal in a splash condenser or as solid zinc oxide in bag filters. During the process, the dust's chlorine and fluorine content also enters the gas phase and condenses as dross in the splash condenser or as salts which contaminate the zinc oxide product. Pretreatment of EAF dust by pyrohydrolysis appeared to offer a means of minimizing the dust's halogen content.

The primary research objectives were to experimentally verify the technical viability of pyrohydrolysis and delineate the process parameters requisite for efficient halogen removal. To this end, representative EAF dust samples were secured and characterized by x-ray diffraction, scanning electron microscopy, and Mössbauer analyses. Specific experimentation focused on the response of halogen extraction to variations in temperature, retention time, water vapor partial pressure, and various additives.

The experimental results proved that 97 to 99 percent chlorine extraction and 80 to 85 percent fluorine extraction are consistently achieved when EAF dusts are blended with silica and roasted at temperatures of 850°C or greater in a furnace atmosphere of 75% steam (diluted by air). A statistical analysis further established that more than a single optimal point exists, thus affording flexibility in parameter selection (e.g., by increasing temperature, the necessary retention time and additive concentrations are decreased).

The findings substantiate the feasibility of pyrohydrolysis roasting in reducing the chlorine and fluorine in the EAF dust to a level that will obviate industrial concerns about halogen contamination. The end-product of the pyrohydrolysis procedure is a de-halogenated, self-fluxing calcine which can be directly charged to a plasma or flame reactor. The paper includes design parameters and flow diagrams illustrating the pyrohydrolysis process integrated with the plasma and flame reactor technologies.

Residues and Effluents - Processing and Environmental Considerations
Edited by Ramana G. Reddy, William P. Imrie and Paul B. Queneau
The Minerals, Metals & Materials Society, 1991

Introduction

During electric arc furnace (EAF) steelmaking, dust-laden gases are evacuated from the furnace atmosphere and filtered in baghouses. The collected particulate matter, known as "EAF dust," is predominantly composed of a mixture of complex metallic oxides. The major elemental constituents are typically iron, zinc, and oxygen, and the minor elements include cadmium, calcium, chlorine, chromium, fluorine, lead, magnesium, and manganese.

In the U.S., EAF steelmakers annually generate about five hundred thousand tons of EAF dust (1). Historically, EAF dust was considered waste and merely discarded in landfills. However, the ban on land disposal of hazardous waste imposed by the Hazardous and Solid Waste Act (HSWA) of 1984 provided an impetus for developing new dust disposal alternatives. Since EAF dust is classified as hazardous waste because of its lead, cadmium, and chromium content, its management and disposal is now strictly regulated (2,3).

EAF Dust Treatment Processes

The hazardous waste designation instigated a search for improved EAF dust treatment and disposal techniques. A desirable dust treatment scheme would enable zinc recovery while rendering a non-hazardous process residue. Several existing pyrometallurgical processes offer the means of accomplishing these objectives. These processes typically effect zinc recovery by the sequential reduction of the oxide, volatilization of the metal, gas phase transport, and product condensation. Lead and cadmium are also reduced and gasified during this process. The carbothermic reduction and volatilization steps are exemplified by the following net chemical reactions:

$$ZnO_{(s)} + C_{(s)} \rightarrow Zn_{(g)} + CO_{(g)} \qquad [1]$$

$$PbO_{(s)} + C_{(s)} \rightarrow Pb_{(g)} + CO_{(g)} \qquad [2]$$

$$CdO_{(s)} + C_{(s)} \rightarrow Cd_{(g)} + CO_{(g)} \qquad [3]$$

The vaporized metals exit the thermal processing unit in the offgas stream. The zinc recovery options include liquid metal in a splash condenser or reoxidation and condensation of a crude oxide powder.

Three prominent commercial pyrometallurgical processes are based on rotary (or *Waelz*) kiln, plasma furnace, and flame reactor technologies. Rotary kiln technology currently accounts for the majority of thermal EAF dust processing capacity in the U.S. Other pyrometallurgical and hydrometallurgical approaches have been proposed and developed, though with few exceptions these processes have not been embraced by the industry.

Research Objectives

Chlorine and fluorine compounds in EAF dust inhibit effective utilization of the pyrometallurgical technologies because they volatilize during thermal processing and later condense during the zinc recovery stage. The condensed salts are problematic because they form accretions in the gas handling systems as well as affect product recovery. For example, if the zinc vapor in the reactor off-gas is condensed as metal in a splash condenser, experience has shown that a chloride dross will form and interfere with efficient splash condenser operation. If zinc is collected as an oxide product in a bag filter, the product can be contaminated by the condensation products of halogen-containing waste gases.

Pretreating EAF dust for halogen removal prior to the zinc recovery process offers a potential solution to the problem. Hydrolyzing the halogens with steam at elevated temperature was projected as a pretreatment method bearing considerable promise. EAF dust characterization studies indicated much of the chlorine and fluorine may be present in the form of alkaline or alkaline earth compounds. Therefore, it seemed plausible that pyrohydrolysis chemistry, similar to that employed in segregation roasting to create gaseous hydrochloric acid, could provide a means of gasifying the chlorine and fluorine in the EAF dust as hydrochloric acid and hydrofluoric acid, respectively (4).

The research program was organized to investigate pyrohydrolysis as a pretreatment method. Specifically, the study was designed to demonstrate the technical feasibility of the process for chlorine and fluorine removal. Other key objectives were confirmation of the procedure's effectiveness on various EAF dust samples and development of a process flowsheet and design criteria.

Theory

A thermodynamic analysis of the proposed process was performed to establish its theoretical plausibility. To facilitate this effort, a computer simulation model was developed to guide experimental parameter selection by indicating the conditions necessary for maximum chlorine and fluorine extraction. The calculations indicated the pyrohydrolytic de-halogenation reactions are thermodynamically feasible, provided adequate temperature and a high stoichiometric excess of water vapor is present. Further, the model indicated the presence of silica in the furnace charge is essential to attaining adequate halide gas (HCl and HF) vapor pressures.

For example, consider pyrohydrolysis at 850°C of two halogen species, calcium chloride and calcium fluoride, suspected to be present in EAF dust. The simple hydrolysis reactions are represented by the following equations:

$$CaCl_{2(s)} + H_2O_{(g)} \rightleftharpoons CaO_{(s)} + 2HCl_{(g)} \qquad [4]$$

$$CaF_{2(s)} + H_2O_{(g)} \rightleftharpoons CaO_{(s)} + 2HF_{(g)} \qquad [5]$$

Assuming activities of unity for the solid species and a nominal 0.75 atmosphere water vapor partial pressure, the (calculated) equilibrium vapor pressures of HCl and HF are only 5.2 and 0.23 Torr, respectively. The practical implication of this analysis is that the (forward) reactions are favored by a high water vapor partial pressure and very low partial pressures of the acid gases. However, it is known that the vapor pressure of the nascent halide gases can be increased by adding silica to the system (4). The pertinent chemical reactions are:

$$CaCl_{2(s)} + H_2O_{(g)} + SiO_{2(s)} \rightleftharpoons CaSiO_{3(s)} + 2HCl_{(g)} \qquad [6]$$

$$CaF_{2(s)} + H_2O_{(g)} + SiO_{2(s)} \rightleftharpoons CaSiO_{3(s)} + 2HF_{(g)} \qquad [7]$$

Since the standard Gibbs free energies of reaction for equations 6 and 7 are substantially lower than their counterparts from equations 4 and 5, the corresponding equilibrium vapor pressures for the HCl and HF gases rise to 550 and 24 Torr, respectively. Therefore, from a thermodynamic standpoint, the silica enhances the halogen gasification by increasing the equilibrium vapor pressure of the halide gases. It is presumed similar behavior will occur with other halides in the dust.

Temperature also has a distinct effect on the reaction equilibria. The equilibrium vapor pressures of HCl and HF generated by reactions 1 and 2 both increase as a function of temperature (Figure 1), albeit the respective vapor pressures remains quite low even at 1050°C. The temperature dependence of reactions 6 and 7 follows a similar pattern (Figure 2), but the equilibrium vapor pressures are substantially higher as a result of the silica addition. Comparison of the figures again indicates the enormous theoretical advantage imparted by adding silica.

While the preceding established the thermodynamic basis for the pyrohydrolysis process, kinetic factors govern the rates at which the reactions take place and, specifically, the rate at which the chlorides and fluorides are converted to gaseous species. A detailed study of the transport and chemical kinetics of the pyrohydrolysis reactions was precluded by the complex chemical nature of the dust and the configuration of the experimental system. However, the two factors most likely to affect the reaction rate, temperature and the reactant concentrations, were experimentally examined in great detail.

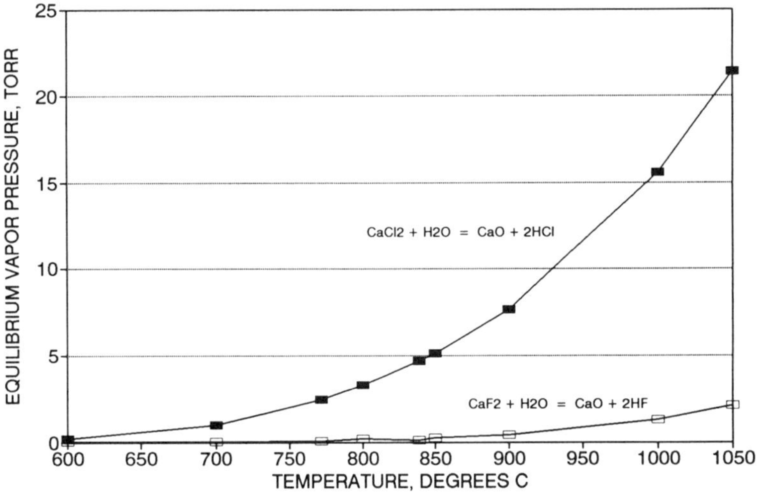

Figure 1. Halide Partial Pressure as a Function of Temperature; P_{H2O} = 0.75 atm., No Silica Addition

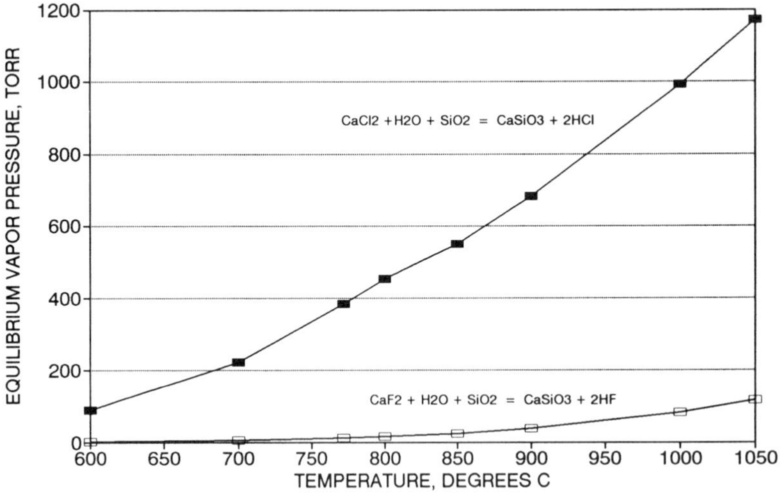

Figure 2. Halide Partial Pressure as a Function of Temperature; P_{H2O} = 0.75 atm., with Silica Addition.

Tungstic oxide is often mentioned in the literature as a catalyst for halogen pyrohydrolysis. Consequently, the research plan included an investigation of the effect of tungstic oxide as well as other additives.

Experimental

An experimental procedure was devised to measure the influence of various parameters on halogen extraction from representative EAF dust samples. The key variables were temperature, solids retention time, the type and concentration of various additives, water vapor partial pressure, and EAF dust composition.

Three bulk samples of electric arc furnace dust, designated "C1", "C4", and "C17", were acquired from industrial sources. The complex and highly variable chemical nature of EAF dust is well-documented in the literature and, accordingly, it was necessary to carefully blend and characterize the dust samples. In Table 1, the chemical analyses of the blends are compared with the average EAF dust analyses from a comprehensive Lehigh University study (5).

Table 1. Chemical Analysis of Composite EAF Dust Samples.

ELEMENT	C1, %	C4, %	C17, %	Lehigh Range, %	Lehigh Avg., %
Cl	2.10	0.81	1.49	0.22-4.54	1.0
F	0.46	1.31	1.30	0.04-6.25	0.9
Ca	6.11	4.35	4.33	0.93-20.8	3.4
Fe	40.12	28.10	38.60	20.6-61.6	35.1
K	---	0.72	0.46	0.06-5.07	0.82
Mg	1.76	2.33	2.00	0.22-5.53	2.27
Mn	4.18	4.17	4.89	0.49-8.16	3.69
Na	---	0.46	1.12	0.27-4.60	0.62
Pb	0.75	1.19	1.48	0.23-3.81	1.5
Si	1.66	1.41	1.69	0.97-7.98	2.5
Zn	10.29	25.70	9.94	0.14-28.0	15.4

Characterization

The chemical analyses revealed the dusts were composed predominantly of iron and zinc oxides, but also contained appreciable concentrations of many other elements. Further characterization of the samples was complicated by their complex chemical nature.

Particle size distributions of EAF dust samples C4 and C17 were measured with a Micromeritics Sedigraph 5000 D Particle Size Analyzer. Both dusts were extremely fine, with over ninety percent of the particles under 10 micron (diameter) and a substantial fraction under 1 micron.

The electron microscopy studies study showed the dusts consisted primarily of spherical particles within the range of 0.25 to 10.0 micron diameter with some larger, irregularly-shaped inclusions. An EDS analysis of individual particles demonstrated their heterogeneity, with at least eleven elements detected in each particle analyzed (excluding oxygen). The particle composition was most consistent among the finest particles (i.e., 0.25 to 1.0 micron diameter), but the larger particles (1.5 and 2.0 micron) were less consistent, primarily with respect to the calcium, chlorine, sodium, manganese, and magnesium concentrations.

X-ray diffraction and Mössbauer analyses established the two major phases present in the dust were zinc oxide (ZnO) and spinel ($MeFe_2O_4$, where "Me" represents various cations such as Fe, Zn, Mg, Mn, et cetera). The spinel phase consisted of magnetic and nonmagnetic fractions, neither of which could be classified as a "pure" compound because of the extensive cation substitution within the lattice.

Attempts to delineate the minor compounds, particularly the halogen-bearing compounds, were less successful. This was due to the relatively low halogen concentrations in the chemically complex dust samples combined with the inherent deficiencies associated with the analytical methods employed (e.g., the overlap of characteristic peaks by more highly-concentrated compounds in the x-ray diffraction analysis).

Apparatus

A schematic drawing of the entire experimental system, including the pyrohydrolysis reactor and the ancillary equipment, is presented in Figure 3.

The pyrohydrolysis experiments were carried out within a cylindrical, quartz reaction vessel measuring approximately 27 centimeters in length and 6.5 centimeters in diameter. The heating source was an Applied Test Systems (ATS) Series 3210 split tube furnace with kanthal A1 heating elements. Furnace temperature was maintained constant (within ±1.5°C) during each experiment by an ATS Series 2010 temperature control system. A chromel-alumel thermocouple extended from the control system to a point directly above the center of the reaction zone.

The reactor was centered within the furnace and held in position by quartz ball joint fittings (Figure 4). Gas-tight seals were maintained by applying a thin coat of high temperature vacuum grease on the ball joints. The kiln was rotated by a detachable sprocket, drive chain, and variable speed drive motor assembly.

The reactor atmosphere was controlled by passing a mixture of oxygen, nitrogen, and water vapor through the reactor at a carefully measured flowrate. Oxygen and nitrogen flowrates were metered by individual capillary flowmeters in proper proportion to form synthetic air. These streams were combined in a common header, mixed, and transferred to the boiler. A combined water vapor and air flowrate of 193.7 milliliters per minute (at 298° Kelvin and one atmosphere) was used for all experiments, except for those specifically designed to examine the influence of the water vapor partial pressure.

The need for a constant water vapor flowrate led to the implementation of the steam generation system shown in Figure 5. The boiler consisted of two concentric 316 stainless steel tubes. The interior tube contained the heating element, a 1000 Watt cartridge heater connected to a variac power supply. A peristaltic pump was installed to deliver water to the boiler at a very low (approximately 0.107 milliliters/minute), constant flowrate through 0.159 centimeter (O.D.) stainless steel tubing. The carrier gas (air) was also introduced to the boiler at this point. This design provided intimate mixing of the carrier gas and the water vapor as well as preheating the gas mixture to a known temperature (480°C). Application of this system eliminated water condensation within the gas transfer system.

Throughout the first two phases of experimentation, the reactor offgas passed through a water-cooled reflux condenser to remove the gaseous halide reaction products. The system was modified for Phase 3 by replacing the reflux condenser with a two stage, packed bed absorption tower (Figure 6). In either configuration, the exit gases passed through a scrubber containing 0.2 molar sodium hydroxide solution before evacuation through the laboratory fume hood.

Procedure

The same general charge preparation procedure was followed for each experiment. Predetermined masses of EAF dust and additives (e.g., silica or tungstic oxide) were measured and manually mixed to consistency prior to being charged to the reactor. The charge was leveled within the reactor to provide an even distribution. The reactor mass was measured before and after charging to verify the actual charge mass.

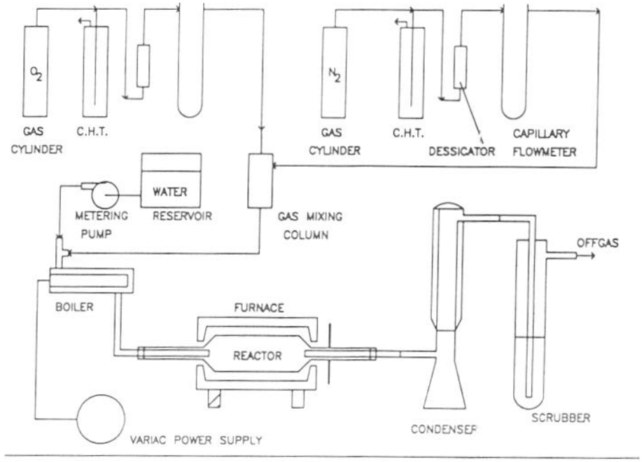

Figure 3. Experimental Apparatus.

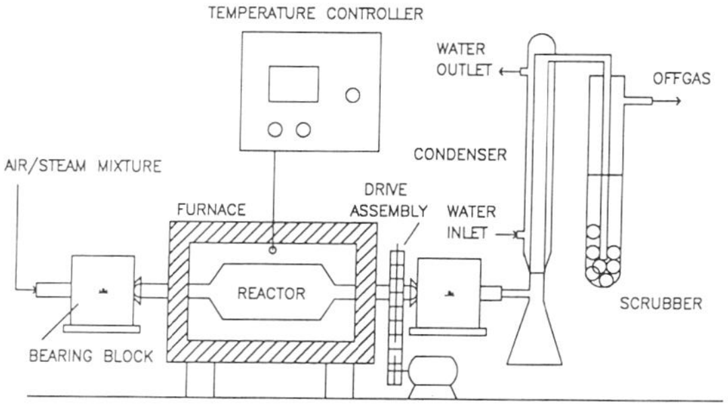

Figure 4. Pyrohydrolysis Reactor and Scrubber System.

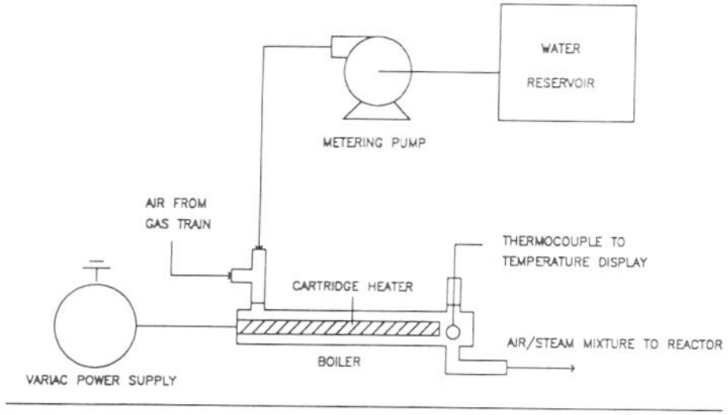

Figure 5. Steam Generation Assembly.

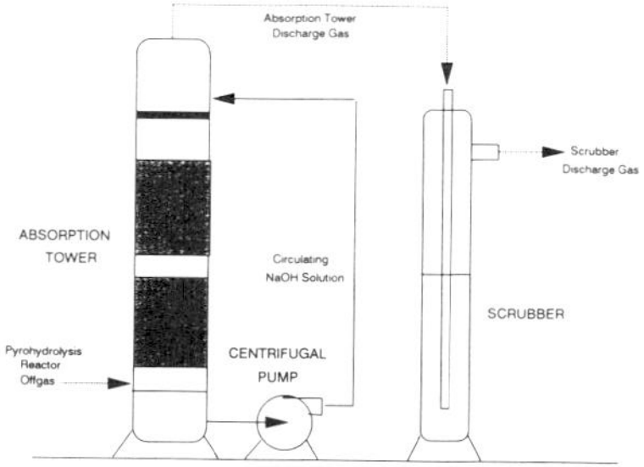

Figure 6. Absorption Tower Assembly.

Once the charge was in place, the drive sprocket was mounted and the reactor positioned in the furnace. The portions of the reactor extending outside the furnace were heavily insulated. The temperature control system was programmed to step up to the desired temperature over a one hour period. Thereafter, another one-half hour to one hour was allowed for the system to reach thermal equilibrium before introducing the steam and synthetic air mixture. The air and water flows were started and allowed to continue for the duration of the experiment.

At the close of each experiment, the steam flow was terminated and the reactor allowed to rotate for an additional one-half hour. The air flow was then terminated and the furnace power disconnected. After the system cooled to near room temperature, the furnace shell was opened and the reactor removed. The drive sprocket was disengaged and the mass of the reactor and calcine carefully measured and recorded. The calcine was usually free-flowing, although some lumpiness was encountered in the experiments conducted at temperatures of 850°C or greater. The recovered calcine was ground to an even consistency with mortar and pestle, then blended for chemical analysis.

The calcine samples were digested via sodium peroxide fusion and the chlorine and fluorine concentrations were determined by ion specific electrode methods. These analytical techniques were selected after careful comparison with alternate methods of sample digestion and analysis. This selection was justified by the accuracy and precision demonstrated by submitting randomly selected duplicate EAF dust samples to separate commercial laboratories. The chlorine and fluorine analyses proved accurate to within 0.02 percent in the calcine samples.

Results and Discussion

The project actually comprised three distinct experimental phases: Phase 1 consisted of a series of scoping experiments to investigate the integrity of the experimental apparatus and evaluate the effectiveness of pyrohydrolysis at relatively low temperatures; Phase 2 was designed to elucidate the effects of the critical process parameters and to determine the conditions necessary for maximum halogen extraction; and, in Phase 3 a statistical method was employed to confirm the process parameters and compare the process applicability on an alternate EAF dust.

Phase 1: Scoping Experiments

The scoping effort encompassed thirteen pyrohydrolysis experiments. The parameters included temperature (300° to 650°C), retention time (three to five hours), the presence of silica and/or tungstic oxide additives (added in ratios of 0.05 to 0.10 to the EAF dust in the charge), and the percentage of excess water vapor in the reactor inlet gas mixture (2390 to 4050 percent).

In general, the attempt to eliminate the halogens at lower temperature was unsuccessful. The maximum chlorine and fluorine extractions realized were only six and eleven percent, respectively, which implied low temperature (i.e., ≤ 600°C) pyrohydrolytic roasting of EAF dust is not particularly effective for gasifying either halogen. However, the results indicated some promise of improving chlorine extraction by elevating the temperature and/or by adding silica and tungstic oxide to the furnace charge.

Phase 2: Parameter Evaluation

The Phase 2 experiments were performed in several suites to gauge the effects of individual variables on halogen extraction. The initial parameter selection was guided by information gleaned from the literature and the outcome of the Phase 1 experiments. The principal variables were furnace temperature (400° to 1050°C), retention time (two to twelve hours), water vapor partial pressure (twenty-five to seventy-five percent, by volume), the type of charge additives (e.g., silica, tungstic oxide, alumina, etc.), and the additive to dust ratio of the charge (zero to 0.30). Except where noted, seventy grams of EAF dust plus any additives were charged to the reactor, the water vapor content of the gas stream entering the reactor was seventy-five percent by volume, and the total gas volume to the reactor was 193.7 milliliters per minute.

Effect of Furnace Temperature and Charge Additives. The initial series of experiments evaluated the influence of temperature on halogen extraction. A secondary objective was to compare the effects of adding silica and tungstic oxide to the furnace charge. In these experiments, retention time was held constant at four hours, the additive to EAF dust mass ratios (when applicable) were maintained at 0.20, and the standard operating conditions were applied to the reactor inlet gas composition and flowrate.

The experiments verified temperature is a dominant variable in the pyrohydrolysis process (Figure 7). At 700°C and below, chlorine extraction did not exceed forty percent regardless of the additive(s) present. However, chlorine extraction approached ninety percent in experiments conducted at 800°C with either silica or silica plus tungstic oxide addition.

The most striking feature is the improvement in extraction efficiency brought about by the silica and/or tungstic oxide addition. Experiments conducted in the absence of the additives generated poor extraction efficiencies (e.g., less than twenty percent extraction at 800°C, and less than ten percent at lower temperatures), but the extraction efficiencies rose dramatically when additives were included. At 700°C for example, the silica addition improved chlorine extraction efficiency by over twenty percent. An additional ten percent extraction was gained when silica and tungsten oxide were present. At 800°C the presence of silica increased chlorine extraction by more than seventy percent over the baseline level.

The 700° and 800°C cases indicate the degree of enhancement caused by tungstic oxide diminished as temperature was increased. The curves representing the silica addition and the silica plus tungstic oxide addition converge near 800°C, where maximum chlorine extractions of ninety percent (silica only) and eighty-eight percent (silica and tungstic oxide) were realized. This phenomenon suggests tungstic oxide is not essential to efficient chloride extraction at 800°C or greater.

The fluorine data (Figure 8) illustrate a strong positive association between fluorine extraction efficiency and temperature, although extraction was consistently lower than the corresponding chlorine extraction. Again, the importance of the additives is conspicuous, particularly the silica. Without any additive, less than fifteen percent extraction was obtained at all temperatures surveyed. Silica addition produced a marked increase in extraction, especially at 700° and 800°C. The inclusion of both tungstic oxide and silica heightened extraction by approximately ten to fifteen percent. In contrast to the chlorine response, the curves do not converge at 800°C.

Effect of Retention Time. The encouraging chlorine and fluorine extraction efficiencies prompted an evaluation of the effect of varying retention time at 800°C. In this experimental series, residence time was varied from four to eight hours, while the other parameters were held constant (i.e., the furnace temperature at 800°C; the standard reactor inlet gas composition and flowrate was maintained; and the charge additive was a blended mixture of silica and tungstic oxide, each compound present at an additive to dust ratio of 0.20).

As expected, the chlorine and fluorine extractions increased as a function of increasing residence time (Figure 9). Chlorine extraction appeared to reach a maximum at approximately ninety-nine percent after six hours. Fluorine extraction rose from fifty-eight percent at four hours to sixty-four percent at eight hours. Fluorine extraction did not reach a maximum, but appeared to increase throughout the temperature range.

Effect of Temperature at Six Hours Retention Time. Another series of experiments was devoted to an evaluation of the effect of temperature on extraction efficiency. This time, the experiments were performed within the temperature range 700° to 925°C, inclusive, and retention time was held constant at six hours. The silica and tungstic oxide additive ratios were 0.20 and the standard conditions were maintained with respect to the reactor inlet gas.

The resultant extraction data (Figure 10) indicate halogen extraction was again highly temperature dependent, similar to previous experiments conducted at four hours retention time. In this case, chlorine extraction was essentially complete at 800°C and above (the calcine contained ≤ 0.01 percent). Fluorine extraction was also highly temperature sensitive, ranging from fifty-

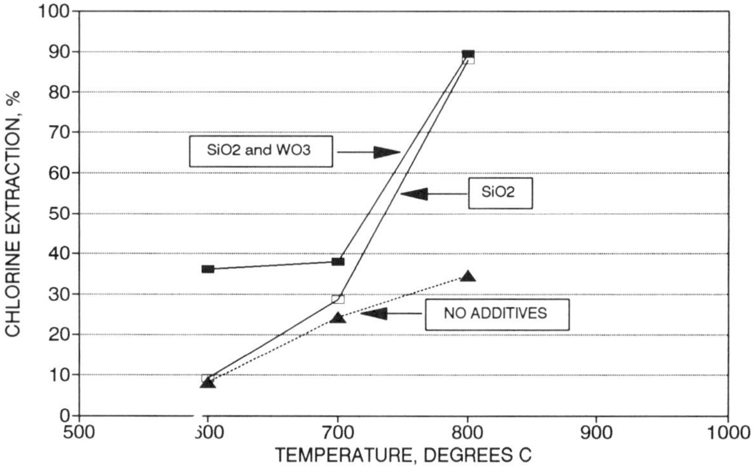

Figure 7. Chlorine Extraction as a Function of Temperature; 4 Hours, 0.20 Additive Ratio.

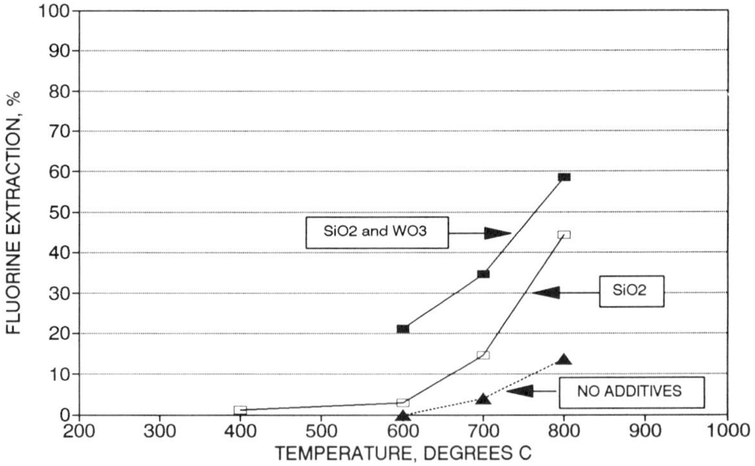

Figure 8. Fluorine Extraction as a Function of Temperature; 4 Hours, 0.20 Additive Ratio.

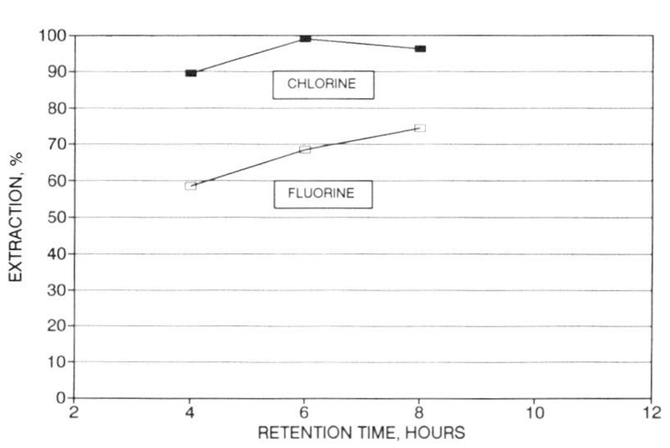

Figure 9. Extraction as a Function of Time; 800°C, 0.20 Additive Ratio.

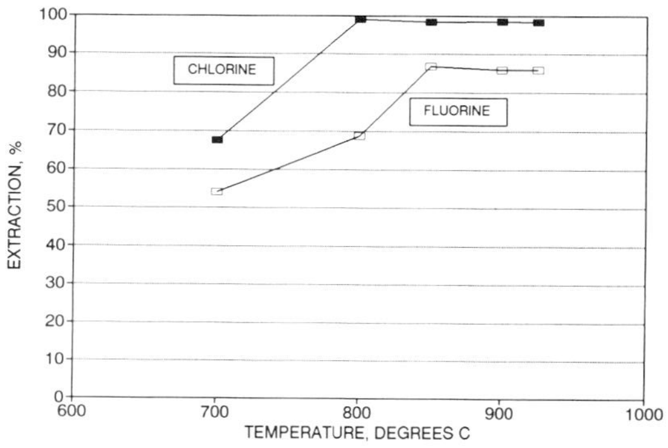

Figure 10. Extraction as a Function of Temperature; 6 Hours, 0.20 Additive Ratio.

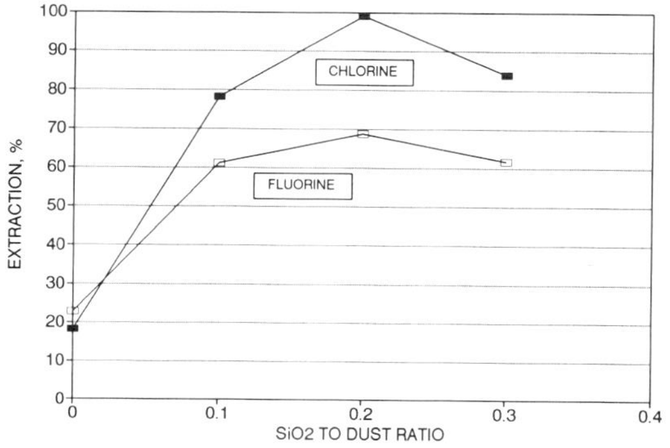

Figure 11. Extraction as a Function of Silica Addition; 800°C, 6 Hours, 0.20 Tungstic Oxide Ratio.

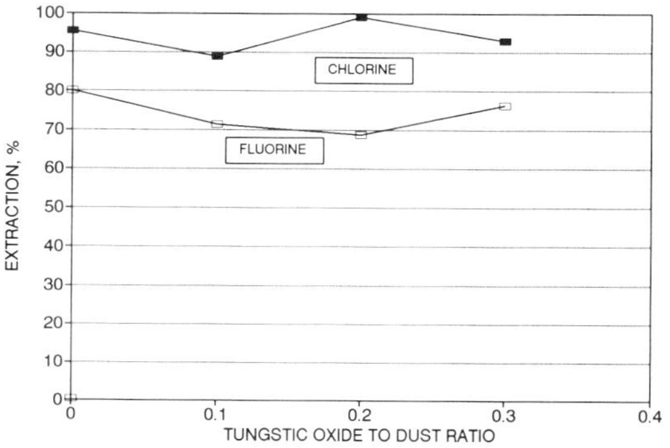

Figure 12. Extraction as a Function of Tungstic Oxide Addition; 800°C, 6 hours, 0.20 Silica Ratio.

three percent at 700° to eighty-seven percent at 850°C. Above 850°C, fluorine extraction became nearly constant at approximately eighty-six percent.

Effects of Additive Type and Concentration. In view of the perceived importance of the silica and tungstic oxide additives, another suite of experiments was designed to more fully investigate the sensitivity of halogen extraction to the additive type and concentration and to evaluate potential substitutes for the tungstic oxide, a relatively expensive reagent. Except where noted, these investigations were performed under conditions of 800°C and six hours retention time with all other parameters held at the predescribed standard conditions.

In the first series of additive experiments, the ratio of silica to EAF dust in the reactor charge was varied between zero and 0.30, while the tungstic oxide to dust ratio was maintained at 0.20. The outcome demonstrated silica is essential to the effective extraction of both chlorine and fluorine (Figure 11). In the absence of silica, poor extractions were obtained for both chlorine (eighteen percent) and fluorine (twenty-three percent). The highest extractions were reached with a 0.20 silica to dust ratio.

Subsequent experiments were executed to certify the effect of the tungstic oxide concentration. These experiments were similar to those described above for silica, except the variable was the tungstic oxide addition, while the silica to dust ratio was constant at 0.20. The halogen extraction efficiencies resulting from these experiments did not display a consistent relationship with the tungstic oxide content of the reactor charge (Figure 12). These results, combined with previous data, indicate tungstic oxide is not necessary for satisfactory extraction at temperatures exceeding the range 800° to 850°C.

Additional experiments were performed to evaluate the effects of other additives on halogen extraction efficiency. Two such suites of experiments were conducted at 800°C and 850°C, respectively. The other parameters were held constant at six hours, seventy five percent water vapor, and a 0.20 silica to dust ratio in the charge.

The extractions obtained at 800°C with three additives (tungstic oxide, calcium oxide, and alumina) and at 700°C with molybdic oxide are compared with the silica-only case (shown as "None") in Figure 13. Tungstic oxide and alumina produced the most favorable results; each compound substantively improved the chlorine and fluorine extraction efficiencies. Alumina actually outperformed tungstic oxide when present at an equal concentration (i.e., an additive to dust ratio of 0.10). Molybdic oxide improved fluorine extraction in comparison with the silica-only case, but the overall chlorine and fluorine extraction efficiencies were not competitive with those obtained using alumina or tungstic oxide. Calcium oxide addition actually appeared to inhibit halogen extraction, decreasing chlorine extraction by approximately nine percent and causing no fluorine extraction.

A similar series of experiments was conducted at 850°C to compared the effects of adding tungstic oxide, nickelous oxide, or alumina with the silica-only case. The results (Figure 14) indicated extremely high chlorine extractions in all cases (including silica-only). Greater than eighty percent fluorine extractions were achieved in three separate experiments: those performed with tungstic oxide (at a ratio of 0.20), nickelous oxide (at a ratio of 0.20), and alumina (at a ratio of 0.10), respectively.

The impressive performance of the alumina additive in the two previous experimental series led to more careful evaluation of its ability to enhance halogen extraction. This series, wherein alumina to dust ratio was the variable, was performed at 850°C, six hours, seventy-five percent water vapor, and a silica to dust ratio of 0.20.

The results (Figure 15) indicated chlorine extraction efficiency is independent of alumina addition (similar to the situation with tungstic oxide). However, the alumina did produce a modest increase in fluorine extraction. It appears the optimum level of alumina addition is approximately five percent of the EAF dust mass. This finding implies fluorine extraction efficiency can be further enhanced by adding a relatively small amount of alumina.

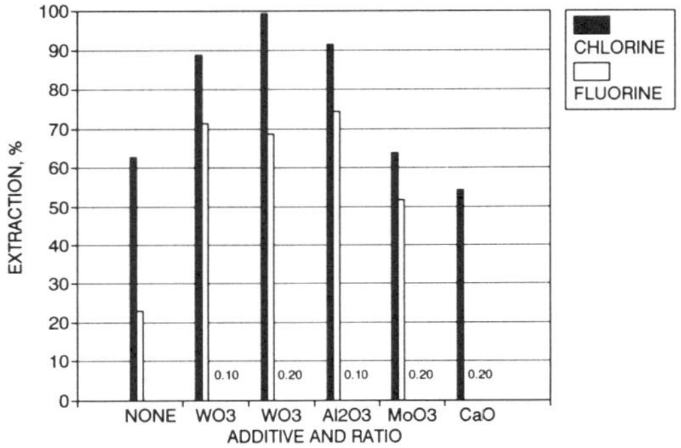

Figure 13. Comparison of Alternate Charge Additives; 800°C, 6 Hours, 0.20 Silica Ratio.

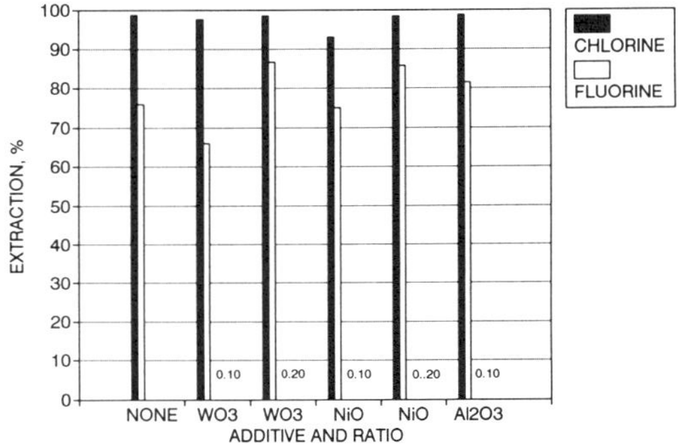

Figure 14. Comparison of Alternate Charge Additives; 850°C, 6 Hours, 0.20 Silica Ratio.

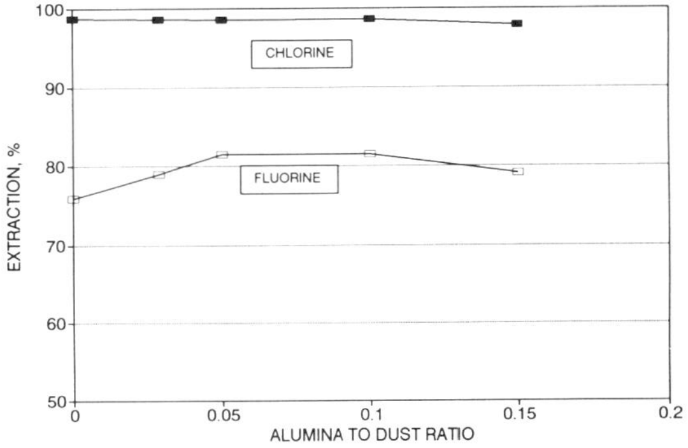

Figure 15. Extraction as a Function of Alumina Addition; 850°C,6 Hours, 0.20 Silica Ratio.

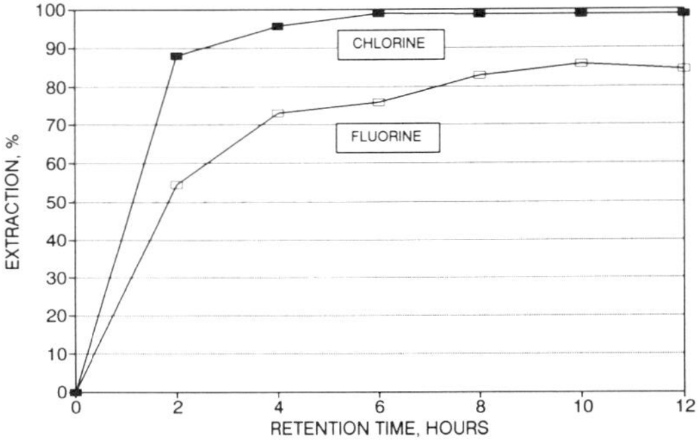

Figure 16. Extraction as a Function of Time; 850°C, 0.20 Silica Ratio.

Silica Additive Only. Previous experimentation had documented the critical role of silica in maintaining high extraction efficiency, but most experiments included tungsten oxide or another additive. Several experimental series' were designed to explore the response to silica in the absence of another additive. In the first series, retention time was varied between two and twelve hours (in two hour increments) at 850°C, seventy-five percent water vapor, and a silica to dust ratio of 0.20.

The halogen extraction results (Figure 16) were similar to those of the previous (800°C) retention time experiments, but were slightly higher at any given retention time because of the fifty degree temperature elevation. Chlorine extraction increased with increased retention time from approximately eighty-eight percent at two hours to an apparent maximum level above ninety-eight percent after six hours. Fluorine extraction was only fifty-four percent at two hours, but increased to a maximum of eighty-six percent after ten hours.

All testwork conducted to this point had utilized minus seventy micron silica gel. For purpose of comparison, a single experiment was performed using minus one hundred and fifty micron sand with the other parameters held constant at 850°C, six hours, and a 0.20 silica to dust ratio.

The result (Figure 17) indicated chlorine extraction is unaffected by the change in particle size; chlorine extraction exceeded ninety-eight percent. Utilization of the coarser silica decreased fluorine extraction by approximately eleven percent (i.e., sixty-eight percent vs. seventy-nine percent for the minus seventy micron silica), indicating the available surface area is an important factor in determining the level of fluorine extraction.

In view of the diminishing effect of the tungstic oxide additive at higher temperatures, experiments were conducted to examine the influence of temperature on halogen extraction with only silica added to the EAF dust charge. Temperature was varied between 800° and 900°C, while retention time (six hours) and water vapor partial pressure (seventy-five percent) were held constant.

A general trend toward increased halogen extraction efficiency as a function of temperature was again evident (Figure 18). In this instance, the increases were relatively minor and amounted to only a few percent. The average extraction efficiencies for both chlorine (approximately ninety-six percent) and fluorine (approximately eighty percent) within this temperature range were very encouraging, confirming tungstic oxide (or other) addition is unnecessary provided the operation is conducted with silica in the range of 800° to 900°C.

Based on this premise, further experimentation was committed to delineating the effect of silica with no other additives present. In these experiments, the silica to dust ratio was varied between zero and 0.20. The other key operating parameters were held constant (i.e., 850°C, six hours, and seventy-five percent water vapor).

The experiments verified the importance of silica in obtaining reasonable extraction efficiencies (Figure 19). An impressive increase in extraction takes place for halogens as the silica ratio was increased from zero to 0.05. Maximum chlorine extraction of approximately ninety-eight percent occurred when the silica addition was between ten and fifteen percent. Conversely, fluorine extraction did not level off, but continued an increasing trend through a 0.20 silica ratio.

Water Vapor Partial Pressure. The effect of varying the water vapor concentration in the gas mixture was measured by adjusting the pertinent flowrates such that volume of water vapor was reduced but the total gas mixture flowrate remained constant. The other parameters were held constants at 850°C, six hours, and a silica to dust ratio of 0.20.

Chlorine extraction was found to be virtually independent of the percentage of water vapor for the range examined (Figure 20). Fluorine extraction increased from sixty-seven to seventy-nine percent as the water vapor content of the inlet gas increased from twenty-five to seventy-five percent. Fluorine extraction appeared dependent on water vapor partial pressure, although the effect above fifty percent water vapor was minor.

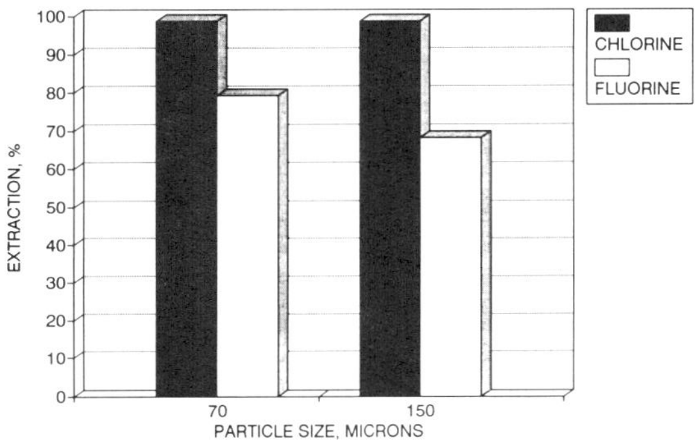

Figure 17. Effect of Silica Additive Size; 850°C, 6 Hours, 0.20 Silica Ratio.

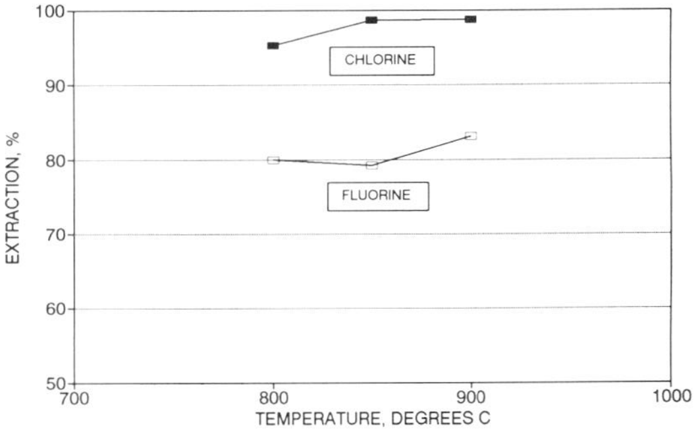

Figure 18. Extraction as a Function of Temperature; 6 Hours, 0.20 Silica Ratio.

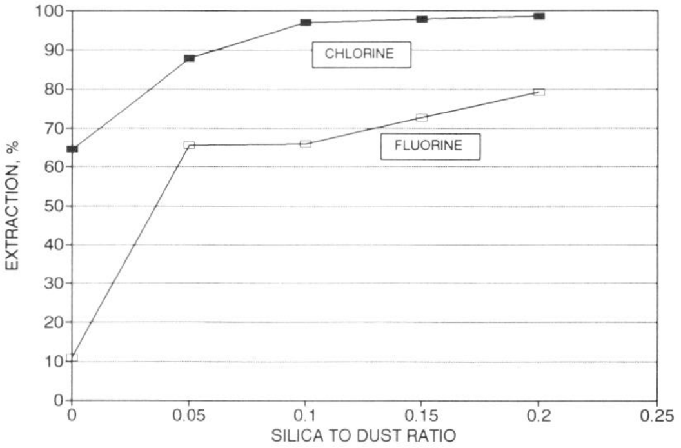

Figure 19. Extraction as a Function of Silica Addition; 850°C, 6 Hours

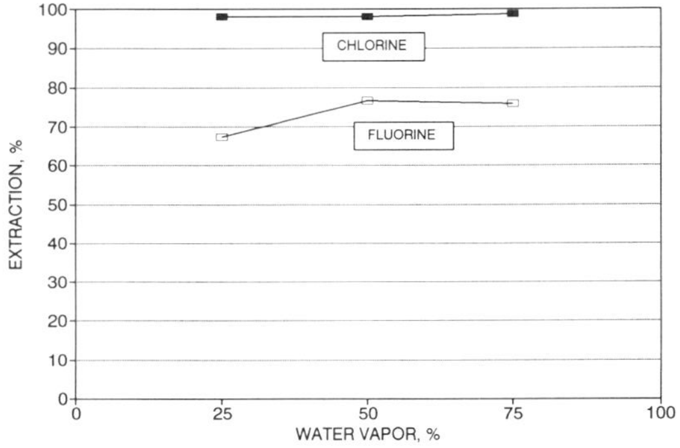

Figure 20. Extraction as a Function of Water Vapor Volume; 850°C, 6 Hours, 0.20 Silica Ratio.

Alternate Dust Compositions. Although the majority of experiments were performed on dust C17, several tests were devoted to EAF dust samples C1 and C4. In summary, the results indicated that the three dusts behaved quite similarly with respect to chlorine (Figure 21). Chlorine extractions exceeding ninety-five percent were achieved when the dusts were roasted at 850°C for six hours with a silica additive (0.20 ratio).

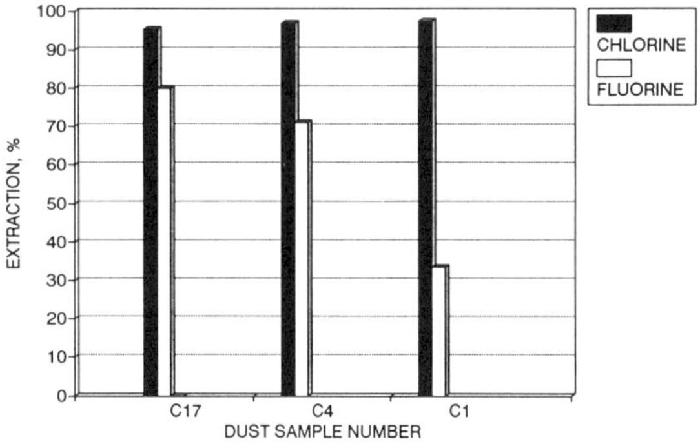

Figure 21. Behavior of Alternate EAF Dusts; 850°C, 6 Hours, 0.20 Silica Ratio.

The fluorine extractions were not as consistent. Under identical conditions, dust C17 yielded the highest extraction (eighty percent), while dusts C4 (seventy-one percent) and C1 (thirty-three percent) proved somewhat more refractory. The poor extraction for dust C1 is partially explained by its relatively low initial fluorine concentration (0.46 percent), which is only about one third the concentration of samples C4 and C17. Therefore, even though the fluorine remaining in the C1 calcine (0.23 percent) was comparable to the best results obtained with dust C17 (between 0.14 and 0.20 percent), the calculated extraction percentage is much lower.

Phase 3: Confirmatory Studies/Statistical Analysis

The Phase 2 experimentation delineated the response of EAF dust sample C17 to an assortment of variables (e.g., temperature, retention time, water vapor partial pressure, and presence of sundry additives). It remained for the parameters to be studied collectively, a necessity because of the possible interaction between parameters.

A statistical technique, commonly known as the *Box-Wilson* method (6), was applied in Phase 3 to systematically evaluate all combinations of variables in a minimum number of experiments. In this method, the response (i.e., halogen extraction) is optimized to an assortment of independent variables. An initial matrix of experiments is employed to establish the response surface, and the process is subsequently optimized through additional experiments conducted along the vector of steepest ascent.

Seven variables were investigated in the Phase 3. The Phase 2 work identified furnace temperature, retention time, and the amount of silica addition as critical variables affecting halogen extraction. The other variables were the silica particle size, the reactor rotation speed, and the amounts of alumina and tungstic oxide added to the furnace charge.

A seven-variable, eight-experiment matrix (Table 2) was applied to EAF dusts C4 and C17 to develop and compare their response surfaces under otherwise identical conditions. The variables (and their respective high and low values) were temperature (825° to 875°C), silica to dust ratio (0.10 to 0.20), solid retention time (five and seven hours), tungstic oxide to dust ratio (zero to 0.03), alumina to dust ratio (zero to 0.03), silica particle size (seventy and one hundred and fifty microns), and kiln rotation (5.5 and 7.7 revolutions per minute).

In general, the average chlorine extractions for the two dusts were very close, about ninety-three percent for dust C17 and ninety-four percent for dust C4 (Table 3). The highest chlorine extractions for dust C17 (ninety-nine percent) and C4 (ninety-eight percent) were each obtained under the experimental conditions designated OC-3.

Without exception, the fluorine extractions were lower than their counterpart chlorine extractions. Dust C4 generally proved more refractory with respect to fluorine extraction; the average extraction from dust C17 (seventy-three percent) significantly exceeded that of dust C4 (fifty-six percent) (Table 4). The highest extraction efficiencies (eighty-five percent for C17 and eighty-seven percent for C4) were again obtained from the OC-3 experiments.

Table 2. Phase 3 Initial Parameter Matrix.

FACTORS STUDIED	Temp. °C	WO_3 g	SiO_2 g	Al_2O_3 g	Kiln rpm	SiO_2 μm	Time hr.
Base Level	850	1.0	10.5	1.0	6.57	110	6.0
Unit	25	1.0	3.5	1.0	1.12	40	1.0
High Level	875	2.0	14.0	2.0	7.69	150	7.0
Low Level	825	0.0	7.0	0.0	5.45	70	5.0
EXPT. OC-1	825	0.0	7.0	0.0	5.45	70	5.0
OC-2	875	2.0	7.0	0.0	7.69	150	5.0
OC-3	875	0.0	14.0	0.0	7.69	70	7.0
OC-4	825	2.0	14.0	0.0	5.45	150	7.0
OC-5	875	0.0	7.0	2.0	5.45	150	7.0
OC-6	825	2.0	7.0	2.0	7.69	70	7.0
OC-7	825	0.0	14.0	2.0	7.69	150	5.0
OC-8	875	2.0	14.0	2.0	5.45	70	5.0

Table 3. Chlorine Extraction Efficiency for Dusts C17 and C4; Initial Optimization Series.

Experiment Number	C17 Calcine Cl, %	C17 Extraction %	C4 Calcine Cl, %	C4 Extraction %
OC-1	0.34	80.5	0.06	93.2
OC-2	0.07	95.8	0.05	94.4
OC-3	0.02	98.7	0.02	97.6
OC-4	0.13	91.5	0.06	92.5
OC-5	0.03	98.2	0.02	97.7
OC-6	0.14	91.4	0.03	94.9
OC-7	0.18	88.1	0.14	82.6
OC-8	0.04	97.4	0.02	97.5
AVERAGE	0.12	92.7	0.05	93.8

Only minimal co-extraction of zinc and iron was observed (Tables 5 and 6). Zinc extraction averaged less than five percent for dust C17 and less than two percent for dust C4. (The zinc data for dust C17 is skewed by an apparently aberrant assay for experiment OC-3; disregarding this experiment, the average zinc extraction was approximately three percent.) The average iron extraction efficiencies were less than one percent for both dusts.

Table 4. Fluorine Extraction Efficiency for Dusts C17 and C4; Initial Optimization Series.

Experiment Number	Dust C17 Calcine Fl, %	C17 Extraction %	C4 Calcine Fl, %	C4 Extraction %
OC-1	0.70	54.8	0.76	43.9
OC-2	0.37	75.0	0.64	52.8
OC-3	0.21	85.1	0.16	87.5
OC-4	0.39	71.2	0.70	43.1
OC-5	0.30	79.8	0.68	49.6
OC-6	0.43	70.3	0.66	50.5
OC-7	0.52	61.5	0.73	40.7
OC-8	0.21	84.4	0.26	78.7
AVERAGE	0.39	72.8	0.57	55.9

Table 5. Zinc Extraction Efficiency for Dusts C17 and C4; Initial Optimization Series.

Experiment Number	C17 Calcine Zn, %	C17 Extraction %	C4 Calcine Zn, %	C4 Extraction %
OC-1	11.2	0.1	24.5	0.0
OC-2	10.2	0.0	22.5	5.8
OC-3	8.25	15.0	22.5	0.8
OC-4	9.00	3.2	21.6	0.7
OC-5	9.70	4.9	23.8	0.0
OC-6	9.45	5.2	24.0	0.0
OC-7	9.25	0.3	21.8	0.0
OC-8	8.40	9.1	20.6	4.6
AVERAGE	9.43	4.7	22.7	1.5

The halogen extraction data were reduced to provide a comparative measurement of the variable parameters (Table 7). The temperature coefficients for both dusts are by far the largest, indicating temperature was the most influential variable examined. For example, the data were interpreted to indicate that for each 25°C increase in temperature, the resultant change in chlorine extraction would be a 4.8 percent increase for dust C17 and a 3.0 percent increase for dust C4. By similar analysis, fluorine extraction would

Table 6. Iron Extraction Efficiency for Dusts C17 and C4; Initial Optimization Series.

Experiment Number	C17 Calcine Fe, %	C17 Extraction %	C4 Calcine Fe, %	C4 Extraction %
OC-1	36.4	0.1	26.7	0.0
OC-2	33.0	0.0	26.2	0.0
OC-3	35.8	0.0	24.7	0.0
OC-4	29.3	2.1	22.7	0.0
OC-5	33.4	2.4	24.2	0.0
OC-6	33.0	0.0	24.7	0.0
OC-7	32.0	0.0	21.3	1.3
OC-8	29.4	1.1	24.4	0.0
AVERAGE	32.8	0.7	24.4	0.2

Table 7. Effect of Seven Independent Variables on Halogen Extraction Efficiency.

Variable and Units	Dust C17 Chlorine Response	Dust C4 Chlorine Response	Dust C17 Fluorine Response	Dust C4 Fluorine Response
Temperature (25°C)	0.048	0.030	0.083	0.113
Retention Time (1 hr.)	0.023	0.019	0.038	0.018
Silica (3.5 g)	0.012	-0.012	0.028	0.067
Tungstic Oxide (1.0 g)	0.013	0.010	0.025	0.004
Alumina (1.0 g)	0.011	-0.006	0.012	-0.010
Silica Size (40 μm)	-0.007	0.020	0.009	0.093
Kiln Rotation (1.12 rpm)	0.008	-0.014	0.002	0.020

be expected to increase by 8.3 percent in dust C17 and by 11.3 percent in dust C4 for the same 25° temperature increase (within the range studied).

The data indicate chlorine extraction was most strongly influenced by temperature and retention time. The other five variables had only a minor effect upon chlorine extraction. The fluorine extraction from each dust was most noticeably affected by temperature, retention time, and the amount of silica in the furnace charge. The impact of these variables on fluorine extraction was much greater than for chlorine.

The responses of dusts C4 and C17 also differed in some respects. For example, the fluorine extraction of dust C17 exhibited a greater dependence on tungstic oxide concentration than did C4. C4 responded more favorably to the finer silica particle size (e.g., fluorine extraction increased by over nine percent when the finer silica particle size was used) and higher kiln rotation speed.

Because fluorine had proved more difficult to remove in all previous experimentation, optimization was based on fluorine extraction efficiency. The steepest ascent vector was calculated from the coefficients generated from the fluorine extraction efficiencies. This was accomplished by multiplying each of the effect coefficients (from Table 7) by the respective units of variation for each parameter studied; the products represented the incremental change for each variable. The best path was decided by successive additions of these increments to the corresponding base levels (from Table 3). Since temperature had proven the most influential variable for both dusts, the path of steepest ascent was keyed to ten degree (Celsius) increments, and the corresponding units for the other six remaining parameters were adjusted accordingly. A few additional experiments were conducted at various points located along the path of best response as shown in Tables 8 and 9.

Table 8. Path of Best Response, EAF Dust C4.

Best Path Experiment	Temp. °C	WO_3 g	SiO_2 g	Al_2O_3 g	Kiln rpm	SiO_2 μm	Time hr.
OC4-09	830	1.0	8.9	1.07	5.45	70	5.87
OC4-10	840	1.0	9.7	1.03	5.45	70	5.94
	850	1.0	10.5	1.00	5.45	70	6.00
OC4-12	860	1.0	11.3	0.97	5.45	70	6.06
	870	1.0	12.1	0.93	5.45	70	6.13
OC4-11	880	1.0	13.0	0.90	5.45	70	6.19
	890	1.1	13.8	0.86	5.45	70	6.26
	900	1.1	14.6	0.83	5.45	70	6.32

Table 9. Path of Best Response, EAF Dust C17.

Best Path Experiment	Temp. °C	WO_3 g	SiO_2 g	Al_2O_3 g	Kiln rpm	SiO_2 μm	Time hr.
OC17-11	830	0.8	9.6	0.88	5.45	70	5.63
OC17-09	840	0.9	10.0	0.94	5.45	70	5.82
	850	1.0	10.5	1.00	5.45	70	6.00
OC17-10	860	1.1	11.0	1.06	5.45	70	6.18
	870	1.2	11.4	1.12	5.45	70	6.37
OC17-12	880	1.4	11.9	1.18	5.45	70	6.55
	890	1.5	12.4	1.24	5.45	70	6.74
	900	1.6	12.8	1.30	5.45	70	6.92

The experiments were performed at the same four temperatures for each dust (i.e., 830°, 840°, 860°, and 880°C). The retention time and the levels of silica, tungstic oxide, and alumina addition were set by the values established by the path of best response for each dust. Kiln speed and silica particle size were held constant at 5.45 rpm and seventy microns, respectively.

The extraction data (Tables 10 and 11) indicate the optimization procedure was effective in establishing a response vector. The extraction efficiencies of fluorine (and also chlorine) increased in the predicted direction without exception.

Table 10. Best Path Results, EAF Dust C4.

Experiment Number	Dust C4 Residual Chlorine %	Dust C4 Chlorine Extraction %	Dust C4 Residual Fluorine %	Dust C4 Fluorine Extraction %
OC4-09	0.02	97.7	0.64	51.8
OC4-10	≤0.02	≥97.7	0.60	54.6
OC4-12	≤0.02	≥97.7	0.42	67.0
OC4-11	≤0.02	≥97.6	0.29	78.0

Table 11. Best Path Results, EAF Dust C17.

Experiment Number	Dust C17 Residual Chlorine %	Dust C17 Chlorine Extraction %	Dust C17 Residual Fluorine %	Dust C17 Fluorine Extraction %
OC17-11	0.17	88.9	0.48	66.8
OC17-09	0.05	96.7	0.45	68.6
OC17-10	0.03	98.0	0.33	76.6
OC17-12	0.02	98.7	0.25	82.1

Excellent chlorine removal efficiency was obtained in seven of the eight experiments. The four experiments performed on dust C4 yielded nearly uniform chlorine extraction efficiencies of approximately ninety-eight percent (equivalent to only 0.02 weight percent chlorine remaining in the calcine).

The experiments conducted with dust C17 also produced high chlorine extraction efficiencies, with three of the four experiments removing better than ninety-seven percent of the chlorine. While the results from dust C4 were nearly constant, there is a discernable gradient in extraction for dust C17 that suggests the chlorine in this dust is more amenable to pyrohydrolysis at higher temperature.

The fluorine extraction efficiencies of both dusts exhibited a marked gradient. Interestingly, the fluorine extraction efficiencies were noticeably higher for dust C17 than for dust C4, the reverse of the chlorine behavior. The extraction efficiencies ranged from approximately fifty-two to seventy-eight percent for dust C4, and from sixty-seven to eighty-two percent for dust C17.

It is also interesting that higher fluorine extraction efficiencies were previously obtained at lower temperatures, which indicates that more than one "peak" or optimal point exists on the response contour. In this case, it is considered likely that the silica additive concentration was insufficient to promote the desired level of fluorine extraction at any given temperature.

In summary, the most impressive halogen extraction efficiencies were produced from experiments conducted at or above 850°C, for six hours or greater, with a 0.20 silica to dust ratio, in an atmosphere composed of seventy-five percent by volume water vapor (diluted by air). Under these conditions, further (minor) improvement in fluorine extraction was found possible by adding a relatively small amount of alumina to the furnace charge. These criteria represent the optimal conditions (for dust sample C17), as established by the research effort. The results of the statistically-designed experimentation on dust sample C4 indicate the process can be effectively applied to other dust types.

The Phase 3 analysis compliments the Phase 2 observations. Phase 3 showed more than one point of optimization exists on the response surface (for dust C17). Instead of precisely reproducing the Phase 2 optimal parameter values, Phase 3 indicated high halogen extraction efficiencies could be obtained by elevating the temperature, with shorter retention time and lower additive concentrations. This result is favorable both from a process and an economic standpoint, because it provides operators with a degree of flexibility and permits selection of process parameters based on the comparative costs of the fuel and necessary reagents.

The differences between Phase 2 and Phase 3 chiefly result from the selection of the base levels and incremental variations incorporated in the Phase 3 matrix. Temperature was the dominant variable, and since such a large range (fifty degrees Celsius) was examined, the temperature effects overshadowed those of other parameters. Nevertheless, the Phase 3 effort established the same critical parameters of temperature, time, and silica concentration.

The Phase 3 experimentation also revealed the response of dusts C17 and C4 were quite similar in most respects. The (expected) variability in response was within reasonable limits. An acceptable level of halogen extraction was demonstrated possible with both dusts with only a minor variation in the process parameters.

Conclusions

The experimental results firmly established pyrohydrolytic roasting can effectively eliminate chlorine and substantially reduce the fluorine content of EAF dust. The pivotal observations are the strong positive associations between halogen extraction and the temperature, retention time, and silica content of the charge. The values selected for these three parameters essentially dictate the effectiveness of the process.

Without exception, the combined Phase 2 and Phase 3 results demonstrated (in contrast to chlorine) the fluorine extraction is much more sensitive to variations in the operating parameters. In addition to the critical parameters of temperature, time, and silica addition, fluorine extraction efficiency was found to be influenced by three other factors: the presence of alternate charge additives (i.e., alumina), water vapor partial pressure, and the particle size of the silica additive.

The experimentation demonstrated the EAF dust chlorine content can be reduced by more than ninety-eight percent and the fluorine content can be lowered by as much as eighty-five percent. These levels of halogen removal should be wholly satisfactory to industrial concerns. The residual fluorine, typically in the range of 0.15 to 0.25 percent, would not be expected to present problems in the thermal zinc recovery processes. In the flame reactor or the plasma furnace technologies, the basic slag could absorb much of the residual fluorine as calcium fluoride.

The zinc and iron co-extraction percentages were minimal. Typically, less than five percent of the zinc and two percent of the iron contained in the EAF dust samples were lost under optimal roasting conditions.

It is concluded that the pyrohydrolysis process offers considerable promise for industrial adoption and that sufficient data has been generated to design, with confidence, a pilot plant scale pyrohydrolysis reactor.

Process Design Criteria

The following general design criteria are proposed to initiate evaluation of the response of various dust compositions:

Table 12. Proposed Process Design Criteria.

Parameter	Recommended	Range
Temperature, °C	850	800 to 900
Solid Retention Time, hrs	8	6 to 10
Water Vapor, volume %	75.0	50.0 to 90.0
Silica to EAF Dust Ratio	0.20	0.15 to 0.25
Silica Size, microns	70	70 to 150
Alumina to EAF Dust Ratio	0.02	0 to 0.10

Although the process was demonstrated to be viable with differing EAF dust types, a universal set of process parameters for all EAF dust compositions was not established. It is believed the process parameters can be fine-tuned to successfully de-halogenate most EAF dusts.

The research revealed the operating parameters are interrelated and more than a single point of optimization occurs on the response contour. In practical terms, this means there is a degree of flexibility that will allow limited parameter adjustment for economic considerations. For example, it may be desired to operate above 850°C to minimize the retention time, additives concentrations, and steam volume necessary to achieve halogen specifications. These criteria are highly dependent on the composition of the dust(s) submitted for pyrohydrolytic treatment. The ceiling temperature for the process is about 900°C; at higher temperatures the furnace charge begins to fuse. In certain cases, such as when the lead oxide (melting point of 886°C) concentration of the dust is fairly high, a lower operating temperature is advisable.

Process Flow Diagram

The pyrohydrolysis process is compatible with existing thermal zinc recovery technologies. Two emerging thermal treatment technologies which could benefit from the application of the pretreatment method are those based on the flame reactor and the plasma furnace technologies. Schematic flow diagrams of the pyrohydrolysis process integrated with these technologies are presented in Figures 22 and 23.

As envisaged, EAF dust is collected and stockpiled in storage bins to permit adequate surge capacity and the requisite chemical analysis for establishing process control. The dust must be well-blended with the silica (and alumina, if necessary) prior to introduction to the pyrohydrolysis reactor.

The pyrohydrolysis reactor could be a direct-fired rotary kiln or a multiple hearth roaster. In either case, the roaster atmosphere is slightly oxidizing through addition of secondary air. The steam for the pyrohydrolysis reaction is produced by a small boiler, preferably using heat recovered from the plasma or flame reactor offgases. The steam, secondary air, and combustion gases move countercurrent to the solids. The offgas is treated by a high temperature electrostatic precipitator in series with a multiple stage absorption tower. The dusts collected by the electrostatic precipitator revert to the EAF dust storage area and are recycled through the dehalogenation process.

The de-dusted gas stream, containing the residual water vapor, oxygen, nitrogen, hydrogen chloride, and hydrogen fluoride gases, advances to the absorption tower, which utilizes a recirculating solution of caustic to remove the acid gases. A stream of the absorbent solution is bled from the tower reservoir to remove excess solution and prevent concentration of chloride and fluoride. The bleed stream is treated in an evaporative crystallizer to remove the halogens as salts of sodium and the carryover base metals.

To use the sensible heat of the hot calcine to full advantage, the calcine is charged directly to the plasma or flame reactor, along with the fuel

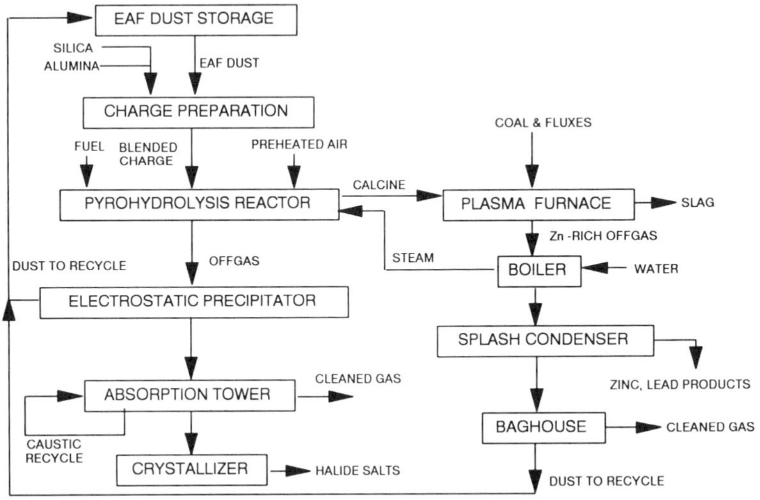

Figure 22. Pyrohydrolysis Process Integrated with Plasma Furnace for Zinc Metal Recovery.

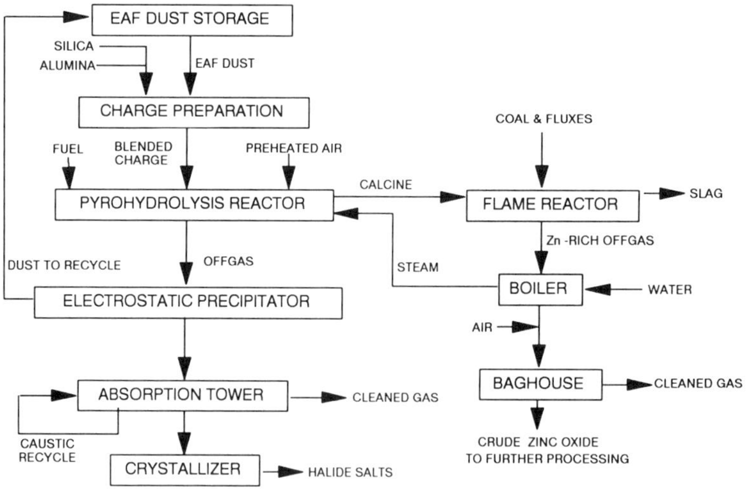

Figure 23. Pyrohydrolysis Process Integrated with Flame Reactor for Zinc Oxide Recovery.

(coal) and any necessary fluxes. The calcine should be self-fluxing with regard to silica, since silica is added for the pyrohydrolysis roast. In the reactor, the oxides of zinc, lead, and cadmium are selectively reduced and volatilized from the EAF dust. The molten furnace product is a non-hazardous iron-rich slag (in certain applications a separate, iron-rich alloy could be produced for recycle to an EAF). The molten product is intermittently tapped and allowed to solidify.

As shown in Figure 22, the furnace exhaust gases pass through a heat exchanger and enter a splash condenser, where the volatile metals are condensed in a lead bath. The gases leaving the condenser are cooled by air injection and pass through a baghouse for final de-dusting prior to release into the atmosphere. Alternatively, the furnace offgases could be cooled through air injection and subsequently cleaned in a baghouse to recover zinc as a crude oxide product. This scenario is depicted in Figure 23.

Acknowledgements

The authors wish to acknowledge the research project was funded by the United States Bureau of Mines through the Generic Minerals Technology Center for Pyrometallurgy, under grants USDI-1175129 and USDI-1105129. Also, thanks are due Mr. John Pusateri of Horsehead Resource Development Corporation, Inc. for providing the EAF dust samples, and Mr. Peter Cowx of Tetronics Research and Development Company, Limited for alerting the authors to the problems caused by chlorine and fluorine in EAF dust.

References

1. Donald R. Mac Rae, "Electric Arc Furnace Dust -- Disposal, Recycle, and Recovery" (Report No. 85-2, Bethlehem Steel Corporation, 1985).

2. "Identification and Listing of Hazardous Waste," *Federal Register* 45 (98); 33084-33124 (19 May 1980).

3. C.W. Avent and J.C. Moriss III, "Environmental Regulation of Electric Arc Furnace Dust: Alternatives and Decisions," *I&SM*, 5 1987: 63-69.

4. I. Iwasaaki, "A Thermodynamic Interpretation of the Segregation Process for Copper and Nickel Ores," *Minerals Science and Engineering*; 4 (2) 1972: 14-23.

5. "Characterization, Recovery, and Recycling of Electric Arc Furnace Dust;" (Report, Lehigh University for U.S. Congress Department of Commerce, 1982).

6. D. M. Himmelblau, *Process Analysis by Statistical Methods*, New York: John Wiley and Sons, 1970, 253-257.

THE COMMERCIAL DEVELOPMENT OF PLASMA TECHNOLOGY:

EAF DUST APPLICATION

David T. Bunney, Mark S. Mazanek, Dr. John. K. Pargeter

In December of 1987, International Mill Service, Inc. signed an exclusive licensing agreement with Tetronics Research & Development Co. Limited of Faringdon, Oxfordshire, England to commercialize plasma technology which had been demonstrated in an EAF dust processing application. The intent of the process is to reduce and volatize the zinc, lead and cadmium oxides inherent in EAF dust and collect these elements as metals in an ISP (Imperial Smelting Process) zinc splash condenser, and to render the resultant slag an inert, non-hazardous material.

With these objectives in mind, IMS started commissioning facilities in 1989 at Florida Steel in Jackson, Tennessee and Nucor-Yamato Steel in Armorel, Arkansas.

This paper will address the commercial development and progress of the technology at these two installations.

Process Description

In the EAF steelmaking process typically the off gas stream is cleaned by some form of baghouse. The particulate collected is known as Electric Arc Furnace dust and has been designated by the U.S. EPA as hazardous waste number K061, once it leaves the collection system. This designation was given to the material due to the high toxic concentrations of lead and cadmium (nickel and chromium for stainless EAF dusts) inherent in the dust.

IMS' method of operation is to locate the facilities on the client mill's property adjacent to the baghouse. The dust is transported directly to IMS via a closed conveyance system. This method enables the technology to be an additional processing step in the steelmaking process. Once the dust has been transported to the IMS facility, it is fed, metered and mixed mechanically in the feed conveyance system with a carbon reductant (typically coke breeze). The carbon reductant is approximately 10% of the dust feed rate. This mixture passes through the feed system and enters the furnace

Residues and Effluents - Processing and Environmental Considerations
Edited by Ramana G. Reddy, William P. Imrie and Paul B. Queneau
The Minerals, Metals & Materials Society, 1991

vessel through two ports in the roof. The mixture is gravity fed and falls into the molten slag bath which is maintained in a temperature range of 1400°-1500°C. The energy required for the operation is supplied by a DC hollow carbon cathode, using nitrogen as a plasma gas, and operated in an extended arc mode. At this temperature the oxides of zinc, lead and cadmium are rapidly reduced and vaporized from the bath into the gas stream. Also occurring is the selective reduction of iron from its ferric state (Fe_2O_3) to its ferrous state (FeO).

Additionally, a small amount of iron is generated which is returned to the mill as scrap.

The gas stream, which consists primarily of carbon monoxide and carbon dioxide along with the metal vapors, passes through the roof of the furnace into refractory lined ductwork, which is connected to an ISP (Imperial Smelting Process) zinc splash condenser. In the condenser a splasher rapidly spins in a molten bath of zinc metal, which is maintained between 550°-600°C, and creates a violent showering effect of small zinc droplets. These droplets act as a condensing medium and as coalescing sites to condense the zinc, lead and cadmium vapors from the gas stream, which has entered at approximately 1100°C.

The metal mixture collected is periodically tapped from the condenser and either liquated and cast, or directly cast into molds for outside sales. The metal vapors not collected in the condenser, along with the carbon monoxide/carbon dioxide gas mixture, pass to an afterburner which oxidizes the remaining metal vapors and burns the carbon monoxide to carbon dioxide.

Air is added to the ductwork downline from the afterburner to facilitate cooling of the oxidized gas stream. The off-gas enters a baghouse where the gases are cleaned and vented to the outside atmosphere. The particulate collected in the baghouse is recycled to the system for further recovery and closing of the loop.

Two additional product streams not previously discussed are produced by the process. The largest volume product produced is a slag which is batch tapped from the furnace approximately every two hours. The slag is an inert, non-hazardous material which has had 97-100% of the toxic elements (i.e. lead, cadmium) removed. Any remaining lead and/or cadmium, as well as any additional toxic elements are tightly bound in an $FeO-SiO_2-CaO$ based crystal lattice. This binding significantly inhibits the leachability of these elements. After cooling, the slag is sized and readied for outside sale for various applications.

The second product produced is known as "SCDR" which stands for "splash condenser dross residue" and was so named by the EPA in its final rule signed on August 8, 1991, dealing with land disposal of K061. This material is produced in the condenser and periodically cleaned from it. This residue is commodity-like, high in zinc metal

and zinc oxide content, and is sold on the secondary market to a thermal processor for further recovery.
See Figure 1 for a process flow diagram.

PLASMA PROCESSING OF EAF DUST

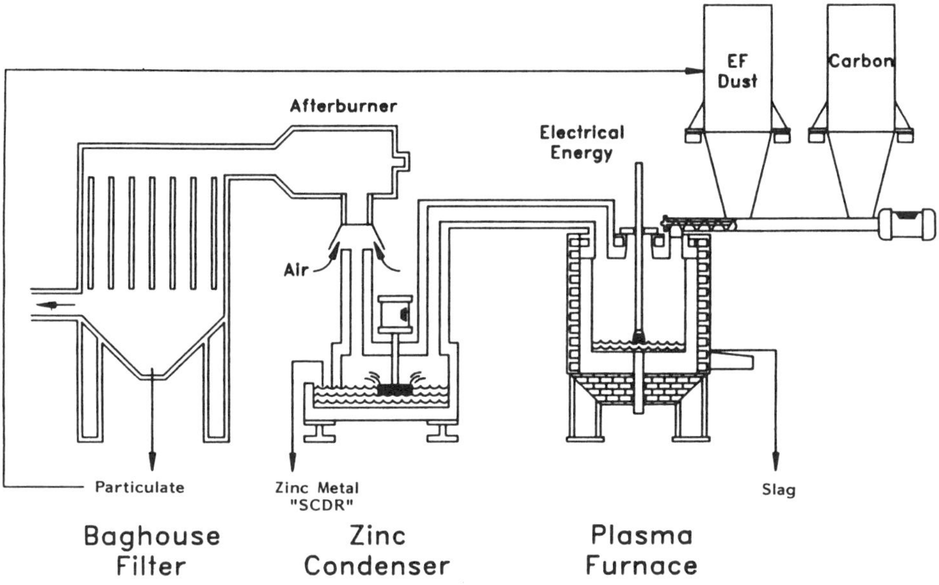

FIGURE 1

Operating Experience

The performance to date with the process has proven to be very typical of a start-up with any new technology. This, of course, means it has been difficult and has required numerous equipment modifications and changes.

The process originally utilized a Tetronics-designed transferred arc, water-cooled plasma torch. Due to the high operating temperature in close proximity to the arc, the torch precessed in a circular motion around the furnace vessel to insure even heating and to hinder instantaneous vaporization of elements intended to remain in the slag bath. Early performance of the torch was extremely poor as premature failure was a common occurrence.

Typical failure involved either pinhole leaks or rupture of the torch with the resultant effect being the introduction of water in the vessel. Water in the vessel causes extremely poor zinc recovery efficiencies as the temperature of the gas stream cools, and combined with increasing carbon dioxide levels and the presence of water vapor, promotes the back reaction of zinc to zinc oxide.

Due to poor torch life, the cost to operate the system with this method was very uneconomical. The best monthly average torch life was 30 hours, which was achieved at the Jackson facility in September of 1989. For this reason, as well as safety concerns, the decision was made to switch to a hollow graphite cathode at both facilities.

On November 15, 1989, operations restarted at the Nucor-Yamato Steel facility with the new cathode system. Argon was then being used as the plasma gas and was introduced through the center of the cathode. After a short trial period it was evident the operation was electrically similar to the operation utilizing the water-cooled plasma torch. Normal operating voltage has been in the range of 310-340 volts with a resultant arc length of 14-18 inches. Observation of the arc via an inspection port indicated the arc pattern was very uniform and left the entire periphery of the 10" diameter cathode and expanded to leave an 18-20" diameter footprint on the slag bath surface.

Though the performance of the new cathode has been very successful, it created a new set of problems to solve. As graphite is conductive and operates at high temperatures and as this phenomenon is not conducive to operating with a refractory roof (which was highly suitable for use with the plasma torch), the temperature of the cathode at the point of entry into the furnace vessel, combined with the harsh environment of the gas stream, promoted the thermal and chemical decomposition of the roof.

The solution to the problem was to raise the roof 20 inches, electrically insulate it from the furnace body, and to install water cooled sections. After these modifications, the system was able to operate at higher production rates. This solution, however, created further problems.

With the original water-cooled plasma torch outer sleeve being non-current carrying, the seal at the penetration hole into the furnace was a simple design and highly effective. This changed drastically with the introduction of the hollow carbon cathode and water-cooled roof, which provided a convenient path to ground. The situation is currently managed to an acceptable level by utilizing an IMS-designed refractory sleeve and sealing system. There is still room for improvement, but process downtime due to this problem is now a relatively minor concern.

Several major benefits were realized as a result of the torch change. First, the operating cost dropped dramatically as the need for expensive torch part changes was no longer required. Secondly, cathode consumption is at a level of 0.6 lb/ton of EAF dust, which is lower than expected. Thirdly, the power requirement to complete the reactions in the furnace was significantly reduced. Normal power consumption is at a rate of approximately 980 KWHr/ton of EAF dust. This reduction is a result of the fact that significant energy was lost to water cooling of the plasma torch.

Additionally, savings were noticed because the torch was operating hot under higher load requirements and the argon flow had to be significantly increased to provide cooling. With the new cathode system, nitrogen was successfully tested as the plasma gas. This conversion to nitrogen provided a large operating cost savings.

The change in plasma systems enabled the operation to achieve higher consistent operating levels. Soon after the operations were stabilized, it became evident both facilities were underpowered to handle the required higher capacity levels of 18,000 TPY and 10,000 TPY at Armorel and Jackson, respectively. In the second quarter of 1991 the power capabilities of the rectifiers were increased at both facilities to 3.6 and 2.6 megawatts, respectively.

With increased operating rates and power levels IMS began to incur problems with the anode section of the furnaces. The anode system design was quite simple and operated fairly well at low power levels. This design consisted of three 4" X 4" steel billets which extended from the bottom of the furnace through the hearth and made contact with the melt, and were cooled externally with water-cooled cheek plates. At increased levels the operation became unstable electrically and current loads on the anodes were usually out of balance. Due to this, a number of runouts along the anodes occurred at both plants. The decision was made to change the bottom design using conductive refractory as the base of the furnace hearth, in contact with redesigned imbedded anodes. This modification has to-date proven successful, and has permitted the process to run at capacity rates under full power.

The major key to the process is the ability to collect zinc as molten metal in the condenser. This task has proven to be more difficult than anyone in this industry has expected. Though the required metallurgical reactions and thermodynamics are quite simple, achieving the process parameters and maintaining them are very difficult. After the furnace vessel-related problems were under control, efforts were concentrated on the condenser.

Due to the proprietary nature of the zinc collection, its importance to the requirements of the process, and the highly competitive nature of the development of metals recovery technology, we are unable to provide any details of the exact problems and solutions which have been met and overcome. Suffice it to say, numerous problems have been encountered, extensive modification and redesign has occurred, and further work is ongoing in order to achieve the original anticipated collection levels.

To date, the maximum average zinc recovery for an individual pass though the process has been in the range of 73-75% against a theoretical maximum of 85%. As is evident, further work is required, but most of the problems have been overcome.

Summary

International Mill Service, Inc. started commissioning and development work of two commercial plasma processing installations in mid-1989. The problems encountered with the technology have been significant and have required numerous, extensive, and highly creative solutions. IMS is confident the remaining challenges can be met and will facilitate the expansion of the technology for use by other customers, as well as other possible applications.

AN UPDATE ON THE START-UP AND OPERATION OF ZIA TECHNOLOGY'S IRRS PROCESS INVOLVING AN ALL-METALLIC REPORT

Andrew A. Fyfe

Vice President International
Zia Technology, Inc.
4172 Rayfield Court
Mississauga, Ontario
Canada L4Z 1E7

Michael D. Lanyi

Manager-Technology Development
Primary Metals Group
Air Products and Chemicals, Inc.
7201 Hamilton Boulevard
Allentown, PA 18195

Abstract

The start-up and operating results obtained at Zia Technology Inc.'s first commercial operating facility at Caldwell, Texas are described. The operation which stated in early 1991 is the first to employ the patented Inclined Rotary Reduction System (IRRS) with the associated all-metallic retort for the recovery of all metallics contained in EAF dust.

Residues and Effluents - Processing and Environmental Considerations
Edited by Ramana G. Reddy, William P. Imrie and Paul B. Queneau
The Minerals, Metals & Materials Society, 1991

3.2 NON-FERROUS FLUE DUSTS

FLUE DUSTS:

AN IDEAL FEED FOR RESOURCE RECOVERY

John E. Litz

Industrial Compliance Technologies, Inc.
1746 Cole Boulevard, Suite 300
Golden, Colorado 80401

Abstract

Metal-bearing flue dusts are produced by some ferrous and non-ferrous industries. Many dusts are being or can be treated for recovery of metal values. Operating and proposed flowsheets for the most common dusts are reviewed. Advantages, disadvantages, product marketability, and cost implications are discussed for each flowsheet.

Residues and Effluents - Processing and Environmental Considerations
Edited by Ramana G. Reddy, William P. Imrie and Paul B. Queneau
The Minerals, Metals & Materials Society, 1991

Introduction

Flue dusts are the curse of the pyrometallurgical industry. In times past flue dusts were shifted between smelters using different processes until the constituents either reported to a saleable product or a disposable slag. As the major smelting companies have closed operations or spun them off, the ability to shift flue dusts became difficult. Another contributing factor is that often the quality of the feed to the smelters has declined, causing the flue dusts to contain larger and larger quantities of deleterious components.

The problems outlined above lead to development of flue dust treatment processes. Superficially flue dusts are an ideal material to process profitably. Typically a valuable component is present in significant quantity and the particle size is fine such that grinding before treatment is not required. However, the quantities of similar dusts be limited, keeping a commercial plant too small to be economical.

Three dusts have been subjected to considerable R & D efforts over the past twenty years. In some cases commercial processes were developed, whereas in other cases the processes have not left the laboratory. This paper reviews processes treating: 1) Electric arc furnace dust, 2) primary copper smelter dust and 3) secondary copper smelter dust.

Electric Arc Furnace Dusts

The rise of the "mini-mill" in the steel industry led to generation of considerable flue dust. These "mini-mills" use electric arc furnaces for recycling steel scrap into pig iron and low grade alloys. Since much of the feed is derived from scrap automobiles, the dust has a high zinc content. Typical analyses of EAF dusts are listed in Table I.

Much of this EAF dust is partially leached to produce zinc sulfate solution. However, aggressive leaching solubilizes significant quantities of impurities. In addition the residues are environmentally unacceptable as determined by TCLP.

Three thermal processes have been operated on a pilot and commercial scales. These are: 1) Horsehead Industries, Flame Reactor; 2) Zia Technology, Inclined Rotary Reduction Process: and 3) Elkem a/s Engineering, Multi-Purpose Electric Furnace Process. All three process produce waste products that are recyclable or pass the TCLP.

TABLE I. - ANALYSES OF TYPICAL EAF DUSTS

Element	A	B	C	D	E	F	G
Iron	39.0	22.9	32.4	28	31.6	30.1	38.6
Zinc	16.7	27.5	18.3	22	15.7	14.7	10.6
Lead	2.5	3.9	2.0	3	3.1	2.0	1.2
Manganese	3.5	2.7	-	-	3.0	3.4	5.0
Copper	0.3	0.2	-	-	-	0.1	0.1
Cadmium	-	-	0.05	0.1	-	0.1	0.02
Chromium	0.1	0.1	0.2	0.8	-	0.1	0.2
Magnesium	1.4	1.2	-	-	2.7	1.7	1.5
Calcium	5.8	8.4	-	-	11.1	10.9	7.4
Silicon	1.6	1.1	-	-	2.1	0.9	1.5
Chlorine	0.8	2.3	-	-	0.7	0.7	1.4

A simplified flowsheet for Horsehead's, Flame Reactor Process[1/] is shown by Figure 1. EAF dust is fed into a coke-fired reducing flame (coke, air and/or oxygen). The flame temperature, >2000°C, reduces and volatilizes the metals in the dust. The non-volatile components form slag which coats the reactor and flows down the walls into the slag separator. The slag-free gases then are combusted in an after-burner forming metal oxides. The oxides are collected in a dust collector and are marketed.

The flowsheet for the Zia Technology' Inclined Rotary Reduction Process[2/] is shown in Figure 2. EAF Dust, pelletized with coal and flux, is fed into an inclined rotary furnace fired. The coal reduces the lead, cadmium and zinc oxides to metals that volatilize. Iron is reduced and remains in the pellets. The pellets are quenched and are suitable for feed to an electric arc furnace.

The metal vapors are oxidized in an afterburner that surrounds the retort. The oxidized gases are cooled and the dusts are collected. The collected dusts then are pelletized with coal and flux, and fed into the retort. In the retort coal reduces the metals and the zinc, cadmium and lead volatilize. The volatile metals are recovered in a splash condenser. Prime Western grade zinc metal and low-zinc lead alloy are recovered by liquation. Non-volatiles from the retort recycle to the rotary furnace. Volatiles from the retort recycle through the afterburner.

Elkem demonstrated their EAF flowsheet, see Figure 3, in a pilot run reported in April, 1987[3/]. EAF dust, pelletized with coke and flux, is fed to an air tight, "Multi-Purpose" electric arc furnace. The charge is heated to 1400°C, reducing and volatilizing the zinc plus some lead. The volatiles are condensed using a splash

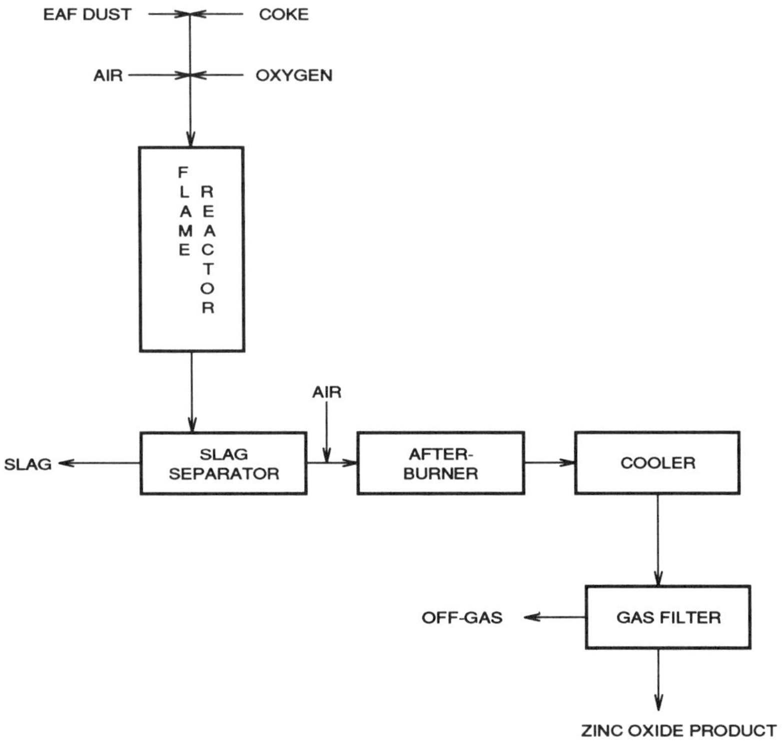

Figure 1 - Horsehead Process for Treating EAF Dust.

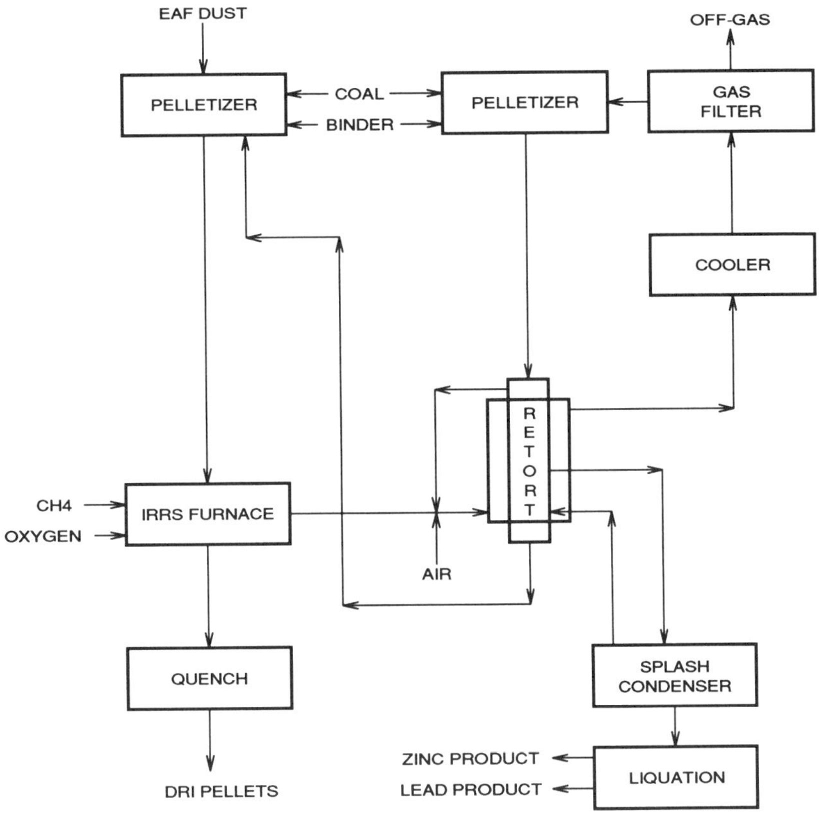

Figure 2 - Zia Technology Process for Treating EAF Dust.

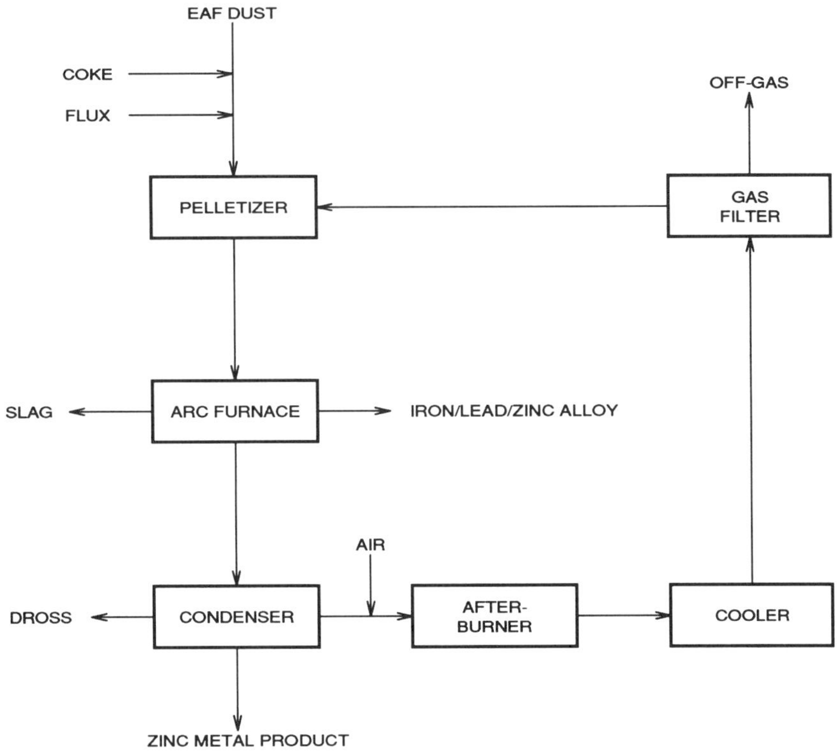

Figure 3 - Elkem Process for Treating EAF Dust.

condenser. The condenser off-gas is cooled and any dust is collected by a dust collector and recycled to the arc furnace.

Table II compares recoveries for the three processes. Only the Horsehead shows high recoveries. The Zia paper does not show the source of their data; whereas the Elkem data are from short, small scale pilot runs that did not come to equilibrium. Both these latter processes are capable of >95% recoveries of zinc.

TABLE II. - RECOVERIES TO PRODUCTS FROM EAF DUSTS

		Recovered, Percent		
Process	Product	Zinc	Lead	Cadmium
Horsehead	Oxide Dust	90	97	99
Zia	Zinc alloy	83	0.1	
	Lead alloy	2	67	
Elkem	Zinc metal	59		
	Furnace metal	2.5	69	

Note: Recoveries are calculated from references or were reported in the reference.

The quality of the products from the three processes are shown in Table III. The Zia and Elkem processes recover the zinc as high quality metal. The Horsehead process recovers low quality, zinc oxide. Since Horsehead is a primary zinc producer, this oxide is suitable for recycling to their primary production. The zinc oxide product may not offer adequate return for a stand-alone treater.

The published reagent and energy costs for the processes are listed in Table IV. Power for pumps, fans, etc. must be added to the costs for all three processes.

When the processes are compared. The Flame Reactor probably has the lowest capital cost, but offers the lowest value product. The Elkem technology probably has the highest capital cost and is very sensitive to power costs. The Zia Technology process appears to offer a moderate capital cost, while producing high value end products.

TABLE III. - PRODUCT QUALITY FROM EAF DUSTS

Process	Product	Contained, Percent Zinc	Lead	Cadmium	Iron
Horsehead	Oxide Dust	60	7.7	0.8	6.4
	Slag	3.7	0.07	-	45.
Zia	Zinc alloy	98.5	1.5	-	-
	Lead alloy	10.	89.	1.0	-
	DRI Product	0.8	0.1	-	46.
Elkem	Zinc metal	99.4	0.33	0.02	0.01
	Furnace metal	1.7	4.7	0.1	88.3
	Slag	0.6	0.02	0.02	23.

TABLE IV. - REAGENT AND ENERGY COSTS

Reagent	Units/t dust	Horsehead	Zia	Elkem
Coal	t		0.3	
Coke	t	0.5		0.07
Oxygen	t	0.5	0.54	
Methane	m^3		200	
Silica	t			0.1
Power	MJ			2000

Primary Copper Smelter Dusts

All primary copper smelters produce significant quantities of dusts. In some cases these dusts were (or are being) treated to recover arsenic as a saleable product. Since arsenic is often the primary impurity, the removal of arsenic allows recycling of the dust residual dust. However, most smelters do not produce dust containing enough arsenic for economic recovery. Therefore, they often just recycle them hoping that the circulating load will not be oppressive and that the impurities will report to the slag faster than to the metal. Typical analyses of copper smelter dusts are listed in Table V.

TABLE V. - ANALYSES OF TYPICAL PRIMARY COPPER SMELTER DUSTS

Element	A	B	C	D	E	F	G
Copper	13.6	13.8	27.0	15.7	15.5	14.1	25.4
Arsenic	13.5	11.2	3.5	8.0	12.6	12.9	1.3
Lead	14.7	17.0	7.2	2.2	3.9	4.9	1.0
Zinc	2.1	1.5	5.4	5.5	7.0	2.8	2.1
Iron	3.8	3.8	20.4	15.4	13.1	18.4	18.6
Molybdenum	0.9	1.6	-	-	-	-	0.6
Cadmium	0.2	-	-	-	-	0.2	0.2
Antimony	0.2	0.4	0.3	0.3	0.4	0.4	-
Bismuth	1.7	1.9	0.1	0.5	0.5	0.6	0.2

Two options are available when considering treating flue dusts. Research groups of Kennecott Copper have conducted many programs on treating dusts over the years. Their process, U. S. Patent 4,149,880,[4] was pilot tested at a leaching scale of 190 L and does not recover the arsenic.[5] Kennecott also has developed processes that recover arsenic. A second group researching flue dust processing is Artech Recovery Systems. Artech has been developing the Cashman Process, U. S. Patent 4,655,829.[6] Artech and ARCO Coal tested the Cashman Process on Anaconda, Montana smelter dusts at a scale of 25 kg per hour.[7]

Figure 4 depicts the Kennecott flowsheet. Flue dust is leached under pressure with sulfuric acid, precipitating the arsenic as scorodite ($FeAsO_4 \cdot 2H_2O$), converting the lead to insoluble lead sulfate, and solubilizing most of the remaining metals. Since the Kennecott dusts contain significant molybdenum, it is removed from the leach solution by solvent extraction and recovered as calcium or ammonium molybdate. The copper is precipitated using cementation onto scrap iron and recycled to the smelter as sponge. Most of the solution is recycled to the leach to provide enough iron for the scorodite. Zinc and cadmium are precipitated from the balance of the solution as hydroxides.

The flowsheet for the Artech/Cashman Process as applied to Anaconda dusts is shown in Figure 5. The dust is leached under pressure with sulfuric acid in a calcium chloride brine. Arsenic is precipitated as scorodite, but the other metals are solubilized. The residues readily pass the TCLP. Lead, bismuth and silver are the first metals removed from the solution. These are hydrolyzed to a low chloride product suitable for recycling by a lead smelter. Copper is recovered by solvent extraction and electrowinning. Cadmium and mercury are recovered by precipitation with zinc dust. A zinc oxide/hydroxide product then is recovered by hydrolysis.

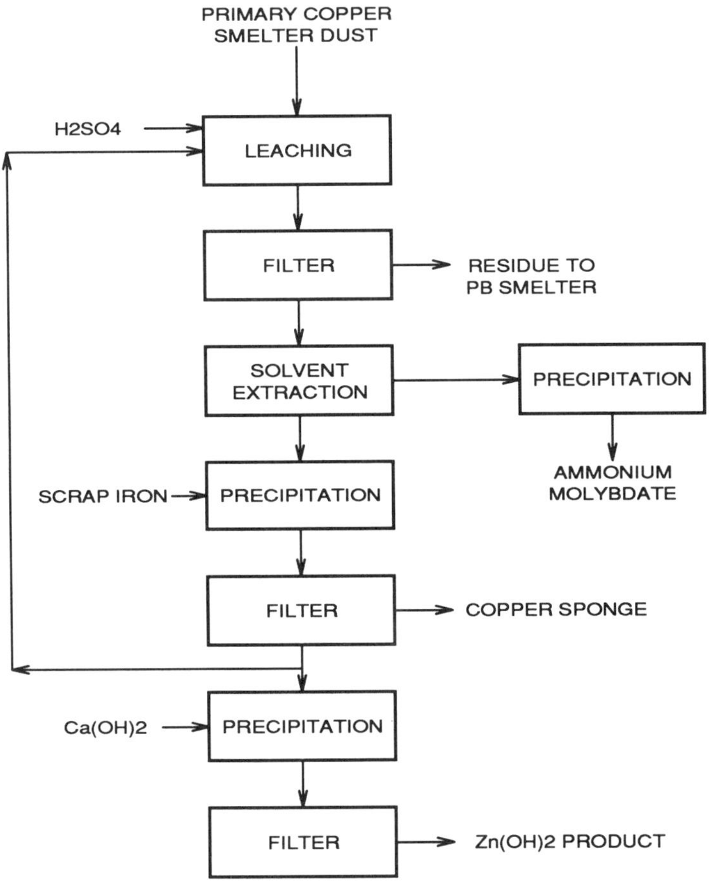

Figure 4 - Kennecott Process for Treating Primary Copper Smelter Dust.

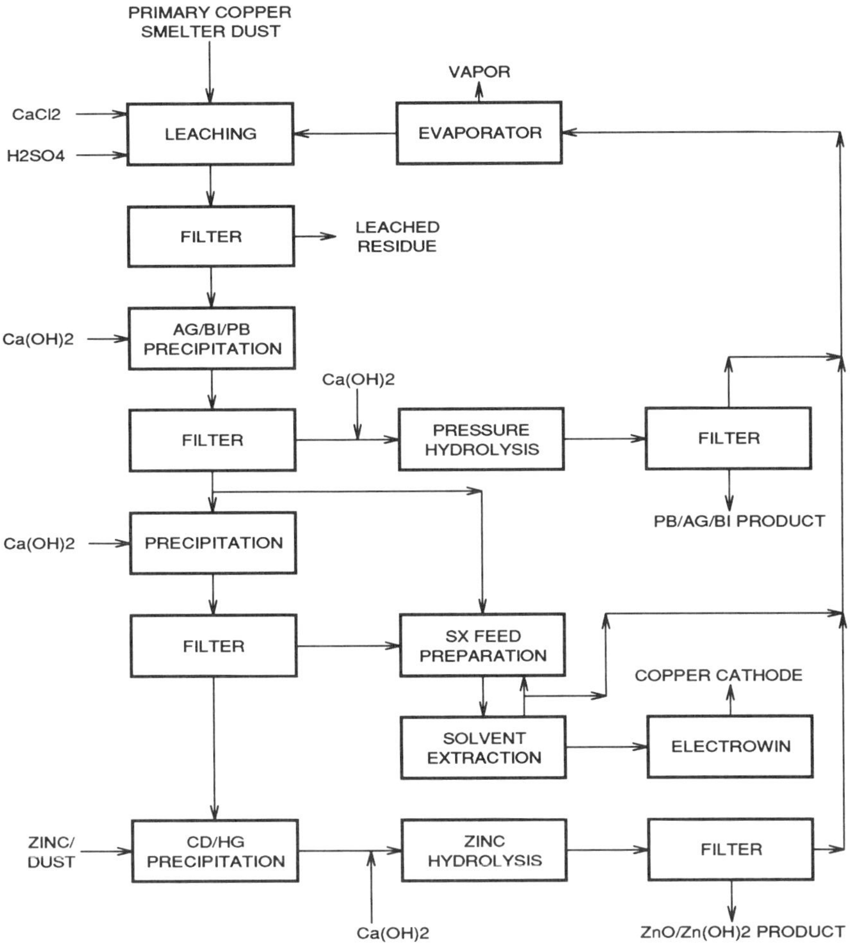

Figure 5 - Artech/Cashman Process for Treating Primary Copper Smelter Dust.

Enough information is not available to make accurate economic comparisons of the two processes. The Kennecott process only produces a molybdenum product for marketing. The cement copper product is recycled to the smelter. Calcium molybdate produced during the pilot plant contained: 42-44% Mo, 20.5-21.5% Ca, 0.01-0.04% S and 0.02% As. As such it should be marketable for direct steel addition. The copper cementation also precipitated 65% of the soluble bismuth. The recycle of this bismuth may ultimately cause problems for the smelter.

The Artech/Cashman Process converts most of the values to saleable products. The copper is recovered as electrowon metal, the lead, bismuth, silver and some copper are converted into a concentrate suitable for recycling to a lead smelter, zinc is recovered as an oxy-hydroxide, and cadmium and mercury are recovered as retortable precipitate. The zinc product contained 66% Zn, 3% Ca, 0.3% Mg, and <0.1% Cl. Mercury can be separated from the cadmium precipitate by retorting. However, the market for cadmium is limited.

In both processes the major costs are sulfuric acid used in leaching, lime used for neutralization plus the cost of recovering the copper - scrap iron or power for electrowinning.

Secondary Copper Smelter Dusts

Dusts generated by secondary copper smelters traditionally have been recycled by shipping to primary copper or lead smelters. As such the return is negligible to negative and much of the values are not recoverable by copper or lead smelters. Hydrometallurgical processing allows for potential recovery of all major components. Analyses of typical secondary copper smelter dusts are listed in Table VI.

Hydrometallurgical processing of secondary copper smelter dusts was proposed by the U. S. Bureau of Mines in 1989, Report of Investigations 9199[8/]. The flowsheet proposed by the Bureau is shown in Figure 6. One potential step is not shown on the flowsheet. If the dust contains considerable chloride, most of the chloride can be removed by washing with hot, dilute sodium carbonate solution. The low-chloride dust then is leached with ammoniacal ammonium carbonate. Most of lead and tin are converted to carbonates and remain with the leach residue. The Bureau tested a low copper dust, 0.9% Cu. Therefore, the solubilized copper plus lead and tin were removed from solution by precipitation using zinc dust. Zinc is then precipitated from the purified solution by adding carbon dioxide to adjust the alkalinity to 7 pH. The zinc precipitate, $ZnO \cdot Zn(NH_3)_2CO_3$, is calcined at 400°C to convert it to ZnO. The Bureau produced zinc oxide containing 80.2% Zn, 0.007% Cu with no lead or tin detected.

TABLE VI. - ANALYSES OF TYPICAL SECONDARY COPPER SMELTER DUSTS

Element	A	B	C	D	E	F	G
Copper	3.2	1.5	0.9	6.2	8.1	4.7	4.4
Arsenic	0.02	0.003	-	-	0.06	0.05	-
Lead	15.2	16.6	16.0	11.7	11.5	14.9	21.2
Zinc	45.6	36.3	40.4	16.5	16.2	29.6	38.9
Iron	0.4	0.1	-	7.1	3.7	2.4	0.8
Cadmium	0.08	0.01	-	-	0.07	0.11	-
Tin	0.8	15.0	4.3	-	1.0	1.0	0.8
Calcium	0.3	0.1	0.1	7.0	6.5	4.2	-
Silicon	1.5	0.3	-	-	1.3	2.6	-
Chlorine	5.2	0.2	0.05	6.5	7.4	4.6	8.1

The author performed a similar study on dust containing higher concentrations of copper and arsenic. In that study copper was recovered by solvent extraction with intent to produce tribasic copper sulfate and the zinc was precipitated by thermal decomposition of the ammonium carbonate solution. Purifying from arsenic in solution was not possible, although the majority of the arsenic reported to the first crop of zinc precipitate. Calcining in a slightly reducing atmosphere at 900°C reduced the arsenic from 700 to 19 ppm. Distribution of values from the two studies are shown in Table VII.

Acid processing of secondary copper smelter dusts also has been the subject of study. Figure 7 depicts two flowsheets having potential for economic processing. Weak sulfuric acid leaching solubilizes up to 90% of the copper, zinc, nickel, manganese, and antimony plus up to 80% of the iron.

The solubilized copper can be precipitated from solution by addition of scrap iron or zinc dust, depending on whether or not addition of iron will affect the zinc recovery steps. If zinc electrowon zinc metal is the goal, the solution can be purified using conventional tank house techniques. If recovery of zinc/iron sulfate is the goal, only soluble nickel and cobalt need to be removed.

Since none of the processes has been tested in a pilot plant, it is not possible to compare capital or operating costs. Acid leaching followed by producing zinc/iron sulfate offers the lowest capital and operating costs. The process also has the lowest potential return. The ammoniacal routes and the zinc electrowinning process have potential to generate high value zinc products, but with significantly higher capital and operating costs.

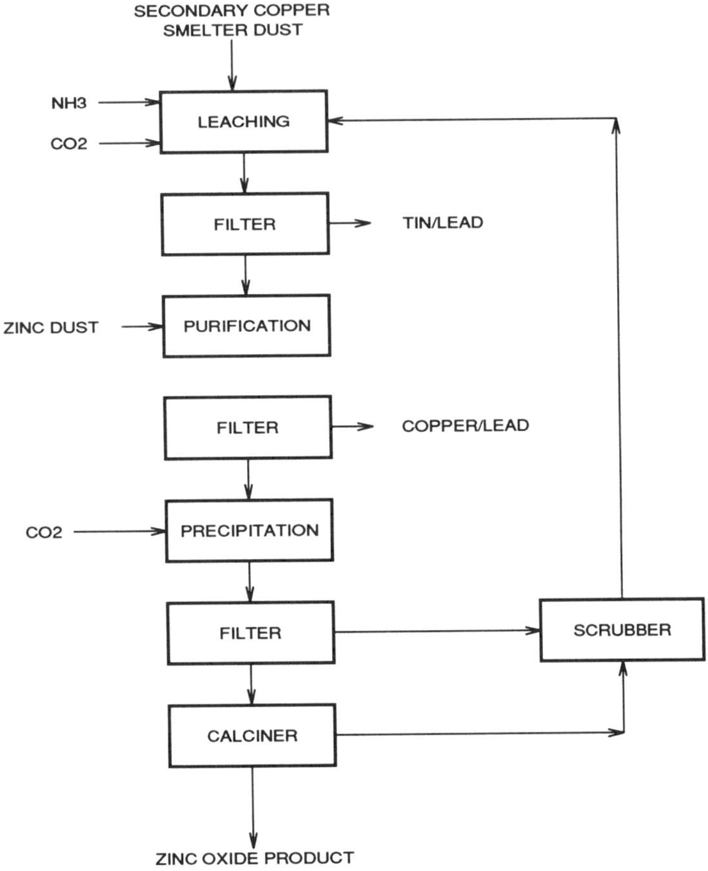

Figure 6 - U. S. Bureau of Mines Process for Treating Secondary Copper Smelter Dust.

TABLE VII. - DISTRIBUTION OF VALUES BY AMMONIACAL LEACHING PROCESSES

Product	Zinc	Copper	Lead	Tin	Arsenic
USBM					
Leach residue	11	52	97	92	-
Cement product		48	3	2	-
Zinc product	94				-
Author					
Leach residue	4	17	99	92	38
Solvent phase	0	83	0	0	0
Cement product		0	1	0	0
First zinc crop	28	0	0	0	31
Zinc precipitate	68	0	0	0	30

1. J. F. Pusateri, C. O. Bounds, and L. W. Lherbier, "Zinc Recovery via the Flame Reactor Process," Journal of Metals, August 1988, 31-35.

2. Norman L. Kotraba, and Michael D. Lanyl, "Metal Recovery from EAF Dust Using the Inclined Rotary Reduction Process," (Paper Presented at the AIME Annual Meeting, Denver, Colorado, February, 1991), 15.

3. Elkem a/s R & D, "EAF Dust Treatment, Demonstration of Technology, Final Report," (Report Elkem a/s Engineering, April, 1987).

4. John D. Prater, and Barry A. Wells, "Recovery of Copper from Arsenic Containing Metallurgical Waste Materials," U. S. Patent 4,149,880, April 17, 1979, 4.

5. John D. Prater, Joseph W. Schlitt, and Kenneth J. Richards, "Kennecott Process for Treating Copper Smelter Flue Dusts," source unknown

6. Joseph B. Cashman, "Metal Recovery Process," U. S. Patent 4,655,829, April 7, 1987, 6.

7. Artech Recovery Systems, "Combining Pyrometallurgy and Hydrometallurgy; An Environmentally Sound Solution," (Paper prepared for 30th Annual Conference of Metallurgists, Ottawa Ontario, August, 1991).

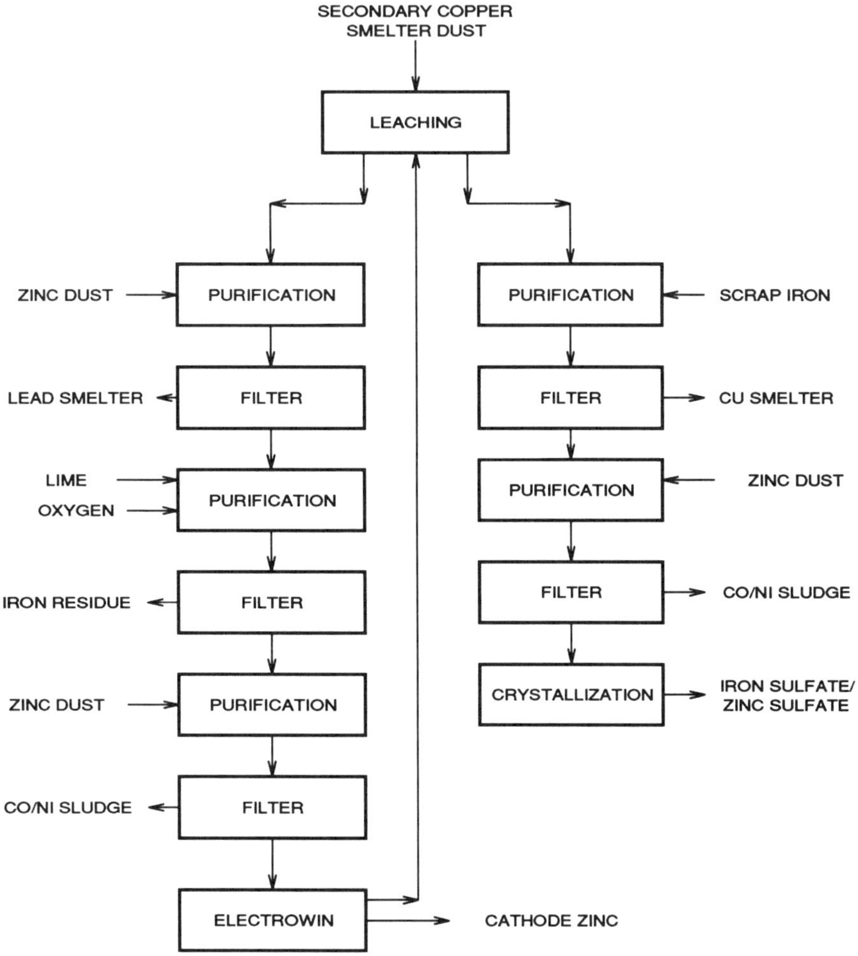

Figure 7 - Sulfuric Acid Processes for Treating Secondary Copper Smelter Dust.

NEW PROCESS FOR PRODUCTION OF A ZN-FE-BASED ANIMAL MICRONUTRIENT FROM SECONDARY-COPPER FLUE DUST

M. P. Bohrer, H. H. Law, and P. A. O'Hara*
AT&T Bell Laboratories
600 Mountain Ave.
Murray Hill, NJ 07974

R. E. Burgess**
AT&T Nassau Metals
P.O. Box 218
Gaston, SC 29053

P. B. Queneau and N. Hazen
Hazen Research, Inc.
4601 Indiana St.
Golden, Co 80403

Abstract

A closed-loop hydrometallurgical process has been developed for treatment of smelter flue dust generated by copper recycling. The process is a semi-continuous operation in which zinc, iron, and copper are first extracted from the dust with a sulfuric acid solution. The lead and precious metals, initially present in the dust, remain as solids and can be recycled to the copper smelting furnace. Cementation with iron and zinc metal purifies the zinc and iron-rich liquid stream prior to spray drying. The spray-dried product, consisting of a mixture of zinc sulfate monohydrate and iron sulfate monohydrate, is a desirable micronutrient additive for use in animal feeds. Treatment of the smelter flue dust in this manner will generate a positive return on investment, as well as mitigate a potential environmental liability.

* Current address: CH2M Hill, 2510 Redhill Ave., Santa Ana, CA 92705

** Current address: Center for Waste Minimization, South Carolina Department of Health and Environmental Control, 2600 Bull St., Columbia, SC 29201

Residues and Effluents - Processing and Environmental Considerations
Edited by Ramana G. Reddy, William P. Imrie and Paul B. Queneau
The Minerals, Metals & Materials Society, 1991

Introduction

Secondary copper recycling generates electrolytic copper from a variety of sources. Typically, furnaces in the copper recycling plant generate flue dusts containing zinc, lead, copper, tin, nickel, iron, precious metals, sulfur, chloride, and many other elements in minor concentrations. These metals are primarily present as oxides but some sulfates and chlorides are also present. The primary values in the flue dusts are zinc and precious metals.

Numerous hydrometallurgical, pyrometallurgical, and electrometallurgical processes have been reported for extracting valuable metals from these furnace dusts and other sources such as low-grade ores, tailings, slags, and sludges (1). In particular, Pearson (2) has reported the use of various acidic and alkaline reagents to leach zinc from secondary sources. In a companion study (3), various extractants were employed to leach zinc from flue dust generated by secondary copper smelting. Treatment with sulfuric acid at an elevated temperature was found to yield the highest zinc recovery, $\sim$90%. Subsequent purification of the Zn solution using Fe and Zn cementation followed by drying, produces a high purity Zn/Fe sulfate salt. This product is ideally suited for use in the animal feed market, where $ZnSO_4$ and $FeSO_4$ are typically combined with other salts to make a "premix" feed supplement (4).

On the basis of lab-scale tests (3) and an engineering feasibility study performed by Hazen Research, the hydrometallurgical process shown schematically in Figure 1 was proposed for treatment of smelter flue dust generated by copper recycling. Briefly, the process consists of mild soda ash leaching to reject chloride in a manner similar to that reported by Salomon-De-Friedberg (5) for leaching of zinc oxide fume; sulfuric acid leaching of the soda ash leach filter cake to solubilize zinc, copper, and part of the iron; removal of the solubilized copper by cementation of the acid leach slurry with scrap iron; rejection of heavy metal impurities, particularly nickel and cadmium, from the leach/cementation filtrate using zinc cementation; and spray drying of the purified liquor to produce a salable zinc/iron sulfate product. The copper-bearing acid leach/cementation filter cake is recycled to the smelter. In this study, a pilot-scale test of this proposed process was performed.

Experimental Procedures and Results

Dust Characterization

Furnace dust generated at the AT&T Nassau Metals facility in Gaston, SC was used for the pilot-scale tests. A single bulk composite sample, consisting of 58% blast furnace dust and 42% converter furnace dust (dry basis), was used for the pilot-scale study. The composition of the dust mixture is given in Table I. The blast furnace dust which had been treated with water and pelletized at the baghouse, was dried for 24 hours at 100 °C, crushed, and then ball milled prior to blending with the untreated converter furnace dust.

Water Leaching

Two hundred kg of the dry dust composite was leached in a 27% solids slurry for 4 hours at 90 °C and pH 9 to 9.5 using Na_2CO_3 for pH adjustment. The water leaching was performed batchwise in a covered, cone-bottom,

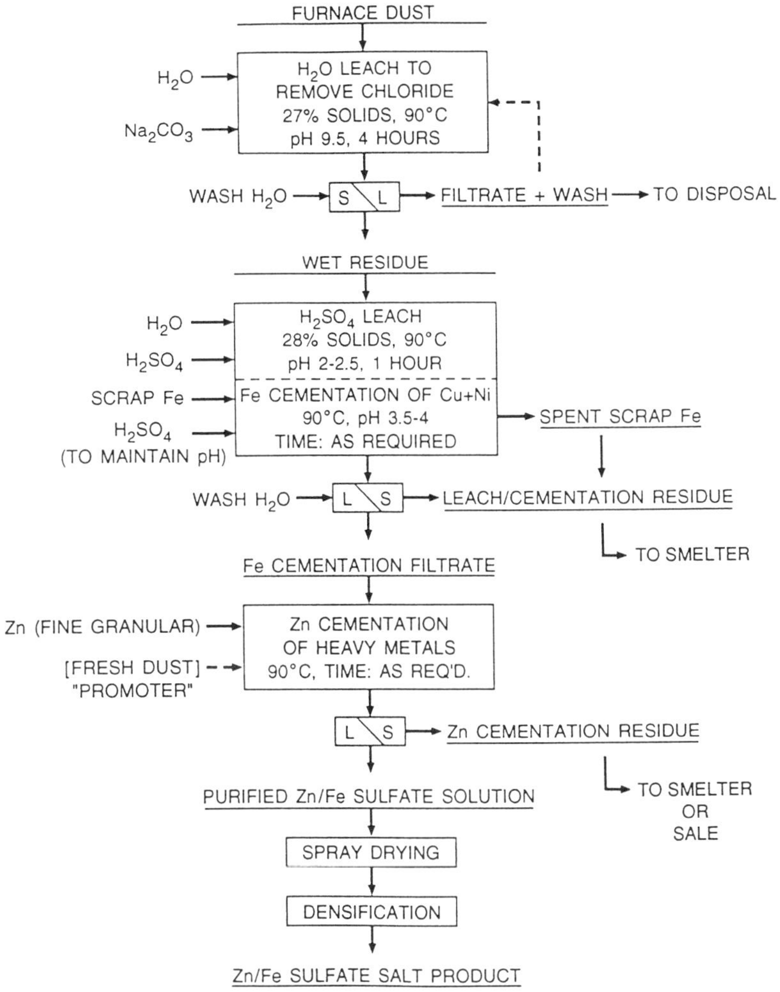

Figure 1 - Schematic diagram of furnace dust pilot plant process.

stainless steel tank agitated with a rubber-covered impeller. A stainless steel steam coil was used to heat the slurry. After 4 hours, the final slurry was vacuum filtered, without settling, through 0.76 m diameter table-top Buchner filters. The cakes were washed with approximately 1.75 CSD's (cake solution displacements) of water. The water leach residue was advanced, wet, to the acid leach/Fe cementation step.

Table I. Chemical Analysis of Composite Dust

Element	Unit	Assay
Zn	%	21.8
Pb	%	8.66
Cu	%	9.53
Fe	%	1.89
Cd	%	0.162
Ni	%	0.116
Co	%	0.014
Ca	%	6.72
Mg	%	0.454
Au	mg/kg	15.4
Ag	mg/kg	450
Pd	mg/kg	7.00
Pt	mg/kg	0.625 (poor repeats)
Rh	mg/kg	0.313
Al	%	2.04
Si	%	1.82
Ba	%	0.062
As	ppm	530
Sb	%	0.224
Sn	%	1.10
Cr	%	0.118
Mn	%	0.055
Se	ppm	1.8
Hg	ppm	2.5
S-total	%	1.46
Cl	%	5.68
F	%	0.40
C-total	%	3.95

Conditions and results of the water leach are presented in Table II. The leach achieved 74% chloride removal from the dust and 7% solids weight loss with negligible extraction of heavy metals. The residue still contained 1.6% (dry basis) chloride, which was dissolved in the acid leach and carried through to the final salt product.

In order to test whether additional washing would improve the chloride rejection, a small portion of the wet residue was repulped at 17% solids in pH 9 water, filtered, and washed well with more water. Overall Cl extraction was ncreased to 84% by this treatment, with the rewashed solids assaying 1.06% Cl. The bulk of the wet residue was not rewashed prior to acid leaching.

Table II. Summary of Pilot Plant Water Leach

Leach Conditions:		Flocculation:	
Dust weight, kg	200	Flocculant	Percol 351 (1 g/l)
Slurry H_2O, l	540	Flocculant volume, l	4
Initial % solids	27	Flocculant dose, ppm	22 (solid basis)
Target pH	9-9.5		
Na_2CO_3 required, kg	12.79		
Temperature, °C	90		
Time, hours	4		
Final slurry pH	9.9		
Filtration and Washing:			
Filtrate volume, l	351		
Filtrate specific gravity	1.03		
Wash volume, l	284		
No. CSD's	1.75		
Wash specific gravity	1.02		
Wet cake weight, kg	347		
Wet cake density, kg/l	1.59		
% Moisture	47		

Element	Product Assay			
	Dust Feed (%)	Residue (%)	Filtrate (g/l)	Wash (g/l)
Zn	21.8	23.9	<0.001	-
Cu	9.53	10.3	<0.0005	-
Fe	1.89	2.12	0.0015	-
Cd	0.162	0.156	-	-
Ni	0.116	0.114	-	-
Pb	8.66	8.14	<0.05	-
Co	0.014	0.017	-	-
Cl	5.68	1.61	17.0	8.02
Mg	0.454	0.476	<0.001	-
Ag	450 mg/kg	453 mg/kg	-	-
Au	15.4 mg/kg	24.0 mg/kg	-	-
Pd	7.00 mg/kg	5.94 mg/kg	-	-
Pt	0.625 mg/kg	0.313 mg/kg	-	-

Acid Leach/Cementation with Scrap Iron

The bulk of the wet filter cake from the water leaching, 346 kg, was reslurried to 28% solids and heated to 90 ° C in the same tank that was used for the water leaching step. Technical grade sulfuric acid was added to maintain a leach pH of 2.0 to 2.5 (751.5 kg H_2SO_4/1000 kg dry solids was required). The acid leach was conducted for one hour, at which time Fe cementation was begun. The Fe cementation was done at 90 ° C to enhance the rate of cementation and reduce iron consumption as reported by Biswas and Reid (6).

During cementation, the slurry pH was allowed to rise to the 3.5 to 4.0 range, but was maintained below pH 4 by addition of 21.5 kg H_2SO_4/1000 kg dry solids feed over the course of the cementation. Initially, 62 kg of shredded scrap iron, contained in four 316 SS screen "baskets", were suspended in the agitating slurry. It was found that the baskets of coarse scrap Fe quickly fouled with a coating of cemented copper and residue solids. The coatings were manually knocked loose periodically, and the scrap Fe was replaced with a fresh 62 kg charge after four hours of operation. The coarse Fe in the baskets was in the slurry for a total of 11 hours at 90 ° C over two days, including an overnight period without heating. Only 85% Cu cementation was achieved in this time, with 6 g/l Cu remaining in solution. Impurity levels in solution (normalized to 100 g Zn/l) versus cementation time are shown graphically in Figure 2, with explanatory notes given in the caption.

Table III. Filtration Data for Pilot Plant Acid Leach/Iron Cementation

Filter press used	R&B, Model HPL300/14
Effective area, m^2	2.0
Filter media	Polypropylene woven filter cloth
Operating pressure, Pa	6.20×10^5 (90 psi)
Batch 3 feed slurry weight, kg	141
Feed slurry % solids	15.6
Filtration rate, solids, kg/hr/m^2	24.4
Filtration rate, solution, l/hr/m^2	85.6 (estimated)

Filtration washes	Rate, l/hr/m^2	Cum CSD's
W-1 (10 l displaced)	53	0.75
W-2 (10 l displaced)	118	1.5
W-3 (10 l displaced)	151	2.25
W-4 (10 l displaced)	159	3.0
W-5 (10 l displaced)	179	3.75

In order to improve cementation efficiency, the baskets of Fe were removed and cementation was continued by adding 24 kg of minus 10-mesh scrap iron directly to the slurry. The minus 10-mesh Fe was in the slurry for a total of 18 hours at the reaction temperature of 85 to 90 ° C. This time was spread over three days, including two overnight periods without heating. As shown in Figure 2, addition of fine iron at 11.9 hours quickly reduced the Cu concentration to the 5 mg/l level, with corresponding drops in Ni and Cd

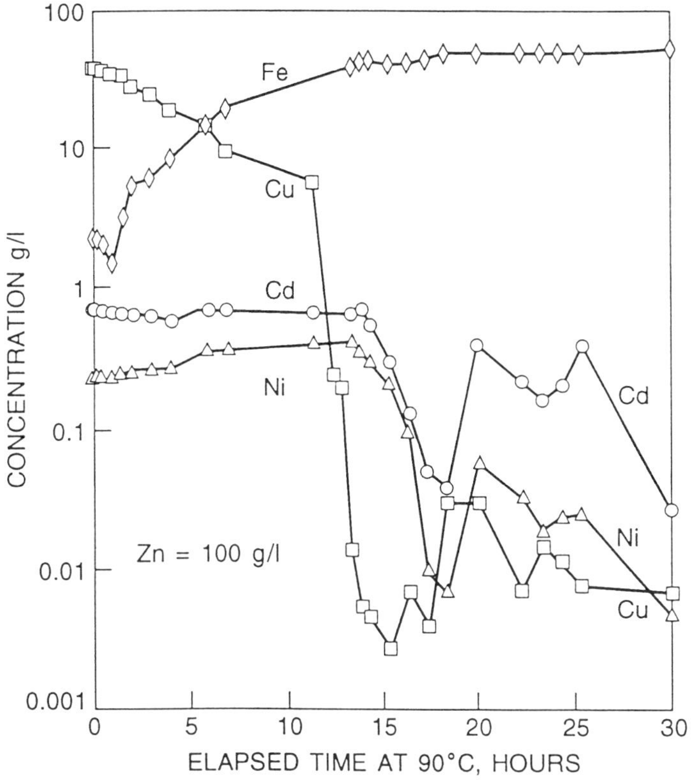

Figure 2 - Impurity and iron concentrationos during Fe cementation versus elapsed time at 90 ° C. Concentrations are normalized to 100 g/l Zn. Scrap iron in baskets added at 1 hour and replaced with fresh scrap iron at 5 hours. Minus 10-mesh scrap iron added to slurry and baskets removed at 11.9 hours.

concentrations. The first overnight period of agitation without heating however, allowed the impurity metal levels to increase again, probably due to air entrainment in the slurry, partially reoxidizing and redissolving the Cu, Ni, and Cd. Cementation was continued the next day at 90 °C and the impurity concentrations decreased again. The second overnight period was passed without slurry agitation to minimize air-slurry contact. On the third day, the final cementation slurry was diluted with water to 85 g/l Zn to improve filtration properties, reheated, and advanced to filtration in a plate-and-frame filter press (R&B, 2.0 m^2 filter area) operated at 6.2×10^5 Pa (90 psi). Five filtration batches were required to process the slurry. Each filter batch was washed with 50 to 60 liters water, with the first 10 liters of displaced solution added to the primary filtrate. Filtration and washing rate data were collected during the third filtration batch and are presented in Table III.

Table IV. Summary of Pilot Plant Acid Leach/Iron Cementation Conditions

H_2SO_4 Leach:	
Wet solids weight, kg	345.6
% solids	53
Dry solids, kg	183
Target pH	2-2.5
H_2SO_4 required, kg	140.4 (93.2% H_2SO_4)
Temperature, °C	90
Time, hours	1 after all reactants added
Iron Cementation:	
Target pH	3.5-4
H_2SO_4 required, kg	8.0 (93.2% H_2SO_4)
Temperature, °C	90
Part 1- Scrap Fe feed, kg	2 x 62
Reaction time, hours	11 at 90 °C
Part 2- Scrap iron feed, kg	24.2 (minus 10 mesh)
Reaction time, hours	18 at 90 °C
Final Slurry Filtration:	
Number of batches	5 yielding ~30 kg wet cake/batch
Filtrate volume, l	410 ±10
Filtrate specific gravity	1.32
Wash volume, l	243
No. CSD's	3.5-4.5 (CSD=Cake Solution Displacement)
Wet cake weight, kg	153
% Moisture	28.9

The filtrate from the five batches were combined to yield approximately 410 liters of solution assaying 85 g/l Zn and 46 g/l Fe with a specific gravity of 1.32. The composite filtrate was advanced to the next stage for heavy metal impurity cementation using zinc. The bulk of the fine cement Fe remained on the bottom of the reactor tank and was collected separately after the residue slurry had been withdrawn from the tank.

An overall summary of conditions and results of the acid leach/Fe cementation are presented in Tables IV and V, respectively. Zinc extraction from the dust was 90%, accompanied by 90% Cl extraction, 91% Mg, and 78% Co extractions. Approximately 13% of the scrap Fe added for cementation was solubilized. Overall dust solids weight loss in the leach/cementation was 40%. Less than 50 mg/l each of Cu, Cd, Ni, and Pb remained in solution.

Table V. Product Analyses for Pilot Plant Acid Leach/Iron Cementation

Element	Assay			
	H_2O Leach Residue (%)	Final Residue (%)	Filtrate (g/l)	Wash Composite (g/l)
Zn	23.9	2.02	84.8	13.0
Cu	10.3	5.57	0.006	0.001
Fe	2.12	14.2	45.6	5.45
Cd	0.156	0.168	0.023	-
Ni	0.114	0.125	0.004	-
Pb	8.14	7.76	0.042	-
Co	0.017	0.015	0.144	-
Cl	1.61	0.06	6.73	0.9
Mg	0.476	0.064	1.64	-
Ag	453 mg/kg	403 mg/kg	-	-
Au	24.0 mg/kg	24.9 mg/kg	-	-
Pd	5.94 mg/kg	7.50 mg/kg	-	-
Pt	0.313 mg/kg	<0.313 mg/kg	-	-

Zinc Cementation

The primary purpose of Zn cementation was to reduce Cd to the 5-10 mg/l level in the Zn/Fe sulfate solution. Cu and Ni already had been reduced to this level in the Fe cementation step. In the first stage of Zn cementation, 410 liters of the Fe cement filtrate was reacted for 3 hours at 81 to 90 °C with the addition of 700 g fine granular Zn metal along with 2000 g of the raw furnace dust. A nitrogen blanket over the slurry was not used during the reaction time. During Zn cementation, the slurry was allowed to evaporate to a target solution Zn concentration of 100 g/l (15-20% volume reduction).

Previous work (3-8) has shown that a number of elements such as Pb, Cu, Sb, Cd, and As either alone or in various combinations can be used as "activators" or "promoters" to improve the rate of zinc cementation of heavy metals. In the companion study (3), Pb and Cu were found to dramatically improve the rate of Ni and Co cementation using Zn metal at 90 °C. Antimony or Cd did not increase the rate of Ni removal. Substitution of 5 g/l of fresh furnace dust which contains substantial amounts of Cu and Pb, was found to promote cementation of heavy metals. The 700 g of Zn metal used in the first stage of Zn cementation represents a 125-fold stoichiometric excess for the Cd in the filtrate; however, the fresh dust introduces heavy metals which, upon dissolution, also require Zn for cementation. If total dissolution of Cu, Cd, and Ni occurred from the added dust, the stoichiometric ratio of cement Zn to impurity metals would only be 3.2. The true excess of Zn metal would be intermediate between 3.2 and 125, since total dissolution from the dust would not occur at the slurry pH of 4.

After 3 hours of cementation in stage 1, Cd concentration remained at 23 mg/l, the same as the feed solution. The final slurry from stage 1 cementation was flocculated with a 200-ppm dose (solids basis) of Percol 1011, settled in the tank, and the bulk of the solution decanted through a cartridge filter to remove suspended fine solids, which consisted primarily of hydrated ferric oxides formed by air oxidation of a small fraction of the Fe^{+2} in solution. The settler underflow was filtered on a pan filter and the filtrate combined with the decant.

The incomplete removal of Cd in the first Zn cementation stage was attributed to an insufficient stoichiometric excess of Zn, due to dissolution of heavy metals from the added fresh dust, and to possible redissolution of cemented Cd by air oxidation in the absence of an inert gas atmosphere over the slurry. Nitrogen blanketing over the head space of the tank was instituted at the beginning of stage 1 slurry settling and was continued through stage 2 cementation and liquid/solids separation.

In the second Zn cementation stage, 2700 g of fresh fine granular Zn was added to the solution and reacted for two hours at 85 to 90 °C under the N_2 blanket. No fresh dust was added, therefore the Zn metal represented a 640-fold excess for the 23 mg/l Cd in solution. The Cd concentration was reduced to 6 mg/l after one hour. Apparently, 1 mg/l Cd was achieved after 0.5 hour reaction, but increased to 6 mg/l at one and three hours.

The final Zn cementation slurry was settled, decanted, and the settled solids filtered as in stage 1. The filtered cement residues from both stages were combined for assay. The 315 liters of final purified Zn/Fe sulfate solution was "preserved" by the addition of 1 g/l formic acid to minimize oxidation of ferrous iron to ferric during subsequent storage and handling. The final solution was assayed for a wide range of impurity species, which are presented in Table VI. The solution contained 102 g/l Zn and 49 g/l Fe with a specific gravity of 1.35. Approximately 4 g/l of orange solids remained suspended in the final solution even after a pass through a cartridge filter. Analysis of a solution sample before and after clarification using a Millipore filter indicated only about 12% of the solution Cd was associated with these solids. Therefore, it was not considered necessary to further clarify the solution. The bulk of the final purified solution was advanced to the next step for product preparation.

Table VI. Product Analyses for Pilot Plant Zinc Cementation

Element	Assay	
	Fe Cement Filtrate (g/l)	Final Filtrate (g/l)
Zn	84.8	102.0
Fe	45.6	48.5
Mg	1.64	1.91
Cl	6.73	8.10
F	0.310	0.220
	(mg/l)	(mg/l)
Cu	6.0	7.9
Cd	23.0	5.7
Ni	4.0	4.2
Pb	42.0	6.4
Co	144.0	121.0
As	-	0.3
Sb	-	<5.0
Cr	0.8	0.3
Mo	-	<50.0
Se	-	<0.001
Ba	-	<1.0
Hg	-	<0.005
Ag	-	<0.5

Preparation of Zn/Fe Sulfate Salt Product

Small samples of the zinc/iron sulfate products were prepared in the laboratory by evaporation and drying (110 °C maximum) portions of the final purified Zn cementation solution. Larger samples were prepared by spray drying the same solution on a pilot scale unit at Niro Atomizer, Inc., Columbia, MD. The spray drying unit consisted of a 120-mm spinning disk inside a 1.52 m diameter, insulated, SS chamber. The chamber was ~2.44 m high, including a conical 1.52 m bottom. A cyclone was attached for removing fine solids and a portion of the solids were recycled to the feed solution. The inlet gas temperature was 400 °C and the outlet gas temperature was 130 °C. The bulk density of the spray dried salts was 669 kg/m^3 (tapped) or 1169 kg/m^3 (pulverized). The salt products were analyzed for the species known to be of concern to commercial animal feedstock producers. All of these data are presented in Table VII.

The highest impurity levels in the Zn/Fe sulfate salt products prepared from the purified Zn cementation solution were the light metal elements and halide anions not affected by cementation. The salts contained 1.7% Cl, 0.4% Mg, 0.3% Na, 0.09% Ca, 0.09% F, and 0.06% Al. Cobalt, which was not cemented by Fe or Zn, was the highest heavy metal impurity at 260 ppm. The other heavy metals were present in concentrations of 5 to 50 ppm.

Table VII. Analyses of Zn/Fe Sulfate Salt Products

Element	Unit	Assay	
		Evaporation	Spray Dried
Zn	%	22.5	21.5
Fe	%	10.8	10.0
Mg	%	0.422	0.430
Cl	%	1.69	1.77
F	%	0.088	-
Cu	ppm	14	86
Cd	ppm	13	12
Ni	ppm	28	50
Pb	ppm	30	17
Co	ppm	266	250
Ca	ppm	870	1180
Na	ppm	3140	3790
Al	ppm	650	500
Te	ppm	6	-
As	ppm	4	<1
Sb	ppm	<50	<50
Cr	ppm	-	12
Mo	ppm	<10	-
Se	ppm	<0.1	-
Bi	ppm	10	-
Tl	ppm	<400	-
Sn	ppm	<500	-

The final solution yielded 434 g of salt per liter upon drying. The product salts have a calculated composition of 59 to 62% $ZnSO_4 \cdot H_2O$ and 30 to 33% $FeSO_4 \cdot H_2O$. The monohydrate form of $ZnSO_4$ was confirmed by X-ray diffraction of the salts prepared by evaporation. The Mg, Ca, Na, Al, and Cl impurities contribute an additional 3 wt% when the metals are expressed as sulfate, yielding a total weight accountability of 92.5 to 97.5%. The balance may be assumed to be additional water of crystallization.

Discussion and Summary

Each step of the proposed dust recovery process has been tested successfully. The process effectively converts zinc and iron in the furnace dust along with scrap iron into a salable zinc/iron sulfate monohydrate product. The only waste streams from the process are a dilute brine solution from the water leaching step and water vapor from spray drying. The spent scrap iron and leach residue from the acid leach/iron cementation step is rich in Cu, Fe, and Pb. These solids which also contain the precious metals originally present in the dust, can be recycled to the copper smelting furnace, leading to a higher steady state level of Pb and precious metals in the furnace. The precious metals and some of the Pb will be recovered in the copper electroplating tank sediments, with the majority of the Pb exiting the system via the furnace slag.

An alternative mode of operation that would avoid the higher levels of Pb in the furnace, would be to separate the acid leach and iron cementation steps. Filtration of the acid leach slurry would generate a precious metals and lead-bearing residue for sale to lead smelters. The high-grade copper cement generated by iron cementation of the acid leach filtrate would be recycled to the copper furnace.

The only other by-product of the process is the cement solids from the zinc cementation step. This solid is a relatively high-grade zinc source that should be marketable to a smelter capable of handling zinc with cadmium, nickel and antimony.

It is interesting to note that zinc cementation of Cu, Cd, and Ni in the second stage was effective in the pilot-scale tests without further addition of "activators" or "promoters". Numerous combinations of "activators" such as Sb (7,9-12), As (7-11), Pb (7,9), Cd (7), and Cu (7-9,11,12) have been reported in the literature for improving zinc cementation of heavy metals. It is possible that the Zn/Fe sulfate solution contains a sufficient amount of these "activators" such that further addition is not required to yield effective removal of these heavy metals. This is in contrast to lab-scale tests (3) where addition of Cu and Pb in small quantities was found to greatly improve the rate of Ni removal. Differences in the mass transport characteristics and degree of oxidation between the lab-scale tests (<100 ml) and pilot-scale tests may account for these observed disparities.

The prevention of oxidation by means of an inert gas blanket over the reaction slurry was found to be critical in order to prevent the redissolution of the cemented metals such as Ni and Cd as well as to prevent the oxidation of ferrous ions to ferric ions. In commercial operations the use of a sealed tank should be sufficient.

Estimates of the profitability of this process for a copper recycling operation that generates 1 million pounds of dust per month indicate a highly favorable return on investment. Capital costs are estimated to be in the range of \$2.5 to 3.5 million with annual operating costs of \$1.0 to 1.5 million. Annual revenues on the order of \$1.5 to 2.5 million are generated from the sale of the Zn/Fe sulfate, the recovery of precious metals, and the recycling of copper. Actual costs and revenues will vary, depending on the actual process design, annual dust production, and composition of the dust.

The primary market for zinc sulfate is in agricultural applications, either as fertilizer micronutrient or as animal feed supplement. The Zn/Fe sulfate product from the furnace dust process described in this paper is targeted to the animal feed market since typical animal feed premixes contain both zinc (as ZnO or $ZnSO_4$) and iron (as $FeSO_4 \cdot 6H_2O$ or $FeSO_4 \cdot H_2O$ or $FeCO_3$) along with other nutrients such as manganese, copper, and iodine (4). The premix is blended with animal feed at a dosage of one pound premix per ton of feed. Premix compositions vary widely, depending on the type of animal and on the preferences of the feedlot owner. Other premix additives include calcium, cobalt, magnesium, phosphorous, potassium, and sulfur. A 1:1 to 3:2 weight ratio of Zn:Fe as mixed monohydrate is the desired range of compositions for the animal feed premix market. The Zn/Fe sulfate product from the pilot-scale tests contained ~2:1 Zn to Fe. Addition of ferrous sulfate would be required to reach the desired composition.

The maximum and preferred impurity levels for various elements in animal feed premixes are listed in Table VIII. These levels are based on discussions with premix producers and distributors. The spray dried product listed in Table VII is found to meet all of the preferred impurity levels with the exception of Cr which was found to be at the 12 ppm level, although no Cr was detected in the evaporated sample.

Table VIII. Standards for Impurities in Animal Feed Premixes[1]

Impurity	Maximum (ppm)	Preferred (ppm)
Pb	250	100
Cd	40	25
As	2	1
Ni	200	50
Cr	10	10
Mg	5,000	1,000
Cl	5,000	1,000

[1]Based on discussion with premix producers and distributors.

In summary, the results of the pilot-scale tests clearly demonstrate the feasibility of the proposed hydrometallurgical dust treatment process. Conversion of secondary copper smelter flue dust in this manner will generate a positive return on investment, as well as mitigate a potential environmental liability.

References

1. R. G. Reddy, "Mineral Waste Treatment and Secondary Recovery," Journal of Metals, 38, (April 1986), 49-55.

2. D. Pearson, "Recovery of Zinc from Metallurgical Dusts and Fumes" (Paper presented at the AIME Annual Meeting, Las Vegas, Nevada, February 1980).

3. H. H. Law, M. P. Bohrer, P. A. O'Hara, J. Sapjeta, and V. Tierney, "Recovery of Metals from Copper Smelter Furnace Flue Dust" (Paper to be presented at the Annual TMS Meeting, San Diego, California, March 1992).

4. J. Nelson, "The Animal Feed Industry as a Possible Market for "Clean" Metal Salts and Compounds from Reclamation Operations," Proc. 2nd Internat. Symp. Recycling of Metals and Engineered Mats., ed. by J. H. L. Van Linden, D. L Stewart, and Y. Sahai, TMS, (1980), 447. TMS Meeting, 1980), 477.

5. H. Salomon-De-Friedberg, "The Leaching of Zinc Oxide Fume with Soda Ash to Control Halides at Cominco" (Paper presented at the CIM Annual Meeting, Edmonton, Canada, 1988).

6. A. K. Biswas and J. G. Reid, "Investigation of the Cementation of Copper on Iron at Elevated Temperatures," Trans. Inst. Min. Metall., 82, (1973).

7. R. W. Adams and R. L. Chapman, "New Developments in the Zinc Dust Purification of Zinc Plant Electrolyte" (Paper presented at the Aus.I.M.M. Melbourne Branch, Symposium on "Extractive Metallurgy", November 1984).

8. S. Fugleberg, "Principles of Cadmium Production and Their Application at the Outokumpu Zinc Plant in Kokkola" (Paper presented in the "Internat. Symp. on Hydrometallurgy" at the Am. Inst. Mining, Metal., Petrol. Engs. Meeting, Chicago, 1973), 1144-1167.

9. F. J. J. Bodson, "Method of Purifying Zinc Sulphate Solutions", U. S. Patent 3,826,648, (1974).

10. A. D. Milkov, M. I. Kolarov, N. T. Kunchev, I. M. Iliev, K. S. Spassov, and G. S. Buyukliev, "Method of Employing Elemental Zinc for the Purification of Aqueous Solutions of Metallic Salts", U. S. Patent 4,078,040, (1978).

11. S. Cammi, A. D'Este, and A. Pescetelli, "Process for Purifying Solutions of Zinc Sulphate", U. S. Patent 4,637,832, (1987).

12. Capps, R. L., and J. M. Boxall, "Zinc Leach Residue Treatment and Solution Purification at the National Zinc Company" (Paper presented at the AIME Annual Meeting, Los Angeles, California, February 1984).

COPPER EXTRACTION FROM SMELTER FLUE DUST BY LIME-ROAST/AMMONIACAL HEAP LEACHING

by

Zhi-biao Yin,, Eric Caba, Lee Barron, Darren Belin, William Morris, Matt Vosika, and Robert Bartlett*

College of Mines and Earth Resources
University of Idaho
Moscow, Idaho 83843
(208) 885-6195*

ABSTRACT

Copper smelter flue dusts often cannot be directly recycled to the smelting process and accumulate as hazardous wastes requiring environmentally acceptable disposal. Because of the limited amount of flue dust, a separate copper extraction process must be simple and require a small plant investment. A flue dust process has been developed, consisting of the following steps: (1) roasting a pelletized mixture of hydrated lime and flue dust to fix arsenic and sulfur as insoluble calcium salts, (2) heap leaching the roasted pellets with a buffered ammonia/ammonium salt solution to extract copper as an ammine complex in a lined cell that is the final repository of the leached pellets, and (3) boiling ammonia from the lixiviant to precipitate copper. Condensed ammonia and the filtered solution are returned to leaching. Heap leaching in the final repository cell lowers the capital investment significantly compared with competing extraction processes. Chemistry with respect to arsenic, sulfur and copper is discussed.

Residues and Effluents - Processing and Environmental Considerations
Edited by Ramana G. Reddy, William P. Imrie and Paul B. Queneau
The Minerals, Metals & Materials Society, 1991

Introduction

Copper smelter flue dusts containing arsenic are hazardous materials requiring environmentally acceptable disposal, preferably with resource recovery under the RCRA and CRCLA regulations. However, the known resource recovery processes entail capital and operating costs greater than can be justified by the revenue of the recovered metal. Consequently, the smelting industry is tending to forego resource recovery in favor of encapsulation and storage in a certified Class I hazardous waste site, or in a dedicated facility located at the source Superfund site. Order of magnitude costs for this option with portland cement encapsulation are estimated at $80 per ton, not including transportation. Industrial adoption of a new resource recovery process requires that its cost, including capital amortization, exceed the cost of encapsulation and storage only by the amount of the offsetting revenue from sale of metal. A lime-roast/ammoniacal heap leaching process will be described that is expected to meet this goal.

Process Elements

The process flow sheet is shown in Figure 1 and consists of the following steps:

1. roasting a pelletized mixture of hydrated ground lime and flue dust to fix arsenic and sulfur; a venturi scrubber removes dust and any trace of escaping arsenic and sulfur.

2. heap or vat leaching the roasted pellets with a solution of ammonia and an ammonium salt in a lined cell that is the final repository for the leached pellets.

3. boiling the resulting copper pregnant liquor to vaporize ammonia and precipitate copper hydroxide, which is filtered and sold. The condensed ammonia and filtrate are combined and recycled as the leaching solution.

No reagents are required other than ground lime, fuel, water and a small amount of ammonia and ammonium salt to make up entrainment losses in the copper filter cake. The ammonium salt must be one with an anion that does not form insoluble calcium salts, e.g. chloride and nitrate. Anions that form insoluble calcium salts such as carbonate will leach copper but also release unacceptable amounts of arsenic as arsenate ion. Sulfate ion is somewhat intermediate.

Environmental Considerations

The lime-roast/ammoniacal leach process recovers copper while converting the flue dust to an acceptable product for permanent disposal in a lined repository. Arsenic is converted to calcium arsenite/arsenate, sulfur is converted to calcium sulfite/sulfate (anhydrite) and the heavy metals are entirely hydrolyzed and in a solid matrix at a pH above 10. Copper is dissolved during leaching because it is complexed with ammonia. Zinc, and some nickel if present, will also be dissolved but precipitated and removed in the basic copper filter cake. Iron and other heavy metals are not complexed or dissolved. The leaching process is a closed system with no water effluent.

Upon completion of the project the permeable heap is washed to remove ammonia, which renders any residual copper, nickel or zinc insoluble. Acid is never generated or used in the process. Sulfur and arsenic remain captured as calcium sulfate and calcium arsenate without being altered during leaching using the basic lixiviant.

Although insoluble under the processing and permanent storage conditions of the repository, calcium arsenate's solubility under the acidic conditions of EPA's Toxicity Characteristic Leaching Procedure (TCLP)[1] exceeds the allowable arsenic limit of 5 ppm. With respect to all other heavy metals air dried leach residues met the TCLP test limits. The copper concentration in TCLP test extract solutions was less than 1 ppm.

Because of the arsenic solubility in the TCLP test, the leached residue must be stored permanently in a lined repository. The flue dust extraction process is designed to permit copper heap leaching to be carried out within the final repository, in order to avoid the additional operating and capital cost involved with a separate leaching facility.

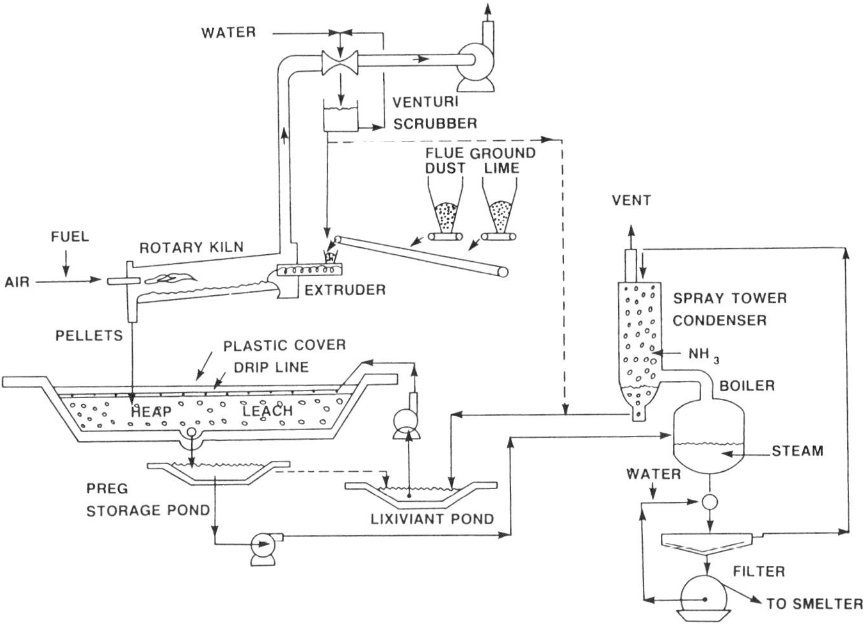

Figure 1. Flowsheet for the Lime-Roast/Ammoniacal Heap Leach Process.

Lime-Roasting

The lime-roasting process was originally invented[2-4] for roasting copper flotation concentrates while simultaneously capturing the oxidized sulfur as calcium sulfate (anhydrite). The roaster calcine was subsequently leached with sulfuric acid to dissolve copper, which was electrowon as copper cathodes. This process was successfully developed and piloted as a continuous industrial process by Kennecott in the mid 1970s. It was an alternative to smelting, in the event that smelting became environmentally proscribed.

More recently, Taylor, et al.,[5] have lime roasted arsenopyrite and shown that arsenic capture with hydrated lime during roasting is even more efficient than sulfur capture. Twidwell and Mehta[6] performed lime roasting experiments on copper smelter flue dust and obtained very high retention of arsenic in the roaster calcine as calcium arsenate and calcium arsenite.

The normal oxidation reactions of residual sulfide minerals in the flue dust will occur during lime-roasting. Excessive temperatures must be avoided to prevent the formation of insoluble copper ferrites. A final roasting temperature in the vicinity of 600C has been found to be optimum in the previous studies, cited above, but good copper extractions have been achieved between 450C and 800C, and the optimum roasting temperature will vary with the specific roaster feed material. 600C is also about the temperature at which hydrated lime decomposes:

$$Ca(OH_2) = CaO + H_2O \qquad (1)$$

Although the sulfide mineral oxidation reactions are exothermic, some additional fuel, such as natural gas, will be required for most flue dusts because they have been substantially oxidized to sulfates during prior smelting.

The formation of calcium sulfite and calcium sulfate from sulfide minerals occurs by release of SO_x gas during oxidation of the mineral followed by localized capture of the SO_x by CaO:

$$Ca(OH)_2(s) + SO_2(g) = CaSO_3(s) + H_2O \qquad (2)$$

$$Ca(OH)_2(s) + SO_3(g) = CaSO_4(s) + H_2O \qquad (3)$$

Thus, it is a solid/gas reaction rather than a solid/solid reaction. Because a gaseous intermediate, SO_x, is formed, an intimate mixture of flue dust and hydrated lime is needed to prevent escape of SO_x into the flue gas stream and obtain a high degree of sulfur capture.

Arsenic is captured by what is believed to be the same mechanism as that for sulfur and in the same temperature range.

$$3Ca(OH)_2(s) + As_2O_3(g) = Ca_3(AsO_3)_2(s) + 3H_2O \qquad (4)$$

$$Ca_3(AsO_3)_2(s) + O_2(g) = Ca_3(AsO_4)_2(s) \qquad (5)$$

Agglomerating or extruding the mixture into a ball or pellet enhances SO_x capture.

Partially roasted lime pellets will show an unreacted core surrounded by a fully-reacted rim; see Figure 2. SO_x generated inside the pellet must diffuse out through the calcine rim, and before a significant distance has been travelled SO_x will have reacted with the hydrated lime.

Forming pellets or agglomerates is also an advantage for subsequent heap leaching. The resulting high uniform permeability of the heap permits unhindered downward percolation of the lixiviant solution. The pellet calcine produced by lime roasting is porous and permits the absorption of leaching solution into its pores by capillarity, and, consequently, the dissolution and extraction of copper.

Hydrating the lime prior to roasting makes the lime more reactive[2,5] and improves the capture efficiency for both sulfur and arsenic. Laboratory roasting experiments with lime-flue dust pellets (about 1 cm diam by 1 cm long) were monitored to detect arsenic in the effluent gas from the roaster and none was detected. Of course, in a commercial system, any traces of sulfur or arsenic escaping into the roaster flue gas will be captured by a bag house or scrubber and recycled to the roaster, as shown in Figure 1. Scrubber liquor can be doped with lime or alkali if necessary to ensure complete capture of arsenic and sulfur.

Sulfates present in the flue dust before lime-roasting will also react with hydrated lime, and sulfates present in the flue dust must be considered in computing the amount of lime needed in the roaster charge.

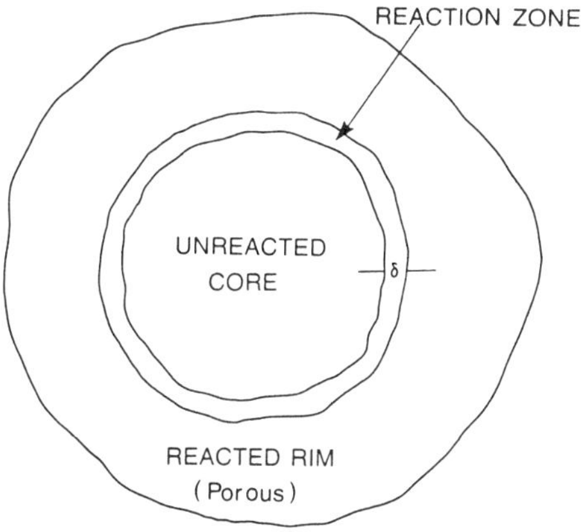

Figure 2. Cross-section Sketch of a partially Roasted Pellet.

Leaching Copper with Ammoniacal Solutions

Various copper materials have been leached with ammine carbonate solutions (ammonia plus ammonium carbonate) and these include copper scrap, cement copper and various forms of oxidized copper such as roaster calcines[7]. A Kansas City plant successfully leached copper scrap with ammine carbonate for many years[8]. More recently, Kock et al.,[9] successfully used ammine carbonate solutions to leach copper from electroplater sludges obtained from precipitating electoplater waste solutions with a large excess of lime. Copper extractions of +97% were obtained.

Oxidized copper is leached by the following reaction, and similar reactions for other oxidized copper minerals,

$$CuO(s) + (NH_4)_2CO_3 + 2NH_3 = Cu(NH_3)_4CO_3 + H_2O \qquad (6)$$

A lixiviant solution composition with a ratio of 2 moles of ammonia to each mole of ammonium carbonate is optimum. This is evident from the chemical reaction stoichiometry, but it has also been experimentally verified[7].

However, ammonium carbonate is unsatisfactory for leaching arsenic containing lime-roasted flue dust because it will partially dissolve arsenic[10] while precipitating calcium carbonate by the following reaction:

$$Ca_3(AsO_4)_2 + 3(NH_4)_2CO_3 = 3CaCO_3(s) + 2(NH_4)_3AsO_4 \qquad (7)$$

Consequently, a different ammonium salt, one that does not form insoluble calcium salts must be used. Experiments have been performed successfully with both ammonium nitrate and ammonium chloride as the ammonium salt used in conjunction with free ammonia. The results of these experiments are summarized in the next section. The leaching reaction with ammonia and ammonium nitrate is,

$$CuO(s) + 2NH_4NO_3 + 2NH_3 = Cu(NH_3)_4(NO_3)_2 + H_2O \qquad (8)$$

And, the leaching reaction with ammonia and ammonium chloride is,

$$CuO(s) + 2NH_4Cl + 2NH_3 = Cu(NH_3)_4Cl_2 + H_2O \qquad (9)$$

Anions other than nitrate and chloride are also expected to be effective, but may be too costly.

It is necessary to have an ammonium salt present with ammonia so that the charge balancing anion for the cupric ammine complex $[Cu(NH_3)_4{}^{++}]$ is not hydroxyl ion, which is what will result if only ammonia is used in the leaching solution. When only ammonia is used, the pH runs above 12 due to the hydroxyl anion and copper is precipitated as the hydroxide rather than dissolved as the ammine complex. With an ammonium salt present in sufficient quantity, the pH is lower, allowing copper to dissolve as the ammine complex. The computed optimum pH range for dissolving copper as the ammine complex is shown in Figure 3, reproduced from Tozawa, et al.,[11].

Arsenic solubility is minimized in the pH range 10 to 11, where it remains as a stable solid, calcium arsenate. Arsenic would be precipitated as calcium arsenate if it were not already in that form as a result of the lime-roast.

Preliminary Leaching Experiments with Lime-Roasted Flue Dust

The chemical analysis of the flue dust, obtained from an Anaconda, Montana site, is shown in Table I. This material was blended with $Ca(OH)_2$ and pressed into cylindrical pellets, 1.27cm x 1.5cm, weighing about 2.5 gram and roasted at 650C for two hours. Air was passed through the tube furnace holding the boat and into an absorbing solution. No detectable amounts of arsenic and sulfur were determined in the gaseous effluent of any of the roaster experiments.

Roasted pellets were crushed and agitation leached for at least 24 hours in solutions containing ammonia and ammonium salts. Leaching conditions and copper and arsenic extractions are summarized in Table II. Extractions were determined from atomic absorption analysis of leach solutions and from wet chemical analysis of leached residues using standard methods.

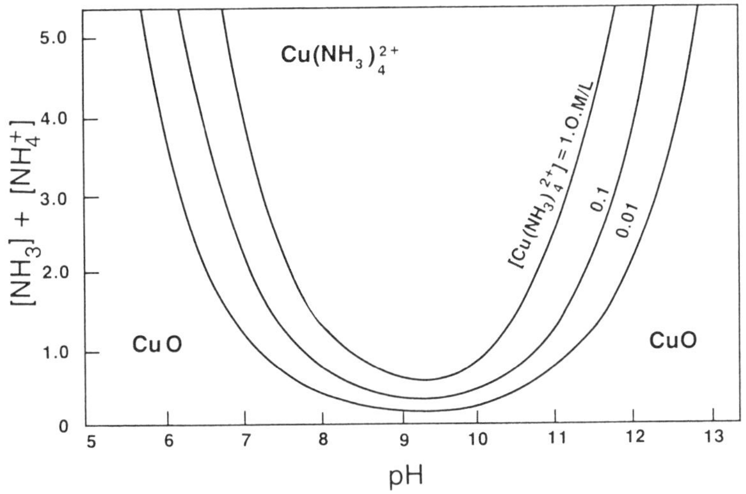

Figure 3. Copper Ammine Complex, $Cu(NH_3)_4^{++}$, Stable Region at 25C[11].

These leaching results provide the following conclusions. Neither copper nor arsenic are appreciably extracted in aqueous ammonia without an ammonium salt. Most of the copper is extracted in all of the solutions containing both ammonia and an ammonium salt ($CO_3^{=}$,NO_3^-,Cl^-,$SO_4^{=}$) in a pH range 10 to 11.

Table I

Chemical Analysis of Flue Dust Sample (CP-T7)

Moisture Content	29.99 Wt. pct
Solids Composition	
Copper	14.48 Wt. pct
Total Sulfur	7.31 Wt. pct
Total Arsenic	7.81 Wt. pct

Ammonium salt is needed to provide anions other than hydroxyl ion, to balance the cupric ammine complex charge without an excessive pH (hydroxyl ion concentration) where the copper ammine complex cation is unstable.

Leaching experiments using ammonium carbonate dissolved half the arsenic. Leaching with ammonium sulfate dissolved some arsenic, 3.3%, and may be environmentally unacceptable.

A leaching solution (lixiviant) with ammonia and either ammonia chloride or ammonium nitrate dissolved most of the copper in the lime-roasted flue dust but very little of the arsenic, less than 0.3%. This much arsenic deposited in the copper precipitate would not be a problem for further treatment at a copper smelter.

Table II

Summary of Preliminary Leaching Experiments with Lime Roasted Flue Dust

	$NH_3/(NH_4)_2CO_3$	$NH_3/(NH_4)_2SO_4$	NH_3/NH_4NO_3	NH_3/NH_4Cl
Lixiviant Solution				
Ammonia (M)	2.2	4.0	4.0	4.0
Ammonium Salt (M)	1.1	1.5	1.1	1.25
pH	10.5	10.5	10.5	10.5
Pregnant Liquor				
Copper (gpl)	3.65	2.83	3.03	3.20
Arsenic (mg/l)	1407.0	83.0	7.9	7.6
pH	10.6	10.5	10.6	10.55
Metal Extraction				
Copper (%)	72.7	56.4	60.4	63.7
Arsenic (%)	55.5	3.3	0.3	0.3

Footnote: The copper extraction in only ammonia was 4.9%.

Heap Leaching in Final Repository Cells

Heap leaching has been extensively developed as a commercial process for the extraction of gold from permeable ore heaps using basic sodium cyanide solutions. The lixiviant solution is distributed evenly on the surface of the heap and solution percolates downward without flooding the heap. The ore particles are wetted by the percolating solutions which penetrate the ore particle pores by capillary action.

While fine particles are deleterious to good percolation, large particles are slow to leach. A pelletized feed of about 1 cm uniform dimension is an ideal size compromise for heap extractions in reasonable residence times -- several weeks to a few months.

Ammoniacal leaching requires that the heap be covered with impervious plastic during leaching to prevent evaporation of ammonia. Consequently, drip irrigation is the preferred method. Flexible plastic solution distribution lines with drip emitters will be spread on the surface of the heap and covered with an impervious sheet of heavy duty plastic. The drip emitters are spaced at about one per square meter. The plastic sheet may be covered with a blanket of earth materials to protect it from wind and sunlight. Drip irrigation, originally developed to conserve agricultural uses of water, has been successfully applied for gold ore leaching. Gold leaching is normally terminated in winter months when the solution freeze, but drip irrigation lines buried beneath a few feet of ore or other earth materials are insulated and gold leaching is now practiced throughout the year at some mines.

The planned heap leaching cell for lime-roasted flue dust should be fully lined and contain a french drain system to collect percolating solution. Also, solutions may be transferred from the pregnant liquor pond to the lixiviant pond and recirculated in the event of a process upset or to increase the pregnant liquor grade before processing for copper recovery. Continuous recycle with partial withdrawal of pregnant liquor for processing is a likely mode of leaching operation, especially as the lime-roasted pellets become more depleted in copper, in order to provide a high grade pregnant liquor to the ammonia boiler for copper recovery.

Vat Leaching in Final Repository Cells

Vat leaching differs from heap leaching in that the cell (vat) holding the roasted flue dust is flooded with the leaching solution. When an adequate copper concentration is achieved pregnant liquor is partially withdrawn for copper recovery and replaced to keep the vat full of leaching solution. Circulation of the leaching solution continues until most of the copper has been extracted, upon which all of the remaining solution is withdrawn.

Both vat and heap leaching can use two or more cells to advance solutions from cell to cell countercurrent to the leached solids. Fresh lixiviant would go first to the most extensively leached (depleted) cell and end in the cell with the least prior extraction.

Vat and Heap Leaching Experiments

Simulated vat and heap leaching experiments were performed on cylindrical pellets (1.27 cm 1 diam x 1.5 cm) of lime-roasted Anaconda flue dust. The amount of $Ca(OH)_2$ used, 30 wt pct, was hyperstoichiometric with the total arsenic and sulfur content. Again, arsenic and sulfur in the tube furnace roaster flue gas were below detectable limits. The lime-roasted pellets had a specific gravity of 1.43 g/cc. Because of their uniform size, their apparent bulk density, in a heap or vat, is only 42 lbs/cu ft (0.67 g/cc).

Vat leaching of uncrushed pellets was conducted in each of the following lixiviants:

$$4M\ NH_3 + 4M\ NH_4CL$$
$$4M\ NH_3 + 4M\ NH_4NO_3$$
$$4M\ NH_3 + 2M\ (NH_4)_2SO_4$$

Approximately 100 grams of pellets were flooded with lixiviant (110 ml) in each experiment and allowed to leach for seven days.

One of the purposes of these experiments was to determine if there were practical solubility limits for copper being dissolved by any of the lixiviants, since adequate copper concentration in the pregnant liquor is required to obtain energy efficient boiling and precipitation of copper from the pregnant liquor.

A solubility limit of about 13 gpl copper was found for the nitrate lixiviant. Copper bearing crystals, precipitated in the lixiviant, could not be positively identified by x-ray diffraction. However, many of the diffraction peaks and lattice spacings were similar to those of $Cu(NH_3)_2(NO_3)$.

Precipitation did not occur with the other two lixiviants. After seven days, when the other lixiviants were decanted the copper concentrations were 35.8 gpl in the chloride lixiviant corresponding to extraction of 40.5% of the copper and 50.8 gpl in the sulfate lixiviant corresponding to extraction of 57.6% of the copper.

Leaching was continued for each lime-roasted pellet sample with a fresh 110 ml of lixiviant. The nitrate sample became saturated again, at about 13 gpl of copper. The samples were leached for a total of 12 days at which time the solution was decanted and a water wash was begun. Cumulative copper extractions from the lime-roasted pellets, based on atomic absorption analysis of the leach solutions, through 12 days are shown in Figure 4. In a commercial application, using lime-roasted pellets of this size, leaching should be continued for a total period of 4 to 6 weeks to obtain high copper extractions.

There was no apparent physical disintegration of the pellets during the leaching period. Arsenic assays, obtained by a wet chemical method from the first pregnant liquor decanted at seven days are shown in Table III.

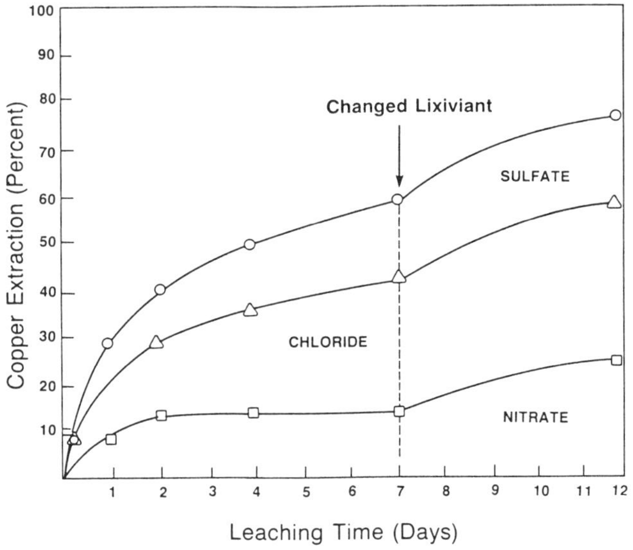

Figure 4. Vat Leaching Copper Extractions from Lime Roasted Cylindrical Flue Dust Pellets.

Table III
Arsenic Concentration in Vat Leaching Pregnant Liquor

Lixiviant	Concentration, gpl	Molarity	pH
Chloride	0.012	1.6×10^{-4}	10.0
Nitrate	0.010	1.4×10^{-4}	10.1
Sulfate	0.061	8.2×10^{-4}	10.5

These results are consistent with our preliminary experiments with these lixiviants, summarized in Table II, and the thermochemical determinations of Robins(10). Somewhat higher arsenic solubilities result with the sulfate lixiviant because of gypsum precipitation and a lowering of the calcium solubility. For the chloride and nitrate lixiviants, the arsenic molarity obtained agrees with thermochemical predictions. Under these environmentally optimum conditions with pH at about 10, calcium arsenate is extremely stable.

Simulated heap leaching was studied by percolation leaching the same sized lime roasted pellets used in vat leaching. Leaching was conducted in a 3 cm diam x 110 cm column using the 4.0 M ammonia/2.0 M ammonium sulfate lixiviant. The lixiviant was very slowly dripped onto the pellet bed at a rate of 2.5 cc/sec per m^2 of ore bed area, which is a typical heap leaching percolation rate as practiced in the gold mining industry. Each day, 36.5 cc of pregnant liquor was withdrawn, representing production sent to copper recovery, and the remaining pregnant liquor, about 153 cc, was recycled with 50 cc of fresh lixiviant solution. The steady state solution inventory, both in and out of the column, was approximately 500 cc.

The cumulative copper extraction (pct) and the pregnant liquor concentrations (gpl) are shown as a function of leaching time in Figure 5. Copper extraction is very rapid, apparently because the roasted pellets are very porous. 70% of the copper is extracted in 12 days, during which time the copper grade in the pregnant liquor remains above 25 gpl. Cumulative copper withdrawal for production in the daily 36.5 cc aliquots in nearly linear, showing that a steady rate of production for most of the copper extraction (70%) can be maintained.

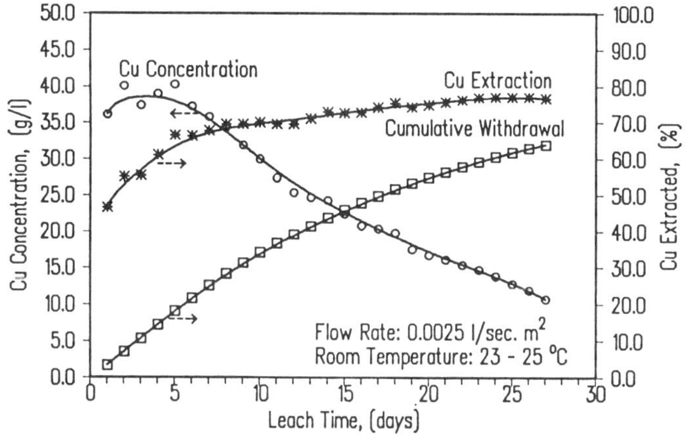

Figure 5. Copper Extraction by Heap Leaching Lime Roasted Flue Dust Cylindrical Pellets (1.27 cm diam x 1.5 cm).

Copper Precipitation by Boiling Ammoniacal Solutions

Boiling the pregnant liquor at atmospheric conditions releases ammonia vapor and is expected to precipitate a copper hydrolysis product. X-ray diffraction of the precipitates yielded by boiling pregnant liquor did not reval an identifiable crystalline copper spectrum for any of the lixiviants used to generate the pregnant iquor.

Boiling and precipitation can occur with steam, submerged burners or any convenient heat source[7,8]. The flowsheet in Figure 1 shows steam, which would be raised from an on site package steam plant burning natural gas or fuel oil. Although generating steam in a waste heat boiler attached to the roaster is possible, it is not recommended for the small scale and modest exhaust temperature of the roaster. Boiling in a spray dryer with incomplete drying may be desirable to avoid scaling of the copper precipitate on walls of the boiler apparatus.

It is desirable to have a good concentration of copper in the pregnant liquor to minimize energy consumption while boiling ammonia. Energy consumption is inversely proportional to copper solution grade. Heap leaching allows a small solution volume relative to the volume of roaster calcine. And, with a high concentration of ammonium ions in the leach solution an adequately low pH will result and permit sufficiently high copper ammine complex concentrations.

Ammonia Condensation-Scrubbing

Ammonia vapor released from the boiler is cooled with circulating lixiviant, e.g., in a packed tower, and condensed to regenerate the lixiviant (leach) solution. With a large lixiviant storage pond in a reasonably cool climate circulating lixiviant through the condenser, additional cooling with a heat exchanger probably will not be required. The condenser is essentially a closed system, but minor venting of accumulated air or other non-condensible gases may be required.

The copper precipitate slurry is filtered. The separated liquid is also recirculated to the lixiviant liquor, as shown in Figure 1.

Stopping the Leaching Process

When the economic limit in extracting copper has been achieved it will be necessary to stop the process to prevent further leaching of small amounts of ammine complexed copper. Although water washing can be used, the amount of water is limited if a water balance on the entire leaching system is maintained. Purging ammonia from the system by removing the heap cover sheet and drying the impounded leached residue, e.g., through a summer season, may be necessary. It may be possible to enhance residue drying by blowing air through the french drains formerly used to withdraw leach solution.

References

1. Title 40-261.24 of the Code of Federal Regulations, in Federal Register, Vol 51, No. 114, June 13, 1986, pp. 21685-21590.

2. R.W. Bartlett and H.H. Haung, "The Lime-Concentrate-Pellet-Roast Process for Treating Copper Sulfide Concentrates", J. of Metals, Vol. 25, No. 12, pp. 2-8 (1973).

3. H.H. Haung and R.W. Bartlett, "Oxidation Kinetics of Lime-Copper Concentrate Pellets", Met. Trans. 7B, pp. 369-374 (1974).

4. F.P. Haver and M.M. Wong, Mining Engineering, 24, No. 6, p. 52 (1972).

5. P.R. Taylor, et al, "Lime Roasting of Refractory Precious Metal Ores", in Precious Metals, ed. by M. Jennings, The Met. Soc., Warrendale, PA (1991).

6. L.G. Twidwell and A.K. Mehta, "Disposal of Arsenic Bearing Copper Smelter Flue Dust", Nuclear and Chem. Waste Management, 5, pp. 297-303 (1985).

7. W. Kunda, et al, "Production of Copper from the Ammine Carbonate System", in Copper Metallurgy, ed. by R.P. Ehrlich, The Met. Soc., Warrendale, PA, pp. 27-69 (1970).

8. J.F. Whitaker, Wire and Wire Products, 36, pp. 1346-8; 1456-7 (Oct. 1961)

9. J.W. Kock, et al, "Leaching Composites of Five Plating Wastes with Ammonium Carbonate", in EPD Congress 90, ed. by D.R. Gaskell, The Met. Soc., Warrendale, PA, pp. 529-538 (1990).

10. R.G. Robins, "The Solubility of Metal Arsenates", Met. Trans.12B, pp. 103-109, (1981).

11. K. Tozawa, Y. Umetsu and K. Sato, "On Chemistry of Ammonia Leaching of Copper Concentrates", in Extractive Metallurgy of Copper, ed. by J.C. Yannopoulos and J.C. Agarwal, The Met. Soc., Warrendale, PA, Chapter 36, (1976).

THE CASHMAN PROCESS TREATMENT OF SMELTER FLUE DUSTS AND RESIDUES

Richard S. Kunter, and William E. Bedal

Artech Recovery Systems, Inc.
433 Park Point Drive, Suite 140
Golden, Colorado 80401

Abstract

Artech Recovery Systems, Inc has developed a low pressure and temperature leach process called the "Cashman Process" to extract metals from arsenical flue dusts and residues and fix the arsenic in an environmentally stable form as ferric arsenate (scorodite). The process was pilot tested at Hazen Research in an integrated plant including continuous recycle from August 1989 to October 1989 during which several tons of flue dust were processed. Based on this pilot program, the process was deemed technically feasible and produced commercially salable products. Residues from this pilot program were subjected to a long-term stability test jointly designed by PTI environmental services and the USEPA. The process, its versatility, and the nature of the products is discussed in some detail.

Residues and Effluents - Processing and Environmental Considerations
Edited by Ramana G. Reddy, William P. Imrie and Paul B. Queneau
The Minerals, Metals & Materials Society, 1991

Introduction

The economic recovery of metals from ores, concentrates and byproducts having a significant arsenic content represents a considerable challenge to the metallurgical industry.

A wholly pyrometallurgical process approach is becoming less acceptable because of ever more stringent regulatory and environmental constraints. In a copper smelter, for example, attempts to keep flue dust in closed circuit result in the buildup of impurities which cannot be tolerated because of the deleterious effects of arsenic, bismuth, antimony, etc. on copper quality. In these smelters such problems can be overcome by utilizing a hydrometallurgical process to treat a portion or all of the resulting flue dust. This would give the pyrometallurgical facility the advantage of lowering or eliminating the flue dust recirculating load, thereby increasing the facilities capacity, and at the same time allowing the facility to process feeds with a greater arsenic content. However, in instances of treating arsenic rich feeds, a wholly hydrometallurgical approach may offer the cleanest solution.

In the selection of a hydrometallurgical process, one of the most important concerns should be the long-term stability of the arsenic containing residue. Ever changing environmental standards should convince the smelting industry that the superior environmental performance of a process has great advantages, both politically and operationally. The arsenical residue disposal problem will grow since many of the future sources of the common base metals (copper, lead, zinc), as well as gold, have arsenic contents that effectively preclude their processing by conventional pyrometallurgical methods alone.

The fume/flue dust or electrostatic precipitator dust that is recovered from the off gases of many smelter operations typically contains arsenic and metals of value, and is often considered hazardous. In some instances, these hazardous flue dusts have been disposed of without regard to their hazardous nature. Because the arsenic component of the flue dust is in a chemically unstable/soluble form, weathering of such flue dust dumps can lead to the contamination of local groundwater systems.

The metallurgical industry is faced with two problems in the treatment of arsenic containing feeds. In addition to the development of economic extraction processes for metals, the industry must also develop processes which address ever increasing environmental concerns. From a combined economic and engineering point of view, there is a great incentive to solve both problems simultaneously. This means developing flow sheets that allow for the economic recovery of the contained metals as well as the stabilization of the arsenic in a residue that will meet current and future environmental regulations. The Cashman Process provides the necessary solution in many cases.

The Cashman Process

The Cashman Process was initially developed to treat arsenic containing base metal sulfide concentrates. Subsequent work showed that by the choice of appropriate operating parameters, the basic principles of the process could be applied to effectively treat arsenical flue dusts.

The Cashman Process is a hydrometallurgical process utilizing a hydrochloric acid leach under somewhat elevated temperature and pressures which solubilizes metals of economic interest such as copper, zinc, lead, silver, gold, nickel, cobalt and bismuth, and simultaneously leaves toxic elements such as arsenic in an environmentally stable leach residue. The Cashman Process uses a patented hydrochloric acid leach process (Patent No. 4,655,829) developed by CSS Management Corporation in Skykomish, Washington. Artech is the licensor of the Cashman Process and is currently conducting efforts to identify and develop commercial applications for the Cashman Process.

To bring about the oxidation of arsenic(III) to arsenic(V) necessary for the precipitation reactions, as well as the oxidation of the base metal sulfides and arsenopyrite, oxygen (sometimes as air) is used. The effectiveness of the Cashman Process depends largely on the leach chemistry as well as efficiency of oxygen dispersion throughout the reaction slurry, which is achieved by use of a low temperature, low pressure autoclave with an efficient mixing and gas injection system.

Materials of construction are always a matter of concern when developing design engineering and operating criteria. When calcium chloride is used as one of the reactants in the Cashman Process, the pregnant leach slurry contains solubilized base metal chlorides which are corrosive toward standard materials of construction. For these reactions, acid resistant brick lined autoclaves are required. Ancillaries such as piping and pumps must also be corrosion resistant. Although materials of construction are a concern, reliable components are readily available.

Reagent additions are determined by the chemistry and mineralogy of the feed. Operating conditions for each feed type are typically of low temperature and pressure. The reactor is heated indirectly with steam. Reaction time depends on grind size, pulp density, and mineralogy and is typically 30-60 minutes for a flue dust that contains only small quantities of sulfides.

Stability of Arsenic Containing Solids

The chemistry of arsenic is complex and much of the early work on the composition and stability of arsenic containing solids precipitated from solution is difficult to interpret. In most hydrometallurgical processes, the arsenic is precipitated as ferric arsenate ($FeAsO_4$). Recent studies have indicated that in order to minimize the solubility of the arsenic in the precipitate, the Fe:As cation ratio in the precipitate should be greater than the inferred stoichiometry of 1:1. This means that excess ferric ions and other cations are co-precipitated with the ferric arsenate. In addition, the studies indicate the temperature and pH (acidity) of precipitation should be controlled to produce a crystalline ferric arsenate product.

These conditions are difficult to maintain in typical commercial hydrometallurgical operations. Many processes designed to recover valuable base metals from arsenic rich sources are also designed to minimize iron dissolution because the recovery of base metals from iron rich solutions is often difficult. This apparent dissolution selectivity is achieved by either actually preventing the iron minerals from dissolving, or causing them to simultaneously dissolve and reprecipitate in a form other than ferric arsenate. This results in an arsenic compound that is unstable and has relatively high solubility.

In summary, most of the proposed hydrometallurgical processes for treating arsenic containing feeds result in residues that are not chemically stable and not of low solubility because (a) the acidity (pH) regime in which the arsenic compound was formed is unsatisfactory and/or (b) there is insufficient cation concentration in solution to "completely" precipitate the arsenate anion.

In order to meet environmental disposal requirements, other processes may require additional processing stages, either to retreat the leach residue, and/or to precipitate the soluble arsenic from the leach slurry before disposal or treatment for recovery of valuable metals. The Cashman Process obviates the need for these additional stages because it permits the precipitation of all the arsenic in the feed in a stable non-soluble form in a single unit operation. Furthermore, Artech is successful at producing residues that better the toxicity tests using stoichiometric iron to arsenic ratios of 1.2 : 1. Selected samples of the leach residues have been chemically and mineralogically characterized and shown to be mixtures of predominantly crystalline ferric arsenate and gypsum.

Long Term Stability Testing

Independent test work conducted in 1989 and 1990 by PTI Environmental Services-[1] of Boulder, Colorado to simulate the long term stability of leach residues derived from treating copper smelter flue dusts by the Cashman Process substantiate the long term stability claims of the Cashman Process leach residues. The U.S. Environmental Protection Agency (EPA) approved the procedures used by the independent test firm for this test work. In the test work, arsenic and cadmium in aqueous solution were extracted from leach columns and their concentrations were monitored over several months. The report on these tests concluded that it would take 4 to 8 million years to gradually leach all of the arsenic from a 3-meter thick pile of Cashman Process residues openly exposed to the environment. Additionally, the suitability of several substrates to mitigate any solubilized arsenic and cadmium were investigated. Some of these tests demonstrated that the leachate which passed through the substrate passed drinking water quality standards for cadmium and arsenic. The long term stability test information from the above report provides a basis for making a technically appropriate choice of disposal sites.

Mandated Stability Testing of Arsenic Containing Residues

Although there are many methods of evaluating the stability of process residues, the most commonly used method in the United States is that devised by the EPA called the Toxicity Characteristic Leaching Procedure test (TCLP)[-2] which replaces their earlier mandated EP Toxicity Test. The TCLP test involves leaching samples of residue under specific conditions and measuring the concentrations of arsenic and other metals in the resulting solution. Although the TCLP test does not necessarily give a true indication of the long term stability of a residue under various residue weathering conditions, U.S. regulations will require that all new residues must pass the TCLP test, or more stringent standards, before disposal is allowed.

Below is a representation of the results of applying the TCLP test to Cashman Process residues from treating a high arsenic content flue dust. As the figure depicts, the process residue contains very little soluble arsenic. The full TCLP results on this Cashman Process residue and unprocessed material are presented on page 10. Table IV contains toxicity test results for several feeds and residues produced during testing of the Cashman Process.

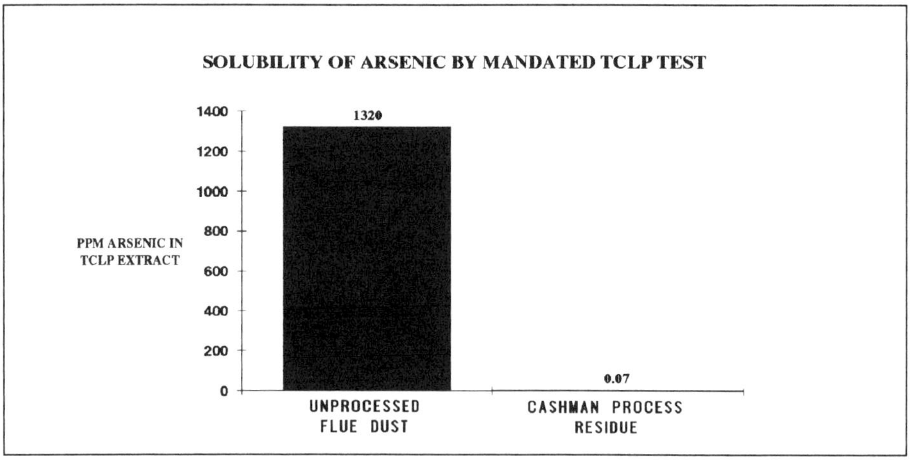

Figure 1. - Graph of Cashman Process Residue and Untreated Material.

Cashman Process Description

Arsenic containing flue dust

In base metal smelters, dusts emanating from roasting, smelting and converting operations typically contain arsenic, bismuth, cadmium and antimony as process impurities as well as valuable base and noble metals. The base metals are present in the fume/flue dust as complex mixtures of oxides, ferrites, etc. while the arsenic is typically present as arsenic(III) oxide. In order to dissolve the base metals from the flue dust, during the Cashman Process a reaction slurry of about 20% solids is made acidic. If the flue dust has a relatively low iron content, an additional iron source is required. Calcium chloride is used so that the base metals are dissolved as chloro complexes. The arsenic is co-precipitated primarily as ferric arsenate but also as a mixture of copper/iron and other complex metal arsenates with gypsum (hydrated calcium sulfate) which are very stable.

Implementing the Cashman Process at an existing smelter site is illustrated in a following conceptual flowsheet (figure 2). After solid/liquid separation of the slurry, the pregnant leach liquor can either be processed by a combination of selective precipitation, solvent extraction/ion exchange and electrowinning techniques to recover the dissolved base metals separately, or in bulk form as a mixed oxide / hydroxide / carbonate base metal precipitate which can be recycled to the smelting furnace. This proprietary conversion reaction (called the metathesis reaction) has been developed by Artech to convert chlorides to oxide products, and to recycle the chloride in the Cashman Process, thereby eliminating residual chloride content in the products which might cause corrosion damage in subsequent unit operations.

Implementing the Cashman Process at a now demolished copper smelter site at Anaconda, Montana, is illustrated in the simplified conceptual flowsheet (figure 3).

Generally a plant utilizing the Cashman process would include the following:

- Feed preparation;
- Autoclave leaching, filtration and disposal of process residue;
- Metals recovery; and
- Barren stream recycle.

Feed Preparation

Mineralogical and metallurgical test work are used to identify the optimum reagent additions required to provide maximum metals recovery and maximum environmental performance. Most feed constituents of less than 85 mesh generally do not require grinding, although screening and wet grinding can be utilized, where necessary. The material to be treated is slurried with recycled calcium chloride brine, sulfuric acid, and a source of iron (if sufficient iron is not contained in the waste) and is then introduced into the autoclave.

Autoclave Leaching, Filtration and Disposal of Residue

Materials are leached in an autoclave under somewhat elevated temperature and pressure conditions for 30-120 minutes. The hot leached slurry is then pressure filtered to produce a solid residue and a leach liquor. The solid residue generally contains sulfates, iron oxides, silica, sulfides and ferric arsenate. The leach liquor may contain dissolved metals such as copper, lead, zinc, silver, mercury, bismuth, cadmium and gold. The solid residue is washed with water prior to disposal. Metals recovery operations are applied to the filtered leach liquor and the barren liquor containing the chloride is then recycled.

CASHMAN PROCESS FLOWSHEET FOR OPERATING SMELTER

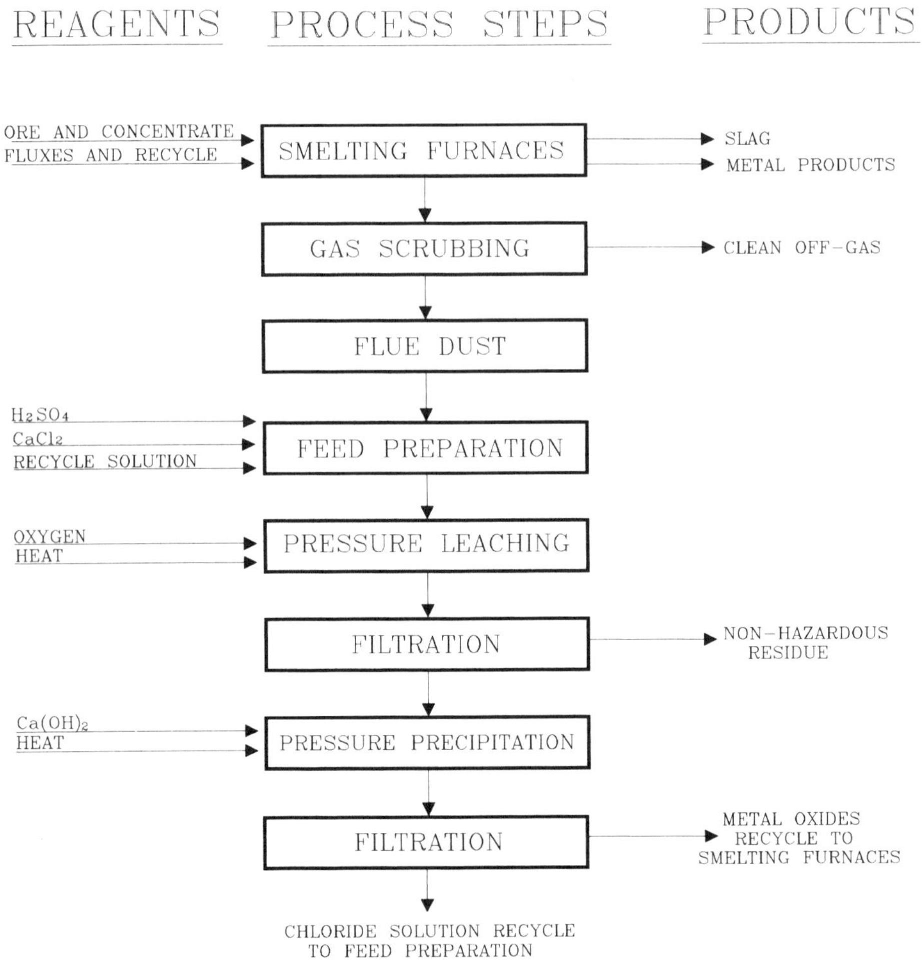

FIGURE 2

CASHMAN PROCESS FLOWSHEET FOR DEMOLISHED SMELTER

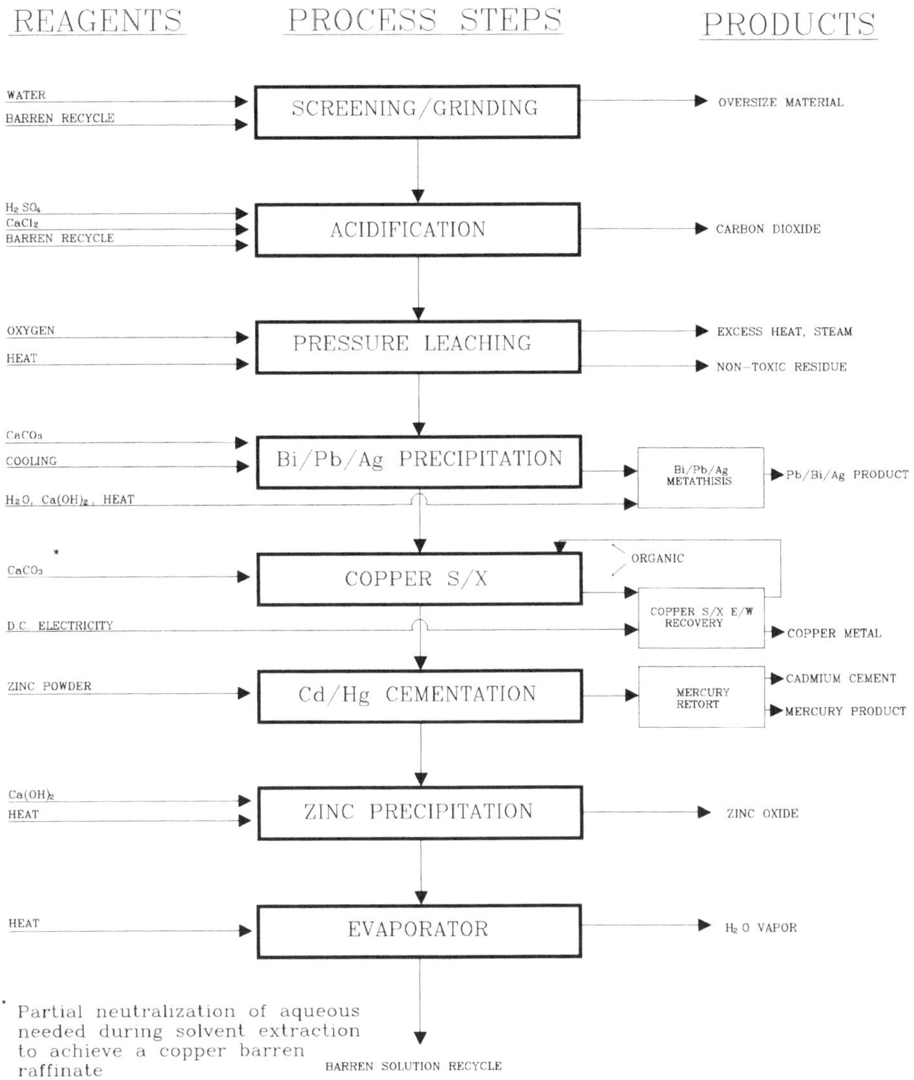

FIGURE 3

Metals Recovery

The best method of recovering metals from the leach liquor will vary, depending upon the quantity and the type of metals present. The following discussion first focuses on the most likely metal recovery methods, but also identifies alternatives developed for Anaconda flue dust which could be employed, if necessary. If the metal being discussed is the last to be extracted, it is assumed that the remaining barren filtrate would be recycled at that point. Not all flue dusts or residues would necessarily have all of these metals present or alternatively may have other metals not mentioned present.

The Cashman process can be utilized at existing smelting facilities to eliminate hazardous flue dust from the smelting process. Treating the flue dust would generate two products: a non-hazardous residue primarily consisting of $CaSO_4$ (calcium sulfate), SiO_2 (silica) and $FeAsO_4 \bullet 2H_20$ (ferric arsenate or scorodite); and a mixed metal oxide/carbonate precipitate that could be recycled to the smelter feed to combine with and, in part, replace the fluxing components.

Artech has piloted a proprietary elevated temperature and pressure precipitation procedure which allows the precipitation of metal oxides and hydroxides from a chloride solution. Using this procedure to precipitate all the metals solubilized during the leaching stage produces a precipitate consisting of metal oxides, metal hydroxides, metal carbonates and CaO. This precipitate could then be either recycled to the smelting furnaces or directed to other metal recovery operations at the facility. For instance, the precipitate could be digested with sulfuric acid and directed to existing solvent extraction circuits.

Alternatively, metals could selectively be recovered as discussed in the following sections:

Bismuth, Lead and Silver Recovery. Following filtration the autoclave leach liquor is cooled and $CaCO_3$ and copper powder are added to precipitate a combined lead/bismuth/silver material. The filtered precipitate is then metathesized at elevated temperatures and pressure with water and hydrated lime $Ca(OH)_2$ to produce a combined lead/silver/bismuth metal product.

If desirable, separate lead, bismuth and silver products could be produced. In this case the autoclave leach liquor is partially neutralized to precipitate bismuth, and cooled to crystallize lead and silver. The filtrate from this operation is passed to the copper extraction step. Bismuth, lead and silver are then recovered using a series of selective leach and precipitation operations. Bismuth is precipitated as bismuth oxychloride (BiOCl). Lead is precipitated as lead sulfate ($PbSO_4$), and silver is cemented using zinc powder. A barren leach cake is produced and recycled to the reactor feed and all filtrates are recycled.

Copper Recovery. If copper is present in the feedstock, the bismuth/lead/silver precipitation filtrate will generally contain cupric chloride. Cathode copper is produced from this solution using a standard solvent extraction/electrowinning (SX/EW) circuit. The filtrate from the SX/EW circuit is advanced to the cadmium cementation if cadmium is present.

In certain circumstances producing cathode copper may not be the most economic. Another alternative developed by CSS and Artech involves recovering copper as an oxide precipitate via a two-step process. First, the filtrate is heated and reacted with limestone and hydrated lime, precipitating the copper as atacamite ($Cu_2(OH)_3Cl$). The atacamite is converted to copper oxide by combining it with water and hydrated lime at elevated temperatures and pressures (metathesis reaction). Filtrate from the copper precipitation is then advanced to a cadmium cementation step if cadmium is present in the feedstock.

Cadmium and Mercury Recovery. The filtrate from the copper solvent extraction circuit could contain cadmium, zinc and mercury, if present in the feedstock. If present, cadmium (with mercury following the cadmium) and mercury are removed through cementation with zinc dust. This cementation process involves the reduction of cadmium to a sponge and the oxidation of zinc metal to zinc ions in solution. The mercury, if present, would be removed from the cadmium sponge by retorting.

Zinc Recovery. The solution from cadmium cementation is combined with hydrated lime at elevated temperatures and pressure to precipitate zinc as a mixture of zinc oxide and zinc hydroxide.

Barren Recycle Stream. Barren liquor from the zinc recovery unit may need to be passed through an evaporation step to remove water before recycling to the initial feed preparation operation.

Table I Indicates product compositions obtained utilizing the above process unit operations:

Artech Product Material Compositions and Characteristics

Element	Copper Oxide	Silver Cement	Bismuth Oxychloride	Cadmium Cement	Lead Sulfate	Zinc Oxide
Cu	66.2	4.3	1.7	1.8	0.06	0.1
Pb	1.6	45.0	6.5	27.7	65.0	0.1
Bi	0.04	0.15	64.0	0.01	0.07	0.01
Zn	1.8	2.5	0.12	32.0	0.02	66.4
Fe	0.16	0.45	1.0	0.03	0.2	0.05
Cd	0.07	0.01	0.01	35.0	0.005	0.01
Ca	3.0	1.1	0.03	1.02	N/A	2.9
Ag	0.0014	21.0	0.22	0.31	0.15	0.0004
As	0.002	0.6	0.1	0.001	0.04	N/A
Sb	0.006	N/A	N/A	N/A	N/A	0.002
Cl	1.0	N/A	N/A	2.1	N/A	0.03
Mg	0.08	0.01	0.02	0.06	N/A	0.3
K	0.02	0.11	0.01	0.07	N/A	0.1
Na	0.03	0.01	0.03	0.06	N/A	0.1
Hg(ppm)	7.5	7.0	0.2	N/A	0.4	3.0
Al	0.65	0.01	0.8	0.06	0.02	0.07
SO_4	1.4	N/A	N/A	0.68	N/A	0.4
Te	N/A	0.03	0.15	N/A	0.01	N/A
Au(ppm)	N/A	0.3	N/A	16.0	0.5	N/A
Mn	N/A	N/A	N/A	N/A	N/A	0.8

* All compositions in percent, except as noted.
* N/A = Not Analyzed

Chemistry of Cashman Process Application to Base Metal Sulfide Ores and Concentrates

This is a variation of the chloride-oxygen process developed by the US Bureau of Mines[-3]. The chemistry is complex with a number of competing reactions taking place at different rates. Some of the reactions that occur can be presented in the following simplistic forms:

$2MS + 4HCl + O_2 \rightarrow 2MCl_2 + 2S + 2H_2O$ (M = Pb,Zn,Fe,Ni,Co,Bi)

$4CuFeS_2 + 20HCl + 5O_2 \rightarrow 4CuCl_2 + 4FeCl_3 + 8S + 10H_2O$

$MS + 2CuCl_2 \rightarrow MCl_2 + 2CuCl + S$ (M = Pb,Zn)

$4FeS_2 + 15O_2 + 8CaCl_2 + 2H_2O \rightarrow 4FeCl_3 + 8CaSO_4 + 4HCl$

$2FeAsS + 7O_2 + 2CaCl_2 + 2H_2O \rightarrow 2FeAsO_4 + 2CaSO_4 + 4HCl$

$MS + 2FeCl_3 \rightarrow MCl_2 + 2FeCl_2 + S$

$4FeCl_2 + O_2 + 6H_2O \rightarrow 4FeO(OH) + 8HCl$

$2S + 2H_2O + 3O_2 \rightarrow 2H_2SO_4$

$CaCl_2 + H_2SO_4 \rightarrow CaSO_4 + 2HCl$

The above equations reflect that some reactions consume acid and some produce acid. During the leaching stage the pH is continuously monitored and adjusted to ensure that the optimum conditions for maximum metal extraction are maintained. At maximum metal extraction the arsenic is precipitated as stable ferric arsenates.

Test Results

To date, major emphasis has been directed towards treatment of stockpiled flue dust derived from the shutdown copper smelter operations of ARCO at Anaconda, Montana, however other feedstocks have been tested (see Table IV and V on the next page). Successful pilot scale tests were conducted by Hazen Research in Golden, Colorado in late 1989 on Anaconda flue dust. Autoclaving has been performed both as a batch operation and as a continuous operation.

The results in the following chart illustrates the chemical stability of a residue produced from processing a high arsenic content flue dust and highlight one of the major benefits of the Cashman Process. This metal recovery treatment process for flue dust yields residues of high chemical stability.

Concentrations in leachate from applying the TCLP test procedures to untreated flue dust and Cashman Process residues are compared below with the general EPA maximum permissible levels.

Table II TCLP test results for treated and non-treated flue dust

ANALYSIS IN ppm			
Element	EPA (max)	Untreated	Treated
Arsenic	5	1320	0.07
Barium	100	.019	0.09
Cadmium	1	36.6	0.05
Chromium	5	.2	0.04
Lead	5	1.69	0.74
Mercury	0.2	.002	0.0001
Selenium	1	.2	0.21
Silver	5	.019	0.01

The following table reflects partially optimized metal extractions from the same high arsenic content flue dust referred to in the previous Table II and Figure 1.

Table III Metal extractions from the flue dust in Table II

METAL EXTRACTIONS	
METAL	PERCENT
COPPER	93.8
CADMIUM	91.6
SILVER	85.1
BISMUTH	95.3
ZINC	89.8
LEAD	94.3
GOLD	85.3

Table IV below shows toxicity results for Cashman Process residues and unprocessed feeds from eight different feedstocks which have been tested.

Table IV

Toxicity Results for Cashman Process Residues and Unprocessed Feed

Material	Type of Test	Extract Number	Toxicity Results (ppm) As	Ag	Ba	Cd	Cr	Hg	Pb	Se
Feed # 1	EP TOX	N/A	52.0000	0.0090	<0.001	11.4000	0.0070	0.0010	0.8000	<0.001
Residue # 1	EP TOX	N/A	0.1300	0.0500	<0.001	0.0700	0.0660	0.0030	2.7000	<0.001
Feed # 2	EP TOX	N/A	23.0000	0.0050	0.0100	26.0000	0.0050	0.0004	1.2800	0.0005
Residue # 2	EP TOX	N/A	0.3500	0.0200	0.0050	0.8000	0.0050	0.0006	2.9400	0.0340
Feed # 3	EP TOX	N/A	28.0000	<0.001	<0.001	33.6900	0.0040	<0.001	2.2000	<0.001
Residue # 3	EP TOX	N/A	0.0030	0.0350	<0.001	0.4870	0.0210	0.0001	1.5000	<0.001
Feed # 4	EP TOX	N/A	0.6000	0.0100	0.4000	10.000	<0.001	0.0004	7.0000	0.0009
Residue # 4	EP TOX	N/A	0.0400	0.0400	0.0100	0.0400	0.0200	0.0005	2.3800	0.0001
Feed # 5	EP TOX	N/A	0.3500	0.0010	<0.0001	33.2200	0.0280	<0.0001	53.4000	<0.0001
Residue # 5	EP TOX	N/A	3.6000	0.0001	<0.0001	0.0660	0.0230	<0.0001	1.3000	<0.0001
Feed # 6	TCLP	2	0.0099	0.0300	0.0990	65.000	0.0400	0.0001	0.6700	0.3500
Residue # 6	TCLP	1	0.0500	0.0100	0.0900	0.160	0.0400	0.0001	0.1400	0.0350
Feed # 7	TCLP	1	1,300.00	0.0190	0.0190	36.600	0.2000	0.0020	1.6900	0.2000
Residue # 7	TCLP	1	1.200	0.0100	0.0100	0.030	0.0100	0.0006	1.5600	0.0560
Feed # 8	TCLP	1	2,100.00	0.0090	0.0600	41.400	0.0090	0.0004	9.2800	0.0580
Residue # 8*	TCLP	1	1.100	0.0100	0.0100	0.169	0.0100	0.0009	15.9000	0.1000

* This sample is in the process testing stage as of 2/25/91 (only two scoping tests performed to date).
NOTE: The EPA has identified feed # 5 as a K064 waste

N/A = Not Applicable

Continuous pilot scale tests on the treatment of the flue dusts have confirmed flow sheets for producing products such as copper oxide or electrolytic copper, zinc oxide, bismuth oxychloride, lead sulfate, silver cement, cadmium sponge and mercury. Both batch and continuous production have been piloted for Artech's proprietary metal oxychloride to metal oxide conversion reaction (metathesis).

Table V contains metal extractions for eight different feedstocks treated by the Cashman Process.

Table V

Metal Extractions for Material using Cashman Process

Material	Gold	Silver	Copper	Lead	Zinc	Bismuth	Cadmium
Feed # 1	93.75%	N/D	80.40%	47.46%	97.84%	N/D	98.78%
Feed # 2	N/D	93.10%	N/D	61.57%	98.11%	N/D	N/D
Feed # 3	95.78%	N/D	89.11%	N/D	74.89%	N/D	95.37%
Feed # 4	69.53%	74.03%	90.71%	80.91%	81.98%	N/D	99.09%
Feed # 5	N/D	N/D	94.45%	42.35%	83.34%	29.95%	98.85%
Feed # 6	90.37%	72.05%	90.56%	45.40%	56.78%	98.30%	91.44%
Feed # 7	49.63%	88.59%	96.44%	86.27%	90.59%	94.71%	98.47%
Feed # 8	64.38%	96.77%	94.41%	61.74%	97.12%	92.63%	99.37%

N/D = Not Determined

The analyses of the eight feedstocks and residues are presented in Table VI.

Table VI

Analytical Results for Cashman Process Residues and Unprocessed Feed

Material	As	Ag	Ba	Cd	Cr	Hg	Pb	Se
	Analytical Results (%)							
Feed # 1	11.6200	0.0110	n/a	0.5340	0.0160	0.0040	2.6800	n/a
Residue # 1	5.4200	0.0010	n/a	0.0065	0.0106	0.0001	1.4080	n/a
Feed # 2	11.2500	0.0160	n/a	n/a	<0.001	<0.001	44.5000	n/a
Residue # 2	7.9400	0.0010	n/a	n/a	n/a	<0.001	14.9500	n/a
Feed # 3	5.3400	0.0190	n/a	0.1656	0.0115	<0.001	1.8980	n/a
Residue # 3	4.9200	0.0010	n/a	0.0070	0.0133	0.0002	1.8960	n/a
Feed # 4	2.9650	0.0090	0.0540	1.0050	<0.001	0.0001	23.7000	0.0004
Residue # 4	3.1300	0.0030	n/a	0.0100	n/a	0.0001	4.9500	n/a
Feed # 5	1.4150	0.0061	n/a	0.1860	0.0069	0.0001	0.4920	0.0001
Residue # 5	1.0100	0.0017	n/a	0.0160	n/a	n/a	0.2700	n/a
Feed # 6	7.7700	0.0163	0.1792	0.1660	0.0110	0.0003	9.5600	0.0002
Residue # 6	8.5750	0.0022	n/a	0.0030	n/a	n/a	1.5500	n/a
Feed # 7	2.0600	0.0160	n/a	0.1560	0.0125	0.0012	1.9700	n/a
Residue # 7	2.0800	n/a	n/a	0.0017	0.0118	0.0030	1.0740	n/a
Feed # 8	22.9000	0.0603	n/a	2.9500	n/a	<0.0001	25.9000	n/a
Residue # 8*	15.9000	0.0015	n/a	0.0140	n/a	n/a	7.4900	n/a

n/a = Not Analyzed

Commercial Development

The concept for the commercial development is a three stage approach: first, laboratory bench scale tests are carried out; second, pilot plant trials are performed; and third, basic design and engineering feasibility studies are conducted. To date this latter work has been performed by Bechtel Corporation in San Francisco.

Bechtel Corporation has completed an engineering design of a Cashman Process plant for a Superfund flue dust project incorporating many of the metal recovery steps discussed in this paper. The Bechtel report states that the *"process plant is designed to be zero discharge and has no water effluent"*[-4].

After signing a confidentiality agreement the client is encouraged to participate in all stages of process development. Artech provides the client with:

- Results from all tests to support the claims of the Process.
- A demonstration that the Process can be run on a suitably sized pilot scale for an acceptable period of time with equipment that exhibits reliability.
- Basic data necessary for the projection of revenues and operating and capital costs.

Capital and Operating Cost Considerations

The capital and operating costs of a Cashman Process application to a flue dust or residue are very material and site specific. Oxygen costs for instance will depend upon the amount of material which requires oxidation. In general, abandoned or demolished sites with little or no infrastructure and a finite amount of material to be processed will have the highest capital and operating costs. Existing smelters, on the other hand, with extensive infrastructures and unit operations provide the greatest opportunity for reduced capital and operating costs as many intermediate products can be integrated into existing circuits. For example copper recovery can be achieved utilizing existing SX/EW plants at many facilities.

As it is difficult to characterize a typical plant including recovery operations, it may be more useful to quantify the cost of a Cashman Process pressure autoclave which would be common to all of the plants. The installed capital cost for a 100 tpd pressure autoclave assuming a "typical" oxidized flue dust containing 6% arsenic and less than 1% sulfides would be 1.6 million dollars[-5].

Advantages of the Cashman Process

Advantages of the Cashman Process Leach Over Other Pressure Leach Systems

- Oxygen partial pressure is not the rate determining factor in the oxidation of iron and arsenic, but oxygen dispersion and pulp agitation are critical.
- Temperatures and pressures are significantly lower than for other autoclave systems.
- During the oxidation leach, the arsenic is precipitated in a highly stable form in a residue that significantly betters the TCLP test requirements. Cashman Process residues, if left exposed will take millions of years to gradually release their potentially hazardous constituents back into the environment.
- The process leaches base and precious metals and precipitates the arsenic in a single stage leach (other processes require two or more stages).

Advantages of the Cashman Process Leach Over Conventional Flue Dust Management or Treatment Processes

- The Cashman Process could replace the usual procedure of recycling flue dusts within the smelter, thereby allowing for increased smelter throughput, increased metal purities and less dust handling problems.
- Smelters could accept ores and concentrates higher in impurity element content, and still achieve product specifications.
- Readily leachable arsenic trioxide or other arsenic complexes are converted to stable ferric arsenate. Arsenic is estimated to leach from these materials over millions of years, which would virtually eliminate loading of groundwater with arsenic.

- Long-term management and potential liabilities associated with discarding waste products can be minimized.

- Waste volumes can be reduced by the removal of metals.

- Natural resources are conserved because otherwise discarded metals are returned to industrial use.

The Future

All experimental work carried out to date has demonstrated the versatility of the Cashman Process in treating hazardous metallurgical dusts and residues to yield arsenic containing residues that significantly better the TCLP toxicity test requirements. The arsenic in the Cashman Process residues are reintroduced to the environment over millions of years, simulating natural processes. This is a substantial breakthrough and is a major advantage of the Cashman Process when compared with alternative processes.

U.S. Superfund law highlights the exposure to potential future liability, regardless of the legality today, if thought isn't given to future environmental standards when disposing of wastes. In the face of mounting international environmental pressures, the controlled precipitation of the arsenic in a stable non-soluble crystalline form will have obvious economic, engineering and environmental advantages which will have to be considered in future commercial smelting and roasting operations.

REFERENCES

1. A. Davis, "Anaconda Smelter Flue Dust Cashman Process Residue Long-Term Stability Study" (Report by PTI Environmental Services, Boulder, Colorado, 1990).

2. Toxicity Leach Procedures (TCLP) (40 CFR, part 268, App 1).

3. Gary A. Smyres et al Recovery of Metal Values from Complex Sulfides (United States Patent # 4,410,496 Oct 18, 1983).

4. "Comprehensive Feasibility Study of the Arsenical Flue Dust Treatment Anaconda, Montana" (Bechtel Corporation, San Francisco, California, March 1990 Vol. I sec. 1, pg 7)

5. Vender Quotes (Koch-Knight # 918-057 BP).

Hydrometallurgical Process of Copper Converter Dust

at the Saganoseki Smelter & Refinery

M. Tomita, M. Higashi, S. Oto, and H. Okamoto

Saganoseki Smelter & Refinery, Nippon Mining Co., Ltd.
Saganoseki, Oita, JAPAN 879-22

Abstract

One of the most essential factors for a custom smelter to strengthen its competitive power is to improve its comparative standing in purchase of concentrates from the world market. It is especially critical for Japanese smelters who depend almost 90% of their raw materials on import. In order to accomplish this breakthrough, the development of the new technology which enables us to treat concentrates containing higher impurities is inevitable.

Nippon Mining Company Ltd. made an intensive research on processing complex ore for three years from 1979, and successfully developed an original hydrometallurgical process to treat converter dust for eliminating impurities from the copper smelting circuit. The commercial plant was constructed at the Saganoseki Smelter & Refinery in 1982, and its nine years operation has been satisfactorily proven the advantages of the technology. The concept of this process and its operational performance are discussed.

Residues and Effluents - Processing and Environmental Considerations
Edited by Ramana G. Reddy, William P. Imrie and Paul B. Queneau
The Minerals, Metals & Materials Society, 1991

Introduction

The Saganoseki Smelter & Refinery of Nippon Mining Company Ltd., which is the largest coastal copper smelter in Japan, presently has the annual production capacity of 330,000 metric tons. The Saganoseki consists of 2 Outokumpu-type flash furnaces, 6 Peirce-Smith converters (two to four converters operating), 2 anode furnaces, one casting wheel, 3 sulfuric acid production lines, 3 tank houses, along with the necessary auxiliary equipment.

The Saganoseki is a custom smelter which imports 100 percent of its concentrates from overseas mines. Figure 1 shows variations in the concentrations of arsenic, antimony, bismuth, lead and zinc impurities in the flash furnace feed at the Saganoseki from 1978 to 1982.

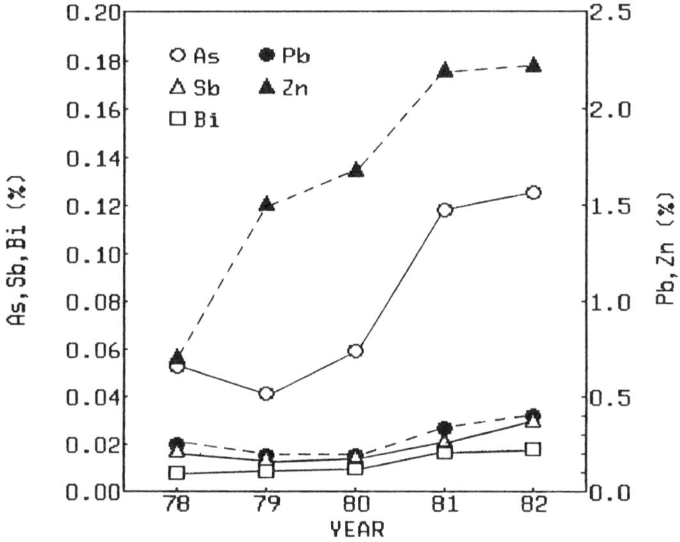

Figure 1 - Variations in the concentrations of impurities in the flash furnace feed

Copper concentrates processed were becoming increasingly impure year after year. Therefore the development of the new technology which enables us to treat concentrates containing higher impurities was indispensable for the Saganoseki.

The Saganoseki conducted an intensive research work on eliminating impurities from its smelting circuit for three years beginning in 1979, and successfully developed an original hydrometallurgical process to treat converter dust for impurity elimination. The commercial plant was constructed in 1982, and has been operating consistently with the satisfactory performance. The concept of the process and its operational results are introduced in this paper.

The development of hydrometallurgical process for dust treatment

The First Stage

The impurities are concentrated in the flash furnace E.P dust & converter E.P dust.
To determine whether the tolerance limit of impurities in the smelting process could be enlarged through treatment of these dusts, investigation of impurities distribution was carried.

The Second Stage

By way of impurities investigation as mentioned in the first stage, it was revealed that flash furnace dust or converter dust treatment was considered the necessesity for improved impurities processing ability. And, hydrometallugical process was adopted because of environmental respect.
Table I shows the typical analyses of flash furnace dust and converter dust.

Table I Typical analyses of flash furnace E.P dust & converter E.P dust

	As %	Bi %	Pb %	Zn %	Cd %	Cu %	Fe %
Flash furnace dust	1.5 ~ 3	0.3 ~ 0.7	3 ~ 5	5 ~ 10	0.3 ~ 0.7	13 ~ 17	8 ~ 12
Converter dust	5 ~ 8	2.5 ~ 3.5	20 ~ 25	10 ~ 16	1.0 ~ 4	4 ~ 6	1 ~ 2

The table clarifies that the impurities content of converter dust is approximately 3-5 times as that of flash furnace dust. While, from the quantitative point of view, when 25,000 tons/month of copper is produced, flash furnace dust is approximately 1,500 tons/month and converter dust is only 500 tons/month, if this much. As a result of studying both the dust amount and the impurities content, it has been determined that converter dust should be independently treated.
In developing the process of converter dust treatment, the concept is as follows:

A. Lead should be separated and recovered as lead sulfate to be processed into lead material.
B. Bismuth and antimony shall be concentrated into the lead sulfate and recovered in the lead smelting process.
C. Zinc and cadmium shall be recovered as hydroxide to be processed into raw material in the zinc smelter.
D. Arsenic shall be separated independently of the other elements as much as possible and recovered as a stable sulfide.

The Third Stage

According to the plan described in the foregoing, the process flow was determind based on beaker test. Subsequently, a bench scale pilot plant was constructed and, after half-a-year of pilot operation, equipment specifications were decided.

Flowsheet of the Converter Dust treatment Process

Figure 2 shows a flowsheet of the hydrometallurgical process for converter dust.

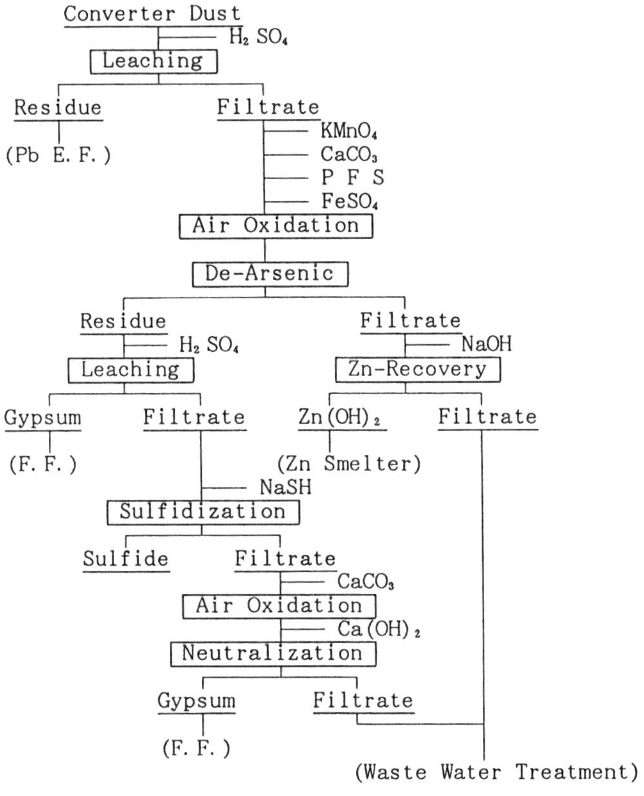

PFS : Poly Ferric Sulfate
PbEF: Electric Furnace for Pb Smelting
FF : Flash Furnace

Figure 2 - Flowsheet of the hydrometallurgical process for converter dust

The process consists of the following six continuous process.

(1) Dust leaching process
(2) De-arsenic process
(3) Iron arsenic compound leaching process
(4) Sulfidization process
(5) Neutralization process
(6) Zinc recovery process

Each process is outlined in the paragraphs to follow:

(1)Dust leaching process

In this process, converter dust is leached with sulfuric acid, separating lead, bismuth and antimony in the residue, and arsenic, copper, zinc and cadmium in the filtrate.
Figure 3 shows the leaching ratio when converter dust is leached with sulfuric acid.

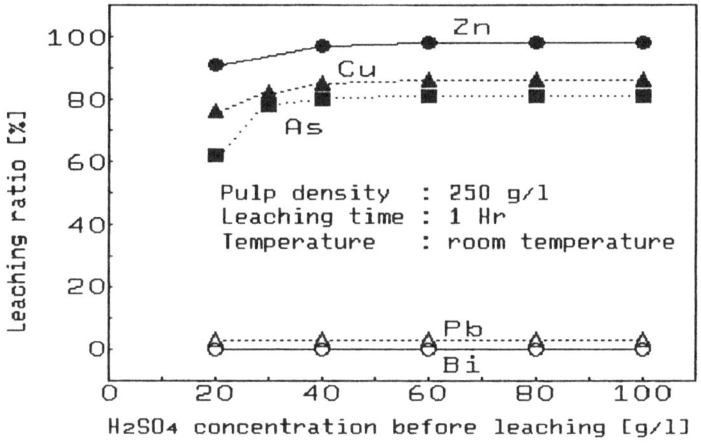

Figure 3 - Leaching ratio vs sulfuric acid concentration

The relationship is shown between the H_2SO_4 concentration before leaching and the leaching ratio under a pulp density of 250 g/l, leaching time of one hour, and room temperature. As the figure shows, lead and bismuth are only negligibly leached while arsenic, copper and zinc are thoroughly leached. When the H_2SO_4 concentration before leaching exceeds 40 g/l, the leaching ratio of arsenic, copper and zinc become nearly saturated. As a result, the H_2SO_4 concentration before leaching was rated 50 g/l. The effect of pulp density is that, up to 250 g/l, the arsenic leaching ratio tended to be lower.
When the bismuth content in the dust is high, the leaching ratio of arsenic decrease because arsenic exists as bismuth arsenic compound, as shown in the morphology analysis in Table II.

Table II Morphology analysis of converter dust

As	$BiAsO_4$, $FeAsO_4$
Cu	$CuSO_4 \cdot H_2O$
Fe	$ZnO \cdot Fe_2O_3$
Zn	$ZnSO_4 \cdot H_2O$, $ZnO \cdot Fe_2O_3$
Cd	$CdSO_4$
Pb	$PbSO_4$
Bi	$BiAsO_4$

(2) De-arsenic process

In this process, arsenic in the filtrate after leaching is independently separated from zinc, cadmium and copper which remain in solution.
A conventional method of removing arsenic from the leach solution is to separate copper at first as copper sulfide or cement copper, then add iron ion and, separate it as iron arsenic compound, while oxidizing with a pH of 4 or higher.
However, this method has the defects of requiring a copper eliminating process, precipitating cadmium in the iron arsenic compound since the pH is made 4 or higher, and lowering the content of the arsenic in iron arsenic compound.
In this process, the following development has been added to compensate for the defects of conventional methods.

A. Arsenic in the converter dust is trivalent(As^{3+}) or pentavalent(As^{5+}). As^{3+} ion, which is hard to precipitate and separate, is oxidized into a As^{5+} ion form with potassium permanganate.
B. Subsequently, ferrous sulfate and poly ferric sulfate are added to deposit arsenic as an iron arsenic compound while oxidizing with air and keeping the pH at around 2.5 with calcium carbonate. If sodium hydroxide is used instead of calcium carbonate, the de-arsenic ratio is lowered.

Figure 4 shows the relationship between the arsenic eliminating ratio and pH when the arsenic/iron mole ratio is 1/1.2.

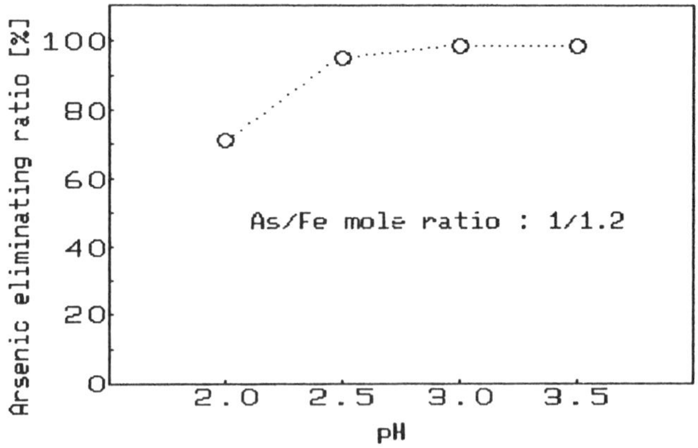

Figure 4 - Arsenic eliminating ratio vs pH

As the figure clarifies, according to this process, the eliminating ratio of arsenic as iron arsenic compound at the pH value 2.5 or higher is high (approx. 95%) and becomes nearly saturated. While, as Figure 5 shows, the relationship between the Cu/As weight ratio and pH tells that, at the pH value of 3 or higher, the precipitating volume of copper into iron arsenic compound becomes large, and so does that of cadmium. Therefore, pH must be strictly controlled at 2.5 or thereabout, and the control must be automatic.

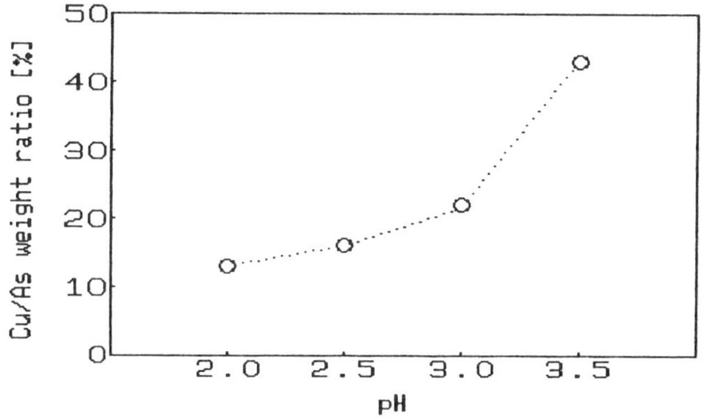

Figure 5 - Cu/As Weight ratio of precipitate vs pH

Using the aforementioned process control, we succeeded in establishing the process that minimizes the precipitation of large amounts of copper, cadmium and zinc while maximizing arsenic precipitation.

(3) Iron arsenic compound leaching process

The arsenic that was fixed as iron arsenic compound in the process described above is leached with sulfuric acid. Figure 6 shows the relationship between the H_2SO_4 concentration and the leaching ratio.

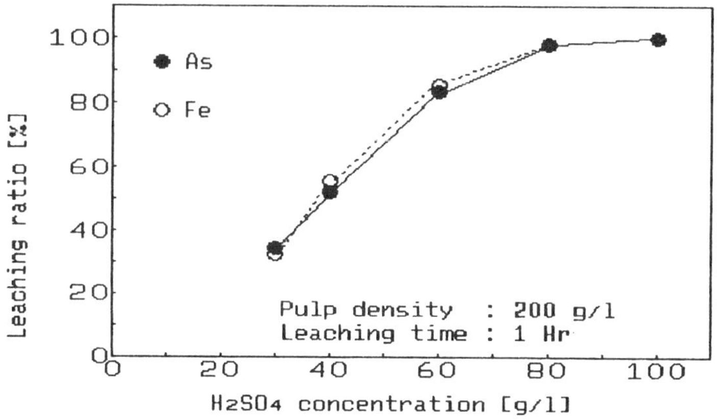

Figure 6 - Leaching ratio of Iron arsenic compound vs sulfuric acid concentration

As the figure shows, at the H_2SO_4 concentration of 80 g/l or more, nearly 100% of the leaching ratio is attained.

(4)Sulfidization process

Arsenic in the filtrate from the iron arsenic compound leaching process is precipitated as arsenic sulfide by sulfidizing with sodium hydrosulfide.

(5) Neutralization process

The filtrate after sulfidization contains a very small amount of arsenic and a large amount of ferrous iron. The arsenic and iron are neutralized with calcium hydroxide to be precipitated and separated together with gypsum. To improve the efficiency of sedimentation of the arsenic and the filterability of the precipitate, ferrous iron is air oxidized into ferric iron at pH 4-5 and neutralized at pH 10-11 with milk of calcium hydroxide.

(6) Zinc recovery process

Zinc, cadmium, etc. contained in the filtrate after the de-arsenic process is neutralized to approximately pH 9-10 with sodium hydroxide, to be separated and recovered as hydroxide.
The filtrate after zinc recovering is processed to the waste water treatment.

Construction of Converter Dust Treatment Plant

Construction of the plant was started early in 1982 and completed in September of the same year. The converter dust treatment equipment is arranged as shown in Figure 7.

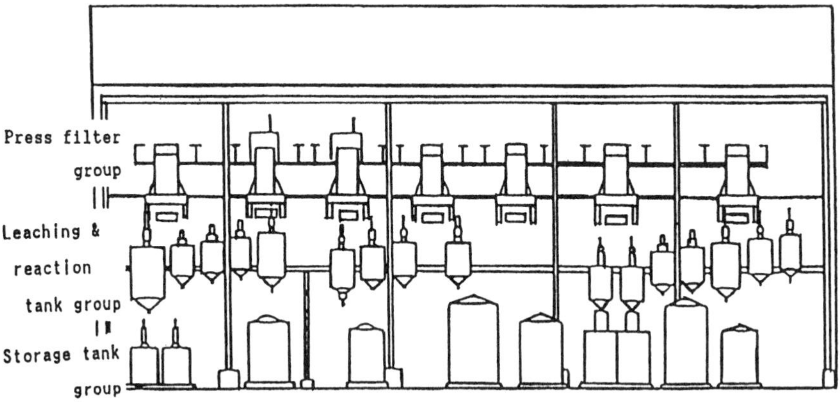

Figure 7 - Equipment layout of dust treatment plant

The converter dust treatment plant consists of a press filter group in the upper floor, leaching and reaction tank group in the middle floor, and a filtrate storage tank group in the first floor.

Table Ⅲ shows specification of main equipment of this plant.

Table Ⅲ Main equipment specifications for the commercial dust treatment plant

Process	Main equipment	Unit	Specifications
Reciving	Dust Bin	1	$160m^3$, Bag Filter $250m^2$ Table Feeder ×1
Dust Leaching	Jet Mixer Leaching Tank Press Filter	1 1 1	0.4mφ×1.4mH, 1.5T/h $10m^3$ FRP Fully-Automatic Compression Type $106m^2$
De-arsenic	Reaction Tank Oxidation Tank Press Filter	3 2 1	$3m^3$×2, $6m^3$×1 FRP $2.5m^3$ SUS 316 Fully-Automatic Compression Type $62m^2$
Zinc Recovery	Reaction Tank Press Filter Container Bag Packing Machine	1 2 1	$3m^3$ FRP Fully-Automatic Compression Type $77m^2$ Semi-Automatic Type 6Bag/h
Iron Arsenic Compound Leaching	Leaching Tank Press Filter	2 1	$3m^3$ Rubber Lined Fully-Automatic Compression Type $49m^2$
Sulfidi-zation	Sulfidization Tank Degassing Tower Absorbing Tower Press Filter Repulping Tank	2 1 1 1 2	$3m^3$ FRP $3m^3$ FRP $6m^3$ Rubber Lined Fully-Automatic Compression Type $98m^2$ $10m^3$ Rubber Lined
Neutrali-zation	Reaction Tank Oxidation Tank Press Filter	2 3 1	$6m^3$ FRP, $10m^3$ SUS $2.5m^3$ Rubber Lined Fully-Automatic Compression Type $77m^2$

Dust Treatment Operational Achievements

Figure 8 shows the recent data of the concentrate impurities content of the flash furnace feed, which has been undergoing a change at a high level since 1982.

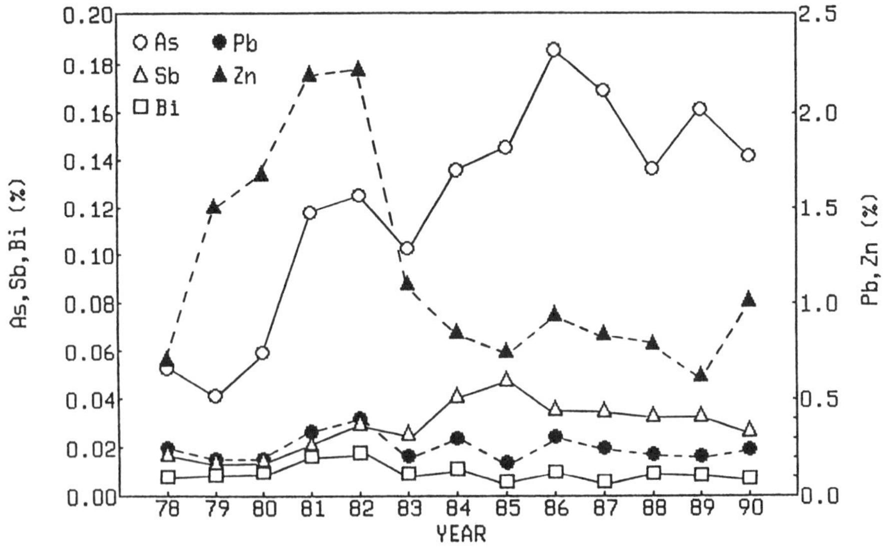

Figure 8 - Variations in the concentrations of impurities in the flash furnace feed

Figure 9 shows the typical material balance of the dust treatment operation.

As the figure shows, 61% of the arsenic is separated as arsenic sulfide. Bismuth, lead and antimony are distributed nearly 100% in lead residue and recovered in lead smelting process.

Zinc and cadmium, almost 90% separated as hydroxide, is sent to a zinc smelter.

The distribution of copper in this process is 4% in arsenic sulfide, 13% in lead sulfate and 82% in zinc hydroxide. Copper in the zinc hydroxide is recovered in the zinc smelting process, to be returned to the copper smelting process.

Main chemical agent consumption when 400 tons/month of converter dust is treated is as follows:

H_2SO_4	160 tons/month
$CaCO_3$	120 tons/month
NaOH	95 tons/month
$[Fe_2(OH)_n(SO_4)_{3-n/2}]_m$ (Fe^{3+} : 11%)	70 tons/month
NaSH	33 tons/month
$FeSO_4 \cdot 7H_2O$	23 tons/month
$Ca(OH)_2$	20 tons/month
$KMnO_4$	3 tons/month

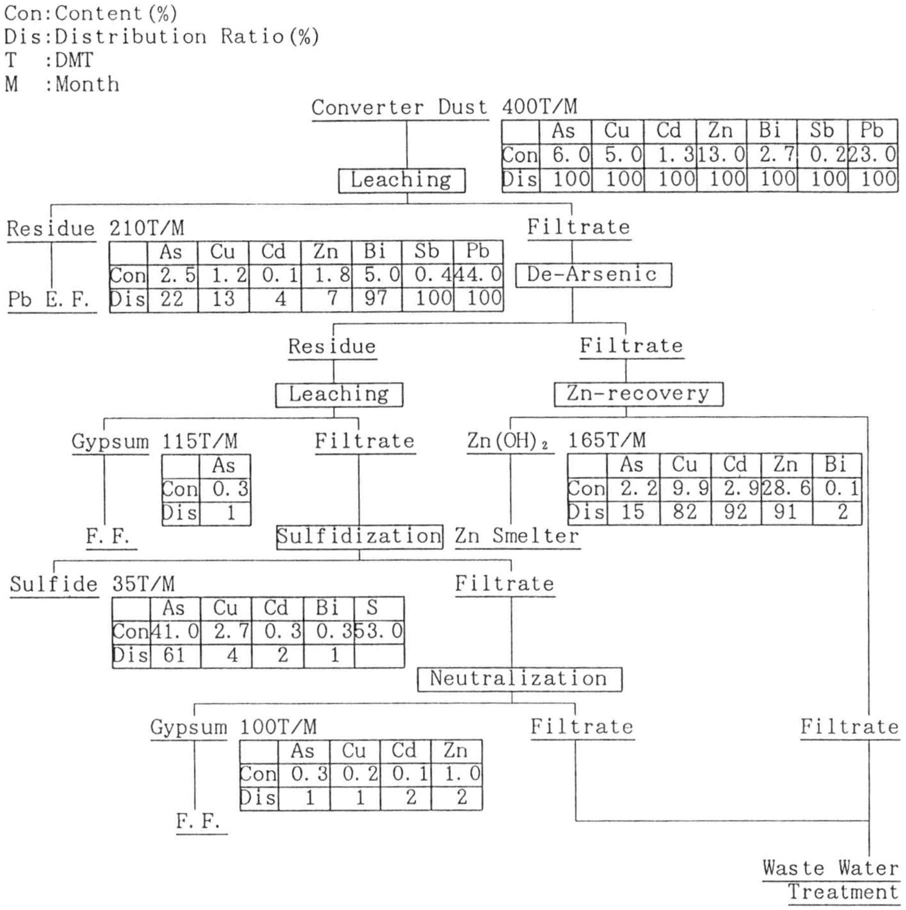

Converter Dust 400T/M

	As	Cu	Cd	Zn	Bi	Sb	Pb
Con	6.0	5.0	1.3	13.0	2.7	0.2	23.0
Dis	100	100	100	100	100	100	100

Residue 210T/M

	As	Cu	Cd	Zn	Bi	Sb	Pb
Con	2.5	1.2	0.1	1.8	5.0	0.4	44.0
Dis	22	13	4	7	97	100	100

Gypsum 115T/M

	As
Con	0.3
Dis	1

$Zn(OH)_2$ 165T/M

	As	Cu	Cd	Zn	Bi
Con	2.2	9.9	2.9	28.6	0.1
Dis	15	82	92	91	2

Sulfide 35T/M

	As	Cu	Cd	Bi	S
Con	41.0	2.7	0.3	0.3	53.0
Dis	61	4	2	1	

Gypsum 100T/M

	As	Cu	Cd	Zn
Con	0.3	0.2	0.1	1.0
Dis	1	1	2	2

Figure 9 - Typical material balance of the dust treatment operation

As indicated, since arsenic, bismuth, antimony, cadmium, lead and zinc have been separated in the copper smelting process, the impurities processing capacity of the copper smelting is significantly improved.

Conclusion

Through hydrometallurgical process for converter dust at the Saganoseki Smelter & Refinery, it has become possible to treat copper material with many impurities, such as arsenic, bismuth, antimony, cadmium, lead and zinc.

RECOVERY OF METALS FROM COPPER SMELTER

FURNACE FLUE DUST

H. H. Law, M. P. Bohrer, P. A. O'Hara*, J. Sapjeta, and V. Tierney**

AT&T Bell Laboratories
600 Mountain Ave.
Murray Hill, NJ 07974

Abstract

Large amounts of waste material in the form of flue dust are generated in the operation of copper smelter furnaces. As a result of increasingly stringent disposal, it has become important to develop methods for treating such metallurgical wastes to extract metals and convert them into salable products. A two-step closed-loop hydrometallurgical process has been developed: (1) extraction with sulfuric acid removes about 90% of zinc and decreases the dust volume by 50%; (2) extraction with hydrochloric acid dissolves the lead and further decrease the dust volume by 75%. A process scheme for generating salable products is proposed.

* Current address: CH2M Hill, Santa Ana, CA 92705

** Current address: AT&T Quality Management Engin., Union, NJ.

Residues and Effluents - Processing and Environmental Considerations
Edited by Ramana G. Reddy, William P. Imrie and Paul B. Queneau
The Minerals, Metals & Materials Society, 1991

Introduction

Copper recycling is generally accomplished through a pyrometallurgical and electrolytic operation in which various metals are separated from the copper resulting in the production of refined copper. In a typical recycling facility, the smelting process (Figure 1) begins with the melting of low-grade copper scrap with coke and limestone in a blast furnace. Metal slag, consisting mainly of iron, silicon, aluminum, and calcium floats to the top, and is then separated. Most of the zinc and lead are vaporized and form fine particles of oxide dust, which are filtered out as flue dust in a baghouse. A higher grade molten copper at the bottom of the furnace is converted into "black copper" pellets and is passed to a converter furnace along with high-grade copper scrap (>25% Cu). In the converter furnace, air is injected to oxidize the impurities which again separate out as slag or dust. The copper is then transferred to the anode furnace, where the oxygen impurities are removed and the copper is cast into anodes. In the tankhouse operation, the copper anodes are electrolytically refined into cathodes that are subsequently cast into salable forms.

The copper recycling process generates or has the potential for generating other resources as well. For instance, precious metals can be recovered from the sediment in the electrolytic tanks and the slag material can be sold for use as construction materials. The flue dust represents another potential source of valuable metals. The furnace dusts that are generated by the copper recycling furnaces contain mostly oxides of volatile metals, such as zinc, lead, and tin. Indeed the baghouse operation is the major outlet for these reactive metals. Table 1 shows typical compositions of dusts from blast and converter furnaces.

Table 1. Flue Dust Composition

Oxide	Blast Furnace Wt %	Converter Furnace Wt %
ZnO	29.4	20.1
CaO	12.0	1.0
PbO	10.2	14.0
SiO_2	7.0	.5
Al_2O_3	6.5	.7
FeO	5.5	1.1
Cu_2O	5.2	3.3
SnO	1.5	8.7
Sb_2O	3.7	.7
NiO	.3	.2
Au	10 ppm	10 ppm
Pd	7 ppm	7 ppm
Ag	35 ppm	35 ppm

Zinc oxide and lead oxide account for about 35-40 wt% of the dust. Zinc represents the most valuable component of the dust, worth about 10 cents/pound of dust while the precious metals add an additional 6 cents/pound of dust. Other potential resources include lead, copper, and tin. Without any recovery process, the dust becomes a

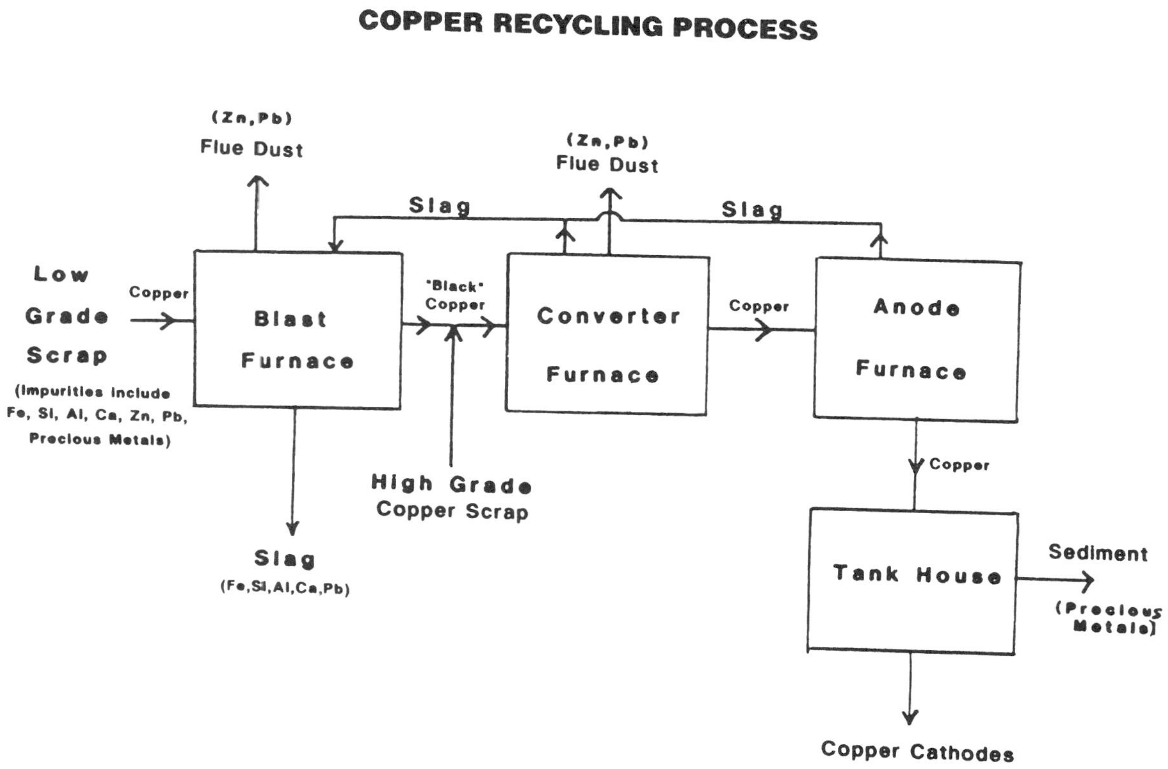

Figure 1. Schematic diagram of a copper recycling process.

include lead, copper, and tin. Without any recovery process, the dust becomes a significant liability since the presence of lead requires that the dust be disposed in a hazardous landfill.

This waste problem is not unique to copper recycling. Numerous other pyrometallurgical processes such as brass smelting and steel production also generate metal-rich flue dusts. As a result of increasingly stringent government regulation of waste disposal, it has become important to utilize metallurgical wastes as sources of metals and other chemicals. Methods have been developed in which metals are extracted from flue dusts and converted into salable products. These methods include pyrometallurgical [1], plasma-based [2], and hydrometallurgical [3-9] processes. Pyrometallurgical and plasma-based processes are energy intensive and require a relatively high grade scrap source or some other economic advantage such as the ability to utilize process heat for other applications to be profitable.

Recovery of zinc by hydrometallurgical processes with sulfuric acid has been reported [3-7]. Powell et al. [3] described a process in which brass smelter flue dusts are treated with sulfuric acid to separate zinc and copper from lead and tin. Zinc is then recovered as zinc sulfate or metallic zinc, and the lead-tin mixture is sold. Datillo et al. [4,5] have developed a process in which flue dust is treated with sulfuric acid to extract zinc. The resultant solutions are then purified and used to electrogalvanize steel wire. Mohri and Yamada [6] report a hydrometallurgical process for recovering copper, zinc, and cadmium from flue dust generated by copper ore flash smelting furnaces. Lead is converted to lead sulfate and recovered in a purified form. Harris et al. [7] describe the development of a process to treat secondary zinc sources such as slags, drosses, tailings, and flue dusts to recover zinc as zinc sulfate monohydrate. Their process consists of washing to remove water solubles such as chloride, acid leaching of zinc and other metals, oxidation to precipitate iron oxide and iron arsenate, purification via cementation with zinc dust, and spray drying to recover the zinc sulfate product. Jha and Duyvesteyn [8] reported a two-stage leaching process for recovery of zinc from steel plant dusts. This process consists of an atmospheric leaching step in which fresh dust is leached with acidic zinc sulfate solution from the following pressure leaching step. The thickened solid phase is sent to an autoclave, where it is treated with sulfuric acid. Iron is precipitated and the solution is returned to the atmospheric leaching step where it is recovered as low-iron zinc sulfate solution.

Another approach was developed by Burrows [9]. Hot ammonium chloride solution is used to remove iron first from flue dusts. Zinc metal dust is then added to the solution to precipitate metallic lead and copper. Upon cooling, the resultant solution yields crystals of zinc oxide.

The objective of this work is to examine various methods of treating copper smelting furnace flue dust in order to decrease or eliminate waste volume and to recover valuable components. The most efficient treatment of the dust would involve a process that could extract the major components and convert them into a salable form. If zinc and lead were removed, it would substantially decrease the amount of dust needing disposal. It would also eliminate the necessity of utilizing hazardous waste facilities, significantly reducing costs.

Experimental Procedures

Solution Extraction of Metals

Blast furnace and converter furnace flue dusts were obtained from the AT&T Nassau Metals copper recycling facility. Blast furnace dust, received in the form of pellets, was crushed, ground, and mixed well. Converter furnace dust (non-pelletized) was used as received. The dust was dried at 100°C for 24 hours to remove water. For extraction rate experiments, the ground dust was fractionated by size using standard screens.

Extraction experiments were carried out by adding a known volume of extracting solvent to a weighed amount of dust, stirring for 30 minutes at a specified temperature, and filtering by suction. For extraction rate experiments using larger dust particle sizes, longer mixing times were necessary. During extraction rate experiments, small samples of the mixture (<1% v/v) were removed periodically and rapidly filtered. The filtrate was analyzed for copper, lead, and zinc using atomic absorption to determine the extent of extraction. In other experiments, the total mixture was filtered after stirring was complete. The residue was washed with a small amount of water, and dried at 100°C for ~24 hours (until no further weight reduction was observed). Any residues to be used for further extractions were ground with a mortar and pestle after drying. The quantity of dust that was extracted into solution was determined by the difference in weight of the dried dust before the experiment and that of the dried residue. Metal concentrations in the residues were determined with an atomic emission method.

Purification of $ZnSO_4$ Solution

A composite sample of dust consisting of 58.3 wt% blast furnace dust and 41.7 wt% converter furnace dust was used for water washing, acid extraction, and solution purification experiments. Water washing of dust samples was accomplished by mixing dust and deionized water in a beaker (0.38 g dust/ml water). Agitation was provided with a four-bladed stirrer, and the solution was heated by means of a hot plate connected to a temperature controller. Sodium carbonate was added as required to maintain the solution pH at 9 to 9.5. After 3 hours, the slurry was suctioned filtered followed by four displacement washes with deionized water. The volume of each wash was equal to ~10% of the total slurry volume. The percent moisture of the cake was determined gravimetrically and the chloride content of the filtrate and washes was measured by silver nitrate titration.

Wet cake from the water washing experiments was mixed with deionized water in a beaker to make a slurry containing 40-60% (w/v) solids. The slurry was heated to 90°C on a hot plate and maintained at this temperature with a temperature controller. Concentrated sulfuric acid (95-98%) was added dropwise to maintain the solution at pH 2 to 2.5. Agitation was provided with an overhead four-bladed stirrer. After 3 hours, the solution was suction filtered and the filter cake was washed with 4 displacement washes (volume/wash ~20% of total slurry volume.)

Filtrate from the acid extraction experiment was used for cementation experiments using scrap iron. Scrap iron, obtained from AT&T Nassau Metals was sized and particles of minus 10-mesh size (Tyler) were used. Samples (100 ml) of filtrate were heated to 90°C with stirring. Scrap iron equal to 3-4 times stoichiometric based on the total amount of Cu+Ni+Cd was added. The solution bubbled and the temperature rose ~10°C when the Fe was added. The solution was sampled periodically over ~4 hours after which time the copper-iron cement was separated by suction filtration.

Filtrate from the iron cementation was further purified through addition of zinc dust (fine powder supplied by AMAX Zinc Co.). The zinc dust was added to 20 ml aliquots of the solution in plastic containers. Experiments were performed with zinc dust alone (30 times stoichiometric for Ni + Cd (30X)), zinc dust (30X) with furnace dust (0.08 g), or zinc dust (30X) with $PbSO_4$, Sb_2O_3, and $CuSO_4$. The amounts of Pb (=1.5% of zinc dust), Sb (=6% of Ni in solution), and Cu (=100% of Ni in solution) were chosen based on the solution purification procedure of the National Zinc Co. [10]. After adding the reagents, each container was partially covered and placed in the water bath (90°C) and agitated with a magnetic stirrer. Samples (0.10 ml) were taken every hour for 4 hours.

Results and Discussion

Gravimetric Separation Processes

The large difference between the density of zinc oxide and lead oxide (5.6 vs 9.1 g/cm^3, respectively) led us to consider the possibility of a physical separation of zinc and lead oxide. Upon close examination of the furnace flue dust with a scanning electron microscope and X-ray analysis, we found that the particle size of most of the dust is smaller than 0.1 μm and the zinc and lead oxides are not physically separated. Thus a mechanical separation of zinc and lead would not be feasible. Subsequent work focused only on hydrometallurgical processing.

Solution Extraction of Metals

Based on the work reported in the literature, both acidic and alkaline solutions were evaluated. For the initial screening experiments, extractions were carried out using three solutions: HCl, H_2SO_4, and 1.2 M NH_4OH with 0.2 M NH_4Cl. The acid concentration ranges were from 1 M to 4 M for H_2SO_4 and from 1 M to 3 M for HCl. The ratio of the volume of the acid solution to the amount of dust was varied from 2 to 10 ml/g of dust for the acid solution and from 10 to 100 ml/g of dust for the alkaline solution.

The percent of zinc dissolved by HCl, H_2SO_4, and the NH_4OH solution is shown in Figure 2. The highest percentage of zinc removed by the alkaline solution was about 50% at a solution volume of 100 ml/g of dust, whereas the values were about 80% and 100% for HCl and H_2SO_4, respectively, at a solution volume of only 10 ml/g of dust. For acid solutions, the most important parameter appears to be the acid content, i.e., moles of acid/g of dust (Figure 3). The acid concentration did not appear to affect the efficiency of extraction. The ratio of dust to solvent volume also had little effect, provided there was enough free acid initially present to ensure complete reaction with the dust. Although HCl only removed about 80% of the zinc, it dissolved a higher percentage of the dust (Figure 4).

Of the three solvents, only HCl dissolved lead oxide to any extent. Figure 5 shows that the percent of lead dissolved is proportional to the initial HCl content. Less than 50 mg/l lead was detected in either the alkaline solution or H_2SO_4.

Based on the results presented, the alkaline solution (1.2 M NH_4OH + 0.2 M NH_4Cl) is not as desirable as an acid solution because of the large volume of the alkaline solution that would be required. Both acids dissolved over 50% of the total dust. The HCl could only extract 70-80% of the zinc, along with a small but significant amount of lead. The H_2SO_4 consistently extracted more than 90% of the zinc, while leaving

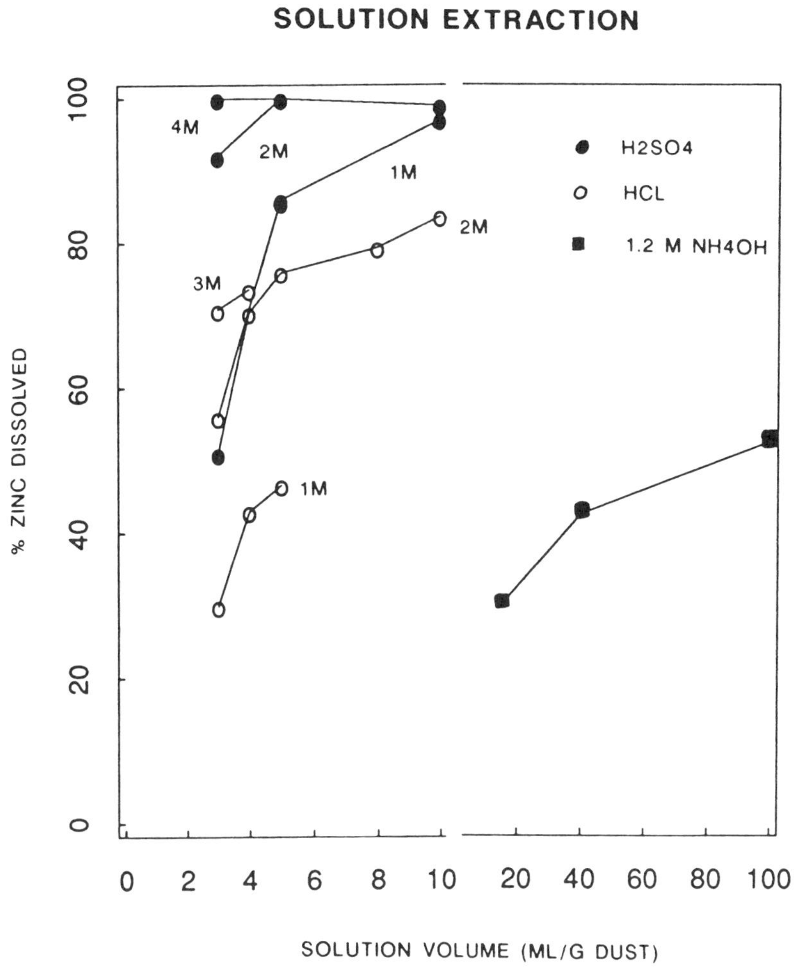

Figure 2. Solution extraction of zinc from blast furnace dust using H_2SO_4 (1 M, 2 M, and 4 M), HCl (1 M, 2 M, and 3 M), and 1.2 M NH_4OH + 0.2 M NH_4Cl.

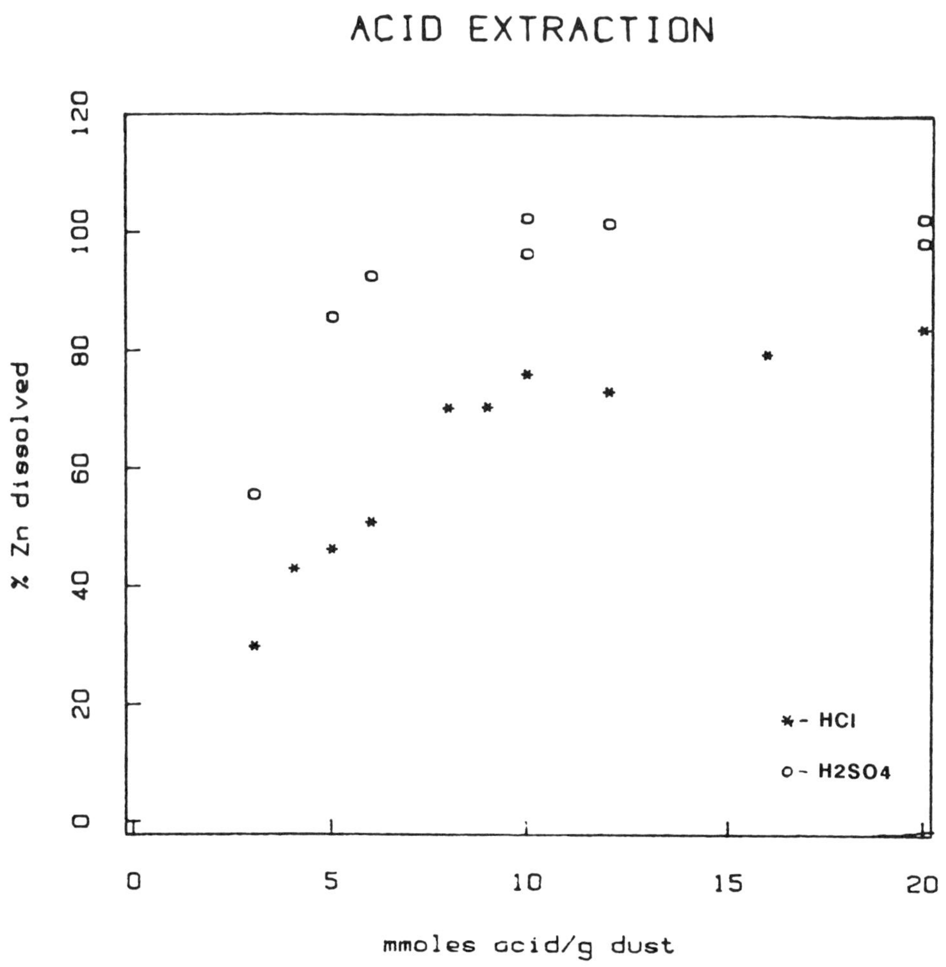

Figure 3. Percentage of zinc dissolved from dust vs. acid content for H_2SO_4 and HCl.

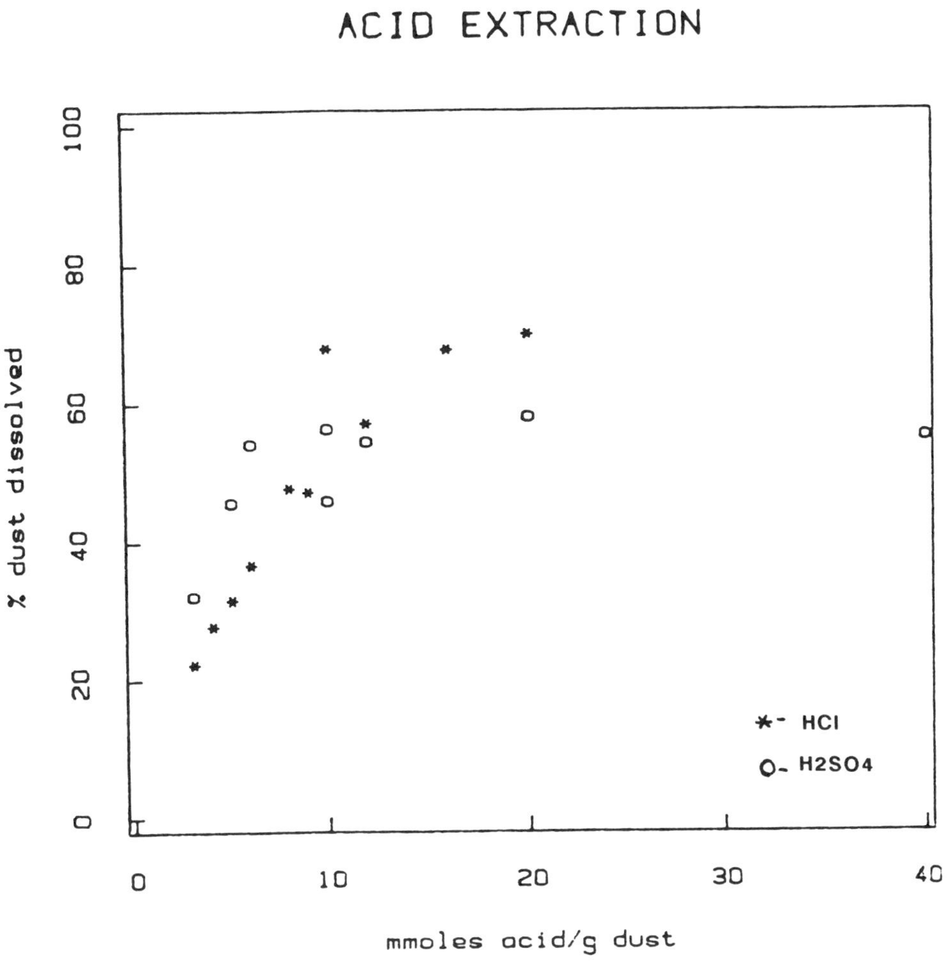

Figure 4. Percentage of dust dissolved vs. acid content for H_2SO_4 and HCl.

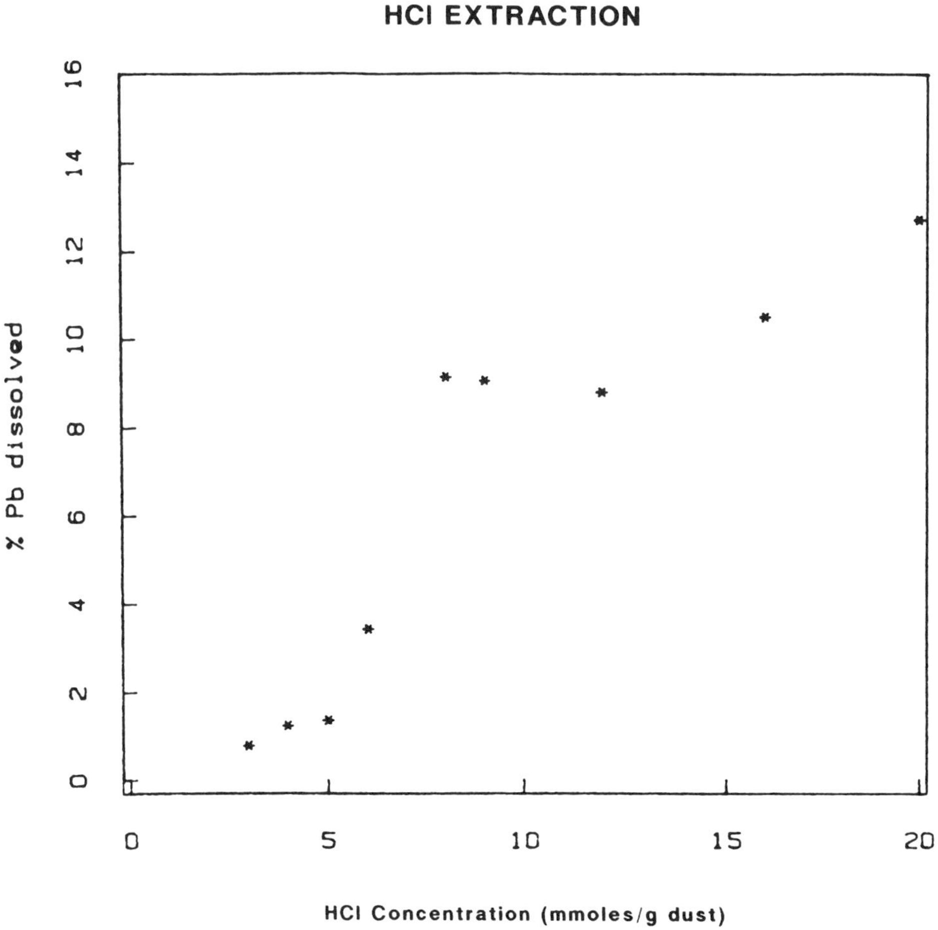

Figure 5. Percentage of Pb dissolved vs. acid content for extraction of residue with HCl.

virtually all of the lead in the insoluble residue. A typical analysis of the dust, filtrate, and residue from an extraction experiment in which 2.0 M H_2SO_4 was used is given in Table 2. The results show that the major components of the acid solution are zinc (43.1 g/l) and copper (8.8 g/l), while the major component of the residue is lead (16.7 wt.%).

A series of extraction experiments were performed on the residue to try and further decrease the volume by removing the lead. Based on results from the preliminary extraction experiments, HCl was chosen as solvent. Acid concentration did not appear to affect the efficiency of extraction: 2, 4, and 6 M HCl all dissolved the same amount of residue. Increasing the temperature from 25 to 80 °C increased the amount of residue dissolved by 60-70%, but still less than half of the residue could be dissolved. Increasing the extraction time had no effect. However, after three consecutive extractions, more than 50% of the lead had been removed from the residue. Also, by decreasing the residue/solvent volume ratio to 1:20, about 75% of the residue could be dissolved.

The results of extraction rate experiments are given in Figure 6. The extent of reaction is calculated as the ratio of the quantity of copper that has been extracted to the total quantity that is in the dust. The extent of copper extraction was found to be equivalent to the extent of zinc extraction, but copper was used because of the relative ease of the measurement. Results show that, in 10 minutes, at least 65% extraction can be achieved with particles smaller than 2.0 mm, and that 70% extraction can be achieved with particles smaller than 0.5 mm. As indicated above, >90% extraction is achieved within 30 minutes at room temperature.

Table 2. Typical Extraction Results

Metal	Dust Wt %	Filtrate* g/L	Residue Wt %
Zn	28.32	43.16	2.7
Pb	8.74	.02	16.68
Cu	5.98	8.76	0.76
Ca	5.45	0.59	9.56
Fe	2.89	2.58	2.27
Si	2.55	0.97	3.32
Al	1.56	1.24	1.78
Sn	0.68	0.46	0.74
Ni	0.42	0.24	0.89
Cd	0.16	0.54	0.17
Sb	0.12	0.07	0.09

* Based on standard extraction ratio of 20 g dust/100 ml acid solution.

Purification of $ZnSO_4$ Solution

One possible market for zinc recovered in this process is the agricultural market. Zinc sulfate monohydrate, a preferred form of zinc sulfate for use as a fertilizer micronutrient or animal feed premix, is easily obtained from the acid extraction filtrate

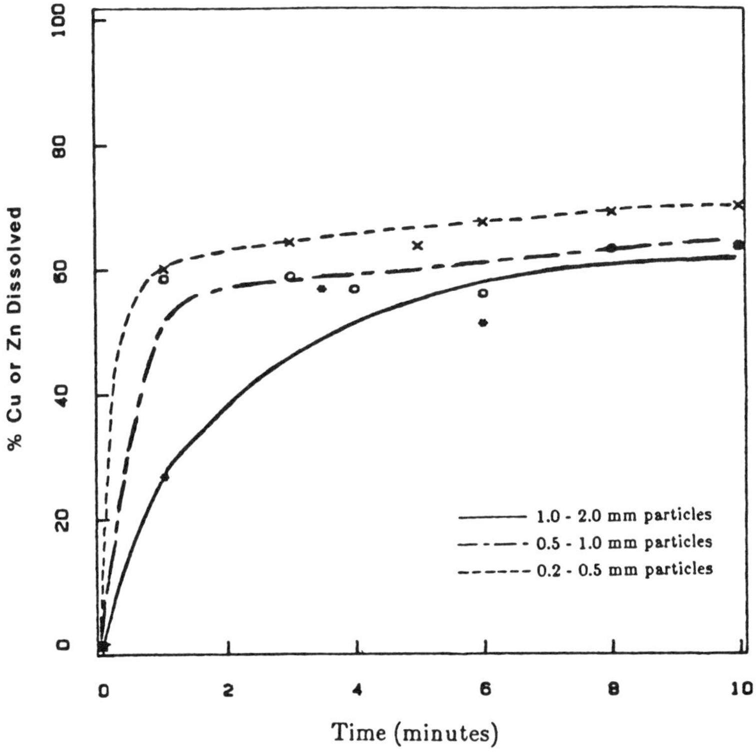

Figure 6. Percentage of Cu or Zn dissolved from blast furnace dust vs. time for several different size fractions of dust. Standard ratio of 20 g dust/100 ml 2 M H_2SO_4 was used.

through spray drying or crystallization. Of the other components in the acidic zinc solution, only Cu, Pb, Ni, and Cd are at a level to be of any concern for the intended application. The more stringent standards are for feed premix where the maximum allowable levels for Pb, Ni, and Cd are 250, 200, and 40 ppm, respectively, while the preferred levels are 100, 50, and 25 ppm, respectively[1].

Since the presence of chloride ions is not desirable, samples of the dust were first water washed to remove chloride and any water soluble metals. Based on chloride analysis of the solids before and after water washing, 76% of the total chloride content of the composite dust sample was removed when washed at 90°C. At 23, 50, and 75°C, the chloride removal was measured to be 73, 82, and 78%, respectively, indicating that there is no advantage to heating the solution for the water washing.

Acid extraction of the composite dust sample was carried out at 90°C using 0.49 ml of concentrated sulfuric acid/g of dust. This is equivalent to mixing the dust with a ~3M H_2SO_4 at a solution volume of 3 ml/g of dust. After 3 hours, ~95% of the Zn had been extracted from the solids, similar to the results presented above for sulfuric acid. The recovered solids consist primarily of $PbSO_4$ (14.9% Pb, dry basis), but also contain Sn (1.7%), Fe (1.7%), Cu (1.4%), Zn (1.2%), Al (0.8%), Sb (0.4%), Si (0.3%), and traces of Ni and Cd.

Purification of the acid extraction filtrate was accomplished through a two step cementation process using first scrap iron, and then zinc dust. Cementation is the reduction of a dissolved metal ion to the elemental state by the action of a more electropositive metal. The cementation of copper with iron is one of the oldest hydrometallurgical reactions. This reaction can be written as:

$$Fe^{o} + Cu^{++} \rightarrow Fe^{++} + Cu^{o}$$

The results of the cementation with scrap iron are given in Table 3.

Table 3. Acid Extraction and Cementation with Iron

	Liquor Concentration (mg/L)	
Element	After Acid Extraction*	After Cementation**
Cu	36300	24
Pb	54	<2
Ni	289	372
Cd	715	762

* Average of two experiments.

** Average of eight experiments.

1. Personal communication with P. B. Queneau based on discussion with premix producers and distributors.

After 4 hours, the Cu and Pb levels are found to be 24 and <2 mg/l, respectively. In fact, these concentrations were reached within the first 60 to 90 minutes. Ni and Cd on the other hand, are not reduced at all, and in most experiments were actually found to rise. Presumably this increase is due to dissolution of these metals from the scrap iron.

As the final purification step, Zn dust is added to the solution after the iron-copper-lead cement is removed by filtration. Previous studies [10,11], have indicated that Ni, Co, and Cd can be effectively removed via cementation by adding an excess of Zn dust (>10 times stoichiometric) along with minor amounts of activators such as antimony, arsenic, copper, and lead. In the present study, the procedure described by Capps and Boxall [10] was employed. They found that Sb, Pb, and Cu aided in the cementation of Ni and Cd. Antimony was added as a Sb_2O_3 slurry in water, equivalent to 6% of the Ni, Pb was added as PbO, equivalent to 1.5% of the zinc dust, and copper is added as a $CuSO_4$ solution, equivalent to the Ni content. The results of the zinc cementation experiments are summarized in Table 4.

Table 4. Zinc Cementation

Final Solution Concentration (mg/L)					
Zn Dust Alone				Zn+Furnace Dust	
Ni	Cd	Ni	Cd	Ni	Cd
19	36	15	55	35	28
88	41	13	119	21	54
105	21	22	51	28	95
75	18	15	70	25	69
45	27	14	106	14	11
73±35	30±10	14±5	70±38	24±7	63±41

In six experiments using a 30 times stoichiometric amount of zinc dust alone, the Ni concentration was reduced from 254±50 to 73±35 mg/l and Cd was reduced from 634±70 to 30±10 mg/l. Somewhat better results are obtained for Ni removal when the Pb, Sb, and Cu reagents are added. The mean of the Cd value is increased to 63 in this case, however, the variability is quite large as evidenced by a standard deviation of 41 mg/l. Equivalent results are obtained when furnace dust is substituted for the Pb, Sb, and Cu reagents. The amount of dust added was chosen to give the equivalent amount of Pb, resulting in an excess of Cu (1.75X) and less Sb (0.25X) than used above. The high variability found in the Cd values may be due to different degrees of oxidation in these relatively small sample volumes. No effort was made in these experiments to reduce oxidation. In spite of this variability, it can be concluded that zinc dust can be used to reduce both Ni and Cd to reasonably low levels.

Process Definition

Based on previous work in the literature and the experimental results of this study, a hydrometallurgical process for treating the copper recycling flue dust is proposed. Initially, the dust is slurried in water to remove chloride and other water-soluble species. The dilute brine solution is sent to a waste treatment facility. The solids are then mixed with sulfuric acid to extract Zn, Cu, and other metals. Filtration is then used to separate the zinc-rich acid filtrate and the solid residue. The solids, composed primarily of $PbSO_4$ can be treated in a number of different ways, including: 1) recycle to the copper blast furnace, 2) treat with HCl to extract Pb, 3) sell to a lead smelter for the value of lead and precious metals, or 4) dispose in a hazardous landfill. The first option is relatively easy to accomplish and is attractive due to the fact that all precious metals are retained in the system. The only drawback is that the increased Pb levels may be detrimental to the operation of the furnace and the lead content of the slag will increase. Alternatively, HCl can be used to extract a portion of the lead before recycling the solids to the furnace. Clearly, the least desirable option, assuming no market can be found for the Pb-rich solids, is to dispose of them as hazardous wastes.

The zinc solution is purified in two cementation steps, first with iron and then with zinc. The solids from the iron cementation step, which contain most of the copper extracted by the sulfuric acid, are recycled to the furnace. The iron that is added to the solution can be removed through oxidation (raise pH and bubble air into solution) or it can be left in solution, producing a final product consisting of zinc and iron sulfate. The solids from the zinc cementation step contain significant amounts of Zn, Ni, and Cd. These can be sold, recycled to the furnace, or disposed. In the final stage, the zinc sulfate (and possibly iron sulfate) is collected by spray drying.

The economics of the proposed process are enhanced by the fact that a large market exists for high quality zinc sulfate. The majority of the zinc sulfate is used for agricultural applications as a fertilizer micronutrient or an animal feed premix. In general, animal feed premixes contain both zinc and iron sulfate along with other compounds. This suggests that iron added during cementation should be left in solution, producing a higher valued product consisting of a mixture of iron sulfate and zinc sulfate. A pilot-scale test of the proposed process has been completed and the results are reported in a separate paper [12].

Acknowledgements

We are grateful to R. Burgess of AT&T Nassau Metals Corporation for providing samples of the furnace dust. We also wish to thank Mr. Ron Burnham (AT&T Nassau Metals) for providing excellent analytical support, AMAX Zinc Co. for providing a sample of zinc dust, and Hazen Research, Inc. (Mr. Nick Hazen, Mr. Al R. Williams, and Mr. Paul B. Queneau) for their guidance and many helpful suggestions.

References

1. Fukubayashi, H. H., and L. W. Higley, "Recovery of Zinc and Lead From Brass Smelter Dust," U. S. Bureau of Mines Report #7880, 1974.
2. Herlitz, H., S. Eriksson, and J. Skogberg, "Getting Metals Out of Steelworks Dust," *MBM,* 67, May 1986.
3. Powell, H. E., H. Fukubayashi, L. W. Higley, and L. L. Smith, "Recovery of Zinc, Copper, and Lead-Tin Mixtures from Brass Smelter Flue Dusts," U. S. Bureau of Mines Report #7637, 1972.
4. Datillo, M., E. R. Cole, Jr., and T. J. O'Keefe, *Conservation and Recycling,* **8,** 399-409, 1985.
5. Datillo, M., E. R. Cole, Jr., T. J. O'Keefe, R. A. Kosher, and F. Chen, "Recycling Zinc From Flue Dust for Electrogalvanizing," *Proceedings of the International Symposium on Recycle and Secondary Recovery of Metals,* 159-172, Fort Lauderdale, FL, December 1985.
6. Mohri, E., and M. Yamada, "Recovery of Metals from the Dusts of Flash Smelting Furnace," *Proceedings of the Joint MMIJ-AIME Meeting,* 520-533, Denver, CO, September 1976.
7. Harris, H. G., R. M. Lattig, D. H. Cranney, W. F. Tucker, and A. U. Blackham, "Hydrometallurgical Production of Zinc Sulfate Monohydrate from By-Product Materials," *Mining Congress J.,* **64,** 43-46, 1978.
8. Jha, M. C., and W. P. C. Duyvesteyn, "A Two-Stage Leaching Process for Selective Recovery of Zinc from Steel Plant Dusts," *Proceedings of the International Symposium on Recycle and Secondary Recovery of Metals,* 143-157, Fort Lauderdale, FL, December 1985.
9. Burrows, W. H., U. S. Patent #3,849,121, November 19, 1974.
10. Capps, R. L., and J. M. Boxall, "Zinc Leach Residue Treatment and Solution Purification at the National Zinc Company," *AIME Annual Meeting,* Los Angeles, February 1984.
11. Adams, R. W., and R. L. Chapman, "New Developments in the Zinc Dust Purification of Zinc Plant Electrolyte," *Aus. I.M.M. Melbourne Branch, Symposium on Extractive Metallurgy,* 169-178, November 1984.
12. M. P. Bohrer, H. H. Law, P. A. O'Hara, R. E. Burgess, P. B. Queneau and N. Hazen, "New Process for Production of a Zn-Fe-Based Animal Micronutrient from Secondary-Copper Flue Dust" (Paper to be presented at the Annual TMS Meeting, San Diego, California, March, 1992.)

3.3
SLAGS, DROSSES AND SLUDGES

AN IMPROVED PYROMETALLURGICAL METHOD FOR THE RECOVERY OF LEAD FROM BATTERY RESIDUE

C.A. Pickles
Dept. of Materials & Metallurgical Engineering
Queen's University, Kingston, Ontario K7L 3N6

and

J.M. Toguri
Dept. of Metallurgy and Materials Science
University of Toronto, Toronto, Ontario M5S 1A4

ABSTRACT

At the present time, pyrometallurgical processes are employed for the recovery of lead from battery scrap. These processes have serious pollution problems, primarily due to the emission of sulphur dioxide and airborne lead in the form of dust and fume. In addition it is necessary to dispose of solid wastes and liquid effluents which contain lead and in some cases sodium sulfide-containing compounds.

In this work an improved process was investigated for the recovery of lead from battery residue. This involved pelletization of the residue with sodium carbonate and moisture. Then, the pellets were melted without carbon and the sodium sulphate salt was removed. Subsequently, the molten lead oxide was reduced with carbon. This process minimizes the sulphur dioxide emissions and leads to very low sulfide contents of the sodium sulfate salt.

INTRODUCTION

At the present time, lead is recovered from lead-acid batteries by pyrometallurgical smelting processes. Usually, the smelting of lead-acid battery residue is performed in two stages. In the first, the drained and de-cased scrap is smelted in a reverberatory or rotary furnace. The charge melts and the major impurities such as antimony are oxidized into a slag phase. The litharge slag which forms contains the majority of the oxidizable impurities (1,2). The lead bullion is relatively pure. In stage two, a lead-antimony-tin alloy is produced by the reduction of the slag. This second stage can either be performed in the same furnace or in a small blast furnace or cupola (3). If the lead content of the final slag is low it can be discarded. This two-stage process generates two products, one of which can be refined to a high purity lead while the alloy product can be modified to produce a marketable antimonial alloy.

These conventional pyrometallurgical processes have a number of environmental problems which can be attributed to the presence of the lead sulphate in the battery residue. If the residue is smelted in the presence of carbon, sulphur dioxide will be liberated. In addition, since the lead oxide is in finely divided form, lead-containing particulates can be

Residues and Effluents - Processing and Environmental Considerations
Edited by Ramana G. Reddy, William P. Imrie and Paul B. Queneau
The Minerals, Metals & Materials Society, 1991

released. Capture of the dust, fume, and gas is both difficult and expensive. Furthermore, slag disposal is a major problem. The sulphur dioxide emissions can be reduced by adding soda ash (Na_2CO_3), for example in the rotary kiln smelting of battery residue (4). However, it is difficult to control the reducing potential in the rotary kiln and thus the final slag contains sulphides and thiosulphates. This type of slag requires special disposal techniques. Another alternative is to pelletize the residue with moisture and sodium carbonate and then smelt the pellets in the presence of carbon (5). Even in this case, it is difficult to control the reducing potential in the system and the slag tends to be high in sulphide sulphur (S^{2-}).

A number of hydrometallurgical processes have been proposed for the treatment of battery residue but have achieved only limited success (6-12). The major disadvantages of leaching the residue are: there are many steps in the process, there are environmental problems associated with the treatment of the waste liquor (e.g. the Na_2SO_4 solution), and the lead carbonate must still be dried and reduced at high temperatures to metallic lead.

The kinetics of conversion of lead sulphate to lead carbonate in a sodium carbonate aqueous solution has recently been reported (13). The exchange reaction

$$PbSO_4 + Na_2CO_3 = PbCO_3 + Na_2SO_4$$

occurred rapidly at room temperature and the lead carbonate recovery was ninety-eight percent. Thus, when the battery residue is pelletized with sodium carbonate in the presence of moisture, the desulphation reaction will proceed essentially to completion.

In this paper, an improved pyrometallurgical process for the recovery of lead from battery residue is described. A flowsheet of the proposed process is shown in Figure 1. The battery residue is mixed with sodium carbonate and moisture, pelletized and sintered. Then, the pellets are melted and molten lead oxide and sodium sulphate are produced. The sodium sulphate is removed and the lead oxide is reduced with carbon. This process has the following advantages over the conventional pyrometallurgical processes:

(1) reduced emissions of lead and lead oxide particulates,
(2) the sulphur dioxide is fixed in a stable form in the sodium sulphate slag,
(3) the sulphide content (S^{2-}) of the sulphate slag is low.

EXPERIMENTAL

Raw Materials

The analysis of the battery residue as-received is shown in Table I. This material was ground in a ball mill and the metallic lead was separated by sieving the ball mill product at 28 mesh (Tyler). The minus 28 mesh material contained 5.92 mass percent metallic lead. The sodium carbonate was minus 200 mesh and was Fisher reagent-grade. Fisher activated charcoal in the size range of 6 to 14 mesh was employed as the reducing agent.

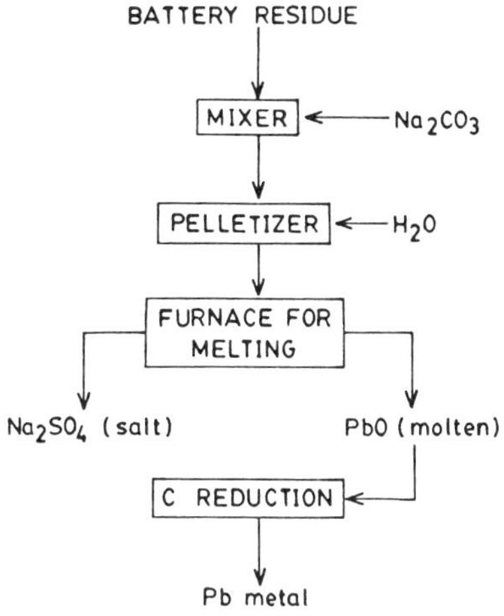

Figure 1: Flowsheet of the proposed process for the recovery of lead from lead-acid battery residue.

	Battery Residue	Minus 28 Mesh Material	Plus 28 Mesh Material
Total Pb	--	71.67	90.39
Metallic Pb	32	5.92	78.60
$PbSO_4$	36	61.38	11.64
PbO	3	--	--
PbO_2	17	--	--
Al	--	0.02	0.05
Sb	--	2.80	0.43
As	--	0.51	0.25
Bi	--	0.12	0.009
Cd	--	0.17	0.002
Cu	--	0.11	0.22
Fe	--	0.08	0.63
Ni	--	0.005	0.006
Ag	--	0.001	0.0007
Na	--	0.01	0.06
Sn	--	0.10	0.35
Zn	--	0.16	0.59
Total S	--	6.53	1.45
Sulphide S	--	0.045	0.22

TABLE I: Analyses of Battery Residue in mass %, Minus 28 Mesh (Tyler) Material and Plus 28 Mesh (Tyler) Material.

Small-Scale Pellet Melting Experiments

The residue was mixed with sodium carbonate and pelletized in the presence of moisture. These pellets were either air dried or sintered and subsequently one hundred grams of the pellets were melted at $1000^{o}C$ in a fireclay crucible in a resistance furnace. The sinter temperature was varied between $25^{o}C$ and $800^{o}C$. The lead and/or lead oxide was separated from the sodium sulphate and the lead and/or lead oxide recovery was determined. The lead metal, lead oxide, and sodium sulphate phases were analysed for antimony, tin and arsenic. The sodium sulphate salt was analysed for lead. For comparison purposes one series of experiments was performed in which carbon was added. In this case the sodium sulphate was analysed for lead and sulphide sulphur (S^{2-}).

Large-Scale Pellet Melting Experiments

A series of large scale pellet melting tests were performed in order to produce significant amounts of lead oxide for the lead oxide reduction experiments. The conditions employed in these tests are summarized in Table II. Tests were performed with both battery residue and pure lead sulphate. The melting experiments were performed in a gas-fired muffle furnace at $1000^{o}C$ in a fireclay crucible. In some tests, samples of molten sodium sulphate were taken with an iron rod. The first sample was taken once the molten salt had formed. Further samples were taken every five minutes. The samples of the sodium sulphate were analysed for total lead content. After melting was completed the crucible was slowly cooled. Then the crucible was broken open and the sodium sulphate was carefully separated from the lead oxide. The sodium sulphate was white in appearance. The lead oxide was crushed to minus 28 mesh before being employed in the reduction experiments.

Large-Scale Pellet Smelting Experiments

In these tests, lead dioxide (PbO_2), sodium sulphate and charcoal were reacted. The purpose of these tests was to investigate the conditions which would be required to avoid matte formation. The reactants were mixed and then smelted at $1000^{o}C$ in a gas-fired muffle furnace in a fireclay crucible. After cooling in the furnace, the crucible was broken up and the lead metal and the sodium sulphate phases were separated. In most experiments an intermediate phase was formed between the sodium sulphate and the lead. At carbon contents below the stoichiometric requirement for the reduction of lead oxide, the intermediate layer was unreduced lead oxide. For carbon contents above the stoichiometric value, the intermediate phase was a matte which resulted from the reduction of metal sulphate. Samples of the three phases were taken for analysis.

Small-Scale Reduction of Lead Oxide Experiments

In order to test the technical feasibility of the proposed process, the lead oxide from the large-scale pellet melting experiments was reduced with carbon in a fireclay crucible at $1070^{o}C$ in a resistance furnace. Fifty grams of lead oxide were employed in each test. The following variables were studied: reduction temperature, time at temperature and carbon addition. Also, the effect of a flux cover, e.g. Na_2SO_4 was investigated. The effect of these variables was quantified by determining the lead recovery. For comparison purposes, tests were also performed with pure reagent grade lead oxide.

Material	Material Mass	Na_2CO_3 Mass	Initial Mass	Mass After Mixing	Mass After Pellet-izing	Mass After Drying	Mass After Sinter-ing	Drying Temp.	Sintering Temp.	Pb Mass	PbO Mass	Na_2SO_4 Mass
	g	g	g	g	g	g	g	^oC	^oC	g	g	g
$PbSO_4$	1840	700	2540	2531	2973	2533	2212	200	600	15	1184	845
$PbSO_4$	1840	700	2540	2539	2871	2529	2268	204	608	0	1114	827
$PbSO_4$	1840	700	2540	2535	2861	2527	----	25	---	14	1172	834
Battery Residue	3000	700	3700	3697	3998	3678	----	25	---	928	1253	863
Battery Residue	3000	700	3700	3699	3998	3682	3283	194	608	0	2434	876

Table II: Experimental Conditions for the Large Scale Pellet Melting Experiments.

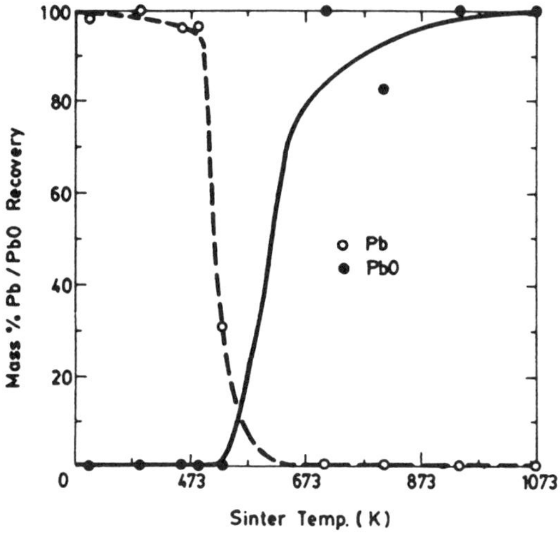

Figure 2: The lead and lead oxide recoveries as a function of sintering temperature. No carbon was added.

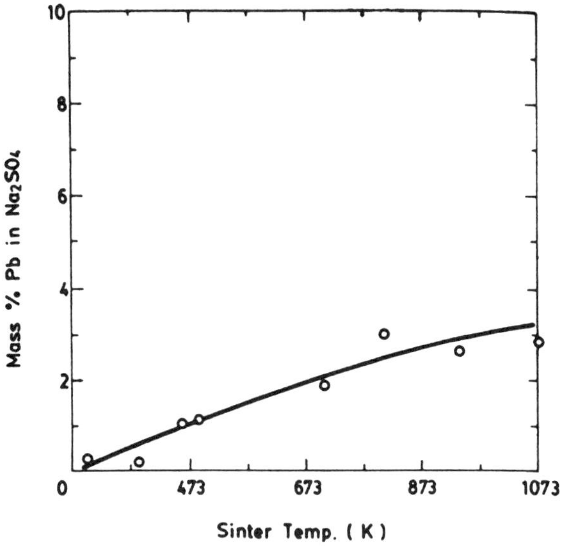

Figure 3: The effect of sintering temperature on the lead content of the sodium sulphate salt from the smelting tests without carbon addition.

RESULTS AND DISCUSSION

Small-Scale Melting Experiments

In these tests, about one hundred grams of the pellets, containing battery residue and sodium carbonate were melted in a fireclay crucible at 1000°C. The sintering temperature of the pellets was varied and since the pellets contained some carbon in the form of organics from the battery casing and separators, the reducing potential in the system was variable. Figure 2 shows the effect of sintering temperature on the lead and lead oxide recoveries. For sintering temperatures below about 250°C the metal recovery was over ninety-five percent while the lead oxide recovery was close to zero. Above this temperature the metal recovery decreased rapidly to zero at about 400°C. This is attributed to the loss of the organics in the residue at sintering temperatures over 200°C. At temperatures above 400°C the lead was recovered as lead oxide with recoveries approaching one hundred percent.

The lead content of the sodium sulphate salt is shown in Figure 3. It can be seen that it increases from very low levels below 200°C to about two or three mass percent at temperatures above 600°C. At low sintering temperatures the reducing potential is high and thus mainly metallic lead is formed. As the sintering temperature increases more lead oxide is formed and this can become entrapped in the sodium sulphate as either lead oxide or more likely lead silicate. The lead silicate will form by reaction of the lead oxide with the fireclay crucible.

The antimony, tin, and arsenic contents in the lead metal, lead oxide, and sodium sulphate salt are shown in Figures 4 to 6, respectively. All three impurities exhibited similar trends. The impurity content of the lead oxide phase increased with sintering temperature. On the other hand, in the lead metal and the sodium sulphate salt phases the impurity content decreased with increasing sintering temperature. Thus, if the pellets are sintered at temperatures above about 400°C the impurities can be concentrated in the lead oxide phase and the sodium sulphate would be relatively clean.

For comparison purposes a series of tests were performed in which carbon was added to the pellets. In these tests the sulphide sulphur (S^{2-}) and lead contents of the salt were determined and are shown in Figure 7. It can be seen that the presence of carbon results in high sulphide sulphur (S^{2-}) contents of the salt. Also, as the carbon content increases, more lead is entrapped in the salt since lead matte begins to form. On a small laboratory scale, the amount of matte is not sufficient to produce a separate layer and remains entrapped in the salt. It is clear from this Figure that if the pellets are simply melted in the absence of carbon then both very low sulphide sulphur (S^{2-}) and lead contents of the salt can be obtained. Without carbon additions the sulphide sulphur content (S^{2-}) of the sodium sulphate was less than 0.01 mass percent.

Large Scale Pellet Melting Experiments

Of particular importance in the proposed process is the lead concentration in the molten sodium sulphate produced in the melting operation. The lead in the salt not only reduces the lead recovery but it is also a contaminant in the sodium sulphate. Thus, the lead content of the salt must be minimized. Figure 8 shows the lead content of the salt as a function of time after salt formation has occurred. Results are

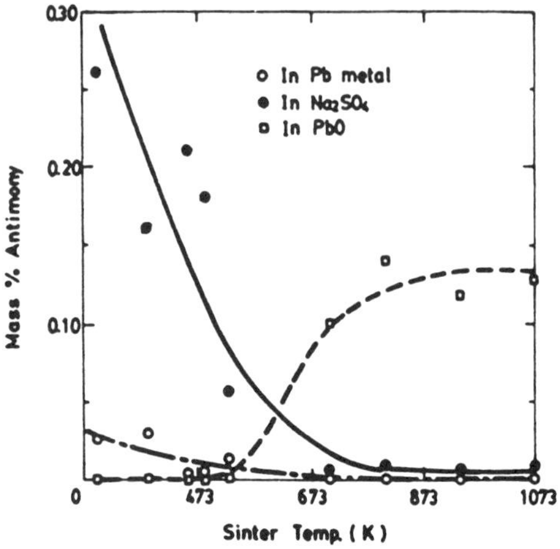

Figure 4: The antimony contents of lead metal, lead oxide, and sodium sulphate from smelting tests without carbon addition.

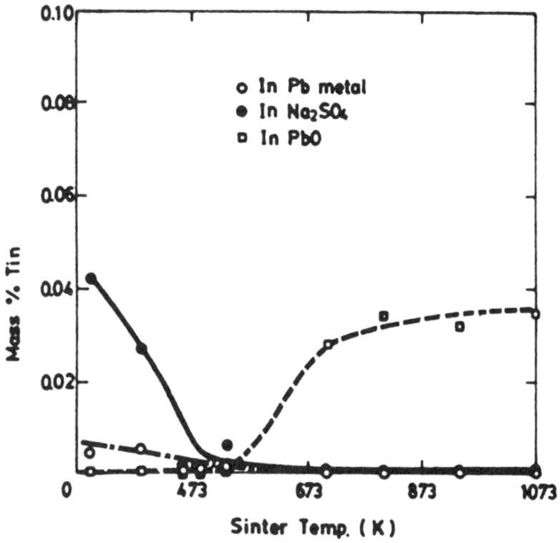

Figure 5: The tin contents of lead metal, lead oxide and sodium sulphate from smelting tests without carbon addition.

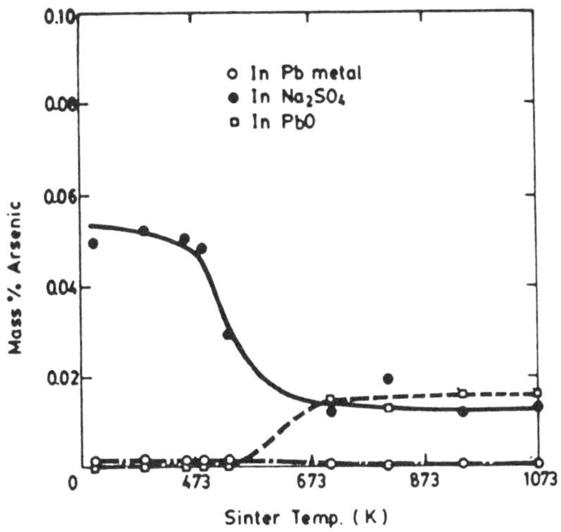

Figure 6: The arsenic contents of lead metal, lead oxide and sodium sulphate from smelting tests without carbon addition.

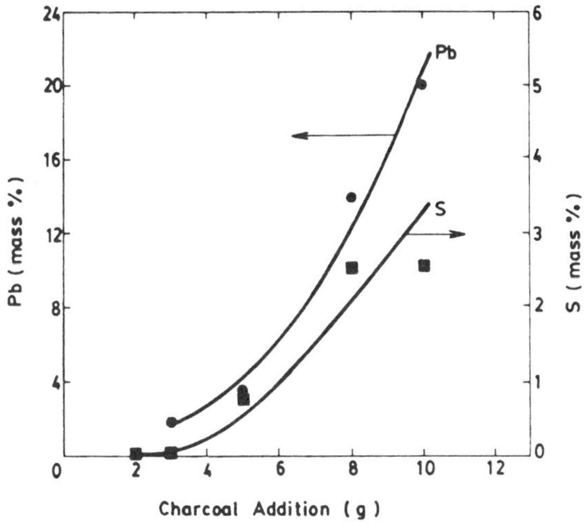

Figure 7: The effect of carbon additions on the lead and sulphur contents of the salt.

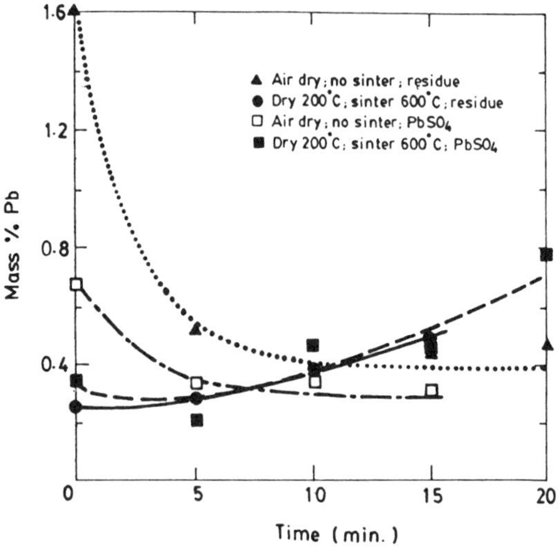

Figure 8: The lead content of the salt as a function of time after molten sodium sulphate formation.

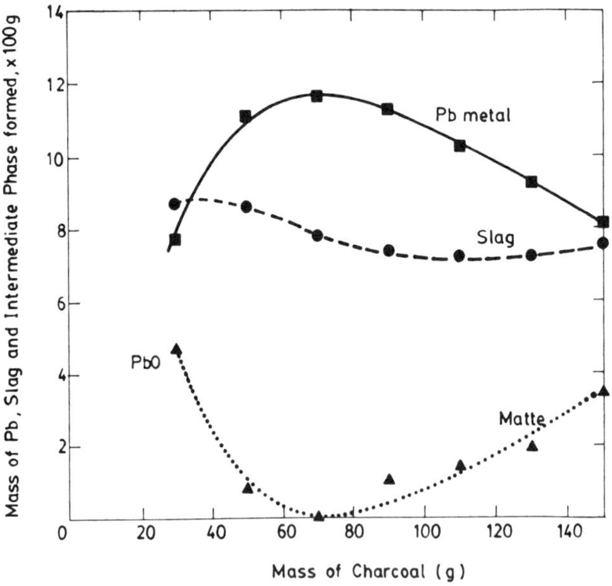

Figure 9: The effect of carbon on the formation of various phases during the smelting of lead dioxide and sodium sulphate.

presented for both pure lead sulphate and battery residue for pellets which were dried in air or were sintered at 600°C. It should be pointed out that the battery residue contains some organics from the plastic components in the battery which will result in the production of some lead during the melting operation. Sintering of the pellets of battery residue at 600°C prior to melting results in the complete removal of these organics and thus no metallic lead is produced. It can be seen that for both battery residue and pure lead sulphate, lower lead contents are achieved more rapidly if the materials are sintered. This could be a result of the formation of some liquid during sintering and thus some separation of the lead sulphate has occurred before the melting operation. The formation of liquid during sintering was observed in a previous publication where it was shown that the strength of the pellets began to increase at about 600°C (5). It is clear that once liquid sodium sulphate forms there is rapid separation of the lead oxide due to the large difference in density of the salt and the oxide. Lead contents in the salt of 0.2 to 0.5 mass percent could be readily achieved. Thus, it would be expected that the equilibrium solubility of lead oxide in sodium sulphate would be less than 0.5 mass percent. In comparison to industrial slags these lead contents would be considered to be very low (14) and are in fact lower than those observed in previous work on the smelting of battery residue in the presence of sodium carbonate and carbon (5). Also, it can be noted that the lead content of the salt increases with time after molten salt formation. This can be attributed to the fact that the experiments were performed in a fireclay crucible and some lead-silicate slag can form, which due to its relatively low density becomes entrapped in the salt. Thus, if fireclay refractories were to be employed it would be necessary to minimize the contact time of the lead oxide with the fireclay. This could be achieved by melting the pellets and allowing only a few minutes for separation before removing the sodium sulphate and reducing the lead oxide to lead. Chemical analysis was performed on the lead oxide and on average the lead oxide both from battery residue and pure lead sulphate contained 2.66 mass percent alumina, 7.42 mass percent silica and 0.22 mass percent iron oxide. Presumably, these impurities came from the fireclay crucible.

Large-Scale Smelting Experiments

The weights of lead metal, salt and the intermediate phase (PbO or matte) are shown in Figure 9. It can be seen that the maximum lead recovery occurs at about 70 grams of charcoal which corresponds to the stoichiometric requirement for the reduction of the lead oxide. Below the stoichiometric value, the lead recovery decreases since there is unreduced lead oxide. Above the stoichiometric requirement, the lead recovery is reduced due to the formation of matte. It is postulated that matte formation occurs according to the following reaction:

$$Na_2SO_4 + PbO + 3C = Na_2CO_3 + PbS + 2CO$$

The lead content, total sulphur content and the sulphide sulphur (S^{2-}) content of the sodium sulphate phase are shown as a function of charcoal addition in Figures 10, 11 and 12 respectively. The minimum lead content of about 0.5 percent lead occurs at the stoichiometric charcoal addition of about 70 grams. Below this value the unreduced lead oxide becomes entrapped in the salt. Above 70 grams of charcoal, matte formation occurs and thus the lead content again increases. The total sulphur content of the salt decreases with increasing charcoal addition as the salt is reduced and matte is produced. On the other hand, as conditions become more reducing, the sulphur (S^{2-}) content of the salt increases.

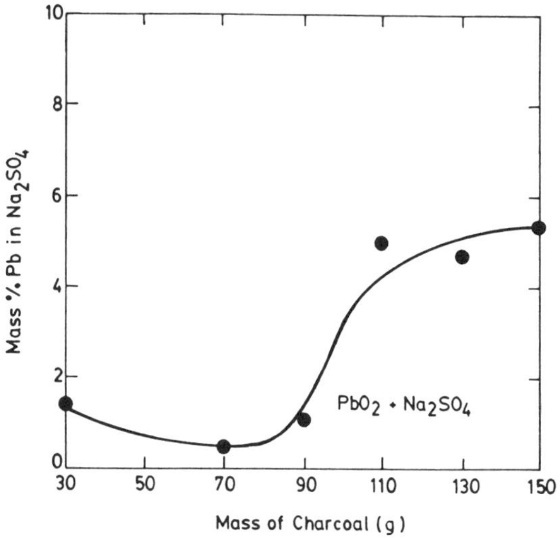

Figure 10: The lead content of the Na_2SO_4 phase as a function of charcoal addition.

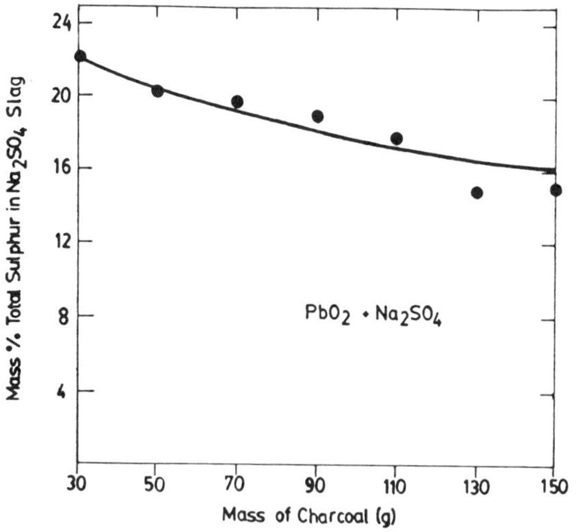

Figure 11: The sulphur content of the Na_2SO_4 phase as a function of charcoal addition.

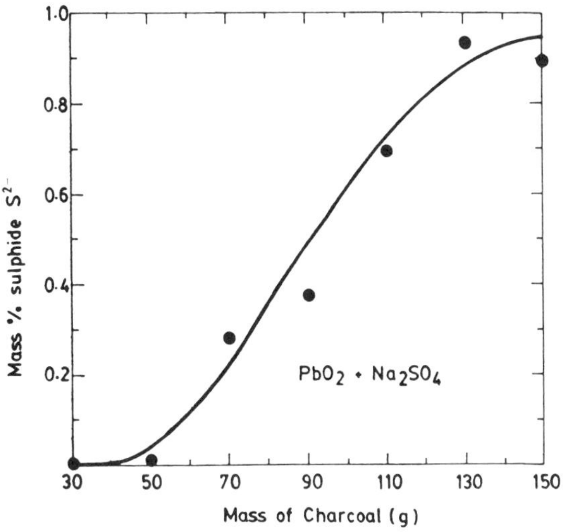

Figure 12: The sulphide sulphur content of the Na_2SO_4 phase as a function of charcoal addition.

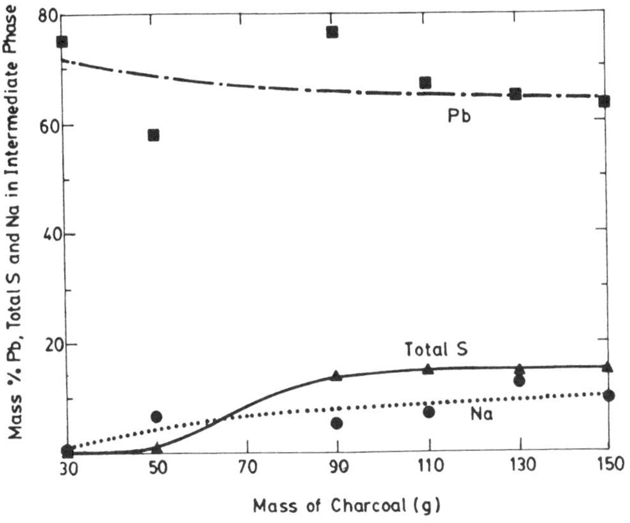

Figure 13: The lead, sodium and total sulphur content of the intermediate phases as a function of charcoal addition.

The lead, sodium and total sulphur contents of the intermediate phases (PbO or matte) are shown as a function of charcoal addition in Figure 13. The lead content of the intermediate phase remains relatively constant since the amount of lead in either lead oxide or lead sulphide is essentially the same. As the charcoal addition increases, the reducing potential increases and thus the sodium content of the intermediate phase increases as matte formation occurs. At low charcoal additions, the sodium would be expected to be present in the intermediate phase as entrapped sodium sulphate. In addition, as the reducing potential increases the total sulphur content of the intermediate phase increases and reaches about fifteen percent at high carbon additions. From the above Figures, it can be seen that the approximate composition of the matte is as follows: 65% lead, 7 to 12% sodium, and 15% total sulphur.

The antimony and sulphide sulphur contents of the lead metal are shown in Figure 14. It is clear that the antimony content of the metal initially increases with carbon and reaches a maximum at about the stoichiometric carbon addition. At higher carbon additions, the antimony content drops to a constant value of 0.002 percent. This can be attributed to the fact that at high carbon additions matte is produced and the antimony forms a sulphide which enters the matte. A similar type of behaviour would be expected for the other impurities which are present in the original battery residue.

The results presented in this section clearly illustrate that the amount of carbon added during the smelting operation is a critical factor. If excess carbon is added then matte formation occurs and the sulfide sulphur (S^{2-}) content of the slag is high. If the carbon addition is below the stoichiometric value then lead oxide forms and again the lead recovery drops. There is relatively narrow range of carbon addition which gives the maximum lead recovery. This effect would be even more pronounced in industrial operations due to the presence of lead sulphate in the feed and also the variability of the feed.

Small-Scale Reduction of Lead Oxide Experiments

Figure 15 shows the effect of carbon additions and the amount of sodium sulphate added, as a flux cover, on the lead recovery. These experiments were performed at 1070°C on the lead oxide from the large scale melting tests on battery residue. About six or seven grams of sodium sulphate were required to obtain recoveries approaching ninety-five percent. The stoichiometric requirement for the reaction

$$PbO + C = Pb + CO$$

for fifty grams of lead oxide is about 2.7 grams. However, as pointed out above, the lead oxide from the battery residue contains about ten mass percent of impurities. This carbon addition gave the maximum recovery of lead. High recoveries were also obtained with a carbon addition of 2.95 grams. At higher carbon additions the lead recovery decreased, presumably due to the formation of matte, which contains both sodium sulphide and lead sulphide. For the same reason, at high carbon additions the recovery decreased with increasing sodium sulphate. Some sodium sulphate is required to act as a flux cover to minimize the loss of lead oxide to the vapour phase. However, both the carbon addition and the sodium sulphate addition have to be selected closely so as to minimize the amount of matte which forms. This matte will be entrapped in the molten salt and thus reduce the lead recovery. The results for pure lead oxide are shown in Figure 16.

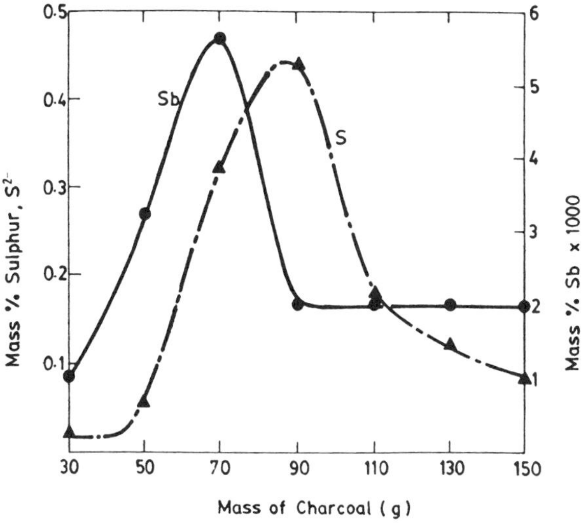

Figure 14: The antimony and sulphide sulphur content in the lead phase as a function of charcoal addition.

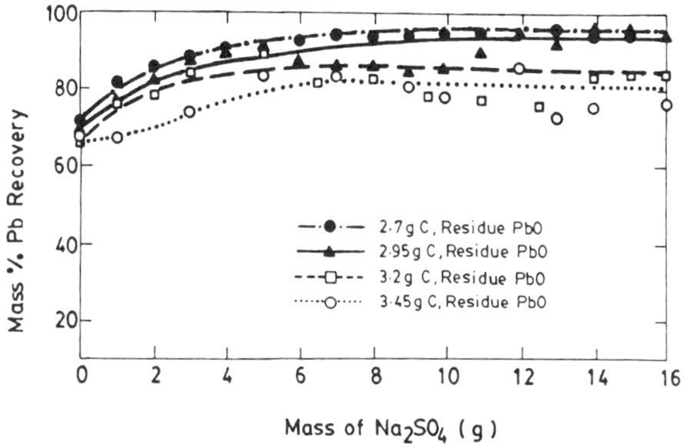

Figure 15: The effect of carbon on the lead recovery at a smelting temperature of 1070°C for lead oxide from battery residue and lead sulphate.

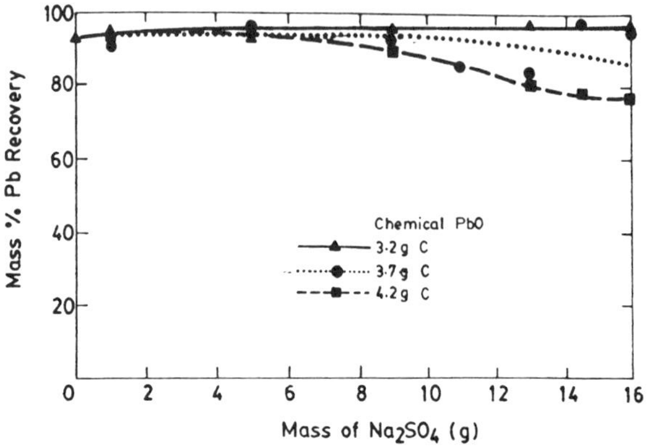

Figure 16: The effect of carbon and sodium sulphate additions on the lead recovery at a smelting temperature of 1070°C for pure lead oxide.

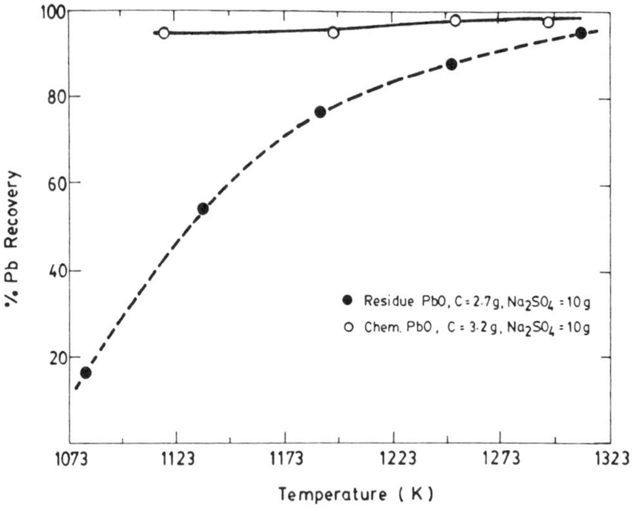

Figure 17: The effect of smelting temperature on the recovery of lead for pure lead oxide and lead oxide from battery residue.

Again, high sodium sulphate additions above about ten grams resulted in decreased recoveries of lead due to matte formation.

The effect of smelting temperature on the recovery of lead from both pure lead oxide and the lead oxide from battery residue is shown in Figure 17. Ten grams of sodium sulphate were added in each case. The lead from the pure lead oxide could be easily recovered at temperatures as low as 850°C. However, for the lead oxide from battery residue, smelting temperatures of nearly 1050°C were required. This can be attributed to the presence of the impurity oxides, such as silica in the lead oxide which decreases its activity.

The effect of reaction time is shown in Figure 18. The furnace sample temperature reached 1050°C after twenty-minutes and then remained constant. Recoveries approaching one-hundred mass percent could be achieved for carbon additions slightly above the stoichiometric value and with ten grams of sodium sulphate. The results in clay-graphite crucibles were marginally higher than in fireclay.

CONCLUSIONS

(1) Pellets of battery residue and sodium carbonate were reacted and melted to produce lead oxide and sodium sulphate. Sulphide sulphur (S^{2-}) levels of less than 0.01 mass percent and lead contents of 0.3 mass percent could be achieved in the sodium sulphate salt phase.

(2) In the absence of a reducing agent the impurities in the battery residue such as antimony and tin will concentrate in the lead oxide. Thus, the sodium sulphate salt phase is relatively pure.

(3) The lead oxide from the melting operation can be reduced with carbon and lead recoveries approaching one hundred percent could be achieved. The carbon addition should be slightly in excess of the stoichiometric and some slag cover, e.g. Na_2SO_4, is required to prevent the vaporization of lead oxide.

ACKNOWLEDGEMENTS

The authors would like to thank the Natural Sciences and Engineering Research Council of Canada (NSERC) for the financial support of this work through the PRAI program. A large amount of the analyses was provided by Canada Metal Co., Toronto, Ontario.

REFERENCES

1. R.D. Prengaman, "Reverberatory-Blast Furnace Smelting at RSR", Lead-Zinc-Tin '80, TMS-AIME World Symposium on Metallurgy and Environmental Control, 1980, pp. 985-1002.
2. S. Bergsoe and N. Gram, "Lead Smelting, Refining and Pollution" Lead-Zinc-Tin '80, TMS-AIME World Symposium on Environmental Control, 1980, pp. 1023-1030.
3. L.A. Lyons and D.B. Gillett, "The New Secondary Lead Smelter of SIMS Metal Pty, Ltd., Brooklyn, Victoria, Australia", Proc. Adv. in Ext. Met., IMM, London, 1977, pp. 105-110.
4. R.C. Egan, M.V. Rao, and K.D. Libsch, "Rotary Kiln Smelting of Secondary Lead", Lead-Zinc-Tin '80, TMS-AIME World Symposium on Metallurgy and Environmental Control, 1980, pp. 953-973.
5. C.A. Pickles and J.M. Toguri, "A Laboratory Investigation of the Soda

Ash Smelting of Lead Acid Battery Residue", Second International Symposium on Recycling of Metals and Engineered Materials, TMS, 1990 pp. 345-360.
6. E.R. Cole, A.Y. Lee and D.L. Paulson, "Update on Recovery of Lead from Scrap Batteries", J. of Metals, Vol. 37, No. 2, 1985, pp. 79-83.
7. M. Kato, M. Imai, K. Arai and T. Izaki, "New Hydrometallurgical Treatment Process of Lead Battery Slime", J. Japan Inst. Metals, Vol. 50, No. 7, (1986), pp. 631-639.
8. D.A. Wilson, "A New Sulfur Dioxide-Free Process for Recovering Lead from Battery Scrap", U.S. Bureau of Mines, R.I. 8123 (1976).
9. F.A. Acoveno and T.A. Fuediger, "Process for the Removal of Sulphur from Battery Wreckers Material Using Ammonium Carbonate Solutions", U.S. Pat. 3383348 (1975).
10. M. Yamada, N. Wakamatsu, and S. Nakano, "Lead Carbonate of High Purity from Crude Lead Sulphate", Japan Tokyo Koho, Japan Pat. 744819 (1974).
11. M.E. Elmore and J.K. Klang, "Process for Recycling Junk Lead-Acid Batteries", U.S. Pat. 4118219 (1978).
12. Diamond Engineering Co., Ltd., "DEC Lead Recovery System from Battery Scrap", (1977).
13. K. Arai and J.M. Toguri, "Leaching of Lead Sulphate in a Sodium Carbonate Solution", Hydrometallurgy, Vol. 12, 1984, pp. 49-59.
14. K.N. Pike, "Secondary Lead Blast Furnace Smelting at East Penn Manufacturing Co., Inc.", Lead-Zinc '90, edited by T.S. Makey and R.D. Prengaman, The Minerals, Metals and Materials Soc., Warrendale PA, 1990 pp. 955-969.

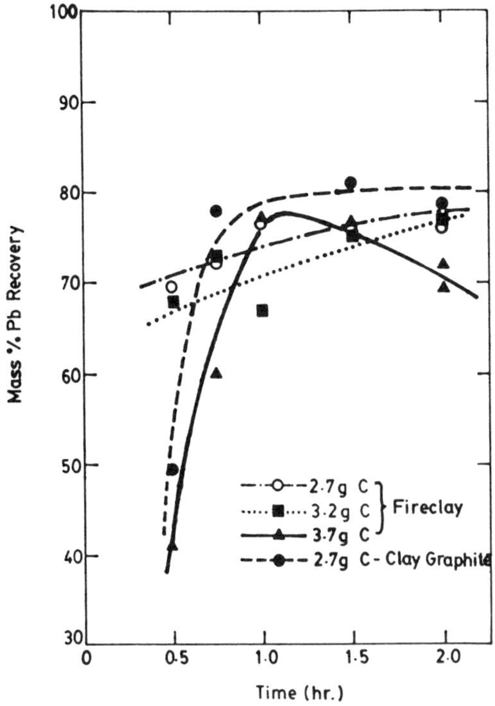

Figure 18: The effect of reaction time on the recovery of lead from lead oxide from battery residue.

SIROSMELT TECHNOLOGY FOR SOLVING

THE LEAD AND ZINC INDUSTRY WASTE PROBLEM

K. R. Robilliard, P. J. King, J. M. Floyd

Ausmelt Pty Ltd
(A.C.N. 005 884 355)
2/13 Kitchen Road
Dandenong 3175, Victoria, Australia

Abstract

An outline is given of the problems facing the lead and zinc industry from the viewpoint of the range of residues currently produced from electrolytic zinc, blast furnace and other processes.

Smelting approaches available using top submerged lancing technology are discussed and analysed. Pilot plant results for a range of lead and zinc-bearing materials are presented.

Residues and Effluents - Processing and Environmental Considerations
Edited by Ramana G. Reddy, William P. Imrie and Paul B. Queneau
The Minerals, Metals & Materials Society, 1991

Introduction

The metallurgical processes involved in the zinc and lead industries must strike a balance between operating cost and recovery of values. It often happens then that in order to keep costs down, recoveries are lessened and process residues and waste materials contain valuable (and often toxic) metal components. Process residues and waste residues which cannot be treated economically are often dumped or stockpiled within the confines of the metallurgical plant.

Table I summarises these residues and waste materials and their main components.

Table I Typical Analyses of Zinc and Lead Wastes

Material	Primary Leach Residue	Jarosite	Goethite	ISF Slag	Pb Blast Furnace Slag	Vertical Retort Residue	Steel Plant Dusts
Zn%	16.9	5	10	6	19.8	3.6	20
Pb%	9.9	5.2	2	0.5	4.1	4.9	3.3
Fe%	28.9	25	40	35.4	12.4	11.2	31.6
S%	12.5	13	4	0.3	0.8	2.2	-
SiO_2%	6.1	3	2.5	17.8	27.2	18.3	-
Cu%	1.1	0.2	1.5	0.1	-	0.8	0.2
Cd ppm	8100	200	500	-	-	-	400
As ppm	1800	600	2000	1000	900	6000	-
Ag ppm	1800	80	80	5	24	100	-
Sb ppm	1700	100	10	10	1200	500	-
C %	0	0	0	0	0	32.2	-

It is clear that significant value is contained in these materials since in many cases they are stockpiled in considerable volume. It is also clear that left untreated, the toxic elements contained in these materials, stored in an unstable condition can create significant environmental problems. This is currently being addressed by legislation in many countries.

Ausmelt, using its 500kg pilot plant facilities in Victoria, Australia, has been active in developing processes using top entry submerged lancing technology to treat these materials [(1)(2)]. These processes have had two principle aims:

- to recover valuable and toxic metals to a fume dust phase.
- to provide an environmentally acceptable slag phase for discard or sale.

In general the process identified for the majority of the materials involves a one or two stage smelting process.

Stage 1: This involves feeding the materials to the submerged land furnace and smelting the materials at temperatures of 1250-1400° C under reducing conditions. The lance is fired with either oil, coal or natural gas at a variety of air to fuel stoichiometries and with lump coal fed with the waste materials to ensure reducing stoichiometries.

Stage 2: In order to increase metal recoveries (from 80 to 98% in Stage 1) to above 98%, a second stage is required. This involves the reduction of the molten slag produced in Stage 1, using the submerged lance as above and lump coal or low air-to-fuel lance flame stoichiometries.

The choice of a one- or two-stage process will depend on the nature of the material to be treated, but in some instances the slag produced from a single stage operation satisfies the environmental guidelines of some countries.

Smelting Waste

Jarosite and Other Zinc Leach Residues

A zinc calcine leaching circuit, forms an integral part of a roast-leach-electrolytic plant. This involves the use of neutral leaching of the calcined concentrate, followed by hot acid leaching of the residue, as required to improve zinc recovery[3]. The residue from such a circuit contains significant levels of zinc, and other metals and this material is stored in residue ponds.

Decades of operation of this type have resulted in millions of tonnes of leach residue in storage ponds in plants around the world. Because the residues contain a variety of metals, they have become of environmental concern and the respective zinc plants are now under some pressure to initiate strategies to not only refrain from producing these waste materials, but also to treat the existing stockpiles. A typical analyses of primary leach residue, jarosite and goethite residues are shown in Table I.

In addition there are further metal values which can be recovered such as gallium, germanium and indium.

Sirosmelt Treatment

Sirosmelt technology [1][2] involves a top entry submerged lance operation. The furnace is cylindrical in shape and it contains feed port, gas offtake and slag handling facilities. A schematic diagram of the furnace is shown in Figure 1.

The tip of the lance is protected by a layer of slag (applied to the lance before submersion) and resides 50 to 300mm below the static slag surface. Fuel (oil, fine coal or natural or liquified gas), air and/or oxygen is injected into the slag via the lance. The combustion zone, adjacent to the lance ensures good mixing of the bath and incoming feed materials (residue and lump coal). Table II summarises the features of the Sirosmelt furnace system.

Sirosmelt offers an economical vehicle for treating these stockpiled waste materials to not only recover the metal values, but to also remove the toxic metals from the materials to produce a slag product which is environmentally benign for discard.

AUSMELT

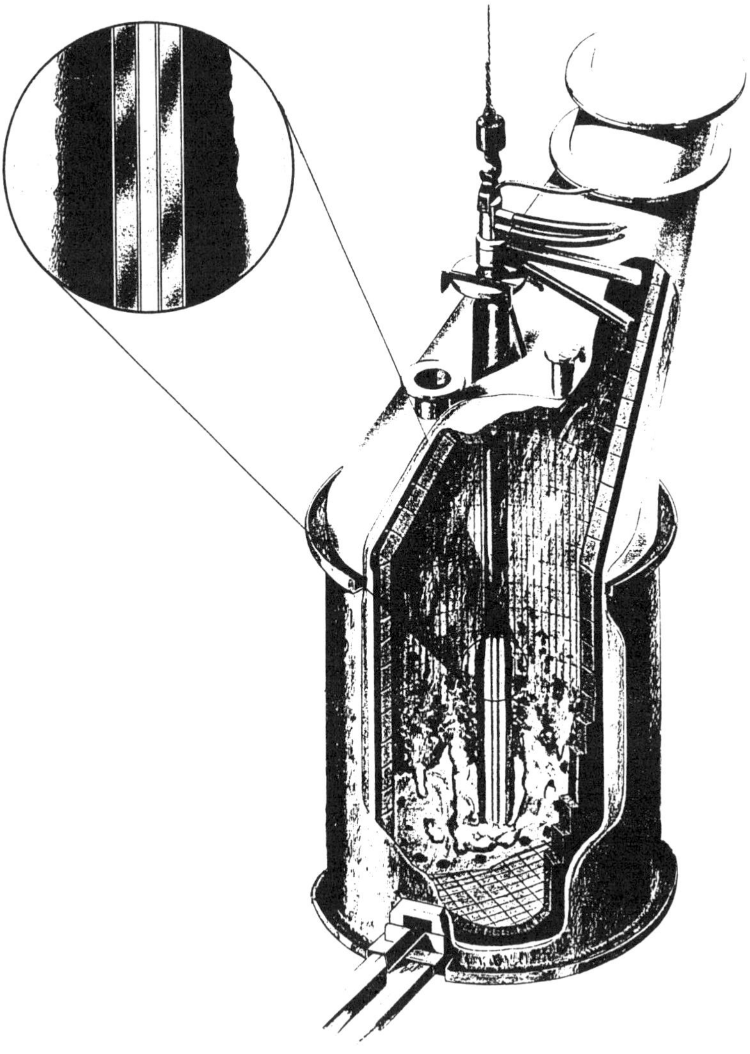

Figure 1 - Schematic Cutaway View of a Sirosmelt Furnace

Based on pilot plant tests, leach residue can be treated in a one-stage or a two-stage operation depending on the zinc content in feed commercial plants are under consideration for smelting rates of as high as 200,000 tpa. The first stage of the operation involves smelting the residue under reducing conditions to produce a slag containing 0.5 to 10% Zn and 0.1 to 5% Pb depending on the feed material in question. The coal, oil or gas-fired submerged lance allows a suitable mechanism for gas-flushing of the bath whereby the volatile leach residue components (zinc, lead, cadmium, bismuth, arsenic and silver) report the gas phase. These metals are reoxidised by afterburning air injected through Ausmelt's patented lance system. The oxide fume is captured in either a standard bag filter or electrostatic precipitator.

Table II Key Features of the Sirosmelt Furnace System

1.	Enclosed refractory-lined furnace facility with top submerged lance technology using slag coated lance cooled by injected air.
2.	Efficient heat and mass transfer from combustion of injected fuel, air/oxygen within a slag bath while feeding materials into the bath.
3.	Low energy cost and small furnace system result from fuel efficiency and rapid smelting rates.
4.	Feed preparation is minimal and feed preparation and handling is inexpensive. (wet or coarse feed suitable).
5.	Any fuel is suitable (Coals, Natural Gas, LPG, Liquid Oil, Heavy Oil, Carbon in feed, Iron Sulphide in feed).
6.	Small carry over of feed material into flue lines results in undiluted volatilised products in fume.
7.	Efficient operation on small scale as well as a large scale.
8.	Multiple reaction zones in the furnace allows for efficient operations.
9.	Oxidising neutral or reducing conditions can be used on call by operator.
10.	Metal or matte phases can be used as collecting agents by control of conditions or by reagent addition.
11.	Discard slags are environmentally acceptable.
12.	Emissions from the furnace are avoided due to small size and tight sealing of ports.

The schematic flowsheets of Figure 2 and Figure 3 illustrate the one and two-stage processes.

The second stage of operation in the Sirosmelt system involves the final reduction of the smelting stage slag, either in the same furnace, or in a second furnace, to produce

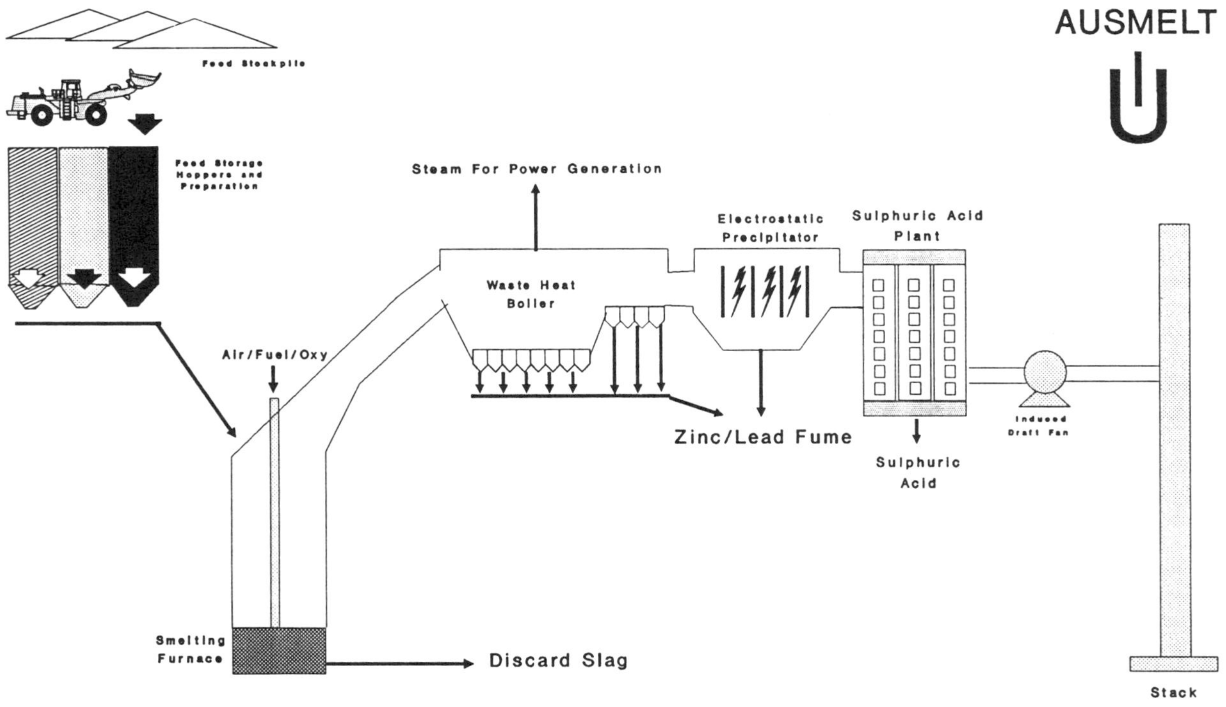

Figure 2: Single Stage Residue Treatment

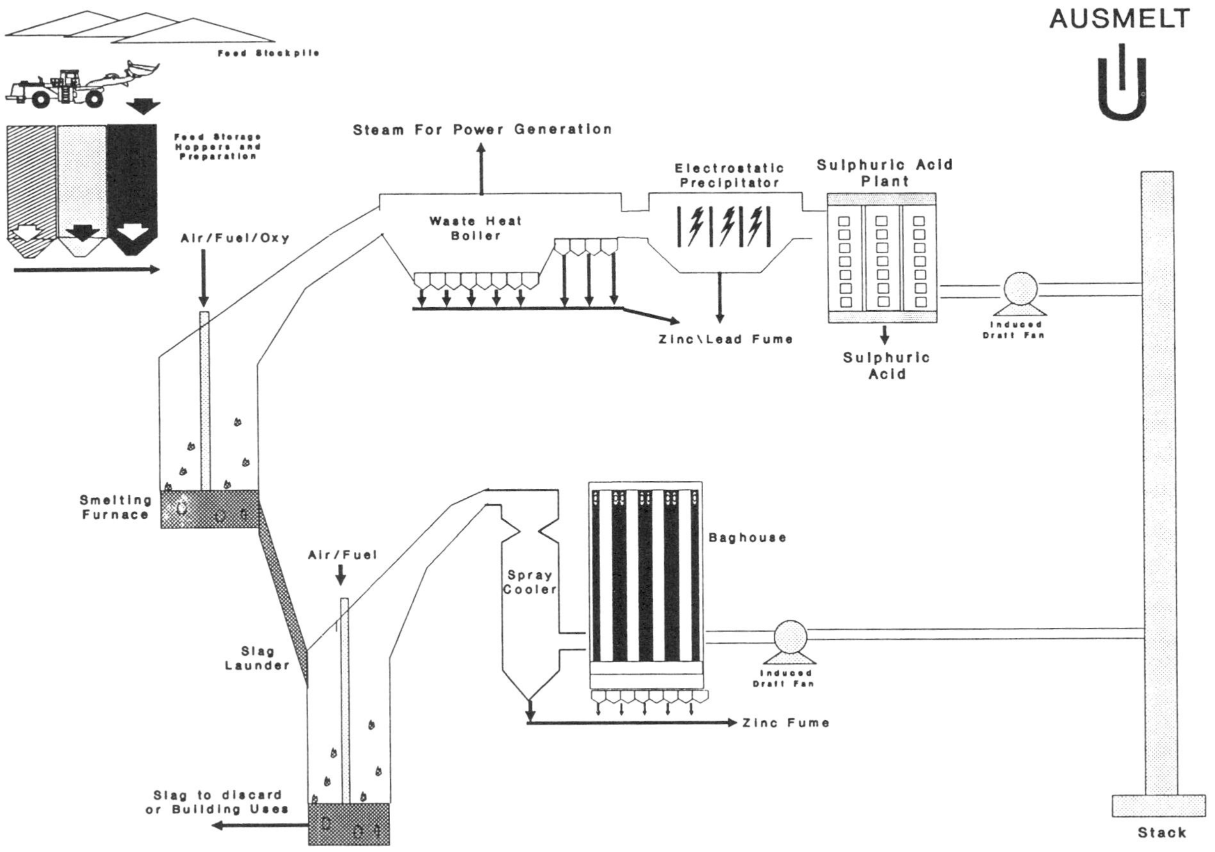

Figure 3: Two Stage Residue Treatment

a discard slag containing low levels of metallic components. Slag products containing 0.5% Zn, 0.01% Pb and insignificant levels of cadmium, arsenic, bismuth and silver are generally produced in the pilot plant. The reduction stage differs from the smelting stage in that, although the submerged lance operates in a similar way, coal only (and not residue) is fed to the bath. Because of the nature of the operation, the fuel requirements are less in the reduction stage than in the smelting stage. From the reduction stage, fume, containing recovered zinc, lead, silver and other metals is produced. This fume differs from the smelting stage fume generally in its higher zinc content. The fume products, from both stages, may be retreated either in a leach circuit or by direct feed in an ISF system, for eventual recovery of zinc, lead and silver (as well as germanium and indium) as metal. Very little carry-over of feed materials occurs in a Sirosmelt system which is designed and operated correctly, and the iron and silica levels in fume are generally less than 2-3%.

The fume product collected separately from the smelting stage of jarosite has the following composition:

44% Pb, 18% Zn, 2.3% As and 2000 ppm Ag.

It is possible to reduce this fume product directly to produce bullion in a Sirosmelt furnace(4). This involves feeding the material with coal to the submerged lancing furnace (using oil, natural gas or coal as fuel) at temperatures of 1150°C to 1200°C. Bullion metal is tapped and the resultant slag may be fumed to recover its zinc in a second stage.

The fume product from this stage and the fume product from the jarosite reduction stage contain high levels of zinc (up to 65%) and this material is suitable for zinc recovery by leaching or ISP operation.

Slag Dumps

The operation of lead blast furnaces and Imperial Smelting Furnaces in various parts of the world has led to the stockpiling of significant volume of zinc-bearing slags. The compositions of typical slags are as shown in Table I.

Because of the valuable metal contents of these materials, the slags cannot be discarded, and in any case, environmental considerations would preclude this.

The Sirosmelt process has been shown to be able to treat these materials to recover the metal values to fume, producing a slag consisting of essentially iron silicate (and other oxidic components). The slag may be introduced to the furnace in either molten or solid (granulated form). In molten form, where liquid slag is introduced to the Sirosmelt furnace continuously via an inlet weir or batch-fed with ladles, there is the advantage of minimising lance fuel requirements during the fuming operation. In solid form, the granulated slag is introduced from a feed chute at the top of the furnace and brought to the molten state using fuel (oil, natural gas or fine coal) burnt at the tip of the submerged lance with air and, if required, oxygen. The extra fuel requirements when using cold feed allow a greater volume of gas to flush the bath, and this allows a greater part of the volatile species to report to the fume phase.

Steel Plant Dusts

In the operation of remelting scrap in electric arc furnaces, where a quantity of the feed charged consists of galvanised scrap, a by product of the process includes a dust material containing significant levels of zinc and other elements. Again because of the high levels of metal components in this substance, disposal is a severe problem and large stockpiles have accumulated at the premises of steel plants.

Steel plant dusts contain typically the components shown in Table I and also have significant levels of chlorine, sodium and potassium.

Using Sirosmelt technology in a process similar to those described above (i.e. a smelting stage to produce a slag with 3-8% Zn and 0.5 to 1% Pb, followed by a reduction stage which removes the remainder of the metal values from the slag), steel plant flue dust may be treated to recover the metal values and to produce a discardable slag.

Of particular concern, unlike other materials is the presence of chlorine. While recovery of zinc and lead to fume has been demonstrated, the amount of chlorine reporting to fume (nearly 100% deportment to the fumes of both stages) would present problems in recycling this material to conventional metallurgical operations. The processing of the feed and product materials for chlorine removal can be handled in a number of ways such as leaching or roasting of the feeds or products. Successful bench scale tests have been carried out by Ausmelt, although discussion of the results are beyond the scope of this paper.

Vertical Retort Residues

A process which has been used in the past for extraction of zinc is the vertical retort process. Zinc oxide (as calcine or sinter) is treated in the retort with coke at 1100° C to 1200° C. Zinc vapour and carbon monoxide are drawn from the top of the retort to a zinc condenser. Spent feed material which consists of gangue and coke ash are withdrawn from the bottom of the retort, and this forms a residue which contains significant levels of metal components. The residue creates a disposal problem. There are large dumps of these materials in existence at present and former smelter sites.

Vertical retort residues have a typical composition as shown in Table I.

The Sirosmelt process is capable of smelting this material to recover the zinc and to produce a discardable slag product. The process used has particular advantages in dealing with the residue in that the carbon content of the material can be utilised as both fuel and reductant.

In a single stage operation, as illustrated in the schematic flowsheet of Figure 2 zinc and lead levels in slag with less than 0.5 and 0.4% respectively can be obtained, whilst recovering 97% and 98% respectively of these metals to two fume phases.

Ideally the process should satisfy the following criteria:

- to achieve good recoveries of metal to fume phases

- to minimise the amount of coke ash remaining with slag
- to minimise the introduced fuel and reductant requirements
- to maximise sulphur retention in the slag.

In achieving combustion of the carbon it is found that sulphur fixation in the slag cannot be achieved, and the optimisation for the other aspects result in more than 80% of the sulphur being eliminated.

For one particular example the optimisation was achieved by using the following operating parameters:

- using natural gas as fuel in the lance at a low air-to-fuel stoichiometry.
- using further air in the lance at a rate to theoretically burn 75 to 90% of the carbon in the feed residue.
- operating at 1300° C to 1350° C.

This operation requires a suitable system for removal of SO_2 from the flue gases. The sulphur can be mostly fixed in the furnace bath under strongly reducing conditions but removal of coke and minimising fuel requirements cannot be achieved.

Combining Feeds

The versatility of the Sirosmelt process has been demonstrated in the treatment of a combination of the above materials. For instance 500kg batches of feed materials were prepared involving zinc leach residue and steel plant dusts in various ratios and these materials were treated in a two-stage operation in the pilot plant, to produce satisfactory slag materials and high recoveries of metal values to fume.

Using Sirosmelt as an Integral part of an Electrolytic Zinc Plant

The neutral leaching step in the electrolytic circuit produces a solid residue containing zinc ferrite, which, as mentioned previously, is now normally fed to a hot acid leaching process for increased zinc recovery.

By taking this product as feed to a Sirosmelt plant the circuit is simplified and the cost of the hot acid leaching and iron precipitation steps can be saved. The same level of recovery of values is achievable as in the Sirosmelt processing of jarosite and goethite.

If a plant already has facilities for hot acid leaching and iron precipitation, then this plant can, in principle, be used to process the fume produced from the Sirosmelt processing of primary leach residue, in a similar fashion to the reprocessing of the zinc fume produced from jarosite processes as shown in Figure 4. Alternatively it can be freed for processing of other feed sources to increase the leaching capacity of the plant if an expansion is needed.

The circuit for smelting primary leach residue to produce zinc fume for recycle and lead bullion is shown in Figure 5. In order to recover lead and silver from the circuit the Sirosmelt fuming furnace is operated to produce a lead-rich fume which is fed to a separate, smaller Sirosmelt furnace system. This furnace is operated to produce lead bullion, and a zinc-rich slag, which is recycled to the Sirosmelt residue smelting furnace. Alternatively the lead residues or fume can be sent to a separate lead

AUSMELT

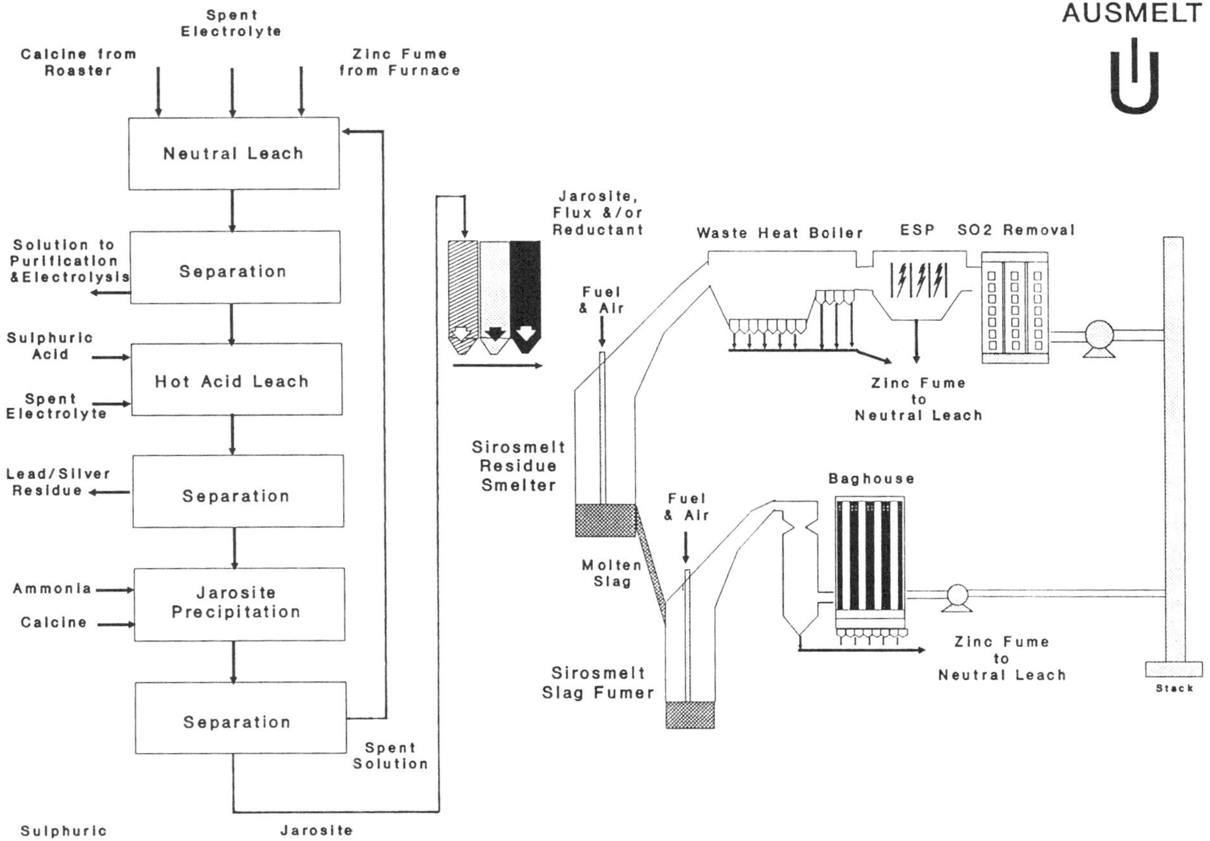

Figure 4: Proposed Ciruit for Sirosmelt Processing of Jarosite

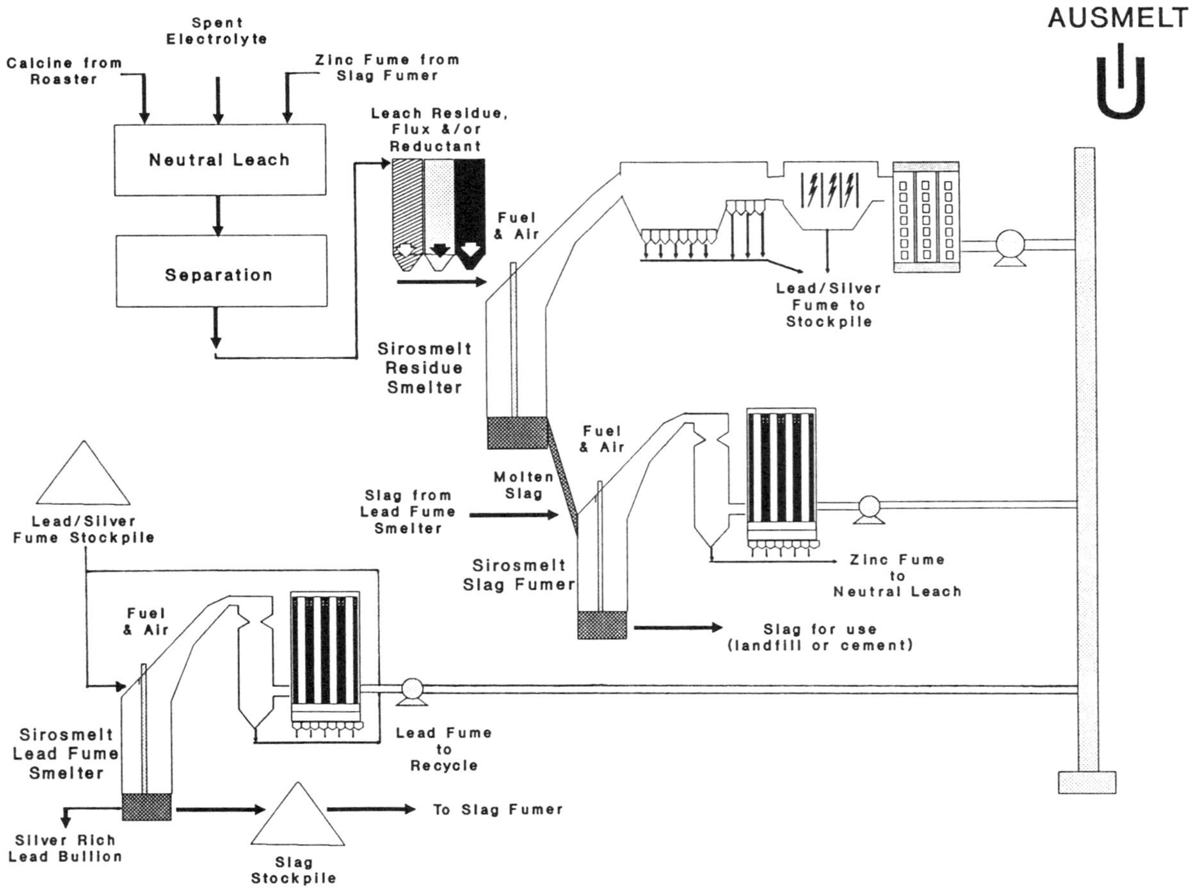

Figure 5: Sirosmelt Processing of Primary Leach Residue

smelter for silver and lead recovery.

Smelting of Ore or Concentrate to Produce Zinc Fume for Feed to an Electrolytic Zinc Plant

The Sirosmelt system can also be used to produce a single fume rich in zinc from high grade zinc concentrate. This fume is then an iron-and sulphur-free-source of zinc for leaching and electrolysis, so that the electrolytic plant does not produce an iron residue.

This circuit is shown in Figure 6.

As mentioned earlier the Sirosmelt process generates very low levels of dust carry over from the feed material into the fume product. When correctly designed and operated the bath acts as a scrubber to prevent dust from being lost to the flue gas stream. Thus fume from smelting zinc-containing materials such as zinc concentrates, vertical retort residues, and steel plant dust will contain very little of the non-volatile elements such as iron, silica and alumina. When this fume is used as the feed for a leaching circuit there is no iron residue produced. The primary leach residue will consist mainly of lead, silver, and other volatile elements which had been concentrated in the fume and which have a low solubility in spent electrolyte. This residue will be a suitable feed for a Sirosmelt plant, and provided the lead level is in excess of 40% it can be smelted directly to bullion, as illustrated in Figure 5.

Plant Design and Economics

Plant Design and Economics - Steel Plant Dust Smelting

In the electric arc furnace steelmaking of scrap carbon steel, a dust is produced, which for this example contains typically 22% Zn, 4% Pb, 33% Fe, 4% SiO_2, 4% Cl and 80 ppm Ag. Steel plant dust has been classified by the U.S. Environmental Protection Agency and similar agencies in many countries as a hazardous waste because of the potential leachability of lead, cadmium and hexavalent chromium in the dust. Therefore, the dust must be converted into a chemically stable and environmentally safe form before disposal.

The Ausmelt furnace system is capable of smelting the steel plant dust to produce a chemically stable slag which can be disposed of safely or used for building materials, and a zinc oxide fume which can be sold for its zinc, lead and silver values. The process has been proved on pilot plant scale and its technical and economic significance has been evaluated.

Smelting using the Ausmelt furnaces does not remove chlorine from zinc. Chlorine can be removed from the steel plant dust before smelting or from the fume products after smelting. The removal of chlorine is not discussed in this paper but the Ausmelt process can be modified to optimise the total capital cost and operating cost of smelting steel plant dust and separation of chlorine from zinc.

The schematic process flow diagram of a Sirosmelt plant for smelting 50,000 tpa of steel plant dust of the above composition and the distribution of major elements is shown in Figure 7.

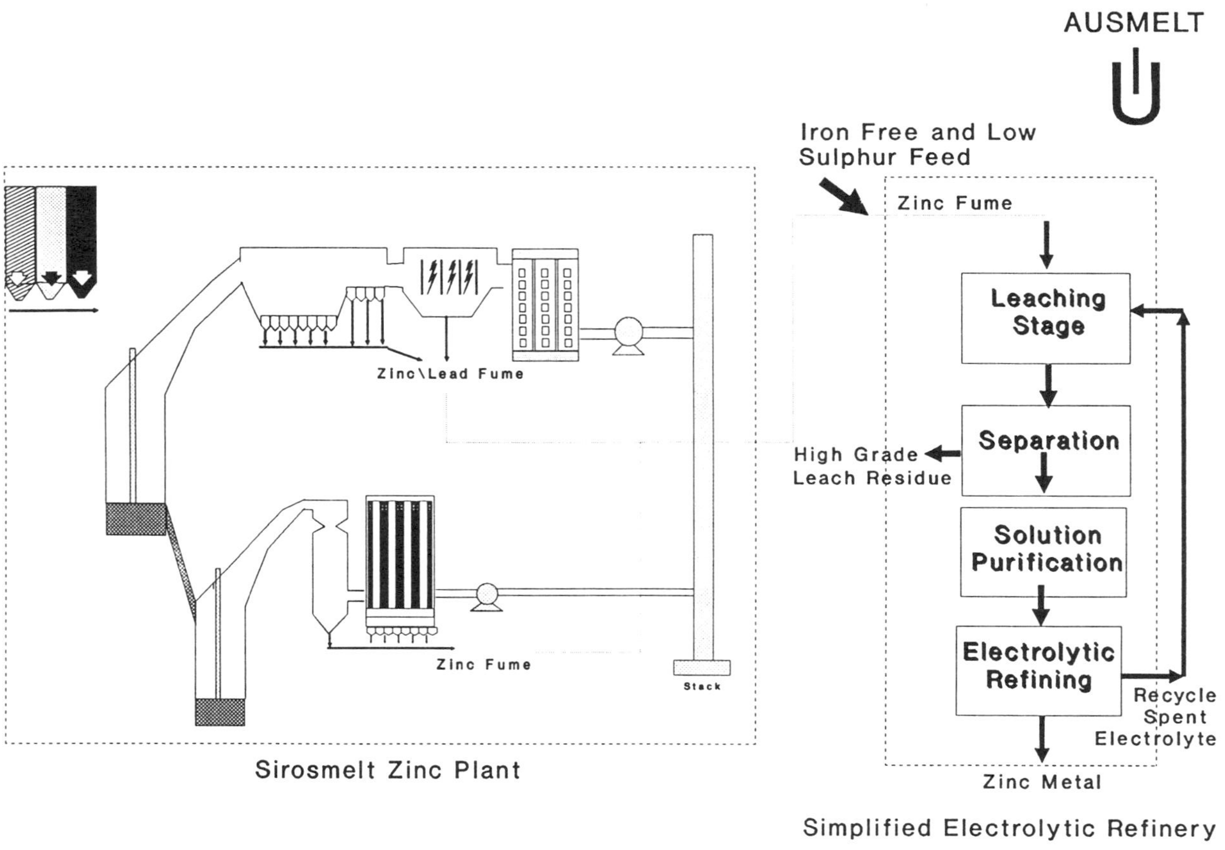

Figure 6: Zinc Concentrate Smelting

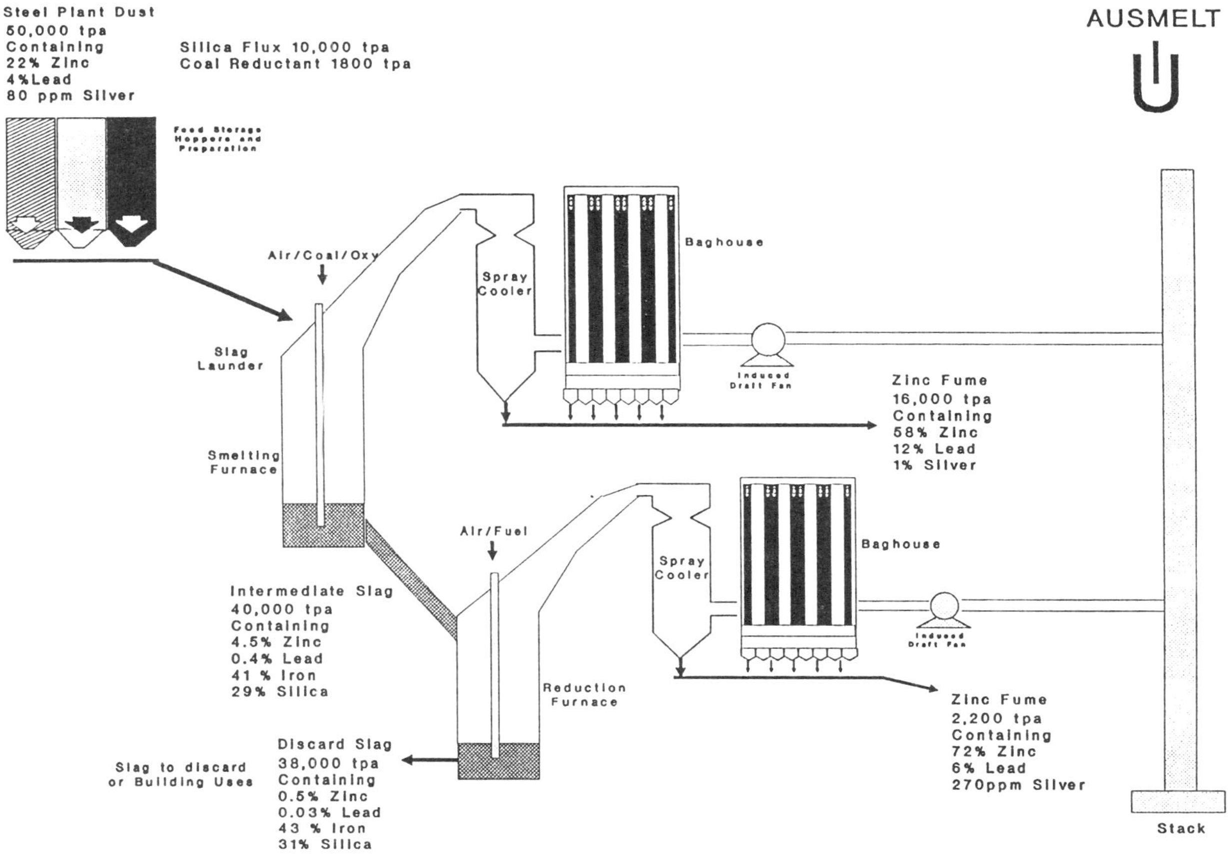

Figure 7: Schematic Process Flowsheet - Steel Plant Dust Treatment

The steel plant dust, silica, sand and lump coal are pugged with water and fed continuously into the first Ausmelt furnace for smelting. Fuel coal and combustion air oxygen enriched to 40% oxygen, are blown into the furnace bath via the lance. The smelting temperature is maintained at 1300° C.

Zinc in the feeding materials is reduced to metal vapour and is re-oxidised in the furnace chamber from afterburning air supplied by Ausmelt's lance system. Lead, silver, arsenic, cadmium, chlorine and some potassium and sodium vaporise also. The furnace gases are drawn to the gas handling train and cooled to 180° C by evaporative cooling. The cooled gases are filtered in a baghouse and discharged to the atmosphere. A first fume product containing typically 58% Zn, 12% Pb and 190 ppm Ag is collected in the baghouse. Depending on the procedure used to handle chlorine the level of this element can be as high as 12%.

A first slag containing typically 4.5% Zn, 0.4% Pb and 25 ppm Ag is transfered continuously from the first furnace via a launder into the second Ausmelt furnace for further zinc fuming from slag.

The second furnace is operated at 1350° C. No oxygen enrichment of the combustion air of the second furnace is required. The handling of the reductant coal, fuel coal, combustion air and furnace gases of the second furnace are similar to those of the first furnace.

A second fume product containing typically 73% Zn, 6% Pb and 270 ppm Ag is produced. The final slag containing 1% Zn or less and 0.03% Pb is tapped continuously from the slag fuming furnace, granulated and discarded.

The total capital cost of the plant including development and other requirement is in the region of $US5 to $US10 million depending on the local site conditions and costs and on whether oxygen enrichment is used. The total cost includes the design and equipment cost, civil works, buildings, services, installation cost, spares, shipping, handling, insurance, commissioning and training, engineering, procurement and construction management. Training and commissioning and all fees are covered. The two furnace system includes the furnaces and structural support, lances, feed and fuel handling systems, evaporative coolers, baghouses, induced draft fans, stack, slag granulation facility, air compressors, water services, electrical and instrumentation and the required buildings.

The operating cost is in the region of $US50-$US100 per tonne of steel plant dust depending on whether oxygen is used for the process and what are the local costs on the particular site. This includes the raw materials, fuel, tonnage oxygen available from an oxygen plant, electricity, water, labour, refractories and maintenance.

Plant Design Variation for Steel Plant Dust

Waste heat can be recovered from the smelting furnace offgases by means of waste heat boilers if it is desired and economically justified. The waste heat boiler replaces the evaporative cooler of the smelting furnace and a smaller baghouse and an induced draft fan for the smelting furnace are required.

When the use of tonnage oxygen is not justified economically because of high costs of oxygen production, the process can be modified for operation without oxygen enrichment. A larger furnace, larger furnace gases handling facilities and more fuel coal, water and electricity are required.

Chlorine and lead can be concentrated in the first fume product leaving most of the zinc in the first slag by smelting the steel plant dust under neutral conditions. The quantity of the first fume product is greatly reduced. This can reduce the requirement of the facilities (which must be corrosion resistant) for any subsequent chlorine removal from the fume product with the trade off of a larger slag fuming furnace. The chlorine concentration in the clean fume is typically 0.5% or less.

The concentration of chlorine in the fume products can be diluted by smelting the steel plant dust with other zinc bearing materials such as zinc leach residue produced in electrolytic zinc plants. The extent of dilution depends upon the zinc concentration in the zinc bearing materials and the ratio of mixing.

Other zinc bearing materials which contain little iron and sulphur but much silica or fluxing materials can also be used instead of silica flux. A possible substitute is zinc silicate ore.

The plant illustrated has two furnaces to recover the maximum amount of zinc, and as shown for most condition would be most economical. If the slag from the smelting furnace is acceptable for disposal or use a single-furnace system can be considered with a small saving in capital and operating cost.

REFERENCES

1. Baldock B R, Floyd J M, Lightfoot B W and Robilliard K R, "Smelting of Metallurgical Waste Materials Containing Iron Compounds and Toxic Elements". International Patent Publication Number WO 91/02824, 7 March 1991.

2. Floyd J M, "Top Submerged Injection with a Shrouded Lance". International Publication Number WO 91/05214, 18 April 1991.

3. A R Gordon, "Zinc Extraction and Refining" in Scientific and Technological Development in Extractive Metallurgy. G K Williams Memorial Volume. The AusIMM, Symposium Series No 43, 1985 pp 87

4. Floyd J M, Lightfoot B W, Robilliard K R and Swayn G P: "Lead Smelting Using the Sirosmelt Furnace System". Lead and Zinc in the 1990's. International Lead and Zinc Group Meeting, Sao Paulo, Brazil. 27 pages, 5-7 February 1991.

PROCESSING OF ANTIMONY AT THE SUNSHINE MINE

Corby G. Anderson, Suzzann M. Nordwick and L. Ernest Krys

Sunshine Precious Metals Inc.
P.O. Box 1080
Big Creek Road
Kellogg, Idaho 83837
208-783-1211
FAX 208-783-1211 EXT. 202

Abstract

Antimony has been utilized by mankind for thousands of years. However, due to recent environmental concerns, antimony is often considered to be an impurity element. Since antimnoy is commonly found in precious metal refractory ore bodies, extraction is essential. The Sunshine Mine in Kellogg, Idaho has recovered antimony from its precious metal tetrahedrite deposits since the 1940's. A plant was initially constructed as a primary antimony source during World War II. This facility also decreased the downstream smelting penalty for the silver-copper-lead concentrate and provided an industrial commodity from a potential environmental nuisance. This paper describes the Sunshine antimony processing facility. Specifically, the relevant process chemistries involved in concentrate leaching, metal electrowinning, and sodium antimonate production will be discussed. In addition, ongoing areas of research to improve unit operations and treatments implemented to meet environmental discharge requirements will be covered.

Residues and Effluents - Processing and Environmental Considerations
Edited by Ramana G. Reddy, William P. Imrie and Paul B. Queneau
The Minerals, Metals & Materials Society, 1991

Introduction

In the realm of metallurgical processing, antimony is quite commonly perceived as an impurity element and environmental menace. This is particularly true in non-ferrous smelting or when treating refractory precious metal ore bodies. The first recorded use of antimony was found in a vase fragment excavated at Tello, Chaldea.[1] The relic was dated to 4000 B.C. Since that time, antimony has found service to man in a variety of applications. It is commonly alloyed with lead to make bullets and storage battery grids. Babbitt metal for bearing material contains 5% to 21% antimony. Other compounds of antimony are also finding great application. Sodium antimonate is commonly used as an additive in the glass industry, and the use of antimony oxide is thriving in flame retardant and paint pigment industries.

In the United States, major deposits of antimony are found in the Couer d'Alene Mining District of Idaho. The Sunshine Mine was discovered within this area in 1884. Well known for its precious metal deposits, it stands as the world's largest individual producer of silver.[2] Presently, Sunshine is also the largest producer of antimony metal in the country. The native ore body at the mine is argentiferrous tetrahedrite ($Cu_2S \cdot Sb_2S_3$), a copper-antimony sulfide mineral containing silver values.

Early in the history of the operation, the copper levels in the mill concentrate indicated that copper smelting would be advantageous. However, the high antimony content was inappropriate. Thus, the concentrates were shipped to lead smelters where the contained antimony, copper and bismuth resulted in an unfavorable treatment schedule. In 1937, Sunshine began a detailed investigation to reduce the antimony in their concentrates. The end result of this effort was the discovery of the Sunshine antimony process. It produces a low antimony concentrate which meets copper smelter specifications. This process is based on patented hydrometallurgical principles invented by Holmes.[3] The operation is counter to most antimony operations where pyrometallurgical (e.g. volatilization smelting, precipitation smelting, sulfide liquation) techniques have been the dominant methods for production since before Christianity.

Additionally, world circumstances enhanced the need for antimony and provided added motivation for implementation of this novel technology by Sunshine. The outbreak of World War II caused disruption of a strategic supply of antimony metal necessary for munitions manufacture. Consequently, the United States government encouraged and authorized creation of the production facility. Construction of the plant started March 1942, and operation began in September.

Overall, the Sunshine facility is a completely integrated mine-to-market operation. Besides the antimony plant, the current operating configuration includes the mine, a mill, a silver-copper refinery, and a fully functional mint. A conceptual flowsheet illustrating this complete integration is shown in Figure 1. The specifics of these operations will not be discussed here. They are separate from the antimony plant, and some have been covered in other articles.[4,5,6,7,8,]

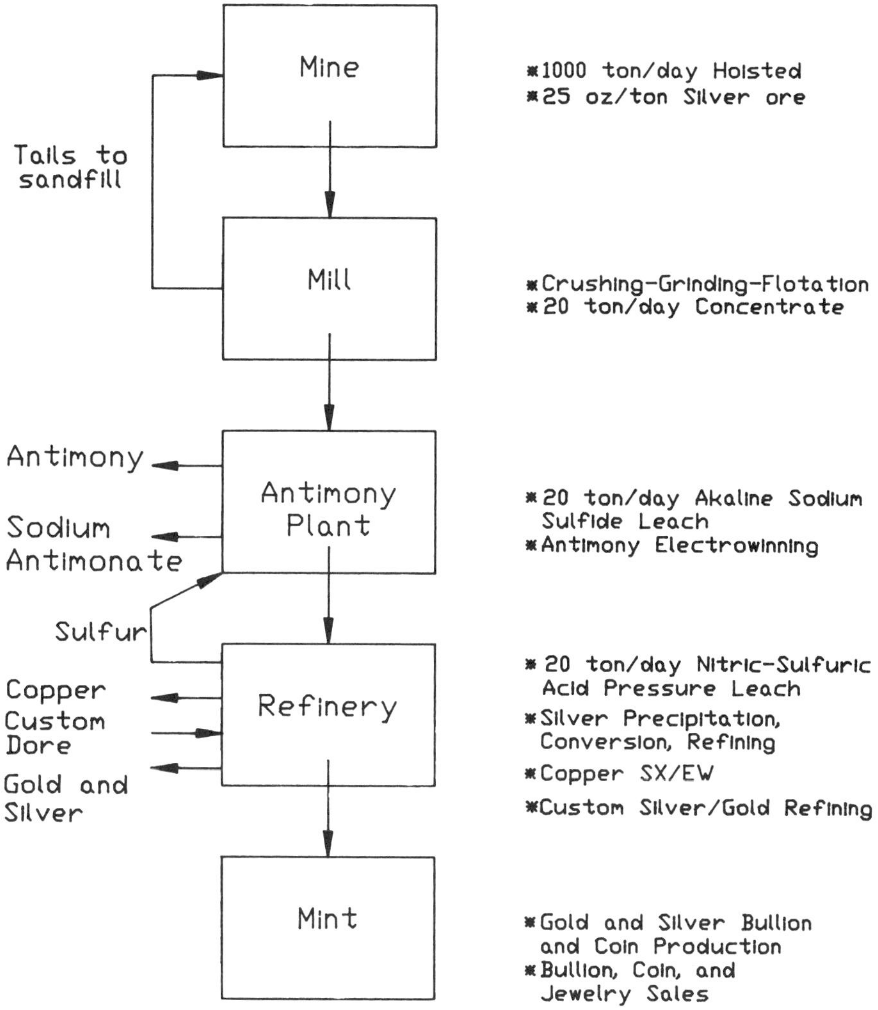

Figure 1 - Sunshine Facilities Flowsheet

In summation, this paper will illustrate the use of novel technology by Sunshine for the production of a functional material from substances commonly classified as wastes. At this juncture, the plant technology has been around for fifty years with little change. However, this will not hold true in the near future. A research effort is now underway to overhaul and upgrade the facility. Many of the technologies from this research effort have matured to the point that they will soon be implemented plant-wide. Specific quantitative results of the studies to develop these technologies will be the subject of subsequent publications. Therefore, this presentation will serve as an overview of the existing plant operations with an introduction to future process improvements.

Antimony Plant Operation

Background

Hydrometallurgical techniques for antimony recovery are fairly new, but they are not uncommon throughout the world. The basis for most of these processes is the knowledge that an alkaline sodium sulfide solution acts as an effective solvent for most antimony compounds. Conversely, most metals are highly insoluble in this solution. The exceptions include antimony, arsenic, tin, and mercury. Thus, a high degree of selective separation of antimony is available. This principle is employed at antimony operations in the USSR, China, and at Sunshine. A potential-pH diagram of the alkaline sodium sulfide leach system is illustrated in Figure 2.

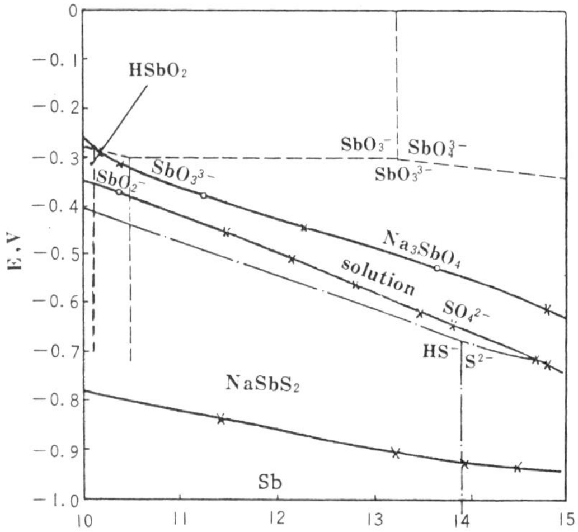

Figure 2 - Potential-pH diagram for the basic negative potential field of the Sb-Na-S-H_2O system at T = 25°C, where: $[Sb]_T$ = 0.5 mole/Kg and $[S]_T$ = 3 mole/Kg.[9]

Currently, the Sunshine antimony operation consists of four major unit operations: leaching, filtration, electrowinning, and tails treatment. A schematic of the entire

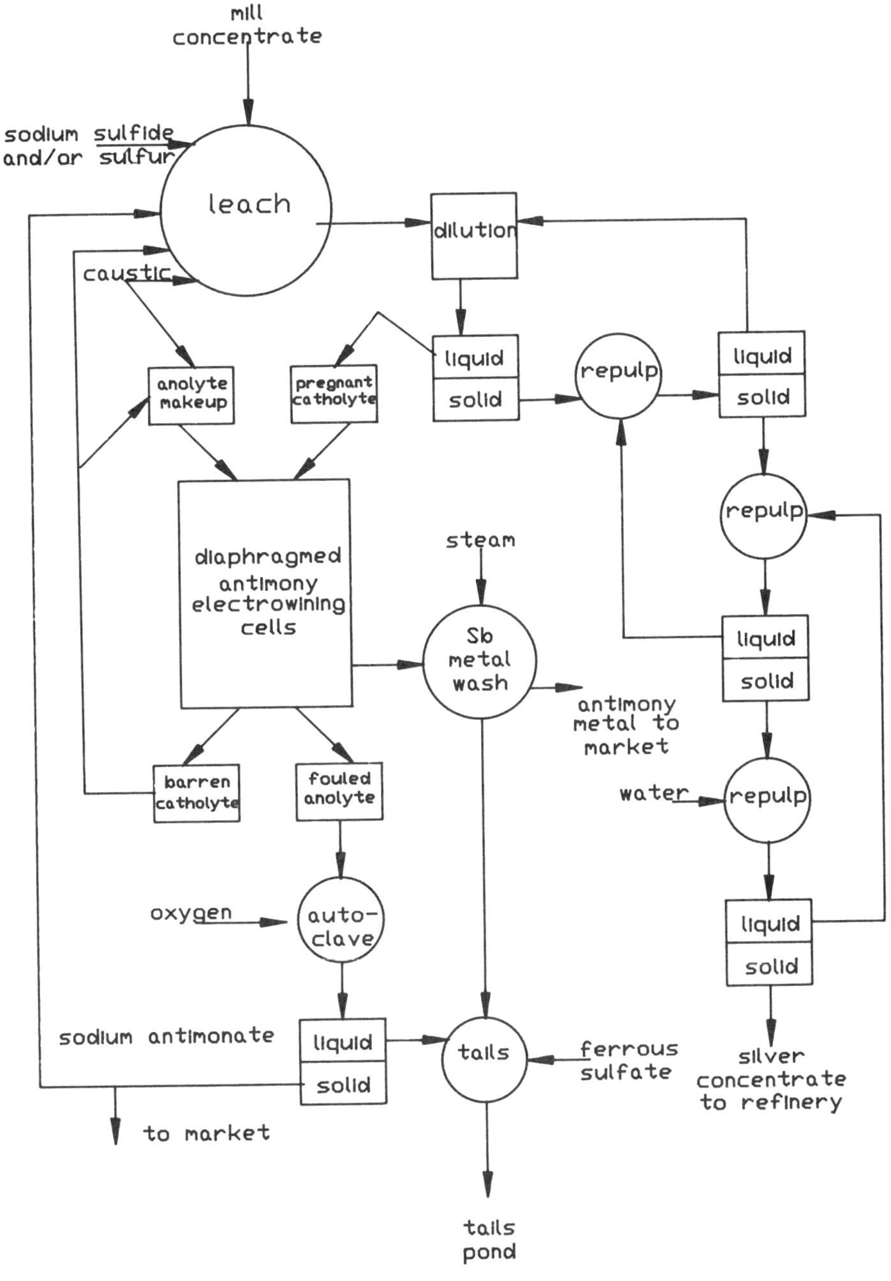

Figure 3 - Sunshine Antimony Plant Flowsheet

facility is presented in Figure 3. Feed for the plant is provided from the upstream concentrator where 1000 tons per day of mine ore is milled to 20 tons. A typical analysis of relevant feed and product streams is shown in Table I.

Table I

Typical Composition of Sunshine Mine Mill and Antimony Plant Products

Product Type	Ag, oz/Ton	Cu, %	Sb, %
Concentrator Feed	25	0.4	0.4
Antimony Plant Feed	1200	22.0	19.0
Silver Refinery Feed	1300	24.0	0.5

Leach

In practice, the antimony containing concentrate is batch leached in a hot alkaline sodium sulfide solution in one of four 9000 gallon mild steel tanks. The leach solution is created with a mixture of depleted electrolyte from the downstream electrowinning process (i.e. barren catholyte), elemental sulfur, sodium sulfide, and sodium hydroxide. The initial sodium sulfide (Na_2S) concentration is equivalent to 300 grams per liter. The initial leach solution also contains about 10 grams per liter antimony and some residual sodium hydroxide. The solution matrix then solubilizes the antimony and any arsenic which is not present as arsenopyrite. The leached antimony species are complexed by sulfide into thio compounds which include $NaSbS_2$, Na_3SbS_3, Na_3SbS_4, and $Na_4Sb_2S_5$.

In the preparation of the leach solution and in other plant solutions assays, the total alkalinity is determined using a simple acid-base titration. This test consists of titrating the sulfide, hydroxide and carbonate species with 0.0624 N sulfuric acid to a methyl orange endpoint. Since this quick procedure does not allow for any delineation of the individual alkaline species, the resultant total titratable alkalinity is expressed in terms of grams per liter of sodium hydroxide. This number can be converted to total pounds of alkalinity. Initially, a leach batch contains around 17,500 pounds or 250 grams per liter of alkalinity expressed as sodium hydroxide.

Elemental sulfur is used in the leach because of favorable economics and the ready supply produced by the downstream silver refinery. However, sulfur use results in the formation of species other than sulfide. Both polysulfide (Na_2S_x) and thiosulfate ($Na_2S_2O_3$) are formed along with sulfide in the alkaline solutions making up the leach batch. This is illustrated simplistically in the following scenario:

$$4S^o + 6NaOH = 2Na_2S + Na_2S_2O_3 + 3H_2O \quad (1)$$

$$(X-1)S^o + Na_2S = Na_2S_X \text{ (where X= 2 to 5)} \quad (2)$$

$$S^o + Na_2SO_3 = Na_2S_2O_3 \quad (3)$$

There are indications that, like sulfide, the oxidized sulfur species can act as effective leaching agents. However, unlike sulfide, they adversely effect the efficiency of downstream electrowinning.

In actual practice, the leach solution is first heated by internal steam coils to a temperature above 95°C. Approximately 8 tons of concentrate are then added. The leaching is carried out to completion in eight to ten hours at 104°C. The endpoint is determined by analyzing the leached residue. Any leach which gives a residue analysis determined as less than 1% antimony is considered finished. The resulting leached mixture contains antimony, arsenic, sulfur, and sodium ions as well as varying amounts of hydroxides and carbonates. The leach reactions thought to occur are:

$$Na_2S_{(Aq)} + Cu_2S \cdot Sb_2S_{3(s)} = Cu_2S_{(s)} + 2NaSbS_{2(Aq)} \quad (4)$$

$$NaSbS_{2(Aq)} + Na_2S_{(Aq)} = Na_3SbS_{3(Aq)} \quad (5)$$

$$3Na_2S_{(Aq)} + Cu_2S \cdot Sb_2S_{3(s)} = Cu_2S_{(s)} + 2Na_3SbS_{3(Aq)} \quad (6)$$

$$Na_2S_{X(Aq)} + (X-1)Na_3SbS_{3(Aq)} = (X-1)Na_3SbS_{4(Aq)} + Na_2S_{(Aq)} \quad (7)$$

$$2Na_2S_{(Aq)} + Cu_2S \cdot Sb_2S_{3(s)} = Cu_2S_{(s)} + Na_4Sb_2S_{5(Aq)} \quad (8)$$

Due to the oxidizing power of polysulfide, the major species in solution is sodium thioantimonate (Na_3SbS_4).

Solid/Liquid Separation

When leaching is complete, the slurry is diluted to a total titratable alkalinity equivalent to 200 grams per liter expressed as sodium hydroxide. This is done to allow enhanced solid separation from the viscous alkaline solutions. Dilution is accomplished using water recovered from downstream repulping and solid/liquid separation stages. The resultant slurry consists of an antimony containing pregnant catholyte solution of about 60 grams per liter antimony and a high grade copper-silver residue.

Solid/liquid separation initially takes place by conventional settling and thickening. Both primary and secondary thickeners are used with a repulp stage in between. Residue from the secondary thickener is repulped a second time and recovered with a disc filter. A final repulping step follows and the final stage of filtering is provided by a drum filter. To maximize separation, the wash water is run counter-current to the residue. Accordingly, fresh water enters the solids recovery process at the third repulp stage. Then following drum filtration, the wash water moves to the second repulping stage. Next, following the disk filter, the wash water is sent to the first repulp stage. Finally, following thickening, the wash water is sent to the leach stage as dilution water.

The solid filter cake contains about 18% to 20% moisture, and it is feed material for the Sunshine silver refinery. Because of its pyrophoric nature, final filtering of the residue is scheduled to occur when it can be quickly trucked to the silver refinery. Otherwise, air oxidation can turn it into a hot red glowing mass which presents downstream difficulty.

The three repulping steps are important because they allow maximum recovery of both the leached antimony and the contained alkalinity. They also provide a copper-silver residue free from alkaline sulfur compounds which would prove troublesome for the downstream nitric-sulfuric acid pressure leach.

Electrowinning

As mentioned before, the leach solution primarily contains antimony in the form of sodium thioantimonate. This solution serves as feed for the antimony electrowinning circuit where antimony metal is recovered on the cathode. Since the sodium thioantimonate does not disassociate solely to ionic antimony, the electrolysis of this sulfur bearing solution is complicated. Thus, a brief discussion of the general chemistry behind electrolysis of alkaline solutions of these thio-salts is a worthwhile endeavor before mentioning the plant operation.

In an ideal electrowinning situation, the thioantimonate ion would be reduced to metal. This would result in an antimony depleted solution enriched in free sulfide ions. No other side reactions would occur and disrupt the process. This liquid could then be recycled for leaching.

As with all industrial processes, this ideality does not happen. Four processes may occur in the catholyte in the electrowinning of alkaline sulfide solutions containing antimony: the deposition of antimony at the cathode, the reduction of electrolyte at the cathode, the reaction of solution species resulting from cathode reduction processes and the evolution of hydrogen. Of these, only the reduction of antimony contributes directly to the Faradaic current efficiency. The mechanism behind reduction of the dipolar anionic antimonate is complex and the subject of much debate.[10,11,12,13,14] For example, one mechanism relies on atomic hydrogen being present at the cathode. It reduces antimony to the metallic state on the cathode plate. This is thought to occur as:

$$H^{+} + e^{-} = H \qquad (9)$$

$$SbS_4^{-3} + H = SbS_3^{-3} + HS^{-} \qquad (10)$$

$$SbS_3^{-3} + 3H = Sb^{o} + 3HS^{-} \qquad (11)$$

Of course, for illustrative purposes, it will be accepted that in some manner the negatively charged antimony containing ions are discharged at the negative cathode plate. All four major cathode processes are illustrated below.

- Antimony reduction at the cathode (contributes to current efficiency):

$$SbS_4^{-3} + 2e^{-} = SbS_3^{-3} + S^{-2} \qquad (12)$$

$$SbS_3^{-3} + 3e^{-} = Sb^{o} + 3S^{-2} \qquad (13)$$

- Oxidized sulfur species reduction at the cathode (reduces current efficiency):

$$S_2O_3^{-2} + 2e^{-} = S^{-2} + SO_3^{-2} \qquad (14)$$

$$S_X^{-2} + 2(X-1)e^{-} = (X)S^{-2} \qquad (15)$$

$$3S_X^{-2} + 2(X-1)\ Sb^{o} = 2SbS_3^{-3} \qquad (16)$$

$$S_x^{-2} + (X-1)SbS_3^{-3} = (X-1)SbS_4^{-3} + S^{-2} \quad (17)$$

- Reaction of species resulting from cathode reduction processes (can effect current efficiency depending on species concentrations):

$$(X-1)SO_3^{-2} + S_x^{-2} = (X-1)S_2O_3^{-2} + S^{-2} \quad (18)$$

$$SbS_4^{-3} + SO_3^{-2} = SbS_3^{-3} + S_2O_3^{-2} \quad (19)$$

- Hydrogen evolution at the cathode (reduces current efficiency):

$$2H_2O + 2e- = H_2 + 2OH^- \quad (20)$$

Of importance is the fact that hydrogen evolution only occurs to an appreciable extent when the antimony concentration is low (i.e. 10 to 20 grams per liter). Thus, this phenomena can be controlled by monitoring the antimony level in the catholyte during electrowinning. Unfortunately, the other deleterious reactions are largely a function of solution makeup and are not readily controlled.

The situation at the anode is even more complex. Essentially the oxidation reactions of hydroxide ion and sulfide ion predominate. The reaction products from these two processes further contribute to an overall complex scenario of solution oxidation. Some of these reactions are illustrated and categorized below.

- Primary anodic reactions:

$$4OH^- = 4e^- + 2H_2O + O_2 \quad (21)$$

$$S^{-2} = 2e^- + S^o \quad (22)$$

- Solution reactions with products of primary anode reactions:

$$2S^{-2} + 3/2O_2 = S_2O_3^{-2} \quad (23)$$

$$S^{-2} + 3/2O_2 = SO_3^{-2} \quad (24)$$

$$SO_3^{-2} + 1/2O_2 = SO_4^{-2} \quad (25)$$

$$S^{-2} + 2O_2 = SO_4^{-2} \quad (26)$$

$$S^o + (X-1)S^{-2} = S_x^{-2} \quad (27)$$

$$S_x^{-2} + (X/4)3O_2 = (X/2)S_2O_3^{-2} \quad (28)$$

$$(X-1)SO_3^{-2} + S_x^{-2} = (X-1)S_2O_3^{-2} + S^{-2} \quad (29)$$

$$SbS_4^{-3} + SO_3^{-2} = SbS_3^{-3} + S_2O_3^{-2} \quad (30)$$

$$S_x^{-2} + (X-1)SbS_3^{-3} = (X-1)SbS_4^{-3} + S^{-2} \quad (31)$$

- Other significant anodic reactions:

$$1/2\ S_2O_3^{-2} + 5OH^- = SO_4^{-2} + 5/2H_2O + 4e^- \quad (32)$$

$$SbS_3^{-3} + S^{-2} = SbS_4^{-3} + 2e^- \quad (33)$$

Of these reactions, the most important are the formation of polysulfide, the formation of thiosulfate, and the oxidation of thioantimonite to thioantimonate (i.e. equations 23, 27, 28, 29, 30, 31 and 33). As previously outlined in the cathode reactions, these species disrupt the reduction processes and ultimately lower current efficiency. As a consequence of using elemental sulfur in the leach at Sunshine, polysulfides and thiosulfates are always initially present in the catholyte (i.e. equations 1, 2 and 3). Moreover, because of anodic processes, more of these species form and can accumulate in the electrolyte unless they are controlled. Therefore, it is important to minimize their interaction with the cathode. This is commonly done by placing canvas diaphragms around the anodes to form separate anolyte and catholyte solutions. In this case, current transfer occurs as part of the sulfur species in the catholyte migrate to and concentrate in the anolyte, while part of the sodium ions migrate from the anolyte to concentrate in the catholyte. This diaphragm cell concept has been employed at Sunshine since plant start-up to control the unwanted diffusion of oxidized sulfur species from the anolyte.

Next, with these general electrowinning principles in mind, the operating plant will be described. It consists of 144 individual electrowinning cells which are divided into two circuits of 72 cells each. The cell house lay out design is 24 rows of 6 cascading mild steel cells. The two solutions are kept separate by placing a heavy canvas diaphragm between the anodes and cathodes. As previously mentioned, the canvas permits the passage of electric current and prevents the mixing of the anolyte and catholyte. The separate anolyte and catholyte solutions are collected in surge tanks at the end of the cell bank. These are pumped back up to head tanks where they recycle continuously by gravity feed through the cellhouse.

With an individual capacity of 275 gallons, the individual cell dimensions are 30 inches deep, 32 inches wide, and 60 inches long. Each cell contains 9 cathodes and 8 anodes spaced evenly at 2.5 inches from center. The cells, anodes and cathodes are all constructed from mild steel. The anodes are 0.125 inches thick and the cathodes are 0.375 inches thick. Each anode is encased in a heavy canvas basket with a mild steel frame. The baskets are interconnected with rubber hoses and are designed for anolyte solution flow across the face of the anode. The cathodes have a submerged area of 3.33 feet squared per side. The anodes are slightly smaller in surface area to lessen the chance of dendrite growth along the cathode edge.

The cellhouse is run on direct current which is furnished by rectifier conversion of alternating line current. Normally, 1750 amps are provided to the cell circuit bus bars at 450 volts. This results in a cathodic current density of 29 amps per square foot and a anodic current density of 33 amps per square foot. The voltage in each cell varies between 3 and 5 volts depending on the alkalinity of the anolyte, the electrical connection integrity and the overall electrolyte solution chemistry. Electrically, the cells are connected in series while the electrodes within the cells are in parallel. The resistive heat from the applied current passing through the cells keeps the solutions at 50°C to 55°C without external heating. Fortunately, this has proven to be an optimum temperature in that all of the species stay in solution, and not much evaporation is experienced. Overall, current efficiency ranges from 25% to 40%.

The catholyte is made from the antimony leach pregnant solution (i.e. pregnant catholyte). This solution is typically diluted with repulp wash water to 60 grams per liter antimony and 200 grams per liter total alkalinity expressed as sodium hydroxide. An electrolysis batch consists of 12,000 gallons of catholyte. During a circuit change, an equivalent amount of barren catholyte is simultaneously withdrawn from the circuit while fresh catholyte is being added. During electrowinning the catholyte density is reduced. Therefore, to promote adequate residence time in the individual cells, the solution is introduced near the bottom at one end of the electrowinning cell and allowed to overflow the other end to the next cell. The catholyte is considered depleted (i.e. barren catholyte) when it has been electrowon down to 20 grams per liter antimony. At this point, the total titratable alkalinity will approximately be 250 grams per liter when expressed as sodium hydroxide.

As previously illustrated, the regeneration of leach solution from the cathode process is a benefit to overall plant economics (i.e. equations 12 and 13). Sulfide ions tied up with antimony ions are freed by the electrowinning cathode process. This allows recycle back to the leach step where the free sulfide solubilizes additional antimony. In actual plant practice at Sunshine, about half the antimony depleted solution is returned to the leach area. The remainder serves as makeup for the anolyte solution.

Fresh anolyte (i.e. anolyte makeup) is usually comprised of a fifty-fifty solution of barren catholyte and sodium hydroxide solution. The initial total titratable alkalinity is usually 225 grams per liter reported as sodium hydroxide. 3500 gallons of anolyte makeup are required to fill the anolyte circuit. It is considered depleted (i.e. fouled anolyte) and subsequently replaced when it reaches a total alkalinity of 175 grams per liter expressed as sodium hydroxide.

The anolyte solution has four functions.[15,16] First, it completes the electrolytic circuit by serving as conducting solution. Second, it reacts with the evolved oxygen to prevent air contamination around the cells. Third, the anolyte provides an ion exchange transfer vehicle to the catholyte for alkalinity. This allows the total alkalinity of the anolyte to decrease during electrolysis while the total alkalinity of the catholyte increases. Last, the anolyte provides an ion exchange receiving vehicle for the excess sulfur ions from the catholyte. Sulfur species in excess set up resistance in the cells to current transmission. So, at high concentrations, they are considered a contaminant. Sulfur ions migrating to the anolyte combine as oxidized sulfur radicals and are discarded from the system.

For convenience, a compilation of typical antimony plant solution assays is presented in Table II. It includes values for all four cellhouse solutions as well as an assay of fouled anolyte which has been treated for antimony removal (i.e. autoclaved anolyte). The production of autoclaved anolyte will be covered under tails treatment.

Antimony metal is deposited on the cathode plates as a brittle non-adherent layer. It is harvested every four days by briefly removing the cathodes from the cells. The cathode is then hit with a small pneumatic hammer which causes the antimony to break free and fall into a collection bin.

Table II

Typical Analysis of Antimony Electrowinning Solutions
(All Values Expressed In Grams Per Liter)

Solution Type	Sb	As	Total Sulfur	OH	CO_3	S^{-2}	S_x	S_2O_3	SO_3	SO_4
Pregnant Catholyte	60.0	5.0	120.0	3.0	40.0	60.0	20.0	55.0	5.0	20.0
Barren Catholyte	20.0	5.0	120.0	1.0	40.0	80.0	5.0	20.0	35.0	30.0
Anolyte Makeup	12.0	3.0	70.0	40.0	25.0	50.0	4.0	12.0	15.0	12.0
Fouled Anolyte	30.0	5.0	120.0	1.0	40.0	30.0	16.0	110.0	5.0	30.0
Autoclaved Anolyte	0.20	5.0	120.0	0.0	40.0	2.0	2.0	175.0	15.0	30.0

The recovered metal is washed by blowing steam into a tank containing hot water and the metallic antimony. The wash removes any solution or soluble solids which may have adhered to the metal during the plating process. This used wash water is sent to tails treatment. The antimony metal is then air dried and packed into 55 gallon drums. It assays about 97 percent antimony and 2 percent arsenic. At full production, the plant will yield well over 1 million pounds per year of cathode antimony. The metal is sold to secondary smelters for use as a lead hardener or for antimony oxide production.

Tails Treatment

Approximately 10 percent of the antimony which initially enters the plant is lost to the fouled anolyte or cathode wash streams. The fouled anolyte solution also contains some residual caustic soda. Therefore, a system was developed to recover these values as well as clean up streams ultimately discharged from the plant. An effective economical first step is sodium antimonate production by use of an oxygen autoclave. This system reduces the antimony which would have to be sent to further tails treatment.

In practice, fouled anolyte is charged to a 1500 gallon stainless steel autoclave with dual agitators. Initially, steam is added to heat the solution to near boiling. Then, an oxygen overpressure of 90 pounds per square inch is introduced into the reactor. The oxidizing reactions are exothermic and the reaction progresses at a controlled temperature of 105°C.

Solid sodium antimonate precipitates from solution while the arsenic and sulfur species are oxidized and remain in solution. The perceived chemistry describing this process is as follows:

$$Na_3SbS_{4(Aq)} + 4O_2 + 2NaOH + 2H_2O = NaSb(OH)_{6(s)} + 2Na_2S_2O_3 \quad (34)$$

$$Na_3AsS_{4(Aq)} + 9/2O_2 + 2NaOH = NaAsO_{4(Aq)} + 2Na_2S_2O_3 + H_2O \quad (35)$$

$$Na_2S_2O_{3(Aq)} + 2O_2 + 2NaOH = 2Na_2SO_{4(Aq)} + H_2O \quad (36)$$

$$Na_2S_{(Aq)} + 2O_2 = Na_2SO_{4(Aq)} \quad (37)$$

$$2Na_2S_{(Aq)} + O_2 + H_2O = Na_2S_2O_{3(Aq)} + 2NaOH \quad (38)$$

Typical analysis of autoclave feed and product solutions are found in Table II. Respectively, these solutions are designated as fouled anolyte and autoclaved anolyte.

The solid sodium antimonate product is removed by sedimentation and decantation. Depending on market conditions, it is either accumulated for sale or recycled back to the leach step for antimony and hydroxide recovery. The filtrate containing the sodium arsenate and residual antimony is treated with solid ferrous sulfate to a pH of 8 to 8.5. Conceivably, precipitated compounds of arsenic and antimony form in solution:

$$2AsO_4^{-3}{}_{(Aq)} + 3FeSO_{4(Aq)} = Fe_3(AsO_4)_{2(s)} + 3SO_4^{-2} \quad (39)$$

$$2SbS_4^{-3}{}_{(Aq)} + 5FeSO_{4(Aq)} = Sb_2S_{3(s)} + 5FeS_{(s)} + 5SO_4^{-2} \quad (40)$$

After pH reduction is accomplished, a small amount of quick lime is added to polish the solutions with hydroxide precipitation of residual ferrous ion in solution. This method was adopted from Chinese industrial practices.(17) It has proved exceptionally effective in removing the potential pollutants from the discharged effluents, and it has been quite easy to incorporate in the plant scenario.

Originally, the antimony plant relied on the use of barium sulfide for tails treatment and solution regeneration. This was part of the technology patented by Holmes.(18) In this process, fouled anolyte was mixed with a solution of barium sulfide. Oxidized sulfur species in the solution, present from anodic reactions, combine to form insoluble precipitates. In turn, this allows sulfide to be released into solution thereby regenerating solutions for leaching. The conceived reaction scenario is:

$$Na_2SO_{4(Aq)} + BaS_{(Aq)} = Na_2S_{(Aq)} + BaSO_{4(s)} \quad (41)$$

$$Na_2SO_{3(Aq)} + BaS_{(Aq)} = Na_2S_{(Aq)} + BaSO_{3(s)} \quad (42)$$

$$Na_2S_2O_{3(Aq)} + BaS_{(Aq)} = Na_2S_{(Aq)} + BaS_2O_{3(s)} \quad (43)$$

$$Na_2CO_{3(Aq)} + BaS_{(Aq)} = Na_2S_{(Aq)} + BaCO_{3(s)} \quad (44)$$

The solid precipitates of barium were then separated from solution and reduced with coal to barium sulfide in a furnace. This solid "black ash" was extracted with hot water to form a barium sulfide solution which was used again to treat fouled anolyte. In theory, this process could have been used on fresh catholyte solutions to aid in raising current efficiency. However, this practice was never implemented. In fact, the entire black ash process was eliminated from the plant operation in the late 1940's.[19]

Conceived Process Improvements

For fifty years, the antimony plant at Sunshine has remained relatively unchanged. This is true for several reasons. First, since it is the only one of its kind in the western world and it was developed in-house, any comparisons to other existing facilities as a measure of performance were next to impossible. Next, the process chemistries have proved to be quite elusive due to the lack of analytical tools for measuring the sulfur species involved. Lastly, unlike other non-ferrous metals, developments in antimony processing have been few and largely unpublicized in the western world. Thus, there have not been resources available to draw upon in order to improve the antimony process used at Sunshine.

Despite these difficulties, it was decided in 1990 to proceed with an intense in-house research program to better understand the process and to encourage new developments. There was a twofold motivation behind this effort. First was the continual process disruption caused by the excessive maintenance requirements of the diaphragmed electrowinning cells. Second was the dower economics of the plant caused by low metal product pricing and high reagent costs. The significant concepts generated from these studies to date are outlined in the following sections.

Assay Procedure Development

The first step in the research was development of accurate qualitative and quantitative assay procedures to define the solution chemistry. This was not easy as most of the sulfur species in the system are metastable. However, techniques were developed around an auto-titrator which successfully allowed delineation of all species in the system. This was important for both research and plant operation. As an example, the developed titration procedures are now used to quickly and accurately measure all alkaline species in the anolyte solution. This has translated into improved plant performance by optimizing use of sodium hydroxide in the anolyte.

Leaching

The alkaline sulfide leach is expected to remain the core process for antimony removal at Sunshine. An investigation is currently being conducted on the characteristics of this process. Results to date indicate the following noteworthy phenomena:

- The reaction kinetics and the resultant recoveries of antimony appear to be sensitive to system temperature. This is to be expected as, commensurate with most rate processes and known antimony solubility data, higher temperatures should directly increase reaction rates and recoveries.

- Sodium thiosulfate appears to be an effective leaching agent for tetrahedrite. At first, it was thought that this product of sodium sulfide oxidation was inert in the leaching process.

- Sodium polysulfide, known to be an effective leaching agent because of its oxidizing prowess, was found not to be effective at all studied concentrations. In fact, an abundance of polysulfide tends to impede reaction velocities and recoveries.

- Sodium carbonate, sodium sulfate and sodium sulfite concentrations appear to have little effect on reaction rate or recovery.

- The presence of excess sodium hydroxide enhances the leaching capabilities of the solution. This is probably due to the effect it has on suppressing the hydrolysis of sodium sulfide to sodium hydrosulfide. Sodium hydrosulfide does not leach tetrahedrite.

At this juncture, work is continuing to completely define these preliminary results.

Single Electrolyte Electrowinning

Other research efforts have indicated that a novel method of diaphragmless electrowinning could be employed. Both lab scale and pilot plant studies are underway at this time. This technology appears to offer current efficiencies comparable to the diaphragm based technology now being employed. Cathode metal quality, as seen in Table III, is much higher. Also, without the use of diaphragms, the single electrolyte system is

Table III

Comparison of New Metal Product With Current Metal Product

Old Sunshine Diaphragm Cell Metal	New Sunshine Diaphragmless Cell Metal
Sb = 97 %	Sb = > 99.0 %
As = 2.0 %	As = 0.02 %
Hg = 0.04 %	Hg = Trace

simpler to operate and maintain while the cell house voltages are much lower. This new operating practice will allow the entire volume of barren electrolyte to be returned to the leach area to utilize the free sulfide content for antimony solubilization. This technology centers around a proprietary purification process developed at Sunshine. Quickly and simply, this novel unit operation provides a separation of antimony from both arsenic and the various oxidized sulfur salts which can disrupt electrowinning.

High Grade Sodium Antimonates

By employing the same purification process which allows for single solution electrowinning, research has also indicated that high grade sodium antimonates can be produced. This will allow for further plant and product diversification. At this point, details concerning this procedure have been restricted from publication. However, the assay of the high grade sodium antimonates along with a compilation of industrial specifications is given in Table IV. These industrial specifications are some of the most stringent applicable to these products. As seen, the new materials exceed the impurity requirements.

Table IV

Impurity Levels of Sunshine Sodium Antimonates Relative to The Most Stringent Available Market Specifications

Sodium Antimonate $NaSbO_3$		Sodium Hydroxy Antimonate $NaSb(OH)_6$ or $NaSbO_3:3H_2O$	
Market Specifications	Sunshine Product	Market Specifications	Sunshine Product
Sb = 60-63 %	Sb = 61.0 %	Sb = 48-49 %	Sb = 49 %
As = < 100 ppm	As = 65 ppm	As = < 100 ppm	As = 62 ppm
Pb = < 1000 ppm	Pb = < 10 ppm	Pb = < 1000 ppm	Pb =<10 ppm
Cu = < 50 ppm	Cu = 4 ppm	Cu = < 50 ppm	Cu = 3 ppm
Fe = < 100 ppm	Fe = 69 ppm	Fe = < 100 ppm	Fe = 28 ppm
Sn = < 500 ppm	Sn = < 5 ppm	Sn = < 500 ppm	Sn =< 5 ppm
Ni = < 50 ppm	Ni = 2 ppm	Ni = < 50 ppm	Ni = 2 ppm

Caustic Production

Today's caustic soda market is volatile in terms of price and supply due to environmental concerns about chlorine by-products from caustic production. Thus, it would be beneficial to be self reliant in this commodity. Work has been done to define how best to implement an on-site caustic generation facility. As such, the facilities at Sunshine are a perfect match for implementation of a modified lime-soda caustic (i.e chemical caustic) manufacturing facility. This technology is well established and is based on the following reaction scheme:

$$CaO_{(s)} + H_2O = Ca(OH)_{2(Aq)} \quad (45)$$

$$Na_2CO_{3(Aq)} + Ca(OH)_{2(Aq)} = CaCO_{3(s)} + 2NaOH_{(Aq)} \quad (46)$$

This process has been widely used for caustic manufacture. In fact, up until 1940 and before the advent of electrochemical techniques, it was the primary method of production in the United States.[20] In practice, the solutions resulting from this process are around 140 to 150 grams per liter sodium hydroxide. Energy intensive and costly triple effect evaporation is used to upgrade to a product of about 50 sodium hydroxide percent by weight. Also, a key step is the calcination of recycled calcium carbonate back to calcium oxide, a costly and energy intensive process which is usually done at 1100°C to 1300°C in a kiln. The reaction of interest is:

$$CaCO_{3(s)} + Heat = CaO_{(s)} + CO_{2(G)} \quad (47)$$

At the Sunshine facility, the nature of the plant processes will eliminate much of the energy use required in traditional lime-soda caustic manufacture. This is true for two reasons. First, the antimony plant does not require highly concentrated caustic solutions. Most, if not all, of the evaporation necessary in the established procedure can be eliminated. Next, the downstream nitric-sulfuric acid pressure leach plant requires neutralization in the copper solvent extraction process. The resultant calcium carbonate from the chemical caustic plant could efficiently replace the calcium hydroxide currently being utilized there. This would allow the product from the silver refinery calcium hydroxide slaking unit to be redirected upstream for primary causticization with soda ash which is the principal step in the lime-soda process.

Summary

The Sunshine antimony plant has successfully produced antimony metal from the orebody of the Sunshine Mine since 1942. Though this is accomplished with state of the art technology, economic and environmental circumstances have changed over the past five decades. Thus, the advances garnered from Sunshine's in-house research efforts will enable the existing facility to further establish its role as a major producer of antimony products from refractory precious metal ore bodies. This will be done with lower operating costs and with increased diversity in process options and marketable products.

Acknowledgments

Thanks is given to Sunshine Precious Metals for making it possible to write and present this paper. Additional appreciation is given to those who successfully designed, built and operated the original Sunshine antimony facility.

References

1. Zhao, T., The Metallurgy of Antimony, Central South University of Technology Press, Changsa, The People's Republic of China, June, 1988.

2. Bennett, E., Idaho Geological Survey, University of Idaho, Moscow, Idaho, May, 1985.

3. W. Church Holmes, Sunshine Mining Co., "Electrolytic Recovery of Metals", U.S. Patent 2,331,395, October 12, 1943.

4. Caldon, F.,Sunshine Mining Co., "Treatment of Metal Bearing Mineral Material", U.S. Patent 4,084,961, April 18, 1978.

5. Ackerman, J. B. and Bucans, C.S. "Plant and Process Startup of the Sunshine Silver Refinery", Minerals and Metallurgical Processing, AIME-SME, Feb. 1986, pp. 20-32.

6. Anderson, C.G. and Harrison, K.D. "Optimization of Nitric-Sulfuric Acid Pressure Leaching of Silver From Refractory Sulfide Concentrates", Precious Metals 1990, D.A. Corrigan, ed. International Precious Metals Institute, San Diego, June, 1990, pp. 119-151.

7. Ackerman J.B., "The Sunshine Mining Company's Silver Refinery", Proceedings of the Fourth Western Regional Conference on Precious Metals and the Environment, AIME, Lead, South Dakota, Sept. 1990, pp. 219-235.

8. Ackerman, J.B., and Anderson, C.G., "Plant-Scale Processing of Silver Chloride", Precious Metals 1991, L. Manziek, ed. International Precious Metals Institute, Naples, Florida, June, 1991, pp. 253-274.

9. Tang, M. and Zhao, T., "A Thermodynamic Study on The Basic And Negative Potential Fields of The Systems of Sb-S-H_2O And Sb-Na-S-H_2O", J. Central-South Inst. Min. Metall., 1988, Vol. 18, pp. 35-43.

10. Levin, A.I., Shestitko, V.S., and Ishchenko, N.V., , "Mechanism of the Electrodeposition of Antimony From Alkali Metal Sulfide Electrolytes", Ural. Politekh. Inst. Sverdlovsk, USSRIzv. Vyssh. Ucheb. Zaved., Tsvet. Met., 1969, 12(3), pp. 74-78.

11. Tvertsyn, V.S., "Scientific Notes. Theory and Practice of Antimony Electrodeposition when Treating Sulfide Ores", No. 5, Knizhnoe Izdatel'stvo, Izhevsk, 1960.

12. Jiang, H., Shu, Y., Zhao-Ruirong, Gong, Z., Yang, S., "The Electrochemical Behavior of Sb Electrowinning in Na_2S-NaOH Solution", Jinshu Xuebao, 1982, Vol. 18, pp. 377-386.

13. Kobozev, N.I., and Monblanova, V.V., ZhPKh XXIX, 1955.

14. Biryukov, N.D., "The Mechanism of Anodic and Cathodic Processes During The Electrolytic Deposition of Antimony from Solutions of Thio-Salts", Journal of Applied Chemistry of the USSR, October, 1963, pp. 2098-2108.

15. Gould, W.D., "The Antimony Plant of The Sunshine Mining Company", Paper Presented at The Northwest Mining Association Meeting, December, 1953.

16. Barr, L. N., "Sunshine Mining Co. - Metallurgy Summary", Internal Report - Sunshine Mining Company, Kellogg, Idaho, 1973.

17. Chen, C., "Treatment of Wastewater from Antimony Hydrometallurgy", Huanjin Wuran Yu Bangzhi, Vol. 7, 1985, pp. 7-11.

18. W. Church Holmes, "How Electrolytic Antimony is Made at The Sunshine Plant", Engineering and Mining Journal, 145, March 1944 , pp. 54-58.

19. Palmer, H., Personal Communication with authors, Sunshine facility, July 11, 1991.

20. Parikh, V.A., "Chemical Caustic Soda Production - Techniques and Economics", Nojes Development Corporation, 1962, pp. 36-39.

THE ENVIRONMENTAL ASPECTS OF THE IN-PLANT COOLING OF

ALUMINUM MELTING FURNACE DROSSES

Richard P. Roberts

SECO/WARWICK Corporation
180 Mercer Street
Meadville, Pennsylvania 16335

Abstract

This paper will discuss the methods for the cooling of aluminum melting furnace dross and their impact on the environment. Also discussed will be a case history of an AROS Dross Processor installation, which is designed to maximize the recovery of metal content of the dross in a manner which is not only environmentally cleaner, but which also prepares the output in such a way as to facilitate downstream processes. Handling of the discharged material will be discussed in regards to the recycling of the preserved metal content and deposition of the unusable materials.

Residues and Effluents - Processing and Environmental Considerations
Edited by Ramana G. Reddy, William P. Imrie and Paul B. Queneau
The Minerals, Metals & Materials Society, 1991

INTRODUCTION

If aluminum furnace dross consisted only of aluminum and aluminum oxide, the task of handling the material would be greatly simplified. In reality, however, dross contains other substances, such as the oxides of alloying metals, salt, kryolite and other flux residues, as well as aluminum carbides and nitrides. Although presently, most drosses will pass leachate testings for landfill, they may also contain varying levels of chromium, cadmium, and lead.

Another factor in the equation is that dross contains valuable metal which can be economically recovered. New technologies are being introduced, but still, by far the most common way to recover the metal content is through the use of rotary salt furnaces. Disposal of the salt flux is not prohibitively difficult at present, although rising landfill fees are making that form of disposal economically less attractive as landfill capacity decreases, and environmental controls tighten.

The aluminum recovery/oxide separation system described here is designed to cool a dross charge quickly to reduce or eliminate the burning of recoverable metal while considering all aspects of environmental protection.

DROSS COOLING METHODS

Over the years, many approaches have been made to the problem of the handling of hot melting furnace dross.

Traditional Methods

The traditional method of dealing with hot furnace skim is simply to spread the material out to allow it to cool. This allows intimate contact with atmospheric oxygen, promoting burning of the free metal and causing fumes and dust which are nearly impossible to control. The dross remains at high temperatures for long periods, which, combined with the fumes, create hazards to personnel in the area.

Approximately half of the metal content of the dross is lost during this process to a thermite reaction, which is converting the aluminum to aluminum oxide.

Various other methods for dross handling have been tried over the years, including water-cooled vibrating conveyors, presses to squeeze out the free metal, and other systems, all of which have more or less fallen out of use. The main purpose of each of these designs was the preservation of the materials metal content.

Current Methods

One popular type of system, still in use, is the rotary drum type dross cooler. These units use a large rotating vessel which is partially immersed in a water bath. Additional water is generally applied to the outside of the drum by a water spray system, or by devices attached to the drum.

Hot dross is admitted to the drum by a feeder device. The tumbling action of the material against the wall of the drum rapidly cools the charge, reducing the temperature below the burning point with far less metal loss. Hoods are placed at the charge and discharge ends of the machine to capture escaping dust and fumes.

An evolution of this type of machine is the AROS (Aluminum Recovery/Oxide Separation) Dross Processor, which has been in use in Europe since 1982, and which has more recently been introduced in to the United States.

AROS Operational Features

The AROS system uses a special refractory lined pan into which the furnace dross is skimmed. Following transport to the machine, the pan is placed on the lifting device at the charge end (Figure 1). The pan is elevated into contact with a mating hood inside the processing unit. The fit between the pan and this inner hood is such that air leaking in is reduced or eliminated, thereby quickly creating an oxygen-starved environment and eliminating escaping dust. A scooping device inside the inner hood places a small amount of dross at a time into a water cooled drum. The scoop is bolted directly to the end of the drum and rotates with it. The drum has a double wall which forms a water jacket completely surrounding it. The water jacket is channeled to produce a parallel flow in the recirculating water.

SECO/WARWICK

Figure 1

Grain Size and Metal Analysis of Cooled and Ground Dross in the "Aros-DP"

Analysis of Cooled and Ground Aluminum Dross		
Grain Size (mm)	% of Each Size By Weight	Analytical Metal Content (%)
>8	15	>95
8 - 4	15	95
4 - 2	12	90
2 - 1	8	80
1 - .5	5	75
.5 - .25	9	65
.25 - .125	14	50
.125 - .063	13	20
<.063	9	5
	100%	

8mm Screen Mesh			
Fraction Type	Grain Size (mm)	% of Each Size By Weight	Analytical Metal Content (%)
Coarse	>8	15	>95
Fine	8 - .2	53	80
Dust (Baghouse)	.2 - 0	32	< 20

AROS Operational Features - Continued

The scoop introduces a small amount of dross at a time into the drum. The small scoops of newly admitted dross are charged onto the bed of previously cooled dross. This is done to reduce the thermal stress on the drum. The short delay in cooling time does not affect the metal recovery from the dross as the elimination of oxygen eliminates additional burning of the metal.

As mentioned earlier, the scoop is mounted directly to the end of the cooling drum, therefore, if the main drive should stop, the scoop would stop and no more hot dross would be admitted. This allows the charge in the drum to be safely cooled. Even if the drum drive and the water supply would fail at the same time, the jacket is designed to contain more than enough water mass to cool the charge.

AROS units are built with and without grinding equipment. In the grinder type models, the dross exits the cooling drum into the grinding chamber (Figure 2). In this chamber, the lumps of solidified aluminum are used in a manner similar to a ball mill. The grinder provides a purifying action by reducing the friable oxides to a dust which is drawn off to the baghouse. Because the charge end of the machine is sealed, the airflow confines itself to the grinding chamber and material discharges only, greatly reducing the airflow requirements. The temperature of the exiting air does not exceed 200°F and is free of sparks making for a relatively simple and inexpensive baghouse design.

The output from the grinding section is fed into a screening system which separates the coarse and fine particles (Figure 3 - 4).

In the designs which do not include the grinding equipment, the oxides, which would be pulverized and drawn into the baghouse, appear in the fine material discharge, increasing it's bulk and reducing it's average metal content.

Without the grinding chamber, the baghouse filter serves only to control the dust at the discharge openings. The baghouse requirements are, therefore, further reduced in these models.

Figure 2

View into grinding chamber of AROS DP

Figure 3

Fine fraction discharge from AROS DP

Figure 4

Coarse fraction discharge from AROS DP

OPERATIONAL CASE HISTORY

The first AROS unit was rated at 6600 pounds per hour and was installed in the Toging, West Germany, plant of V.A.W. The unit has been in continuous service since the summer of 1983.

Machine Output

The unit is charged with dross averaging 60 weight percent aluminum content at an average temperature of 1600°F. The V.A.W. machine was designed without the ability to drain the free-run metal within the machine during the charging cycle. Following transport to the area of the dross processor, a taphole in the pan is opened, allowing the metal which has collected in the bottom of the pan to run out.

Newer processors are designed with a mold incorporated into the lift platform which receives the molten metal as it drains from the taphole. The stirring action of the scoop charger greatly increases the volume of free-run metal. The output of the machine averages as follows:

1. 7.5% liquid aluminum drained from the troughs prior to processing (the V.A.W. installation was designed without the ability to drain the liquid metal within the machine during the charging cycle).

2. 21.5% coarse fraction, approximate analytical aluminum content 90%+.

3. 34% fine fraction, approximate analytical aluminum content 75%.

4. 37% dust fraction, approximate analytical aluminum content 22%.

The coarse fraction is remelted in an induction furnace within the plant with a recovery rate of approximately 80% of the metal content. The fine fraction is sent to a sister company at Grevenbrioch, West Germany, where it is remelted in a rotary salt furnace, with a recovery rate of 75%. Approximately 1/3 of the dust is sold for the production of exothermic salts for use in the iron foundry industry. The remainder of the dust is deposited in a landfill.

Table one indicates the output of the machine and the analytical metal content of each particle size. The very large lumps of dross have metal contents in excess of 95%.

The right hand section of the table indicates the effect of the screening system on the output of the machine. The unit uses a 4 mm mesh screen to do the separation. This results in the metal contents indicated. The screening system is generally designed to suit downstream processes. If the coarse fraction is going to be recharged into a traditional reverberatory melting furnace, a larger mesh size would be chosen. This is done because small particles tend to float on the molten aluminum bath only to be skimmed off again. If a submerging device is used with the reclaimer, a smaller mesh size would be chosen, allowing a greater percentage of the output to be kept in-house.

The metal content in the baghouse dust was originally thought to be a result of the grinding process. Closer examination revealed that the bulk of this aluminum was the result of fine particles of metal in the dross which were being preserved by the system. These fine particles are the first to be burned when exposed to oxygen. The rapid quenching and low-oxygen atmosphere results in the preservation of particles that are fine enough to be entrained in the airflow, thereby being drawn off into the baghouse.

Maintenance History

The wear areas of the machine fall into three (3) main areas. The first is the lining within the grinding drum. This lining consists of abrasion resistant steel plate and iron castings which are bolted to the shell of the drum. These plates were replaced after approximately 10,000 hours of operation.

Another wear area is the charging scoop assembly. The leading edge of the scoop is rebuilt approximately every six months. A spare scoop is kept ready for change-out and replacement requires approximately four to six hours. In other installations, scoop chargers have been operated for up to two years without attention, however, the six month interval is recommended as scooping efficiency falls off if the operating period is lengthened.

Another area of maintenance is the refractory lining in the dross pans. The pans are taken out of service every six weeks. Typically, the lining is repaired on the first outage and replaced on the second outage.

Manpower requirements, including operating time, routine maintenance, and maintenance of the dross pan linings is 1850 manhours per year. The unit is operated using untreated river water with an average consumption of approximately 40 gpm (during the actual cooling phase only). The device uses a closed system and, therefore, no water is evaporated.

DOWNSTREAM PROCESSING

Recovery methods using salt fluxes in rotary furnaces are coming under close scrutiny by environmental agencies. Even without tighter regulation, landfill deposition of salt will become increasingly expensive, as available space becomes filled. This trend is causing interest in salt reclamation processes and especially in salt-free metal recovery systems. These systems are doing much to reduce the volume of residue but, to date, no one has succeeded in completely closing the waste loop in the processing of aluminum dross.

Producer's Viewpoint

The operator of the casthouse is charged with maintaining a certain level of production with minimum metal loss. At the same time, he must adhere to certain safety and environmental standards.

Current dross cooling technologies not only reduce metal loss; they also allow the dust and fumes to be far more easily contained, simplifying their control.

Some users of this equipment mix the baghouse dust back into the balance of the cooler's output. The lower metal content of the dust lowers the overall metal content of the end product, but it allows all of the material to be sent to the reclaimer, eliminating the dust as a potential disposal problem.

Reclaimer's Viewpoint

If the average metal content in the dross is increased, the amount of flux used in rotary reclaiming furnaces can be reduced. If the reclaimer is using one of the newer saltless systems, the amount of non-metallic product is reduced. Either way, the amount of residue for disposal is reduced.

CONCLUSION

The waste materials of industry are being placed under tighter and tighter scrutiny every day. Aluminum furnace dross not only represents a potential disposal problem, it can represent an economic loss as well, in the form of recoverable aluminum, which is lost by traditional methods. The installation of dross cooling equipment is one of the few areas where a device which can offer many environmental advantages also has a very well defined amortization period. For this reason, much more attention is being paid to the way dross is handled in the typical aluminum plant.

SLAG HANDLING

IN THE IRONMAKING INDUSTRY

Arthur W. Cooper, P. Eng., M.B.A.

Manager of Engineering Services,
Paul Wurth Ltd.,
3310 South Service Road,
Burlington, Ontario, Canada
L7N 3M6

ABSTRACT

This paper contains a review of the development of blast furnace slag handling and processing technology in the Ironmaking industry over the past 30 years, with an emphasis on economic and environmental factors. The requirements and practice of air-cooling slag for use as aggregate and of pelletizing and granulating slag for use in cement manufacture are described. The operation of slag pits, slag pots or ladles, the slag pelletizer and early granulation systems are compared with the new continuous granulation, dewatering and handling systems developed in the past decade.

Residues and Effluents - Processing and Environmental Considerations
Edited by Ramana G. Reddy, William P. Imrie and Paul B. Queneau
The Minerals, Metals & Materials Society, 1991

INTRODUCTION

The traditional ironmaking process has changed very little since 1910.[1] Technological advances have increased productivity, and new processes such as Direct Reduction are looming on the horizon, but most of the world's iron is still produced by blast furnaces. As a by-product of the ironmaking process, a slag composed of 30-50% lime (CaO), 8-24% Alumina (Al_2O_3), 28-38% Silica (SiO_2), 1-18% Magnesia (MgO) and 0.5-2.5% Sulphur, along with some trace compounds, is produced in the approximate ratio of 300 kg slag per ton of hot metal.[2] This means that a furnace producing 3000 tons per day of pig iron will produce about 900 Tons of slag. With a specific gravity of 2, the slag has a volume (molten) of 15,900 cu. ft. By comparison, the iron (specific gravity 7.3) has a volume of only 14,500 cu. ft., so, from a handling standpoint, slag is the greater problem.

Unlike most non-ferrous and electric arc furnace processes, ironmaking slag is tapped with the iron. The slag and iron leave the furnace through the same taphole as an emulsion and are separated on the casting floor in the main iron trough, which is dimensioned to provide sufficient time for complete separation to occur (Fig. 1). Since the slag is less dense, it rises to the top of the iron pool and is directed to the slag runner system by means of a 'skimmer' set in the trough.

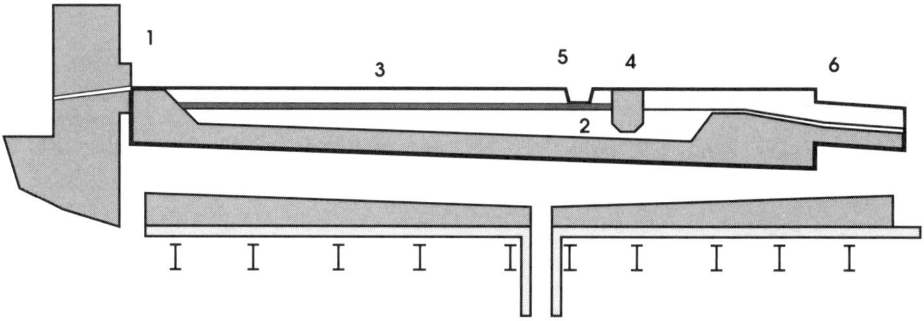

FIGURE 1: TYPICAL MAIN IRON TROUGH

1) Tap-hole
2) Iron Pool
3) Slag Layer
4) Skimmer
5) Slag Runner Takeoff
6) Iron Runner Takeoff

The iron flows under the skimmer to the iron runner system and is loaded into hot metal cars ('torpedo' ladles) to be transported to the steelmaking shop, while the slag is used on-site in land reclamation projects, or is processed and sold as a by-product.

Fortunately, blast furnace slag can be processed to provide a variety of commercial products. There are generally considered to be four distinct types of blast furnace slag, resulting from different methods of cooling. Air-cooled slag is slag which has been allowed to solidify under ambient conditions, possibly with surface application of cooling water sprays. 'Expanded' or 'foamed' slag has controlled amounts of water, air or steam injected to make it foam up prior to solidifying. 'Granulated' slag is solidified by quenching with water, normally high pressure sprays, to produce granules with the size and consistency of beach sand. 'Pelletized' slag is quenched with air and a small amount of water after being flung through the air. Before the slag solidifies, surface tension forms the slag into smooth pellets of roughly spherical shape.[3]

SLAG PIT OPERATION

Slag pit operation consists of pouring the molten slag into an open pit adjacent to the furnace, allowing it to solidify, and then excavating and hauling the solid slag off-site (Fig. 2). To keep the furnace in production, sufficient pit area must be available so that the entire slag production can be handled during the cycle time (cooling plus excavating) for one slag pit. Most North

Figure 2: SLAG PITS AT ARBED BELVAL, LUXEMBOURG. The pit in the left foreground is empty, the center pit is full of hardened slag and the pit in the background is being sprayed with water (note the steam emissions).

American furnaces have two pits of approximately 3,000 to 5,000 ft^2 each (effective size is less) but some large, multi-taphole furnaces have more (Fig. 3).

FIGURE 3: CASTHOUSE LAYOUT for a large blast furnace showing four tapholes and six slag pits.

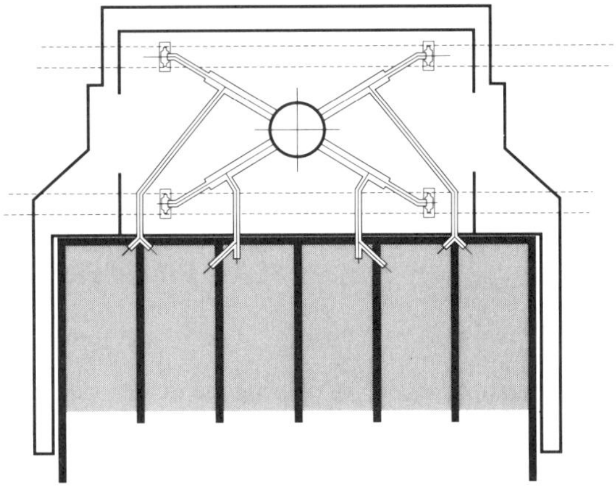

Water is sprayed on the molten slag after it is cast to speed up the cooling process, but this practice causes pollution in the form of sulphur-laden acidic steam (see Fig. 2). Damage to surrounding structures and electric and control circuits have been estimated as high as $750,000 per year at some blast furnaces from the corrosive effects of slag pit emissions. This low-level, uncontained steam also causes visibility problems with local traffic. Although slag pits are not a point source for emissions and have escaped official scrutiny so far, they are clearly a source of considerable pollution.[4] Because of the area involved, it is difficult to provide effective pollution abatement equipment for open slag pits.

SLAG POT OPERATION

Slag pots are used widely in non-ferrous plants where slag volumes are relatively small and flow rates are low. In ironmaking, slag pots are normally found on older, single taphole furnaces (Fig. 4). Most operators are converting to pit operations if the space is available, due to the expense of maintaining slag ladles and tying up a large proportion of the in-plant railroad system to handle slag. There is also a significant cost involved in maintaining the complicated slag runner system necessary to fill five or six slag pots each cast (Fig. 5). Estimates of total cost for slag pot operations are as high as $10.00 per ton of slag.

Once the slag is in the pots, it must still be dumped somewhere on the property, with the result that the problems described in the section under slag pit operation still exist, but at a different location.

FIGURE 4: SLAG POTS IN USE.

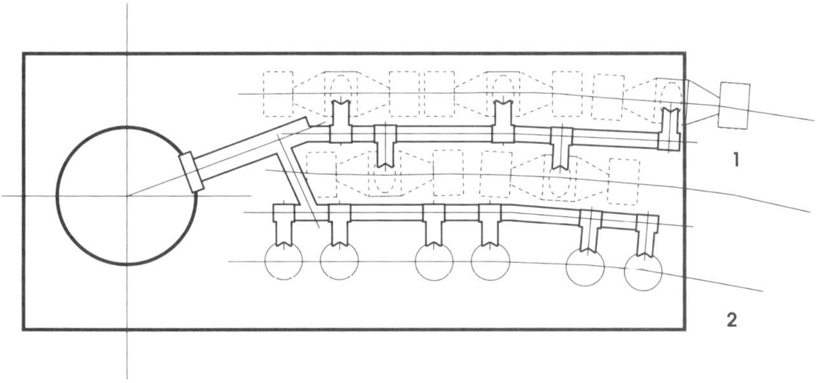

Figure 5: TYPICAL CASTHOUSE LAYOUT USING SLAG POTS FOR SLAG HANDLING.

1) Hot Metal 'Torpedo' Car Tracks
2) Slag Pot Track

In North America, most iron slag is air-cooled and used as an aggregate in railway and road ballast, granular base and engineered fill, for which it is ideally suited. The trend in Europe and Japan is toward granulation with water to produce a vitrified (glassy) slag for use in the manufacture of slag cement. A small percentage of slag is expanded for use as an aggregate in lightweight and semi-lightweight structural concrete. Pelletized slag is screened, and the larger, light pellets are also used in the production of lightweight concrete and concrete blocks. The smaller, dense pellets are normally suitable for grinding for use in cement manufacture. Slag cement is also the primary end use for granulated slag. There is sufficient lime in iron slag to give it cementitious properties if the slag is vitrified by rapid quenching to prevent crystallization. Ground granulated blast furnace slag is substituted for Portland cement in ratios as high as 80% slag/20% Portland in some European countries,[5] but typically in North America, the cement replacement is limited to 50%.

The replacement of Portland cement by ground granulated blast furnace slag results in a net energy savings, due to the substantial energy invested in the slag, during the ironmaking process. Wet process Portland cement requires about 6,700 MJ/Tonne (5.77×10^6 BTU/ton) to manufacture, compared to about 1700 MJ/Tonne (1.46×10^6 BTU/ton) for ground slag.[6]

"...Only four blast furnaces in North America are currently granulating their slag, or have granulators under construction."

In addition, slag cement is a premium product in terms of cement color, sulphate resistance and ultimate strength. Given this favorable energy saving and the possibility of adding considerable value to the raw slag, it seems surprising that only four blast furnaces in North America are currently granulating their slag, or have granulators under construction.

SLAG PROCESSING SYSTEMS

PIT GRANULATION

The granulation process is remarkably simple in principle. Molten slag is poured, either from a slag ladle or directly from the furnace, into a pit filled with water, or is hit with sprays of water to rapidly quench it. Blast furnace slag forms granules of the approximate consistency and color of beach sand, which have a glassy or vitrified structure because the rapid quenching does not allow the molecules to align to form crystals. This vitrification of the slag is essential to its use in cement, since crystalline structures of the same composition do not exhibit hydraulic behavior.[7] The granulated slag must then be separated from the granulation water and transported to the drying and grinding plant.

In the simplest granulation processes, the slag is granulated in a pit, the water is allowed to drain off, and the slag sand is excavated by loader (Fig. 6). By the mid 1960s more sophisticated systems were being developed to accomplish such goals as water recirculation, automatic slag unloading and remote processing. These have come to be known as slag granulation systems, when, in fact, they are granulation, dewatering and handling systems.

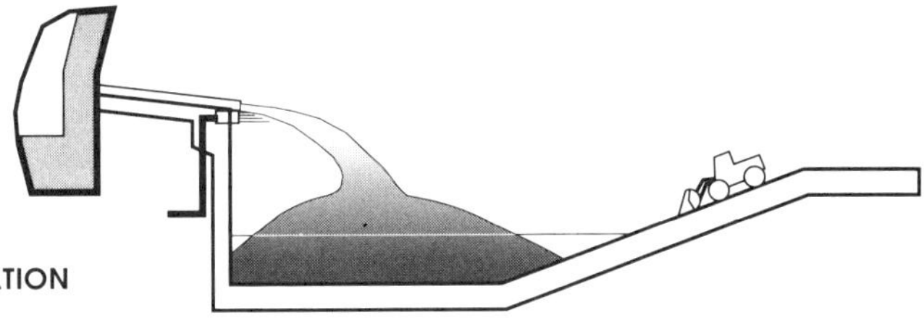

FIGURE 6: PIT GRANULATION SYSTEM.

GRAVEL LAYER FILTER BED SYSTEM

The gravel layer filter bed system is the ultimate development of the pit granulation system (Fig. 7). The slag is granulated at the furnace in a granulation head or 'blowing box' (Fig. 8) by high pressure water sprays. The slag/water slurry flows down a runner to an

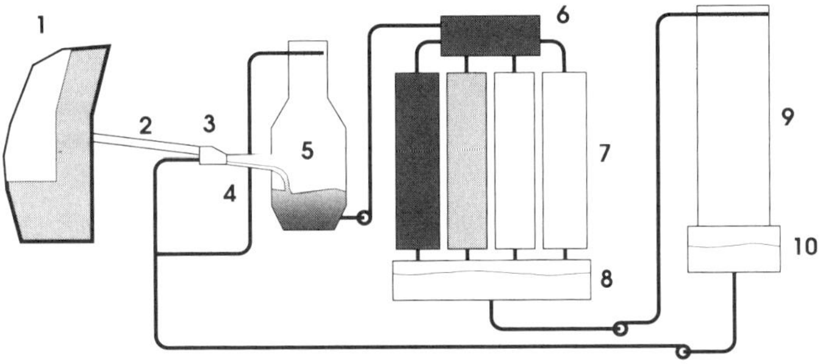

FIGURE 7: GRAVEL LAYER FILTER BED SYSTEM

1) Blast Furnace
2) Slag Runner
3) Granulation Head
4) Cold Runner
5) Agitating Tank
6) Distributor
7) Filter Beds
8) Hot Water Reservoir
9) Cooling Tower
10) Cold Water Reservoir

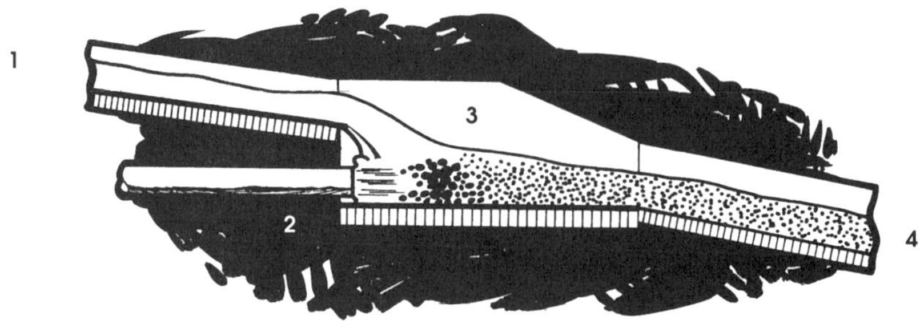

FIGURE 8: GRANULATION HEAD

1) Slag Runner
2) Water Sprays
3) Slag Slurry
4) Cold Runner

agitating tank (Fig. 9) where water is injected to keep the slag in suspension. The slurry is then pumped to the remote filter bed installation consisting of four or five concrete dewatering pits (Fig. 10). A distributor directs the slurry to the next pit in the programmed dewatering sequence. A number of casts are poured into the same pit during the sequence.

FIGURE 9: AGITATING TANK

1) Cold Runner
2) Condensation Water
3) Agitating Water
4) Slurry Pump

The water filters through the slag and the gravel and is drained to a sump through perforated pipes buried in the gravel. The slag layer acts as a sand filter and traps the majority of the fines which can cause plugging of the filter bed and abrasion and wear in the water circuits. The time necessary to drain the water is a function of the pit area (i.e., filter surface) so installations for large furnaces occupy a lot of plant space and are capital intensive.

FIGURE 10: GRAVEL LAYER FILTER BEDS AT ARBED, BELVAL IN LUXEMBOURG. Three pits are visible. The bucket crane and the movable pit covers are positioned above the center pit.

Once the slag is dewatered, a computer-controlled bucket crane excavates the pit (Fig. 11). Slag produced during this phase is directed to another pit. After a pit has been excavated, the filter layer is 'regenerated' by filling the pit with water and back-flushing with compressed air to free up any slag fines which might be lodged in the gravel or the pipes.

Control of atmospheric emissions is achieved by condensing the steam formed during granulation in a condensation tower above the agitating tank. Large, movable covers on the filter beds contain the residual steam coming from the hot slurry as it drains. The closed-loop water circuit eliminates water pollution concerns.

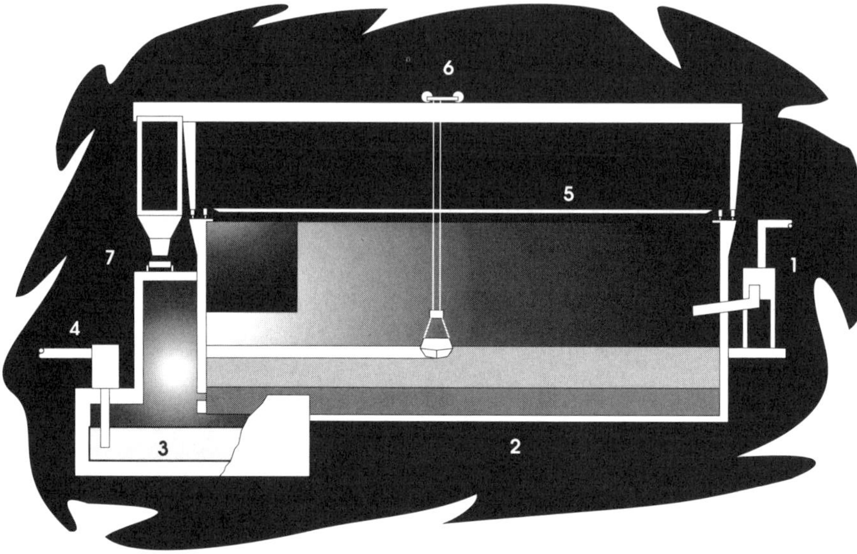

FIGURE 11: FILTER BED CROSS-SECTION

1) Slurry Distribution Box
2) Filter Layer
3) Hot Water Reservoir
4) Hot Water Pumps
5) Movable Pit Covers
6) Programmable Bucket Crane
7) Unloading Conveyor

The gravel layer system is reliable, produces a good product and is not maintenance intensive. However, its large size and capital cost are significant disadvantages. Because of the plant area required, very few gravel layer systems can be located adjacent to the blast furnace, so pumping the slag slurry to a remote dewatering station is normally required. The extreme abrasiveness of slag makes it necessary to use lined pipes and special slurry pumps, increasing the capital and maintenance costs. However, until the early 1980s, the gravel layer filter bed system was considered the system of choice where space constraints would allow it. The plant built by Atlantic Cement in 1981 at Sparrows Point is a gravel layer system and has been operating successfully for over ten years.

FILTER HOPPER SYSTEMS

The filter hopper or filter bin systems (Fig. 12) developed by AJO-STAHLBAU in conjunction with Hoesch Hüttenwerke in Europe, and the Rasa Trading Co. in Japan in the mid 1960s, attempted to address the large plant area required for the gravel layer and other pit based system. The filter hopper systems use large dewatering silos lined with screens to dry the slag, essentially extending the filtering system upward to conserve plant space. For these systems, pumping the slurry is an absolute requirement, and because the filtering is somewhat less efficient, more fines escape to the process water, making separate recirculating systems necessary to trap the fines.[8,9]

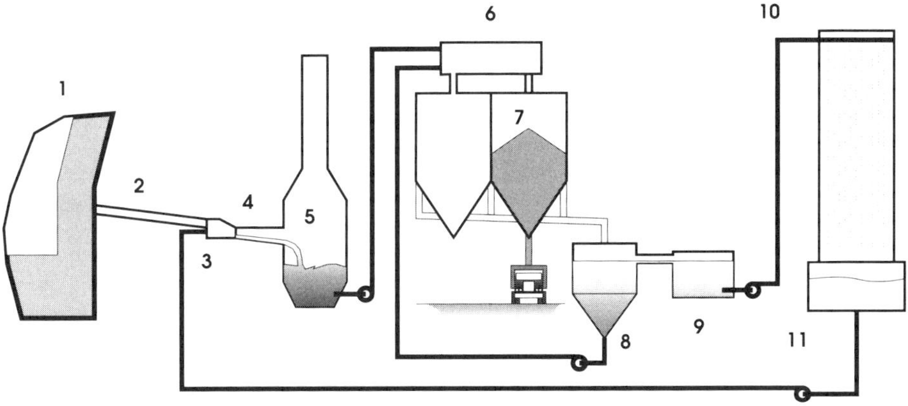

FIGURE 12: FILTER HOPPER SYSTEM

1) Blast Furnace
2) Slag Runner
3) Granulation Head
4) Cold Runner
5) Agitating Tank
6) Distributor
7) Filter Hoppers
8) Fines Settling Tank
9) Hot Water Overflow Tank
10) Cooling Tower
11) Cold Water Reservoir

Although they occupy slightly less area, filter hoppers are otherwise similar in performance to gravel layer systems. However, the dewatering hoppers have proven to be more maintenance intensive than the filter layer system, the internal screens and water circuits having a tendency to plug up and the system suffering from the effects of corrosion.[10] The RASA system was successful in Japan, with more than thirty installations. The filter hopper system was less successful in Europe.

THE NATIONAL SLAG PELLETIZER

National Slag Limited has been producing expanded slag for sale as lightweight aggregate since 1953 in Hamilton, Ontario, site of STELCO and DOFASCO, Canada's two largest steelmakers. Prior to 1968, a variety of processes were used, based on injecting water, air or steam into the slag. During the 1960s, prior to the advent of external iron desulphurizing, the ironmakers increased the sulphur content in the slag to about 2.5% in order to lower the sulphur content in the iron. As a result, National Slag's expanding slag process released large quantities (4,000 ppm) of Hydrogen Sulphide (H_2S) into the atmosphere. The resulting official and public pressures resulted in the development of the National Slag Pelletizer.[11]

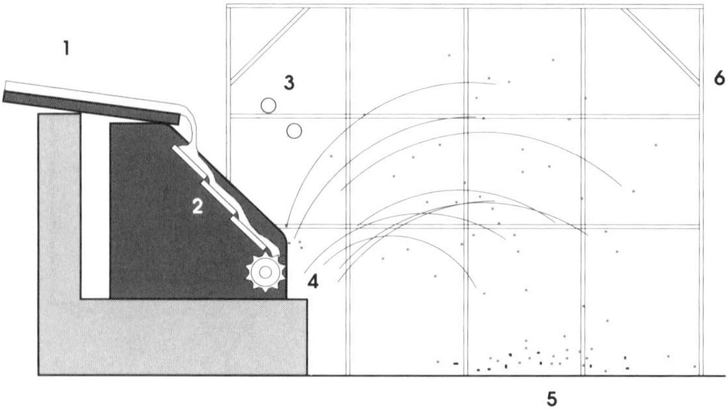

FIGURE 13: NATIONAL SLAG PELLETIZER

1) Slag Runner	4) Pelletizing Drum
2) Foaming Tables	5) Pellet Landing Area
3) Foaming Water Sprays	6) Chicken-wire Enclosure

The pelletizer (Fig. 13) consists of a set of water sprays to start the molten slag foaming, vibrating feed plates (foaming tables) to feed the expanding slag to the pelletizing drum, and the drum itself, which spins at about 300 rpm. The vanes on the drum splatter the slag into droplets and fling it through the air. Surface tension causes the droplets to form roughly spherical pellets while the flight throughout the air cools them quickly enough to form the glassy structure necessary for cement manufacture. The pellets vary in size from less than 1/8" to larger than 1/2", a much wider range than granulated slag. The particles are excavated by loader; those larger than 3/16" are used as lightweight aggregate for concrete while the smaller particles are separately ground by a local cement company for use in blended slag cement.

The pelletizer was originally fed from slag ladles, but by 1975, a pelletizer was successfully installed directly at the end of the slag runner at DOFASCO's No. 4 Blast Furnace. Normally, the pelletizer is limited to a maximum slag flow of three tons per minute, so most blast furnace installations require two pelletizers (Fig. 14). The pelletizer has enjoyed considerable success, with at least 27 installations in eight countries by 1986.[12]

FIGURE 14: TWIN PELLETIZER INSTALLATION AT ARBED, BELVAL IN LUXEMBOURG showing foaming tables and pelletizing drum. The right-hand machine is covered with slag wool.

Among its advantages are a lower installed cost than typical water granulation systems and the ability to produce lightweight aggregate. However, the pelletizer is limited in the amount of slag which it can handle and the equipment is susceptible to damage by unexpected flows of iron. Although it was originally developed to reduce atmospheric emissions, the pelletizer has no method for controlling the steam, noise and remaining H_2S (10 ppm) which are produced during operation. (H_2S is detectable by smell at concentrations as low as 13 ppb).

Slag wool is captured to a certain extent by a large chicken wire canopy surrounding the pellet landing zone, but the large space required for flinging the pellets and the airflow necessary for cooling make further pollution control enhancements difficult. The requirement for a loader and operator add to the operating and capital cost, and in terms of maintenance, the pelletizer is not as robust as most water granulation systems. In Europe, the lightweight aggregate produced by the pelletizer commands a premium price, about twice that of cement-grade slag, so ARBED, a major Luxembourg steel producer made significant but unsuccessful efforts to address these concerns. A large concrete hood was installed to control noise and emissions, and vibrating screens/feeders were arranged in the pellet landing area to extract and pre-size-size the pellets.

CONTINUOUS SLAG GRANULATION AND DEWATERING SYSTEMS

Increased iron production and the acceptance of slag cement in Japan and Europe created a demand for slag granulation as a means of handling the large quantities of slag produced and adding value to the product. The rebuilding and upgrading of older existing facilities which had originally been designed for much lower production rates resulted in severe space constraints. There was no plant area available near the furnace to incorporate or enlarge slag pits, slag pots were prohibitively expensive to maintain and the older plants were so crowded that remote pit or hopper dewatering systems could not be installed. The solution arrived at in the late 1970s, which appears obvious in retrospect, was to replace the static filtering of the pits and hoppers with dynamic dewatering machines.

BUCKET WHEEL DEWATERING SYSTEM

Developed by AJO-STAHLBAU in the late 1970s, the system consists of a bucket wheel rotating in a trough (Fig. 15). The slag is granulated and fed into the trough where it is scooped up by the bucket wheel. The bottoms of the buckets have steel or plastic screens to allow the

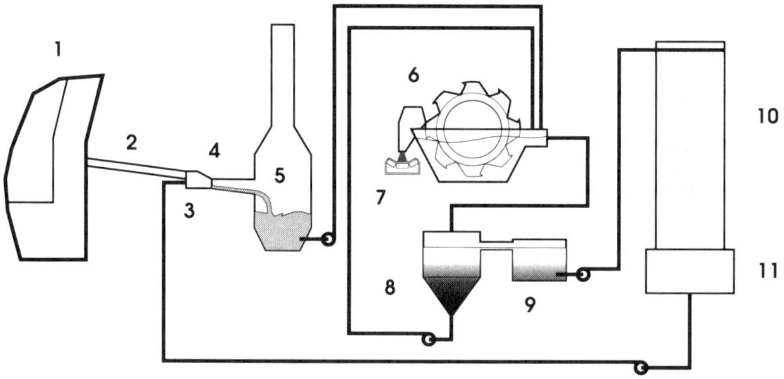

FIGURE 15: BUCKET WHEEL SYSTEM

1) Blast Furnace
2) Slag Runner
3) Granulation Head
4) Cold Runner
5) Agitating Tank
6) Bucket Wheels (2)
7) Unloading Conveyor
8) Fines Settling Tank
9) Hot Water Overflow Tank
10) Cooling Tower
11) Cold Water Reservoir

water to drain from the slag. The dewatered slag is then deposited on a conveyor belt for transport to a truck loading silo or stockpile.[13] A bucket wheel system was installed at Algoma Steel's No. 7 Blast Furnace in 1985 along with a grinding plant to make slag cement, primarily for the mine backfill market.[14]

SCREW CLASSIFIER DEWATERING SYSTEM

The RASA Trading Co. of Japan developed a system of dewatering based on the use of a series of screw classifies to remove slag granules from settling tanks and deposit them on a conveyor (Fig. 16). The slag is granulated and fed into an angled trough. The screw conveys the slag from the first trough to a secondary set of screws for final dewatering. From here, the slag falls on a belt conveyor for removal from the furnace area.[15]

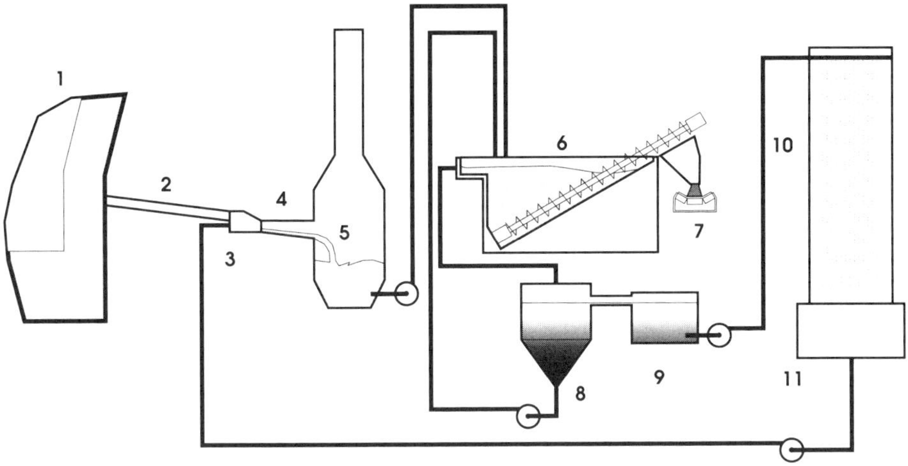

FIGURE 16: SCREW CLASSIFIER SYSTEM

1) Blast Furnace
2) Slag Runner
3) Granulation Head
4) Cold Runner
5) Agitating Tank
6) Screw Classifier
7) Unloading Conveyor
8) Fines Settling Tank
9) Hot Water Overflow Tank
10) Cooling Tower
11) Cold Water Reservoir

'INBA' FILTER DRUM DEWATERING SYSTEM

In 1974, SIDMAR, a Belgian steelmaker installed a RASA remote filter hopper system with agitating tank on blast furnace 'A' due to lack of available space, operational problems with filter bed systems on their other furnace, confidence in Japanese technology and the necessity to abandon open-loop water circuits for environmental reasons. Citing a number of disadvantages, including maintenance and energy costs, SIDMAR proceeded with the design of their own dewatering system, first as a scale model and then in a full-scale pilot operation, commissioned in March, 1981 (Fig. 17).[16]

FIGURE 17: INBA SYSTEM PILOT INSTALLATION AT SIDMAR, BELGIUM, commissioned in March, 1981. The system is seen in operation, the dewatered slag is visible on the conveyor belt coming from the drum. All emissions from the granulation process are captured.

The pilot plant made use of the agitating tank from the existing filter hopper system on Blast Furnace 'B' as a receiving hopper for the granulated slag slurry. From here, the slag feeds into the dewatering drum through a distribution box which spreads the slag along the entire length of the drum (Fig. 18). The water filters through the slag layer and the drum screens, and is collected in the water tank below the drum (Fig. 19). As the drum rotates, the internal vanes lift the slag out of the water and deposits it on the conveyor inside the drum. The conveyor removes the dewatered slag granules from the furnace area.

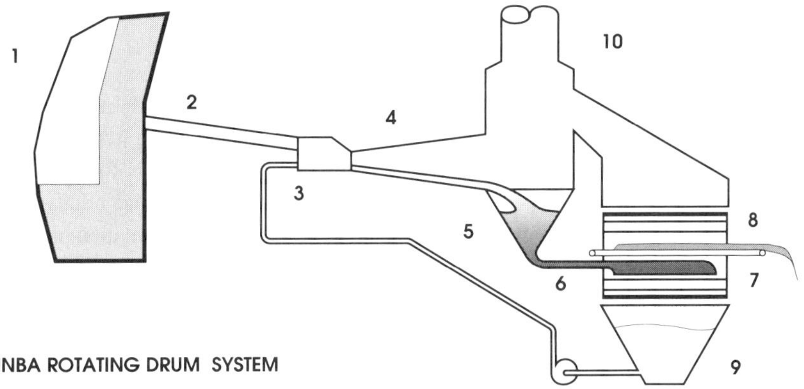

FIGURE 18: INBA ROTATING DRUM SYSTEM

1) Blast Furnace
2) Slag Runner
3) Granulation Head
4) Cold Runner
5) Receiving Hopper
6) Distributor
7) Rotating Drum
8) Unloading Conveyor
9) Water Tank
10) Condensation Tower

The filtered water is pumped back to the granulation head (hot water system) or to cooling towers (cold water system). High pressure water and compressed air nozzles continually clean the drum screens during operation. The PLC control system monitors the instantaneous slag flow rate and adjusts the drum speed accordingly. As in the filter bed system, the steam is collected from the granulation process and condensed in the tower above the

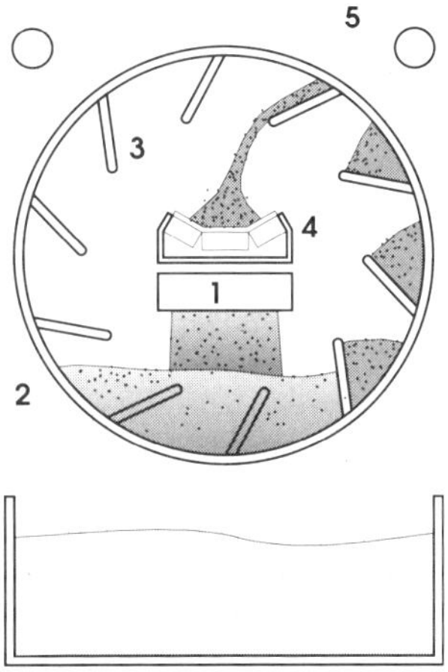

FIGURE 19: INBA DRUM CROSS-SECTION

1) Distributor
2) Slag Layer
3) Axial Vanes
4) Unloading Conveyor
5) Purging Air and Water Sprays

receiving hopper. The filter drum is completely enclosed so residual steam emissions are eliminated.

The compact size permits the dewatering/handling portion of the plant to be located adjacent to the furnace so pumping of the slurry is not normally required. The two most recent granulation systems to be installed in North America have been INBA's at Weirton Steel and USX Gary Works.

COMPARISON OF CONTINUOUS SYSTEMS

Each of these processes make it possible to granulate, dewater and transport the slag away from the furnace at normal tapping rates. Each is compact enough to allow direct gravity- fed operation from most furnaces and each is significantly less expensive to build and operate than previous static dewatering systems. Due to their compact size and relatively concentrated emissions, capture and treatment of the steam generated in the granulation process is relatively easy.

The screw classifier system suffers major wear problems due to the extreme abrasiveness of the granulated slag. The bucket wheel also experiences wear problems, but to a lesser extent because there is less direct grinding action with the slag granules. Since the process water does not get filtered during the process, both systems permit high concentrations of fines in the process water.[17] This leads to abrasion of components and necessitates additional settling circuits for these systems. The INBA system deposits the slag directly into the drum, so there is no abrasion from trying to dig through a pile of slag. The incoming slag falls on the layer of slag already in the drum, so the drum structure and the filter screens are protected from the direct impact. The water filters through the slag layer as well as through the screens. This 'self-filtering' effect makes the INBA work like a gravel layer filter system and traps fines which would otherwise escape into the process water. The other systems decant the process water from the slag receiving tank, whereas the INBA separates the slag from the water before the water is reused.

The self-filtering effect of the slag layer in the INBA drum traps fines which would otherwise escape into the process water

These process differences result in increased plant availability and significant savings in operating and maintenance costs for the INBA System. For this reason, it is currently the most popular granulation system, with over thirty-five installations built since 1981, including 12 in Japan and two in the U.S. In addition to iron blast furnace slag processing, the INBA technology is being applied to the granulation of steelmaking slags and nonferrous slags. The granulation of nonferrous matte is also under investigation.

ECONOMIC FACTORS

A number of factors have delayed the application of granulation technology in North America, while encouraging it in Europe and Japan. Until recently, the economic advantage of slag cement was insufficient to overcome the capital start-up costs associated with what would essentially be a new industry.

MARKET FORCES

Furnace operators in North America have not had an established market for slag cement, even though slag was first used for Portland cement in the U.S. in 1896.[18] To them, slag has always been a problem for disposal. For this reason, most North American ironmakers air-cool the slag and subcontract their slag handling to third parties, sometimes receiving a small payment for the raw material. The contractor normally excavates the slag and sells it as construction bulk aggregate. The size of the North American market has absorbed all the air-cooled slag which could be produced,[19] so no need was seen to 'improve' the product by the aggregate suppliers.

The use of ground slag as a replacement for Portland cement means marketing granulated or ground slag to the cement companies. For the cement producers to substitute ground slag for up to 50% of their Portland cement would require that they keep the slag separate from the Portland for addition at the mixer (separately ground, blended slag cement), or mix the granulated slag with the cement clinker before grinding (inter-ground blast furnace slag cement). Separate grinding requires separate plant facilities for drying, grinding, handling and storage. On the other hand, intergrinding can give varying results because the slag, which must be more finely ground, is normally harder than the cement clinker. The clinker is, therefore, selectively ground more finely than the slag, with potentially adverse effects on the strength of the product. The trend in North America is toward separately ground slag as a replacement for Portland cement.[20]

The ability of ground slag to replace Portland cement allows cement producers to increase their production without increasing plant capacity. However, in the period from 1930 to 1940, North America was in the Great Depression and from 1941 to 1945 the Second World War took its toll on the economy with the result that cement producers 'enjoyed' 16 years of excess capacity. During this period, there was no incentive to invest in slag granulation and grinding equipment.

The following 12 years saw a shortage of cement capacity, but the technology to produce ground slag on a commercial scale had not been developed , and then, as now, North America benefited from extensive natural resources and cheap energy, so the energy saving inherent in blast furnace slag was not particularly attractive.

Due to the shortage of clinker capacity, the cement industry increased capacity between 1957 to 1960, to the point that clinker capacity became over-built. Once again, this excess capacity became an economic barrier to the use of ground slag for cement. With the facilities in place and the capital dollars spent, the marginal cost to produce and grind clinker is low compared to the marginal cost to purchase and grind slag.[21]

Table 1: KEY EVENTS IN BLAST FURNACE SLAG HISTORY

1822	GERMANY	Granulated slag and lime used in cement
1830	USA	Slag used for road construction
1892	GERMANY	Commercial production of Portland blast furnace slag cement
1896	USA	Granulated slag and slaked lime cement production started
1902	USA	First Slag Cement Specification by U.S. Corps of Engineers
1910	JAPAN	First blast furnace slag cement produced in Japan
1938	USA	First Portland blast furnace slag cement produced in the U.S.
1946	USA	ASTM C205 - first U.S. Portland blast furnace slag cement specification
1968	CANADA	National Slag Pelletizer invented
1978	GERMANY	First continuous granulation system developed (bucket wheel)
1981	BELGIUM	INBA Rotating Drum granulation system developed
1982	USA	First modern North American granulation system, (filter beds) built by Atlantic Cement (now Blue Circle)
1986	CANADA	Bucket wheel system built at Algoma Steel by Reiss Lime (now Koch Carbon)
1990	USA	INBA system built at Weirton Steel by Koch Carbon
1991	USA	INBA system built at USX Gary by Koch Carbon

STANDARDS

Another barrier to the development of the separately ground slag market was the lack of applicable standards. It was not until 1982 that ASTM C989, "Standard Specification for Ground Iron Blast Furnace Slag for use in Concrete and Mortars" was published.[22] Previous specifications in the U.S. covered slag cement (grinding together granulated blast furnace slag with slaked lime - 1902) and Portland blast furnace cement (inter-ground slag and clinker - C205 - 1946, now type IS in C595)[23] but the consensus in the industry favors separately ground, blended slag cements both in terms of performance and the ability to tailor the blend to the end user's application.[24]

QUALITY CONTROL

Early slag granulation systems had significant problems with pollution and because slag chemistry was not well understood, and blast furnace operation was a highly variable 'art', the quality of the slag product was often inferior. This led to a decline in production until, by the early 1920s, only one slag cement plant still operated.

The commercial production of Portland blast furnace slag cement is generally considered to have begun in Germany in 1892. It spread rapidly through Europe, whose limited available cement and centralized steel industry led naturally to this method of enhancing production.

In the U.S. Portland blast furnace slag cement (type IS) was first manufactured in 1938 and by the mid-1950s, 12 cement plants were producing it.[25] However, the increase in kiln capacity from 1957 to 1960 diminished the market for Portland slag cement. Compared to European production, where blast furnace slag cement has comprised up to 60% of total cement production (Netherlands),[26] U.S. production has never exceeded 1.5% of total cement production.[27]

In contrast to North America, Europe and Japan had a number of economic and logistic incentives to granulate slag for cement. These include lack of naturally occurring raw materials, serious space constraints and large steel industries.

IRONMAKING TECHNOLOGY

Another factor which influenced the growth of slag granulation technology was the Second World War. In Japan and Continental Europe, much of the steel industry was destroyed. Subsequent rebuilding after 1945 was undertaken with a conscious effort to modernize and rationalize iron production with the result that large gains in productivity ensued.[28]

In 1910, the dawn of the modern ironmaking age, the blast furnace had developed to the point that it would be recognizable to a modern furnace operator. The best blast furnaces could produce up to 500 tons of iron per day. By 1930, designs had been developed for a 1000 ton per day furnace, essentially scaling up the earlier furnace and making some mechanical and structural improvements. This style of furnace was in widespread use in the U.S. by the start of World War II. After the war, the design was enlarged again to produce 1200 to 1500 tons per day and although the design is now more than 40 years old, this furnace is still prevalent in the North American steel industry, albeit modified considerably. Today, this type of furnace will produce 3000 tons per day and in some cases, with extensive modernization up to 5000 tons per day.[29]

Partially as a result of a conscious effort to maximize efficiency, the Japanese and Europeans began to build much larger blast furnaces, starting in the late 1960s and early 1970s. The largest blast furnaces are capable of 10,000 tons per day or more, a quantum leap from the previous generation of furnace. The impact of blast furnace technology on slag production is also significant. In the mid 1950s, the slag produced per ton of pig iron in the U.S. using high-grade ores averaged about 1000 lb. Therefore, for a production rate of 1500 tons of iron, 750 tons of slag would be produced. In Europe, where lower quality ores were used, the ratio could reach 1:1, or a production of 1500 tons of slag per ton of iron.[30]

Slag production as a function of iron production has decreased dramatically since then, but the increase in furnace size has meant that slag handling is a bigger problem than ever. A furnace producing 8000 tons per day of iron will produce about 2400 tons per day of slag, or about 42,000 ft^3. The time and space necessary to air-cool and excavate such a large quantity of slag is a considerable incentive to granulate, purely as a means of simplified slag handling.

Table 2: ECONOMIC FACTORS IN THE SLAG MARKET

	NORTH AMERICA	EUROPE	JAPAN
CEMENT MARKET	- Excess Cement Capacity - Cheap Energy - Available Raw Materials	- Limited Cement Capacity - Limited Energy - Limited Raw Materials	- Limited Cement Capacity - Limited Energy - Limited Raw Materials
AGGREGATE MARKET	- Large Market - 100% Use of Slag	- Large Market - Substitutes Available	- Large Market - Substitutes Available
IRONMAKING TECHNOLOGY	- Leading Technology Before WWII - Aging Plants, Many Small to Midsize Furnaces - No Disposal Problem - Less-onerous Emission Control Laws	- Rebuilt after WWII - High Production - Slag Disposal Problem - Emission Control Laws	- Rebuilt after WWII - High Production - Slag Disposal Problem - Emission Control Laws - Scarce Land Area
DISTRIBUTION CHANNELS	- Large Territory - Transportation Network Undeveloped	- High Population Density - Well-developed Transportation Network	- High Population Density - Well-developed Transportation Network

DISTRIBUTION CHANNELS

Another influence on the growth of slag granulation and slag cement is the transportation infrastructure. Many European and Japanese steel mills are located on navigable waterways. Both Europe and Japan have well-developed transportation systems and high population densities. The short distances and inexpensive transportation costs, in contrast to North America, are favorable for the market development of a bulk product like slag cement.[31] The first modern slag cement facility in the U.S. was the Atlantic Cement (now Blue Circle) plant at Bethlehem Steel in Sparrows Point, Maryland, built in 1981. One of the keys to its success has been this plant's access to the entire eastern seaboard by water transportation.

To summarize, factors which have limited growth of the slag cement (and, therefore, slag granulation) market in North America, include unfavourable early experiences with the product, excess capacity in the cement industry, long distances to markets, full use of air-cooled slag by the aggregate industry and a certain lack of interest in slag as a value-added product by steelmakers.

EMISSION CONTROL SYSTEMS

Emission control has increasingly been a concern in the design of new and retrofit installations. During the 1970's, heightened awareness and stricter legislation resulted in the addition of pollution control systems to slag processing plants.

The primary pollutants from slag processing are sulphur species, specifically, SO_2 and H_2S, resulting from the sulphur in the slag coming into contact with air and water at elevated temperatures. The sulphur originates in the metallurgical coke used as fuel for the blast furnace. The rapid quenching which occurs in the granulation process is beneficial in trapping the sulphur within the slag to a certain extent. The emissions which are generated are relatively easy to contain, since the granulation head acts as a point source. This allows the emissions to be treated directly by means of a condensation tower.

The primary pollutants from slag processing are SO_2 and H_2S...

Early non-continuous systems like the gravel layer filter bed system still had difficulty to deal with fugitive emissions from the filter beds themselves, so large, movable covers were added to the beds.

COLD WATER VERSUS HOT WATER

Granulation with recirculated water reduces the potential for pollution by overflow from the slag granulation process. However, the quantity of slag produced during a typical furnace cast is sufficient to raise the temperature of the water considerably. This has the effect of increasing the amount of steam generated, thereby increasing the required design capacity of the condensation tower. Additional process water is required to completely condense the steam, so an overflow of process water must be handled within the plant.

Incorporating a cooling tower in the water circuit, with the attendant reservoirs and pumping systems, allows the make-up water entering the circuit to be better balanced with the process requirements while still condensing all the steam. This minimizes the overflow from the system and reduces the cost of the condensation equipment.

Granulating with cold (35° C) water was originally felt to be a requirement for the production of cement-grade slag. However, recent research into slag cement strengths and new techniques for measuring the glass content necessary for good slag cement indicate that hot water (80° C) granulation will produce good quality granulated slag. Cold water granulation has been shown to result in lower absolute levels of SO_2, apparently due to improved solubility of SO_2 in cold water. Where maximum capture of emissions is required, the trend is to cold water granulation.

The condensed steam, H_2S and SO_2 are mixed with the rest of the water in the granulation circuit. Since it is a closed system, dissociation of the ions in the slag occurs to a certain extent until a stable concentration is reached. The large quantity of lime in the slag results in a basic solution (average pH of 8-9) within the water circuit. Residual H_2S and SO_2 is neutralized and any precipitate is recirculated to the granulation head and then is filtered out in the slag.

DISPOSAL OF WASTE OR BLOWDOWN WATER

It is relatively common practice to use blowdown water from the blast furnace gas cleaning plant to cool the slag in slag pit operations, thereby disposing of water which would otherwise require expensive treatment to remove heavy metals such as zinc and lead, ammonia, cyanides, phenols and fluorides. The evaporation of this blowdown water on air-cooled slag forms the EPA BAT-1 treatment model as of 1982.[32] It appears that blowdown water could be used as make-up water for slag granulation systems without affecting slag quality, since a wide range of water qualities have been used in the granulation of slag for use as cement. Pollutants which might be trapped in the slag are at such low concentrations that cement quality would not be affected.

CONCLUSIONS

In contrast to Europe and Japan, the development of the slag cement market in North America was inhibited for a number of economic reasons, including the geographic distances involved in delivering the product to the market, the initial capital investment required to enter the market, the existence of a developed, alternative market for primary raw material and the lack of demand caused by overcapacity in the Portland cement industry. The relative indifference to the use of slag as a cement substitute has restricted growth of slag processing in the North American ironmaking industry until recently.

Table 3: BLAST FURNACE SLAG HANDLING SYSTEMS			
TIME FRAME	**SYSTEM**	**ADVANTAGES**	**DISADVANTAGES**
1700's	**SLAG PITS**	-Simple solution	-Emissions hard to control -Low value product, -Expensive loader -Large plant area needed
1800's	**PIT GRANULATION**	-Simple solution -Cement-grade product	-Emissions hard to control -Dewatering difficult -Handling difficult
1900's	**SLAG POTS**	-Compact arrangement	-Expensive railroad costs -Slag must still be dumped -Low value product
1960's	**GRAVEL LAYER FILTER BED SYSTEM**	-Cement-grade product -Reliable -Can handle large slag amounts	-Capital-intensive -Large plant area -Normally, abrasive slag must be pumped
1968	**FILTER HOPPER SYSTEM**	-Cement-grade product -Requires less space than Filter Bed System	-Capital-intensive -Large plant area -High wear and plugging due to fines, wool
1968	**PELLETIZER**	-Compact -Inexpensive -Can produce semi-light aggregate, cement slag	-Emissions hard to control -Limited slag capacity, -Affected by slag quality
1978	**BUCKET WHEEL SYSTEM**	-Compact, continuous -Cement-grade product -Good emission control	-High wear and plugging due to digging action and fines
1979	**SCREW CLASSIFIER SYSTEM**	-Compact, continuous -Cement-grade product	-High wear and plugging due to digging action and fines
1981	**INBA ROTATING DRUM SYSTEM**	-Compact, continuous -Cement-grade product -Good emission control -Low maintenance	

Over the past thirty years, slag granulation, dewatering and handling systems have undergone considerable development in Europe and Japan, with the result that the cost per ton of slag handled has decreased, capital costs for the installations have decreased, the required installations can now be retrofitted to even the most congested plants and complete treatment of emissions is possible.

In North America, increased awareness of pollution control, recycling and energy conservation, coupled with economic opportunities in the use of blast furnace slag have resulted in granulation systems at Bethlehem Steel (Atlantic Cement, 1979), Algoma Steel (Reiss Lime, 1985), Weirton Steel and USX Gary (both Koch Carbon, 1991).

REFERENCES

1. McLean, A., "Historical Development and Principles of the Iron Blast Furnace" (Lecture Presented at the Blast Furnace Ironmaking Course, McMaster University, Hamilton, Ontario, May, 1987) P. 15

2. Lea, F. M., The Chemistry of Cement and Concrete, Third Edition, (Edward Arnold (Publishers) Ltd., 1970) P. 455

3. Emery, J. J., "Slag Utilization in Pavement Construction" Extending Aggregate Resources, (American Society for Testing and Materials, 1982) P. 99

4. Stephenson, R. L., "Availability and Uses of Blast Furnace Slag" (Paper presented at the University of Alabama in Birmingham Slag Cement Seminar, April 30-May 1, 1981) P. 7

5. Bintner, E., et al, "Paul Wurth's Granulating Process" (Paper presented at the University of Alabama in Birmingham Slag Cement Seminar, April 30-May 1, 1981) P. 2

6. Emery, J. J., "Slag Utilization in Pavement Construction" Extending Aggregate Resources, (American Society for Testing and Materials, 1982) P.102-103

7. Hooton, R. D., "The Reactivity and Hydration Products of Blast Furnace Slag" Supplementary Cementing Materials for Concrete (ed. Malhotra, V. M., Canadian Centre for Mineral and Energy Technology, 1987) P. 254-255

8. Fieser, A. H., "Slag and Slag Cement Practices in Japan" (Paper presented at the University of Alabama in Birmingham Slag Cement Seminar, April 30-May 1, 1981) P. 17

9. Mueller, K., "Hoesch/Ajo Slag Granulation System" (Paper presented at the University of Alabama in Birmingham Slag Cement Seminar, February 18-19, 1982) P.7

10. Monteyne, G., "Sidmar's Blast Furnace Granulation Systems" (Paper presented at the University of Alabama in Birmingham Slag Cement Seminar, February 18-19, 1982) P.4

11. Cotsworth, R. P., "National Slag's Pelletizing Process" (Paper presented at the University of Alabama in Birmingham Slag Cement Seminar, April 30-May 1, 1981) P. 2

12. Hooton, R. D., "The Reactivity and Hydration Products of Blast Furnace Slag" Supplementary Cementing Materials for Concrete (ed. Malhotra, V. M., Canadian Centre for Mineral and Energy Technology, 1987) P. 254

13. Mueller, K., "Hoesch/Ajo Slag Granulation System" (Paper presented at the University of Alabama in Birmingham Slag Cement Seminar, February 18-19, 1982) P. 11-13

14. Kirkpatrick, E., "Production of a New Granulated Blast Furnace Slag at Reiss Lime Company of Canada, Ltd." (Paper presented at the International Workshop on Granulated Blast Furnace Slag in Concrete, Toronto, Ontario, October 22-23, 1987) P. 426-427

15. Fieser, A. H., "Slag and Slag Cement Practices in Japan" (Paper presented at the University of Alabama in Birmingham Slag Cement Seminar, April 30-May 1, 1981) P. 24-25

16. Monteyne, G., "Sidmar's Blast Furnace Granulation Systems" (Paper presented at the University of Alabama in Birmingham Slag Cement Seminar, February 18-19, 1982) P.4

17. Kirkpatrick, E., "Production of a New Granulated Blast Furnace Slag at Reiss Lime Company of Canada, Ltd." (Paper presented at the International Workshop on Granulated Blast Furnace Slag in Concrete, Toronto, Ontario, October 22-23, 1987) P. 429-430

18. Lewis, D., "History of Slag Cements" (Paper presented at the University of Alabama in Birmingham Slag Cement Seminar, April 30-May 1, 1981) P. 3

19. Hooton, R. D., "The Reactivity and Hydration Products of Blast Furnace Slag" Supplementary Cementing Materials for Concrete (ed. Malhotra, V. M., Canadian Centre for Mineral and Energy Technology, 1987) P. 250

20. Spellman, L. U., "Advantages for slag Cement in the North American Market" (Paper presented at the University of Alabama in Birmingham Slag Cement Seminar, April 30-May 1, 1981) P. 1

21. Hauser, K., "Slag Cement, Cement Slag and Their Specification" (Paper presented at the University of Alabama in Birmingham Slag Cement Seminar, February 18-19, 1982) P. 3-4

22. ASTM C989, Standard Specification for Ground Iron Blast Furnace slag for Use in Concrete and Mortars. (American Society for Testing and Materials, 1982)

23. Lewis, D., "History of Slag Cements" (Paper presented at the University of Alabama in Birmingham Slag Cement Seminar, April 30-May 1, 1981) P. 5

24. Spellman, L. U., "Advantages for Slag Cement in the North American Market" (Paper presented at the University of Alabama in Birmingham Slag Cement Seminar, April 30-May 1, 1981) P. 1

25. Lewis, D., "History of Slag Cements" (Paper presented at the University of Alabama in Birmingham Slag Cement Seminar, April 30-May 1, 1981) P. 3-5

26. Wiebenga, J. G., "Blast Furnace Slag Cement in the Netherlands" (Paper presented at the International Workshop on Granulated Blast Furnace Slag in Concrete, Toronto, Ontario, October 22-23, 1987) P. 24

27. Lewis, D., "History of Slag Cements" (Paper presented at the University of Alabama in Birmingham Slag Cement Seminar, April 30-May 1, 1981) P. 5

28. McLean, A., "Historical Development and Principles of the Iron Blast Furnace" (Lecture Presented at the Blast Furnace Ironmaking Course, McMaster University, Hamilton, Ontario, May, 1987) P. 31

29. McLean, A., "Historical Development and Principles of the Iron Blast Furnace" (Lecture Presented at the Blast Furnace Ironmaking Course, McMaster University, Hamilton, Ontario, May, 1987) P. 15, 31, 34

30. Stephenson, R. L., "Availability and Uses of Blast Furnace Slag" (Paper presented at the University of Alabama in Birmingham Slag Cement Seminar, April 30-May 1, 1981) P. 2-4

31. Pierson, C. U., "Realities of Producing and Marketing Blended Cements" (Paper presented at the University of Alabama in Birmingham Slag Cement Seminar, April 30-May 1, 1981) P. 3

32. U. S. Environmental Protection Agency (EPA), Development Document for Effluent Limitations: Guidelines and Standards for the Iron and Steel Manufacturing Point Source Category (Vol. II, Final), Ironmaking Subcategory, EPA 440/1-82/024/ (Washington, D. C.: EPA, Effluent Guidelines Division, May, 1982) Pp. 348, 353, 375, 376, 397, 400, 402, 411-418

Treating Zinc Crust by Vacuum Distillation

Dai yongnian

Research Institute of Vacuum Metallurgy
Kuuming Institute of Technology
Kunming 650093, Yunnan, P.R.China

Abstract

The methods of treating zinc crust and its development in the last hundred years are compared and reviewed. There are better effects from normal pressure metallurgy turned into vacuum metallurgy. The traditional methods dezinc only. But we must consider that delead on further step. From lab tests to industrial scale experiment are researched. The first step-dezinc, second-delead, crude metal of zinc, lead and silver are produced. The flowsheet of treating zinc crust are simplified, material consumption reduced, work condition improved, environment protect mended. This method is pointed out that its application can be expanded.

Residues and Effluents - Processing and Environmental Considerations
Edited by Ramana G. Reddy, William P. Imrie and Paul B. Queneau
The Minerals, Metals & Materials Society, 1991

1.Foreward

About 80% crude lead of the world is refined by pyrometallurgy processes. The major step of it is desilverization by adding zinc. The precious metals in the crude lead are extracted into the zinc crust. Then silver and gold are recovered from zinc crust. Since a major part of silver of the world comes from the crude lead, the treatment of zinc crust becomes a principal process in the pyrorefining lead smelters.

In recent years, the productive equipment of pyrorefining lead are built up in our country. So, zinc crust must be treating. From 1986, this project is researched by Research Institute of Vacuum Metallurgy, Kunming Institute of Technology. They suggest a program to dezinc, to delead and to get crude silver. In 1987, the institute undertook the research work. Then crude silver (contains~ 70%Ag)(2)(8), crude lead and crude zinc are produced. And a new process of crystallization to extract silver was tested. All the experiments were at industrial scale, which were successful. Its flowsheet is short and equipments are few. These achievements of getting three crude metals had not reached in many fireign lead smelters.

Many foreign methods of treating zinc crust are introduced in this paper, and those compared with our process for selecting optimum plan of our smelter. Really, this process may be used to reform the lead smelters abroad.

2. The traditional methods of treating zinc crust

The ancient process in which silver was separated from crude lead is called "Cupellation". In the process lead is oxidized and oxide dross is formed, the residue was metal silver. Partial crystallization process was discovered by Pattinson in 1833. After many times of crystallization, lead contains 8.5~ 14g/t silver and an alloy contains 1.8% Ag is gotten. Oxidization of large amount of lead is kept away from the flowsheet. In 1850, the Parkes process was suggested. A certain amount zinc is added into molten lead, then a Zn-Ag(Au)-Pb alloy is formed and floated on the lead surface. It is called zinc crust. Nearly all silver (gold) in the lead is caught in the zinc crust. This method is used in all pyrometallurgy lead smelter, till now.

The zinc crust compositions of smelters are different as table 1.

Table 1. Zinc crust compositions of many plants,%(5)(6)

	Ag	Zn	Pb	Au	Cu
N plant	10.69	10.7	76.54		3.3
N plant	3.4	2.2	75.5	0.015	
Friberge	4.0	40	54	0.015	2.0
Trifolberge	2.6	20.0	60		
Presse crust	10	30	60		
Continuous desilver crust	15~20	60~65	15~20	0.016 ~0.019	0.4
Liquation under salt layer crust	25	65	8~10		

The data of table 1 shows that major components of zinc crust is Ag-Zn-Pb alloy, another little part is many other elements. So the purpose of treating zinc crust is to separate the three metals and to produce single metal for use.

After Liquation of zinc crust, so much lead and a rich silver alloy are produced. Then through dezincification and deleadization processes, crude silver is obtained.

Dezincifization in the early time retort distillation was used, it is reported in 1898. The zinc crust is put in the retort, fuel burns outside the retort to 1200°C, in the retort 1000~1200°C. Zinc evaporated, Zinc vapor diffused through the opening of zinc is 65-90%. By-products amount: blue powder 1%, dross 5-10%, and alloy is produced. This alloy contains 4-10% Ag, 80-85% Pb, 9.5-10% Zn.

In this process, zinc separation from zinc crust is accomplished. In recent years the equipment has some improvement. The retort is enlarged, the fuel from coal has changed to gas or liquid fuel, and the process is mechanized, Till now this process and its equipment is used widely, German Stoberge lead smelter use it, and according to the scale of our smelter German Specialist suggest to use it.

In U.S.S.R, from 1954 distilling the zinc crust in arc furnace is used. Zinc vapor is condensed in spray condenser. The furnace power was 400KVA, area of furnace bottom-2.9 m^2, inside height of the furnace-1.8m, three graphite electrodes diameter 200mm, revolutions of condenser's graphite rotor 500-900 rpm. Volume of condenser $2.7 \times 1.0 \times 0.5 m^3$. The composition of forming slag: SiO_2 40-45%, CaO 20-30% Na_2O 25-35%,1-2%. Coke breeze is added into the raw material. Temperature under the furnace roof 1000-1100°C, phase voltage 100-120V, electricity 1500-2000A. Depth of the hearth 450-500 mm, which contains 150-200mm lead layer. Inside furnace, gas pressure 10-59 pa, at time of charging 200-246 pa larger than atmosphere, the slag is replaced per day, furnace capacity 3-4 t/m^2, electric power consumption 600 KWH/t, electrode use up 5 Kg/t. The compositions of raw materials and products are show in table 2.

Table 2. Composition of raw materials and products of electric furnace smelting zinc crust

	Rate	Composition(C) and Recovery(R)							
	%	Pb		Zn		Au		Ag	
		C,%	R,%	C,%	R,%	C,g/t	R,%	C,g/t	R,%
Charged	100	65.8	100	22.69	100	457	100	86187	100
Zinc crust	2.2								
Coke breeze									
Obtained									
Pb-Ag alloy	74.4	86.7	98.81	1.53	5.0	604.5	98.83	114505	99.28
Condensed zinc	19.7	1.52	0.46	98.22	85.5	25.8	1.12	2185	0.5
Oxide	1.5	4.46	0.10	7.43	5.0	15.0	0.05	1803	0.01
Dust, loss	6.3	----	0.63	----	4.5	----	----	----	0.19

These two normal pressure distillation processes are good for vaporization of zinc. Zinc is obtained and Pb-Ag alloy is obtained as residue.

The first industrial vacuum furnace for distillation of Zinc crust is called Lefferrer Vacuum furnace which is published in 1957. In the furnace, graphite rod resistance are used for heating. The process temperature 750-800°C, electric power 100 KVA, residue gas pressure 1333pa, treating charge amount 850-2000 Kg/d, duration of batch operation 14hr. Condenser's temperature 450°C, power consumption 850-850 KWH/t. Compositions of charge and products are shown in Table 3. Metal loss: silver 1.3%, zinc 10%.

After this furnace was used in many smelters, the superiority of vacuum distillation for zinc crust was shown.

The induction vacuum furnace for treating zinc crust is published in 1985[10], named Hoboken vacuum retort, which is a electric induction furnace 450 KVA, lined with graphite brick. Its condenser is lined by fire brick. A rotating Vane pump for drawing gas. Working pressure 1000 pa. Charge batch 3.5 t high silver content zinc crust (about 28-30%Ag), process duration: melting/ distillation/discharge 12-14 hr. Productions: zinc and silver alloy (contains 65-70% Ag). The material in the furnace is agitated by electro-magnetic force, it is useful for distillation.

Table 3 Treating zinc crust in furnace

	Amount	Pb		Ag		Zn	
	Kg	C,%	R,%	C,%	R,%	C,%	R,%
Charged							
Zinc crust	1000	8	100	25	100	65	100
Lead	300						
Produce							
Pb-Ag alloy	650	57	97.5	38	98.8	1.75	0.17
Zinc	630	1.8	2.98	0.15	0.38	98	94.95
Blue powder	20						
dross		3			0.15	94	

Table4 shows; dezinc and producing zinc, Pb-Ag alloy is produced in the four furnace. From normal pressure turned to vacuum process, many results are improved: working temperature reduced from 1000-1100°C to 750-800°C; silver content in alloys raised, then in the next step lead oxide from cupellation reduced, from 0.622 t to 0.1865 t per t of crust.

On comparing the four processes we could arrive at the following conclusion: vacuum furnace is better than the normal pressure processes, and vacuum electric induction furnace are good one. But German specialist said that vacuum induction

Data of these all four furnaces compared in table 4.

Table 4. Data of distilling zinc crust by the four processes

Terms	Normal pressure ditillation	Electric thermal	Vacuumeelectric resistance furnace	Vacuum electric induction furnace
Process style	Batch process	Simecontinuous	Batch process	Batch process
Working pressure, pa	Normal pressure	Normal pressure	1333	1000
Working temperature, ℃	Outside 1200 ° inside 1000–1100 °	inside 1000 ° –1100 °	inside 750–800 °	
Charged crust, Kg	4660	4560	1950 +490 pb	1950 (contains 25–30% Ag)
Batch charge,t	0.5–0.7	4t / m^2.d	0.85–2 t	3.5
Produced pb–Ag alloy, Kg	3230	3040	1120	660
Ag% in pb–Ag alloy	4–10	11.4	38	65–70
Lead oxide produced by alloy cupellation, Kg	2900	2700	900	360
Lead oxide per t crust, t	0.6223	0.592	0.3688	0.186
Energy consumption	heavy oil 180 –270 l / t charge	600 KWH / t Charge	800–850 KWH / t Charge	

furnace is more expensive and its scale is larger than others.

We think that zinc crust dezinc in vacuum distillation is quick, complete and does not oxidize. So vacuum furnace must be used instead of normal pressure equipments as retort and arc furnace. And now, for treating the zinc crust, vacuum furnace must work out in its full advantages, after dezinc to delead as metal vapor. There is no need to oxidize lead forming lead oxide, reducing reduction lead oxide process to get lead metal. So, the flowsheet of treating zinc crust is shortened, metal loss is reduced, production price is decreased.

From our experience of research work and production in many years. We could separate lead from silver in vacuum furnace, and it is economical.

3. The experiments in our Institute

At first, a series of laboratory tests were carried out, then expanding it to industrial scale. Because of the lack of pyro-refining crude lead plant in our country, the test raw material is artificial. The zinc crust is put back into crude lead many times. So it is oxidized deeply. The crust contains a great many oxide, large amount dross in the metal ingot. So before the crust put into research, it must be melted under salt layer. Got alloy 71% by weight, oxide dross 29%. The compositions of alloy are:

	Ag	Pb	Zn
%	1.69	93.34	3.99
	1.08	91.74	2.82

its amount 3650Kg.

A. Vacuum distillation dezinc

Many types of vacuum furnaces are worked with, then a furnace with good effect was selected for smelting the alloy.

The vacuum furnace worked continuously. Test temperature, 700-950°C. Residue pressure in the furnace 666.3-1333 pa. The raw material in the charging furnace is converted with salt layer.

Working time 13 hr, 1781 Kg alloy is treated, it corresponds to 3288 Kg/d. Energy consumption-1470 KWH/t alloy. Salt used up 5Kg/t.

Metal recovery: silver > 99%, zinc > 98.8%

Composition of produced crude zinc: Zn 98.5%, Pb 1.48%

Composition of produced Pb-Ag Alloy; Ag 1.7%, Pb 96%, Zn trace to 0.25%

Dezinc rate: 99%

So, dezinc process is successful product Ag-Pb alloy for next step of treatment.

B. Vacuum distillation delead

A type of vacuum furnace is tested for separating Pb-Ag alloy. 4 t lead/day evaporated. It is built up in July 1988. Continuous operation. The composition of the charge:

	Pb	Zn	Ag
%	97.2	0.05	1.65

process temperature 1020°C. Residue pressure in the furnace 27-53 pa. Charge rate 110-150 Kg/hr. Products amount shown in Table 5.

Table 5. The Amount of products

Charge	Products, Kg		
Kg	Condensed Lead	Residue alloy	dross
3054	2948	43.31	30.90

Composition of products:

	Pb %	Ag %	Au g/t
Condensed lead	99.16	0.19	4.5
Residue alloy	16.40	68.97	1860
Dross	60.22	2.12	48

Real Capacity of the furnace 3.3 t/d
Energy consumption: 725 KWH/T
Recovery of metals:(%) Pb 99.45; Ag 99.1; Au 96.7

Condensed lead can be redistilled, or back to add zinc desilver process, the silver and gold in it can be recovered.

The dross is metal covered with a layer of oxide. After reduction smelting the product lead back to the adding zinc desilver process.

As seen in this process, the lead can be separated as metal not oxide from silver. It is distilled two times, the contents of Ag, Au and Fe in the lead could reach the level of refined lead.

4. Conclusions

Zinc crust treatment, after Parkes process is discovered, becomes an important part of crude lead refinement. In last hundred years, the method have some improvement, the major one is vacuum distillation dezinc instead of normal pressure distillation, which lowers the process temperature from 1000-1100°C to 750-800°C, reduced the production of lead oxide from 0.592-0.6223t to 0.186-0.3688 T/t crust, and improved some data of process. So it instead of normal pressure distillation step by step. But these improvement not concerned the vaporization separating lead of the crust, conserved lead oxidizing process (cupellation) also.

Because the matter evaporates in vacuum more easily than in air. Kunming Inst. of Tech. had been investigating zinc crust distillation. The first step distilled zinc, second process evaporated lead. Crude silver, lead and zinc are obtained. The experiments from the laboratory test to industrial scale all are successful. Crude lead and zinc can be returned to add silver and gold are produced.

The industrial test equipments and results are more better than the traditional process. It can be used in production. The advantages of it: the productive flowsheet is simple, consumption of matters is little, the data of technique and economy is better than the traditional process, working condition is good, and it can be protecting environment good.

It could be predicting that if this method is used in the smelter, the investment of equipment and technique will be lower than importing them from abroad very much. And production cost will be economy also. By using technique, the electrode lead of battery and cable lead are produced by crude lead vacuum distillation in some plant. The economical profit is good. So, for the height valuable zinc crust using these processes will have better profit than traditional methods.

If the continuous process vacuum furnace will be changed to batch process for the high melting point materials, some work of the furnace construction will be done. But on the experiences of ours, it will not be difficult to carry it out.

5. Rerences

1.Dai Youngnian, Discussion of Treating methods for zinc crust, June 1989.

2.Huang Zhikai et al. Zinc crust vacuum distillation to dalead, Industrial experiment Report, June 1989

3.He ZiKai et al. Zinc crust vacuum distillation to delead, Industrial experiment Report, June 1989

4.Li Manlin et al. The primary research of Liquation separating Pb-Ag alloys, June 1989

5.D.M. Chikhnov, Metallurgy of zinc. 1944

6.Teach and Reasearch Division of Base Metal Metallurgy, North-Bast Tech. Inst, Metallurgy of lead, 1960 7.H.O.H Metallurgy of lead, 1898

8.M.P. Smirnov, Refinement of Lead and Treatment of by-products, 1977. pp.164-169

9.V.F. Leferrer: Vacuum Dezincing of Parkes process zinc crust. Trans. AIMEV Nov. 1957. 1450-1460.

10.K.R.Barrett, R.P.Knight. Lead Bullion Reining and Precious Metal Recovery, in <<Extraction Metallurgy 85>> pp.683-708

11.He Zikai et al. Research of vacuum Distillation of Pb-Ag alloy. <<Non-ferrous Metallurgy>>, 1984, No.4.pp16-20

12.Luo Yixiang et al. <<Science and Technology of Kunming Inst. of Tech.>> 1982, No.2 pp 1-8

13.<<Nonferrous Metal>> (U.S.S.R.),1983, No.7. pp.17-18

INTEGRATING JAROSITE RESIDUE PROCESSING IN HYDROMETALLURGICAL ZINC WINNING - COMPARISON OF FIVE POTENTIAL PROCESSES

F.Elgersma

Delft University of Technology
Laboratory for Process Equipment
Leeghwaterstraat 44, 2628 CA DELFT
THE NETHERLANDS

and

T.W. Zegers

Delft University of Technology
Faculty of Mining and Petroleum Engineering
Mijnbouwstraat 120, 2628 RX DELFT
THE NETHERLANDS

Abstract

Hydrometallurgical zinc winning processes are frequently based on iron containing concentrates which yield a goethite, jarosite or hematite residue. Jarosite or goethite are mostly stored in waste ponds which are lined with foil. Water percolating through the pond is withdrawn from the deposit and decontaminated before discharge. Despite these measures, more and more pressure is put upon zinc companies to treat their (historical) residues and to produce environmentally acceptable residues.
The criterion which determines this acceptability is the potentiality of leaching out hazardous elements by natural processes. This paper provides the comparison of five different process designs for integrated jarosite treatment in a hydrometallurgical zinc winning plant leading to environmentally acceptable residues. The processes are designed for producing 200 ktons/yr zinc and for treating 80 ktons/yr historical jarosite. Mass and energy balances are used to determine parameters such as:

- the energy use per ton waste treated and
- the amount and composition of final residues.

The conclusion of the work is that pyrometallurgical jarosite treatment integrated in hydrometallurgical zinc winning is preferable to hydrometallurgical jarosite treatment. Integrated pyrometallurgical processes lead to a smaller quantity of remaining hazardous residues and require less energy.

Residues and Effluents - Processing and Environmental Considerations
Edited by Ramana G. Reddy, William P. Imrie and Paul B. Queneau
The Minerals, Metals & Materials Society, 1991

1. Introduction and Aim

The winning of primary non-ferrous metals from concentrates inevitably leads to the formation of residue materials, since naturally occurring undesirable impurities and valuable metals are part of the same concentrate. Since metal users impose strict specifications on their raw material, the primary metal industry has to separate the desired metal from the impurities. The metal is sold, but impurities frequently remain close to the metal winning plant in a storage facility. Fortunately many 'impurities' are no longer regarded as such, since process technology has become available to win more and more metals economically as by-product. However, in particular iron and arsenic, which are found in many concentrates, are still ending up in residues [1,2].

This paper limits its scope to hydrometallurgical zinc winning. Worldwide, an annual production of about 6 Mton of zinc is realized. About 80 % of this quantity is produced in hydrometallurgical zinc winning plants which consist of a roasting, leaching and electrowinning step. The problem of the dissolution of iron in the leach circuit is solved by precipitating either jarosite, goethite or hematite from the leach solution. Precipitating these iron containing residues also prevents the built-up of some of the other impurities (e.g. Ge, As) which accompany zinc in the concentrate, in the electrolyte. Most hydrometallurgical zinc winning plants separately produce a lead-silver residue which may be commercially used for lead and/or silver winning and a gypsum residue which is formed after bleeding spent acid in order to prevent the accumulation of a.o. magnesium and manganese.
The quantity of iron residue formed during zinc winning depends on the iron content of the concentrate and on the particular iron residue that is precipitated. For a given iron content the overall mass of jarosite residue exceeds the mass of goethite and of hematite. However, operating a jarosite precipitation step in a plant is relatively easy and gives the highest zinc recovery, which has led to the current situation in which jarosite forming plants dominate the zinc winning industry. For this reason this paper only deals with jarosite. The jarosite residue formed during historical and current zinc winning finds no application and is therefore either disposed off in storage ponds, discharged in sea or stored in artificial caves in granite rock. The situation with respect to goethite is similar. Hematite is claimed to find application in cement industries [3].

In The Netherlands storage ponds containing jarosite are lined with HDPE-foils. Percolating water is withdrawn from the deposit and decontaminated before discharge. Despite such measures, more and more pressure is exerted on zinc companies to treat their (historical) residues and to produce environmentally acceptable residues. The criterion which determines this acceptability is the potentiality of leaching out hazardous elements by natural processes. Since current jarosite does not fulfill this criterion, it is regarded as hazardous chemical waste [4]. Society's criticism on (further) storage of hazardous chemical waste has increased, since it is conflicting with the aim to develop a sustainable world economy [5].
In The Netherlands three geographical factors add to the concern on hazardous waste. The first factor is the high population density (± 450 inhabitants/km^2) which means that space for whatever purpose is limited. The second factor is that almost 15 million people are depending on drinking water which to a large extent is extracted from ground water resources. The third factor is the groundwater table (freatic surface), which is very near the surface. Some aqueous pollutants (e.g. nitrate) have already been shown to percolate to these resources as a consequence of human (agricultural) activities. In order to prevent the storage of jarosite residues in The Netherlands from remaining a potential hazard to drinking water resources and, of course, to the soil

itself, it has been decided that further storage of jarosite will be prohibited in the near future. Moreover, current storage facilities are to be emptied. The jarosite residue will have to be converted into an iron product which does no longer possess unacceptable leaching properties. This conversion process is preferentially combined with the primary zinc winning plant.

The aim of this paper is to compare five hydrometallurgical zinc winning processes for integrated residue treatment on residue quantity and composition as well as on energy requirements. The choice for the individual options has been made as follows. Option I considers a treatment unit which is almost entirely complementary to a conventional primary zinc winning plant. A single recycle stream returns the zinc recovered from the jarosite residue to the primary plant. The second option interferes with the primary plant, since two characteristic unit operations in the primary plant are abandoned. The third and fourth option introduce further changes in the primary plant. The fifth option however, utilizes the same primary plant as the first option, but introduces a principally different treatment method. The five options together cover only a small range of possible flow sheets for primary zinc winning plants with integrated residue treatment, but include basically new developments which are worthwhile to be discussed in more detail. Due to the fact that the technology presented in this study has never been built on an industrial scale, it is not possible to include a section on capital costs with any significant accuracy.
This paper might also be useful for zinc winning companies in other countries which face a comparable residue problem, but have to cope with other constraints (e.g. geographical factors, energy prices, residue specifications etc.).

2. Methodology

In the subsequent sections the five processes are discussed as a combination of (unit) operations for which mass and energy balances are calculated.
Mass balances are based on chemical reaction equations (see appendix I) which are supposed to have a conversion of 100% (apart from one well defined exception). This assumption is justified by the fact that recycling flows which are required to cope with conversions below 100%, do not influence the overall mass balance of the process. Solid/gas phase separation devices are assumed to be ideal, i.e. solids are completely separated from the gases. In order to calculate the required energy for drying residues, a residual moisture content after liquid filtration of 40 % (weight) is assumed. In the hydrometallurgical treatment process all filter and washing devices are assumed to yield a filter cake with a 40 % (weight) moisture content. For the hydrometallurgical treatment plant the fluxes of the recycle flows were calculated because these partly determined the required amounts of sulfuric acid, sodium hydroxide and energy.
For calculating energy balances some of the pyrometallurgical equipment was sized, because furnace sizes, operation temperatures and residence times are major parameters determining the overall process energy requirements.
All processes are designed to produce 200 ktons/yr zinc and simultaneously treat the residue due to this production as well as 80 ktons/yr historical jarosite residue. In order to facilitate the calculations a constant concentrate and residue composition is assumed.

3. Raw Materials and Energy Sources

Hydrometallurgical zinc winning processes use sphalerite concentrates which contain a number of impurities. The assumed composition of the concentrate is given in table 1. The presence of impurity elements such as e.g. germanium, cobalt, nickel, aluminum and indium is neglected. This is justified because these elements are not of major importance for calculating the overall mass and energy balances. The elements lead, zinc, copper, cadmium and iron are assumed to be present as sulfides. The other elements are assumed to be present as oxides.

Table 1 The assumed composition of historical jarosite and zinc concentrate

element	% in conc.	mass/year (ton)	% in hist. jarosite	mass/year (ton)
Zn	54	205200	1	800
Fe	8	30400	26	20800
Pb	1.5	5700	4	3200
Cd	0.16	608	0.045	36
Cu	0.25	950	0.075	60
Ba	0.6	2280	2	1600
Ca	0.6	2280	1.5	1200
Si	0.6	2280	2	1600
As	0.165	628	0.2	160
Mg	0.032	122	0.1	80
Mn	0.025	95	0.08	64
H	--	--	1.91	1562
O	1.068	4058	46.38	37137
N	--	--	1.74	1392
S	33	125400	12.97	10384
Total	100	380001	100	80075

The composition of historical jarosite residues is determined by the composition of the concentrate but also by the process lay-out of the zinc winning plant. Differences between jarosite residues obtained from different zinc winning plants are caused by the fact that the lead/silver residue is not always separated from the jarosite residue. This paper uses a worst case scenario in which the lead/silver residue is part of the jarosite residue. The composition of the historical jarosite residue is also given in table 1. In historical jarosite 80% of the iron is present as ammonium and 20% as hydronium jarosite. Other metals are assumed to be present as (hydrated) sulfates. Arsenic is assumed to be present as arsenic(V)oxide and 25% of the zinc load as (unreacted) sphalerite.

Additional essential raw materials are zinc dust, ammonia, lime, oxygen, sand, sulfuric acid, ferric sulfate and sodium hydroxide. All these raw materials are assumed to be 100% pure except oxygen. Sand, required for slag formation, is assumed to consist of 100% silicium dioxyde. Oxygen is assumed to be technically pure (95 %), the remainder being nitrogen. The overall energy requirements for the integrated zinc winning and residue treatment operations are split up in electricity requirements and natural gas consumption. The electricity is generated in a natural gas fired power station. Natural gas is only assumed to be the energy source because this reflects the Dutch situation best. Natural gas and electricity are compared by using the following conversion factor which includes energy losses in power generation and power transport: 3.0 kWh/Nm^3 gas or 10.8 $MJ_{elec.}$/Nm^3 gas. The heat of combustion of Dutch ('Groningen') natural gas is 35.2 MJ/Nm^3 gas. Thus, an overall conversion efficiency of 30.7% is assumed. The composition of Groningen natural gas is given in appendix II.

4. Description of the processes

4.1 Process option I

Figure 1 presents a simplified flow-sheet of an integrated zinc winning and residue treatment plant [6,7]. The primary zinc winning plant conventionally starts with an oxidative fluid bed roaster which converts concentrate into calcine. The relevant chemical reactions are presented per block in appendix I. The obtained calcine is leached at 70°C and pH = 3-5 ('neutral leach') to yield a solution of zinc, cadmium and copper sulfate.
The remaining solids still contain about 10% of the zinc present in the concentrate as zinc ferrite. The solids are subsequently leached cocurrently at a temperature between 90 and 95°C and a pH ranging from 1.5 to 1.0. At the lower pH zinc ferrite is dissolved while at the higher pH iron is precipitated as (ammonium) jarosite. The zinc content of zinc ferrite is only to a minor extent incorporated in jarosite and is returned to the neutral leach. The neutral leach solution is purified by the addition of zinc dust in order to cement out copper and cadmium.
In practice, cobalt is also cemented out but due to the simplified concentrate and jarosite residue composition the need for cobalt-related additives and equipment is ignored. After this purification the clean electrolyte is used for zinc winning. The copper-zinc-cadmium cement is partly redissolved to yield a cadmium/zinc solution and a copper cement. Cadmium is electrolyzed after further refining and the remaining acidic zinc solution is returned to the leach circuit. Spent acid from the electrolysis is recycled to the neutral leach, the acid leach and to the fly-ash leaching process in the treatment unit. The sulfuric acid production during electrolysis can not fulfill the demand in the leaching processes. Therefore fresh sulfuric acid is added to the acid leach. This results in excess water in the leach-electrowin circuit which necessitates a bleed stream. The bleed stream consists of spent acid, which has a low zinc concentration. This minimizes zinc losses and simultaneously prevents the accumulation of manganese and magnesium in the circuit. Lime is added to the bleed in order to precipitate metal hydroxides. This operation yields a contaminated gypsum residue and a purified neutral aqueous solution which, after passing further waste water treatment units, is discharged without environmental risks.
The gypsum residue, together with the solid residue that remains after the acid leach (jarosite and lead/silver residue) and the historical jarosite residue are dried and subsequently fed into a flash furnace. The purpose of this first treatment operation is to produce an iron-silicate slag with a low heavy metal content and a negligible leachability of those heavy metals that remain in the slag. Sand and natural gas are essential for obtaining the optimum slag formation conditions (± 1300°C). In order to cope with the demands of environmental legislation, secondly a batch slag fuming is operated at ± 1300°C, during which coal is used as reductant and lances are inserted in the slag bath for enhancing the mass transfer between the liquid slag and the gas phase. In this step the slag is fumed to further decrease the volatile heavy metal content. The two operations together yield an iron-silica slag which also contains calcium, barium, magnesium and manganese and fly-ashes which contain lead, cadmium, arsenic, copper and zinc.

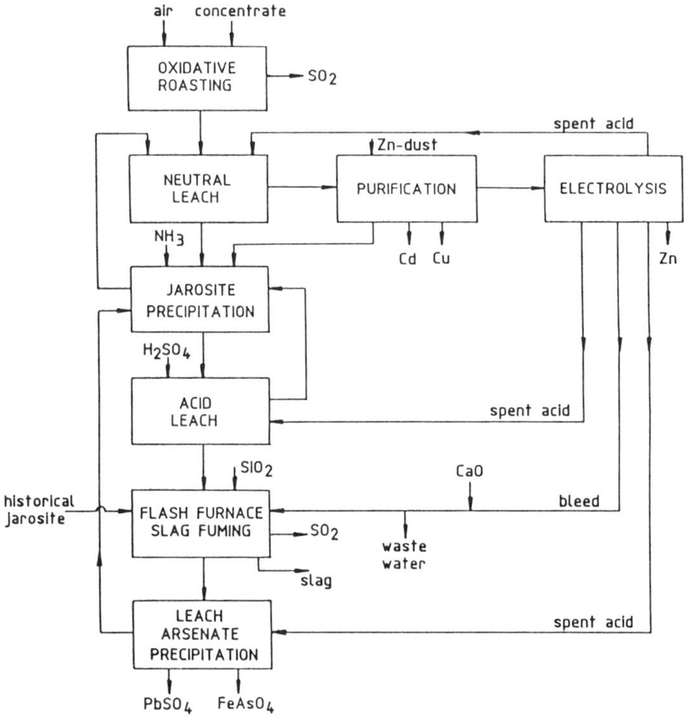

Figure 1: Simplified flow-sheet showing the option I integrated jarosite treatment process.

The fly-ashes are leached in a complicated set of unit operations which prevent the accumulation of arsenic, yield a saleable lead sulfate residue and enable the recovery of cadmium, copper and zinc. A more detailed discussion of this leach unit is beyond the scope of this paper. Finally, a ferric arsenate residue is precipitated simultaneously with gypsum in the leach unit. This residue does not have any application and has to be stored in a suitable hazardous chemical waste pond. However, the volume of hazardous chemical waste that has to be stored is reduced to 5% of the original jarosite waste, since the slags produced in this process are suitable for civil engineering purposes.

4.2 Process option II

Figure 2 presents a simplified flow-sheet of an alternative process for integrated zinc winning and residue treatment. This option only differs from option I in that no jarosite precipitation and no acid leach plant are incorporated. The elimination of these steps is justified by the fact that the zinc content of the neutral leach residue will, after passing the residue treatment process, finally end up in the neutral leach in a dissolved form. Thereby equipment and maintenance costs are reduced and the energy requirements in the primary plant are decreased.

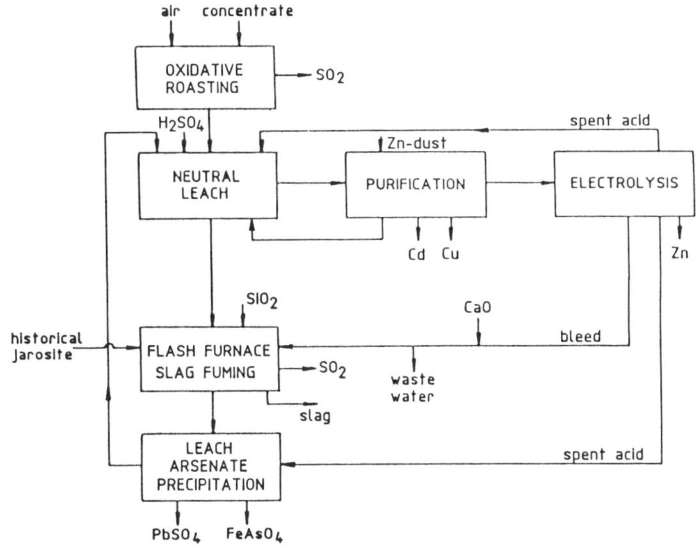

Figure 2: Simplified flow-sheet showing process option II

4.3 Process option III

Figure 3 represents a second alternative process. This process resembles process option II, but is more complicated. In order to avoid the major increase of the initial zinc load to the flash furnace this process operates a reductive roasting step. This step reduces the zinc ferrite content of calcine by exposing the calcine to ± 850°C in a reductive gas atmosphere. Zinc ferrite is then converted to zinc oxide which is easily soluble in the neutral leach and to magnetite which is very slightly soluble [8,9]. Simultaneously, a fly ash is formed, which is leached in the fly-ash leach in the treatment unit. The advantage of reducing the zinc ferrite content in calcine is gained by introducing an extra process step which increases the energy requirements.
Another difference between the options II and III regards the gypsum bleed. In this option the neutral leach solution which contains a lot of zinc at a low acidity is partly bled off. Lime is added to the bleed in order to precipitate waste gypsum which contains the metals as hydroxides. All zinc contained in the gypsum residue is recovered in the flash furnace and the slag fumer. Compared to option II, savings are made on fresh sulfuric acid inputs, because the amount of acid neutralized by lime has decreased.

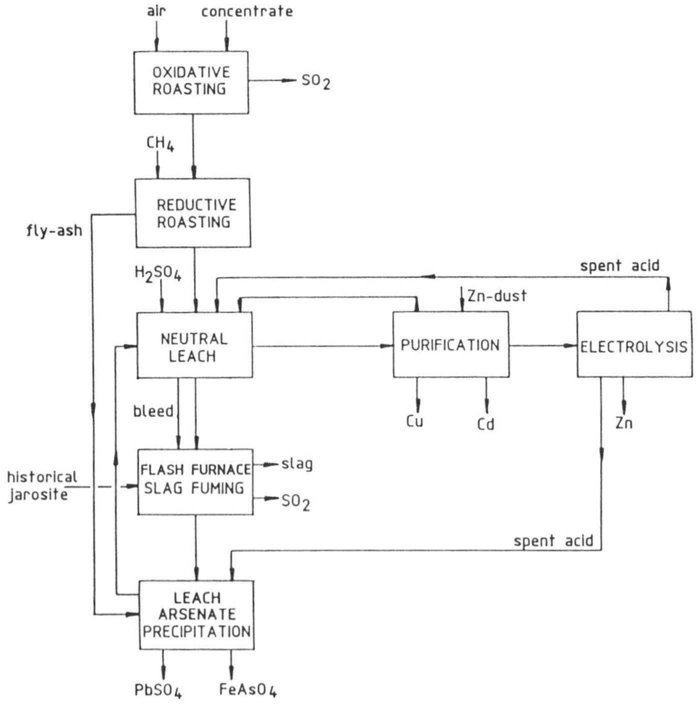

Figure 3: Simplified flow-sheet showing process option III

4.4 Process option IV

Figure 4 shows the simplified flow sheet for the fourth option. This option closely resembles option III. The only difference is that in option III historical jarosite residue is fed into the flash furnace, whereas in option IV this residue is fed into the oxidative roaster. Neglecting the technical difficulties which may arise from feeding a very fine ($d_{50,vol.}$ < 20 µm) pulverized material into a fluidized bed roaster, it is easily demonstrated that the energy required for decomposing 80 ktons/yr historical jarosite residue is much smaller than the energy obtained by roasting 380 ktons/yr of sphalerite concentrate. By feeding the jarosite residue to the oxidative roaster the combustion energy of the concentrate is utilized efficiently because the heat transfer in fluidized beds is excellent. Combining the jarosite residue and the concentrate does decrease the steam production in the waste heat boiler following the oxidative roaster, but the main application of this steam, heating the acid leach and jarosite precipitation vessels, has also vanished.

By adding historical jarosite residue to the concentrate it is essentially enriched in iron, leading to a considerable increase in zinc ferrite production. However, the oxidative roasting step is followed by a reductive roasting step which eliminates the major part of the zinc ferrite. The flash furnace is thus fed with a smaller quantity of residue.

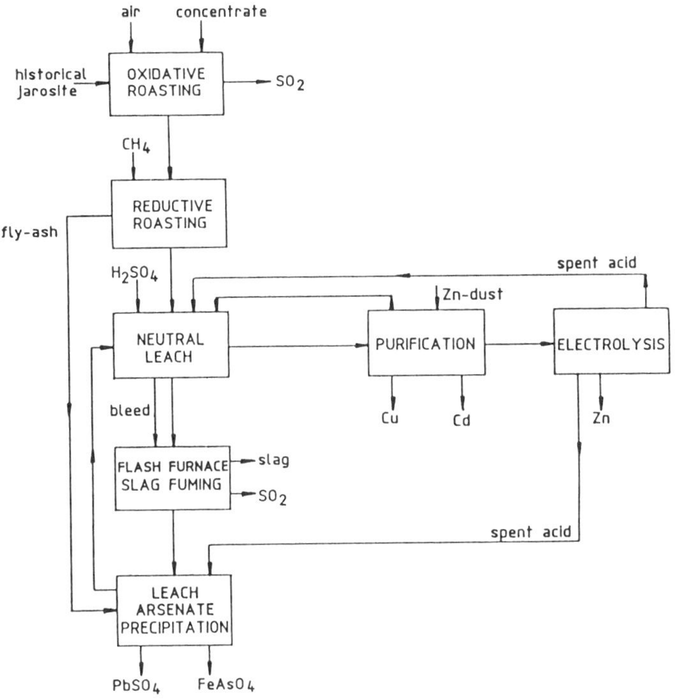

Figure 4: Simplified flow-sheet showing process option IV

4.5 Process option V

Option V is schematically presented in figure 5. This option includes a hydrometallurgical treatment unit. The primary winning unit resembles the unit described for option I and leads a.o. to a combined jarosite and lead/silver residue as well as to a bleed gypsum residue. The fresh jarosite residue combined with historical jarosite is leached under reductive conditions by adding sulfur dioxyde [10]. This yields a solid residue containing lead, barium, calcium, arsenic and silica. The obtained solution contains zinc, copper, cadmium, ammonium, magnesium, manganese and, most of all, iron(II). Ferrous sulfate is then crystallized as smozolkonite (monohydrate) in a fractional crystallization unit. The unit as such is described in appendix III. It is operated at 160 °C in order to improve crystal purity, since the relative solubility of ferrous sulfate monohydrate, compared to zinc sulfate monohydrate is small at elevated temperatures (140-160°C). The partition coefficient for the incorporation of zinc in ferrous sulfate is defined as follows:

$$D=\frac{[Zn]_s/[Fe]_s}{[Zn]_l/[Fe]_l} \qquad (1)$$

in which l represents the liquid phase and s the solid phase.

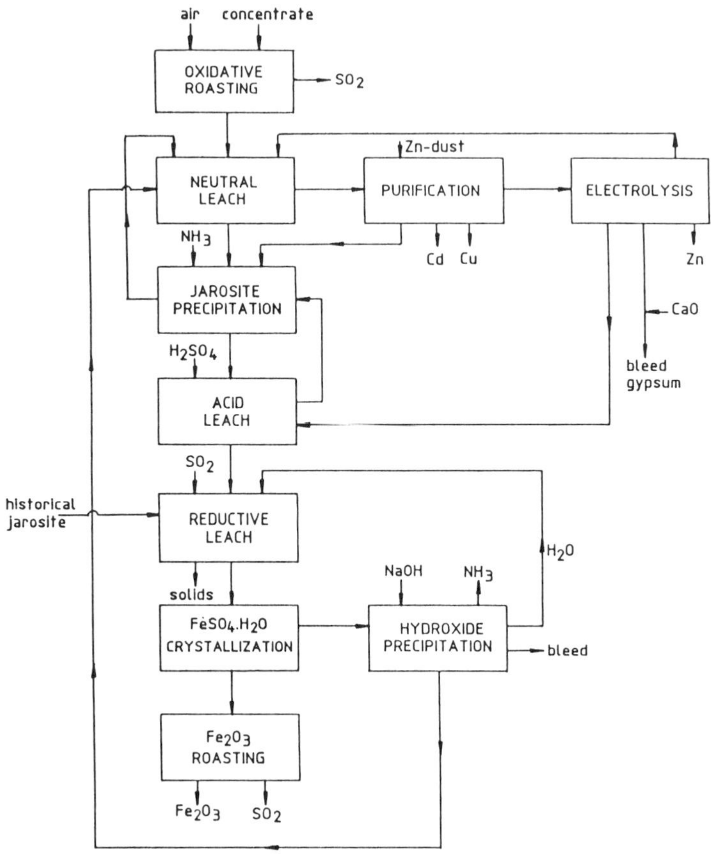

Figure 5: Simplified flow-sheet showing process option V

In this study this coefficient is assumed to be constant. Its value is assumed to be equal to 0.2. This is still far beyond the thermodynamically feasible value at 160°C which, based on the ratio of the solubility products, equals about 0.05. The value of 0.2 is based on experimental results obtained by one of the authors [11]. Zinc is chosen as the impurity which determines the effectiveness of this operation for three reasons:

i) The site is a zinc winning plant which benefits from selling zinc instead of losing it in residues.
ii) The potential application of hematite in the primary iron and steel industry sets very strict specifications.
iii) Zinc is the most abundant 'impurity' while producing clean iron residues. Other impurities like copper and cadmium will behave similar to zinc, but their quantity is much smaller and therefore the incorporation is neglected.

The fractional crystallization unit yields a relatively pure ferrous sulfate which is roasted at 600 - 800°C to yield hematite and sulfur dioxyde. The so-produced hematite has to meet the standards for use as a raw material in the steel industry. The remaining liquid phase contains, after oxidation, iron(III), zinc, cadmium, copper,

manganese, ammonium and magnesium. The metals are precipitated by adding caustic soda, which removes ammonium as ammonia. The precipitate is returned to the neutral leach in the primary plant, since zinc, cadmium and copper are valuable. Magnesium and manganese are finally removed with the bleed gypsum. The sodium sulfate solution remaining after the hydroxide precipitation is recycled for its heat content. However, a bleed is required to limit the accumulation of sodium and sulfate ions. This bleed may be treated with lime to yield a relatively pure gypsum which will find application in the building industry and a diluted caustic soda solution which may be utilized in waste water neutralization.
This option does not include processes which are operated at very high temperatures. Therefore, it is interesting to calculate whether the energy consumption per unit waste will be lower.
The remaining residues, however, are assumed to have quite different properties from those that have experienced a pyrometallurgical treatment. The impurity content of hematite is determined by the final composition of ferrous sulfate monohydrate. This composition is determined by the partition coefficient as given in equation (1). Since the partition coefficient and the aqueous ferrous concentration are constant for a given temperature, the uptake of zinc in the crystals is influenced by the slurry density, which is defined as the quantity of ferrous sulfate monohydrate crystals per 1000 kg of water. In this study three different values are chosen for the slurry density i.e. 50, 100 and 200 kg/1000 kg water. Since the output of produced ferrous sulfate monohydrate is constant, regardless of the slurry density, the differences in slurry density are realized by changing the quantity of water, i.e. by varying the dilution of the pregnant solution.

5. Mass balances; results

Appendix I provides all reaction equations which were used to calculate the mass balances for the different processes. The overall results for the pyrometallurgical treatment process are given in table 2 and for the hydrometallurgical treatment process in table 3. Since the raw materials are the same for all processes, the options I to IV yield about the same amount of products. Minor differences occur in lime and sulfuric acid use and sulfur dioxyde and slag production. Major differences, however, occur between the options I to IV and option V in residue composition and amount. Obviously, the hydrometallurgical option produces a larger quantity of residues. The Pb/As/Ba-residue contains only 20 % lead by weight, which probably inhibits its application as raw material for lead winning.
Table III shows that with a decrease of the slurry density the amounts of chemicals required for operating this process increase drastically. The chemical use at low slurry densities clearly reaches unacceptably high levels. The aqueous bleed containing sodium sulfate rises as a consequence of this. The reason for the increased chemical use is lying in the fact that with a lower slurry density the fraction dissolved iron over crystallized iron is enhanced and therefore the recycling of ferric ions increases. Each recycled ferric ion is precipitated as ferric hydroxide, consuming caustic soda, and redissolved with sulfuric acid. Redissolved ferric hydroxide is then precipitated as jarosite, consuming extra ammonia. These mass balances also confirm the decreasing zinc content in the hematite product with decreasing slurry density. However, even at a slurry density of 50 kg/1000 kg H_2O the hematite product still contains 229 ppm of zinc, which exceeds the steel industry's limit of 120 ppm Zn in pig iron [12], which corresponds to 81 ppm in hematite, assuming pig iron to contain 4 % carbon. In case a lower value of the partition coefficient D is practically feasible, the corresponding final contamination of hematite also decreases and enables the utilization of hematite in the steel industry.

Table 2 : Results obtained from mass balance calculations for four pyrometallurgical integrated jarosite treatment options (all figures in kton/yr)

	option I	option II	option III	option IV
raw materials				
concentrate	380.0	380.0	380.0	380.0
zinc dust	2.7	2.7	2.7	2.7
hist. jarosite	80.1	80.1	80.1	80.1
ammonia	2.5			
lime	3.8	3.8	6.2	6.5
sulfuric acid	51.0	18.9	19.5	25.5
sand	25.2	25.2	26.8	27.0
ferric sulfate	2.7	2.7	2.7	2.7
coal	2.0	2.0	2.1	2.1
products				
zinc	208.7	208.7	208.7	208.7
cadmium	0.6	0.6	0.6	0.6
copper	1.0	1.0	1.0	1.0
lead sulfate	13.0	13.0	13.0	13.0
slag	111.4	111.4	116.1	116.5
sulfur dioxyde	303.4	280.2	283.2	287.0
residues				
arsenate/ gypsum	4.8	4.8	4.8	4.8

The residues obtained with the hydrometallurgical process consist next to the Pb/As/Ba-residue of waste gypsum which has to be stored because its heavy metal content inhibits applications.

Table 3 : Results obtained from mass balance calculations for the hydrometallurgical integrated treatment process for three different values of the slurry density in the fractional crystallizers (all figures in kton/yr)

	M_t =50 kg/ton H_2O	M_t =100 kg/ton H_2O	M_t =200 kg/ton H_2O
Raw materials			
concentrate	380.0	380.0	380.0
zinc dust	2.7	2.7	2.7
hist. jarosite	80.1	80.1	80.1
ammonia	12.8	5.7	3.7
lime	2.3	3.2	3.4
sulfuric acid	206.4	99.7	70.3
caustic soda	326.8	111.5	51.5
Products			
zinc	208.1	207.8	207.2
cadmium	0.6	0.6	0.6
copper	1.0	1.0	1.0
hematite	73.2	73.2	73.2
zinc in hematite	0.02	0.15	0.7
sulfur dioxyde	206.4	257.0	271.0
Residues			
Pb/As/Ba-residue	44.2	44.2	44.2
waste gypsum	8.8	11.3	11.9
zinc in gypsum	0.4	0.5	0.6
aqueous bleed	4,186.9	1,056.2	499.5
dissolved Na_2SO_4	579.8	197.9	91.4

6. Energy balances: results

Estimates were made of the energy requirements of all five options described in section 4. Table 4[a,b,c,d] provides the obtained results for the options I to IV. The energy requirements of the primary winning plant are based on literature [13]. The relations used for estimating the energy use of 'classical' unit operations in zinc hydrometallurgy are given in appendix IV. Appendix V discusses the estimates of the energy use of the reductive roasting step. Appendix VI provides the calculations of the energy requirements for the residue dryers. The design aspects and energy requirements of the flash furnaces are discussed in appendix VII. The batch slag fumers are discussed briefly in appendix VIII.

In options I to IV the flash furnace and the batch slag fumer yield large quantities of hot combustion gases, which are suitable for power generation. In the options III and IV the hot gases leaving the reductive roaster and in the options II to IV part of the hot gases obtained during oxidative roasting are also used for this purpose. The gases are utilized to produce saturated steam at 100 bar (312 °C). This steam is subsequently superheated in a separate unit which utilizes natural gas as fuel. The energy requirements of the superheater and the resulting power production are described in appendix IX. Appendix X discusses the energy requirements of the fly-ash leaching in the options I to IV.

Table 4[a] The electrical energy requirements for the options I to IV (all figures in kWh/ton SHG zinc, negative signs indicate energy releases)

step	option I	option II	option III	option IV
roasting	110	110	110	183
sulfuric acid production	300	278	281	285
reductive roasting	-	-	-	100
leaching	179	72	72	72
electrolysis	3,373	3,373	3,373	3,373
melting and casting	127	127	127	127
others	71	71	71	71
residue drying	36	32	32	14 + 26
flash furnace	140	86	99	59
electrostatic precipitator	0.7	0.5	0.5	0.3
slag fumer	132	132	137	137
fly-ash leach & miscellaneous	100	100	100	100
on-site power generation	-1,477	-1,366	-1,503	-1,161
net use	3,092	3,016	3,000	3,386

Table 4[a] indicates the importance of on-site power generation. The main differences in the power consumption are related to the flash furnace. This is explained by the fact that the flash furnace requires technical oxygen, for which a power consumption of 400 kWh/ton technical oxygen is assumed. The flash furnace in option I has the largest oxygen demand and consequently the largest power consumption.
The natural gas consumption, which is given in table 4[b], increases significantly due to the on-site power generation.

Table 4[b] The natural gas requirements for the options I to IV (all figures in Nm^3/ton SHG zinc)

step	option I	option II	option III	option IV
reductive roasting	-	-	31	33
residue drying	90	81	81	36 + 65
flash furnace	226	142	154	99
slag fumer	125	125	130	131
superheater	94	87	95	74
net use	535	435	491	438

Table 4^c The low and high pressure (saturated) steam balance for the options I to IV (all figures in ton steam/ton SHG zinc, negative signs indicate steam releases).

step	option I	option II	option III	option IV
low pressure (T = 130 °C)				
oxidative roasting	-1.92	-0.25	-0.25	-0.14
leaching	2.23	0.45	0.45	0.45
residue drying	-0.44	-0.40	-0.40	-0.51
fly-ash leaching	0.2	0.2	0.2	0.2
deficiency	0.07	0	0	0
high pressure (T = 312 °C)				
oxidative roasting	-	-1.28	-1.28	-1.22
reductive roasting	-	-	-0.24	-0.24
flash furnace	-3.72	-2.06	-2.22	-1.16
fumer	-1.13	-1.13	-1.18	-1.18
superheater	4.85	4.47	4.92	3.80

Table 4^c shows that option I has a small low pressure steam deficiency. This steam defiency is converted into Nm^3 natural gas equivalents by assuming an energetical efficiency of 85 % during steam production and additionally 10 % heat losses to the environment during the utilization of the steam. Thus, 1 ton of steam (130 °C, H=2.72 MJ/kg) is produced by burning 111 Nm^3 of natural gas (heat of combustion 32 MJ/kg). The options II to IV generate just as much low pressure steam during the oxidative roasting as is required for the leaching processes. The remaining energy is used to generate high pressure steam which is utilized, after superheating, for power generation.
In order to compare the overall energy requirements for the options I to IV, the Nm^3 natural gas equivalent is defined. The equivalent of 1 MJ electric energy is 2.963 MJ heat obtained by combustion of natural gas. This relates to an energetic efficiency of 30.7 % during power generation. Table 4^d provides the overall results of the energy requirements.

Table 4^d The overall energy requirements for the options I to IV (in Nm^3 natural gas equivalents/ton SHG zinc)

energy carrier	option I	option II	option III	option IV
gas	535	435	491	438
electricity	1,031	1,005	1,000	1,129
steam	8	-	-	-
overall use	1,574	1,440	1,491	1,567

Table 4^d shows the minor differences in energy consumption between the four options. Option I is consuming more energy than option II, because option I maintains jarosite precipitation and hot acid leach steps in the primary plant, whereas these two steps

are abandoned in option II. Introducing a reductive roaster in option II, which yields option III, increases the energy consumption and makes the plant somewhat more complex. Option IV, which is equivalent to option III in all respects but the location of the historical jarosite feed, shows that this alternative even further increases the energy consumption.
Before conclusions are drawn it is essential to emphasize that the differences in energy consumption are relatively small. It is even doubtful whether based on the accuracy of the estimates which were used, it is justified to conclude that the differences reported on the bottom line of table 4^d are significant. It is, however, evident that the energy consumption of the options I to IV equals in average $1.5*10^3$ Nm^3/ton SHG zinc.

The energy calculations for option V are given in the tables $5^{a,b,c,d}$. The rotary furnace which is utilized for roasting ferrous sulfate is described in appendix XI. Three different types of steam are used in the treatment plant. The medium pressure steam (180 °C or 200 °C) is used for heating the (feed) streams to the fractional crystallizer. The low pressure steam is used for the leaching steps in the primary winning units, as well as for heating the reductive leach. The off-gases of the residue dryer and the rotary furnace are used for generating medium and low pressure steam. Steam is generated by combustion of natural gas with an assumed energetical efficiency of 85 %. The heat losses to the environment while using steam are assumed to be 10 %. As far as reasonably possible, waste heat is used for heating purposes and water is recycled. The steam circuits are closed loops between the crystallizers and the boilers, thus utilizing the heat content of the condensate as efficient as possible. The heat content of the waste water stream which leaves the hydroxide precipitation is not lost since this stream is partly recycled as feed wash stream to the fractional crystallization unit and the reductive leach.
Table 5^a shows the power requirements for the options V. Comparing table 5^a with table 4^a it becomes evident that the fact that no power is generated by using waste heat, increases the net power consumption for the options V considerably. Table 5^b gives the steam consumption for the hydrometallurgical option. For all values of the slurry density there is a steam deficiency. However, in particular for lower values of the slurry density, this steam deficiency becomes very large. This has an important effect on the natural gas requirements which is given in table 5^c. It is also obvious that the influence of whether or not dissolution is required in the fractional crystallizer is not of influence on the energy requirements for low values of the slurry density.

Table 5^a Power requirements for option V for different values of the slurry density M_t (in kg $FeSO_4.H_2O$/1000 kg H_2O) (all figures in kWh/ton SHG zinc)

step	M_t = 50 kg/ton H_2O	M_t =100 kg/ton H_2O	M_t =200 kg/ton H_2O
oxidative roasting	110	110	110
acid production	213	255	269
leaching	751	365	247
electrolysis	3,373	3,373	3,373
melting and casting	127	127	127
others	71	71	71
residue dryer	29	29	29
rotary furnace	4	4	4
fractional crystallizer and reductive leach	77	77	77
sum	4,755	4,411	4,307

Table 5[b] Steam requirements in option V for different values of the slurry density M_t (in kg $FeSO_4.H_2O$/1000 kg H_2O) and the redissolution percentage RD (all figures in ton steam/ton SHG zinc, negative signs indicate steam releases).

	M_t = 50 RD = 0	M_t = 50 RD =100	M_t =100 RD = 0	M_t =100 RD =100	M_t =200 RD = 0	M_t =200 RD =100
low pressure *(T = 130°C)*						
oxi. roaster	-1.92	-1.92	-1.92	-1.92	-1.92	-1.92
leaching	9.31	9.31	4.42	4.42	3.07	3.07
reduct. leach	-	-	0.10	-	0.07	-
residue dryer	-	-	-	-	-0.07	-
rotary furnace	-	-	-	-	-0.08	-
deficiency	7.39	7.39	2.60	2.50	1.07	1.15
med. pressure *(T = 180°C)*						
fractional crystallizer	3.40	3.42	0.70	1.10	0.09	0.58
residue dryer	-0.59	-0.59	-0.59	-0.59	-0.09	-0.58
rotary furnace	-0.08	-0.08	-0.08	-0.08	-	-
deficiency	2.73	2.75	0.03	0.43	0	0
med. pressure *(T = 200°C)*						
fractional crystallizer	1.41	1.41	0.60	0.60	0.37	0.37
residue dryer	-	-	-	-	-0.37	-
rotary furnace	-	-	-	-	-	-0.06
deficiency	1.41	1.41	0.60	0.60	0	0.31

Table 5[c] The natural gas consumption for option V for different values of the slurry density M_t (in kg $FeSO_4.H_2O$/1000 kg H_2O) and the redissolution percentage RD (all figures in Nm^3 natural gas/ton SHG zinc).

	M_t = 50 RD = 0	M_t = 50 RD =100	M_t = 100 RD = 0	M_t =100 RD =100	M_t =200 RD = 0	M_t =200 RD =100
residue dryer	72.1	72.1	72.1	72.1	72.1	72.1
rotary furnace	38.3	38.3	38.3	38.3	38.3	38.3
steam production	1,179.2	1,180.9	342.3	366.1	119.0	154.0
sum	1,289.6	1,291.3	452.7	476.5	229.4	264.4

The results in table 5[c] are added to the results in table 5[a] by using the same conversion factor as used for the options I to IV.

Table 5^d The overall energy requirements for the option V (in Nm^3 natural gas equivalents/ton SHG zinc)

energy carrier	M_t = 50 RD = 0	M_t = 50 RD =100	M_t =100 RD = 0	M_t =100 RD =100	M_t =200 RD = 0	M_t =200 RD =100
gas	1,289.6	1,291.3	452.7	476.5	229.4	264.4
power	1,585.0	1,585.0	1,470.3	1,470.3	1,435.7	1,435.7
overall use	2,874.6	2,876.3	1,923.0	1,946.8	1,665.1	1,700.1

Table 5^d clearly shows the steep rise of the natural gas requirements of this option with decreasing slurry density. Comparing the results presented in table 4^d and table 5^d it is evident that option V requires more energy than the options I to IV despite the fact that the process temperatures in option V are much lower.
This demonstrates the advantage of high temperature processes; waste heat may be used as a source of energy with a high intrinsic value, which in this case is used for power generation. The hydrometallurgical process uses energy of a lower quality since the applied temperatures are much lower, but because of that, its waste heat can not be used for power generation.

7. Conclusions

The following conclusions are drawn:

- The hydrometallurgical zinc winning processes with integrated pyrometallurgical jarosite treatment processes as discussed in this paper (options I to IV) reduce the quantity of residue to be stored to less than 5 % by weight of the original amount of jarosite residue.

- The options I to IV show relatively small differences in energy requirement per unit zinc produced. The most energy-efficient process appears to be option II. Its energy consumption is below 1500 Nm^3 natural gas equivalents/ton SHG zinc.

- The hydrometallurgical zinc winning process with integrated jarosite treatment process as discussed in this paper as option V produces two residue streams which have no application: waste gypsum and Pb/Ba/As-residue. The two streams combined amount about 25 - 30 % by weight of the original amount of jarosite residue. In order to apply the produced hematite in the steel industry the zinc content should be very low, which complicates the process and requires large amounts of chemicals and energy. The process parameter which determines the incorporation of zinc in ferrous sulfate monohydrate is the partition coefficient D. The relatively high value of this partition coefficient inhibits the practical application of an integrated hydrometallurgical treatment process, which may change when lower values of the partition coefficient D appear to be feasible in industrial practice.

- Hydrometallurgical integrated jarosite treatment with fractional crystallization of ferrous sulfate monohydrate is an interesting technique, but given the final residue quantity and the high energy requirements it is not competitive with a pyrometallurgical treatment process, which offers better opportunities for waste-heat use and yields a smaller amount of unapplicable residue.

Acknowledgements

The Dutch Ministry of Housing, Physical Planning and the Environment is gratefully acknowledged for financially supporting this research study. The authors are indebted to prof. G.M. van Rosmalen for her support and stimulating comments. J.H. Versteegh M.Sc. is acknowledged for carefully reading of the manuscript.

Literature

[1] J.E. Dutrizac & A.J. Monhemius (Eds.), Iron Control in Hydrometallurgy, Ellis Horwood Publishers, Chichester (GB) (1986)

[2] R.G. Reddy, J.L. Hendrix and P.B. Queneau (Eds.), Arsenic Metallurgy, Fundamentals and Applications, TMS-AIME Publications, Warrendale (PA), (1988)

[3] ibid [1], A. von Röpenack, Hematite - the solution to a disposal problem - an example from the zinc industry, pp. 730 - 741

[4] Staatscourant (Dutch State Journal), Building Materials Directive, published 26-06-1991

[5] The World Commission on Environment and Development, G.H. Brundtland (chairman), Our Common Future, Oxford University Press (1987)

[6] Outokompu OY, Process for treating jarosite residue in an outokompu flash furnace, patents pending

[7] J.H. Versteegh, Response to the key issues facing the zinc industry, paper presented at the GDMB-meeting, march (1991), Freiberg

[8] R. Püllenberg, Hydrometallurgische Verarbeitung eisenreicher Zinkkonzentraten nach thermischer Teilreduktion des Röstgutes, Ph. D. Thesis, RWTH Aachen, Germany (1981)

[9] J. Krüger and R. Püllenberg, Überlegungen zum Eisenproblem bei der hydrometallurgische Zinkgewinnung, Erzmetall, Band 34, no. 7/8 (1981) pp. 380 - 387

[10] A.J.A. Konings, Reduktive Auflösung von Jarosit, paper presented at the GDMB-meeting, may (1988), Antwerp

[11] F. Elgersma, O.S.L. Bruinsma and G.M. van Rosmalen, The incorporation of zinc during high temperature ferrous sulfate crystallization, unpublished results (1991)

[12] Hoogovens Groep B.V., Het hoogovenproces, internal report (1989) p.59

[13] H.H. Kellogg, Energy use in zinc extraction, in: Lead, Zinc & Tin '80, J.M. Cigan, T.S. Mackey and T.J. O'Keefe (Eds.), TMS-Warrendale (1979), pp. 28 - 47

[14] G. Bruhn, J. Gerlach and F. Pawlek, Löslichkeit von Salzen und Gasen in Wasser, Z. Allg. Anorg. Chemie 337 (1965) pp. 68 - 79

[15] J.R. Boldt, Jr. and P. Queneau, The winning of Nickel, Methuen and Co. Ltd. (1967) pp. 320 - 322

[16] J.M. Coulson and J.F. Richardson, Chemical Engineering, Vol. II (SI-units), Pergamon Press, Oxford (1980) p. 749

[17] A.K. Biswas and W.G. Davenport, Extractive Metallurgy of Copper, 2nd Ed. Int. Series on Materials Science and Technology, V32, Pergamon Press (1980) p.169

[18] G.B. Gill, Non-ferrous Extractive Metallurgy, John Wiley and Sons (1988) p.45

[19] R.H. Perry (Ed.), Chemical Engineers' Handbook, 6th Ed., McGraw-Hill, New York (1984), p. 20-119

[20] T.A.A. Quarm, Slag fuming - kinetic or thermodynamic ?, Engineering and Mining Journal, january (1968) pp. 92 - 93

[21] Budelco B.V., Milieu-Effect-Rapportage ten behoeve van de vergunningverlening voor de bouw van een jarosietverwerkingsfabriek (Request for permission for building a pyrometallurgical integrated jarosite treatment plant), june (1991), The Netherlands

[22] I. Barin and D. Knacke, Thermodynamic properties of inorganic substances, Springer Verlag, Berlin (1973)

[23] T. Rosenquist, Principles of Extractive Metallurgy, 2nd Ed., McGraw-Hill, Singapore (1983)

Appendix I Chemical reaction equations used for calculating the mass and energy balances

1. Oxidative roasting (all options)

concentrate

$2\ ZnS + 3\ O_2 \rightarrow 2\ ZnO + 2\ SO_2$ (ditto for Pb and Cd)

$4\ CuS + 5\ O_2 \rightarrow 2\ Cu_2O + 4\ SO_2$

$ZnS + 2\ FeS + 5\ O_2 \rightarrow ZnO.Fe_2O_3 + 3\ SO_2$

$ZnS + 2\ FeS_2 + 7\ O_2 \rightarrow ZnO.Fe_2O_3 + 5\ SO_2$

historical jarosite (option IV)

$4\ NH_4Fe_3(SO_4)_2(OH)_6 \rightarrow 6\ Fe_2O_3 + 8\ SO_2 + O_2 + 2\ N_2 + 20\ H_2O$

$2\ H_3OFe_3(SO_4)_2(OH)_6 \rightarrow 3\ Fe_2O_3 + 4\ SO_2 + 2\ O_2 + 9\ H_2O$

$2\ PbSO_4 \rightarrow 2\ PbO + 2\ SO_2 + O_2$

$2\ ZnSO_4.7H_2O \rightarrow 2\ ZnO + 2\ SO_2 + O_2 + 14\ H_2O$ (ditto for Mg)

$2\ 3(CdSO_4).8H_2O \rightarrow 6\ CdO + 6\ SO_2 + 3\ O_2 + 16\ H_2O$

$4\ CuSO_4.5H_2O \rightarrow 2\ Cu_2O + 4\ SO_2 + 20\ H_2O + 3\ O_2$

$2\ MnSO_4.H_2O \rightarrow 2\ MnO + 2\ SO_2 + O_2 + 2\ H_2O$

$2\ CaSO_4.2H_2O \rightarrow 2\ CaO + 2\ SO_2 + O_2 + 4\ H_2O$

$ZnO + Fe_2O_3 \rightarrow ZnO.Fe_2O_3$

2. Reductive roasting (options III and IV)

$12\ ZnO.Fe_2O_3 + CH_4 \rightarrow 12\ ZnO + 8\ Fe_3O_4 + CO_2 + 2\ H_2O$

$CH_4 + CO_2 \rightarrow 2\ CO + 2\ H_2$

After the reductive roasting step, regardless of the zinc ferrite content of the calcine, 96 % of the zinc metal is assumed to be zinc oxide and 4 % has remained zinc ferrite. Additionally 0.5-1.0% of the concentrate ends up in a fly-ash. The fly-ash is reported to contain: Pb 52%, Cd 8.1% and Zn 5.5%. These elements are assumed to be present as oxides, which makes them contribute 72.1% to the fly-ash weight. The remainder is assumed to be all of the arsenic(III) oxide present in the calcine. The total amount of fly-ash then equals 3729 ton/a which indeed lies in the range of 0.5-1.0% of the concentrate mass. The figures are based on experimental work by Püllenberg [8].

3. Neutral leach (all options)

calcine

$ZnO + 2 H^+ \rightarrow Zn^{2+} + H_2O$ (ditto for Cd, Pb, Mg, Ba, Ca and Mn*)

* Manganese is supposed to be present as MnO; Mn_3O_4 or Mn_2O_3 may also occur, but are neglected

$Cu_2O + 2 H^+ \rightarrow 2 Cu^+ + H_2O$

$4 Cu^+ + O_2 + 4 H^+ \rightarrow 4 Cu^{2+} + 2 H_2O$

$Pb^{2+} + SO_4^{2-} \rightarrow PbSO_{4(s)}$ (ditto for Ba)

$Ca^{2+} + SO_4^{2-} + 2 H_2O \rightarrow CaSO_4.2H_2O_{(s)}$

magnetite (options III and IV)

Despite the fact that magnetite is slightly soluble in the neutral leach solution, its solubility is neglected. Any dissolved Fe^{2+} will be oxidized to Fe^{3+} and then precipitate as ferric hydroxide by aerating the neutral leach solution.

hydroxide precipitate (option V)

$Zn(OH)_2 + 2 H^+ \rightarrow Zn^{2+} + 2 H_2O$ (ditto for Cd, Cu, Mn and Mg)

$Fe(OH)_3 + 3 H^+ \rightarrow Fe^{3+} + 3 H_2O$

4. Jarosite precipitation (options I and V)

$NH_4^+ + 3 Fe^{3+} + 2 SO_4^{2-} + 6 H_2O \rightarrow NH_4Fe_3(SO_4)_2(OH)_{6(s)} + 6 H^+$

$3 Fe^{3+} + 2 SO_4^{2-} + 7 H_2O \rightarrow H_3OFe_3(SO_4)_2(OH)_{6(s)} + 5 H^+$

$NH_3 + H^+ \rightarrow NH_4^+$

5. Acid Leach (options I and V)

$ZnO.Fe_2O_3 + 8 H^+ \rightarrow Zn^{2+} + 2 Fe^{3+} + 4 H_2O$

$H_2SO_4 \rightarrow 2 H^+ + SO_4^{2-}$

6. Purification (all options)

$$Cu^{2+} + Zn \rightarrow Cu + Zn^{2+}$$ (ditto for Cd)

In order to get an effective cementation process a 50% molar excess of zinc dust is added to the first step. After separating the Cd-Cu-Zn-cement from the solution, cadmium is selectively redissolved. Subsequently, cadmium is once more cemented with 50% molar excess zinc dust. The zinc dust used in the cementation process is recycled to either the neutral leach or the jarosite precipitation as an acidic zinc solution.

7. Electrolysis (all options)

$$Zn^{2+} + 2\,H_2O \rightarrow 2\,Zn + 4\,H^+ + O_{2(g)}$$

In the options I, II and V the bleed to remove excess water and to prevent the accumulation of magnesium and manganese consists of spent acid, which is formed during the electrolysis process. Therefore, the reactions used to describe the gypsum precipitation in the bleed stream are discussed in this section.

$$Zn^{2+} + 2\,OH^- \rightarrow Zn(OH)_{2(s)}$$

(ditto for Mg^{2+} and Mn^{2+} (all options) and Cd^{2+} and Cu^{2+} (options III and IV))

$$CaO + 2\,H^+ \rightarrow Ca^{2+} + H_2O$$

$$CaO + H_2O \rightarrow Ca^{2+} + 2\,OH^-$$

$$Ca^{2+} + SO_4^{2-} + 2\,H_2O \rightarrow CaSO_4.2H_2O_{(s)}$$

8. Flash Furnace and Slag Fuming (options I to IV)

waste gypsum

$$2\,CaSO_4.2H_2O \rightarrow 2\,CaO + 2\,SO_2 + 4\,H_2O + O_2$$

$$Mg(OH)_2 \rightarrow MgO + H_2O$$ (ditto for Zn and Mn and Cd and Cu (option II and III))

historical jarosite (options I, II and III)
fresh jarosite (option I)

$$4\,NH_4Fe_3(SO_4)_2(OH)_6 \rightarrow 6\,Fe_2O_3 + 8\,SO_2 + O_2 + 2\,N_2 + 20\,H_2O$$

$$2\,H_3OFe_3(SO_4)_2(OH)_6 \rightarrow 3\,Fe_2O_3 + 4\,SO_2 + 2\,O_2 + 9\,H_2O$$

$$2\,PbSO_4 \rightarrow 2\,PbO + 2\,SO_2 + O_2$$

$$2\ ZnSO_4.7H_2O \rightarrow 2\ ZnO + 2\ SO_2 + 14\ H_2O + O_2 \quad \text{(ditto for Mg)}$$

$$2\ 3(CdSO_4).8H_2O \rightarrow 6\ CdO + 6\ SO_2 + 16\ H_2O + 3\ O_2$$

$$4\ CuSO_4.5H_2O \rightarrow 2\ Cu_2O + 4\ SO_2 + 20\ H_2O + 3\ O_2$$

$$2\ MnSO_4.H_2O \rightarrow 2\ MnO + 2\ SO_2 + 2\ H_2O + O_2$$

$$2\ CaSO_4.2H_2O \rightarrow 2\ CaO + 2\ SO_2 + 4\ H_2O + O_2$$

$$2\ ZnS + 3\ O_2 \rightarrow 2\ ZnO + 2\ SO_2$$

$$As_2O_5 \rightarrow As_2O_3 + O_2$$

$$4\ Cu_2O + CH_4 \rightarrow 8\ Cu + CO_2 + 2\ H_2O$$

$$4\ ZnO + CH_4 \rightarrow 4\ Zn + CO_2 + 2\ H_2O \quad \text{(ditto for Pb and Cd)}$$

$$2\ MgO + SiO_2 \rightarrow Mg_2SiO_4 \quad \text{(ditto for Ca)}$$

$$MnO + SiO_2 \rightarrow MnSiO_3 \quad \text{(ditto for Ba)}$$

$$2\ Fe_2O_3 + 2\ SiO_2 \rightarrow 2\ Fe_2SiO_4 + O_2$$

specific for options II, III and IV:

$$2\ ZnO.Fe_2O_3 + 2\ SiO_2 \rightarrow 2\ ZnO + 2\ Fe_2SiO_4 + O_2$$

specific for options III and IV:

$$2\ Fe_3O_4 + 3\ SiO_2 \rightarrow 3\ Fe_2SiO_4 + O_2$$

The so-obtained Cu,Cd,Zn,Pb,As - fly-ash reoxidizes during cooling:

$$2\ Zn + O_2 \rightarrow 2\ ZnO \quad \text{(ditto for Pb and Cd)}$$

$$4\ Cu + O_2 \rightarrow 2\ Cu_2O$$

9. Fly-Ash leaching unit (options I to IV)

$$ZnO + 2\ H^+ \rightarrow Zn^{2+} + H_2O \quad \text{(ditto for Cd, Pb and Ca)}$$

$$Pb^{2+} + SO_4^{2-} \rightarrow PbSO_{4(s)}$$

$$Ca^{2+} + SO_4^{2-} + 2\ H_2O \rightarrow CaSO_4.2H_2O_{(s)}$$

$$Cu_2O + 2\ H^+ \rightarrow 2\ Cu^+ + H_2O$$

$$4\ Cu^+ + O_2 + 4\ H^+ \rightarrow 4\ Cu^{2+} + 2\ H_2O$$

$Fe_2(SO_4)_3.6H_2O \rightarrow 2\ Fe^{3+} + 3\ SO_4^{2-} + 6\ H_2O$

$2\ As_2O_3 + 2\ O_2 + 4\ Fe^{3+} + 6\ H_2O \rightarrow 4\ FeAsO_4 + 12\ H^+$

10. Reductive leaching (option V)

$NH_4Fe_3(SO_4)_2(OH)_6 + 6\ H^+ \rightarrow NH_4^+ + 3\ Fe^{3+} + 2\ SO_4^{2-} + 6\ H_2O$

$H_3OFe_3(SO_4)_2(OH)_6 + 5\ H^+ \rightarrow 3\ Fe^{3+} + 2\ SO_4^{2-} + 7\ H_2O$

$2\ Fe^{3+} + SO_2 + 2\ H_2O \rightarrow 2\ Fe^{2+} + 4\ H^+ + SO_4^{2-}$

$ZnSO_4.7H_2O \rightarrow Zn^{2+} + SO_4^{2-} + 7\ H_2O$ (ditto for Mg)

$3(CdSO_4).8H_2O \rightarrow 3\ Cd^{2+} + 3\ SO_4^{2-} + 8\ H_2O$

$CuSO_4.5H_2O \rightarrow Cu^{2+} + SO_4^{2-} + 5\ H_2O$

$MnSO_4.H_2O \rightarrow Mn^{2+} + SO_4^{2-} + H_2O$

11. Fractional ferrous sulfate crystallization (option V)

$Fe^{2+} + SO_4^{2-} + H_2O \rightarrow FeSO_4.H_2O$

12. Ferrous sulfate roasting (option V)

$4\ FeSO_4.H_2O \rightarrow 2\ Fe_2O_3 + 4\ SO_2 + O_2 + 4\ H_2O$

13. Hydroxide precipitation (option V)

$Zn^{2+} + 2\ OH^- \rightarrow Zn(OH)_2$ (ditto for Cd, Cu, Mg and Mn)

$NH_4^+ + OH^- \rightarrow NH_3 + H_2O$

$H^+ + OH^- \rightarrow H_2O$

$Fe^{3+} + OH^- \rightarrow Fe(OH)_3$

Appendix II The composition of Groningen natural gas in vol. %

CH_4	81.29
C_2H_6	2.87
C_3H_8	0.38
C_4H_{10}	0.15
C_5H_{12}	0.04
C_6H_{14}	0.05
N_2	14.32
O_2	0.01
CO_2	0.89

Appendix III Details on the hydrometallurgical treatment process

In order to obtain the (model) results presented in table III the following design assumptions have been made.

1. Zinc content of process jarosite

Process jarosite consists of ammonium and hydronium jarosite and of an insoluble residue which consists of lead sulfate, barium sulfate, arsenic(III)oxide, gypsum and silicium dioxyde. The jarosite residue still contains water in which zinc sulfate remains dissolved. It is assumed that in spite of the washing procedure operated in the primary winning plant 3 % of the dry weight of jarosite residue consists of (soluble) zinc sulfate (which equals 1.21 % (weight) zinc). This assumption leads to an increasing zinc load to the reductive jarosite leach with increasing jarosite production. Given the same concentrate intake, an increasing jarosite production is the consequence of a larger recycle from the treatment unit.

2. Zinc content of historical jarosite

Historical jarosite has a lower zinc content (1 % weight), since due to water percolation through the ponds, part of the zinc content has been 'washed' out and been recovered by the drainage system.

3. Reductive leach

The reductive leach is operated at 100 °C. Sulfur dioxyde is first dissolved in water (200 kg/m^3) at 15 °C. The slurry density of the feed to the reductive leach equals 300 g/kg H_2O. After the reductive leach the remaining solids are filtered and subsequently washed with recycled hot sodium sulfate solution. The wash-solution is added to the ferrous sulfate solution produced in the reductive leach and fed into the first crystallizer of the fractional crystallization unit.

4. Description of the fractional crystallization unit

Crystallization processes are used because of their potentiality of producing high purity solids with only limited energy requirements. In the case of zinc and ferrous sulfate the selectivity for producing relatively pure ferrous sulfate crystals is determined by the ratio of the solubility of zinc sulfate over ferrous sulfate. The higher this ratio, the higher the selectivity. The solubility diagrams of ferrous and zinc sulfate are given in figure 6 [14]. The solubility of zinc sulfate is still substantial above 140 °C, whereas the solubility of ferrous sulfate is low in that temperature range. For this reason the unit which is aimed at crystallizing 'clean' ferrous sulfate is designed to operate at 160 °C.

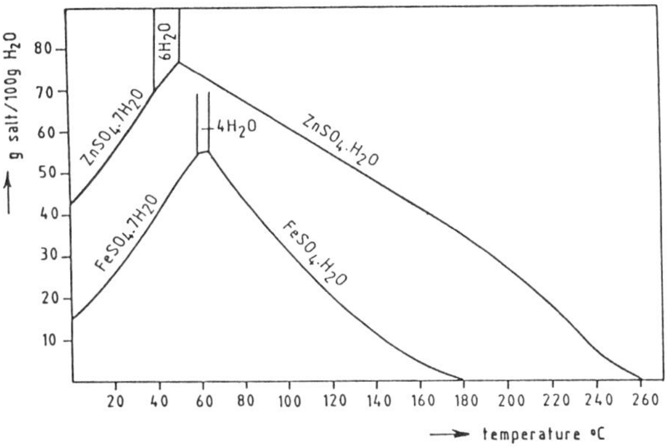

Figure 6 **The solubility diagram of zinc and ferrous sulfate as a function of temperature (g anhydrous salt/100 g H_2O)**

The proportionality constant between the composition of the liquid phase and the resulting composition of the crystalline phase is called the partition coefficient and is defined in section 4.5. Lowering the ratio of zinc over iron in the liquid phase obviously decreases the zinc content of the obtained crystals. Thus, a unit operation is designed in which the ratio of zinc over iron in the liquid phase gradually decreases towards the point where the crystals leave the unit. This technique is called 'fractional crystallization'. The fractional crystallization concept is frequently used in organic crystallization.

The operation principle is as follows. The pregnant solution enters the first of a group of five subsequent crystallizers and yields relatively contaminated crystals, which were grown at a high supersaturation. The latter factor stimulates the formation of highly agglomerated crystals with inclusions of contaminated liquid. These crystals are filtered. Subsequently, they are washed with the filtrate originating from the second crystallizer. The filtrate and the wash liquid leave the unit and are used to preheat the feed streams before they enter the hydroxide precipitation step. After the crystals from the first step are washed they enter the second step. Two extremes should now be distinguished.

The first extreme is that no cooling is required in order to obtain a crystalline phase in equilibrium with its surrounding solution, which has a lower zinc over iron ratio. The driving force for this process is the discrepancy between the feed crystal composition and the crystal composition in equilibrium with the solution present in the second crystallizer. The process described above may thermodynamically be feasible, but still be unpractical for kinetical reasons. In that case, the kinetics of the process may be speeded up by redissolving the crystalline phase at a lower temperature. One unit of a fractional crystallizer with complete redissolution is shown in figure 7 and forms the second extreme. In figure 7 a redissolution step at 120 °C is used. This temperature is applicable for slurry densities up to 100 kg $FeSO_4.H_2O$/1000 kg H_2O. For a slurry density of 200 kg $FeSO_4.H_2O$/1000 kg H_2O a redissolution temperature of 100 °C is required.

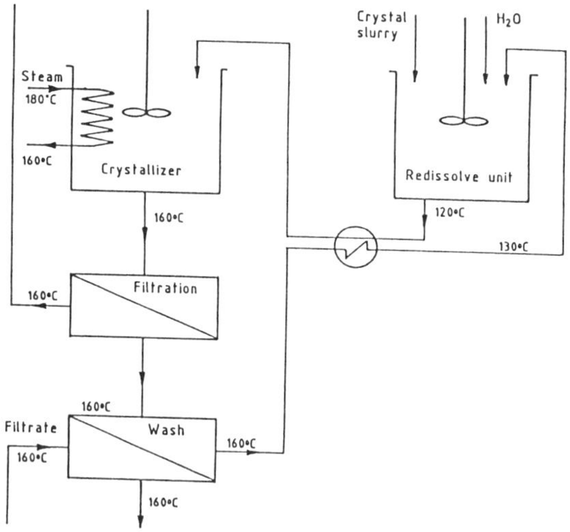

Figure 7 **One crystallizer in a fractional crystallization unit, in this case with complete redissolution**

For this particular application no research results are available which enable to predict the extent to which redissolution is required in a practical process. For this reason the energy consumption of the designed unit is presented as a range. The lower value is valid when redissolution is not required, the higher value includes 100 % redissolution.
The second step in the crystallization unit is equivalent to the third and fourth step. The fifth step utilizes water (instead of filtrate) to wash the crystals. In order to prevent redissolution during washing, the wash water temperature is designed to be higher than 175 °C. The justification for choosing five crystallizers is given in figure 8, which shows the residual zinc content of ferrous sulfate monohydrate for different numbers of crystallizers and different values of the slurry density. Obviously, the slow increase in crystal purity with an increasing number of crystallizers limits its number at about five.
Figure 8 is valid for any redissolution percentage between zero and 100, because for every step it is assumed that equilibrium between the solid and the liquid phase is reached. Additional water is added in step 2, 3 and 4 in order to limit the slurry

density to the desired value, i.e. 50, 100 or 200 kg $FeSO_4.H_2O$/1000 kg H_2O. Figure 8 reveals the importance of this slurry density for obtaining ferrous sulfate monohydrate of sufficient purity. Despite the fact that a slurry density of 50 kg $FeSO_4.H_2O$/1000 kg H_2O yields a low residual zinc content, it is evident from table 3 in section 5 that the chemicals demands for this low value of the slurry density rise to an unacceptable high level. This is due to the high recycle rate which is required for obtaining high purity crystals.

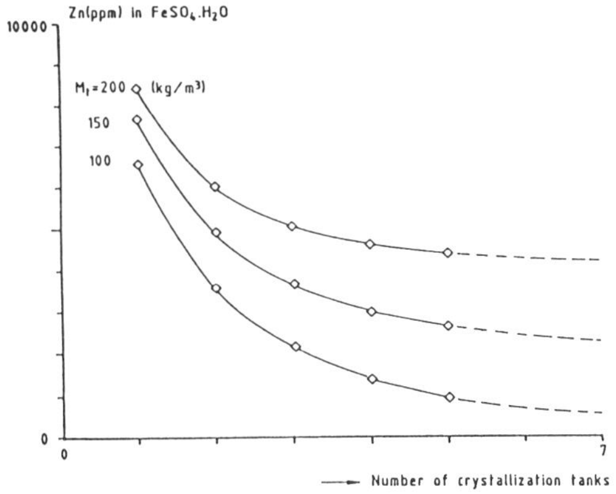

Figure 8 **The zinc load in ferrous sulfate monohydrate after a different number of crystallizers for T = 160°C and D = 0.2. M_t is given in (kg crystal/1000 kg H_2O) which is approximately equal to (kg/m³ H_2O).**

Figure 9 shows the importance of the partition coefficient for the feasibility of this process. The lower the value of the partition coefficient, the more promising this process is likely to be. For this particular application the hydrometallurgical option can not reasonably reach the desired purity of ferrous sulfate monohydrate, because the value of the partition coefficient is too high.

5. Drying and roasting

The resulting ferrous sulfate monohydrate crystals are obtained as a 60 % (weight) slurry, which is dried in a pneumatic dryer. This yields a dried crystalline product and a hot off-gas which contains over 50 % (mole) of water vapour. The off-gas is used to produce steam (180°C, 10 bar) and is subsequently partly condensed in a heat exchanger, heating the feed to the reductive leach. The solids are fed to a rotary furnace which converts ferrous sulfate to hematite and sulfur dioxyde. The solids heat recovery in this step is limited to heating of the incoming combustion air. The hot gas which is rich in sulfur dioxyde, is used to produce steam (180 °C, 10 bar) and is subsequently fed to a sulfuric acid plant.

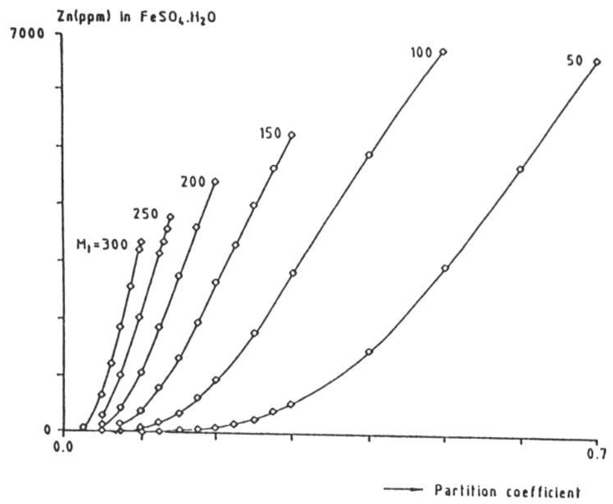

Figure 9 **The resulting zinc load in ferrous sulfate monohydrate using five crystallizers operated at T = 160°C for different values of the slurry density M_t (kg crystal/1000 kg H_2O) and the partition coefficient D.**

Appendix IV The energy use in primary electrolytic zinc winning operations

In order to estimate the energy use in the 'classical' electrolytic zinc winning operations the results of a survey published by Kellogg [13] were used. The concentrate used in Kellogg's calculations resembles the concentrate assumed in this paper. This justifies the use of his figures and assumptions. This study uses the following energy use relations (per unit operation).

1. Roasting

Roasting is an exothermic process, which yields 1.05 ton steam/ton concentrate in option I and V. The obtained steam has an energy content of 2.72 MJ/kg at a temperature of 130°C. The options II, III and IV produce steam which is used for heating (130°C) and steam which, after being superheated, is used for power generation. The balance between the two products is determined by the demand for 'heating steam' in the different options. The surplus of energy is used for power generation, as is shown in table 4[c].
The roasting process includes the electrostatic precipitators, filters, fluid bed blowers etc. The power requirements of the roasting process are estimated at 60 kWh/ton concentrate. In option IV historical jarosite is fed into the roaster, which reduces the steam production, due to its endothermic decomposition. The power requirements for option IV are estimated at 100 kWh/ton concentrate.

2. Sulfuric acid production

For the production of sulfuric acid a double contact process is assumed. The electrical power requirements for this process are given as 135 kWh/ton acid (100 %). The sulfur dioxyde formed during the concentrate roasting as well as during the thermal jarosite or ferrous sulfate treatment is assumed to be fed into the sulfuric acid plant.

3. Leaching, precipitation and purification

Included in the operations leaching, precipitation and purification are the pumps, filters, stirring devices etc. The electrical power requirements are given as 177 kWh/ton zinc. However, in this 'ton zinc', drosses and zinc dust required in the purification are included. For each ton of Special High Grade (or SHG, 99.995 %) zinc which may be sold, R tons of zinc are required, in which R represents the ratio of (SHG zinc + dross + zinc powder) over SHG zinc. In this study the value of R is chosen as 1.0129, which is slightly lower than the value reported by Kellogg. Thus, a slight increase of the energy efficiency since the publication of Kellogg's survey (1980) is taken into account. For the leaching, precipitation and purification section in option I the value of 177 kWh/ton zinc is multiplied by R to yield 179.3 kWh/ton SHG zinc.
In the options II, III and IV however, no jarosite precipitation and hot acid leach are operated. Arbitrarily it is assumed that the electrical energy requirements for the neutral leach and purification section amount 40 % or 71.7 kWh/ton SHG zinc.
Option V does operate a jarosite precipitation and hot acid leach step, but in option V, due to the large recycle of iron, the relationship between a ton SHG zinc and the precipitation and hot acid leaching of iron is no longer valid. In order to compensate for the increased electrical energy requirements the literature value of (177*R) kWh/ton SHG zinc is multiplied by the ratio P which equals (ton jarosite residue formed) over (ton jarosite residue formed in option I). This ratio probably overestimates the increase in the leach-precipitate section, because the neutral leach and purification flows are not significantly higher. In order to compensate for the overestimated value here, no correction for the magnitude of the recycle is made in the hydrometallurgical treatment plant's electrical energy requirements.
The steam requirements reported in literature vary considerably. This study uses the value of (2.2*R) ton steam/ton SHG zinc, which equals 2.23 ton steam/ton SHG zinc. This value is multiplied by 0.2 for the options II, III and IV which yields 0.45 ton steam/ton SHG zinc. This correction is made because these options do not include hot acid leach and jarosite precipitation which are the major steam consuming steps in this section. For option V the value of the steam requirement equals (2.2*R*P) ton steam/ton SHG zinc.

4. Electrolysis

Included in the electrical energy requirements for the electrolysis are a.o. pumps. The stripping of the cathodes is also included. It is reported that the overall requirement equals 3330 kWh/ton zinc or (3330*R=) 3373 kWh/ton SHG zinc.

5. Melting and casting

This procedure requires 125 kWh/ton zinc or (125*R=) 126.6 kWh/ton SHG zinc.

6. Others

The 'others' include laboratory facilities, offices, heating etc. They require 70 kWh/ton zinc or (70*R=) 70.9 kWh/ton SHG zinc.

Appendix V Energy requirements for reductive roasting

The reductive roaster is operated to convert the calcine which is obtained in the oxidative roaster into a calcine low in zinc ferrite. The net reaction equation for this process is given in appendix I, section 2. For calculating the thermal and electrical energy requirements for this step more detailed assumptions were required.
In order to obtain a gas atmosphere with sufficient reductive potential a CO_2/CO-ratio of 1.5 is taken, based on [8] and references therein. This ratio is almost comparable to the one used in the nickel industry for reduction of nickeliferrous iron oxide to magnetite and Ni/Fe-metal [15].
The roaster is operated at 800 °C. At that temperature and for a CO_2/CO-ratio of 1.5 the heat of combustion of Groningen natural gas (see appendix II) equals 433 kJ/mole. Heat is required for:

i) the decomposition of zinc ferrite into zinc oxide and hematite,
ii) the conversion of hematite into magnetite,
iii) the reduction of arsenic(V)oxide to arsenic(III)oxide,
iv) the heat of vaporization of lead oxide, arsenic(III)oxide, cadmium and zinc and
v) the heat losses due to radiation and convection.

The conversion of cadmium and zinc to its metallic form is not included, since these metals volatilize and re-oxidize inside the roaster which consumes no net energy for the respective chemical reations.
The process is assumed to be carried out in a rotary kiln. The heat transfer from the gas to the solids is assumed to take place in the first 33 % of the length of the kiln. The solids which originate from the oxidative roaster were assumed to have a temperature of 600 °C at the entrance and have to be heated up to 800 °C by a gas which reaches a final temperature of 850 °C. The heat losses are estimated at 3 kW/m^2.The heat transfer coefficient is estimated at 50 W/m^2.K and the logarithmic mean temperature equals 440 K, given a combustion temperature of natural gas of 1800 °C for this CO_2/CO-ratio. Assuming the thermal isolation to have a thickness of 0.20 m and the kiln to have a length/diameter-ratio of 20, the following data are obtained:

Option III:	natural gas consumption	31.2 Nm^3 gas/ton SHG zinc
	kiln length	47 m
	kiln diameter	2.33 m
Option IV:	natural gas consumption	32.7 Nm^3 gas/ton SHG zinc
	kiln length	50 m
	kiln diameter	2.46 m

Electrical energy requirements are assumed to be equal for the kilns, and are estimated to be 100 kWh/ton SHG zinc.
The heat content of the off-gases is used to generate steam (312 °C, 100 bar). This steam is superheated in a separate heater and used for power generation.

Appendix VI Energy requirements for residue drying

Wet solids are dried before they are fed into pyrometallurgical processes like roasting or slag production. This is especially important in the latter case, because the gas volume produced, determines the equipment size.
In this paper pneumatic dryers were selected as equipment. For pneumatic dryers the following electrical and thermal energy requirement estimates are available in literature [16]:

i)	electrical energy requirement	0.2 MJ/kg vaporized water
ii)	thermal energy requirement	4.5 MJ/kg vaporized water

Thus, the thermal energy efficiency is equal to about 50 %, since the heat of vaporization of water equals 2.26 MJ/kg at 100 °C. The dryer off-gases are used to generate saturated steam (130 °C) in the options I to IV and 180°C in option V. They are also used in option V for preheating the feed to the reductive leach.
The residual moisture content in the solids is 1 % (weight) for all dryers except the one in option V. In option V a negligible water content of the solids is essential in order to avoid the formation of $FeSO_4OH$ during the roasting of ferrous sulfate. $FeSO_4OH$ is a relatively stable intermediate which is formed in the presence of water. The formation of $FeSO_4OH$ increases the sulfur content of hematite which is undesirable.
In option IV two pneumatic dryers are applied. The first dryer is used to evaporate water from historical jarosite residue which is fed into the oxidative roaster. The second dryer is required to dry the feed of the flash furnace.

Appendix VII The flash furnace: design aspects and energy requirements

The flash furnace is chosen as a first process step towards reducing the content of cadmium, zinc, lead, arsenic and to some extent copper in the final slag residue. The technology of flash furnaces in itself is not new, but they have not found application in low-grade residue treatment.
In a flash furnace a liquid slag bath is obtained at temperatures ranging from 1200 - 1400 °C, depending on the melting point of the slag. During its continuous operation the relatively volatile heavy metals evaporate during suspension smelting and are collected as their oxides or sulfates in the fly-ash.

1. The specific gas flow

Flash furnaces are commonly used for smelting copper and copper/nickel sulfide concentrates. Their capacities are reported on the basis of tons of solid feed treated per unit bath area or per unit furnace volume per day [17,18]. This is not a solid design basis because the corresponding gas flow varies considerably with the degree of oxygen enrichment of the combustion air.
Commercial furnaces vary in gas flow from less than 500 m^3/m^2 bath.hour (INCO, Copper Cliff) to more than 22OO m^3/m^2 bath.hour for Outokompu-type furnaces using pre-heated air only for combustion of fuel. This leads to corresponding variations in dust carry-over. Traditional reverberating furnaces have gasflows close to 1000 m^3/m^2 bath.hour, with a dust load of about 1 % of the charge. Since this is about intermediate for the range calculated for flash furnaces we have chosen a gas flow of 1000 m^3/m^2 bath.hour as basis for our design.

The gases originate from the combustion of natural gas and from the partial decomposition of the residue during the process. It is economical to limit the required gas flow to a minimum, since the bath size is linearly proportional to it and so are the equipment costs and the energy losses. This aim is achieved by using oxygen-enriched air. However, using technically pure oxygen as a fuel is expensive, so an optimum should be chosen. In this study, without further optimization, it is assumed that 67 % of the oxygen demand which is not fulfilled by the oxygen liberated by the decomposition of the residue, is supplied as technically pure oxygen (95 % O_2, 5% N_2), the remaining 33 % originating from air.

2. Energy supply and demand

Groningen natural gas (see appendix II) is used as fuel. Its heat of combustion is estimated at 700 kJ/mole gas. The combustion energy is used for:

i) endothermic decomposition reactions;
ii) heating the resulting gases, slag and fly-ash;
iii) coping with the heat losses to the environment.

Based on the composition of the slag which for all pyrometallurgical options predominantly consists of a mixture of FeO, SiO_2 and CaO, melting points of the slags were obtained by using ternary phase diagrams. For each option the operation temperature of the flash furnace was chosen 100 °C above the so-determined slag melting point. In order to ensure good heat transfer, the gas was assumed to have a temperature of 200 °C above the slag melting point. The heat content of the slag at 1200 °C was estimated at 1600 kJ/kg, for 1300 °C it was 1700 kJ/kg and for 1400 °C it was 1800 kJ/kg. The specific heat content of the fly-ash was assumed to be equal to the specific heat content of the slag.

The energy required for decomposition reactions is obtained by calculating the heat effect of the chemical reactions, mentioned in appendix I for the options I to IV. This also yields the heat content of the product gases. The differences in the zinc feed content in different options have no influence on the composition of the slag leaving the flash furnace, since the reduction potential of the gas atmosphere is for each option adjusted to the feed zinc load of the furnace.
The energy lost to the environment is obtained assuming the following:

i) the thickness of the isolation of the furnace is 0.30 m;
ii) the thermal conductivity of the isolation is 2 W/m.K;
iii) the outer temperature of the furnace is 125 °C;
iv) the outer surface area of the apparatus equals five times the slag bath surface area;
v) water cooling is applied in order to prevent hot-spots.

The energy losses to the environment constitute between 10 and 16 % of the overall energy inputs for the different options.
Having determined the entire energy requirement, the overall gas flow, the natural gas consumption and the slag bath size are calculated. Table 6 summarizes the results for the options I to IV.

Table 6 Calculation results for the flash furnace application as low-grade residue convertor

option no.	feed ton/hour[1]	bathtemp °C	techn. O_2 Nm^3/hour	air Nm^3/hour	nat. gas Nm^3/hour	bath area m^2
I	32.57	1,400	7,436	16,801	6,721	277
II	29.50	1,300	4,566	10,317	4,205	162
III	29.78	1,300	5,239	11,834	4,579	178
IV	25.29	1,200	3,146	7,110	2,924	97.5

1 A year is assumed to include 7008 productive hours; thus an availability of 80% is assumed

After the flash furnace the hot gases are used to produce steam (312 °C, 100 bar) which, after additional heating, is used for power generation. After the waste heat boiler, the fly-ashes are collected in electrostatic precipitators. The energy requirements of electrostatic precipitators are described in literature [19]. For this application a specific energy consumption of 0.85 kW/m^3.s is assumed, given a gas temperature of 312 °C.
The technically pure oxygen is produced in a cryogenic turn-key air separation unit, which will not be discussed here. The energy requirement for obtaining technically pure oxygen is estimated at 400 kWh/ton O_2 (technically pure).

Appendix VIII Calculation of the energy requirement of the batch slag fumer

The batch slag fumer is used to produce a slag which meets the environmental standards with respect to the leaching of hazardous metals by natural processes.
Pilot plant experiments, carried out the Pasminco Research Centre under assignment of Budelco B.V. have confirmed that the composition and the leaching properties of slags produced from jarosite residue fulfil the strict requirements of the Dutch Government for use in building materials. The contents and background of the Dutch 'building materials directive' are beyond the scope of this paper [4]. It should be noted that this directive demands a residual antimony concentration below 100 ppm, before a material is accepted for use as natural gravel substitute in concrete. Achieving this limit results in a zinc concentration below 1000 ppm.

In order to reach these very low residual zinc concentrations in the slag, a high CO/CO_2-ratio is required. This is obtained by using natural gas as fuel and coal as reductant. The carbon requirements are about 800 kg/ton zinc which is to be removed [20]. The zinc load to the slag fumer equals about 1.25 % of the quantity of slag. The slag leaving the fumer contains about 0.1 % of zinc. This means that the net zinc removal equals about 1300 tons/year. This requires about 2000 tons of coal/year, assuming the carbon content of coal to be 50 %, which is reasonable for a coal containing a medium amount of volatiles.
For this paper the quintessence of the pilot plant tests is the quantity of energy required for slag fuming. This equals 233 Nm^3 natural gas/ton slag [21]. The data

given in the tables in section 6 are obtained by relating the zinc and the slag production.
The electrical energy is required for producing technical oxygen (95 % O_2, 5 % N_2) which is needed for obtaining a bath temperature of 1300 °C at that particular CO/CO_2-ratio.

Appendix IX The superheater and the on-site power station

In the options I to IV the hot combustion gases resulting from the flash furnace and the batch slag fumer are suitable for use in power generation. In the options III and IV the off-gases of the reductive roaster are also used. After these high temperature steps waste heat boilers are used to produce saturated steam at 312 °C (100 bar). Steam, which is useful for power generation has to be superheated in order to prevent condensation on turbine blades. The final temperature of superheated steam prepared for power generation is typically 560 °C. In order to heat steam from 312 °C to 560 °C, 0.528 MJ/kg is required. This energy is obtained by burning natural gas, with a supposed energetic efficiency of 85 %. Since for the options I to IV the amount of 100 bar steam is known, the natural gas consumption is easily calculated.
The superheater produces steam which is used for generating power with an efficiency of 30.7 %. Thus, the power production equals 1.10 $MJ_{elec.}$/kg steam (100 bar, 560 °C).

Appendix X Fly-ash leaching unit and miscellaneous

The fly-ash leaching unit uses energy for redissolving the fly-ash and for selectively precipitating ferric arsenate. The overall energy requirements for the fly-ash leaching unit are assumed to be:

i) 0.2 ton steam (low pressure, 130 °C)/ton SHG zinc
ii) 50 kWh/ton SHG zinc

The options I to IV have similar fly-ash leaching units, so for all options the same estimates are used. For the options I to IV a power consumption of 50 kWh/ton SHG zinc is assumed for miscellaneous purposes.

Appendix XI The rotary furnace for ferrous sulfate roasting

The feed of this furnace is assumed to consist of anhydrous ferrous sulfate which decomposes to hematite and sulfur dioxyde. The heat effect of this reaction is 195 kJ/mol $FeSO_4$ at 670°C [22]. This temperature is chosen because the higher the temperature, the faster the reaction proceeds. However, above 670 °C a phase transition of hematite occurs which increases the required heating energy, without increasing the reaction speed [23].
The rotary furnace is operated countercurrently. Its calculated length equals 60 m. The internal diameter equals 2.6 m and the wall thickness is assumed to be 0.2 m. Given a heat transfer coefficient of 50 $W/m^2.K$, the overall heat losses to the environment are calculated at 6 %. The energy required for operating the furnace in option V equals 38.3 Nm^3 natural gas/ton SHG zinc. The hot gases leaving this unit are used for generating medium pressure steam.

PROPOSED TREATMENT OF NEUTRAL LEACH RESIDUE

AT BIG RIVER ZINC

Michael Altepeter and Stephen James

Big River Zinc Corporation
Route 3 and Monsanto Avenue
Sauget, Illinois 62201

Big River Zinc Corporation, located in Sauget, Illinois, operates a neutral leach residue circuit. The residue produced in the circuit is sold to lead smelters for subsequent recovery of valuable metals. A research program was initiated several years ago to find a means of treating the residue at Big River Zinc to recover the valuable metals. The Ausmelt submerged lance reactor was selected as the best alternative for the first step of the process. Pilot scale tests were conducted at Ausmelt's facilities in Melbourne, Australia. Separations were excellent. 99%+ of the Zn, Pb, Cd, and Ag contained in the residue reported to the oxide fume. 99%+ of the SiO_2, Fe, Mg, and Ca in the residue reported to the slag which passed the EPA TCLP test. Subsequent leaching of the oxide fumes and purification of the leach solutions using conventional procedures produced solutions suitable for zinc electrowinning and an upgraded lead sulfate cake for sale.

Residues and Effluents - Processing and Environmental Considerations
Edited by Ramana G. Reddy, William P. Imrie and Paul B. Queneau
The Minerals, Metals & Materials Society, 1991

Introduction

Historically, the Big River Zinc (BRZ) plant in Sauget, Illinois, has sold the neutral leach residue produced at the plant to lead refineries for the contained Pb and Ag. Typical assays of the residue are as follows:

% Zn	% Fe	% Pb	Oz/T Ag	
16.0	10.0	9.0	40	(1)

To produce such a residue the average assays of the blended concentrate feed to the roasters has been

% Zn	% Fe	% Pb	Oz/T Ag	
59.5	1.4	1.3	6.0	(2)

Higher iron concentrates are not used as feed for the plant because they produce lower grade residue with correspondingly higher zinc losses from undissolved ferrite and produce a much lower value neutral leach residue. Installation of a Goethite or Jarosite processing circuit increases zinc recovery but yields iron wastes which pose serious disposal problems. (1)

A research program was initiated to find an alternative means of treating higher iron neutral leach residues. This program identified the Ausmelt submerged lance technology as the best means of separating the valuable materials from the residue and producing a non-hazardous slag which passes the EPA TCLP test.

This paper will discuss results of the following testwork:

1. Pilot scale tests of the Ausmelt submerged lance technology using BRZ neutral leach residue.

2. TCLP tests and mineralogical characterization of the slag.

3. Bench scale leach tests with the oxide fumes.

4. Purification and electrolysis of the subsequent leach solution.

Figure 1 shows a simplified flowsheet of the proposed neutral leach recovery process utilizing Ausmelt's submerged lance technology.

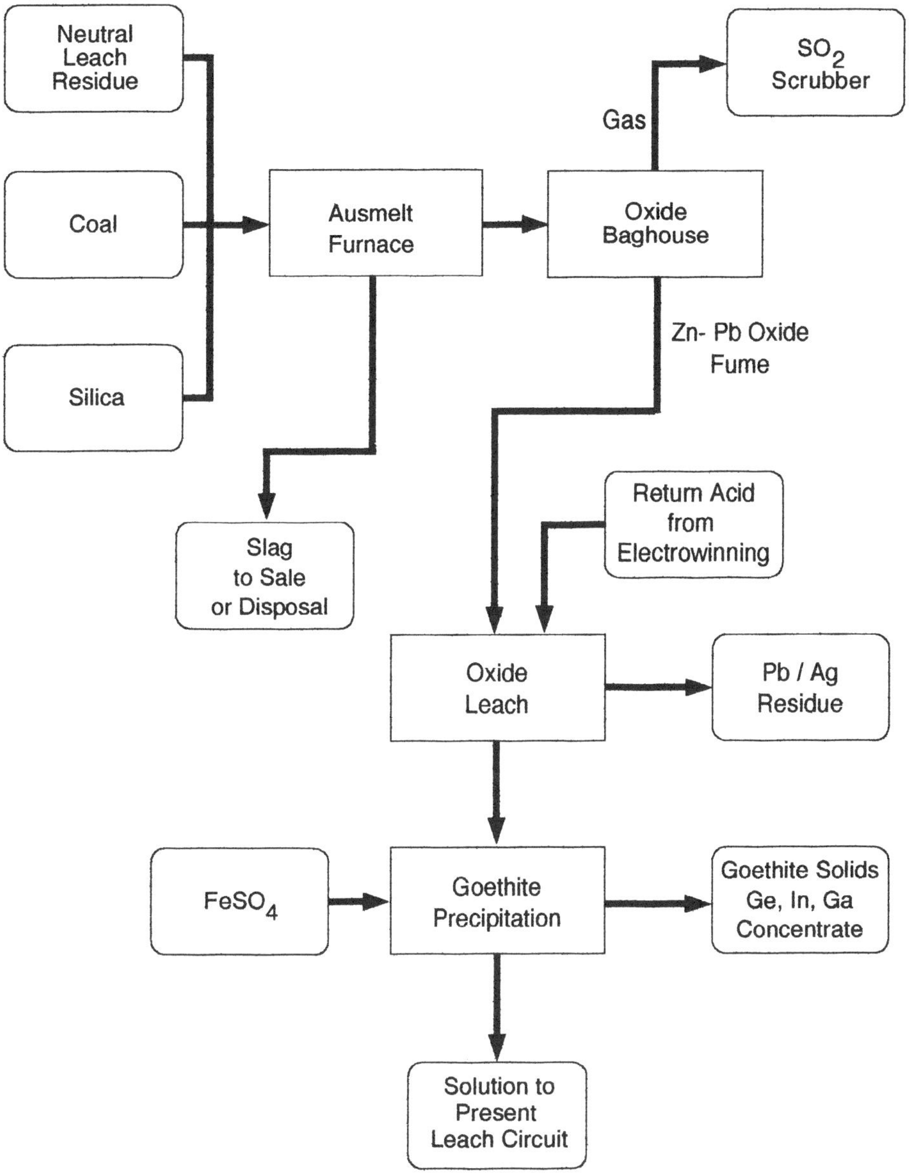

Figure 1
Simplified Flow Sheet of the
Big River Zinc Neutral Leach Residue Treatment
Utilizing Ausmelt Submerged Lance Technology

Ausmelt Pilot Scale Tests

The details of Ausmelt's submerged lance technology will not be described here because it has been discussed in the Literature (2), (3), (4). Pilot scale tests using BRZ neutral leach residue as feed were conducted at Ausmelt's facilities in Dandenong, Victoria, Australia. Nine trials were conducted using 500 Kg of residue for each trial. Fuming was accomplished in two steps, Smelting and Reducing. During the Smelting step, moist residue along with silica and lump coal was fed for 200 minutes into the agitated slag bath maintained at 1250 to 1300 C. Smelting produced a lead-rich oxide fume and removed 90% of the sulfur from the batch. During the Reducing step, only lump coal was added to the slag bath for 90 minutes and the temperature was raised to 1400 C. The Reduction step produced a zinc-rich fume and removed another 5% of the sulfur originally in the residue. No matte was produced during the pilot tests, only slag and fume.

Separations of the elements in the residue were excellent. (Table I).

Table I Trial 5 Assays and Distributions

		ASSAYS				DISTRIBUTIONS	
		Neutral Leach Residue	Smelting Fume	Reducing Fume	Final Slag	Combined Fumes	Final Slag
Zn	%	16.2	31.5	59.1	0.05	99.5	0.5
Fe	%	10.0	0.2	0.1	27.2	0.2	99.8
Pb	%	8.6	34.0	14.4	0.001	100.0	0.0
Cd	%	.59	2.45	0.78	0.003	99.4	0.6
Cu	%	0.82	0.35	0.39	0.95	10.2	89.8
As	%	0.08	0.33	0.14	0.004	94.6	5.4
Ag	oz/t	42.4	169.5	54.0	0.2	99.3	0.7
CaO	%	9.7	0.1	0.1	16.8	0.2	99.8
Al2O3	%	4.1	1.5	0.6	6.1	4.9	95.1
MgO	%	0.6	0.1	0.2	8.1	0.4	99.6
SiO2	%	8.0	0.1	0.1	18.8	0.1	99.1
In	ppm	80	190	330	5	92.7	7.3
Ga	ppm	270	30	212	440	5.4	94.6
Ge	ppm	180	175	575	5	94.4	5.6
S	%	13.9	7.1	2.7	0.01	7.1*	0.1

* The remaining S reports to the scrubber as SO_2.

All distributions are expressed in %; assays are expressed as noted.

Mineralogy

Mineralogical Characterization of the Slag

The slag consists of glassy Ca-Fe-(Mg)-Silicates. Some of the silicates are well crystallized and the crystals correspond to the mineral Kirschsteinite. Spinel-type minerals are embedded in the silicate matrix as either dendrites or octahedral particulates. The spinels generally are Fe-Al-Mg oxides which contain the trace elements: zinc, lead, cadmium, copper, nickel, manganese, titanium, calcium, and silica.

Mineralogical Characterization of the Fumes

The fumes consist of ultrafine dust which contains the following minerals:

- ZnO - Zincite
- $PbSO_4$ - Anglesite
- $Pb_2O(SO_4)$ - Lanarkite
- $CdSO_4$
- Pb_2O_3
- Elemental sulfur

The typical size of the individual oxide particle is 0.5 micron and has a sperical shape. However, many of the particles form agglomerates up to 20 microns in size.

Slag

EPA TCLP Results

The slags produced during the Ausmelt trials easily passed the EPA TCLP test (Table II) and could be sent to a sanitary landfill. However, BRZ is exploring the possible sale of the slag for building material.

Table II TCLP Test Results Ausmelt Slags (mg/l)

	Ag	As	Ba	Cd	Cr	Hg	Pb	Se
Trial 5	<.005	<.005	4.31	.051	<.020	.0007	.23	.011
Trial 6	<.005	<.005	5.51	.062	<.020	<.0005	.23	.017
Trial 8	<.005	<.005	6.22	.020	<.020	<.0005	.31	.007
Standard	<5.0	<5.0	<100	<1.0	<5.0	<0.2	<5.0	<1.0

Oxide Fume Treatment

Leaching of the Fume

The first step in recovering the zinc and cadmium from the oxide to leach it with H_2SO_4. Spent cell acid from the zinc electrowinning circuit at BRZ will be used as the leaching solvent because the leach filtrate will be incorporated into the existing BRZ plant circuit.

Experimental. Samples of smelting and reducing fume from a single Ausmelt trial were mixed before leaching in proportion to the weight of each generated during the trial. Then the fume was slurried in the minimum volume of a mixture of water and plant cell acid to wet the solids. The slurry was then transferred to a larger leach vessel. A pH controller metered cell acid into the slurry to maintain pH at 2.0 during the leach. Temperature was maintained at 55 C. The solids were flocculated and allowed to settle, the clear liquor was decanted, and the settled solids were filtered.

Results. Extractions of zinc and cadmium during leaching were good (Table III). The greatest difficulty encountered was ensuring that the solids were thoroughly wetted because the dry powdery oxides tended to "ball up" when contacted with water or acid. The use of a homogenizer to thoroughly mix the oxide with the solution improved zinc extraction considerably in test L10.

Table III Leach Results
Fraction Of Element Extracted From Solids (%)
(pH 2, 55 C, 3 Hours)

TEST	HOMOGENIZER USED?	Zn	Fe	Cd	Cu	As	Ge	Sn	Sb	In	Ga
L4	NO	97.0	17.0	89.7	93.3	24.6	99.0	–	–	–	–
L9	NO	95.8	21.0	86.7	90.1	40.1	99.9	–	62.4	9.8	–
L10	YES	99.3	26.0	89.0	97.5	75.2	99.9	42.8	85.9	39.1	64.2

Leaching the fumes produced a $PbSO_4$ concentrate rich in Ag (Table IV).

Table IV Leach Results
Assays of Pb/Ag Solids

	%Zn	%Pb	%Cd	%Cu	%As	oz/t Ag	%SO4
L4	2.85	60.3	0.40	0.04	0.54	323	33.0
L9	3.95	60.9	0.52	0.06	0.43	364	32.4
L10	0.75	65.0	0.46	0.02	0.19	388	31.8

The concentrations of zinc, iron, copper, and nickel were approximately the same as typical BRZ plant neutral leach liquor. The concentration of cobalt is lower. However, the concentrations of other elements listed on Table V are far higher than those normally encountered in the zinc plant's neutral leach liquor because

1. they precipitate during neutral leaching of calcine,
2. they fume along with zinc during the Ausmelt process,
3. they leach readily from the oxide fumes.

Table V Leach Results
Leach Liquor Assays
(mg/l or g/l as indicated)
(pH 2.0, 55 C, 3-5 Hour)

	Zn g	Fe mg	Cd g	Cu mg	Co mg	Ni mg	As mg	Ge mg	Sb mg	In mg	Ga mg	Sn mg	Se mg	Tl mg	Te mg
L4	169	232	4.0	1000	–	19	432	140	37	–	–	0.4	–	–	–
L6	158	160	3.9	940	0.7	17	375	154	22	12	15	0.1	0.5	0.9	0.3
L7	169	180	4.4	1016	1.0	19	333	170	–	10	12	–	0.4	0.9	0.3
L8	160	118	3.9	470	3.8	12	305	150	–	–	7	–	0.4	–	–
L9	149	122	3.7	800	1.0	18	148	136	7	–	.9	.1	–	0.8	0.1
L10	165	95	4.0	860	0.9	16	390	140	10	20	23	0.1	0.2	0.9	0.6

Purification of the Leach Filtrate

Efficient zinc electrowinning requires extremely pure electrolytes [5]. Therefore all the impurity elements listed in Table V must be removed from the solution before electrowinning. Testwork confirmed that the iron, cadmium, copper, cobalt, and nickel in the leach filtrate can be removed in BRZ's present Leach and Purification circuit. The concentrations of the other elements far exceed those typically experienced at BRZ. Therefore these elements must be removed from the solution before it can be returned to the existing Leach and Purification circuit.

Iron purification removed the impurities quite successfully. The ensuing filtration was slow and the solids were difficult to handle. Goethite precipitation also successfully removed the impurities from the solution and generated easily filtered solids (6), (7), (8).

Experimental. Reagent $FeSO_4$ was added to a sample of leach filtrate to bring the ferrous Fe concentration to the required level. The solution was heated to 80 C with agitation and was sparged with air as either ZnO or calcine was added to the slurry to reach and maintain a pH of 4.0 until the Fe concentration was $\leq$ 10 mg/l. Finally the slurry was filtered and the solids washed.

Results. As, Ge, Sn, Sb, In, Ga, Se, and Te were successfully precipitated from solution in the Goethite solids (Table VI).

Table VI Leach Filtrate Goethite Purification
Solution Assays Before and After Purification (mg/l)

		As	Ge	Sn	Sb	In	Ga	Se	Tl	Te
P6	Before Purification	375	154	.12	22	12	15	.47	.9	.27
	After Purification	.12	1.1	.02	.07	.04	.18	.06	.7	.06
P7	Before Purification	333	170	.25	-	10	12	.41	.9	.26
	After Purification	.04	.14	.05	.03	.04	.25	.21	-	.15
P8	Before Purification	146	136	.04	7	.51	.91	-	.8	.14
	After Purification	.07	.50	.004	.05	.02	.23	.10	.9	.13
P9	Before Purification	148	136	.05	7	.51	.91	-	.8	.14
	After Purification	.07	.50	.004	.05	.02	.23	.10	.8	.13
P10	Before Purification	390	140	.11	10	20	23	.24	.9	.35
	After Purification	.24	.63	.004	.08	.04	.19	.05	.9	.07

Germanium concentrated in the Goethite solids (Table VII). The BRZ neutral leach residue program will focus on determining methods of recovering the Ge or selling the solids in the future. The concentration of Ge is higher in test P8 because the final pH before filtration was 3.5 rather than 4.0 so some of the unreacted ZnO was leached from the solids. The concentration of Pb was higher in P9 because calcine was used to neutralize the solution rather than reagent grade zinc oxide.

Table VII Leach Filtrate Goethite Purification
Goethite Solids Assays (%)

	Zn	Fe	Pb	Cd	Cu	As	Ge	SO4
P6	28.0	11.0	0.17	0.39	2.72	1.7	0.67	29.2
P7	15.5	37.2	-	0.25	1.55	1.3	0.57	14.6
P8	5.4	42.5	0.34	0.06	3.10	5.0	1.94	-
P9	17.0	31.1	2.40	0.29	0.31	1.5	1.22	-

The concentration of ferrous iron in solution before Goethite precipitation effects the final concentration of the impurities (Figures 2, 3, 4). 90% of the arsenic, germanium, tin, tellurium, antimony, indium, and gallium were precipitated with 0.5 g/l iron; however, before the solution can be recycled to the plant, the germanium concentration should be less than 1.0 mg/l. 2-3 g/l iron are required to precipitate germanium to less than 1 mg/l.

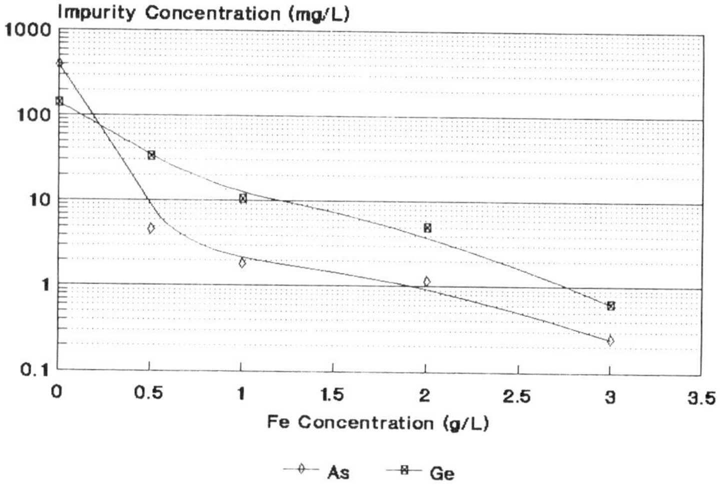

Figure 2 Effect of Iron Concentration on As, Ge

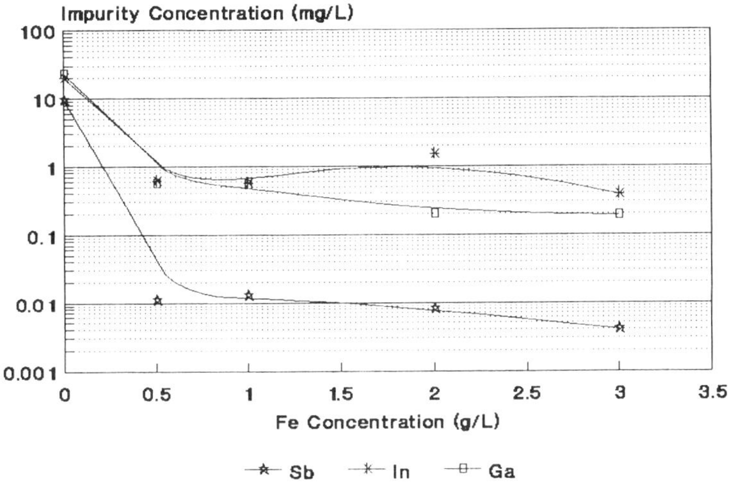

Figure 3 Effect of Iron Concentration on Sb, In, Ga

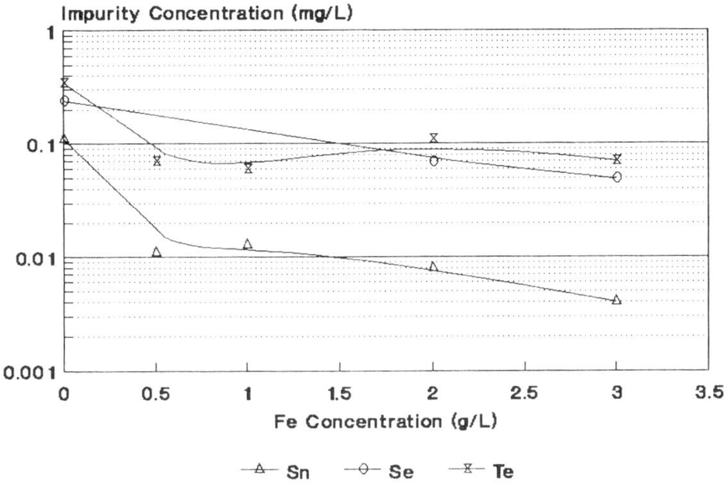

Figure 4 Effect of Iron concentration on Sn, Se, Te

Electrolysis

After Goethite precipitation the solutions were subjected to typical BRZ copper/arsenic purification and cadmium precipitation. Cyclic voltametry tests were performed (9) on the resulting "purified" solutions and the results were positive. When larger volumes of solution are generated during the pilot scale Leach and Purification tests, the resulting purified solutions will be used as feed for test cells which produce 1500 g of zinc per day to determine the suitability of the solution for electrowinning.

Acknowledgment

The authors wish to acknowledge Dr. Guy W. Lussiez for his participation in evaluating pyrometallurgical methods of treating neutral leach residue. They also wish to thank Dr. T. J. O'Keefe of the University of Missouri - Rolla for testing and evaluating the suitability of the "purified" $ZnSO_4$ solutions for electrowinning.

References

1. J. O. Buckle and H. M. Freeman "A Review Of The Environmental Considerations Of The Disposal Of Residues From Selected Hydrometallurgical Processes," Iron Control in Hydrometallurgy, 1986, 753-772.

2. J. M. Floyd, P. King, G. Guorgi, "Sirosmelt Technology For Environmental Compliance In Zinc Plants," The International Lead and Zinc Study Group - Geneva, October 1990.

3. J. M. Floyd, B. W. Lightfoot, "Sirosmelt And The Wide World Of Opportunity," Engineering and Mining Journal, June 1985, 52-56.

4. M. Waladon, G. R. Firkin, J. Bultitude-Paull, B. W. Lightfoot, J. M. Floyd, "Top Submerged Lancing For Recovery Of Zinc From ISF Slag," Lead-Zinc '90, 607-628.

5. C. J. Kreuss, "Effects Of Minor Elements On The Production Of Electrolytic Zinc From Sulfide Concentrates," Zinc '85, 753-772.

6. J. A. Andre and N. J. J. Mason, "The Goethite Process In Retreating Zinc Leach Residues," (paper presented at the 102nd AIME Annual Meeting, Chicago, Illinois 1973).

7. P. T. Davey and T. R. Scott, "Removal Of Iron From Leach Liquors By The Goethite Process," Hydrometallurgy, 2, (1976), 25-33.

8. J. M. Boxall and S. E. James, "Experience With The Goethite Process At National Zinc," Iron Control in Hydrometallurgy, (1986), 676-686.

9. Discussion with Dr. Thomas O'Keefe, Materials Research Center, University of Missouri-Rolla, on tests conducted by Dr. O'Keefe on the BRZ/Ausmelt purified solutions.

History of Effluent and Residue Treatment

at Tin Processing Corp, 1942 - 1946

John Dasher - San Francisco[1]

ABSTRACT

One of our War Plants processed Bolivian concentrates as low as 17% Sn by the Arnhem process (pressure leach with HCL) and sent the waste acid down a ditch to Galveston Bay where the chlorides, mostly iron, hydrolyzed, Oyster men, and Texas, demanded impoundment until a disposal process could be developed. A recycle process found by lab work was recommended, piloted and a plant built. This involved cementation of As Sb, Cu, Bi and Ag on light moving scrap iron and of Sn & Pb on Al turnings (then not recyclable), evaporation and crystallization of hydrous $FeCl_2$, roasting crystals in a muffle furnace and absorbing HCl for reuse. The residue from leaching the low grade concentrates was floated to remove undissolved sulfides, classified and passed over sand and slimes tables to get a smeltable concentrate, a discardable sand tail and a slimes tail worth treating. After desliming at 2 micro meters cassiterite, oxide flotation gave good, lab results, but commercial equipment wasn't adequate then. Chloride volatilization, with or without smelting gave high recoveries of Sn as $SnCl_2$, whose marketing was questionable. I have heard this was finally fluxed and smelted.

[1]John Dasher retired as a principal engineer from Bechtel in 1982. He is a Senior Member of TMS, and of SME and, of the AIME Legion of Honor. He has fifty patents and publications, including, from his hobby, "Using the Modern Metric System" (1). During the period of this paper, he was Chief, Research at Tin Processing Corp.

Residues and Effluents - Processing and Environmental Considerations
Edited by Ramana G. Reddy, William P. Imrie and Paul B. Queneau
The Minerals, Metals & Materials Society, 1991

Introduction

At the start of WWII, Metals Reserve of Reconstruction Finance moved to obtain an adequate supply of tin, a strategic metal, 100% imported. Contracts with Bolivian producers obtained all their high (+55% Sn), medium (+35% Sn) and low grade (+17% Sn) concentrates. The +35% material had previously been smelted in England, and the low grade was only treated at Arnhem by a Billiton pilot plant using a Schmidt Unax kiln, a spherical paper-mill digester lined with rubber and brick for batch leaching under pressure with HCl, a Nuetsch filter and shaking tables to obtain feed for a small regenerative reverberatory furnace.

Metals Reserve contracted with Mining Equipment Corp., a Billiton subsidiary, to design and operate a facility near Texas City, based on Arnhem technology and to supply alluvial concentrates from their operations in the East Indies.

The microfilmed drawings of Arnhem equipment were used for facility design. Stage I had 6 kilns, 6 digesters, 6 filters and 6 reverbs. Before these were erected Stage II came with an increase to 12 units for roasting and leaching and 10 reverbs. Other facilities included ore storage, a dressing plant, Cottrells for smelting and roasting, shops, product storage, offices, labs and cafeteria.

The Arnhem Superintendent escaped to help in design, but was killed in a plane crash. The staff were Dutch with experience in dredging, Americans with experience in Bolivia, smelting copper, umpire analysis or leaching low grade material that was mostly sinderite ($FeCO_3$) with SO_2 while moonlighting at a Bureau of Mines Nonmetallic Division Station.

When I arrived in November 1942, the facility was operating. The waste acid flowed down a ditch into Galveston Bay. It typically contained, in g/L, 30 - HCl, 25 - Fe^{+3}, 75 - Fe^{+2}, 8 total Sb, As, Ag, Cu and Bi, 8 total Sn and Pb and 5 total Ca, Mg, Al and Zn. It was about 75°C leaving the plant. In the bay, oxidation and hydrolysis formed a lot of $Fe(OH)_3$, which rendered oysters unsalable. Tin Processing was ordered to impound this solution until a treatment method could be found and installed.

Development of Waste Acid Treatment Process

Neutralization with hydrated lime gave a thick mud from which little $CaCl_2$ solution could be extracted and would require more storage volume than the waste acid. When waste acid was flowed over crushed limestone, the free acid was reduced to where ferric hydrolyzed and coated the lumps so well that less than half the $CaCO_3$ was used at minus 15 mm, and less than 15% at minus 100 mm. No Fe^{+2} was precipitated on $CaCO_3$.

The leach plant needed about a quarter of the U.S. HCl production. Recycle of these chlorides as HCl might have promise. The potential value of Sb Ag Bi Cu and Sn was considerable. It was known that heating ferrous chloride in air gave $FeCl_3$ vapor, but when steam was present, the product was hematite (Fe_2O_3) and HCl, to be absorbed for recycle.

Waste acid is a very corrosive brew. There was no metal then to hold it. The leach plant was nonmetallic--acid brick, Korez mortar, Haveg pipes, rubber covering. Each liter will dissolve 70 g of iron (or stainless or Hastalloy a bit slower).

Experiments showed "tin" cans dissolved 99% without agitation (soldered rings remained). Reduction of Fe^{+3} and neutralization of free acid were main reactions. The cement coating formed from less aggressive solution needs agitation to keep the iron surfaces clean. This step was tested in the lab with a Haveg drum 1 m x 0.1 m at 0.25 to 0.4 Hz (15-24 rev/min) containing 50 x 8 mm iron rods. With input at 50°C or higher and 10 min retention, cementation of Sb, As, Ag, and Bi is complete and 95% of Cu is cemented. Temperature is steady. With shorter retention, exit temperature is higher and product has free acid.

Free acid after cementation is undesirable and can be removed by a tank containing heavy scrap. The gas from cementation is mainly moist H_2. It also contains a trace of arsine (AsH_3) and stibine (SbH_3). These account for less than 0.1 mol% of the As and Sb cemented, but are extremely toxic. Gas from reducer must be collected, burned and treated for removal of particulates which are also toxic.

The ferrous chloride solution from cementation has some Sn, Pb, Ca, Mg and Zn impurities. At that time aluminum turnings and borings from the monstrous aircraft industry were not recyclable. These would, however, cement the Sn + Pb from this solution along with the trace of Cu and some Fe. It was easy to deposit a sponge of Sn + Pb electrolytically, but not to get it off the sheet iron cathode. Al cementation appeared preferable.

The next step was to evaporate the solution and form $FeCl_2 \cdot 4H_2O$ crystals,. The solution is still corrosive. Rubber lined evaporators were satisfactory only to 70°C where the $4H_2O$ phase is stable. Tubes of monel, inconel and 316L stainless were indicated by tests to be adequate if Sn + Pb were removed, but if not, only Hastalloy C and Ilium G were adequate. (Ti and Zr were not yet.) These crystals have more H_2O than needed for hydrolysis. Lab tests in pyrex made $FeCl_2 \cdot 2H_2O$ above 70°C, but when temperature dropped in filtering, these reacted with the mother liquor to form $FeCl_2 \cdot 4H_2O$ and a solid cake. Making $FeCl_2 \cdot 2H_2O$ was judged to be impractical. It was also found that at least 2/3rds of the iron could be quite pure. It is possible that fractional crystallization and separate roasting could give a Fe_2O_3 product that could be marketed as a pigment, for producing iron powder or, after agglomeration, a premium ore. As multiple units would be needed, fractional crystallization has little, if any, extra cost. Wescott and Comstock who had been engaged to study the chloride roasting[(2)], suggested a conveyor furnace be used. Our lab roasting tests on the crystals showed dehydration to $FeCl_2$ occurred first. With air present, $FeCl_3$ evolved at 450°C, but with air and steam, HCl evolved above 300°C and the Fe_2O_3 residue had no soluble iron. We recommended a muffled, multihearth roaster, in the hope that steam from dehydration on upper hearths could supply that needed for hydrolysis on the lower ones.

While this development work proceeded, the waste acid was impounded. The local soil contained much clay and was rather impervious except for where some limestone dissolved to make seepage channels. Dilution with water to the SRL pump glands was excessive until flow restrictors were installed. Tests on 1 m^2 ponds of water and waste acid showed evaporation was doubled, presumably by the absorption of sunlight by Fe^{+3}. The Weather Bureau data for Texas City showed evaporation a bit less than rainfall (0.43 m in one hurricane day) but impounded waste acid became more concentrated except for its free acid and some of the heavy metals that hydrolyzed. It seemed reasonable to assume any treatment facility would treat 80% current waste acid and 20% from the ponds. This would provide all the HCl needed for operations, if at least 80% recovery of HCl was obtained.

Our report on this technology and the proposed waste acid recycle facility was sent to Metals Reserve November 17, 1944. A simplified flow sheet is in Figure 1. Estimates of capital and operating costs and value of marketed products were included. These costs and values are by now at least 10X greater. The proposal indicated a quick payoff for doing what had to be done to keep producing our tin requirements. The report appended the contributions of Dr. C.L. Mantell on recovery of Sn and Pb and of Dr. Lincoln T. Work on fractional crystallization.

Metals Reserve then sent Shambra and Shigley from the Dow MgO plant at Freeport and then Sam Strelzoff of Chemical Construction to Tin Processing to evaluate our proposal.

Their reports must have been favorable because soon thereafter Mike Maurer of Chemico arrived to pilot the process to get information for design and construction of the Waste Acid Treatment facility. Chemico first tried a rotary hearth furnace and then a heated ball furnace before settling for muffled multihearths. The war ended with 2 atom bombs and Prof. Antoine Gaudin invited me to MIT to work under a contract from the Army Engineers on ores of an element he couldn't mention.

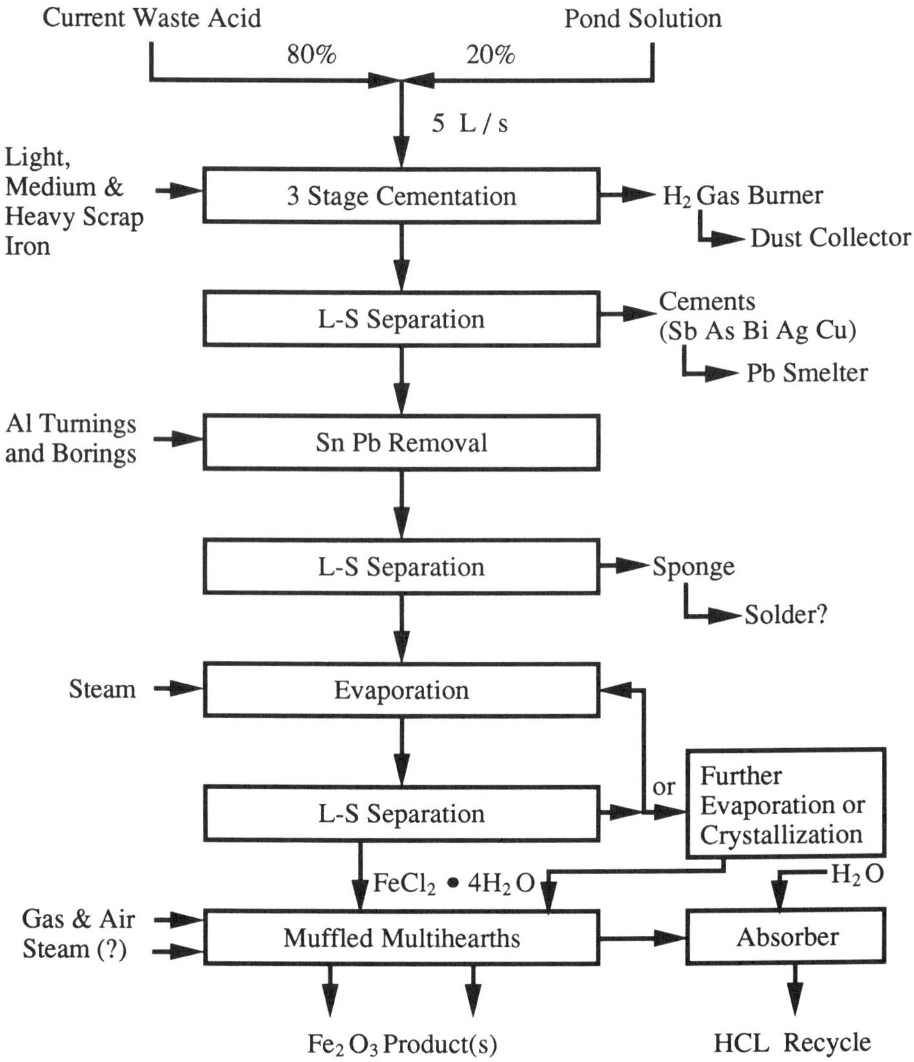

Figure 1 - Proposed Flow Sheet

Induction melting in Ar gave value to Al turnings, and marketing the Fe_2O_3 was deferred. I have heard that cements were sold to a lead smelter with payment for Sb Bi Ag and Cu, and that there was trouble with HCl escaping the roasters which was fixed before the building dissolved in moist Gulf Coast climate.

Treatment of Residues

As a residue is what is undissolved in leaching, at Tin Processing, those from treating medium and high grade concentrates were washed on filters and smelted in two stages to make, after drossing and blending, grade A, 99.8% pure Longhorn 3 star pigs for essential uses.

The residue from low grade material was filtered, washed and sent to the dressing plant where it was classified with sands and slimes tabeled separately.

The Nuetsch filters were a bottleneck. Leach practice was altered. The digesters were stopped for six minutes and the waste acid with slimes decanted. After repulping with washwater, the residue was dumped into the filters and washed further by displacement. There was only 7.6 m from leach building columns to the plant road. Five thickeners of acid brick and rubber covered mechanisms were built in a row in this space to separate slimes and waste acid. Underflow from four thickeners fed the 5th whose underflow was filtered, washed and pumped to a dressing plant. This contained 22 to 30 g Cl^{-1}/kg (dry) and had to be neutralized before contacting mild steel equipment. This was done with hydrated lime.

In the dressing plant, sulfide flotation was used to remove and recycle sulfides which survived leaching, plus elemental sulfur formed therein, then the nonfloast was classified and tabeled.

In January and February 1946, the dressing plant results are given in Table I.

Table 1

DRESSING PLANT BALANCE - JANUARY & FEBRUARY 1946

	Dry Solids		Tin		Distribution
	kt	mass%	kt	mass%	of Sn, %
Feed	6.24	100	3.27	52.4	100
Concs.	4.53	77.6	2.85	62.8	87.1
S Froth	0.085	1.4	0.008	9.3	0.24
To Ponds	1.63	26.0	0.42	25.5	12.7

In this impounded material, tin is present as cassiterite, (SnO_2), 78.8% Sn, 7.0 t/m^3 and nearly all minus 150 μm. Recovery on slime tables was good at least to 37 μm, yet slimes were 30% cassiterite and 1/8th the total Sn. Definitely worth treating.

A patent by Tainton claimed that a neutral roast of slimes with pyrite would volatilize tin as a sulfide. At Tin Processing, we had the flotation froth in which half the S was elemental. The rest was a variety of sulfides of Cu, Fe, Bi, Sb, etc. Lab roasts of froth and slimes volatilized less than half the Sn present. My reports gave no datum. Pommier reports the Bolivian plant using this process didn't work well either (3).

The next approach was to roast slimes with $FeO_2 \cdot 4H_2O$ crystals. With 25% crystals and 5% coal, 90 to 95% of the Sn volatilized at 800°C in an hour. This was condensed as $SnCl_2$ or caught in a packed tower with $Ca(OH)_2$ slurry. The tower showed signs of plugging. The product slurry was hardly filterable. Electrolysis of pure $SnCl_2$, molten, gave Sn at cathode and a mix of Cl_2 and $SnCl_4$ at anode. A need for much development work was apparent. Chloride volatization also worked well on granulated slag. Even more of the Sn was removed when the slimes, crystals and coal were smelted at 1200°C in a pot furnace, whose combustion gases were not treated.

Froth flotation with oleic acid was tried at ambient temperature and at 40°C. Selectivity was negligible. Assuming this was due to dirty surfaces, current samples were dispersed, scrubbed and de-slimed by settling for an hour in a shallow pan and decanting. The decanted slurry contained only cassiterites finer than 2 μm. Floating the settled material with oleic acid at pH 7, gave froths at 40% + Sn and tails under 1% Sn. When this procedure was tried on samples from the ponds, results were poor.

Note that the filters washing the dressing plant feed left 33 g Cl^-/kg solids. Lime addition to protect the steel equipment precipitated Fe^3, Fe^2, Sb^{+5}, As^{+5}, Ag^{+1} and Bi^{+3} on the solids. The oxide flotation feed had been given a good wash, neutralized with Na_2CO_3, dispersed with 3.5 kg/t "N" sodium silicate, de-slimed and adjusted to pH 7 before adding oleic acid [(4)].

The difference between current and impounded slimes was the use of lime and coating by its precipitates. When impounded slimes were leached with waste acid, washed well and treated like current slimes had been, comparable results were obtained.

Equipment was ordered and installed for oxide flotation including filters, thickeners, scrubber, float cells, conditioner and a scroll centrifuge for de-sliming.

There were 2 problems on start=up which had not been resolved when I left for MIT:

(a) The scrubber was my custom design [(5)]. Stator blades were bad castings and broke. Rubber lining did not adhere to pot. Energy absorbed was well below design.

(b) Scroll centrifuge recovered only half the cassiterite of +3 μm vs. 2 μm theoretical. This was not serious but put 17% of the tin in the overflow vs. 12.7% expected. The main problem was the high rate of wear on the scroll with the best hardfacing then available. A new scroll design was promised.

I've heard that some of the impounded slimes containing about 15% Sn were fluxed with oyster shell and iron oxide and smelted after the plant became private. I've never returned.

Bibliography

(1) John Dasher, Minining Engineering, February 1981, P-147.

(2) E. W. Westestt, Letter to Mining to Equipment Corp, 8/17/42.

(3) L. W. Pommier, JOM, November 1979, P-55.

(4) John Dasher & J. E. Norman, US Pat 2471414, 5/31/49.

(5) John Dasher & O. C. Ralston, Bull. Am. Cer. Soc. V 20, #6, June 1941, P-187.

TREATMENT OF CHROMATE RESIDUE BY

DIRECT ELECTROLYSIS IN MOLTEN OXIDES

Donald R. Sadoway

Department of Materials Science and Engineering
Massachusetts Institute of Technology
Cambridge, Massachusetts 02139-4307

Abstract

The viability of treating certain forms of hazardous waste by direct electrolysis in molten oxides is under investigation. Secondary residue or waste mud produced in the course of chromate chemical production is the focus of attention in this study. The goal is to reduce the water soluble hexavalent chromium either to insoluble trivalent chromium or to chromium metal. Electrolysis tests have been conducted in a high temperature cell containing chromate sludge dissolved in an electrolyte composed of Al_2O_3, SiO_2, CaO, and MgO. Oxygen production on a carbon-free anode has been confirmed.

Residues and Effluents - Processing and Environmental Considerations
Edited by Ramana G. Reddy, William P. Imrie and Paul B. Queneau
The Minerals, Metals & Materials Society, 1991

Introduction

Direct electrolysis of molten oxides is the electrolytic decomposition of a metal oxide into molten metal and oxygen gas (1,2). With reference to resource recovery, direct electrolysis has enormous potential. The metal oxide feedstock can be so-called "concentrate" derived from ore or hazardous waste such as chromate sludge. The process in its various embodiments by necessity operates at high temperatures, hence the term, pyroelectrolysis. The process uses carbon-free electrodes; this both avoids the need for energy intensive carbon electrode manufacture and guarantees the absence of greenhouse gas emissions as the by-product of the metal recovery step. Feeding the electrolysis cell and harvesting the products can be done in such a way as to allow continuous operation. The concept applies to a variety of chemistries including titanium, ferroalloys (including stainless steel), rare earth metals, and uranium. This paper is restricted to a discussion of hazardous waste treatment, specifically, secondary residue produced during the course of chromate chemical production.

Chromate residue, like much of the solid industrial waste in this country, is in the form of oxidized metal, some of which is water soluble and thus poses a threat to the environment. Treating this stuff to make it chemically inert is an enormous task. Processes doing so must not only be energy efficient but also have acceptable productivity, i.e., space/time yields. Chromate sludge contains water soluble hexavalent chromium. Pyroelectrolysis in molten oxides has the potential to reduce hexavalent chromium and recover the iron and chromium metal contents of this material. In special circumstances the value of the metal so produced may offset the cost of remediation.

Direct electrolysis in molten oxides has been tried in the past but relied on the use of carbon anodes (3). What is new and unique in the present concept is the absence of carbon. This is critical to the successful implementation of the concept. Materials selection derives from a methodology developed by the author in connection with the problem of finding a carbon-free nonconsumable anode for use in aluminum electrolysis cells (4,5). The methodology is based upon a systems view of the process and therefore takes into full consideration the types of chemical reactions that occur between the key elements of the cell (cathode, anode, sidewall) and the electrolyte and cell products. A fresh look at the use of molten oxides as electrolytic media for metal recovery in the light of the above cited methodology gave birth to the concept of carbon-free direct electrolysis. Among the foundations of the new concept is the recognition that at high temperatures certain materials previously ignored for use in the construction of electrolysis cells have properties that make them suitable as anodes. Furthermore, there is the appreciation of the fact that oxide melts can be modified through changes in solvent composition to form multicomponent solutions that are ionic liquids and therefore capable of sustaining faradaic processes.

These are the key issues in the relevant high temperature physical chemistry as they apply to treatment of chromate residue. Are chromate bearing melts ionic or is there an unacceptably high level of electronic conduction, i.e., what is the electrical conductivity of multicomponent chromate melts as a function of temperature and composition? The answer helps assess the technical viability of the concept (including the energy consumption) and leads directly to the design of the industrial electrolyte. Do the selection criteria call out a suitable material that will function as an oxygen-evolving anode, i.e., in the candidate electrolyte what is the emf series and what are the overvoltages associated with various suspected anodic processes?

Experimental

Tests have begun to examine the technical feasibility of direct electrolysis of chromate sludge. A laboratory scale electrolysis cell was designed and constructed. The cathode consisted of a heel of molten metal; copper was one choice. Electrically insulated from the electrolyte by means of alumina tubing, a tungsten rod was immersed in the molten metal and acted as the current collector. The anode was platinum mesh with an area of 15 cm^2 connected to four platinum wires acting as current leads. A closed-one-end alumina tube, 48 mm O.D., served both as reactor wall and crucible. Temperature was measured by a thermocouple, ASTM Type B, immersed in the melt.

The upper part of the conductivity cell was enclosed in heavy-walled fused quartz tubing, 5 cm O.D. $\times$ 20 cm long, which extended from just above the top of the molybdenum crucible to the stainless steel cell cap. In doing so the tubing passed through the cover plate of the furnace chamber. Vacuum tight connections were made between the fused quartz tubing and both the cell cap and furnace cover by means of O-ring compression fittings. The cell cap had four ports: three compression fittings on the top for the two electrodes and the thermocouple and one tube fitting out the side used for inert gas feed. The cell was heated by a carbon-block resistance furnace with a cylindrical graphite heating element. Machined out of a premium grade of graphite rod the heating element measured 24 cm long, 7.5 cm I.D., 11.5 cm O.D. and had eight 16.5 cm grooves cut into it. Its resistance at room temperature was about 0.008 Ω. Graphite studs connected the heating element to water-cooled current feeds made of heavy-walled copper tubing. Radiation shielding made of molybdenum sheet surrounded the heater. Electrical power at a line voltage of 208 V was stepped down to 10 V with a transformer located between the heating element and the power controller. Temperature was controlled by means of a combination of a proportional controller with integral and derivative control and phase angle fired SCR power controller. The maximum attainable temperature was 1650°C. In previous studies measurements had shown that the vertical temperature zone was 3 cm long, over which the variation in temperature was less than ± 1 K at 1500°C. The furnace was contained in a stainless steel chamber in which it was possible to maintain either an inert atmosphere or vacuum.

The electrolyte was prepared by mixing 38 wt % sludge, 10 wt % SiO_2 and 52 wt % Al_2O_3 which for the particular sludge samples employed in this study gave the following composition:

Fe_2O_3[1]	12.7%	MgO	6.9%
SiO_2	15.0%	CaO	7.4%
Al_2O_3	56.5%	CrO_3	1.5%

After the components of the electrolyte were charged into the crucible and the electrodes and thermocouple were positioned in their respective feedthroughs, the cell was assembled, filled with argon, and heated at a rate of 100 K hr^{-1} to the process temperature of 1550°C. During this time water flashed off the charge. Once the temperature had reached setpoint the electrodes were polarized and lowered into the electrolyte.

---

[1] The figure for iron oxide represents an average of its various forms.

Results and Discussion

Electrolysis was conducted at a cathodic current density of 100 mA cm^{-2}. The cell voltage measured 23 V, and as electrolysis proceeded the voltage slowly rose in accordance with the depletion of cell feed. Argon was directed through the cell in such as way as to exit via a tube located just above the anode. This effectively swept away any gas products. An electronic oxygen sensor with a range spanning 0.1% to 25% monitored the oxygen content of the gas leaving the cell. It took about 20 minutes for the oxygen level in the exit stream to rise above the limit of detection of the sensor. After 2 hours the oxygen level reached 1.2% and hovered at 1.2% - 1.3% which appeared to be the steady state value for the conditions in the cell, i.e., argon gas flow rate and anodic current density.

After about four hours electrolysis stopped, and the cell was cooled to room temperature. The electrolyte and the cathode metal heel were subjected to energy dispersive x-ray analysis. There was no evidence of iron or chromium in the metal heel. At the same time, owing to the relatively small scale of the experiment, i.e., size of cell and duration of experiment, it was not possible to perform an accurate mass balance on the electrolyte. What is clear is that the oxide melt was indeed an ionic conductor as evidenced by the presence of oxygen which was detected only in response to the passage of electric current. When the melt was simply heated to 1550°C and held at this temperature for several hours, no oxygen was detected. Thus, electric current passing through the melt drives faradaic processes. This conclusion is extremely important, as many molten oxides containing substantial amounts of transition-metal oxide, e.g., ferrites and chromites, are known to be electronic conductors. In such melts electrochemical processing is impossible. This experiment then demonstrates that by judicious choice of melt composition one can dissolve transition-metal oxides and retain ionicity.

Possibly, the cathodic the reaction was the reduction of iron and chromium from higher to lower valent forms, as for example,

$$Cr^{6+} + 3e^{-} \rightarrow Cr^{3+} \qquad \text{and} \qquad Fe^{3+} + e^{-} \rightarrow Fe^{2+}.$$

More work is necessary to discover the nature of the cathodic reaction. Linear sweep voltammetry and chronopotentiometry are among the electrokinetic techniques that will be employed. If the reactions proposed above in fact occur, then this represents immobilization of the toxic hexavalent chromium ion. Clearly, from a commercial standpoint it would be desirable to be able to extract the chromium as metal. Perhaps by changes in electrolyte chemistry this goal can be reached. Work continues.

Acknowledgements

The author acknowledges Mr. Tom McKee and Mr. Paul Richter of the New Jersey Department of Environmental Protection for providing not only the samples of chromate sludge but also much of the background information as well. The efforts of Dr. John Hryn, Ms. Susan Schiefelbein, and Ms. Isabel Wang in assisting with the design and construction of the electrolysis cell and helping perform the experiments are recognized.

References

1. "Electrochemical Extraction of Metals," The MIT Report, 19 (1) (1991), 1-2.

2. D.R. Sadoway, "Metallurgical Electrochemistry in Nonaqueous Media," in Proceedings of the Elliott Symposium on Chemical Process Metallurgy, P.J. Koros and G.R. St. Pierre, editors, ISS, Warrendale PA, 1991, pp. 189-196.

3. R. H. Aiken, "Process of Making Iron from the Ore," U.S. patent no. 816,142, March 27, 1906.

4. D.R. Sadoway, "A Materials Systems Approach to Selection and Testing of Nonconsumable Anodes for the Hall Cell," in Light Metals 1990, C.A. Bickert, editor, TMS, Warrendale PA, 1990, pp. 403-407.

5. D.R. Sadoway, inventor, Massachusetts Institute of Technology, assignee, "Apparatus and Method for the Electrolytic Production of Metals," U.S. patent no. 4,999,097, March 12, 1991.

DECONTAMINATION OF SOLUTIONS CONTAINING HEXAVALENT CHROMIUM USING MODIFIED BARKS

GABALLAH I. GOY D. and KILBERTUS G.*

Mineral Processing and Environmental Engineering, CRVM,
CNRS UA 235, INPL, BP 40, 54501 Vandœuvre, France
* ESSTIB, Allée des Hêtres, 88000 Epinal, France

Abstract

Decontamination of synthetic $K_2Cr_2O_7$ solutions containing 10 to 1000 ppm of hexavalent chromium using chemically treated barks have been studied at a laboratory scale. Metal removal from solutions depends on the pH, the initial concentration and the bark species.

The use of a pulp density of 2 % in an agitated flask at the optimum pH for 2 hours leads to a percentage chromium removal of about 95 % of the initial metallic ions content. Other trials conducted by column percolation permit the determination of the maximum retention capacity of the treated bark. Depending on the bark species, oak or pine, the rate of saturation is 110 and 215 mg of Cr per gram of bark respectively. The incineration of the metal charged barks gives recyclable ashes containing 36 and 54 % of chromium metal for the oak and pine barks respectively.

Experiments of Cr^{6+} removal from industrial effluents containing up to 1800 ppm have been carried out on the semi-pilot plant scale. They have lead to the elimination of about 90 % of hexavalent chromium from this industrial waste solution. Further studies are conducted in order to define the hexavalent chromium binding mechanisms on bark constituents.

Residues and Effluents - Processing and Environmental Considerations
Edited by Ramana G. Reddy, William P. Imrie and Paul B. Queneau
The Minerals, Metals & Materials Society, 1991

Statement of the problem

Many industries, especially those of electroplating, discharge waste solutions containing heavy metals such as Cr, Ni, Cu, Zn, Sn, Cd... The conventional methods used at present to remove metal ions from these solutions include, for instance, precipitation, electrochemical reduction, ion exchange, adsorption or bioadsorption, etc. But most of these methods do not present a satisfactory depollution solution with respect to the technology itself or its capital and/or operational costs, i.e. the solubility product in the case of precipitation which may exceeds the actual standard for a specific element or group of elements. It would therefore seem profitable to develop new and inexpensive technologies to achieve the decontamination of industrial waste solutions bearing heavy metal ions.

The wood industry uses the fungicide properties of several heavy metals cations such as Cr, Cu, Sn, ... for wood preservation. The mechanisms of metal ions binding on wood constituents have been studied (1 - 4). It is supposed that metal ions react with the hydroxyl groups of the phenolic constituents of wood, namely lignins and tannins. Wood and barks have almost the same chemical components. The bark chemical composition differs from wood by the presence of a higher proportion of polyphenols such as lignins and tannins (5, 6). Because of its chemical composition, it is possible for bark to bind heavy metal ions of industrial waste solutions.

The annual forestry exploitation of the European community produces up to 105 million m^3 of wood, and 15 million m^3 of bark (7). Wood industries generate an important amount of by-products which represent, for France, about 63% of the total volume of wood consumption (8). About half of these by-products is recycled either as wood pulp, as raw material for boards making or as fuel wood. However bark is considered as a waste and is generally stocked or burned. Only a small proportion of bark is recycled as compost. For the European community, up to 30 million m^3/year of wood and up to 15 million m^3/year of bark are considered as useless by-products of no economic value and which present a disposal problem for the wood industries.

The aim of this research is to use the bark of the European wood industries to depollute industrial waste solutions bearing heavy metal ions. The latter are bound to the former. The resulting material could be incinerated and the ashes used as raw material for the recycling of metals. This paper presents the first step of this research programme. It concerns the removal of hexavalent chromium from synthetic solutions and from an industrial effluent by oak and pine bark which are common, unused and largely available by-products of the European wood industries.

Literature review

Since the seventies a few authors have studied the removal of metal ions from solution by various natural or chemically modified lignocellulosic materials, especially the wastes from wood industries (bark and sawdust). Studies on decontamination of waste solutions rich in heavy metals, by bark and/or sawdust, reported in literature, concern American (9 - 13), Russian (14, 15), Indian (16 - 23) and Japanese (24 - 26) tree species, but the metal binding efficiency of the commonly available European species doesn't seem to have been studied.

The general experimental method employed (9 - 26) for removing heavy metal ions from solutions include both equilibrium and column experiments. The equilibrium studies were conducted by shaking 1 to 10 g of ground bark or ground wood with a solution volume of 100 to 1000 ml containing 10 to 200 ppm of the metal ions in stoppered flasks. The mixture was then filtered and the solution analyzed for the studied metal ion content. The column experiments were usually conducted using glass tubes. 10 to 50 g of ground bark or ground wood was placed into the column. The column was tapped gently to promote even distribution of the packing. The heavy metal solution containing a known concentration of metal ions was passed through the column. Solid adsorbent samples were then analyzed to determine its metal content.

Removal of chromium from synthetic solutions by wood industries wastes was the subject of few studies (11, 16, 22, 23). Some authors used raw wood or bark. Others applied chemical treatment in order to improve the material ability to bind the chromium ions. Tables I and II summarize the results of these researches obtained by using equilibrium or column techniques respectively.

Table I : Literature review of results of chromium removal from waste solutions by wood and/or bark using equilibrium experiments.

Salt	Substrate	Treatment	pH_i	pH_f	C_i	% pulp density	% metal removed	ref.
$K_2Cr_2O_7$	*Accacia arbica* bark	HCHO	6.5	4.0	50	1	86	23
	Terminalia tomentosa bark		5.1	5.6	50	1	96	
	Techtona grandis bark		5.1	5.4	50	1	59	
N.A.	sawdust mixture of various species	none	-	-	50	1	77	16
			-	-	50	5	86	
			-	-	10	1	74	
			-	-	10	5	73	
$Cr(NO_3)_3$	*Sequoia sempervirens* bark	none	-	-	63	1	35	11
$K_2Cr_2O_7$	sawdust mixture of various species	none	2.8	8.5	5	1	83	22
			5.4	9.0	5	1	76	
			6.3	9.0	5	1	76	
			-	-	25	5	82	
			-	-	25	2.5	68	
			-	-	25	1	40	

N.A. Non available, *i* : initial, *f* : final, C : concentration, ref. = reference

Table II : idem table I using column technique.

Salt	Substrate	Treatment	Maximum adsorption capacity mg of metal/g of substrate	ref.
$K_2Cr_2O_7$	*Accacia arbica* bark	HCHO	29.29	23
	Terminalia tomentosa bark		19.50	
	Techtona grandis bark		5.46	

Materials and methods

Characteristics of the barks

The bark of the pedonculate oak (*Quercus pedonculata*) and the scots pine (*Pinus sylvestris*) which are the main oak and pine species exploited in the east of France, were obtained from local sawmills. This material was taken directly from a mechanical debarker. It is mainly constituted of bark and contains less than 10 % wood. This material which will be called bark to simplify the text, was air dried at room temperature for a month, then grounded and sieved to less than 1 mm. Afterwards, the sample was homogenized.

Metal analysis

The analysis of chromium in solution was performed with a Perkin Elmer Atomic Absorption Spectrophotometer. Initial, intermediate and final metal concentrations of the experimented solution were measured.

Equilibrium experiments at laboratory scale

Equilibrium experiments were carried out in stirred and stoppered flasks (Figure 1). The ground raw or modified bark and the potassium dichromate solution ($K_2Cr_2O_7$) or industrial effluent were shaken at room temperature. The chosen pH of the solution was adjusted by the addition of HCl or NaOH. At the end of the experiment the flask mixture was immediately centrifuged. The solution was recovered and analyzed in order to define its chromium concentration.

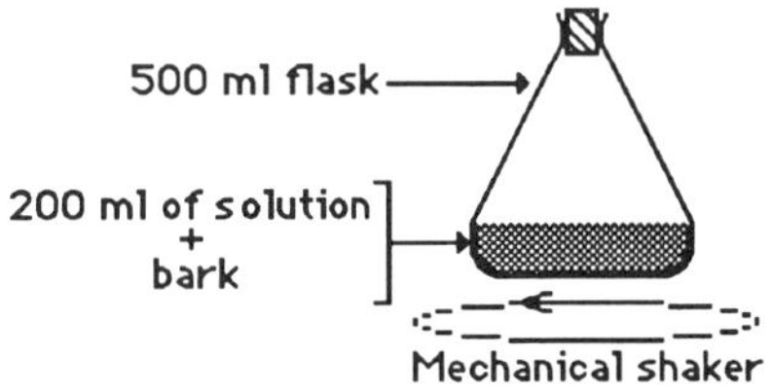

Figure 1
Equilibrium experimental method.

Column experiments at laboratory scale

Column experiments were carried out in order to determine the maximum chromium retention capacity of barks. Tests have been conducted using a glass tube of 25 mm internal diameter (Figure 2). 20 g of modified bark were put into the tube. Glass wool was placed at the top and the bottom of the bed to prevent the dispatching of bark particles. 20 liters of a $K_2Cr_2O_7$ solution were continuously pumped through the substrate at a flow rate of 25 l/h until the concentration of the solution approached the equilibrium which is determined by chemical analysis. A solid sample of the bark was analyzed so as to define its metal content.

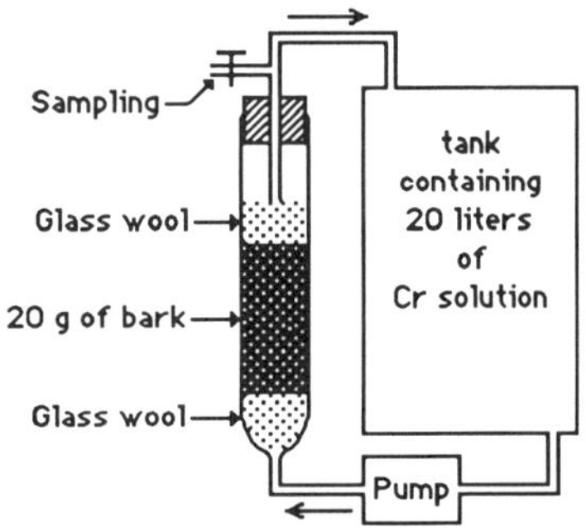

Figure 2
Column experimental method.

Experiments at pilot plant scale

A semi pilot plant apparatus of 500 liters is used for the industrial effluent treatment. As shown by Figure 3, this apparatus permits the experimentation in shaken batch or percolation through columns. The industrial effluent pH is adjusted by addition of HCl or NaOH. Samples are taken from the tank in the case of agitated batch trials, or at the outlet of the column in the case of percolation.

Laboratory scale chromium removal from a synthetic $K_2Cr_2O_7$ solution

Effect of bark treatment on Cr^{6+} removal from solution

Raw barks contain hydrosoluble compounds such as tannins which give a brown colored solution. In order to prevent this leaching, authors have applied different treatments to the barks. They have used for example reagents such as HCOH (27, 28), $(NH_4)_2S_5$ (29), Na_2S_5 (29), HNO_3 (30) A special chemical treatment is applied in this research (31). This treatment seems to enhance the bark's ability to remove hexavalent chromium as shown by Figure 4. In addition, these treated barks give clear final solutions.

Figure 4 shows also the influence of pH on hexavalent chromium removal by barks. It passes through a maximum at pH of about 3.80. At higher or lower pH values the percentage metal removal decreases.

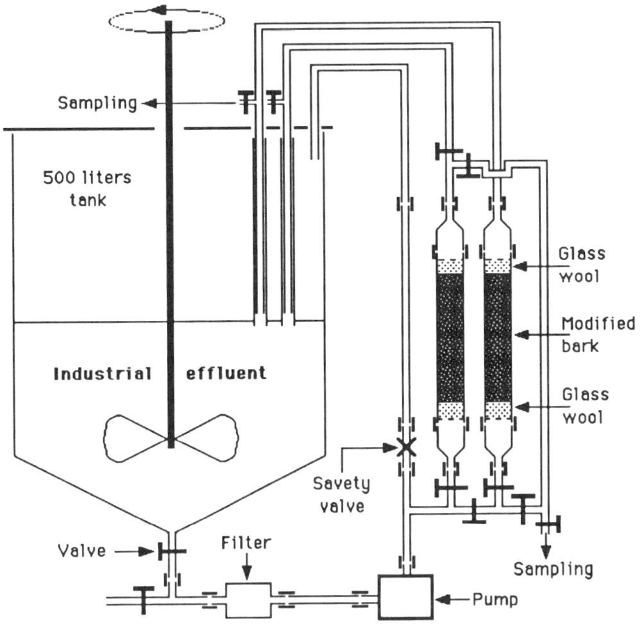

Figure 3 : Pilot plant scale apparatus.

Table III indicates that the chromium removal efficiency of the modified barks is higher than untreated ones, at any pH value. At pH 3.80 percentage removal by modified barks is 98 % of the initial chromium concentration. In the same experimental conditions, this percentage is 80 and 95 % with oak and pine bark respectively.

Laboratory scale trial in shaken batch, $K_2Cr_2O_7$ solution, initial Cr concentration = 100 ppm, solid/liquid ratio = 2 %, shaking time = 2 h, T = 22 °C

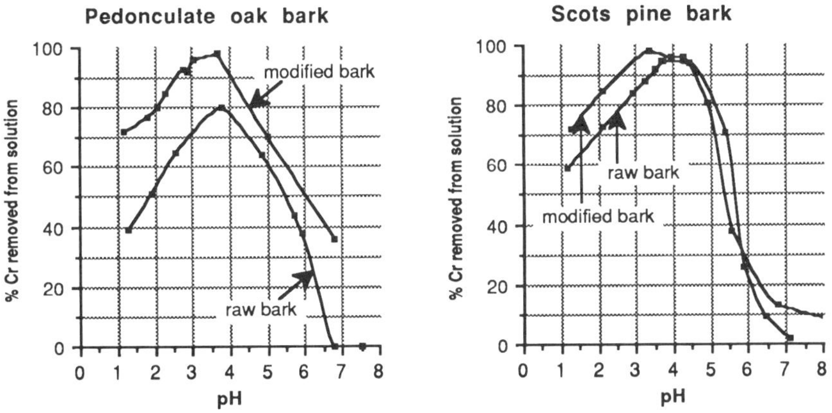

Figure 4 : Comparison between raw and modified barks concerning the hexavalent chromium removal efficiency.

Table III : Comparison between raw and modified barks concerning the Cr^{6+} removal efficiency.

Substrate	% Cr removed from solution		
	pH 2.00	pH 3.80	pH 6.50
Raw oak bark	55	80	10
Modified oak bark	80	98	42
Raw pine bark	72	95	10
Modified pine bark	85	98	18

Kinetics of Cr^{6+} removal from solution

The kinetics of chromium removal from solution by modified scots pine bark were studied at different pH values. Figure 5 indicates that at pH 1 maximum metal removal is attained after only 5 minutes of shaking. Afterwards, the chromium concentration in the solution increases slightly in function of time. This phenomena could be explained by the acid's partial dissolution of the chromium bound with bark.

Laboratory scale trial in shaken batch, $K_2Cr_2O_7$ solution, initial Cr concentration = 100 ppm, solid/liquid ratio = 2 %, temperature. = 22 °C, modified pine bark.

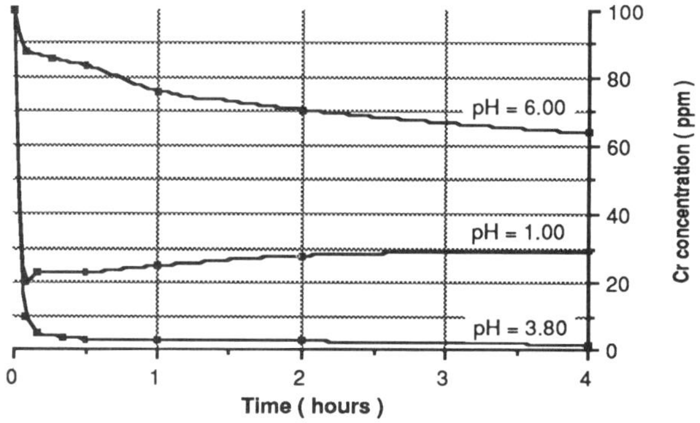

Figure 5 : Kinetics of Cr^{6+} removal from solution by using modified pine bark.

At pH 3.80, the extraction of hexavalent chromium from the solution appears to be very rapid. Up to 95 % of the Cr^{6+} present in a 100 ppm initial solution was removed after 10 minutes of shaking. Equilibrium seems to be reached after 30 minutes. At pH 6.00, the kinetics of chromium removal are slower and chromium concentration in the solution does not seem to reach equilibrium even after 4 hours shaking.

As mentioned before, these results confirm that hexavalent chromium removal depends on the pH of the solution. The optimum pH is about 3.80 for solutions having an initial concentration of 100 ppm. According to experiments of Pizzi A. (2), these results could be explained by the presence of three anionic forms of hexavalent chromium as follows :

$$Cr_2O_7^{2-} + H_2O \rightleftharpoons 2\ HCrO_4^- \rightleftharpoons 2\ CrO_4^{2-} + 2\ H^+$$

Depending on pH conditions and chromium concentration in the solution, one of these anions will predominate (2). It has been proved experimentally (2 - 4) that hexavalent chromium forms stable, insoluble, polymeric complexes with lignins and tannins. $HCrO_4^-$ is the main complexing species (2). $Cr_2O_7^{2-}$ and CrO_4^{2-} exhibit lower complexing properties.

Effect of concentration on the removal of Cr^{6+} using modified scots pine bark

The effect of the solution's initial concentration on the removal of hexavalent chromium by modified pine bark was studied and the results are summarized in Figure 6.

Laboratory scale trial in agitated batch, $K_2Cr_2O_7$ solution, solid/liquid ratio = 2 %, shaking time = 2 h, temperature = 22 °C, modified scots pine bark

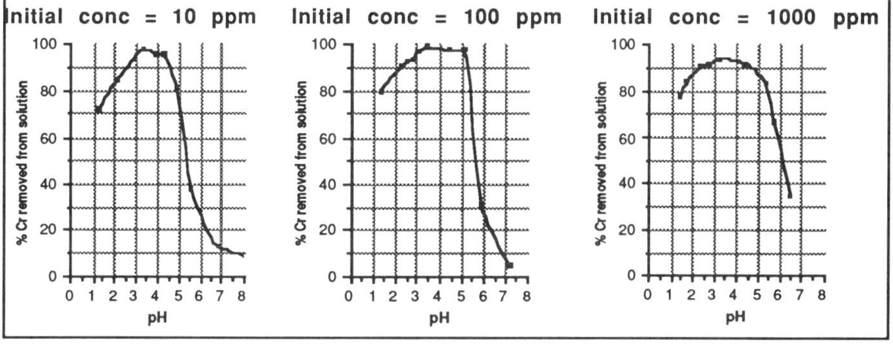

Figure 6 : Cr^{6+} removal from solution by modified pine bark at different initial concentrations.

The percentage chromium removal by the modified pine bark passes through a maximum of about 3.8 as a function of the pH and is independent of the solution's initial concentration. Maximum percentage metal removals are 92, 98 and 98 % for initial concentrations of 1000, 100 and 10 ppm respectively.

Hexavalent chromium retention capacity of modified barks

Column experiments were carried out in order to determine the modified oak and pine barks' chromium retention capacity. The solution was kept in a 20 liters capacity container and continuously pumped through 20 grams of modified bark at a flow rate of 25 l/h (see Figure 2), until the measured concentration of the solution approached the equilibrium. Figure 7 shows that the solution's chromium concentration decreases in function of time. The difference between the solution's initial and final chromium concentration gives the amount of chromium bound with the modified bark. Table IV summarizes the results obtained. The bark's chromium content calculated from the solution analysis is in good agreement with that obtained from the solid analysis. It should be stressed that barks' ashes chromium content is high. These ashes could be useed for chromium recycling.

Table IV : Hexavalent chromium retention capacity of modified oak and pine barks.

Bark	C_i	C_f	Solution analysis Cr bounded with bark (mg of Cr per g of bark)	Solid analysis Cr bounded with bark (mg of Cr per g of bark)	Ash Analysis % Cr metal content
Oak	975	865	110	109	36.2
Pine	990	775	215	206	54.3

C : concentration (ppm), *i* : initial, *f* : final,

Laboratory scale chromium removal from an industrial effluent

The removal of chromium present in an industrial effluent from the electroplating units was studied with modified scots pine bark. Compositions of the effluent used are summarized in table V. The effluent is rich in Cr^{6+} and Zn^{2+} and relatively poor in Cr^{3+}.

Laboratory scale essays, percolation in column, $K_2Cr_2O_7$ solution, solution volume = 20 l, flow rate = 25 l/h, bark weight = 20 g, temperature = 22 °C.

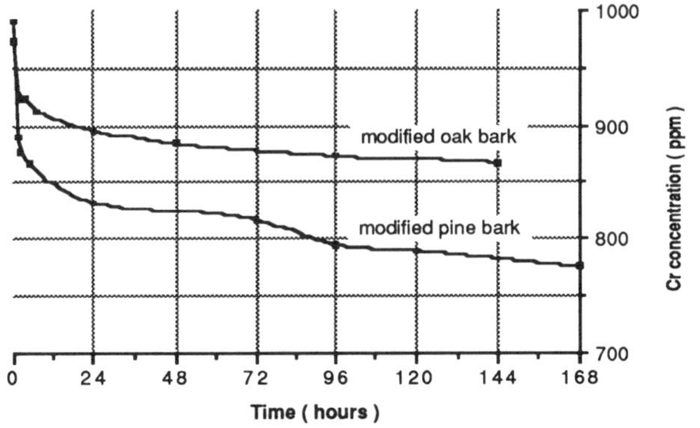

Figure 7 : Hexavalent chromium saturation trial using modified oak and pine barks.

Table V : Average chemical composition of the industrial effluent.

anions	conc. (ppm)	cations	conc. (ppm)
Cl	95	Al	< 10
SO_4	610	Cd	4.5
PO_4	ND	total Cr	1820
NO_3	185	Cr^{6+}	1800
		Cu	<2
		Fe	<2
		Mn	<1
		Ni	2.3
		Pb	<5
		Zn	870

ND = non detected

Effect of pH and pulp density on chromium removal from industrial effluent

Figures 8 shows the influences of pH and pulp density on chromium removal from the industrial effluent by modified scots pine bark at laboratory scale. The optimum pH for chromium removal is about 3.80. This result agrees with experiments carried with synthetic potassium dichromate solutions.

Chromium removal from the effluent depends on pulp density. Applying the same experimental conditions, the chromium maxima percentage removals are 55, 70 and 85 % for pulp density of 1, 2 and 3 % respectively.

Multistep removal of chromium present in an industrial effluent

Chromium removal from this effluent is carried out by means of a four-step treatment using a pulp density of 1 % and a pH of 3.80. This protocol gives a final solution containing 180 ppm of chromium ions as shown on Figure 9 and table VI.

Cr^{6+} can be eliminated from industrial effluent after 4 steps. Concentration of Cr^{3+} seems to increase in the treated effluent. This is probably due to the reduction of Cr^{6+} by the carbohydrate constituent of barks such as cellulose and hemicelluloses as described by Pizzi A. (32).

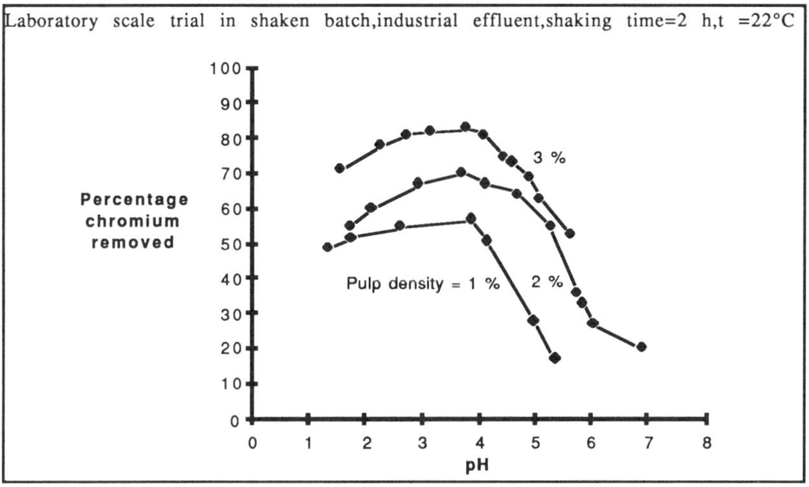

Figure 8 : Chromium removal from industrial effluent by modified scots pine bark, versus pH and pulp density.

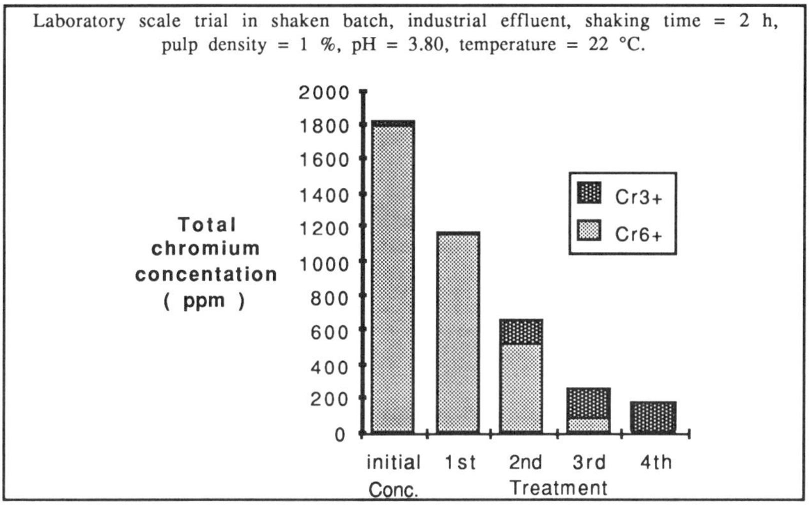

Figure 9 : Multistep removal of Cr from the industrial effluent by modified pine bark.

Table VI : Multistep removal of chromium from the industrial effluent by modified pine bark.

Treatment step	Concentration (ppm)		
	Total Cr	Cr^{6+}	Cr^{3+}
1st	1180	1160	20
2nd	670	530	140
3rd	268	93	175
4th	183	8	175

Pilot plant scale chromium removal from the industrial effluent

Kinetics of chromium removal from industrial effluent at pilot plant scale

Chromium removal kinetics of the industrial effluent by batch technique using a volume of 100 liters are shown by Figure 10. Chromium concentration in the solutions approaches equilibrium after 2 hours. As expected, increasing the pulp density from 1 to 3 % increases the chromium removal from about 40 to 85 % of the initial concentration for a treatment time of 6 h.

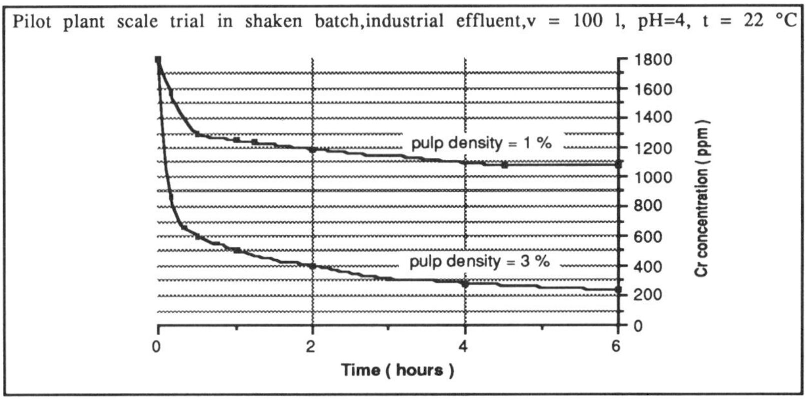

Figure 10 : Kinetics of Cr removal from industrial effluent by modified scots pine.bark, using shaken batch technic at pilot plant scale.

Chromium retention trial of modified scots pine bark at pilot plant scale

The saturation rate of modified bark using chromium of industrial effluent, containing 1880 ppm of Cr, is tested on the pilot plant scale. Two columns connected in series (see Figure 3) are filled with 2 x 250 g of modified scots pine bark. Figure 11 shows the evolution of the chromium concentration measured at the outlet of each column.

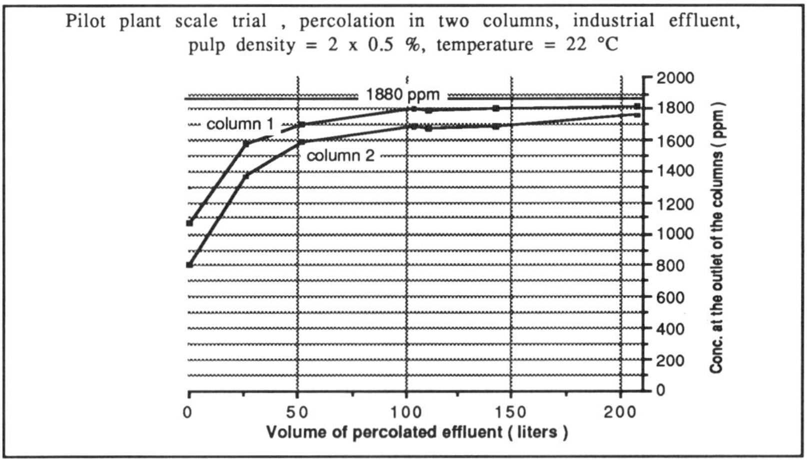

Figure 11 : Evaluation of the modified scots pine bark chromium retention rate at pilot plant scale. Evolution of the chromium concentration versus volume of effluent percolated.

The areas under the horizontal line passing by 1880 ppm and the two curves of chromium concentration evolution permit the calculation of the amount of chromium bound to the bark of each column. Table VII shows that the retention capacity of the first and second columns are respectively 14 and 10 % of the dry weight of the bark.

Table VII : Proportion of Cr bounded on modified scots pine bark.

Weight of bark (g)		Volume of solution (l)		Initial conc (ppm)		Equilibrium conc (ppm)		Weight of Cr removed		% of Cr bounded on bark (% weight)	
column		column		column		column		column		column	
1	2	1	2	1	2	1	2	1	2	1	2
250	250	210	210	1880	-	1820	1760	35 g	26 g	14 %	10 %

Conclusions

Modified oak and pine barks were found to be effective scavengers for hexavalent chromium. Simple and inexpensive treatment appeared to improve the bark's ability to bind chromium and to prevent the leaching of tannins in the solution.

The chromium bound to the bark could be recovered after the incineration of the substrate and the solid residue will give ashes with a chromium content of about 50 %. This material could be considered as raw material for chromium extractive metallurgy.

The results of the research on the different parameters influencing the removal of Cr^{6+} by untreated or modified barks from a synthetic solution permit the following conclusions :

1. the leaching of tannins from bark by the solution can be prevented by pretreating the bark,
2. the modified barks remove a higher percentage of Cr^{6+} than untreated ones (Figure 4),
3. the percentage chromium removal by barks depends on the pH (Figures 4) and to a lesser extent on the solution's initial concentration (Figure 6),
4. the kinetics of chromium removal from the solution by modified pine bark are rapid and pass through a maximum as a function of the pH value (Figure 5),
5. the retention capacity of modified oak and pine barks is about 11.0 and 21.5 % of the barks' dry weight respectively,
6. the incineration of the above saturated oak and pine barks gives ashes containing about 53 and 80 % of Cr_2O_3 respectively.

The results of hexavalent chromium removal by modified pine barks from an industrial effluent containing about 1820 ppm of chromium ions show that :

1. with pulp densities of 1 and 3 %, the Cr^{6+} percentage removal is about 40 and 85 % respectively,
2. the maximum Cr retention capacity of modified pine barks seems to be about 12 % of its dried weight (Figure 11). However in the case of the shaken batch, the chromium retention capacity of the bark is about 6 %,
3. the kinetics of chromium removal from the effluent by modified pine bark in the shaken batch are rapid and Cr concentration approaches the equilibrium after 2 hours (Figure 10),
4. partial reduction of Cr^{6+} to Cr^{3+} is performed by the modified pine bark (Figure 9).

Further studies are focused on the removal of other metal ions such as Cd, Cu, Ni, Zn, etc by raw or treated barks from synthetic or industrial solutions.

Acknowledgment

The authors would like to thank the sawmills Vallar (Chavelot, Vosges, France) and Allencombe (Badonviller, Meurthe et Moselle, France) for providing the bark samples. Thanks to Mrs L. Carpier for the corrections of this paper.

References

1 . S.E. Dahlgren and W.H. Hartford, "Kinetics and mechanism of fixation of copper - chromium arsenic wood preservatives. I. pH behavior and general aspects of fixation", Holzforschung , 26, (2), 1972, 62 - 69.

2 . A. Pizzi, "The chemistry and kinetic behavior of copper-chromium-arsenic/boron wood preservatives. I. Fixation of chromium on wood", J. Polym. Sci. Chem. Ed. , 19, (12), 1981, 3093-3121.

3 . A. Pizzi, W.E. Conradie and A. Jansen. "Polyflavonoide tannins - a main cause of soft-rot failure in CCA-treated timber", Wood Sci. Technol. , 20, (1), 1986, 71-81.

4 . J.G. Ostmeyer et al., "Spectroscopic analysis of southern pine treated with chromated copper arsenate. I : x-ray photoelectron spectroscopy (XPS)", J. Wood Chem. Technol. , 8, (3), 1988, 413-439.

5 . D. Fengel and G. Wegner, Wood : chemistry, ultrastructure, reactions (Berlin, Ed Walter de Gruyter, 1984), 613.

6 . R.W. Hemingway, "Bark : its chemistry and prospect for chemical utilization", 1981, Organic Chemistry and biomass , (Boca Raton, ed I.S. Goldstein CRC, 1981), p.189.

7 . Anonymous, "EUROSTAT, statistiques forestières, tables 1980-1984", Office des publications officielles des communautés européennes, 1987.

8 . R. Brunet, "La valorisation des dechets dans les industries du bois", Le bois national, (17), may 1981, 12-16.

9 . A.C. Waiss Jr. et al., "Adsorption of mercuric cation by tannins in agricultural residues", J. Environ. Quality, 2, (3), 1973, 369-371.

10 . J. Freer et al.,"Removal and recovery of uranium by modified *Pinus radiata* D. Don bark", J. Chem. Tech. Biotechnol., 46, 1989, 41-48.

11 . J.M. Randall et al., "Use of bark to remove heavy metal ions from waste solutions", For. Prod. J. , 24, (9), 1974, 80-84.

12 . J.M. Randall et al.,"Modified barks as scavengers for heavy metal ions",.For. Prod. J. , 26, (8), 1976, 46-50.

13 . J.M. Randall, "Variations in effectiveness of barks as scavengers for heavy metal ions", For. Prod. J. , 27, (11), 1977, 51-56.

14 . L.K. Korovin, M.A. Evilevich and G.A. Voronchikhin, "Biological treatment of wastewater at the Ust-Ilim Lumber Industry Complex", Bum. Prom.-st. ,(7),1982, 27-28.

15 . R.R. Sharifov, F. Yu Ibadullaev and L.A. Mamed'yarova, "Use of industrial wastes for treatment of chromium-containing wastewaters", Izv. VUZOV. Stroit. Arkhit. , (6), 1988, 106-109.

16 . D.S. Bhargava, M.S. Gupta and B.S. Varshney, "Use of sawdust for the adsorption of heavy metals",.Asian Environ , 9, (4), 1987, 29-37.

17 . A.M. Deshkar and S.D. Dara, "Sorption of mercury by *Techtona grandis* bark", Asian Environ , 10, (4), 1988, 3-11.

18 . P. Kumar and S.S. Dara, "Studies on binding of copper ions by some polymeric materials", Chem. era , 15, (11), 1979, 20-23.

19 . P. Kumar and S.S. Dara, "Removal of toxic heavy metal ions from wastwaters using modified agricultural waste materials", Prog. Wat. Tech., 13, 1980, 353-361.

20 . P. Kumar and S.S. Dara, "Modified barks for scavenging toxic heavy metal ions", Indian J. Environ. Health , 22, (3), 1980, 196-202.

21 . V. Kumar and R.S. Sindhu, "Removal of lead ions from its solution by the bark of *Adina cordifolia*", Proc. Nat. Acad. Sci. India. Sect. B , 55, (1), 1985, 94-95.

22 . H.C.P. Srivastava, R.P. Mathur and I. Mehrotra, "Removal of chromium from industrial effluents by adsorption on sawdust", Environ. Technol.Lett. , 7, (1), 1986, 55-63.

23 . Kumar Pawan and S.S. Dara, "Utilisation of agricultural wastes for decontaminating industrial / domestic wastewaters from toxic metals", Agricultural wastes , 4, 1982, 213-223.

24 . M. Morita, M. Higuchi and I. Sakata, "Binding of heavy metal ions by chemically modified woods", J. Appl. Polym. Sci. , 34, (3), 1987, 1013-1023.

25 . R. Suemitsu, M. Osako and N. Tagiri, "Use of dye-treated sawdusts for removal of heavy metals from wastewaters", Sci. Eng. Doshisha Univ. , 27, (1), 1986, 41-48.

26 . M. Fujii, S.I. Shioya and A. Ito, "Chemically modified coniferous wood barks as scavangers of uranium from sea water", Holzforschung, 42, (5), 1988, 295-298.

27 . J.M. Randall et al.,"Removing heavy metal ions from water",U.S. 3925192 (Cl. 210-38, B01D) 09 Dec 1975, Appl. 495 018, 05 Aug 1974, 4 pp Division of U.S. 3925192.

28 . J.M. Randall et al.,"Removing heavy metal ions from water",U.S. 3944415 (Cl. 75-101BE, C22B) 16 Mar 1976, Appl. 495 018, 05 Aug 1974, 4 pp Division of U.S. 3925192.

29 . B.V. Pilat et al., "Removal of heavy metal ion from wastwater",PCT Int. Appl. WO 90 00,156 (Cl. C02F1/28) 11 Jan 1990, SU Appl. 4452988, 01 Jul 1988, 10 pp.

30 . E. Fujii, "Heavy metal and uranium recovery from waters by pine barks containing catechol", Jpn. Kokai Tokkyo Koho JP 60172348 [85172348] (Cl. B01J20/24) 05 Sep 1985, Appl. 84/26883, 14 Fev 1984, 3 pp.

31 . I. Gaballah, D. Goy, G. Kilbertus and J. Thauront, "Support végétal pour l'adsorption de contaminants en milieu aqueux", Demande de brevet d'invention N° 9102758, 07 Mar 1991.

32 . A. Pizzi, "Chromium interactions in CCA/CCB wood preservatives - Part I. Interactions with wood carbohydrates", Holzforschung, 44, (5), 1990, 373-380.

A FLUIDISED BED ION EXCHANGE SYSTEM FOR TREATMENT OF EFFLUENT WATER OF COKE OVEN AND BY PRODUCT PLANT

H.D. Pandey*, A. Gupta*, S. Bhattacharya*, S.R. Mediratta*,

B.N. Das**, K.V.K. Rao*** and J.S. Murty***

* R & D Centre for Iron & Steel, SAIL, Ranchi-834002, India
** Rourkela Steel Plant, SAIL, Rourkela-769011, India
*** Regional Research Laboratory, Bhubaneswar-751013, India

Abstract

The paper describes a fluidised bed ion exchange method for simultaneous removal of phenols and cyanide from the effluent. Use of an eluant enables regeneration of the resin in the continuous operation mode. The breakthrough experiments conducted both on the laboratory scale and on a pilot plant scale have shown that while the level of phenol can be brought down from 462 ppm to about 6 ppm, the free cyanide is almost totally removed. An examination of data indicates that the method can serve the twin objectives of pollution control and phenolic compounds recovery.

Residues and Effluents - Processing and Environmental Considerations
Edited by Ramana G. Reddy, William P. Imrie and Paul B. Queneau
The Minerals, Metals & Materials Society, 1991

Introduction

Treatment of effluent water from coke oven and by-product plants prior to their discharge to the contiguous water source presents a major challenge as the waste water contains highly toxic pollutants such as phenols, cyanides, thiocyanates, ammonia, tar and oils, acids and suspended solids.

The various methods commonly used for treatment of effluents can be categorized mainly into two types, viz: (i) concentration processes whereby the polluting material is removed from the effluent by concentration into a small volume of waste that itself requires some form of disposal and (ii) oxidation processes whereby the polluting material is oxidized to relatively harmless end products by either chemical or biochemical means.

Biological oxidation processes like activated sludge process, percolating filter process suffer from some inherent problems. These processes are extremely sensitive to physico-chemical conditions like dissolved oxygen which unless present in the required quantity leads to negative consequences[1]. Though the process has been used for treatment of effluent from a variety of industries including pickling and plating lines[2], shock loads can kill the bacteria. Furthermore, it is difficult by the biochemical oxidation process to degrade the cyanide sufficiently so that the treated effluent is able to meet the pollution control norms[3].

The discharge norms for coke oven and by-product effluents are given in Table I from which it can be seen that they are rather stringent particularly for cyanides.

Table I Specific Tolerances for Effluents of Coke Ovens

Sl No	Characteristics		Tolerance Limit[4]
1.	Phenolic compounds (as C_6H_5OH)	mg/l	5.0
2.	Cyanides (as CN^-)	mg/l	0.2
3.	Oil and grease	mg/l	10.0
4.	pH value		5.5 - 9.0
5.	Suspended Solids	mg/l	100.0
6.	Biochemical Oxygen Demand (20 °C for 5 days)	mg/l	30.0

Chemical oxidation processes such as chlorination or ozonisation also may prove to be environmentally more hazardous unless controlled very carefully.

The ion exchange process has been used in the past as a polishing step for effluent treatment, and no serious attempt seems to have been made to use it commercially as a primary effluent treatment process. This is inspite of the fact that ion-exchange resins are safe and more economical to handle.

The present paper outlines the results of an attempt to use the ion exchange process under fluidised bed conditions as a primary and exclusive process to treat the effluent from coke ovens of Rourkela Steel Plant. It is observed that the process not only helps in bringing down the phenols to a considerably low levels but also removes cyanides almost totally. The economics of the process is further improved by recovery of phenol which is a costly chemical.

Earlier, some developmental work was carried out at the Regional Research Laboratory at Bhubaneshwar employing a continuous counter current fluidised bed ion exchange system for effective removal of heavy metals like Zn, Pb, Cd, Ni, Cu, Co from metal waste[5,6]. Laboratory scale studies were conducted using 15 mm diameter column[7]. The present work is an extension of the process to treat effluents from coke oven and by product plant for removal of phenols and cyanides.

Experimental

Laboratory Studies

Sixteen resins were tested[8] for determining their capacity to adsorb phenols and cyanide. The resins tested were in three categories cationic, anionic and non-ionic. The anionic resin Indion FF-IP in OH^- form was found to give the best result. In an earlier study, conducted for measuring the effect of pH on anionic resin uptake it is indicated that strong basic resins in OH^- form have higher uptake of phenol in neutral and basic effluents. The laboratory studies showed that phenol removal is primarily by sorption and in addition cyanide and ammonia also get removed wholly or partially.

The ratio of phenol, cyanide and ammonia uptake were approximately in the range of 30 : 1:2 with a single type of resin which could be used to remove all these constituents at the same time. Using different bed porosities and different feed concentrations breakthrough curves were obtained.

Kinetic study of the sorption of phenol by the resin showed that the process is fast and high throughput can be achieved in a continuous system. Based on the data generated by laboratory scale studies, a pilot plant was set up at Rourkela Steel Plant to treat 1000 l/hr of effluent.

Pilot Plant Studies

The flow sheet of the pilot plant is given in Fig. 1. It consists of tar and oil removing system, main extraction, elution and wash columns and a post treatment system for cyanide removal. The fluidised bed ion exchange system consists of an extraction column, elution column and wash column of 380 mm diameter and 2000 mm height[7] each. As phenol fetches a high price in India ($1000/Ton) and Rourkela Steel Plant already has a recovery system based on sodium phenolate, sodium hydroxide was chosen as an eluant.

Removal of Tar and Oil. Laboratory studies showed that addition of ferrous sulfate to the effluent improves the sorption of cyanides because of ferri-cyanide complex formation. Moreover, ferrous sulfate can also be used as a coagulant for removal of oils and tars. With this in view, the effect of ferrous sulfate on the removal of tar and oil was studied. Ferrous sulfate was varied between 150 ppm to 1300 ppm. Some runs were also carried out with 200-500 ppm alum and mixture of ferrous sulfate and alum. From the result it is seen that ferrous sulfate is as affective as alum in removing tar and oils. A mixture of ferrous sulfate and alum was found not to be superior to either of the material individually. It is also observed that the removal efficiency for tar and oil is strongly dependent on the feed concentrations. If tar and oil is low in feed the removal efficiency is reduced even with more addition of ferrous sulfate. Stirring has no effect on removal of tar and oil after 30 minutes. The optimum conditions

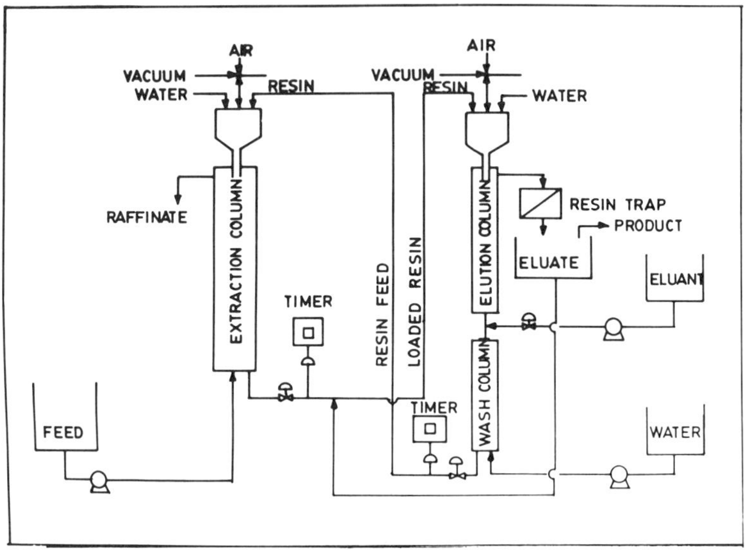

Figure 1 - Flow sheet of the pilot plant.

for tar and oil removal are 30 minutes stirring time and 440 ppm ferrous sulfate for a feed concentration of about 37.4-100 ppm tar and oil. Addition of ferrous sulfate aids in removal of cyanide in the effluent. About 80-90% reduction of cyanides was obtained during tar and oil treatment. The results are shown in Table II.

Table II Data on Tar and Oil Removal

Sl.No	$FeSO_4$ ppm	Tar & oil in feed ppm	Tar & oil on treatment ppm	% Removal
1	150	27.4	5.0	74.45
2	205	27.4	7.0	72.78
3	205	37.4	1.6	95.72
4	397	72.4	9.6	86.67
5	440	100.0	3.6	96.40
6	500	2.0	0.85	57.50

Breakthrough Curves. Four breakthrough curves were made for different bed voidages (ϵ = 0.485, 0.525, 0.535 and 0.630). A typical breakthrough plot for a bed voidage of ϵ = 0.485 is shown in Fig. 2. From Fig. 3 it is seen that breakthrough curves characteristics remain similar to that of laboratory data even after scale up. For example, in Fig 3 the laboratory data breakthrough for a voidage 0.535 with a flow rate of 0.667 l/hr compares well with pilot plant data for the same breakthrough for a flow rate

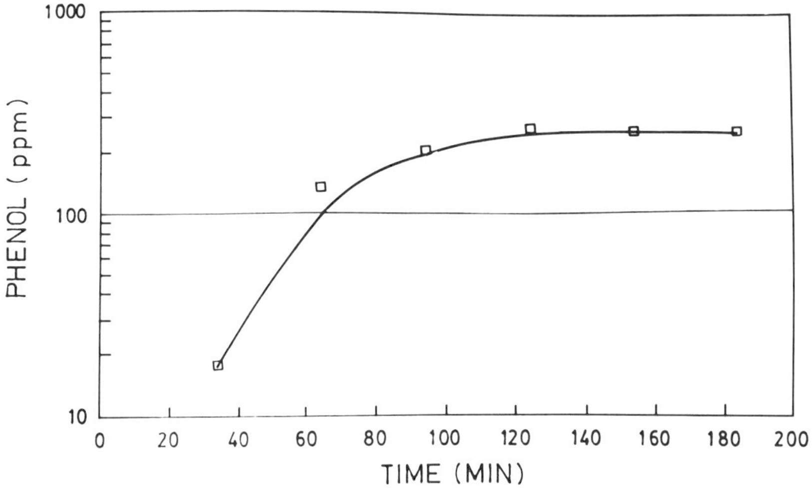

Figure 2 - Breakthrough curve for a bed voidage of 0.485.

of 758 l/hr with the same voidage. Since the data match closely it can be concluded that the data generated in this pilot plant can be used for designing a commercial plant.

Elution. Regeneration of the resin was done using different concentrations of the eluant (NaOH). They are shown in Figs. 4 & 5. Regeneration done with 0.5N NaOH in laboratory scale with 5ml resin, is compared with regeneration done in pilot plant with 80l of resin and the trends are found to be identical in both cases.

Since the data generated on the basic characteristics of the resin Indion FF-IP both in the laboratory and on the pilot plant were found to be in conformity with each other, the pilot plant was used to carry out trials in semi-continuous and continuous modes. This enabled further evaluation of

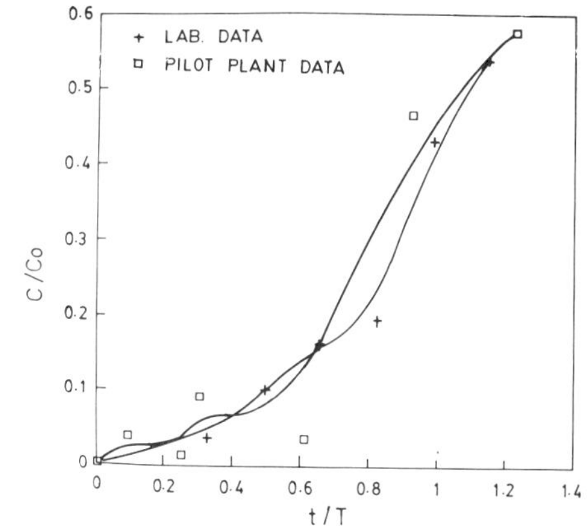

Figure 3 - Comparison of breakthrough data generated during the laboratory and pilot plant studies.

the process for its effectiveness to remove the pollutants drawn directly from the main effluentstream. The results are discussed in the following sections.

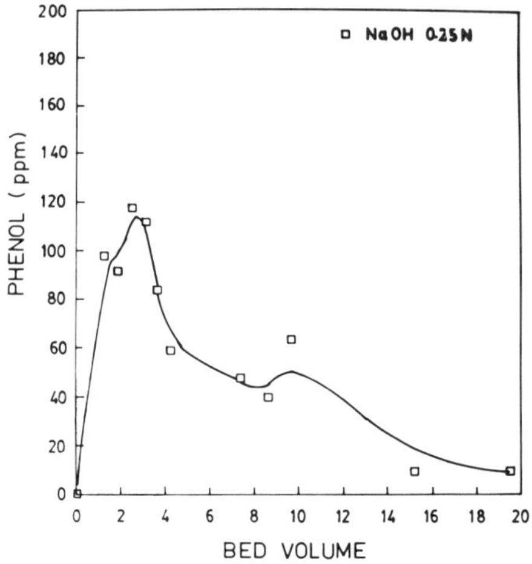

Figure 4 - Regeneration data for the loaded resin with 0.25N NaOH as eluant.

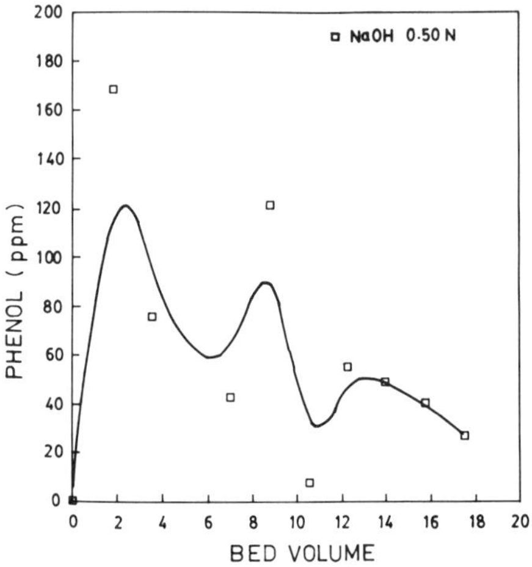

Figure 5 - Regeneration data for the loaded resin with 0.5N NaOH as eluant.

Results and Discussion

In the semicontinous run, the resin was changed at the rate of about 11 litres every 15 minutes. It is seen that the concentrations of phenols and cyanide coming out of the system were drastically reduced thus showing the success of the system for treatment of coke oven effluents. In this run, effluent feed had a phenol concentration of 460 ppm. The treated effluent was phenol free in the beginning and after some time it was stabilized at around 120 ppm. This shows a recovery of about 72%. However, the recovery can be further improved by increasing the bed height thereby enhancing the resin volume.

The data on continuous run given in Table III shows a considerable reduction in both phenol and cyanide concentrations.

Table III Continuous Run Data of FF-IP Resin at a Flow Rate of 410 l/hr
Feed condition: Phenol 462 ppm, Free cyanide 2.08 ppm, Ammonia 156 ppm

SL No	Time min	Phenol conc ppm	Free cyanide conc ppm	Ammonia conc ppm
1	15	6.27	0.019	116.8
2	30	7.84	0.055	95.2
3	45	6.27	0.054	106.4
4	60	6.27	0.064	112.0
5	75	6.27	0.052	106.0
6	90	9.40	0.050	106.0

While the reduction in phenol concentration is observed to be from 462 ppm to about 6 ppm amounting to a recovery of over 98%, the free cyanide is almost totally removed. The system also appears to be marginally effective in bringing down ammonia level.

The reason obviously is the use of anionic resin in the present study. Ammonia can be treated separately either by using a cationic resin or simply by adsorption on some adsorbing media. Attempts were made to separate phenol and ferri cyanide from sodium phenolate solution coming from the elution step but no significant quantity of ferri-cyanide was observed.

Conclusion

1. Fluidised bed ion exchanger is an effective process for treatment of coke oven effluents and can be used as a primary step.

2. Ferrous sulfate is an efficient coagulant which reduces the cyanide content while removing tar and oil and offers an effective pretreatment for cyanide removal prior to ion exchange process.

3. The anionic resin FF-IP in OH^- form is able to reduce phenol and cyanide but is effective only marginally in reducing the ammonia content.

4. Close correlation was found between laboratory and pilot scale studies both for breakthrough curves and the regeneration of the resin.

5. High degree of phenol recovery is possible which should make the process commercially attractive.

Acknowledgments

The authors are grateful to Mr. S.K.Roy, Director and Dr. V. Ramaswamy Executive Director, RDCIS ,SAIL,Ranchi,India for their keen interest and encouragement in the work and permission to publish the paper.

The authors are also thankful to the Director RRL,Bhubaneswar for continuous support and encouragement during the course of the studies. The help received from various laboratories of Rourkela Steel Plant are also gratefully acknowledged. The analytical support rendered by Mr.S.N.Singh,Mr K.D.Ojha and Mr P.K.Sinha of EWML Group,RDCIS are also appreciated.

References

1. G.I.Papkov, "Basic Principles of Design of Biochemical Waste Water Treatment Plants" Koks-i-Khimiya,No.4 (1985), 53-56.

2. Masahiro Fujii et al.," Research on Treatment by Thiobacillus Ferrooxidans of Pickling and Plating Waste Waters at Steelworks" (Technical Report No 37, Nippon Steel , 1988).

3. M.Eliot,ed.,Chemistry of Coal Utilization (New York, John Wiley & Sons ,1981),1139.

4. " Tolerance Limits for Industrial Effluents Discharged into Inland Surface Waters" (Report No IS:2490 Part VII-coke ovens,Indian Standards Institution,New Delhi, 1976).

5. " Design and Development of Continuous Deep Fluidised Bed Ion Exchange System" (RRL Bhubaneswar, 1983).

6. " Studies on Fluidised Bed Ion Exchange System for Metal Recovery and Effluent Treatment" (RRL Bhubaneswar, 1984).

7. " Treatment of Effluent Water Using Fluidised Bed Ion Exchanger-Laboratory Studies" (Report No. R & D.65.10.746.87, R & D Centre , SAIL,1987)

8. B.N.Das et al.," Development of a Fluidised Bed Ion Exchange Process" Proceedings of the Seminar on Impact of Industry on Environment,Ranchi(India).June 5-6,1987.IV.5-1 - IV.5-23.

9. H.D.Pandey et al.," Treatment of Effluent Water Using Fluidised bed Ion Exchanger - Pilot Plant Studies" (Report No.R & D 65:02:1774:91, R & D Centre, SAIL, 1991).

3.4
SCRAP AND OTHER WASTES

RECYCLING OF METALLURGICAL RESIDUES AND EFFLUENTS

Pragna N.H. Bhakta

U. S. Department of the Interior
Bureau of Mines
2401 E. St., N.W.
Washington D.C. 20241

Abstract

The minerals industry is under political and economic pressure to reduce its processing wastes and minimize their impact on the environment. Recycling is one of the means by which the minerals industry can reduce the quantities of effluents and residues that are discharged to the environment. The wise use of available resources is not only an environmental necessity but also an economic benefit as natural mineral resources are depleted and/or become more complex. The purpose of this paper is to review the technical and regulatory issues that will help shape the future of minerals recycling over the next ten years. Research and development to address technical issues will be discussed with emphasis on the Bureau of Mines recycling research program.

Residues and Effluents - Processing and Environmental Considerations
Edited by Ramana G. Reddy, William P. Imrie and Paul B. Queneau
The Minerals, Metals & Materials Society, 1991

INTRODUCTION

The mining industry is faced with the challenge of extracting metal values from lower grade and more complex ores at a time when the overall push is to reduce the industry's impact on the environment. Mining activities over the years have resulted in the accumulation of over 50 billion tons of nonfuel minerals related wastes, and current nonfuel mineral operations add between 1 to 3 billion tons a year. The copper, precious metals, phosphate, and iron industries account for the majority of these wastes. For example, the production of a ton of copper from 0.7 percent copper ore results in approximately 435 tons of wastes in the form of overburden, waste rock, tailings, and slag.

The lack of landfill space and regulations that restrict the disposal of many industrial residues in landfills are realities that the minerals industry has to consider in devising a strategy for waste management. A belief that recycling is one of the means by which society can continue to enjoy its present lifestyle with minimal adverse effect on the environment has pushed recycling into the forefront. The use of recycling as a major component in an overall waste management solution for the minerals industry is discussed in this paper.

The benefits of recycling include resource and energy conservation. For example, in 1988, 2.2 million tons of secondary aluminum was processed in the United States with only 5% of the energy required to recover the same amount of aluminum from imported bauxite ores (1). An additional benefit of recycling is the alleviation of future environmental contamination. For example, recycling lead from automotive batteries not only accounts for over half of the nation's requirements for the material but also reduces the potential for lead contamination.

Regulatory Issues

The 70's and 80's saw the enactment of environmental legislation in the form of the Clean Air and Water Acts, the Resource Conservation and Recovery Act (RCRA, 1976), and the Comprehensive Environmental Response, Compensation, and Liability Act (CERCLA, 1980), known more commonly as Superfund. The Clean Water Act and RCRA are up for reauthorization in Congress during 1992. Reauthorization of RCRA has varied implications for the mineral processing industry. Potentially, all mineral processing wastes that exhibit hazardous characteristics or are listed as hazardous wastes will be subject to regulation under Subtitle C. Nonhazardous (large volume, low toxicity) wastes such as all mining wastes (overburden and waste rock) and the 18-20 mineral processing wastes specifically listed by the Environmental Protection Agency (EPA) in its report to Congress are regulated under Subtitle D. During RCRA reauthorization, the EPA may attempt to extend its authority to implement the current State-run Subtitle D program and may also try to obtain authority for regulating some processes.

In the past, legislation has pressured the minerals industry to focus on minimizing its long-term impact on the environment and streamlining operations. To meet the prescribed requirements, operators have introduced measures which minimize the quantity of regulated materials generated, resulting in more efficient operations. For example, a manufacturer that supplies the aerospace industry decided to minimize the quantity of effluents and residues from its operation for economic reasons. By maximizing resource utilization through improved operating practices in just one unit operation-- painting airplane parts-- it not only saved money by decreasing the volume of effluents requiring disposal, but also decreased the number of personnel and the quantity of raw materials that were needed to complete the job. The legislative trend suggests that this type of innovation will become absolutely necessary in order for the minerals industry to exist in a pro-environmental society.

Recycling Technology

Technology to safely and profitably recycle residues and effluents is needed to practically implement recycling in mineral processing operations. There are four areas within recycling that deserve special attention in the next decade: the design of processes to produce residues and effluents as recyclable products, the development of technology for the recycle of advanced materials, the recovery of metals from currently discarded residues and effluents, and the control of impurities during metals recycling. Although there are other areas that require attention, these four are particularly pressing and are discussed below.

Design for Recycle

The idea of design for recycle is currently being investigated by manufacturers of consumer products such as the automobile and by the chemicals processing industry. The concept of design for recycle is equally applicable to the mineral processing industry because it is clearly more advantageous to design a residue that can serve as a feedstock than to generate a residue that has to be disposed of. The concept has been used to some degree in the minerals industry. For example, the closed loop use of process water results in the "regeneration and recycle" of water around one unit operation or within the process. However, the idea needs to be used more often and innovative applications are required to make an impact on the way process residues and effluents are dealt with.

Recycle of Advanced Materials

The rapid growth of markets for advanced materials makes it imperative that recycling technology for these materials be developed to avoid the loss of valuable resources. The class of advanced materials is a mix of materials ranging from traditional alloys with minor new alloying elements to a material that contains ceramic and metallic components. Recycling technology is required for scrap as well as for residues from the processing and manufacturing of these materials. For example, during the manufacture of neodymium-iron-boron magnets, up to 15 percent of the original material ends up as scrap. In order to maximize the utilization of this valuable material, different manufacturing techniques which generate less scrap need to be investigated or a process to recycle the scrap currently generated needs to be developed.

Minerals from Currently Discarded Materials

The residues and effluents from mineral processing operations contain value in the form of low concentration of metals and/or reagents. The metals and reagents are currently discarded because either the economic considerations are not favorable or the value of the material is not recognized by the generator. The economics of reprocessing some of the residues and effluents might become favorable compared to disposal as changes in regulatory classification of the discarded materials occur. The development of technology which can economically recover these currently lost values would decrease the amounts of wastes that have to be disposed of and would decrease the future liability from environmental contamination.

The Control of Impurities During Recycling

The control of impurities during recycle of scrap, residues, or effluents poses different challenges than conventional processing of ores. The impurity elements are often different and are usually present at different concentrations than in ores. For example, iron ores contain phosphorous and sulfur as impurity elements which have to be removed to specified levels for various uses. Ferrous scrap, on the other hand, contains copper and tin as impurities which have to be within specified levels for various products. In the case of residues and effluents, the impurities might be in the same concentration range as the metal of interest. This provides a challenge for those interested in developing technology for the recycle of low-grade material.

Bureau of Mines Research on Residues and Effluents

The Bureau of Mines has been developing technology for use in the minerals industry since 1910 and has worked in the field of minerals recycling for over three decades. The following discussion presents a few examples of Bureau research under the Minerals and Materials Science Recycling program and the Environmental Technology research program that relate to the processing of residues and effluents.

Recycling Research

The objective of research conducted in the area of recycling is to recover metallic values that are currently lost. The application of technology developed under this program has the added benefit of eliminating future environmental contamination problems. The current program includes work in three of the four areas discussed above: recycle of advanced materials, recycling of residues and effluents, and control of impurities during recycling. Work in the area of design for recycle will be initiated in fiscal year 1993.

Stainless Steel Pickling Liquors: Acid pickling of stainless steels generates approximately 30 million gallons of depleted nitric and hydrofluoric acid solutions. Pickling acids are discarded when the buildup of iron renders the acid bath ineffective. Spent pickle liquors generated by steel finishing operations are listed as a hazardous waste by EPA, K062 in RCRA (2), because of their acidic nature which poses a threat to the environment. Consequently, the cost of their disposal is significantly higher than for a waste that is considered nonhazardous. Currently, these solutions are neutralized to precipitate the heavy metals, and the resulting sludge is disposed of at a secure landfill. The effluent from the neutralization must meet concentration limits (3) on heavy metals and salts in compliance with National Pollutant Discharge Elimination System (NPDES) permits before being discharged to a sewage system or an open waterway. Shortage of secure landfill space and problems with compliance for fluoride and nitrate salts have decreased the cost effectiveness of current disposal practices.

Regeneration and reuse of acid pickling solutions for stainless steel has been under investigation by the Bureau (4,5). In the past, Bureau research has led to a better understanding of the complex pickle bath chemistry (6). New analytical techniques were also developed to better control bath chemistry in order to reduce the volumes of acid requiring disposal.

Based on prior Bureau work, an electro-regeneration process for spent stainless steel pickle liquors has been investigated to further reduce the volume of pickle acid that is disposed of (7). The process involves the extraction of dissolved metals from the solution using an electrodialysis cell shown in Figure 1 (8). The pickling solution is regenerated by the removal of metal ions and the replenishment of hydrogen ions through electrolysis. The choice of the membrane determines the effectiveness of the system as the movement of ions is regulated by the membranes. Bureau research has shown that Nafion[1] perfluorosulfonic acid and ESC 7000 styrene-divinyl benzene cation selective membranes are suitable for this application (8).

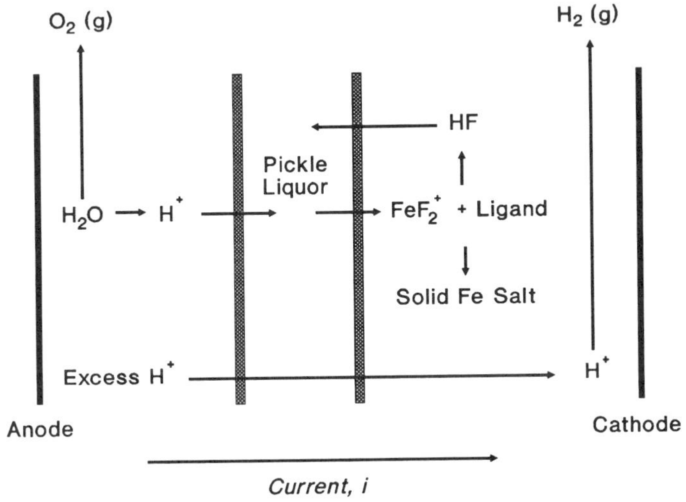

Figure 1. Schematic of Electrodialysis Cell.

Scandium From Tantalum Residues: Scandium is a scarce element used as an oxide in high-energy lasers, nuclear fusion systems, and etching semiconductors. It is used in the metallic form in the manufacture of high-intensity mercury vapor lamps. Sources of scandium have included copper tailings at Bingham canyon, Utah, a fluorite deposit in Montana, and wolframite processing residues in Colorado. The use of scandium in other applications, such as the patented superplastic aluminum alloys by ALCOA, and an increased demand in traditional uses require that new sources of the element be identified and processed. Scandium in tantalum tailings is a new source of the metal recently identified by Bureau researchers (9).

[1]Reference to specific products does not imply endorsement by the U.S. Bureau of Mines.

This source contains approximately 0.25% scandium and over 1.5% each of tantalum and niobium. The process developed by the Bureau to recover scandium is shown in Figure 2. The process to recover scandium involves the sulfation of tantalum tailings at 200°C for 1 to 2 hours, followed by leaching with water at 90°C for 2 hours resulting in over 95% scandium extraction. The leach solution containing scandium, tantalum, niobium, titanium, and zirconium is passed through a solvent extraction step to selectively recover over 99% of the scandium using a DEHPA/ kerosene mixture. The stripping of scandium from the DEHPA fraction is under investigation and work is ongoing to recover the tantalum and niobium values (9).

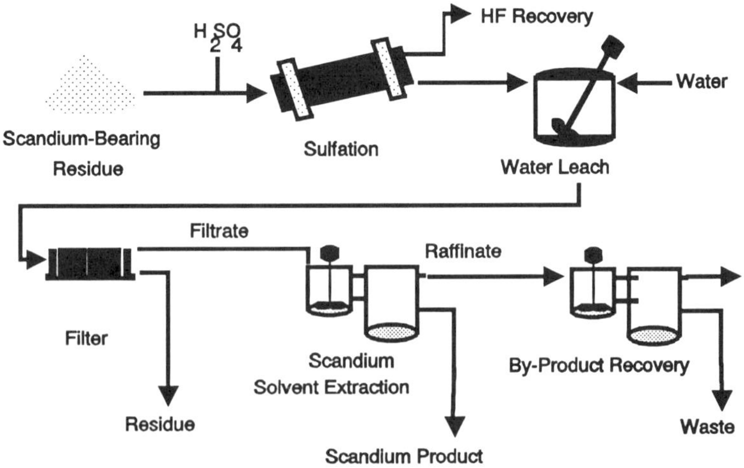

Figure 2. Schematic of Proposed Process for the Recovery of Scandium From Tantalum Residues

Flake Graphite From Steelmaking Kish: Graphite is primarily used in refractories and crucibles for steelmaking, in lubricants, and in electric motor brushes and batteries. The United States currently meets the demand for natural flake graphite by imports from China, Brazil, and Madagascar. A domestic source of flake graphite, which contains a high portion of the size fraction required by users and has been overlooked until now, is steelmaking kish. Kish is a waste material produced as the hot metal cools and consists of iron, graphite, and slag, which is currently disposed of in landfills. The amount of flake graphite that could be recovered from kish is estimated to be more than adequate for supplying all the flake graphite needs of the United States. The Bureau has demonstrated the feasibility of recovering a high-quality product from kish (10). The process involves a concentration, a leaching, and a waste treatment stage. The concentration stage includes dry and wet screening and hydraulic classification. The concentrate from this stage is 80% solids with 71% graphite on a dry basis and is 90% plus 100- mesh material. The concentrate is leached in 20% hydrochloric acid with a 90 minute retention time in each stage of the three- stage leach to remove metallic iron, metal sulfides, and basic silicate slag. The solids from the last stage of leaching are centrifuged and washed. The overflow solution from the last leach stage is pumped to a calcium chloride recovery section where the solution is purified for recycle. The graphite product after the leach is 95% graphite on a dry basis. This product can be further purified, in a two-stage hydrofluoric leach, to 99% graphite on a dry basis. This process is being demonstrated on a pilot scale in cooperation with members of the steel and graphite user industry at a site in Indiana (10).

Environmental Research

One of the objectives of research performed under the Division of Environmental Technology is to prevent the negative impacts of residues and effluents on the environment. The recovery of mineral values is not the primary driving force behind this research. However, it is included here to discuss an alternative approach that could be used when recycling is not practical.

Gold Processing Wastes: A major concern in the precious metals industry is the environmental impact of effluents and residues from heap leaching operations after closure of processing activities. The spent leaching solutions and heaps have residual cyanide that has to be removed or stabilized to prevent its potential escape into the environment. Selenium has also been found in levels much higher than the current drinking water standard of 0.01 ppm. Bioremediation of cyanide and selenium from precious metal process solutions is being investigated in order to reduce total cyanide levels from over 200 ppm to 1 ppm and to reduce selenium from 30 ppm to 0.01 ppm. Bacterial strains isolated from silver tailings pond water have been used to lower the cyanide and selenium concentrations present in precious metal process waters. The bacterial strains Pseudomonas pseudoalcaligenes and Pseudomonas alcaligenes have been successful in reducing the cyanide to 1 ppm and the selenium to 0.01 ppm respectively (11,12). To reduce residual cyanide levels in spent heaps, the use of water as a rinse media is being investigated. Results thus far show water can successfully reduce the residual cyanide concentration to acceptable levels. The rinse water bearing the washed cyanide would then be treated to comply with water standards (13).

CONCLUSION

The treatment of metallurgical residues and effluents has become important in the minerals industry because of the environmental ramifications of neglecting the backend of the minerals life cycle. The future of the minerals industry will be shaped to a large extent by how the industry reacts to the societal concern for the environment. The current approach of waste minimization and recycling of residues and effluents appears to be an appropriate solution to the environmental concerns of society and the economic concerns of industry.

REFERENCES

1. P. A. Plunkert, Minerals Yearbook, Volume 1, Metals and Minerals (U. S. Government Printing Office, Washington, 1990), 113.

2. RCRA Regulations and Keyword Index, 1990 Edition (McCoy and Associates, Inc., Lakewood, Colorado, 1990), 47.

3. D. H. Desy, Iron and Steel, Chapter in Mineral Facts and Problems, 1980 Edition, (BuMines B 671m 1981) 676-681.

4. B. S. Covino, Jr., J. V. Scalera, and P. M. Fabis, "Pickling of Stainless Steels--A Review," (BuMines IC 8985, 1984)

5. J. B. Stephenson, J. C. Hogan, R. S. Kaplan, "Recycling and Metal Recovery Technology for Stainless Steel Pickling Liquors," Environmental Progress, 3 (1984), 50-53.

6. B. S. Covino, Jr., "Fundamentals of Stainless Steel Acid Pickling Processes," in New Steelmaking Technology From the Bureau of Mines, (BuMines IC 9195, 1988), 39-44.

7. G. L. Horter and J. B. Stephenson, "Decreased Acid Consumption in Stainless Steel Pickling Through Acid Recovery," in New Steelmaking Technology From the Bureau of Mines, (BuMines IC 9195, 1988), 45-49.

8. Current Work at Bureau of Mines, Rolla Research Center, Glenn L. Horter, Principal Investigator.

9. D. D. Harbuck and G. R. Palmer, "Scandium Recovery from a Tantalum Waste Residue: A Status Report," To be Presented at The Rare Earths: Sources, Science, Technology, and Applications Symposium, TMS Annual Meeting, March 1-5, 1992, San Diego, CA.

10. L. J. Nicks, P. D. Laverty, and L. A. Walters, "Recovery of Flake Graphite from Steelmaking Kish," to be published as BuMines RI in 1992 series.

11. R. H. Lien, B. E. Dinsdale, and P. B. Altringer, "Biological and Chemical Cyanide Destruction From Heap Leachates and Residues," Chapter 18 in Proceedings of Environmental Management in the 1990's, SME Annual Meeting 1991.

12. P. B. Altringer, R. H. Lien, K. R. Gardner, "Biological and Chemical Selenium Removal From Precious Metal Solutions," Chapter 19 in Proceedings of Environmental Management in the 1990's, SME Annual Meeting 1991.

13. Current Work at Bureau of Mines, Reno Research Center, Paul Comba, Principal Investigator.

BY-PRODUCTS RECYCLING AT ASARCO:

PROCESSING OF DROSSES, SLAGS, DUSTS

G. Archer, B. Bunn, F. Ojebuoboh

ASARCO Technical Services Center
Salt Lake City, UT 84119

ABSTRACT

ASARCO, a large and diverse producer of nonferrous metals, is involved primarily in the smelting and refining of lead and copper. Lead-antimony alloys, antimony trioxide, bismuth, cadmium oxide, selenium, tellurium, silver and gold are among products which result from processing by-products of smelting and refining. The complex mixture of concentrates at some smelters and the high impurity loads at some refineries result in various drosses, sludges, dusts and slags. In this paper, the complex interplant and intraplant transfer of by-products is addressed. These by-products result in an exchange of metals between the lead circuit and the copper circuit. Examples of the transfer of by-products and the prospects for the recycle of valuable products from wastes generated by customers and unrelated processors are also discussed.

Residues and Effluents - Processing and Environmental Considerations
Edited by Ramana G. Reddy, William P. Imrie and Paul B. Queneau
The Minerals, Metals & Materials Society, 1991

INTRODUCTION

Asarco is among the leading base metal smelters and refiners in North America. Products include copper, lead, lead-antimony alloys, zinc oxide, antimony trioxide, bismuth, cadmium oxide, selenium, tellurium, silver and gold. Most of these commodities result from the secondary processing of by-product streams which are generated in Asarco's primary smelter operations - lead and copper. Processing of these by-product streams is facilitated by both interplant and intraplant transfer of residues and wastes.

Interplant recycle includes the transport of copper speiss and copper-sodium mattes from the lead circuit to copper smelting plants. The lead plants in turn process leady dusts generated primarily in the copper facilities. Lead bullion, containing arsenic, antimony, copper, bismuth, tellurium, and silver is transferred to the lead refinery where its processing results in the production of lead, but also antimony trioxide, bismuth, Pb-Sb alloys, dore bars, and a low grade copper matte. The copper matte generated in the refining circuit is returned to the lead smelters for further processing. Lead blast furnace dust, rich in cadmium, is sent to a separate facility where it is processed to produce a rich cadmium fume. The cadmium fume generated is subsequently shipped to a high purity chemicals plant where it is refined into high grade cadmium oxide. Other residues resulting from the processing of the lead blast furnace dusts are recycled to the lead smelters.

Intraplant recycle includes the processing of copper refinery slimes to produce silver, gold, selenium and tellurium; processing of lead refinery residues to recover lead, antimony, bismuth and zinc; recycle of water processing plant sludges to the smelters; and the recycle of various dusts streams within the plants. Table 1 lists the different plants, the products they produce as well as the quantity of non-Asarco materials they process.

Figure 1 is a flow diagram of the various byproduct streams and the plants to which they are shipped for processing.

Recycle scrap generated by customers, as well as unrelated processors, is frequently evaluated for processing by

Asarco. Tellurium and antimony scrap are recycled in our specialty chemical facility. Also, several other lead and precious metal bearing wastes are processed in the lead smelters. Some examples of these non-traditional materials are identified in Table 2, which also shows the most likely Asarco Processing Facility for such materials. Figure 2 gives the location of the various Asarco processing facilities and the different areas of the continent in which non-traditional materials processed in Asarco's smelters are generated.

GENERATION OF INTERMEDIATE PRODUCTS BY-PLANT

Lead Smelters

Asarco presently operates two primary lead smelters. They are located in East Helena, Montana and Glover, Missouri. Both smelters use conventional sinter plant-blast furnace technology to process sulfide concentrates.

Due to the nature of the concentrates and residues processed at the East Helena facility, several by-products are generated and processed in conjunction with the lead bullion. Additional bullion is generated from the processing of a tetrahedrite type concentrate ($AgS+Cu_2S+PbS+Sb_2S_3$) in a kettle process independent of the main sinter-blast operation.

Crude lead bullion is continuously tapped from the lead blast furnace and is transferred to kettles in the drossing plant. Here, cooling of the crude bullion results in formation of a dross. This dross is skimmed and transferred to a drossing furnace for further processing. At this stage the dross contains Cu arsenides, mechanically entrained lead, some PbS, etc. The skimmed bullion is further cooled, and when still molten, sulfur crystals are added and the entire bath mechanically stirred. The addition of sulfur contributes to the further removal of copper from the bullion as a Cu_2S-PbS matte. By controlling the S/Cu ratio, excess copper in the bullion reacts with arsenic and some antimony in the bullion to form arsenides and antimonides (speiss) respectively. The speiss and matte are both combined in a dross, which is intermittently skimmed and transferred to the drossing furnace. Here soda ash is added. This promotes an exchange reaction between PbS and Na, resulting in the formation of a Cu_2S-Na_2S matte, speiss (Cu_3As + Cu_3Sb) and bullion. Matte and speiss are tapped at separate ends of the furnace, while bullion is periodically removed at the bottom of the furnace. The lead is returned to the first drossing kettle, the matte is cast into blocks, and the speiss granulated before shipping.

In addition to the soda matte described above, an intermediate sodium matte is produced as a result of the tetrahedrite processing. This process involves the addition of dried tetrahedrite concentrate and liquid sodium to molten bullion. After agitation, lead, silver and antimony from the tetrahedrite are dissolved in the bullion; whereas a copper-sodium matte is skimmed and transferred to the drossing furnace.

Another by-product at East Helena and at most other Asarco smelting facilities, is sulfuric acid. An on-site acid plant converts SO_2 rich sinter plant gas to sulfuric acid. This acid is marketed by Asarco.

Other by-products produced at East Helena include blast furnace dust, sinter plant baghouse dust and water processing sludge.

The concentrates processed at the Glover plant are much "cleaner" (Missouri concentrate with low As, Sb) than the concentrates processed at East Helena. Consequently, the by-products generated from this facility are fewer than that from East Helena. Dross from the bullion is processed in kettles by sodium addition to produce bullion and a Cu_2S-Na_2S matte. The copper matte is shipped to a copper smelter for copper recovery. The clean concentrates result in the production of a relatively soft bullion which requires little softening. A silver-zinc crust resulting from the desilverization of the bullion is generated and shipped to Omaha for Ag and Zn recovery.

Lead Refinery

Asarco's operations at the Omaha plant are concerned primarily with refining lead produced at the East Helena smelter and custom smelters. The impurity load in East Helena bullion is typically diverse and sometimes high. Bismuth, antimony, tin, silver, and copper are some of the impurity metals found in the bullion. The refining operation, therefore, presents a model for a plant with significant generation of by-products that are processed within the plant or transferred to another Asarco plant.

Operations at the Asarco lead refinery at Omaha are divided into the main lead refinery, antimony oxide production, dore production, and bismuth production. An additional department, the Residue Department, processes assorted intraplant slags and skims prior to transfer to a specific department or another plant. Details of the conventional operations will not be addressed here; however, processes will be examined to the extent that they enhance description of the procedures which generate specific intermediate products - bismuth dross, tellurium slags, softener skims.

East Helena bullion is received at the Omaha refinery in the form of bullion blocks and after remelting, the bullion is softened with oxygen and litharge in a reverberatory furnace. Antimony is recovered directly in softener skims and in the reverb fumes (dust analyzing 1.5% Sb). The skims and dusts are transferred to the Residue Department. The softener skims, which are high antimony slags, are processed in a cupola (blast) furnace. A high antimony Pb metal is recovered. Asarco processs this metal at the Omaha plant for special grades of antimonial lead and, more often, for production of high purity antimony oxide powder.

Further refining of the softened lead is conducted in kettles. An initial kettle refining operation is used to remove tellurium and copper as a soda slag and copper dross respectively. The tellurium-bearing slag is dried and shipped to the Amarillo Copper Refinery for recovery of tellurium metal. The copper dross is processed in the reverberatory furnace in the Residue Department for recovery of contained lead and copper matte.

Two additional steps in the refining of lead at the Omaha plant are the processes of desilverizing and debismuthizing. The drosses generated from both steps are processed within the plant. Desilverizing is performed by the Parkes Process whereby zinc is added to the bullion, thereby generating a silver-zinc intermetallic crust. A resulting dross is processed in liquation kettles to upgrade the crust, then vaccum dezinced for Zn recovery. The remaining lead is removed in the cupels. The dore processing facility at Omaha processes various recycled by-products for the recovery of precious metals. This facility has been recently modernized and uses Britannia technology for the production of dore. This modernization has resulted in much cleaner plant operations and less dore inventory.

Debismuthizing is accomplished by kettle refining using the Kroll-Betterton process developed by Asarco. In this process, calcium and magnesium are added to desilverized, dezinced bullion, and the intermetallic compound, calcium dimagnesium bismuthide, is obtained. The dross which results is skimmed for bismuth recovery in the Bismuth Department of the plant. In the Bismuth Department, lead containing high bismuth contents, is deleaded by chlorine addition, resulting in the production of high-purity metallic bismuth and lead chloride. All by-product lead chloride generated is re-used within the plant.

Copper Smelters

Asarco operates two copper smelters, one at El Paso, Texas and the other at Hayden, Arizona. At the El Paso facility, copper concentrates are presently processed by the roast-reverb method. However, this plant is currently in the process of installing the Contop copper smelting technology. At Hayden, Inco flash smelting followed by conventional (Peirce-Smith) matte converting, is employed. Both smelters employ acid plants for sulfur fixation.

As a result of the matte converting carried out at both plants, leady converter dusts are generated. The converter dust at Hayden is combined with the flash furnace dust and recycled to the flash furnace. The El Paso leady dust is shipped to East Helena for lead recovery. At El Paso, slag from the converting operation is recycled to the reverb, while reverb slag is discarded. Hayden converter slag is recycled to the flash smelter, while flash furnace slag is cleaned in an electric arc furnace for discard.

Both plants generate and recycle water treatment plant and scrubber blowdown sludges.

Copper Refinery

Asarco's Amarillo plant primarily refines anodes produced at Hayden and El Paso. However, several by-products resulting from the electrolytic refining of these anodes are also produced. These include silver, gold, selenium, tellurium, nickel sulfate and platinum group metals. In addition, furnace bricks containing various amounts of precious metals are also processed.

By-Product Flows

As discussed earlier, materials recycled at Asarco can be broadly classified as Interplant Recycle and Intraplant Recycle.

Intraplant recycle and processing include the re-use of various dusts and residue streams resulting from the plant's main smelting and/or refining circuits. These include sinter plant baghouse dusts which are recycled to the sinter machine at both East Helena and Glover; and lead blast furnace dross which is processed at East Helena to produce speiss, matte and bullion. While Figure 1 showed a general picture of interplant recycle at Asarco, Figures 3, 4 and 5 show flowsheets for the processing of some selected intraplant recyclable byproducts.

As a result of Asarco's unique position as a primary smelter of both lead and copper, as well as the nature of the by-products generated by these facilities, processing of residues and products from a specific unit is easily achieved in another. For example, copper matte produced at East Helena, Glover and Omaha is readily processed at the copper smelters which produce copper matte as an intermediate product. On the other hand, leady dusts generated from the converting operations at the El Paso copper smelter, is readily accepted by the lead smelters for the recovery of the lead. Similarly, copper speiss generated in the lead circuit, as a means of removing arsenic from the bullion, is processed at the copper plants for recovery of copper and subsequently for the recovery of partitioned precious metals.

However, interplant recycle is not always due to the ability or the desire of the receiving plant to process a given material. In some instances, material is sent to another facility for processing simply because of the availability of equipment and processing capacity at the receiving plant. One example of this is the roasting of lead blast furnace dust at El Paso. Several years ago, the recycle of this material to El Paso was natural since Asarco also had a lead plant at this facility. However, with the closure of this part of the El Paso operation, the leady residue resulting from the CdO fume recovery is now returned to East Helena.

The flowsheet for the processing of this particular dust is shown in Figure 6. The flow diagram for the recycle of tellurium slag from Omaha through Amarillo is shown in Figure 7.

An ongoing program in which materials suited to our smelters and refineries are processed for the recovery of their contained values, results in the processing of wastes from almost every corner of North America.

Summary

Almost all by-products resulting from the production of Asarco's primary products are recycled through its smelters and refineries. In several instances, the generation of these materials and their processing have resulted in the production and sale of many secondary products such as silver, gold, selenium, tellurium, bismuth, etc. Such recycle of materials, despite the small tonnages it contributes to the primary product output, makes a significant contribution to the economics of the various plants due to the value added nature of the products. In addition, recycle and processing of these by-products within the Asarco network have resulted in a net reduction in the overall tonnages of non-treatable waste generated by the company.

The processing of 'foreign' waste by the traditional metal smelters also contribute to the nation-wide reduction of material to be landfilled. This represents a growing phenomenon in which large metallurgical companies such as Asarco will be of increasing use as outlets for metal containing wastes from small specialty refineries and laboratories.

ACKNOWLEDGEMENTS

We are grateful to ASARCO Inc. for permission to publish this paper. The information involves practice at several plants. The cooperation of plant personnel made the task of preparing the paper possible.

Additional information and suggestions from Dr. V. Ramachandran is gratefully acknowledged. Finally, we thank Mrs. Phyllis Russell for typing the final manuscript.

Table 1

PRODUCT AND MATERIALS SUMMARIES FOR ASARCO Processing FACILITIES

Omaha Lead Refinery

Primary Product:	Refined Lead Bullion			
% From Interplant Transfer:	100%			
Secondary Products:	Dore	Bismuth	Antimony Oxide	Pb-Sb Alloys
% From Byproduct Processing:	4	100	73	100
Non-Asarco Recyclables Trt:	PM Bearing Slag, Film Ash			
Recycled Products Generated:	Ag/Zn Crust, Softener Skims, Cu Matte, Liquation Blocks Lead Chloride, Residue Baghouse Dusts, Antimony Crystals, Tellurium Slag Furnace Refractory			

East Helena Lead Smelter

Primary Product:	Crude Lead Bullion
% From Interplant Transfer:	5%
Secondary Products:	Sulfuric Acid
% From Byproduct Processing:	0
Non-Asarco Recyclables Trt:	Scrap Pb,Ag, Cupels, Ag Borax Slags, Lab Clean-up
Recycled Products Generated:	B.F. Baghouse Dust, Speiss, Cu/Na Matte Acid Plant Dust, Acid Plant Mud

Amarillo Copper Refinery

Primary Product:	Refined Copper			
% From Interplant Transfer:	100%			
Secondary Products:	Selenium	Tellurium	Precious Metals	$NiSO_4$
% From Byproduct Processing:	100	100	100	100
Non-Asarco Recyclables Trt:	PM Sweeps, PM Bearing Slag			
Recycled Products Generated:	Sharp Slag, PM Refractory, Anode Slag, Foul Cathodes, Arsenical Sludge, Various Dusts			

El Paso Copper Smelter

Primary Product:	Anode Copper
% From Interplant Transfer:	6%
Secondary Products:	Sulfuric Acid
% From Byproduct Processing:	100
Non-Asarco Recyclables Trt:	Copper-bearing Foundry Dusts, Electronic Scrap
Recycled Products Generated:	Converter Dust, Roaster Dust, Furnace Dust Scrubber Blowdown CdO Fume, Cadmium Calcine

Table 1, continued:

Hayden Copper Smelter

Primary Product:	Anode Copper
% From Interplant Transfer:	Negligible
Secondary Products:	H_2SO_4
% From Byproduct Processing:	100
Non-Asarco Recyclables Trt:	0
Recycled Products Generated:	Quench Tower Dust, Converter Accretions Scrubber Blowdown, Water Treatment Sludge

Glover Lead Smelter

Primary Product:	Lead Bullion
% From Interplant Transfer:	0
Secondary Products:	None
% From Byproduct Processing:	0
Non-Asarco Recyclables Trt:	Film Blanks, Lead Press Cake, Lens Grindings, Foundry Sand
Recycled Products Generated:	Water Treatment Plant Sludge, Ag-Zn Crust, Cu/Na Matte

Globe Smelter & High Purity Metals

Primary Product:	CdO, Cd, PbO
% From Interplant Transfer:	95 95 100
Secondary Products:	Se, Te, Sb, In
% From Byproduct Processing:	0 0 0 0
Non-Asarco Recyclables Trt:	None
Recycled Products Generated:	$PbSO_4$ Residue

TABLE 2

MATERIALS WHICH ARE RECYCLED BY ASARCO AND WHICH ORIGINATE OUTSIDE THE COMPANY

Material and Origin	Receiving Plant	Disposition of Material
Copper bearing foundry dust, brass foundry	El Paso	Feed to the reverb for copper recovery and flux
Fire assay cupels and slag, analytical labs	E. Helena	Blast furnace feed for lead recovery
Ballistic soil, rifle ranges	Glover	Sinter plant SiO_2 flux
Optical lens grindings, optical lens producers	Glover	Sinter plant SiO_2 flux and Pb & Ag recovery
Tetra-ethyl lead residue, tetra-ethyl lead producer	Glover	Blast furnace feed for lead recovery
Foundry sand, brass foundries	Glover	Sinter plant SiO_2 flux
Spent solid caustic, ss steel foundries	Glover	Refinery, reagent in Pb refining
Lead sulfate, chemical producer	Glover	Sinter plant for Pb recovery
Pb contaminated soil, closed battery plants	Glover	Sinter plant SiO_2 flux

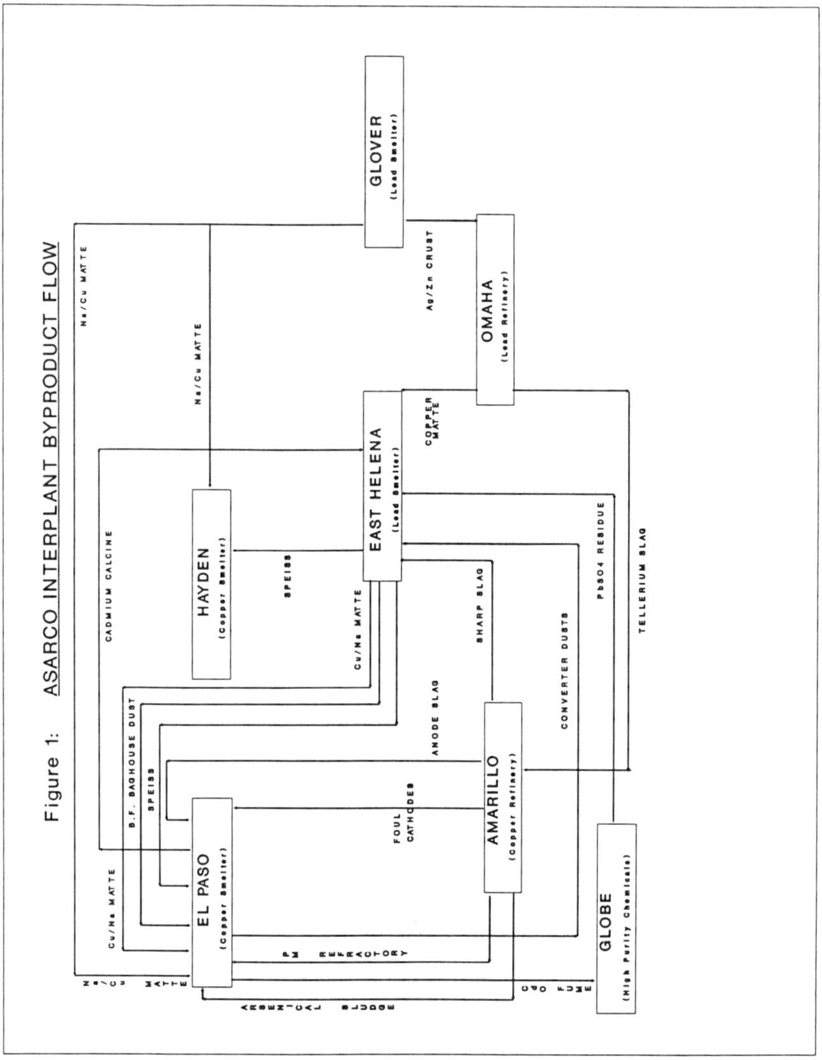

Figure 1: ASARCO INTERPLANT BYPRODUCT FLOW

Figure 2

ASARCO'S NETWORK OF PLANTS AND BYPRODUCT SOURCES

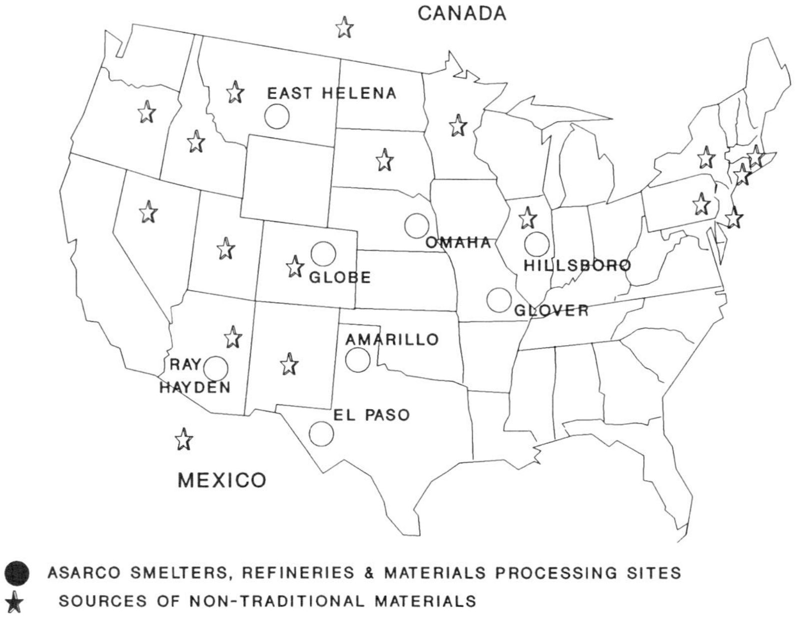

Figure 3.
PROCESSING OF SOFTENER SKIMS AT OMAHA LEAD REFINERY

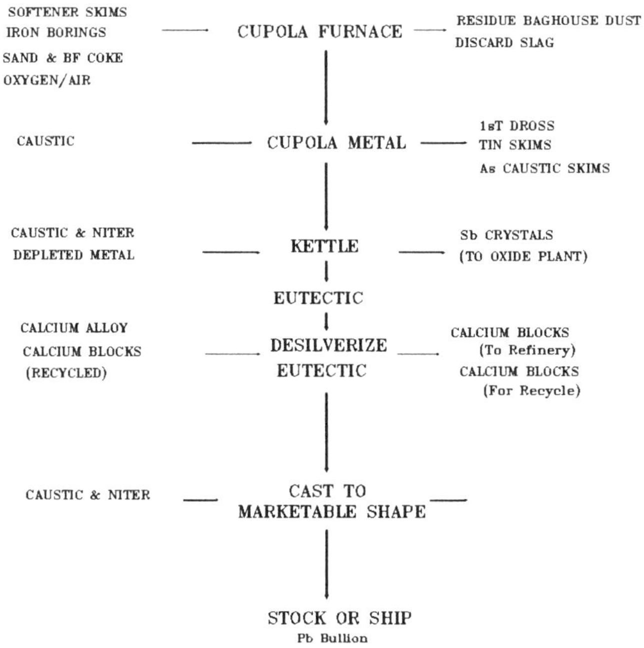

Figure 4

PROCESSING OF DUSTS, DROSSES & CAUSTIC SKIMS AT OMAHA REFINERY

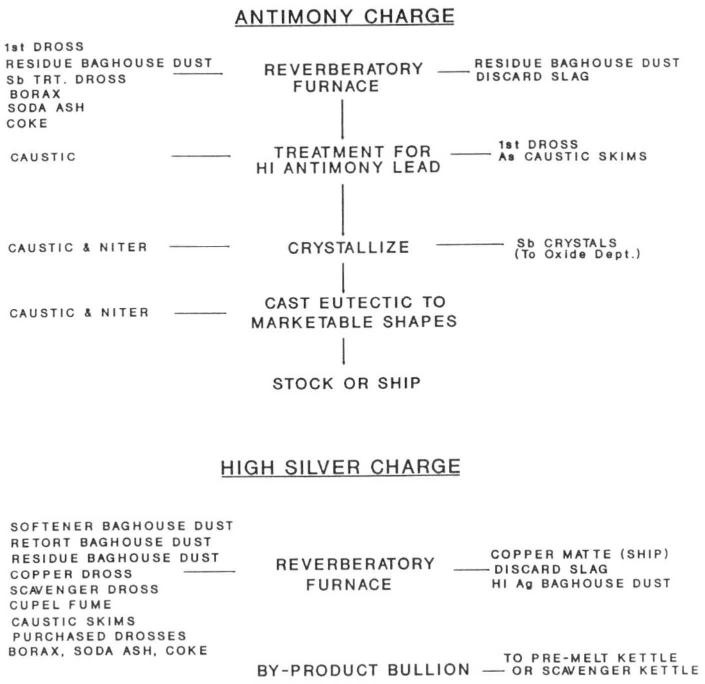

Figure 5

SELENIUM & TELLERIUM RECOVERY FROM TANKHOUSE SLIMES AT AMARILLO COPPER REFINERY

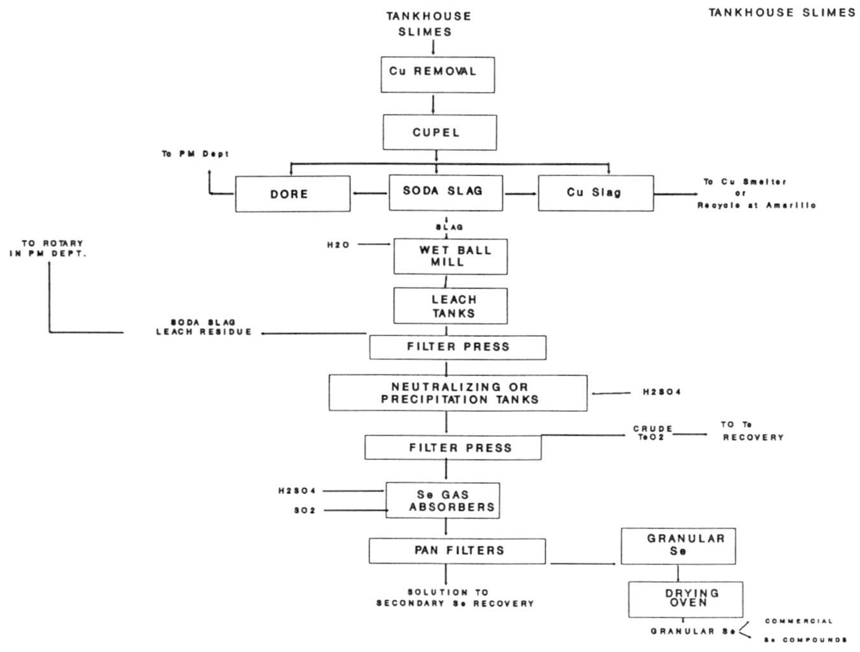

Figure 6

PROCESSING OF LEAD BLAST FURNACE DUST FOR Cd RECOV. AT ELPASO AND GLOBE

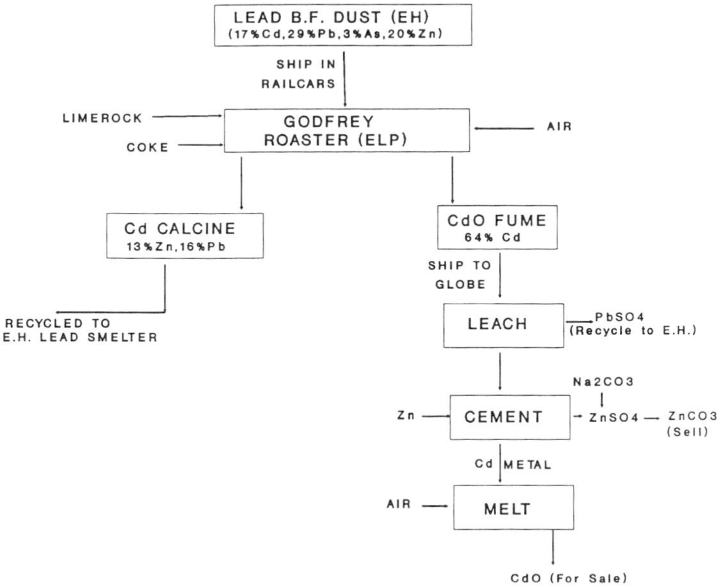

Figure 7

Te RECOVERY FROM OMAHA Te SLAG AT AMARILLO COPPER REFINERY

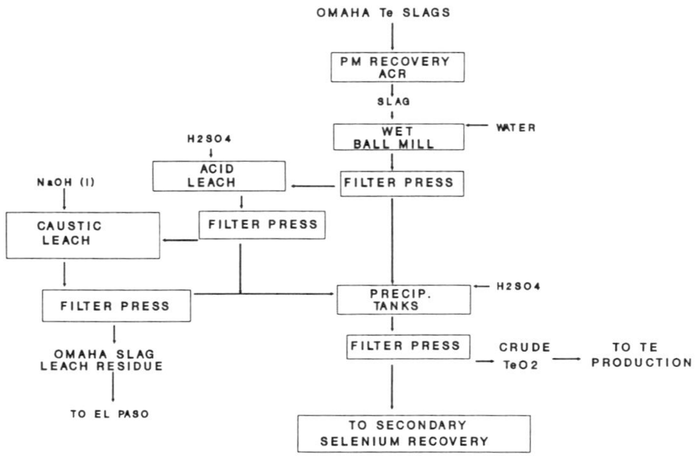

RECYCLE OF TANKHOUSE SOLUTIONS AT CERRO COPPER PRODUCTS

ELECTROLYTIC COPPER REFINERY IN SAUGET, ILLINOIS

John L. Sundstrom

CERRO COPPER PRODUCTS COMPANY
P. O. Box 66800
St. Louis, Missouri 63166-6800

Abstract

Complete systems were designed and are in operation to recycle the effluent streams and to remove impurities from tankhouse electrolyte. Heavy emphasis was placed on making usable products from as many streams as possible at minimum capital and operating costs. Almost complete recycling and reuse of water and sulfuric acid were paramount. Techniques used include solvent extraction - electrowinning, evaporation - crystallization and filtration.

Residues and Effluents - Processing and Environmental Considerations
Edited by Ramana G. Reddy, William P. Imrie and Paul B. Queneau
The Minerals, Metals & Materials Society, 1991

Introduction

Cerro Copper Products Company operates an electrolytic copper refinery of approximately 50,000 tons per year annual capacity at Sauget, Illinois. Refinery operations began in 1929. Anodes to feed the refinery are made within the Sauget Plant from fire refined secondary copper. No. 2 copper and light copper grades of scrap constitute the majority of the feed material. Most of the cathode product is currently sold to the copper tube mill also located within the Sauget Plant. Some cathode is also sold as grade 1 and shipped.

Cementation on iron was the process used to recover copper from tankhouse effluent streams for many years. Effluent streams treated by cementation included all spillage from the cell room, liquors remaining from processing anode slimes, and excess water from washing cathodes.

Cementation was effective at recovering a high percentage of the copper, but had numerous problems and high costs. A low grade cement copper product was produced which was too low in grade (about 30% Cu) to be economically smelted by Cerro's anode furnace, so it had to be sold or tolled through another smelter. The greatest costs of running the cementation operation were associated with treatment of the resulting iron sulfate bearing solution. This solution was treated in the local industrial pretreatment plant by the lime and settle process. Large amounts of lime were required to neutralize the sulfuric acid present, and the iron and other metals precipitated to form large quantities of sludge which were expensive to dispose of. Cement copper was also a difficult product to dry and transport. These problems and regulatory requirements provided the incentives to begin this project.

Flowsheet Development

Starting in about 1982, a search was initiated for better processes to replace cementation. Economic justification of these processes was required.

Direct electrowinning of the copper was tried first on a pilot scale. Dilute solutions gave poor copper recoveries and low current efficiencies. Solutions remaining from leaching slimes were found to be contaminated with tellurium, which co-plated at the cathode. Cathodic material plated from such leach solutions was in the form of a black sludge which rapidly filled the plating cell and contained tellurium up to several percent. Direct electrowinning was found to work acceptably only on uncontaminated refinery electrolyte.

Removing copper from slimes leach liquors by ion exchange resins was tried next. Copper removal using ion exchange resins was acceptable, however, satisfactory solution purity was not obtained without the use of unacceptable amounts of scrub water. Ion exchange on resin beads was abandoned.

Liquid-liquid ion exchange, or solvent extraction was tried and was found to work well in the laboratory. Pilot scale tests followed. Three stage extract, three stage strip, with full continuous countercurrent flow and an electrowinning cell were included in the pilot setup. The extractant used was one part by volume Lix-622 reagent from Henkel Corporation diluted in three parts of KerMac 470-B kerosene. This mixture was recommended by Dr. Kordosky of Henkel Corp. Results were excellent with high copper recoveries and good quality cathodes.

Based on the test results, a solvent extraction - electrowinning plant was designed and built. Full scale solvent extraction - electrowinning began August 1, 1986. The original design was based on simple replacement of cementation tanks with solvent extraction. The vast majority of raffinate or copper depleted solution was still discharged. Some raffinate was used to leach slimes, replacing the water and fresh sulfuric acid makeup previously required. Justification for the SX-EW plant was on strictly economic terms. Payback was based on savings from effluent treatment costs, sulfuric acid purchases, and the much higher grade copper produced. It should be noted that the cathode copper recovered by the SX-EW plant has always been of acceptable quality for direct melting in Cerro's extrusion billet furnaces. Further details of the commercial SX-EW plant operation will be presented later under process control.

Successful operation of the SX-EW plant continued until early 1988. At that time it was deemed desirable to attempt a design that would eliminate discharge of raffinate. Just as at most other copper refineries, nickel is the most abundant and valuable soluble impurity present in the electrolyte. Nickel recovery is virtually a requirement of most discharge reduction plans and is usually recovered by evaporation of water from decopperized electrolyte which forces crystallization of nickel sulfate. Nickel sulfate crystals are then separated from the remaining strong sulfuric acid solution by filtration. A short investigation into other methods of nickel recovery indicated that this classic approach was by far the most appropriate for use at Cerro. Management further specified that design capacity of the nickel recovery system should be 130% of normal nickel inflow under full production conditions while operating the recovery plant five days per week.

Test work conducted previously indicated that copper could be safely plated out of solution in a cell with insoluble (normally lead) anodes down to a copper level of about five grams per liter. It was desired to achieve a nickel to copper ratio of at least ten to one to produce an acceptable grade of nickel sulfate. As the desired maximum nickel content of the refinery electrolyte was thirteen g/l, a copper content of around one g/l was required. A cell design from BEWT Water Engineers called a Chemlec cell was found that appeared to be applicable to removing copper down to the one g/l range. A fluidized bed of glass beads is used in this cell to continuously scrub cathode surfaces. This scrubbing action breaks up the diffusion layer and allows reasonable and adherent plating to proceed at such low copper concentrations with current densities of nearly ten amperes per square foot. Chemlec cells were selected as the final decopperizing cells for the design.

A complete material balance was compiled for the refinery at that time which included water, sulfates, copper, and nickel. A summary of this material balance as prepared to match the selected unit processes is attached as Table One. Preparation of the material balance required numerous flow measurements and laboratory analyses and took several months to complete. In particular, knowledge of the materials entering the by-product recovery area formed the basis of the design, and a high degree of confidence in these numbers was required.

From the material balance, it was obvious that the plant's greatest function would be water re-utilization. Sulfate re-utilization was next in importance. This is somewhat contrary to normal thinking when designing copper and nickel recovery equipment. In fact, control of sulfates has proven to be the most difficult part of controlling operation of the system.

Development of a usable flowsheet block diagram was the next step. Previously identified unit processes (SX-EW, Chemlec cells, evaporation, filtration) were shown as blocks on the diagram. Items identified on the material balance were shown as inputs or outputs on the diagram. Interconnecting plumbing was then shown as lines on the block diagram. Note that complete recycle of black acid, the strong acid remaining after filtering out nickel sulfate crystals, to tankhouse makeup is shown. Interconnecting plumbing was designed to give optimum results and maximize flexibility. Interconnecting plumbing rapidly became very complex. Due to the complexity, it was necessary to break down the entire by-product recovery area into four separate subsystems for presentation here. Block diagrams for these subsystems are given as figures I through IV. As complex as these diagrams are, it should be realized that none of the non-contact flows, such as cooling water, are shown. Emergency storage facilities are also not shown, but were included in the mechanical design. The four subsystems shown are as follows:

I. The tankhouse makeup subsystem.

II. The anode slimes decopperization subsystem.

III. The solvent extraction - electrowinning subsystem.

IV. The liberator and evaporator subsystem.

Sizing of process equipment including the evaporator, the number and size of liberator cells and the storage tanks capacities was done based on the block diagrams and the material balance. Sizing of pumps, piping and other mechanical equipment was then turned over to engineers well versed in this discipline.

Table I Summary of Material Balance for Electrolytic Refinery

Annual Basis Considering Water - Sulfates - Copper - Nickel

Incoming Items	Water Content Gallons/Yr	Total Sulfate Content Lbs/Yr	Copper Content Lbs/Yr	Nickel Content Lbs/Yr
Anodes - 119,959,086 lbs. @ 99.4% Cu	0	0	119,239,331	60,000
City Water to Tankhouse	1,648,895	0	0	0
City Water to Slimes	200,000	0	0	0
Steam to Tankhouse	4,380,000	0	0	0
Steam to Slimes	1,000,000	0	0	0
Sulfuric Acid (23,700 gallons)	Nil	340,000	0	0
HCl to Slimes	4,200	0	0	0
TOTAL INCOMING	7,233,095	340,000	119,239,331	60,000

Outgoing Items	Water Content Gallons/Yr	Total Sulfate Content Lbs/Yr	Copper Content Lbs/Yr	Nickel Content Lbs/Yr
Commercial Cathodes	0	Nil	100,000,000	250
Sx Cathode	0	Nil	234,511	Nil
Liberator Cathode	0	Nil	34,000	Nil
Anode Scrap (19,077,358)	0	0	18,962,894	9,794
Regular Slimes (480,000 lbs/yr)	4,000	90,000	4,800	700
AgCl (24,000 lbs/yr)	460	Nil	126	Nil
$NiSO_4$ Product (300,000 lbs/yr)	3,600	150,000	3,000	49,256
Tankhouse Evaporation	4,745,000	0	0	0
Slimes Evaporation	194,035	100,000	0	0
Condensate Return to Boiler	2,286,000	0	0	0
TOTAL OUTGOING	7,233,095	340,000	119,239,331	60,000

Note: Solids shown as weight of dry solids - moisture shown is extra weight usually lost by drying.

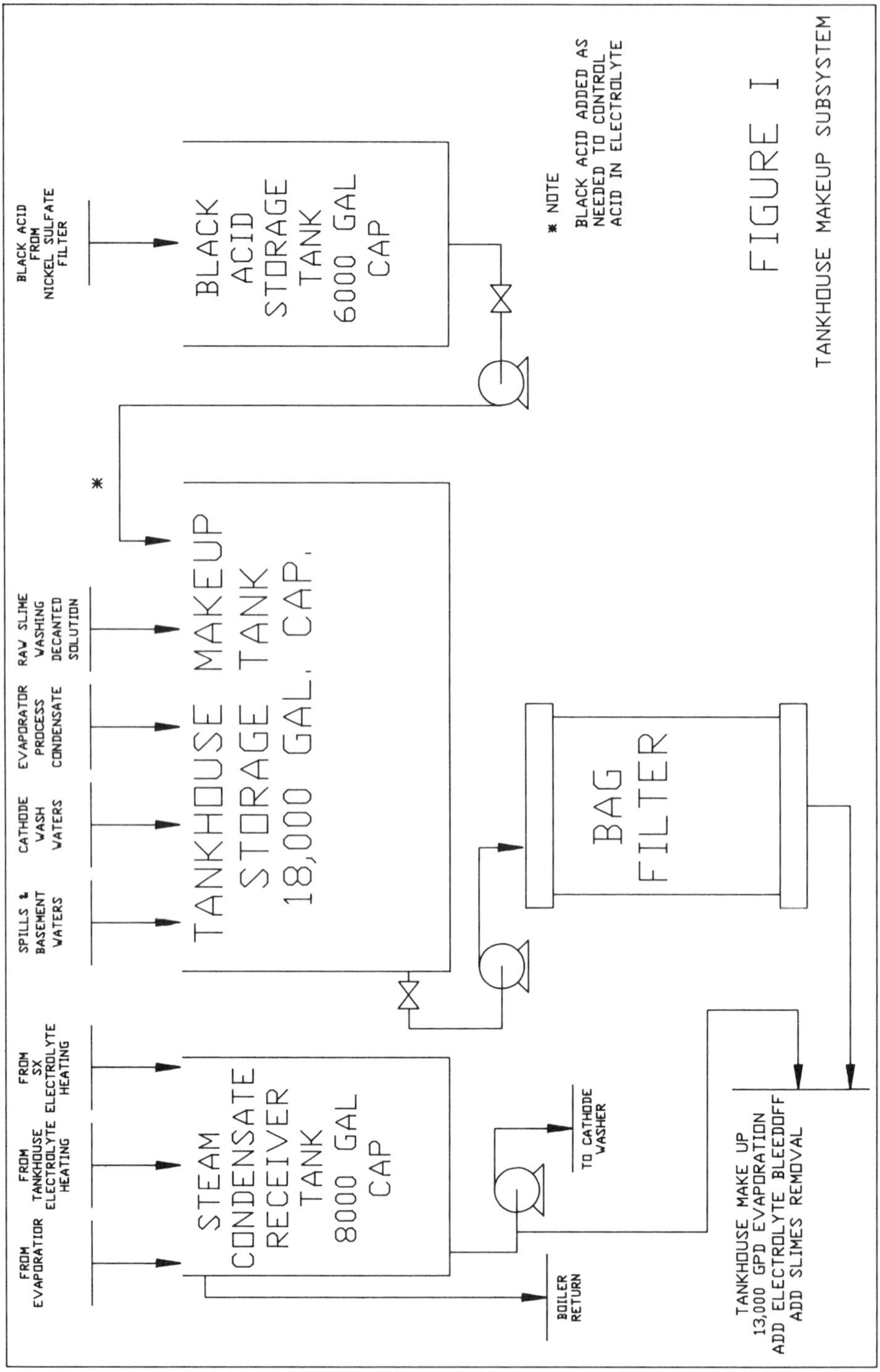

FIGURE I

TANKHOUSE MAKEUP SUBSYSTEM

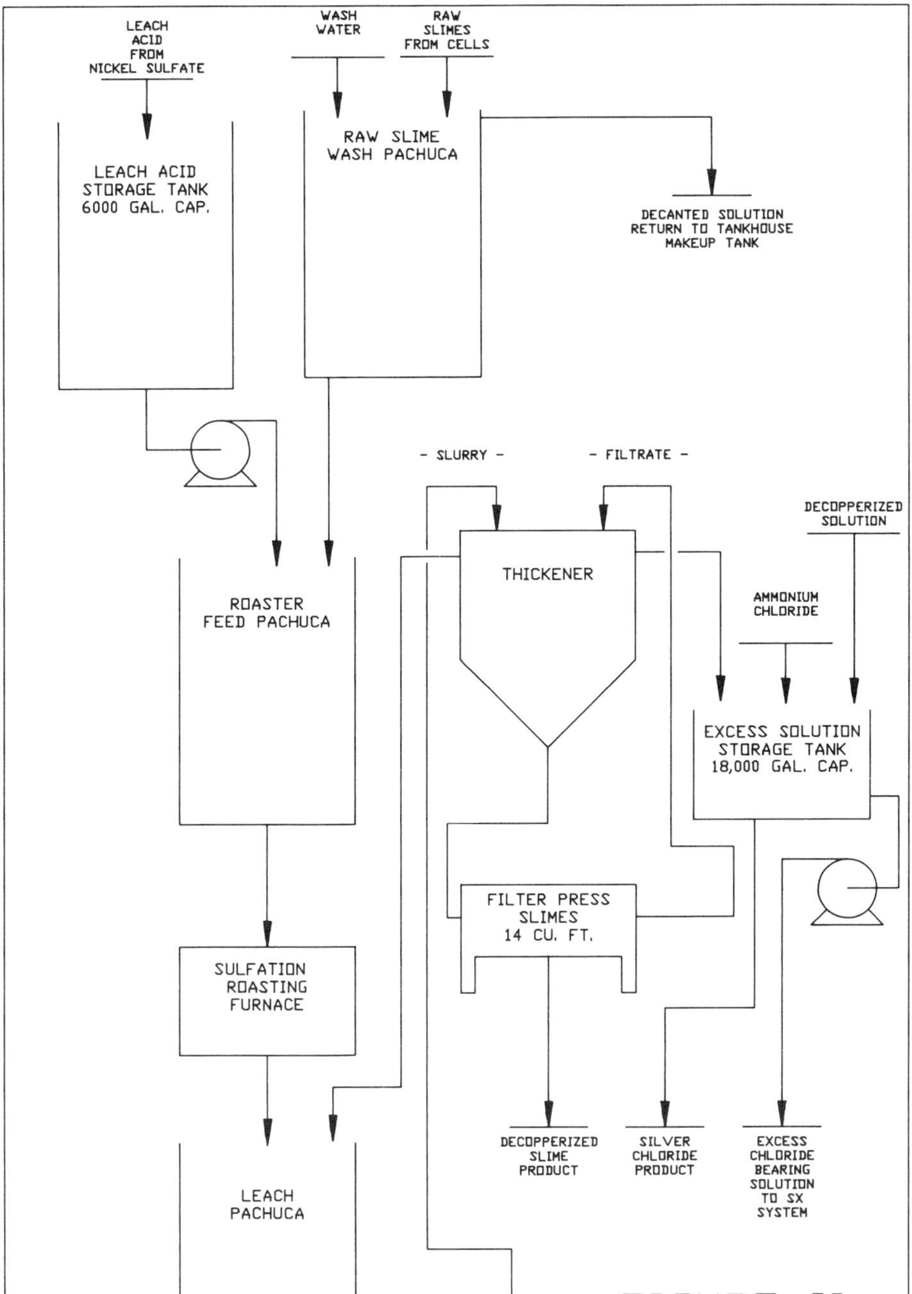

FIGURE II

ANODE SLIME DECOPPERIZATION SUBSYSTEM

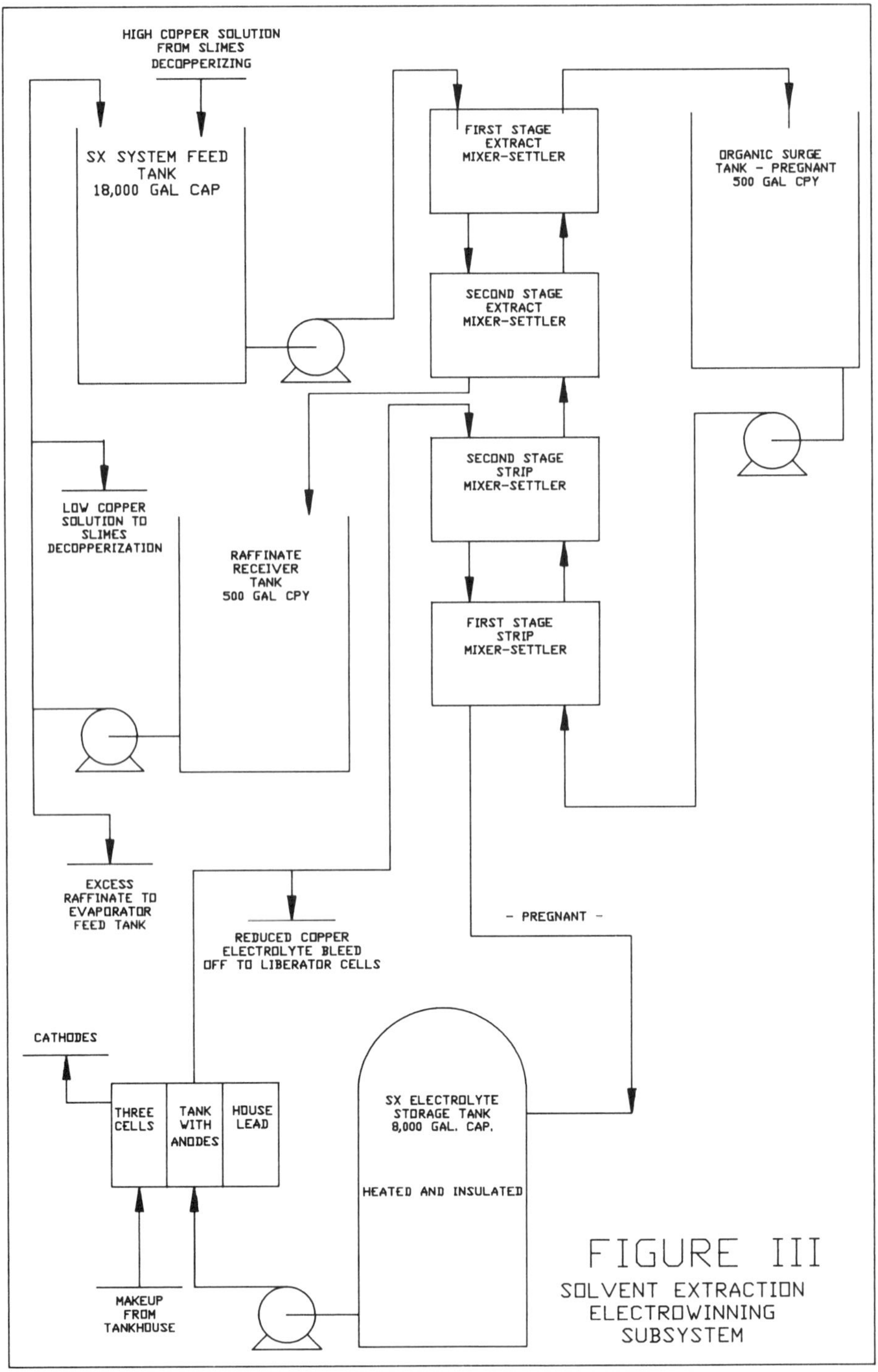

FIGURE III
SOLVENT EXTRACTION ELECTROWINNING SUBSYSTEM

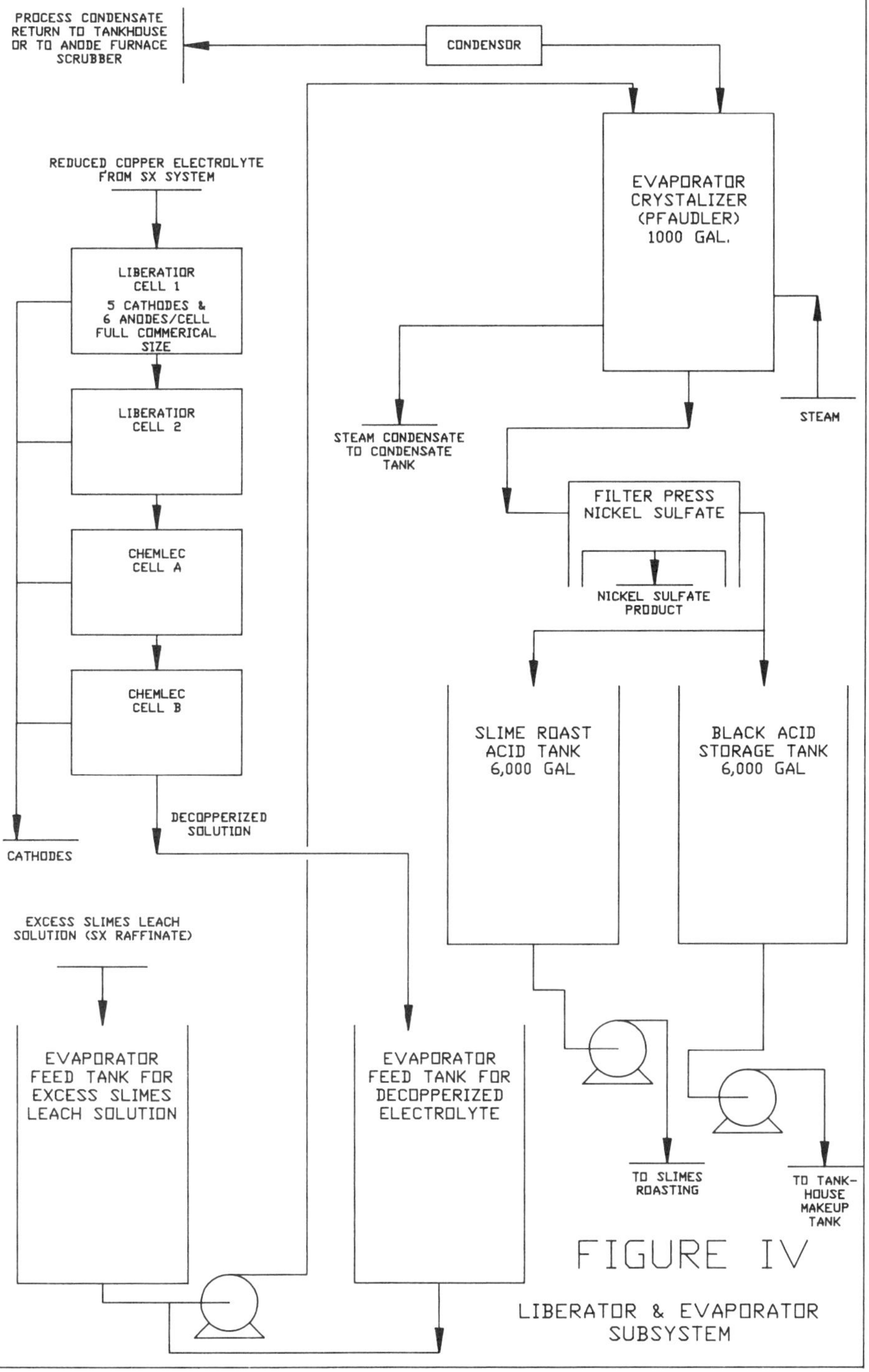

FIGURE IV

LIBERATOR & EVAPORATOR SUBSYSTEM

Mechanical Design and Construction

Additional housing was built in the area previously occupied by the cementation tanks. Approximately 4,200 square feet of new floor space was required. A 40 ft. X 60 ft. segment of this area is serviced by a five ton overhead crane, used primarily to pull cathodes.

Due to the very complex and costly plumbing installed in the building, the building is heated to prevent freezing. Steam heating was used as it was desired to avoid sources of ignition, especially in the solvent extraction area. Heating was specified to maintain 50 degrees F. on the coldest days.

Some site work was required to prevent meteoric water from entering the building. Solutions spilled inside the building are collected in ditches lined with red shale acid brick and furan mortar. All ditch solutions are returned to process at an appropriate point. Flooring in areas where solution contact normally occurs are also paved with acid brick. Flooring in the cellroom basement is plasticized PVC sheeting. This was done to prevent concrete corrosion. It works well and provides a means of collecting spills without contaminating them with concrete debris. A considerable amount of undesirable calcium can dissolve into solutions contacting concrete directly.

Ventilation is provided by a gravity roof vent. Forced ventilation above the liberator and Chemlec cells is provided by hooding, ducting and a fan. Arsine gas detection is provided in the duct and automatically shuts down the rectifier. The arsine detector works, but is sensitive also to hydrogen which appears before arsine.

Emergency tankage was provided. Two 8,000 gallon and one 5,000 gallon fiberglass vertical tanks were included and have proven to be adequate. Most of the other tankage is also fiberglass. Polyproplene has also been used with success. PVC was used for one tank which failed due to softening of the PVC with temperature.

Piping throughout the system is CPVC or 316L stainless where high temperatures and stresses occur.

A Pfaudler 1000 gallon glass lined steel evaporator was chosen. This evaporator was picked due to lowest capital cost. A similar application using a glass lined steel vessel was also seen in another plant which was successful. The Pfaudler vessel is steam jacketed and water can also be applied to the jacket for cooling. Boiling is induced by vacuum. The vessel has a mixer paddle which is also glass lined steel.

Most pumps selected are of the sealless magnetic drive type. Several of the magnetic drive pumps used in the SX system are five years old and are still working. Diaphram type air operated pumps are used to feed the filter presses.

JWI supplied the new filter press for nickel sulfate. This press has a capacity of 20 cubic feet of cake which is nearly adequate to accept a full 1000 gallon batch from the evaporator at once. It is possible to move the tail plate forward, however, and press less than a full press. Filling the press is necessary to obtain adequately dry cakes. Air blowing through the cakes is used to obtain maximum dryness. The press frame is steel, but the 32 inch plates are polypropylene. Piping around the press is all stainless steel due to the high feed pressures, warm temperatures and shock from the diaphram pump.

Mechanical Design and Construction - continued

Six thousand gallon vertical fiberglass tanks were selected for black acid and leach acid storage. These tanks were selected as they can hold an entire tank truck load if it ever becomes necessary to remove acid from the system.

Direct current electrical power is supplied to the liberator and Chemlec cells by a Rapid Electric rectifier unit. Maximum ratings are 1,500 amperes at 15 volts. This is adequate to drive the current density up to 15 a/sf. The rectifier was supplied with both local and remote automatic voltage or current control. Remote automatic current control via a 4 to 20 milliaampere loop is used. All liberator and Chemlec cells are in series and a total voltage drop of around 9 volts is experienced under normal operation. Rolled calcium-tin alloyed lead is used for the anodes in the liberator cells. Platinized titanium mesh anodes are used in the Chemlec cells. Regular tankhouse starter sheets are used in the liberator cells, and starter sheets with extra long loops are used in the Chemlec cells. Sheets are changed once weekly or more often as required.

Safety is an important consideration in a plant of this type. Showers and eye wash stations are provided in several locations due to the rather strong sulfuric acid present in the system. A water fog type fire supression system is installed in the solvent extraction area. In the event one of the fog nozzles opens up, an alarm is automatically sounded and processing is shut down. A berm was placed around the solvent extraction system to contain any spilled organic.

Process Control

A distributed type process control system was selected and installed during installation of the SX-EW. This system works very well, and operator attention is seldom required. Solvent extraction normally operates without an operator present. An alarm sounds inside the cell room on off shifts when something goes wrong. Tankhouse operators then deal with the problem. The distributed control system was expanded to operate the liberator and Chemlec cells.

Control strategy for solvent extraction is to provide constant flows. Feed rates and electrolyte circulation rates are normally about 10 gallons per minute. Organic is circulated at around 22 gallons per minute. These rates maintain the organic as the continuous phase in the mixers. Organic continuous reduces the amount of organic carried out by the raffinate and pregnant electrolyte. Raffinate is normally returned to the SX feed tank until the copper is low enough to feed the evaporator, or the solution is needed to leach slimes. After several passes through solvent extraction, free sulfuric acid in the raffinate runs quite high, sometimes as high as 80 g/l free sulfuric. Copper extraction from such solutions is difficult and slow. A reagent change to LIX-860 which is a stronger extractant helps. Keeping the free acid higher (above 200 g/l) in the strip circuit also helps. The SX-EW system copper recovery capacity is about 1,400 pounds per day and is considerably greater than required.

Control strategy in the liberator and Chemlec cell system is to hold a constant reference voltage by varying the rectifier current. Reference voltage is measured between the cathode buss bar and a copper electrode placed in each cell. A rapid rise in this reference voltage is experienced as the current is raised to the point where ions other than copper begin to discharge at the cathode. Although this reference voltage

Process Control - continued

setup is not perfect, it is adequate to run control. Reasonable quality adherent cathodes are produced with proper settings under automatic control. The reference voltage setting that seems to work best is around .27 volts. The controller is programmed to automatically select the highest of the individual cell reference voltages as the process variable. A second level of control is also used. Current is held reasonably constant by controlled feeding of fresh high copper electrolyte to the cells. This control strategy works surprisingly well. At a current set point of around 700 amperes, and a reference voltage set point of .27 volts, electrolyte is consistently decopperized to around 1 gram per liter. With four cell series connected and a feed copper of 15 grams per liter, 1200 gallons of decopperized electrolyte per day are produced. This is an adequate amount. Feed at 15 grams per liter is available from the electrowinning cells associated with solvent extraction. Copper in these cells is controlled manually by valving cells either to tankhouse circulation or to SX circulation. The SX electrowinning system may contain none, one, two, or three full tankhouse size electrowinning cells. Automatic electrolyte makeup to the SX system is also provided from the tankhouse circulation system. Electrolyte is analyzed daily by the laboratory.

Alarm functions and logging are done automatically by the control system. Flash reports on tank levels, temperatures, flows, alarms, and other system parameters are always available. A remote screen located in the plant gives the operator all necessary information including alarm identities.

Automatic control of the Pfaudler evaporator liquid level is provided. Most other evaporator functions are controlled manually.

The control system has proven adequate to allow one man to operate the by-product recovery system when all equipment is running. A helper is provided when pulling cathodes and when operating the filter presses. Supervision is assigned in the by-product plant on day shift only.

Operating Experience

Commissioning of the electrolyte purification - nickel recovery subsystems began in October of 1990. Many items of interest have arisen during the first ten months of operation.

Crude nickel sulfate produced by the process also serves as the vehicle for removal impurities other than nickel. In Cerro's refinery, these are primarily zinc, iron and calcium. Crystallization percentages of these metals as well as copper and nickel have been 80% higher. Typical nickel sulfate analyses have been as follows:

ITEM	HIGHEST	LOWEST
Nickel	19%	14%
Copper	4%	1%
Zinc	4.5%	2.5%
Iron	2%	1%
Calcium	2%	0.4%
Free sulfuric acid	25%	14%
Moisture (water)	14%	8%

Operating Experience - continued

Ninety evaporator batches were filtered during the first ten months of operation. Filter cake was loaded into jumbo bags holding around 1500 pounds each. 213,000 net pounds of crude nickel sulfate were shipped from these 90 batches. The product was reasonably well received, although zinc content is higher than preferred. A fluxing practice change has been instituted at the anode furnace to reduce incoming zinc.

No unusual impurity buildups in the refinery have appeared. Nickel in the commercial electrolyte system has dropped from about 21 g/l to near 15.5 g/l since recovery operations began. The goal for nickel is 13 g/l. Zinc and iron have remained the same or dropped slightly. Calcium has stayed higher than desired. Calcium sulfate precipitation has caused plugging problems in lines, pumps and valves. Calcium is the reason that a steam condensate system is included on the tankhouse makeukp flowsheet. During the last six months, several uses for hard city water have been converted to steam condensate to further reduce calcium inflow. Arsenic, which is a problem for other refineries, is present only at very low levels in Cerro anodes. Arsenic in the electrolyte after ten months of re-using black acid in the tankhouse is still substantially below 0.5 g/l. Testing indicates that aluminum is not crystallizing into the nickel sulfate and is building up somewhat. Aluminum is entering the system from the mold wash used in anode casting. Aluminum is also still substantially below 0.5 g/l. Organic buildup from addition agents in the electrolyte is viewed as a potential problem, howwever, cathode quality does not appear to have suffered yet.

Several problems with equipment and materials have been experienced. Problems with the pumps used to fluidize the glass beads in the Chemlec cells appeared first. The pumps supplied were horizontal centrifugals with shaft seals. Seal leaks eventually resulted in replacement of these pumps with magnetic drive sealless pumps. Problems were also experienced with glass beads flowing back into the pumps and causing stalling. This problem was cured by installing an inverted trap between the pumps and the cells and by installing bead settlers between the cell overflows and the reservoir tanks.

Problems were experienced with PVC materials softening as a result of higher than expected temperatures in the liberator and Chemlec systems. Structural pieces of PVC are being replaced with polypropylene.

A corrosion problem in the condenser unit of the evaporator occurred. This unit was made of 316L stainless steel. Corrosion occurred in the vapor inlet section, probably as a result of hot chlorides. although only a small amount of chloride is present. A replacement condenser of alloy C276 has been ordered at considerable expense.

Conclusions

Process design goals have been met by the system as designed. No solution bearing copper, nickel or sulfuric acid has been discharged from the refinery area during the first eight months of 1991. Impurity removal rates equalled design capacity. Nickel concentration in the commercial electrolyte has been reduced about 25% in 10 months of full operation. Cathode quality has been maintained.

The system is considered to be fully successful from technical, operational and environmental viewpoints. Savings are being achieved from the sale of nickel sulfate, improved copper recoveries and reduced acid purchases, but payback of the investment will be longer than for most projects.

Acknowledgments

1. Mr. Henry Schweich, President of Cerro Copper Products Company, is acknowledged for permission to publish this paper.

2. Dr. Gary Kordosky of Henkel Corporation, 1844 West Grant Rd., Tucson, AZ, 85745, is acknowledged for providing assistance during both design and operation of the solvent extraction - electrowinning equipment. Henkel Corp. is the current supplier of extractant.

3. Mr. Jack Bradstreet of Pfaudler Division of Sybron Corp., 1000 West Ave., Rochester, NY, 14603, is acknowledged for providing design information on evaporation and crystallization equipment.

4. Ms. Denise Tomlinson of BEWT Ltd., Arden Forest Industrial Estate, Alcester, Warwickshire, B49 6ES, England, is acknowledged for providing design information on the Chemlec cells.

RECOVERY OF CALCIUM FROM THE EFFLUENT

OF DIRECT OXIDE REDUCTION PROCESS

P.D. Ferro, B. Mishra, W.A. Averill*, J.J. Moore, D.L. Olson

Kroll Institute for Extractive Metallurgy,
Colorado School of Mines, Golden, CO 80401
*EG&G Rocky Flats Inc., P.O. Box 464, Golden, CO 80401

Abstract

The direct oxide reduction (DOR) of plutonium oxide generates significant amounts of contaminated waste. This waste primarily consists of calcium oxide-saturated calcium chloride salt. This investigation explores the possibility of recycling the salt by the electroreduction of the calcium oxide. Fused salt electrolysis of a simulated salt mix ($CaCl_2$ + ~10 wt. pct. CaO) is being carried out to electrowin calcium, and the process is being optimized in terms of the calcium solubility, cell temperature, current density and cell design to maximize the current efficiency. This work identifies the major concerns in the development of a process to electrowin calcium from this simulated salt mix.

Residues and Effluents - Processing and Environmental Considerations
Edited by Ramana G. Reddy, William P. Imrie and Paul B. Queneau
The Minerals, Metals & Materials Society, 1991

Introduction

The direct oxide reduction (DOR) process is currently being used as the primary step for commercial production of plutonium metal, according to the following reaction [1]:

$$PuO_2 + 2Ca \rightleftharpoons Pu + 2CaO \quad (1)$$

Plutonium oxide is reacted with calcium to produce plutonium metal and calcium oxide. The calcium oxide is dissolved by molten calcium chloride as it forms. This process is operated at temperatures between 850 and 900° C. The material flow is shown in Figure. 1.

The two principal byproducts of the DOR process, calcium oxide-saturated calcium chloride and spent crucibles, are contaminated with radioactive material as a result of the process. In the current practice the used salt is stored as a contaminated waste. The present investigation focuses on separating the salt into calcium metal and calcium oxide-free calcium chloride, both of which can be recirculated into the process stream.

In 1954, Threadgill [2] developed a process for electroreducing calcium from calcium chloride/ calcium oxide electrolytes. Calcium oxide levels studied in this early investigation were below two wt. pct., and calcium electrowinning was achieved with the rising cathode technique. This technique was based upon continually pulling an electrolyte-covered calcium 'carrot' out of the molten salt. Threadgill reported that calcium electrowinning at calcium oxide concentrations higher than two wt. pct. was not possible because the viscosity of the salt decreases with increased calcium oxide levels. With lower viscosity levels, electrolyzed calcium does not remain covered with frozen electrolyte and ignites in air. Also, lower viscosity levels produce convection currents which were believed to wash calcium back into the bath. The optimal electrowinning conditions, for this system, therefore are calcium oxide content between 0.5 and 1 wt. pct., and electrolyte temperature between 788° C and 816° C.

Lukasko and Murphy [3] have recently developed a method for electrolytically producing calcium metal using an electrowinning/electrorefining sequence. In the initial electrowinning step, a calcium-

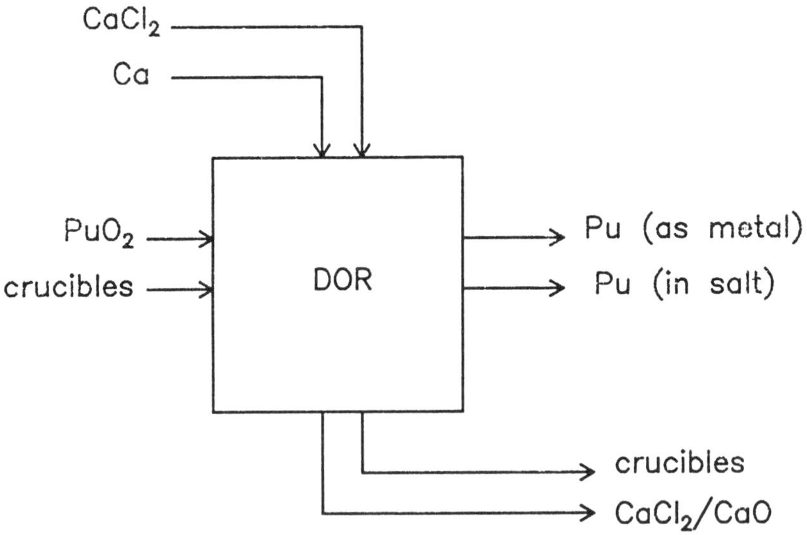

Figure 1. Schematic diagram indicating input and output materials for the direct oxide reduction process.

tin alloy is produced in a molten cathode type of cell using a potassium chloride/calcium chloride electrolyte at 650° C. Tin was chosen as the cathode material due to its high solubility for calcium. The calcium-tin product (~7.5 wt. pct. Ca) of electrowinning is then used as the anode in the subsequent electrorefining step. Electrorefining was done in a calcium chloride/calcium fluoride electrolyte at 850° C. One of the advantages of producing calcium by this two-step method is a lower energy requirement, compared with present commercial production methods [3]. The major advantage of using a liquid tin cathode is that free calcium is not available to reoxidize or form other compounds in the system, thus improving the efficiency and enabling the use of graphite as an anode.

Mulcahey, Poa, and Johnson [4] have reported the results of calcium oxide dissociation experimentation from calcium chloride electrolyte with fifteen wt. pct. calcium fluoride, using a liquid zinc cathode and a consumable graphite anode at 785° C. Calcium fluoride is present in certain plutonium oxide reduction fluxes, since it lowers the melting point of calcium chloride and thereby lowers the necessary operating temperature. Preliminary results indicate that calcium oxide dissolved in molten calcium chloride/calcium fluoride can be electrolyzed to produce calcium dissolved in a liquid zinc cathode. Oxygen gas reacts with the consumable graphite anodes, and is driven out of the cell as carbon dioxide and carbon monoxide. Final calcium weight percentages in the zinc cathode range between 0.27 and 2.28 wt. pct.

Kipouros and Sharma [5] have studied electrolytic regeneration of neodymium oxide (Nd_2O_3) reduction salt, and have reported successful calcium oxide dissociation with calcium recovery in the range of ten to twenty wt. pct. The investigation involved the use of calcium chloride with calcium fluoride levels near fifteen wt. pct.. Calcium oxide levels ranged between 4.74 and 8.38 wt. pct. According to the investigators, low calcium recovery levels were due to calcium carbonate and calcium carbide formation at the anode.

Steele [6] has reported the results of calcium oxide electroreduction using a calcium chloride-based electrolyte with 35.4 wt. pct. lithium chloride (LiCl) and 8.4 wt. pct. calcium oxide. Steele concludes that although this electrolyte has several advantages over pure $CaCl_2$, as a flux for a calcium electroreducing cell, the advantages do not outweigh the major disadvantage of anodic corrosion. The advantages of calcium chloride/lithium chloride-type electrolytes, relative to pure calcium chloride, include lower calcium solubility levels, slower reaction rates for recombination, and lower vapor pressure levels for calcium chloride above the melt.

The solubility of calcium in calcium chloride-based systems at temperatures between 760 and 950° C has been well characterized. A compilation of these investigations, from Axler [7], shows that reported calcium solubility at these temperatures ranges between 2.1 and 16 wt. pct. Peterson and Hinkelbein [8], Sharma [9], Dworkin, et al.[10], Shaw and Perry [11] and Staffanson [12], however, report a maximum solubility level of about three to four wt. pct., indicating that the early reported value of 16 wt. pct by Cubiciotti and Thurmond [13] is high due to the influence of other calcium products in the salt.

A comprehensive survey of the calcium electrowinning work done in the past and their limitations have been described [14]. The analysis of this earlier work indicates that two systems have been successfully used: (1) rising cathode technology, and (2) liquid cathode technology. Both of these methods suggest that calcium should leave the system either through a physical withdrawal, or by using a molten metal cathode. The attempts to deposit calcium in the elemental form within the system, using graphite anodes, have met with little success in the past [6], essentially due to its reactive nature and readiness to form very stable compounds with other components in the system.

One of the goals of the present investigation is to develop a technology to allow for simultaneous calcium electrowinning and plutonium oxide reduction. In such a scenario, plutonium oxide would be chemically reduced to plutonium metal by calcium in a cell which electrolyzes the resulting calcium oxide back into calcium metal and oxygen gas. The cell can be batch charged at a determined rate with plutonium oxide in a molten calcium-calcium chloride bath.

Experimental

The following section describes the materials, apparatus, and procedures used in seven calcium electrowinning trials. The objective of these trials was to determine the operating parameters (temperature, voltage, current densities, electrolyte composition) necessary to dissociate calcium oxide, and to determine problem areas in calcium recovery.

Materials

The calcium chloride and calcium oxide used in the experiments were received in pellet and powder form, respectively. During trial 1 of this investigation and in some pre-investigation melt runs, excessive moisture was present in the starting materials. For runs 2 through 7, to ensure that no moisture was present, calcium chloride and calcium oxide were dried at 250°C and 900°C respectively, both for twelve hours under a nitrogen cover. A vacuum system was developed for periodic removal of collected water during drying. After drying, these materials were kept in an inert atmosphere until ready for use.

Consumable graphite anodes [6.15 mm dia. x 305 mm length] with six-nines purity and a cylindrically shaped stainless steel cathode [102 mm dia. x 1.6 mm thickness] have been used for electrolysis. The immersion depth in the salt for the cathode ranged between 32 mm and 108 mm. Magnesium oxide crucibles of 130 mm diameter, 305 mm height and 3 mm wall thickness were used to contain the molten salt.

Apparatus

The furnace and the cell were contained within an argon atmosphere glovebox with a slightly positive argon pressure. The atmosphere was circulated and kept within acceptable purity limits with a VAC Systems model MO40-H Dri Train. Figure 2 shows a typical cell configurations, with a straight anode sheath.

The temperature of the resistance furnace was controlled with a temperature controller, and a current limiting SCR power controller. Furnace temperature was measured with a chromel-alumel thermocouple placed next to the elements. The temperature of the salt was also monitored with a chromel-alumel thermocouple immersed in the salt bath and contained within an alumina sheath. The graphite anode and stainless steel cathode were connected to a 10 V-50 A max. DC power supply. A stainless steel backup liner was used to protect the furnace well in the event of a crucible failure. An aqueous solution of sodium carbonate was used as the scrubbing medium for the anodic gas to trap any discharge of chlorine gas at the anode. The anodically evolved gases were extracted by creating a natural draft on the exhaust line of the cell.

Dense open-ended alumina sheaths were used to shield the anode on trials 1 and 4-7. The sheaths had an outside diameter of 22 mm, wall thickness of 2.6 mm, and the lengths ranged between 229 mm and 413 mm. For trials 6 and 7, a flared-end version of the tube was tried to enhance oxygen gas collection at the anode. The maximum outside diameter of the flared end of these sheaths was 35 mm.

Figure 3 shows a schematic diagram of the hydrogen evolution apparatus used to determine the amount of calcium present in a salt sample. Dissolving a known quantity of salt in water and measuring the volume of hydrogen gas evolved allows for calculation of the molar quantity of calcium present in the known amount of salt, according to the following reaction:

$$Ca + 2H_2O \rightleftarrows Ca(OH)_2 + H_2 \tag{2}$$

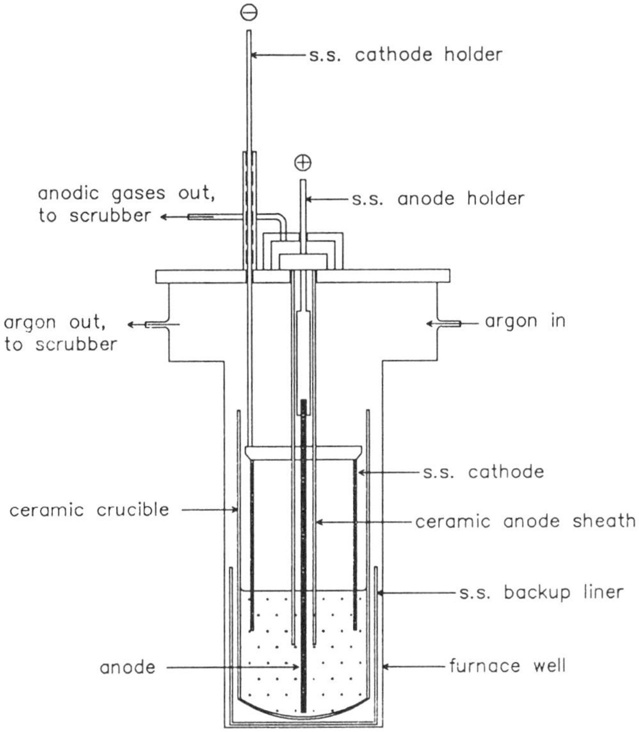

Figure 2. Schematic diagram of the cell configuration with a ceramic anode sheath.

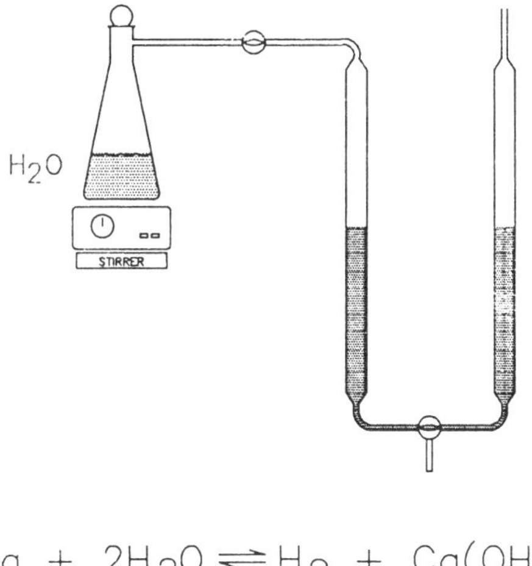

$$Ca + 2H_2O \rightleftharpoons H_2 + Ca(OH)_2$$

Figure 3: Schematic diagram of the hydrogen evolution apparatus for the determination of calcium dissolved in the salt.

To determine the amount of calcium oxide present in a salt sample, the sample was dissolved in distilled water and titrated with hydrochloric acid. The quantity detected by acid titration accounts for both the calcium and calcium oxide present in the salt by converting them into calcium hydroxide according to reactions (2) and (3):

$$CaO + H_2O \rightleftharpoons Ca(OH)_2 \qquad (3)$$

Therefore, to detect the total amount of undissociated and reoxidized calcium oxide in the product salt, a difference was taken between the titrated value and the equivalent oxide of calicum detected by the hydrogen evolution test. It is important to note that any calcium carbide present in the salt also would form calcium hydroxide along with acetylene gas, when dissolved in water and, thus affect the results of both the tests for calcium and calcium oxide determination. The amount of calicum chloride was determined by the weight difference. Since calcium carbonate and free carbon from the graphite anode are the remaining possible components of the product salt and are insoluble in water, these were precipitated out.

Procedure

All operations were performed in an argon atmosphere glovebox. Dried calcium chloride and calcium oxide were mixed in a magnesium oxide crucible, to make a simulated DOR salt mixture. The dry salt mixture was melted in the crucible, and then held at the desired electrolysis operating temperature. During trials 1 and 3, calcium oxide was not sufficiently dissolved, as evidenced by calcium oxide-rich regions at the bottom of the solidified salt. For trials 2 and 4-7, the salt was held for a minimum period of twelve hours at 1000°C, to dissolve all of the calcium oxide present.

As a check on the time-temperature conditions necessary to dissolve ten wt. pct. calcium oxide in calcium chloride, a solubility test was run, using 50 g of calcium chloride and 5 g of calcium oxide in a small alumina crucible. The mixture was heated to 1000°C, and held at temperature for twenty four hours. Post test analysis showed that the calcium oxide level in the final sample was uniform, indicating that calcium oxide dissolution in calcium chloride can be achieved with sufficient time at 1000°C.

Approximately one hour prior to electrolysis, for trials 4-7, the molten salt was manually stirred for one minute, to ensure complete dissolution of calcium oxide.

Cathodic current density ranged between 0.40 KA m^{-2} and 1.44 KA m^{-2} for the seven electrolysis trials conducted. Current was held constant at 25 amperes during these trials, while the voltage varied between 5 and 25 volts. The voltage was dependent upon the amount of exposed anode surface area in the salt and the consumption of the graphite.

Electrowinning times ranged between 3.5 and 10.6 hours, and integrated current for the trials ranged between 107.5 A·h and 265.8 A·h. Average anode life was 84 minutes, and the average number of anodes required per electrowinning trial was 5.1. An increase in the cell voltage to over fifteen volts was used as the criterion for the replacement of anodes during the cell operation, in order to maintain a constant current level of twenty-five amperes. Cell operating data for seven electrolysis trials have been listed in Table I.

After electrolysis, salt samples from different regions of the solidified salt were analyzed for calcium and calcium oxide by the techniques descibed earlier.

Figure 4 shows a plot of theoretical calcium expected vs. the electrowinning time, for a three kg

Table I. Cell Operating Data Calcium Electrowinning Trials.

<u>Electrolyte</u>	
$CaCl_2$, 0-10 wt. pct. CaO	
<u>Cell</u>	
atmosphere:	argon
crucible:	magnesium oxide
operating temp.:	850-970° C
current:	25 A
voltage:	5 - 25 V
time:	3.5 - 10.6 h
<u>Anode</u>	
material:	graphite
size, cm:	6.4 mm dia x 305 mm length
active area:	0.0005 - 0.0011 m^2
current density:	20 - 50 KA m^{-2}
<u>Cathode</u>	
material:	stainless steel
size, cm:	273 x 152 mm
active area:	0.0173 - 0.0312 m^2
current density:	0.40 - 1.44 KA m^{-2}

charge with ten wt. pct. calcium oxide at 850° C. The upper line shows the total won calcium, including calcium dissolved in the calcium chloride, assuming 100 pct. cell efficiency. The lower line shows the expected calcium, beyond the solubility limit.

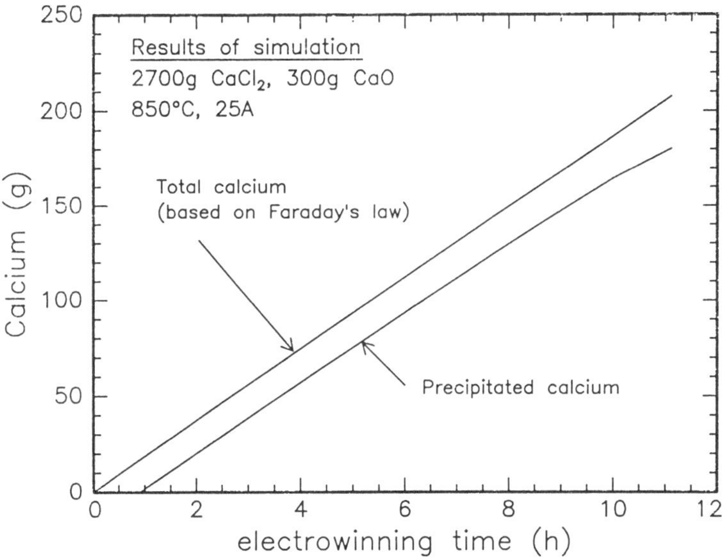

Figure 4. Theoretically predicted recovery of calcium, based on Faraday's law and its solubility in the salt.

In order to construct the diagram shown in figure 4, it was necessary to extrapolate the solubility data [7] in the range of ten wt. pct. calcium oxide. Assuming that the solubility function may be modeled as a power-law function, a logarithmic extrapolation can be used to yield an expression of the following form for maximum calcium solubility (wt. pct.) in calcium chloride at 850°C, as a function of calcium oxide wt. pct.:

$$\%Ca = A + B \cdot (\%CaO)^n \tag{4}$$

From Figure 4 it is possible to determine the minimum length of time between the start of electrolysis and evidence of calcium precipitation beyond the solubility limit. For example, the plot in Figure 4 shows a minimum electrowinning time of about one hour before calcium begins to exceed the solubility limit, for the given conditions.

Results and Discussion

Seven electrowinning trials were performed to determine the parameters necessary to dissociate calcium oxide and to win calcium metal. The results indicate from currents and time used that calcium was present at the cathode during electrolysis for several of the trials, but was not recovered. Calcium at the cathode may have reacted with oxygen, carbon dioxide, or carbon to form calcium carbonate or calcium carbide. Table II shows the attempted electrowinning conditions for individual electrolytic trials.

In the first trial, moisture was present initially in the salt and electrolysis during this trial probably only resulted in the reduction of water and the evolution of hydrogen. For trials 2-7, the salt was dried prior to electrolysis to remove most, if not all, of the moisture present in the calcium chloride.

Trials 2 and 3 were run without an alumina anode sheath, to ascertain that the cell would pass current. Calcium that was won during these two trials probably reacted with oxygen or carbon dioxide at the anode almost immediately. Trial 2 was run at a cell temperature of 880°C, although the salt had been melted and held for approximately twelve hours at 1000°C prior to electrolysis, since the solubility of calcium in the salt is lower at lower temperatures. Since sufficient solubility exists for the oxide in calcium chloride even at 850° C, a higher temperature is initially required only to enhance the dissolution kinetics. Trial 3, on the other hand, was run at a cell temperature of 870°C and was not held at a higher temperature prior to electrolysis. The solidified salt from trial 3 indicated insufficient dissolution of the calcium oxide, due to insufficient time at molten temperatures prior to electrolysis.

Trials 4-7 were run at 970°C, with the salt held at 1000°C for a minimum of twelve hours prior to electrolysis. Trials 4 and 5 were run with straight anode sheaths, and trials 6 and 7 with flared-end anode sheaths.

In trial 4, the orientation of the anode to the anode sheath was such that the graphite anode extended approximately 40 mm below the end of the sheath in to the bath. The final solidified salt mass had a carbon-rich layer on the top, indicating that the anode extension beyond the end of the sheath was excessive. Graphite from the anode disintegrated into the bath and collected at the bath surface during electrolysis. A previous investigation [5] has mentioned graphite on the bath surface as a possible problem, if a current bridge between anode and cathode is formed. Another problem with the anode-to-anode sheath orientation of trial 4 was that the nominal gap between the anode surface and the inside diameter of the sheath was less than 5 mm, which may be insufficient for complete evacuation of the oxygen gas generated at the anode.

To avoid the problem of excessive anode dissolution for trial 5, the anode and anode sheath were immersed to the same depth in the salt, each approximately 75 mm below the bath surface. During electrolysis, maintenance of a constant current level (25 A) was difficult, since the voltage required

Table II. Parameters of Calcium Electrowinning Trials

Trial No.	Electrolyte Temperature °C	Cathodic current density, KAm^{-2}	Anode sheath type	Integrated current, A.h	Theo. calcium expected, g
1	850	0.40	open-ended	200.0	149.5
2	880	0.42	none	107.5	80.4
3	870	0.72	none	115.8	86.6
4	970	1.44	open-ended	265.8	198.7
5	970	1.44	open-ended	188.8	141.1
6	970	1.44	open-ended fluted	215.0	160.7
7	970	1.37	open-ended fluted	168.8	126.2

was relatively high (6-25 V) and fluctuating. The final salt differed from that observed from trial 4 in that a large, carbon-rich layer atop the surface was absent. The cathode had a very thin, whitish salt layer attached which appeared to be high in calcium oxide content. Since the gap between the anode surface and inside diameter of the sheath was relatively small (<5 mm), oxygen generated at the anode may have been liberated into the salt causing a reaction with calcium at the cathode.

In trials 6 and 7, flared-end anode sheaths were utilized, to minimize the possibility of insufficient oxygen evacuation from straight anode sheaths. The orientation between the anode and sheaths, for trials 6 and 7, was such that the anode extended approximately 12 mm below the (flared) end of the sheath. Carbon-rich layers were present on the final solidified salt surfaces for both trials, once again due to graphite disintegration into the salt.

The major result of the experiments discussed above is that calcium metal was not recovered, with several variations of anode sheath and salt temperature. The results also indicate that the cell is operating during each electrolysis trial, in the sense that current is passed through the electrolyte. Because calcium metal was not recovered, it was speculated that one of two electrochemical possibilities accounted for the current that was passed:

1) The current reduced the calcium oxide, but the won calcium reacted with either oxygen and/or carbon dioxide to form calcium oxide, calcium carbonate, calcium carbide or another product;
2) The current reduced chemically-bound water in the salt, forming hydrogen and oxygen gases.

To rule out the second possibility, an experiment was run using an iron anode to see if iron ions would be transported across the electrolyte. Since hydrolysis is energetically favorable compared with iron reduction, an electrolysis test with an iron anode provides information on whether or not chemically-bound water is present in the salt. The experiment was run without an anode sheath, at 970°C using calcium chloride with ten wt. pct. calcium oxide.

The product of the iron anode experiment was an iron-rich deposit at the cathode, which indicated that water reduction was not the dominant cathodic process. Thus, it is expected that calcium was won and accounts for the current that was passed in trials 2 through 7. The evidence of calcium deposit in the iron anode experiment was not conclusive, although one analysis of the salt suggested that calcium was cathodically deposited in the early stages of electrowinning,i.e, before sufficient concentration of iron is built-up in the salt.

Since it was determined that calcium is getting deposited on the cathode and subsequently participating in back reactions, three distinct possibilities existed besides dissolution in the salt, viz. calcium oxide, calcium carbonate and calcium carbide. Therefore, qualitative tests were performed to see if calcium carbonate was present. Calcium carbonate may form through a reaction between calcium oxide and carbon dioxide. In one test, salt (from trial 7) was dissolved in excess water to separate the carbonate and the residue was filtered. Effervescence was observed when this residue was reacted with 0.1 N hydrochloric acid, indicating the possible existence of calcium carbonate in the salt. In a separate test, a sample of the residue was stirred into 0.1 N silver nitrate and allowed to settle. A gray precipitate was obtained which is another positive indication of calcium carbonate in the salt. There is a known solubility of carbon dioxide [15] in similar salt systems, which supports at least the concept of calcium carbonate formation. Also, in the absence of hydrogen evolution on the cathode, the presence of calcium carbonate confirms the dissociation of calcium oxide and, therefore, the winning of calcium.

Besides calcium oxide and calcium carbonate, it is possible that calcium and carbon are reacting to form calcium carbide. Although the salt was not specifically analyzed for calcium carbide, previous investigations have shown that this compound may form under similar conditions [5].

Final calcium oxide weight percentages in the bulk salt for these trials ranged between 6.2 wt. pct. and 9.2 wt. pct. However, insufficient dissolution of the calcium oxide was evidenced in trials 1 and 3 by a calcium-oxide rich region at the bottom of the solidified salt. For trials 1 and 3, the combination of relatively low cell temperature and short time at holding temperature resulted in insufficient dissolution of the calcium oxide.

For trials 2 and 4-7, the calculated calcium oxide percentage was less than the starting percentage indicating that other back reaction products likely exists. Table III lists starting and final calcium oxide weight percentages.

Table III. Initial and Final Calcium Oxide Levels

Run no.	Init. CaO wt. pct.	Final CaO wt. pct.
1	5.0	0.4, 38.0*
2	9.1	6.2
3	9.1	7.1, 23.7*
4	10.0	9.2
5	10.0	8.1
6	10.0	7.3
7	10.0	8.1

* indicates CaO content in CaO-rich region at bottom of solidified salt.

It was observed that alumina, from the thermocouple protection tube and from the anode sheath, dissolves in the salt in small quantities. X-ray diffraction analysis of the salt has been done to confirm the presence of aluminum. The present x-ray diffraction results do not identify aluminum compounds in the salt conclusively, due to the small quantities of alumina involved.

Future Directions

To recover calcium, under DOR conditions with ten wt. pct. calcium oxide, it is apparently necessary to remove electrowon calcium away from the system immediately after its formation. Since calcium has a strong thermodynamic driving force to react with other species in the system, it is necessary to design the cell and operating parameters to allow for winning the calcium faster than it is being lost. Possible ways of overcoming the fast kinetics for back reactions include using a high cathode current density or a high active anode area, or both. For example, arranging the anode and cathode in such a manner as to achieve a high anode area to cathode area ratio, as well as operating at a relatively high current level, may result in a net accumulation of calcium. Also an evaluation of the closed-end porous anode sheaths and reduction in the cell operating temperature needs to be investigated.

A proposed in situ calcium electrowinning/plutonium DOR cell is feasible, if it can be shown that it is possible to continuously and simultaneously a) reduce plutonium oxide with calcium, b) dissociate calcium oxide electrolytically and c) recover plutonium metal. The present results suggest that the in situ cell is feasible, since electrolytic dissociation of calcium oxide is observed. Cell improvements are necessary, however, to design a system which allows for calcium recovery.

Conclusions

It is possible to electrowin calcium from calcium oxide/calcium chloride, although the residence time of calcium in the salt is short if reaction conditions are not controlled. If the deposited elemental calcium is collected at the cathode within the salt, it has the opportunity to reoxidize or form carbonate and carbide in the presence of carbon. The waste problem created by a contaminated DOR process effluent salt can be alleviated by controlling the back reactions and achieving a net deposition rate of calcium beyond its solubility limit in the salt. Possibility also exists for the use of calcium as an insitu reductant for plutonium oxide.

Acknowledgement

The authors appreciate and acknowledge the support of the U.S. Department of Energy and EG&G Rocky Flats, Inc.

References

1. W.Z. Wade and T. Wolf, "The Production of Plutonium Metal by the Direct Reduction of the Oxide" (Report UCRL 50403, University of California Radiation laboratory, Berkeley, CA, 1968).

2. W.D. Threadgill, "An Oxide Process for Metallic Calcium" (Ph.D. thesis, University of Missouri, 1954).

3. J.J. Lukasko, J.E Murphy, "Electrolytic Production of Calcium Metal",(Report of Investigations, RI 9315, United States Bureau of Mines, 1990).

4. D.S. Poa, T.P. Mulcahey, G.K. Johnson, "Electrolytic Reduction of Calcium Oxide to Calcium Metal", (Report, Argonne National Laboratory, 1990).

5. G.J. Kipouros and R.A. Sharma, "Electrolytic Regeneration of the Neodymium Oxide Reduction-Spent Salt, J. Electrochem. Soc., 137, (11), (1990), 3333-3338.

6. R.J. Steele, "An Equilibrium and Electrochemical Study of Oxides in Molten Chlorides" (Ph.D. thesis, Univ. of New South Wales, Australia, 1971).

7. K.M. Axler and G.L. Depoorter, "Solubility studies of the Ca-CaO-$CaCl_2$ system",Molten Salt Chemistry and Technology, ed. M. Chemla and D. Devilliers, Materials Sc. Forum, Vol. 73-75, (1991), 19-24.

8. D.T. Peterson and J.A. Hinkelbein, J. Phys. Chem., 63, (1959), 1360.

9. R.A. Sharma, "The Solubilities of Calcium in Liquid Calcium Chloride in Equilibrium with Calcium-Copper Alloy", J. Phys. Chem., 74, (1970), 3897.

10. A.S. Dworkin et al., Disc. Faraday Soc., 32, (1961), 188.

11. G.S. Perry, S.J. Shaw, "Solubility of Calcium in CaO-$CaCl_2$", (AWE Report O-16/90, AWE Aldermaston, Reading, 1990).

12. L.I. Staffanson, "The Physical Chemistry of Metals in Their Molten Halides", (Ph.D. thesis, London Univ., 1959).

13. D.D. Cubiciotti, C.D. Thurmond, J. Amer. Chem. Soc., 71, (1949), 2149.

14. P.D. Ferro, B. Mishra, W.A Averill, "Electroreduction of Calcium Oxide", Light Metals 1991, ed. E. Rooy, The Metallurgical Society, Warrendale, PA, (1991), 1197-1204.

15. M. Masafumi, A. McLean, "Dissolution of Carbon Dioxide in CaO-$CaCl_2$ Melts", Trans. Iron Steel Soc., 8, (1987), 23-27.

SENSIBLE HANDLING AND APPLICATION OF LIME IN EFFLUENT TREATMENT

C. J. Lewis

Continental Lime, Inc.
670 E. 3900 S.
Salt Lake City, Utah 84107

Abstract

This presentation concerns calcium oxide-better known as high calcium quicklime.High calcium quicklime is produced by calcining high calcium limestone, mostly in the many rotary kilns located throughout the United States and Canada. This paper presents suggestions on the handling of lime based on field experience. The subjects covered are chemical analyses, transportation, lime storage, lime slaking to form a water slurry of calcium hydroxide solids, slurry transportation,slurry viscosity,and scaling of lime slurries. It is hoped that the comments on the foregoing areas will be of use to those designing or operating lime systems.

Residues and Effluents - Processing and Environmental Considerations
Edited by Ramana G. Reddy, William P. Imrie and Paul B. Queneau
The Minerals, Metals & Materials Society, 1991

There are two kinds of commercial lime, the one being a high calcium lime (CaO) and the other a dolomitic lime (CaO·MgO). This presentation will discuss high calcium lime only. Most of the following comments have already been published. It is our purpose to bring most of the practical suggestions on the use of lime into one paper. These involve chemical analyses, transportation, lime storage, lime slaking, lime slurry transportation, slurry viscosity, and the scaling of lime slurries.

Most lime is produced in rotary kilns and the product issuing from the lime kiln will vary in chemical analysis due to the size of the limestone being burned, its source, kiln temperature and other factors. In addition, calcium oxide (quicklime) is perishable and, after cooling, it will pick up moisture from the air and then convert to calcium carbonate, depending upon humidity and exposure to the air. For such reasons, samples of kiln product will not necessarily analyze the same as the lime received by the customer. If the discrepancies must be resolved within a percent or two, an elaborate sampling system must be used. Lime producers and lime users frequently exchange samples when determining the chemical analysis.

It is common practice to determine available calcium oxide by the so-called ASTM sugar procedure. While this analysis is important in some applications where only calcium oxide is of interest, it can be misleading when evaluating lime for acid neutralization. Most limes contain some magnesium which can act as a neutralizing reagent up to a pH of around nine. Also, lime may contain some calcium oxide which is sufficiently overburned to prevent its reaction with the sugar solution but which, nevertheless, can react with acids, particularly if the lime slaker is a ball mill which is grinding the lime very fine. Some years ago, at the Mellon Institute of Industrial Research in Pittsburgh, we developed a procedure for determining the total acid neutralizing value of lime. This procedure is used today and is referred to as the "Basicity Factor" method.

When unloading lime from cars or trucks for transport to a lime storage bin, it is wise to avoid horizontal runs when air pressure is being used to move the lime, since lime tends to settle out of the airstream and plug the horizontal transport pipe. This is also true at elbows. The use of a flexible hose for moving the lime through a horizontal run is becoming more popular. Also, where the transporting system must change direction, sweep elbows are strongly recommended. Most lime storage tanks have a dust collecting system on the top. Sometimes, this system plugs and the tank goes under pressure and ruptures. It is prudent to have a well-anchored blowout port on top of the tank to relieve pressure. Finally, it is considered good practice to have a live bin bottom on the lime storage tank which can vibrate the cone-shaped bottom to assure that the lime does not hang up. Air spargers can be used if the air is dry and so can mechanically-operated hammers.

The discharge of lime from the cone-shaped bottom of the storage tank can be by star valve,moving belt or screw conveyor. The system conveying the lime to the slaker should be engineered so that water vapors and steam from the slaker cannot work backward into the lime-feeding system. Screw conveyors and belt feeders can be set to feed a constant amount of lime per unit time and the slaking conditions can then be adjusted by controlling the water input to the slaker. In some cases, a weigh-belt feeder can be used to advantage. A day tank may also be needed to simplify the lime-feeding system if the quantity of lime used is appreciable. There is a variety of lime slakers ranging from simple mixing tanks to more complicated models. The vendors of this equipment have ample information to recommend the proper slaker.

If slaked lime is placed in a lime slurry storage tank, the tank should be kept gently agitated so the calcium hydroxide solids do not settle to the bottom and compact. If the tank is protected with a lid, this minimizes pick-up of carbon dioxide from the air, although this pick-up is rarely sufficient to be important unless the lime slurry is stored for many days.

There are several important considerations involving the handling of a water slurry of high calcium hydroxide. In the case of some slaked limes, the water slurry containing more than 15% calcium hydroxide solids by weight begins to exhibit a viscosity which renders the subsequent screening or pumping of the slurry difficult. At a solids concentration much above 25% calcium hydroxide solids by weight, some slurries become so viscous that pumping is impossible.This viscosity problem can be alleviated by several procedures. One is to add about two percent of pulverized naturally-occurring gypsum to the slurry. This has been found to break the viscosity and render the slurry pumpable. In some cases, the addition of about one percent sulfuric acid by weight of calcium hydroxide solids accomplishes the same purpose. Indeed, at some coal-burning power plants where a water slurry of slaked lime is used in wet or dry scrubbing, the use of scrubber pond water containing some calcium sulfate in the slaking water for the lime controls the problem of viscosity up to about 22% calcium hydroxide solids by weight.

Another problem involves the transportation of slaked lime slurry where a temperature increase along the transport line occurs. Lime has an inverse solubility and, as temperature increases, the calcium hydroxide precipitates as a solid on the pipe walls, sometimes to a point where the transport system is rendered inoperable. Effort should be made to assure that the temperature in the transport system decreases, rather than increases. Another solution to the problem is the use of flexible hose which can be deformed as necessary to break loose the calcium hydroxide scale.

Still another problem in a slurry transport system is that of diluting the slurry on its way to point of use. If the dilution water is hard, the calcium carbonate in the hard water is precipitated by the lime and will scale the transport system. Finally, calcium hydroxide solids tend to settle, however slowly, and then finally compact. Take-off valves on a continuous loop handling slurry should be in a vertical position above the loop so that when settling occurs, it does not plug the take-off valves. Likewise, if a loop is to become inactive for more than a few days or so, it should be flushed-out with water, or otherwise the calcium hydroxide slurry in the loop may settle out and compact, making it difficult to restart the loop. Finally, it is axiomatic that when pumping lime slurry, air should be prevented from leaking into the pump, because the carbon dioxide in the air will react with the calcium hydroxide to form calcium carbonate, which can scale the system.

Now, the foregoing comments concerning possible problems when using lime are not to be considered serious if the suggested remedial actions are taken. The technology of lime is both physical and chemical. Unless the aforementioned characteristics are appreciated, a lime system can become troublesome and result in engineering problems. However, if properly designed and operated, a lime system can operate indefinitely, troublefree. It is hoped that the foregoing remarks will be of some use to those engineers, chemists and plant operators involved with high calcium quicklime.

EXPERIENCES IN

PROCESSING OF CATHODE LINING FROM ALUMINIUM INDUSTRY

T.A. VENUGOPALAN

HINDALCO INDUSTRIES LIMITED
P.O. Renukoot 231 217
District Sonbhadra
Uttar Pradesh
India

Abstract

The 'Hall-Heroult' Process of reduction of Alumina to Aluminium has remainded basically the same for over 100 years, but the technology has been undergoing drastic changes with respect to selection of materials for Cathode and Anode and also energy consumption. The latest generation pots have a working life of about 1800-2000 days normally. After the failure of the pots, the Cathode Lining is dug out and disposed of on land. Since the lining contains objectional materials causing pollution, various processes have been tried for re-processing of this waste. However, most of the processes have been unsuccessful from the techno-economic point of view.

At Hindalco a new chemical process for converting the fluorine values of the dug out material to Cryolite was developed in 1988 and a one ton per day Cryolite production plant has been operating since 1989. This process is especially applicable where the integrated Alumina and Smelter Plants exist side by side. The experience in operating this Plant is discussed in detail in the paper.

Residues and Effluents - Processing and Environmental Considerations
Edited by Ramana G. Reddy, William P. Imrie and Paul B. Queneau
The Minerals, Metals & Materials Society, 1991

Introduction

Hindalco Industries Limited is operating an integrated Aluminium Industry at Renukoot, India, having a present installed capacity of 300,000 Tones per Annum of Alumina. The Plant started operation in the year 1962 with a capacity of 20,000 Tones per Annum metal only and was subsequently and progressively expanded to the present level. The Aluminium Smelter located at Renukoot has Prebaked Cells operating at 60 - 63 KA with old technology. Initially up to 1985 the reduction pots were lined in monolithic lining which have now been subsequently replaced with cathode blocks under a modernization scheme.

The life of the pots has been varying from 1200 to 1800 days and on failing, the Cathode lining is dug out and disposed on land.

The dug out material from the failed pots have fluorine values which can cause pollution to land underground water. Also because of the presence of carbon and absorption of nitrogen, certain cynides are also formed under specific conditions.

Research Work

A lot of work has been done on recovering the fluorine value economically from the dug out material. The original work done in a small pilot plant based on CO_2, during the years 1971-73 had to be abandoned due to poor economics. After the year 1980, we had a lot of correspondence with M/s VAW of West Germany, who have developed a process for recovering the fluorine value and converting to aluminium fluoride. After detailed study of this method it was found that practical problems like high cost and uncertain availability of required raw material in this specific process, and due to poor economics, this project also had to be dropped. M/s VAW and M/s LURGI of West Germany subsequently evolved a new process of Pyrohydrolysis combined with dry scrubbing by which the fluorine or value could be recirculated and cryolite or aluminium fluoride could be manufactured in the process. But this method is apparently an expensive proposal.

Hindalco's efforts

Meanwhile Hindalco's Research & Development Department which has been conducting various laboratory experiments on this project came out with a successful and technically and economically viable project wherein the capital investment is also much less. Further to the laboratory trials a mini pilot plant trial was also done and the results were studied in detail and it was found that the project was feasible.

Process

The dug out material is crushed in a primary crusher to about less than 2 inch size and then fed to a grinding mill, where measured quantity of cold water was added and wet grinding done. The purpose of the wet grinding is two fold, that is, together with grinding of the material the leaching of the material with cold water will also be achieved. Leaching time is controlled. Discharge from the cold mill was taken to a settling tank where the solids were settled and a clear liquid decanted. The solids from the settler were fed to a centrifuge/filter where all the liquor was decanted. The decanted liquor from the settler and centrifuge will contain the recovered caustic, sodium aluminate and sodium carbonate, and this is pumped to the Alumina Plant at an appropriate point in the system.

The residue from the centrifuge is leached with hot water and caustic at 90°C for a specific time. After leaching, the slurry is again centrifuged in a second centrifuge to separate the residue which is waste carbon (this can incidently be used with pulverized coal in the process steam boiler). The filtrate from this centrifuge is taken to a reactor tank and aluminium sulphate solution, which is prepared from alumina hydrate and sulphuric acid is added for bringing down the pH of the solution. Temperature, pH control and retention time is very important.

The main reaction is:

$$12\ NaF + Al_2(SO_4)_3 \quad \ldots\ 2\ Na_3AlF_6\ \text{(Cryolite)} + 3\ Na_2SO_4$$

Cryolite is precipitated in this tank and the slurry is again sent to a disc or drum filter for recovering the cryolite. The filtrate from this filter (which is mainly sodium sulphate) is sent to the Effluent Treatment Plant. Alternatively, sodium sulphate and Glauber's salt may also be prepared by evaporating this liquor. The filtrate cryolite is dried and calcined at 600°C to give product cryolite. The drier/calciner is provided with adequate pollution control facilities.

The advantage of the process is that practically all the constituents of the dug out material are recovered and utilized. This process is specifically advantageous for integrated Aluminium Industry like Hindalco.

The Cryolite Recovery Plant is operated for 300 days in a year.

The typical analyses of dug out material are as follows:

	Monolithic lining	Cathode block lining
	(- major constituents only -)	
F_2	13 - 15%	7 - 9%
Na_2O	34 - 35%	29 - 30%
Alumina	8 - 10%	6 - 7%
Carbon	Balance	Balance

Problems

i. The entire operation is batch operation and there being wide variation in the fluorine content of the dug out material from batch to batch, the process controls have to be changed from batch to batch at every stage like caustic addition, cryolite precipitation, Alum quantity, etc. No methods of Continuous on line analysis are available. This in turn affects the quality of the end product to some extent. Mainly the composition of dug out material, the fluorine content varies from 20 to 30 per cent.

ii. The Plant has been experiencing reasonably high corrosion rate at certain locations. This problem is being tackled by proper selection of materials of construction.

iii. Cathode dug out material coming out of the Cathode Block Lined pots are too hard to crush and the wear of crushers are heavy. Per ton of cryolite, the quantity of block lined material is much higher than the monolithic lined material and, therefore, this affects production.

iv. It has been found that the temperature of the hot leach, which is an important part in the efficiency of leachability is not properly controlled, there is possibility of some fluorine getting recycled to the Alumina Plant.

v. Selection of the filter and filter cloth for the cryolite is very critical so as to avoid loss of cryolite in the filtrated and at the same time maintaining production.

Advantages

This process is especially advantageous for an integrated Aluminium Complex. There is practically no requirement of disposal site and the process simplifies re-cycling of Hazardous waste. Presently, at Hindalco, all the fresh dug out material is processed directly and at the same time reclamation of old dug out material is also continuing.

Conclusion

Though the process and plant are still under stabilization, this has been a break-through involving a workable process for recycling of an industrial waste. Most of the problems have been tackled and the process is established.

ELECTRIC ARC FURNACE PROCESSING OF SOLID WASTES

J. Kevin Cotchen and Howard F. Davis

Lectromelt Corporation, Pittsburgh, PA

ABSTRACT

The results of "submerged arc" electric furnace processing of a variety of materials is discussed with respect to the process and products, including the characteristics of the products (e.g. TCLP or Toxicity Characteristics Leaching Procedure). The materials smelted include slags and wastes containing various metal oxides including copper, tin, lead, zinc, chrome, nickel and other associated elements. The findings of these tests are compared to those reported in the literature for other operations treating similar "wastes".

Residues and Effluents - Processing and Environmental Considerations
Edited by Ramana G. Reddy, William P. Imrie and Paul B. Queneau
The Minerals, Metals & Materials Society, 1991

INTRODUCTION

Although not generally known, the electric furnace is now successfully being used to process a wide variety of waste streams. Generally these wastes are a mixture of metals and/or metallic compounds including oxides, carbides, sulfides, sulfates and the like, and more often than not they are understandably the products of metal processing operations. However, as will be discussed later, an arc furnace is not limited to the processing of non-organics, in fact, appreciable amounts of organic compounds can be tolerated in the furnace charge as is commonly practiced in smelting operations. An example of this are the wood chip additions made to ferrosilicon furnaces to add bulk to the charge.

The comparatively wide acceptance of arc furnaces partly stems from its long history. Furnaces have been used commercially since 1899, with the first successful installation being in France. Later, in 1906, the first commercial arc furnace was operated in the United States. At this time it wasn't surprising that these early applications centered on steel production, but it was soon found that the intense heat generated by the arc at the foot of the electrode could be used to melt, or decompose, <u>any</u> material. An example of this is the use of arc furnaces to produce fused magnesia refractories as well as high alumina abrasives and fibers. Both of these materials exhibit comparatively high melting points. Additionally, it was found that this heat could be used to force highly endothermic (reduction) reactions to completion. Examples of this are the production of ferroalloys, silicon metal and aluminium. In fact, prior to the invention of the Hall (and Heroult) Cell in 1886, the standard of today's aluminum industry, aluminum was made by chemical means. The difficulty of producing aluminum by wet chemistry is evidenced by the fact that before the U.S. Civil War aluminum was more precious than gold.[1] Aluminum's current price and availability graphically demonstrates one of many successes in utilizing electro-thermic processing to mass produce metals.

Coupled with the long history of arc furnaces is the large volume of technical information that has been generated on their use. Much of this information is still used, but some of it has been forgotten. It is probably best that part is forgotten, such as ill-conceived experiments akin to attempts to turn lead into gold, but there was also much testing performed by companies like Strategic Materials Corporation (Strategic-Udy Processes, Inc.) of Niagara Falls, New York. Around 1960 they performed a wide variety of experiments relating to the separation and reduction of a wide variety of ores utlizing numerous pieces of process equipment including the arc furnace. Currently many of the same techniques used 30 years ago are being reinvented and used to process effluents, some of which date back to the same period.

Unfortunately, the same foundation and history that supports the application of the AC arc furnace also serves to detract from its attractiveness. Recently DC arc furnaces and plasma torch technology have come into the limelight. It is interesting to note that DC furnaces are not new since most early experimentation utilized DC power. However, as the utility systems standardized on the generation and transmission of 60 Hz power, so did the arc furnace. Only recently have large silicon rectifiers been more cost competitive, albeit somewhat more complex, since a great number of thyristors are required for high power systems. Likewise plasma torches with their thoriated tungsten electrodes are being more widely applied.

Some of the purported benefits of using DC is electrode savings and higher metal recoveries due to the electrical effect on migrating ions called cathophoresis. Although data supports claims of lower electrode consumptions, the evidence supporting the existance of the cathophoresis effect has not been conclusive.

The reasons for the popularity of these newer furnaces, given their balanced advantages and disadvantages, is not very clear. The primary reason may be that since high-power DC furnaces and plasma torches are "high tech" they seem more glamorous and "avant garde". Thus they appeal to man's search for new gadgets and a "high tech" solution to what is perceived to be a "high tech" hazardous waste problem. This, however, is not necessarily the case.

Fortunately for AC furnace builders, experimenters and users of plasma torch equipment are quickly discovering its shortcomings. Inherently long arcs from torches are understandably being found to be less energy efficient that a carbon arc, and water cooled electrodes are being found to have a propensity for catastrophic failure. This has led at least one user to replace a torch with a simple graphite electrode. Herein lies the distinct advantage of AC furnaces, their simplicity and hence dependable functionality. This is not to say that torches and DC furnaces do not have their application, in fact, in some cases they are preferred, but a particular application should be scrutinized and tested before use.

Almost every pyrometallurgical application has been, and can be performed using an AC furnace. The following is meant to provide an overview of the authors' knowledge of furnace applications particularly as they relate to processing of solid wastes.

GENERAL

Unfortunately, waste streams are typically a mixture of various elements and compounds. As such their treatment becomes less straight-forward than the treatment of purer materials, particularly if one wishes to recover a saleable product. One such example might be a mixture of metallic oxides, sulfides and carbonates. As will be discussed later, such feeds require multiple processing steps in order to take best advantage of an arc furnace's capability. A particular process may include a slag cleaning furnace step and a smelting/reduction furnace step and possibly even feed pretreatment such as calcination or roasting. In this way one can avoid the common mistake of trying to develop a single piece of process equipment to perform multiple tasks that would best be handled by distinct processing steps. When one attempts to perform numerous tasks in a single vessel, very often the result is a complex piece of equipment that doesn't perform as well as seperate pieces of equipment. Thus one must be sure to critically evaluate the process in terms of performance operability and maintainability.

Some waste streams, although a mixture of oxides, fortunately conform to commonly encountered amalgams such that simple smelting techniques yield a non-hazardous slag product. Generally a discard slag is the desired end result of such operations (neglecting metallic product). Most often (depending upon the quantity of leachable elements) this slag is vitreous with a basicity of less than 1.5, and its glass-like structure results in its low leachability as measured by the TCLP.

Thus toxic metals can be tightly bound in a glass matrix in much the same way that lead is bound in lead crystal. Hence it appears that it is no accident that iron blast furnace slags conform to this general description. However, one must be cautious since some slags (glasses) exhibit appreciable leachability and other slags, that readily degrade to a fine powder upon cooling, surprisingly exhibit low leachability. Therefore until a better understanding of slags is gained each process's slag must be individually characterized. However, in general, one can state that "a glassy slag is a "good" slag".

As far as slag melting furnaces are concerned, there is a common misunderstanding that such furnaces operate with a distinct arc in much the same way as metal melting furnaces. The fact is that these furnaces exhibit high resistance characteristics, the result of which is that the heat is developed in the melt with the vast majority of the heat being generated directly underneath the electrode tip. Thus an electric furnace's high efficiency is a result of the method by which heat is generated, and by proper selection of furnace design and operating parameters these furnaces can be made to perform in a variety of ways. Also misunderstood is the fact that primary control of the melt (slag) temperature is via its composition, with slag temperatures generally equilibrating at 100°C above liquidus under stable conditions. Thus by consulting the appropriate phase diagrams the slag's characteristics can be altered to give the results desired.

With the above in mind, following is a brief summary of various arc furnace operations that

- characterizes the types of "wastes" that have been, and can be processed,
- demonstrates the flexibility of the smelting process, and
- enlightens possible users on the wide applicability of the arc furnace.

COPPER SMELTING AND SLAG CLEANING

One of the largest copper smelters in the world at Anaconda, Montana, is a prime example of the capability of an arc furnace to produce a non-hazarous, vitreous slag. Although the furnace hasn't operated for years at one time the 29 foot by 104 foot long furnace with six in-line electrodes produced over 1000 tons of slag per day with a general composition (although variable) as given in Figure 1.[2] What is most interesting about this slag is that at times the zinc oxide levels were in excess of fifteen percent. However, what is more interesting is the fact that this material is not hazardous. It is now used as a sand blasting grit and according to its Material Safety Data Sheet it is classified as a nuisance dust. Anyone who is cognizant of copper metallurgy can appreciate this fact since copper ores, and hence copper smelter slags, are generally known to contain elements that can be quite toxic (such as lead, zinc, arsenic, selenium, etc).
This is a prime example of potentially hazardous elements being tightly bound in the glass matrix. Also interesting is the fact that, with fluxing, K061 electric furnace steelmaking dust could be melted to produce a slag of similar composition.

Anaconda's operation is not unusual in that there are numerous other similarly-sized operating copper smelters producing a very similar product.

COMPANY (REF)	SOURCE	NOMINAL COMPOSITION (WT. %)						
		CaO	MgO	SiO_2	Al_2O_3	FeO	S	OTHER
ANACONDA [2]	Copper Smelting	7.8	10.9	43.0	5.2	20.	1.1	Zn-8.0, Pb,As,Cu, etc.
STILLWATER MNG. [3]	Slag Cleaning	4.2	14.8	45.0	7.4	17.4	-	
INMETCO [11]	SS Mill Wastes	29.0	13.0	22.0	10.0	-	-	Cr-3.7, Mn-4.8
RSR [•]	Lead Smelter	29.4	-	27.8	-	9.0	4.0	Pb+Sb+As+Sn < 4.5
HOBOKEN [10]	Lead Smelter	24.0	-	28.9	-	21.9	0.16	Zn-1.0, Pb-0.5, Sn-0.36
KAWASAKI CITY [18]	Sewage Sludge	22.5	1.4	10.9	5.2	0.8	-	ORGANIC-34.1, H_2O-22.0, OTHER-3.1
PILOT TEST [18]	Mixed Municipal Incinerator Residue	9.8	0.6	46.8	11.9	4.7	0.2	NOT GIVEN, APPROX. 26.0
MIWO [19]	Blast Furnace Slag & Rock	22.8	11.2	40.8	10.4	8.1	-	K_2O-0.5, Na_2O-2.2
TYPICAL [11]	EAF Dust (K061)	32.4	11.9	26.1	8.5	5.1	-	Zn-1.0, Pb-0.5, Sn-0.36

•EXTRAPOLATED FROM REFERENCE 8

FIGURE 1: COMPOSITIONS OF VARIOUS SLAGS RESULTING FROM ELECTRIC FURNACE PROCESSING

More recently a slag cleaning furnace is being operated by Stillwater Mining in Columbus, Montana. Although this furnace performs a different function by being used to concentrate platinum group metals in a complex iron-nickel-copper sulfide matte, it, like Anaconda, produces a non-hazardous glassy slag which is being used for road bed and general fill. Its typical composition is also given in Figure 1. [3]

Stillwater's furnace is only one of a number of analogous furnaces used for slag cleaning, and although slag cleaning furnaces are typically used to increase metal recoveries from primary smelting operations, this wide variety of furnace applications evidences the flexibility of the arc furnace.

The last application of arc furnace technology for processing of copper containing slags relates to a lab test performed by LECTROMELT for a brass and bronze ingot producer. This producer has accumulated a pile of mixed slag from their continued operations. It has been found that this slag does not pass the TCLP test adopted by the EPA. As a result they have developed a process to recover metal and produce a non-hazardous vitreous slag by processing it in a rotary reverberatory furnace. Lab tests using a single-phase AC laboratory furnace produced a copper alloy containing 7% Sn, 36% Pb, 5% Zn, 2.5% Ni and 1% Fe as well as a glassy slag that passed the TCLP test even though it contained appreciable amounts of zinc. Its composition was somewhat similar to that produced at Stillwater. This is a prime example of waste treatment for metal recovery and detoxification.

NICKEL AND CHROME
(Ferronickel, Cupronickel and Nickel-Chrome Processing)

The arc furnace processing of ferronickel and cupronickel is fairly well known and is being practiced by companies such as INCO, Falconbridge and Cerro Matosa. As in the processing of copper, metal/slag separations are achieved and, as such, it evidences another furnace application.

Probably the most interesting of INCO's operations and most germane to the topic of this syposium is the INMETCO Plant in Elwood City, PA. This plant which began operations in 1978 is based upon pilot tests performed in Port Colborne, Ontario. Although the plant operation remains true to its original concept of processing nickel- and chrome-containing wastes from the stainless steel industry (e.g. baghouse dust, grinding swarf, pickling sludges, etc.) it has more recently been processing other classified toxic wastes such as chrome-magnesia refractory and nickel-cadmium batteries. The products of this operation are stainless master-alloy pigs which are sold back to producers, a non-hazardous slag, and a baghouse dust high in zinc and cadmium content that is shipped to appropriate processors.[4]
A typical slag anaglsis is given in Figure 1.[5] The most relevent step in the process which includes blending, pelletizing and prereduction in a rotary-hearth furnace, is the charging of hot pellets and the final reduction of these pellets in a "submerged arc" electric furnace.

INMETCO's operation is a prime example of a multi-step process used to treat a mixture of hazardous wastes to take best advantage of the capabilities of each piece of equipment so that the electric furnace is not unduly burdened.

TIN

The application of electric furnaces to smelt tin ore concentrates has been known for years and has been practiced in Canada, France, England, West Germany, Portugal, Zaire, and Rwanda.[6,7] Less known is the fact that a smelter for the purpose of tin recovery has been operating in the United States since 1950 at Weirton Steel. This facility has evolved considerably over the years to its present configuration. They currently process tin mill wastes (primarily sludges from Weirton as well as two other mills) as well as tin recovered from an on-site recycling/detinning operation. These tin-rich sludges are processed by hydrometallugical techniques to produce a tin-rich hydroxide which is subsequently, dried and smelted in a 200 kva smelter. The recovered metallic tin is recycled back into the tin plating process, and the drosses created are accordingly disposed.

LEAD

With the large amount of publicity about the long term effects of lead exposure, it is not surprising that both primary and secondary lead producers have been under close scrutiny in recent years. In response, companies such as East Penn Manufacturing in Lyon Station, PA, a secondary lead smelter operator, have invested large amounts of money to safeguard both their employees and the environment. Employee hygiene training, company supplied safety equipment, extensive pollution control systems, on-site sewage treatment facilities, regular ground water testing, and weekly blood tests are all part of their concerted effort. In addition, lead producers are always searching for process improvement and an electric furnace presents one such possibility.

Conventional lead smelting utilizes process equipment such as flash smelters, reverberatory furnaces, and blast furnaces (a.k.a. cupolas) all of which produce large amounts of offgas and hence require extensive fume handling systems. Electric furnaces, on the other hand, create comparatively little fume. For this reason some lead producers have adopted this technology. One domestic producer is RSR Corporation, who from 1983 through 1988 went through a developmental program which included an experimental 1.0 MVA furnace and later a production unit of 2.5 MVA which was later upgraded to 4.0 MVA.[8] This furnace processes the "second run" slag from their reverberatory furnaces. "Reverbs" continue to be the primary means of smelting primary feed stock, which is lead bearing materials obtained after crushing lead-acid batteries. The products of the arc furnace include a bullion containing high antimony, tin and arsenic which is used to produce antimonial lead alloys and a depleted slag which is non-hazardous and disposed of in a landfill. The approximate composition is given in Figure 1.

This is not the first commercial application of an electric furnace to smelt lead bearing waste in the United States. In 1979 AMAX in Boss, Missouri had installed an electric furnace to treat copper, nickel and cobalt-containing lead refinery dross. The electric furnace was chosen because of continued problems with a reverberatory furnace.[9] A good summary of the testing that preceeded the installation of the production furnace is given in the references. Unfortunately, although it is known that the production furnace operated as projected in the reference given herein, no specific operating data is available.

As far as lead processing is concerned, another noteworthy user of electric furnaces is Metallurgie Hoboken-Overpelt.

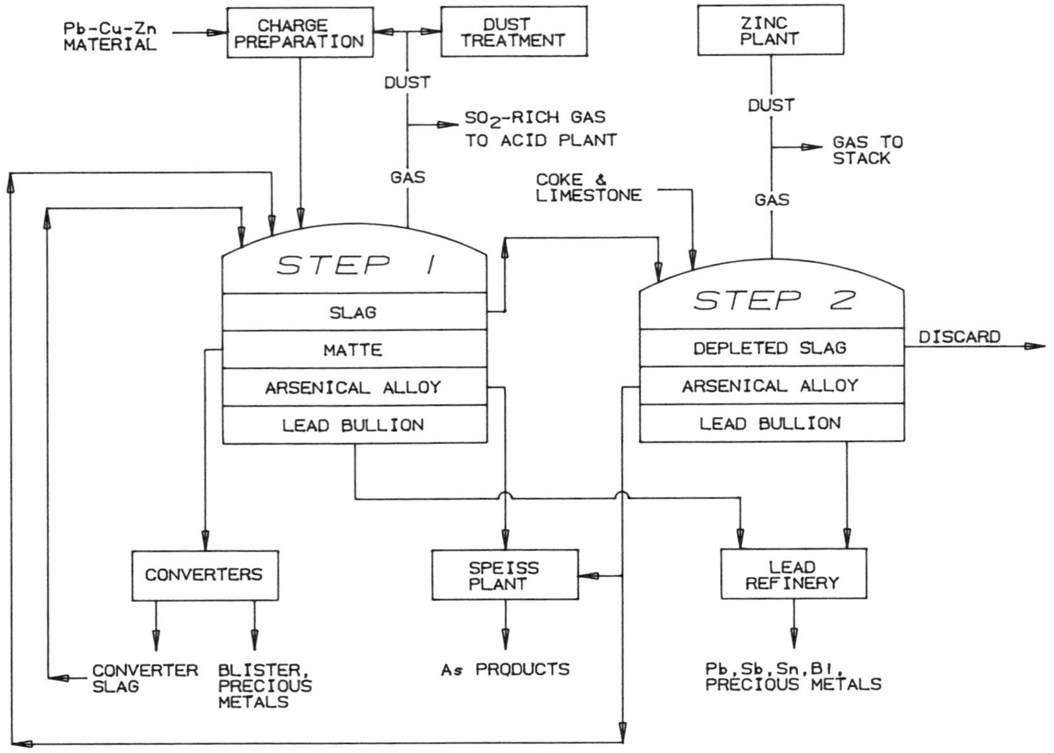

FIGURE 2: FLOW-SHEET OF A TWO-STEP SMELTING PROCESS APPLIED TO COMPLEX Pb-Cu-Zn MATERIALS. (FROM REFERENCE 10)

From a simple lead smelter, whose beginnings can be traced back to the beginning of the century, the facility has grown into a processor of numerous primary and secondary materials. This development is partially due to changes in the primary ores, but it is also due to an increase in the processing of by-products and secondary materials, or what could be considered wastes. Interestingly this development has been such that the majority of the feed is now wastes or recycled material while at the same time the plant has been meeting stricter environmental regulations.

An excellent account of electric furnace pilot testing conducted by Hoboken was given in 1980. This testing was performed on complex Pb-Cu-Zn material wherein a two-step furnacing procedure was performed. The two steps included:

- first, a separation of oxides and sulfides under oxidizing conditions to product lead bullion, a copper matte, an arsenial speiss, and a lead-rich slag.
- Second, a reduction of oxides to produce a bullion, a speiss, a high-zinc dust, and clean slag.

A general flow sheet of this process is given in Figure 2.

This is a prime example in support of earlier statements endorsing multiple-step processing as well as purporting the versatility of the electric furnace. It is best summarized by the authors where they state, "Considering the results of the pilot tests, the combination of the two smelting steps constitutes an attractive process, both economically and technically. The operation of the electric furnaces is stable, control of the operating parameters is easy, and the whole offers a great flexibility with regard to feed materials of varied physical nature and chemical composition. Furthermore, a good control of pollution problems, by means of sealing techniques which have now become usual on commercial electric furnaces, should be emphasized."[10]

ZINC AND STEELMAKING BAGHOUSE DUSTS

There currently exist a number of furnaces, that are being offered commercially which are capable of processing wastes that contain zinc oxide. The primary waste of this type is electric furnace steelmaking dust (K061 waste). This dust has been divided into two classifications, low zinc (less then 15% total zinc) and high zinc (greater than or equal to 15% total zinc) The EPA has dictated that the high zinc dust must be subject to zinc recovery by high temperature resmelting.
Although the mandated recovery of zinc and even the somewhat arbitrary delineation between low and high zinc dusts can be viewed as technical misinterpretations of a number of reports on the subject, the fact is that recovery does slow down the depletion of natural resources and serves to partially offset processing costs. [11,12,13] Nevertheless, the EPA mandate is in affect with (we hope) all steelmakers handling their dust accordingly to preserve our environment.

The thermal processes available includes equipment such as kilns, flame reactors, plasma torches and arc furnaces. The use of arcs is not surprising since the dust was created by arc processing in the first place. Furthermore, whether the arc comes from an electrode or a torch does not appear to make a difference. This relates to an earlier statement alluding to the fact that International Mill Service (IMS) has replaced a plasma torch with a graphite electrode at both their Blytheville, Arkansas and Jackson, Tennessee operations, and the units have been operating well as conventional DC furnaces.[14].

Still others have proposed the use of a more conventional, AC furnace which they call a "Multi-Purpose Electric Furnace". This may be somewhat of a misnomer since a single design cannot cover all applications, however the intent to use an AC furnace for this application (zinc fuming) is clear and certainly feasible. As a matter of fact LECTROMELT Furnaces were installed at Hudson Bay Mining in 1948 to fume off zinc as an oxide. This application is only a variation of this wherein metallic zinc may be collected. The present day viability of this process is presented in a study that shows the costs associated with an AC furnace are much lower than other processes.[15]

SILVER AND PRECIOUS METALS

It is not surprising that one of the world's largest recyclers and processors of silver containing wastes is KODAK, since silver is used extensively in photographic film. However, it may be surprising to learn that KODAK operates an electric arc furnace to recover nearly pure silver for reuse in their film-making process. This furnace has been successfully operating since 1978. Prior to this time a reverberatory furnace was used to melt the silver-containing ash as well as fluxes. Unfortunately the reverb was problematic in that the high thermal conductivity and low specific heat of silver resulted in a continual battle to prevent skulling, but the use of an arc furnace, with its high efficiency and capability to heat the melt directly via resistance, greatly alleviated this problem.

Other electric furnace users that are processing precious metals include the previously mentioned Stillwater Mining and Hoboken, as well as others, particularly Western Platinum LTD. of South Africa.[16] Additionally it is known that in some cases incinerated sewage sludge contains appreciable amount of precious metals. This is an area in which the authors have performed some studies and will be discussed next.

SEWAGE SLUDGE AND MUNICIPAL INCINERATOR FLYASH

The impetus to process municipal wastes in a furnace is obvious when one considers both the detoxification and reduction in volume; particularly when the alternative is secure landfills which do not absolve the producer from liability for this waste.

Although LECTROMELT has had numerous discussions with potential U.S. customers in the past few years, to the author's knowledge there has yet to be a domestic, commercial installation. However, there has been laboratory testing performed by the United States Bureau of Mines (USBM) for the American Society of Mechanical Engineers (ASME) in Rolla, Missouri. Based upon these tests a pilot plant furnace is being built in Albany, Oregon with testing due to begin in October 1991.[17]

In Japan, where the impetus to operate such facilities is greater, data from commercial furnace installations has been published. Furthermore the slags created in these furnaces have been reported to be non-hazardous.[18] The results reported are given in Figure 1. What is very interesting to note is the fact that the dried sewage sludge that is being processed in a furnace contains over 50% organics and water. Thus, as stated earlier, a furnace is capable of handling organics, given the proper fume collection system. Based upon the results already reported and the results of the ASME/USBM testing, it is believed that a greater number of these furnaces may appear in the near future.

OTHER REFERENCES

As mentioned earlier arc furnaces are regularly used to melt refractory oxides such as magnesia and high-purity alumina whose melting points are 2800°C and 2980°C, respectively. The capability to melt and subsequently process such high tempreature materials is a testament to the arc furnace's capability. Similarly, arc furnaces are being used to create oxide melts for spinning into fibers (mineral wool) for use in building products. Because of the low cost nature of these products, the feed materials for these furnaces are typically "wastes" such as blast furnace slag or copper smelter slags, as described elsewhere.[19] Furthermore, although the primary means of producing these oxide melts is in a coke-fired cupola, pressure on these mineral wool producers to reduce their air pollution emissions has been such that electric furnace use is becoming more common.

Another proposed use of an arc furnace is to process wastes from elemental phosphorus production. This is understandable since elemental phosphorus is made in an arc furnace. The general procedure proposed is discussed in U.S. patents 4,919,906 and 4,968,499.

Lastly, in addition to all of the above mentioned applications, arc furnaces are regularly and almost exclusively used to smelt the metal, alloy or synthetic mineral listed below.

- Aluminum Oxide
- Calcium Carbide
- Ferroboron
- Ferrochromium (Silicon Chrome)
- Ferro Columbium
- Ferro Manganese (Silicon Manganese)
- Ferro Molybdenum
- Ferro Silicon
- Ferro Titanium
- Ferro Tungsten
- Mullite
- Pig Iron/Steel
- Silicon Alloys
- Silicon Carbide
- Titanium Oxide
- Zirconia-Alumina
- Zirconium Alloys

CONCLUSION

It is hoped that the preceeding adequately outlines the capability of the AC electric arc furnace. If one considers the variety of feedstocks and the number of applications for which these furnaces are being used, it is fairly obvious that these furnaces exhibit extreme versatility and flexibility such that in nearly all cases the primary waste stream can successfully be processsed given the proper addition of fluxes and/or reductants. Unfortunately this is not to say that the arc furnace is the lowest cost available process. This can only be determined after testing the performance of a study. However in many cases the electric furnace is not only lowest cost, but it is also preferred since the high temperatures developed insure detoxification and/or destruction of the toxins.

In conclusion, although the use of electric furnaces is already extremely diverse, it is believed that their limitations have yet to be reached. One aspect that warrants further attention is sulfide metallurgy. Where presently furnace processing is somewhat restricted to oxide metallurgy, a large number of natural ores occur as sulfur compounds. Since sulfur appears to be nature's collector of many metals we may be able to learn from this example and utilize sulfide metallurgy to help us in solving the environmental problems that we now face.

ACKNOWLEDGEMENTS

The Authors wish to thank our Company for allowing us to dedicate company time to writing this paper, as well as our Secretary Kathy Currie and our CAD Manager, Jim Gormley without whom this paper may not have become a reality.

REFERENCES

1) Morral, F.R., The Twentieth Century Magic Metal-Aluminum, CIM Bulletin, December, 1986.

2) McCoy, J.B., Smelting Practice at Anaconda, Montana, Mining Congress Journal, October 1977.

3) Hodges, G.J. et al, "Slag Cleaning Furnace for Concentration of Platinum Group Metals (PGM's)", Electric Furnace Proceedings Vol. 49, Toronto, 1991.

4) Hanewald, R. and Dombrowski, D.E., "Recovery of Metals from Steelwastes and Production of DRI by the INMETCO Process", I&SE, March 1985.

5) Dick Hanewald, Personnal Communication, INMETCO, 20 August, 1991.

6) Strachan, A.M. and Lonton, I.A., "Electric Smelting of Tin Concentrates at Copper Pass Limited", Lead-Zinc '90, Edited by T.S. Mackey and R.D. Prengaman, TMS, 1990.

7) Internal Communications, "Report on Startup of the Somirwa 500 KVA Tin Smelting Furnaces", LECTROMELT Corporation, February, 1982

8) Eby, D., "Electric Arc Smelting at RSR Corporation", Lead-Zinc '90; Edited by T.S. Mackey and R.D. Prengaman, TMS, 1990.

9) Opie, W.R., Rajcevic H.P. and Major, F.S., "Development of a Submerged Arc Electric Furnace for Treating Lead Refinery Dross", Paper Presented at the Fourth Joint Meeting MMIJ-AIME, Tokyo, 1980.

10) Fontainas, L.M. and Maes, R.H., "A two Step Process for Smelting Complex Pb-Cu-Zn Materials", Lead-Zinc"80, Las Vegas, 1980.

11) MacRae, D.R., "Electric Arc Furnace Dust-Disposal Recycle and Recovery", (Report No. 85-2, Center for Metals Production; Carnegie Mellon University, Pittsburgh, 1985).

12) MacRae, D.R. and Cowx, P.M., "Plasma Furnace Treatment of Electric Arc Furnace Dust as Demonstrated by Bethleham-Tetronics", (Report No 88-2, Center for Metals Production, Carnegie Mellon University, Pittsburgh, 1988).

13) Lherbier, L.W.; "Flame Reactor Process for Electric Arc Furnace Dust", (Report No. 88-1, Center for Metals Production, 1988).

14) David Bunney, Personnal Communication, IMS, 22 August, 1991.

15) Author Unknown, "Electric Arc Furnace Dust Treatment with ELKEM's Multi-Purpose Electric Furnace", Cundall and Associates, Dunwoody, Georgia, October, 1986.

16) Ward, V.C., "Electric Smelting - Some Process Control and Design Aspects: The Smelter, Western Platinum Ltd., Mavikana, Transvaal, South Africa", TMS paper No. A78-28.

17) Internal LECTROMELT Correspondence as contributor to ASME/ISBM Vitrification Project, Albany, Oregon, 1990.

18) Furukawa, T. Inagaki, E. and Shimura, S., "Application of Electric Arc Heating for Melting Treatment of Sewage Sludge and Municipal Incinerator Residue", Presented at the XI Congreso Internacional de Electrothermia, Union Internationale D'Electrothermie, Malaga, Spain, 1988.

19) Cotchen, J.K. and Davis H.F., A Successful Slag Vitrification Arc Furnace", Electric Furnace Proceedings, Vol. 48, New Orleans, 1990.

MELTING AND FLUID PROPERTIES OF SEWAGE SLUDGE

Takashi Fujii*, Takeshi Tsunemi* and Yutaka Shiraishi**

* Osaka Gas Co., Ltd. 19-9, 6-chome, Torishima
Konohana-ku, Osaka, 554 JAPAN

** Tohoku University 1-1, 2-chome, Katahira
Aoba-ku, Sendai, 980, JAPAN

Abstract

In the OG (Osaka Gas) process which includes a fusion process of sewage disposal sludge in coke bed, knowledge of molten and fluid properties of the sludge are most important to controll the process in good condition. Thus, the softening and melting temperatures and viscosity were measured by the sessile drop technique and a rotating viscometer. Obtained results are summarized as follows:
(1) Melting character of sludge slag is varied according to their composition, but closely related to CaO/SiO_2 ratio in ordinary sludge and to $CaO/(Cr_2O_3+Al_2O_3+P_2O_5)$ ratio in Cr_2O_3 containing sludge.
(2) Fluid property of sludge slag has a correlation with optical basicity or Riboud parameter. However, the correlation is poorer than that of metallurgical slags. Especially, chromium oxide containing sludge shows high viscosity and behaves somewhat different manner from ordinary sludge.

Introduction

With the increase in the percent of the population served by the sewerage system, there has been a corresponding increase in the volume of sludge generated from sewage treatment plants. As a result, disposal of sewage sludge (hereinafter referred to as "sludge") has become a grave social problem. Conventionally, the landfill method has been used to dispose of sludge. The reduction of sludge volume by incineration has been employed primarily for maximal use of reclamation sites. The present incineration method faces such problems as the dispersion of incinerated ashes, elution of heavy metals, and the continuing need for landfill sites. At Osaka Gas Co., Ltd., we have developed a coke bed type sewage sludge melting process. In this new process, sludge and a small quantity of coke are fed into a vertical shaft furnace, where combustible matter in the sludge is incinerated on an extremely high temperature coke bed while incombustible matter (ash) is melted to convert a pollution-free and environmentally-safe slag which can be recycled for paving roads or concrete aggregates. The most critical operations including in this process are the melting of non-combustible matter in sludge on the coke bed and the discharging of the resultant matter from the furnace by the adjustment

Residues and Effluents - Processing and Environmental Considerations
Edited by Ramana G. Reddy, William P. Imrie and Paul B. Queneau
The Minerals, Metals & Materials Society, 1991

of the viscosity of the melts. Recycling slag derived from the furnace is one of the main tasks we tackle. This study was aimed to search an appropriate melting and a fluid characteristics of molten sludge, which are essential for stable and economical sludge melting operation, and are also important for generating the good physical properties required for recycling. This paper consists of three parts concerning with 1, melting characteristic, 2, viscosity of molten sludge and 3, crystallization tendency of the sludge.

1. Sludge Melting Characteristics

One of the basic problem on actual operations in the sludge melting process is the melting of non-combustible matter in sludge on the coke bed. In order to facilitate this operation, it is necessary to understand the melting characteristics of the sludge. The idea of lowering the melting point by basicity adjustment, which is used in the steel making industry, has been utilized conventionally in general sludge treatment. This section describes the results of the effects of basicity adjustment, and also reports a new guideline that we have proposed for leather sludge, a new type of sludge that requires disposal. The leather sludge referred to here is sludge generated when treating wastewater containing a high-concentration of chromium oxide discharged from the leather industry. The leather sludge contains an extremely high concentration of Cr_2O_3.

1.1 Experimental Method

We determined melting characteristics in conformance with the ash melting test described among the Methods for Testing Cokes specified by JIS (Japan Industrial Standards), and performed automatic measurements using a Leitz heating microscope that we developed. Changes in sample shape at different temperatures were measured, and three points - softening, melting and pouring points - were measured, as shown in Figure 1.

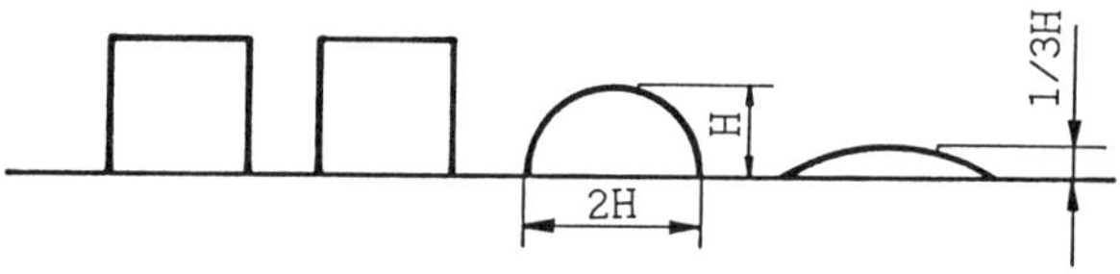

Figure 1 - Deformation of cylinder of sample

1.2 Experimental Results and Discussion

General sludge melting characteristics. The major component of the sludge ashes are CaO and SiO_2, which are considered to affect the fluidity or viscosity of the melt. The ratio of these two components is referred to the basicity as in the steel making industry. In order to determine whether or not this guideline is applicable to sludge, we examined the correlation between basicity and melting characteristics. According to the relationship between basicity and pouring point in Figure 2, the pouring point was the lowest at a basicity being approximately 1.

Leather sludge melting characteristics. As for general sludge, an attempt was made to use basicity as a guideline to determine the leather sludge melting characteristic.

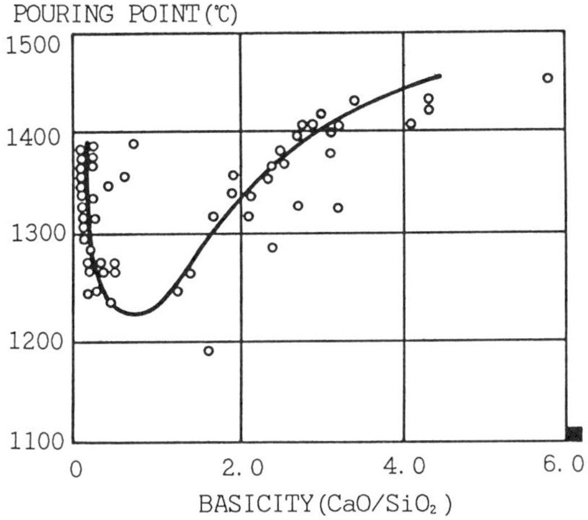

Figure 2 - Relationship between basicity and pouring point

However, it has been found that this method is not applicable to leather sludge, because the ash composition in leather sludge is remarkably different from that in general sludge.
As shown in Table I, leather sludge is characterized by a large amount of chromium oxide content. As a result, basicity is not valid as a guideline in this case, on the contrary in melting of general sludge.

Table I Sludge Ash Composition

Unit:Wt%

Sludge Type	Leather Sludge				General Sludge	
Ash	F	A	B	S	K	C	T
Al_2O_3	3.3	4.2	9.3	3.2	1.3	7.0	13.2
CaO	56.5	41.3	9.4	47.6	45.1	37.8	7.6
Cr_2O_3	3.3	3.8	11.2	11.3	10.7	<0.1	<0.1
Fe_2O_3	5.1	18.7	16.9	3.1	8.4	14.9	18.3
MgO	2.8	1.6	2.1	3.3	4.5	2.4	4.1
P_2O_5	8.8	5.7	13.9	2.3	1.6	6.2	11.3
SiO_2	5.1	12.0	23.0	10.6	5.7	15.9	31.1
Others	15.1	12.7	14.2	18.6	22.7	15.8	14.4

Accordingly, a new guideline was investigated based on a multivariate analysis, which gave a following expression (1).

$$\text{Pouring point}(^\circ C) = 1172.74 + 21.66\times(Cr_2O_3) + 10.70\times(Al_2O_3) + 7.13\times(P_2O_5) + 3.33\times(MgO) - 2.04\times(CaO) - 1.07\times(SiO_2) - 0.60\times(Fe_2O_3) \quad (1)$$

Symbols of compounds (in parentheses) represent the mass percent of the components.

The pouring points determined based on a multivariate analysis have an extremely high correlation with the measured pouring points, as shown in Figure 3. Consequently, it has been found that the melting characteristic can be estimated by identifying ash composition in sludge. Based on the knowledge thus obtained, we studied other guidelines having the same concept as basicity, and consequently obtained next expression (2). Here, the percent of, CaO, a basic oxide, is divided by those of acid oxides Cr_2O_3, Al_2O_3 and P_2O_5. Accordingly, it was found that the melting characteristic can be lowered by increasing the ratio to greater than 1.5.

$$CaO/(Cr_2O_3 + Al_2O_3 + P_2O_5) > 1.5 \qquad (2)$$

CaO, Cr_2O_3, Al_2O_3, and P_2O_5 represent the mass percent of the components.

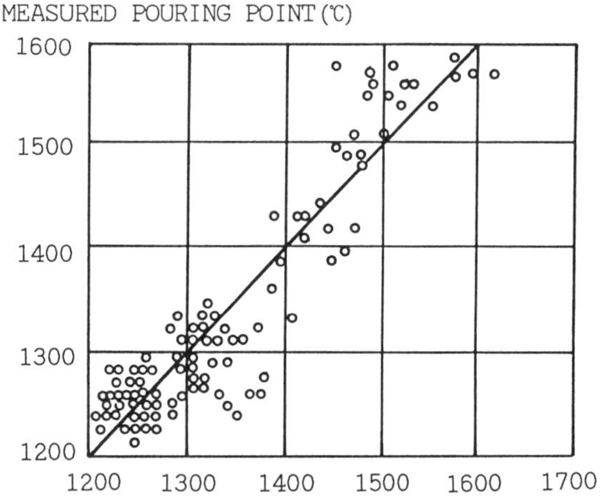

Figure 3 - Relationship between calculated and measured pouring points

2. Sewage Sludge Viscosity

The most critical operation of the sewage sludge melting system involves the melting of non-combustible matter derived from sludge and removing of the resulted melt from the system. Although melting characteristics etc. have been conventionally used as yardsticks for slag discharge operations, the measurement of fluidity is indispensable in determining the melt flow phenomenon. Accordingly, viscosity was measured using the viscometer, Rotovisko employing inner cylinder revolution.

2.1 Experimental Method

The devise used in the Rotovisko is shown in Figure 4. In the Rotovisko, the torque affecting a rotor due to the viscosity of liquid is detected via the twist of a spring, and this is indicated as change in electrical resistance. Viscosity(Pa·s) is given by expression (3).

$$\eta = K \cdot M / N \tag{3}$$

where η = viscosity, M = torque, N = revolution speed, and K = device constant.

Since the coke bed furnace is supposed to have a reductive atmosphere, graphite was used for the crucible and rotor. After the sample was melted down, changes in viscosity were measured by varying the temperature sequentially.

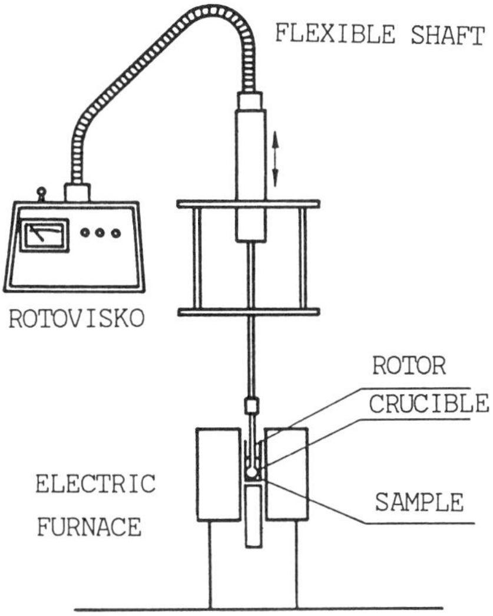

Figure 4 - Experimental device (Rotovisco)

2.2 Experimental Results and Discussion

Results of viscosity measurement. The viscosity of leather sludge, general sludge, and slag was measured. The experimental results can be expressed by the Andrade equation(4).

$$\eta = A\eta \cdot \exp (E\eta/RT) \tag{4}$$

where η = viscosity [Pa·s]
$A\eta$ = frequency coefficient(constant) [Pa·s]
$E\eta$ = viscous fluidity activated energy [kJ/mol]
R = gas constant [kJ/mol-K]
T = absolute temperature [K].

Since the measured data indicated linear relations between log η and 1/T, Arrhenian type equation has been found to apply. As shown in Figure 5, general sludge and leather sludge obviously differ in viscosity characteristics. To obtain the viscosity of 20 dPa·s, at which slag can be discharged, general sludge must be heated to 1,329℃, while leather sludge must be heated much highter temperatures by 50℃ or 100℃. Consequently, it has been found that leather sludge requires high-temperature operation.

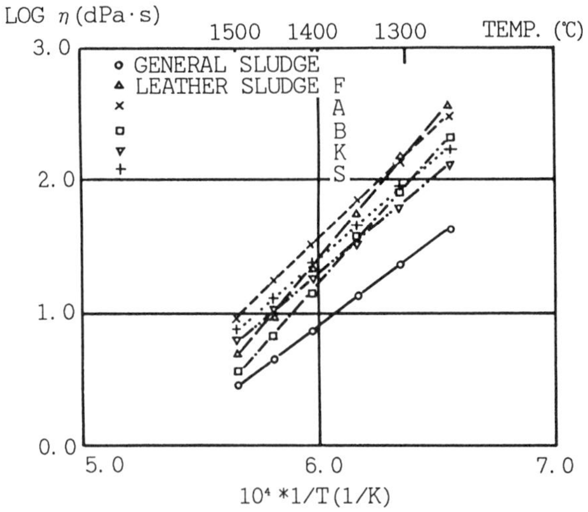

Figure 5 - Viscosity of leather sludge

Influence of chromium oxide on viscosity. Viscosity is generally determined by the chemical composition and temperature. Since leather sludge contains chromium oxide in its ash content, the influence of chromium oxide on viscosity was studied. Viscosity was measured by adding chromium oxide reagents to general sludge ashes. As shown in Figure 6, when 5% and 10% chromium oxide reagents were added to general sludge ashes, which do not include chromium oxide, the viscosity increased and the sludge presented almost the same behavior as that of leather sludge.

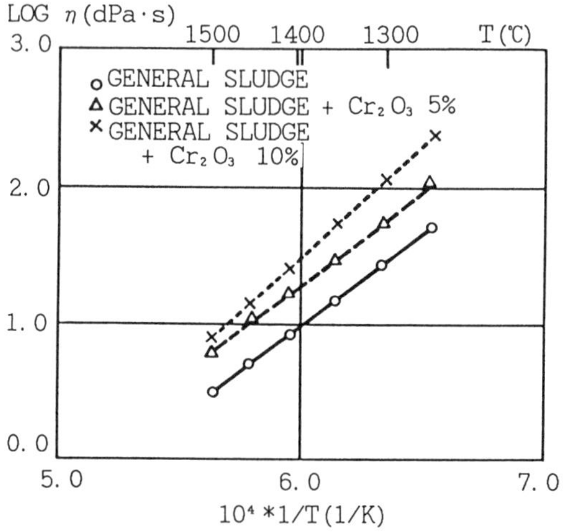

Figure 6 - Effect of chromium oxide addition on viscosity

As shown in Figure 7, viscosity is directly proportional to the chromium oxide content.

Chromium oxide has an extremely high melting point due to its high binding energy and exists as suspended particles even in a multi-component melt. It is therefore considered that the high viscosity is attributable to the presence of non-dissolved chromium oxide particle. Although simple comparisons cannot be made due to the differences in chemical composition, the values for viscosity that we obtained were smaller than those shown in Figure 7. This is because measurements were made in a reducing atmosphere where chromium oxide included in sludge is partially reduced to a metal. As a result, the chromium oxide concentration declines in the course of measurement, and measured viscosity is consequently considered to be lowered. From the above discussion on the effects of chromium oxide on viscosity, it can be concluded that the reducing condition decreases the chromium oxide concentration and then lowers the viscosity of molten sludge.

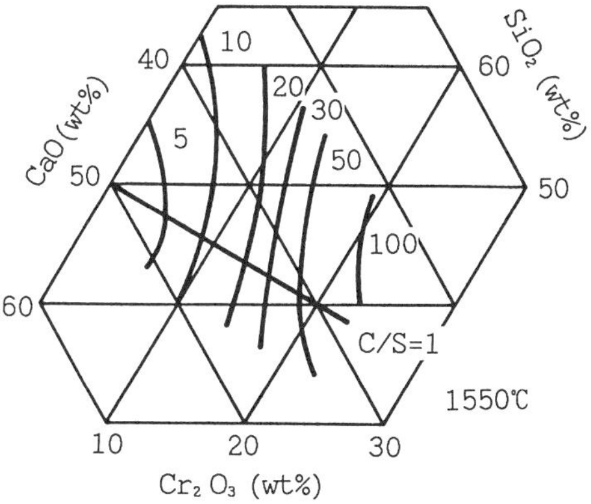

Figure 7 - CaO-SiO_2-Cr_2O_3 iso-viscosity curve at 1550℃

<u>Establishment of viscosity calculation models.</u> Various methods have been attempted for estimating the viscosity of multi-component substances. These methods can be roughly classified into two groups: estimation based on theory and estimation by compiling measured values. This report describes optical basicity as a theoretical estimation method and the Riboud model as a data compilation method.

(I) Optical basicity

Duffy and Ingram (1976) proposed optical basicity, which has recently found numerous advocates in the steel making industry. In this method, viscosity is determined using optical basicity as a guideline. Optical basicity is considered to be applicable to multi-component substances. The optical basicity is given by expressions (5) and (6).

$$\Lambda = \Sigma \Lambda i \cdot Xi \tag{5}$$

$$Xi = (Vi/2) \cdot ni \cdot Ni / \Sigma (Vi/2) \cdot ni \cdot Ni \tag{6}$$

where Λ = optical basicity for multi-component substance,
Xi = equivalent cation fraction
Λi = theoretical optical basicity of component i
Vi = anion charge of component i
ni = number of anion in component i
Ni = mole fraction of component i

These expressions were used to determine the relationship between measured viscosity and the calculated optical basicity of the system. As shown in Figure 8, the relationship tends to show a good correlation.

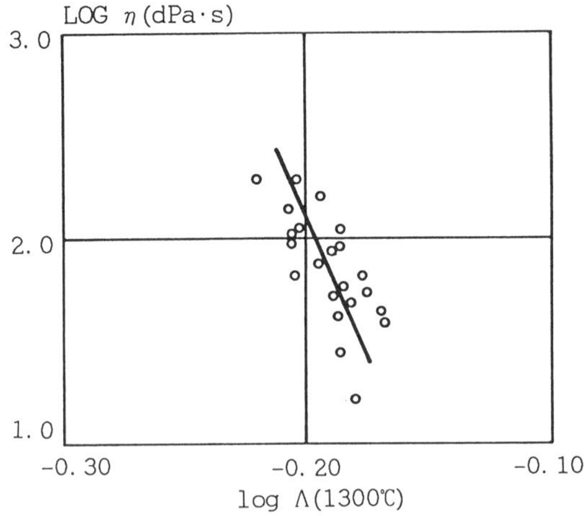

Figure 8 - Relationship between optical basicity and measured viscosity

(II) Riboud model

Several empirical model expressions have been proposed for estimating the slag viscosity. Among these model expressions, the Riboud model was used to examine the measured viscosity values. The Riboud model is given by expressions (7), (8), and (9).

$$\eta = A \cdot T \cdot \exp\ (B/T) \qquad (7)$$

$$A = \exp\{-19.81 + 1.73x(\text{"CaO"}) + 5.82x(CaF_2) + 7.02x(\text{"}Na_2O\text{"}) - 33.76x(\text{"}Al_2O_3\text{"})\} \qquad (8)$$

$$B = 31140 - 23896x(\text{"CaO"}) - 46356x(CaF_2) - 39159x(\text{"}Na_2O\text{"}) + 68833\ x\ (\text{"}Al_2O_3\text{"}) \qquad (9)$$

where X = mole fraction.
η = viscosity [Pa·s].
T = absolute temperature [K].

Using these expressions, the relationship between the measured viscosity values and the viscosity values calculated from the Riboud model was determined. As shown in Figure 9, there was a relatively good correlation as seen in the aforementioned optical basicity.

3. Crystallization of Sewage Sludge Slag

Slag generated by melting sludge can be recycled in various form, depending on the treatment method, e.g. rock-like air-cooled slag and sand-like water-cooled slag and so on. Air-cooled slag can be used as construction material. Since it is imperative to enhance mechanical strength of glass-like air-cooled slag by crystallization, the cooling speed, which plays the major role in generating a crystallized slag, was studied.

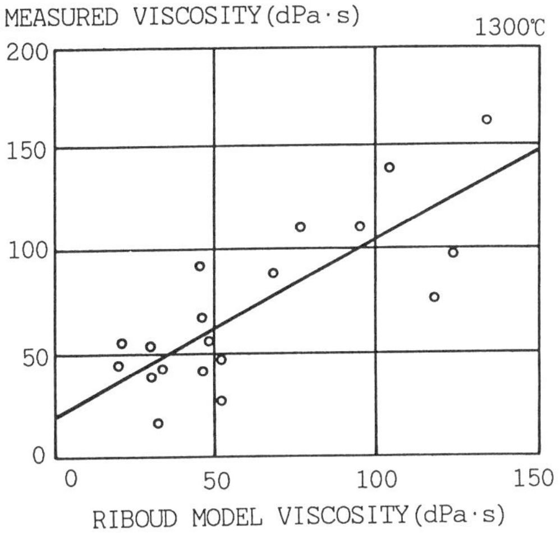

Figure 9 - Viscosity calculate from the Riboud model

3.1 Experimental Method

As a plant test, two vessels ,as shown in Figure 10, were used to control the cooling speeds of the molten slag to measure the crystal and non-crystal portions of molten slag. Based on the data thus obtained, experiments using an electric furnace which produces various cooling patterns were performed in order to determine the quantitative relationship between cooling speed and crystallization. In this analysis, an X-ray diffraction method was used to evaluate a crystallinity.

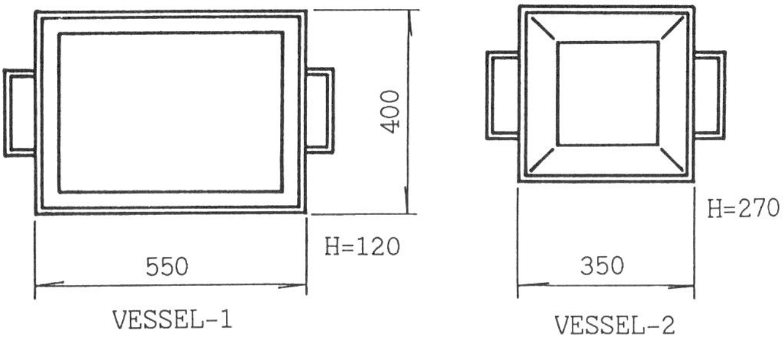

Figure 10 - Vessel for air-cooled slag

3.2 Experimental Results and Discussion

Plant test. As shown in Figure 11, cooling speed was measured at two points on each of the two types of vessels, or at total of four measurement points. Regardless of the type of vessel, crystallization took place at the center, while no crystallization took place on the side wall. The crystallized form found at the center of vessel is referred to as Gehlenite.

Laboratory experiment (electric furnace experiment). Based on the cooling speeds determined at a plant test, crystallinity at 15 different cooling speeds was determined by a use of controlled electric furnace. Degree of crystallization of "Crystal slag" was specified to have more than 40% crystallinity.

(I) Relationship between cooling speed and crystallinity.
The relations between cooling speed and crystallinity at various temperatures after cooling down from 1,400℃ are described as follows: The relations in the range from 900℃ to 1,400℃ showed the best correlations.

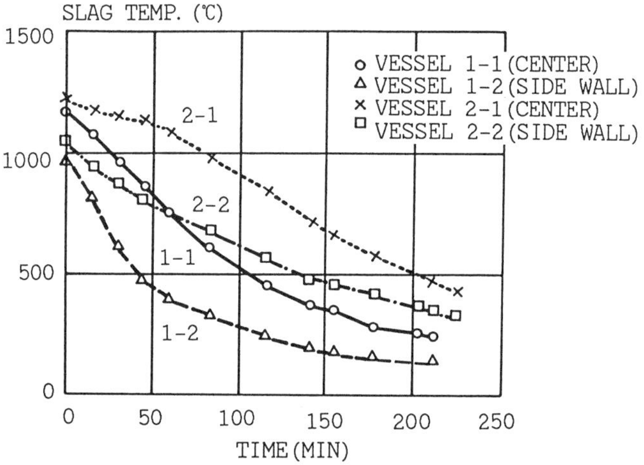

Figure 11 - Slag cooling speed in plant test

Crystal slag was generated by setting the cooling speed at 10℃/min., as shown in Figure 12.

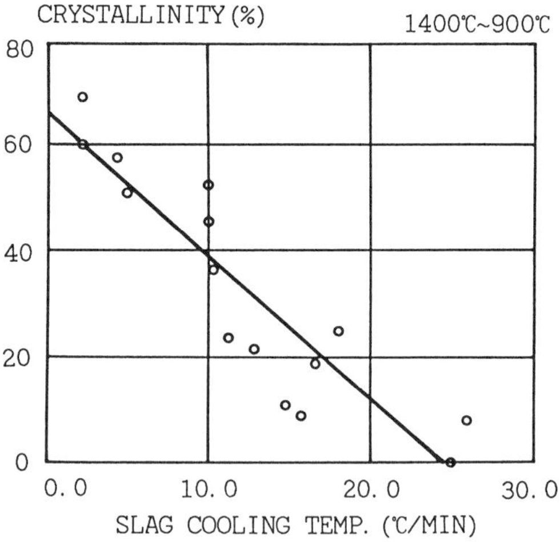

Figure 12 - Relationship between cooling speed and crystallinity

(II) Relationship between cooling speed and crystallinity in each temperature range.

When the relationship between cooling speed and crystallinity in various temperature ranges were compared, the relationship in the range of 900℃ to 1100℃ had the best correlation, as shown in Table II. In this range, the rate of cooling must be controlled at no more than 7℃/min to 10℃/min. When a solid substance is heated continuously , the viscosity decreases drastically at some tempretures, and the substance consequently melts while the temperature remains constant. This temperature is referred to as the melting point, and the viscosity of polymerized substances is approximately 10^1 and 10^2 dPa·s at this tempreture.

Table II Relationship between cooling speed and crystallinity

	1300℃ ~1200℃	1200℃ ~1100℃	1100℃ ~1000℃	1000℃ ~900℃	900℃ ~800℃	800℃ ~700℃	700℃ ~600℃
Correlation Value	-0.673	-0.880	-0.809	-0.819	-0.696	-0.595	-0.672

According to the general sludge viscosity results mentioned in section 2 above, the temperature is approximately 1,300℃ at a viscosity of 10^1 - 10^2 dPa·s. It is well-known in glass engineering that Tc (temperature when crystallization starts) is generally greater than Tg (glass transition temperature). In terms of absolute temperature, Tg ≈ 1/2 Tm ~ 2/3 Tm (Tm = liquid phase temperature); therefore, crystallization takes place at a temperature slightly higher than Tm = 1/2 - 2/3. In other words, crystallization takes place at temperatures slightly higher than 600℃ to 800℃ these results correspond well with the temperatures of crystallization observed in this study.

Conclusion

This study concerns the physical properties of molten slag, which are fundamental in the development of the OG type sewage sludge melting process. The results are summarized below.

1. Leather sludge melting characteristics, which are affected by chromium oxide concentration, cannot be controlled using conventional basicity. We have proposed guidelines to replace basicity and have introduced expressions for estimating melting characteristics.
2. Leather sludge undergoes a temperature rise 50℃ to 100℃ higher than general sludge, even when both sludges have the same viscosity. It has been proven that reduction of chromium oxide is effective in lowering the viscosity , and thus that the coke bed method is superior. It has also been found optical basicity and the Riboud model can be used to estimate viscosity.
3. To crystallize slag, it is necessary to set the cooling rate at 7℃/min to 10℃/min in the temperature range from 1400℃ to 900℃, and especially between 1100℃ and 900℃.

References

1. T. Fujii, T. Kitagawa, "Automatic Measurement Method for Fusibility of Coal and Coke Ash," Coke Circular, 37, 4 (1988) (Japanese)

2. JIS K 2151 Methods for Testing of Cokes (Japanese)

3. JIS M 8801 Methods for Testing of Coals (Japanese)

4. K. Saito, K. Saeki, "Viscosity of Quaternary System of $CaO-SiO_2-Cr_2O_3-FeO$ and the Effect of the Addition of CaF and TiO_2," Japan Soc. Promotion Science. 19th Committee, Report No, 7906, 1965. (Japanese)

5. Makoto Kato , Measurement of Slag and Metal Physical Properties at High Temperatures (Konpass Publishing Co., Ltd. , 1987), 29 (Japanese)

6. Handbook of Physical Properties of Molten Iron and Slag (The Iron and Steel Institute of Japan, 1972), 33 (Japanese)

7. T. Nakamura et al., "Critical review of optical basicity on metallurgical application," 3rd International Coference on Molten Slag and Fluxes 146 (1986)

8. Riboud et al., Fachber. Huttenprax. Metallweiterverarb., 19, 859 (1981)

9. Sumio Sakka , Science of Glass Non-crystalline (Uchida Rokakudo Publishing Co., Ltd. , 1983), 14 (Japanese)

MF PROCESS FOR RECYCLING MATERIALS TREATMENT

Yasuyuki Murayama

Miike Smelting Co., Ltd.,
2-1, Shinkai-Cyo, Omuta-City,
Fukuoka, Japan, 836

Abstract

MF is the semi-blast furnace which has been developed at Mitsui Miike Smelter in the 1960's to treat vertical retort residue. The MF has also been tested for treatment of various recycling materials and wastes. Now various secondaries and wastes (steel dust, zinc leaching residue, Cu-Ag sludge, spent catalyst and spent rubber, etc.) are mainly treated. Powder materials are briquetted with reductant before being fed to the furnace. Products are crude zinc oxide, matte, non-hazardous slag and steam. Zinc and lead are recovered in oxide dust, and copper and silver are recovered in the matte. The MF furnace can be widely applied to many kinds of materials which contain such non-ferrous metal-valuables. In addition, the improvement in operation and technology has effectively made the unit capacity much larger. The MF furnace now has many advantages for these treatment processes.

Residues and Effluents - Processing and Environmental Considerations
Edited by Ramana G. Reddy, William P. Imrie and Paul B. Queneau
The Minerals, Metals & Materials Society, 1991

Introduction

According to our company's reserch, the amount of EAF dust produced in Japan is approximately 450,000 tons per year. There are six recycling treatment companies of EAF dust in our country, including Miike Smelting Co., which are treating 300,000 tons per year. Figure 1 shows the producing conditions of EAF dust and the location of such recycling treatment companies.

Miike Smelting Co. is recycling 90,000 tons of waste materials per year, mostly EAF dust, by its own technology using the half shaft blast furnace called MF Process.

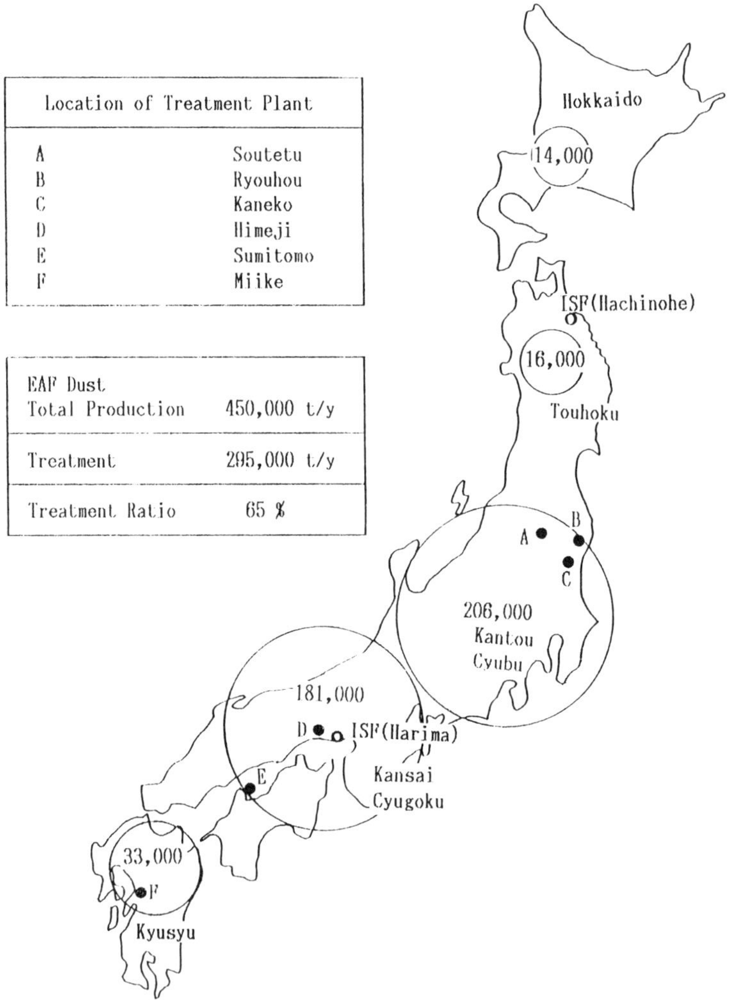

Location of Treatment Plant	
A	Soutetu
B	Ryouhou
C	Kaneko
D	Himeji
E	Sumitomo
F	Miike

EAF Dust Total Production	450,000 t/y
Treatment	295,000 t/y
Treatment Ratio	65 %

Figure 1 EAF Dust Production and Recycling Treatment in Japan

Outline of MF Process

History of Residue Treatment Process at Miike Smelter

Since it started zinc smelting using horizontal retort furnaces in 1913, the Miike Smelter of Mitsui Mining and Smelting Company has equipped electrolysis plants and vertical retort furnaces in turn and produced 108,000 tons of zinc per year at its peak. The residue produced by these processes was partly treated in fuming furnaces; however, for the purpose of treating increasing residue and former piled-up residue, MF Process was developed, and the first MF furnace started operations for practical use in 1965.

During the 20 years from 1965 to 1985, the residue treated by MF Process reached 2,100,000 tons, and the treatment of almost all piled-up residue was completed.

After the production of zinc metal was stopped at Miike by the down-sizing of its zinc smelting department in 1985, it was decided to treat the recycle material outside our company, only small amount of which had been treated there before. At present, only one MF furnace is treating 90,000 tons of waste material per year.

Table I shows the history of zinc production process and residue treatment process.

Table I Process Changing in MIIKE SMELTER

Year	Zinc Production Process	Residue Treatment Process
1913	Horizontal Retort Process start	
1917	Sulfide Acid Plant start	
1936	Electrolysis Plant start	
1952		Fuming Furnace start
1954	Vertical Retort Process start	
1965		No.1 MF Furnace(80t/d) start
1966	Horizontal Retort Process stop	No.2 MF Furnace(80t/d) start
1969		No.3 MF Furnace(120t/d)start
1970	Vertical Retort 32 Furnaces complete	
1973		No.4 MF Furnace(150t/d)start
1974		No.1,2 MF stop No.5 MF Furnace(200t/d)start
1975	Electrolysis Plant stop	No.6 MF Furnace(200t/d)start
1976		Fuming Furnace stop
1979		No.5 MF stop
1982	Vertical Retort from 32 Furnaces to 12 Furnaces	No.3 MF stop
1985	Vertical Retort Process stop	No.4 MF stop

Features of MF Process

Today MF is treating mostly EAF dust, but it is also treating the zinc leaching residue produced by other smelters, and copper and silver recycling materials. Table II shows the typical construction of MF materials.

Table II MF Feed Material (t/year)

EAF Dust	Zn Leaching Residue	Cu & Ag Residue	Vertical R. Residue	Total
55,000	20,000	10,000	5,000	90,000

In Figure 2 is shown the flowsheet of MF Process. In MF Process, lump materials are fed straight to thr MF furnace, and powder materials after being briquette.

In the briquetting process, such materials as EAF dust and leaching residue, silica as flux and coal are constantly fed and transmitted to the dryer, where sulfite liquor as binder is added, blended with the materials, and the moisture is regulated. This mixture, after being milled and mixed in the rod mill, is briquetted and constantly fed to the furnace.

In the MF furnace, zinc and lead oxide dust, matte and slag are produced. This slag, stable and non-toxic, is sold as the material for cement. The oxide dust, carried out from the furnace with the exhaust gas, is collected in the boiler, cooler and the bag. The exhaust gas, after the dust has been collected, contains approximately 400ppm of SOx, which is decreased by desulfurization into less than 5ppm and emitted into the air. The oxide dust, after being washed, filtered and dried in order to remove chlorine and fluorine, is transmitted to ISF at Hachinohe Smelter Co. as crude zinc oxide.

From the MF furnace slag and matte are taken out constantly; the matte, separated and recovered in the settler by its specific gravity, is sold as the material for copper.

Figure 3 shows the conceptual figure of MF furnace.

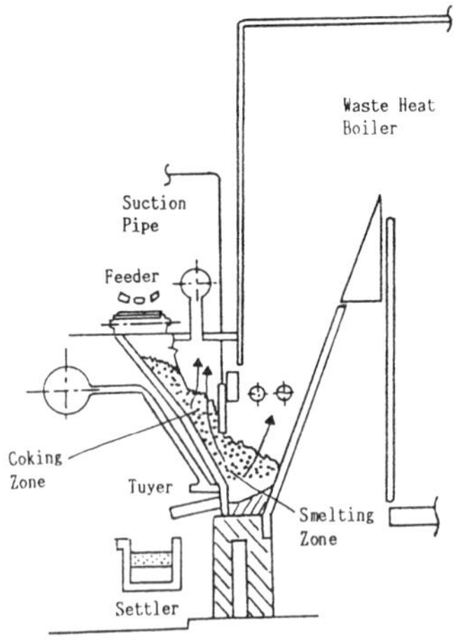

Figure 3 Conceptual Figure of MF Furnace

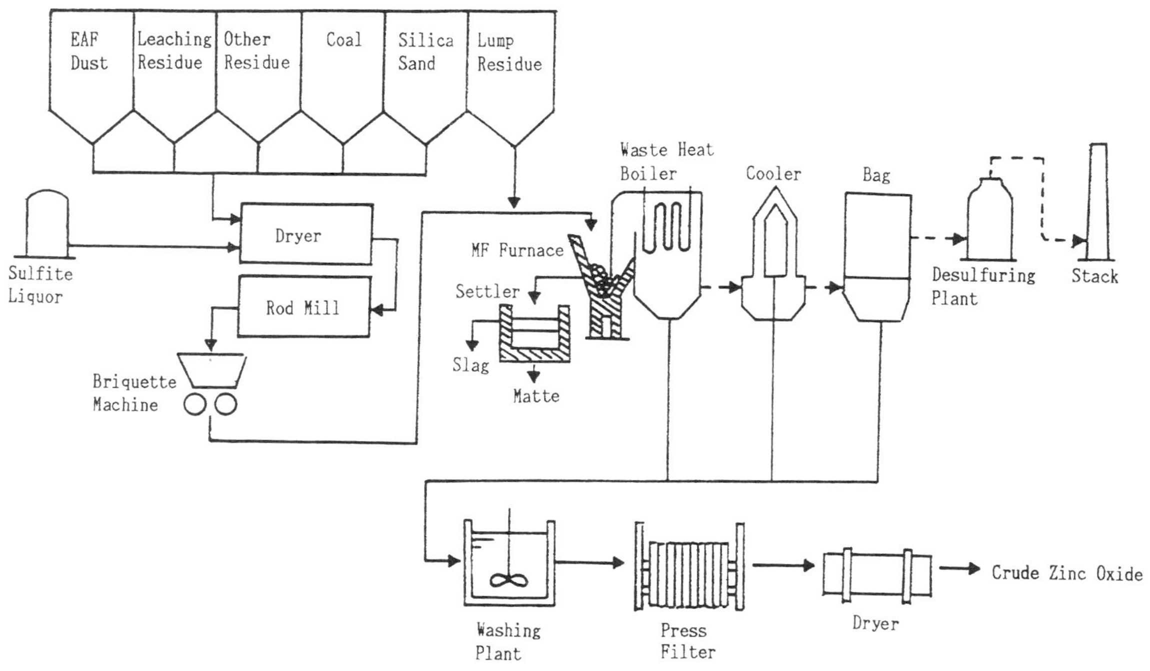

Figure 2 Flowsheet of MF Process

The briquettes fed from the feeding shoot, after being dried and preheated in the coking zone, enter the smelting zone. Zinc and lead are deoxidized there, and after reoxidization turned into oxide dust; copper and precious metals are concentrated in the matte. It is an impotant feature of MF furnace that these metals can be recovered in it at the same time.

Fed into the furnace, 85-90% of the zinc, 90-95% of the lead and 15% of the silver are recovered as oxide dust, and 85% of the copper and 70% of the silver as matte.

Table III shows the typical grade of each product.

In addition, the furnace and the waste heat boiler are constructed in one unit so that efficient recovery of heat is possible.

Table III Typical Assay of Product

	Zn %	Pb %	Cu %	Ag g/t	Fe %
Crude Zinc Oxide	56	11	0.1	120	0.8
Matte	2	0.3	55	1,900	20
Slag	3.5	0.3	0.5	30	35

Technological Progress in MF Process

Changes in MF Materials and the Steps Taken

As mentioned before, MF Process was developed for the purpose of treating the residue in a vertical retort. Later on, by the technological improvement of the vertical retort, the zinc grade contained in the residue was gradually lowered, and the trouble frequently occured, resulting from the generation of a large amount of furnace iron produced by the excessive carbon caused when only the residue was fed. For the purpose of coping with this trouble and using effectively the excessive carbon, zinc leaching residue or some other residue containing zinc and the powder part of the said residue were mixed and briquetted, and then this briquette was treated together with the briquetted residue; thus proper Zn/C grade was maintained.

When the vertical retort process was stopped in 1985, the rate of treating the waste materials such as EAF dust outside our company increased, which previously had been treated in a small amount; thus kinds of MF materials greatly changed and such a change caused a problem. As the rate of feeding EAF dust increased, MF boiler tubes were increasingly corroded by the chlorine contained in it. For this reason, the step was taken to wash the EAF dust in a pellet form by a sprinkler and dechlorinate it in a strage bin and then feed it to the MF furnace. Figure 4 shows the conceptual figure of dechlorination of EAF dust, and table IV, its effect. Figure 5 shows the efect of temperature and time at laboratory test.

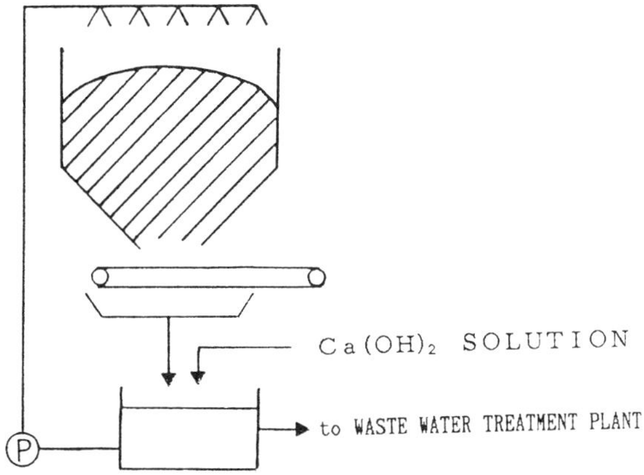

Figure 4 Conceptual Figure of Dechlonation of EAF Dust

Table IV Chroline Removal Treatment of EAF Dust

	Cl %	Zn %	Moisture %
Before Treatment	4.5	20	<1
After Treatment	<0.5	20	15

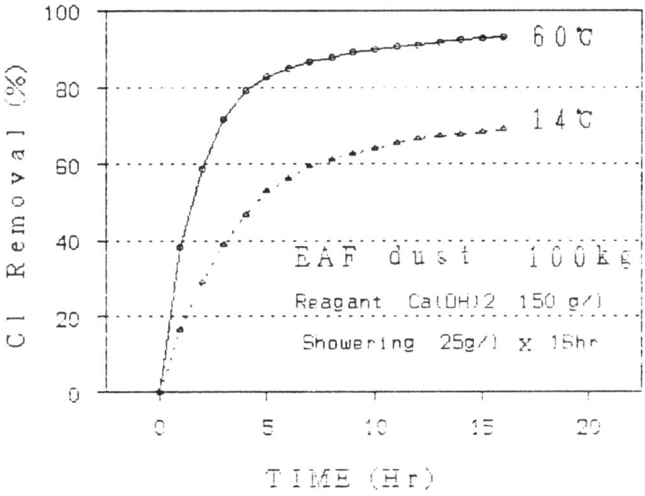

Figure 5 Effect of Time and Temperature on Dechlorination

In addition, the increase of iron by EAF dust made the Fe/SiO2 ratio of the slag higher and raised its melting point and gravity; as a result, the separation of slag and matte became difficult. For this reason, silica sand was added to the materials as flux to maitain the proper composition of slag. Again, copper and silver residue was increased to make copper and silver grade higher, which will be described later. Table V shows the changes of the MF materials' feeding ratio.

Table V MF Row Materials

	Composition (% , g/t)						Feed Ratio (%)				
	Zn	Pb	Ag	Cu	Fe	SiO2	'80	'82	'88	'89	'90
EAF Dust	20	4	80	0.2	30	4	20.1	33.5	49.8	46.6	52.6
Leaching Residue	18	4	200	0.8	30	7	49.0	18.4	10.8	15.2	14.8
Vertical R. Residue	6	5	100	1	19	14	25.5	39.4	16.4	14.9	0.6
Cu & Ag Residue	17	13	230	17	5	3	0.2	0.9	5.1	5.4	6.3
Others (*1)	-	-	-	-	-	-	5.2	7.8	7.4	6.8	12.2
Silica Sand	-	-	-	-	-	95	0	0	10.5	11.1	13.4
FeO/SiO2 Ratio in Slag							2.15	1.61	1.61	1.61	1.62

(*1) Bentnite was used as Flux in 1982.

Reinforcement and Improvement of the Boiler

As chlorine load in MF increased, the corrosion of the boiler tubes worsened, especially that of the Super Heater part exposed to high temperatures. Accordingly, the first improvement was introduced to decrease the heat load of the S.H. part by shifting the front half of the S.H. into a generating tube, decreasing its heating area and lowering the steam temperature, and at the same time to increase the amount of evaporation. By this improvement the amount of evaporation increased from 14.6t/hr to 16.3t/hr.

Figure 6 shows the structure of MF boiler.

The second improvement was to try spray coating nickel chrome alloy on the bending part of the lower part of the tube, as the damage of the S.H.concentrated on that part. As a result, little damage occurred to the spray coated part.

As the amount treated by MF reached 60,000t/y and the evaporating amount of the boiler drew near 16.3t/hr, its maximum power, the evaporating power of the boiler had to be controlled by controlling the amount of the blast air so as not to exceed its maximum power; this fact interfered with the increase of treatment by MF Process. So in 1989 the third improvement was introduced to reinforce the maximum evaporating power of the boiler into 23.3t/hr.

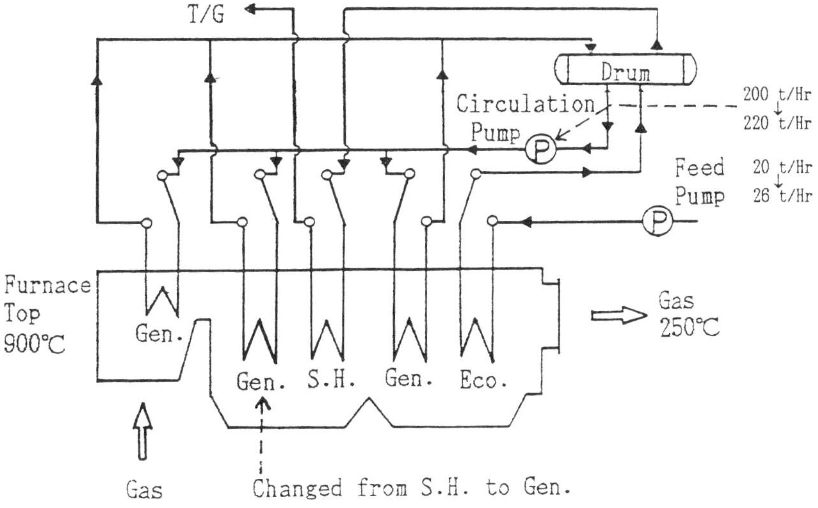

Figure 6 Structure of MF Boiler

Waste tires and waste plastic can be fed straight into MF furnace, and the combustion heat of these materials is also recovered as steam. Table VI shows the heat balance of MF furnace. Approximately 60% of the amount of heat fed into the furnace is recovered as steam, which is used for private power generation.

Table VI MF Heat Balance

Inlet			Outlet		
Object	1,000 kcal/h	Ratio	Object	1,000 kcal/h	Ratio
Coal	15,500	71	Steam	12,800	59
Air	1,800	8	Slag	1,800	8
Additional Waste (Waste Fuel)	4,500	21	Matte	50	0
			Dust	50	0
			Waste Gas	2,600	12
			Cooling Water	2,300	11
			Radiation Loss	2,200	10
Total	21,800	100	Total	21,800	100

Production of High Grade Matte

When the operation started, MF matte was in such a low grade as to contain Cu 3-4% and Ag 500-600g/t, so silver and copper residue was increased to obtain its higher grade. As Figure 7 shows, the copper grade in MF matte depends greatly on the Cu/S ratio in the materials.

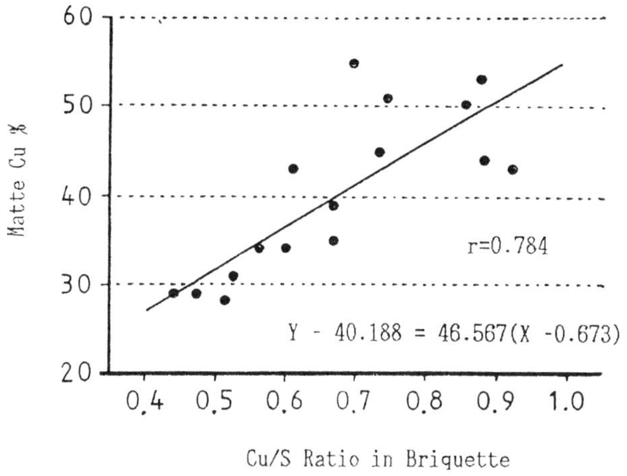

Figure 7 Matte Cu% v.s. Cu/S Ratio in Briquette

In addition, to make the separation of slag and matte easier, the forehearth of MF was altered from a former ladle of slanting type into a large-sized settler in a matte tapping method to make their separation more effective and decrease the loss of Cu snd Ag into the slag. Figure 8 shows the improved effect by the introduction of the settler.

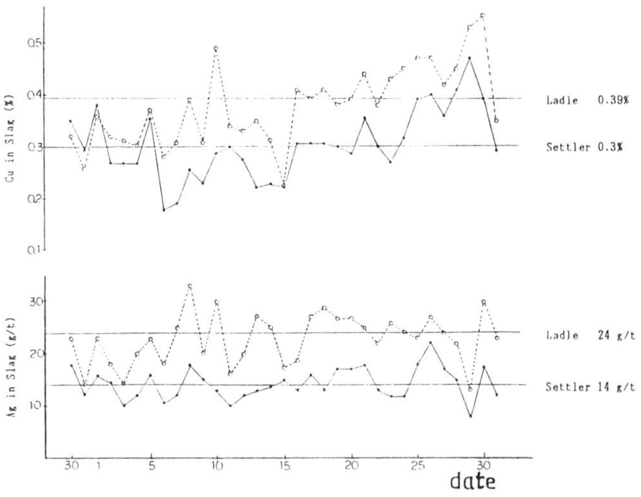

Figure 8 Effect of Introduction of Settler

By the lowering of S grade resulting from the decrease of the feeding ratio of vertical retort residue or zinc leaching residue and by the increase of the feeding amount of copper and silver residue, matte containing Cu 50-60% was produced in 1990. Figure 9 shows the changes of the copper grade in MF matte.

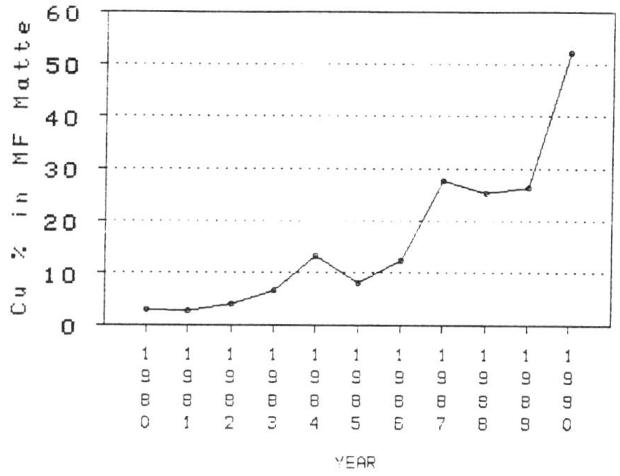

Figure 9 Changes of Cu% in MF Matte

Contrary to the rise of the copper grade, the S grade in the matte was lowered to 5-15%, its properties changing as well. Figure 10 shows the comparison of the former matte and the latest matte which is higher in copper grade based on the results analysed in their forms by EPMA; in the former matte Cu exists as sulfide, while in the higher grade matte most of Cu or Fe is metalized. For this reason, as a result of more uncrushable matte being produced, there are some problems arising from its treatment at Cu smelters and the measures to be taken are under examination.

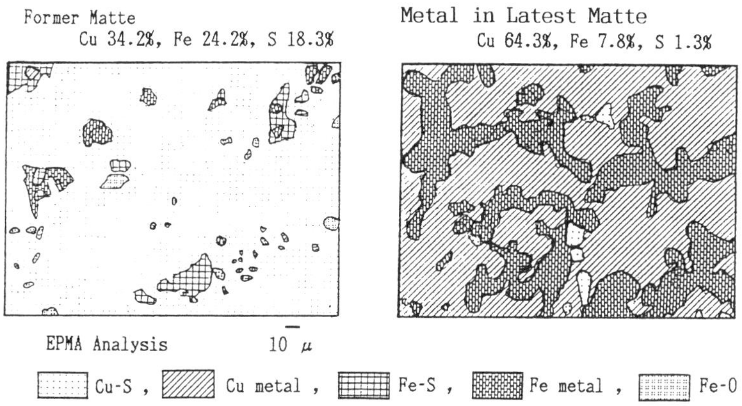

Figure 10 Comparison of the Former Matte and the Latest Matte

Increase of MF Treatment

Since 1985, only one MF furnace has been in operation, and improvements have been made to increase its treating capacity.

Our briquetting plant, which previously was feeding briquettes to four MF furnaces, is supposed to have enough producing capacity to cope with the increase of MF treatment. So improvements were made to feed at a stable rate homogeneous and strong briquettes to MF furnace. In 1989 the computer system was renewed, and the monitoring and controlling system of the plant was strengthened. As a result, MF materials of no less than 11 kinds of waste materials, coal and silica were fed steadily from the bin at a fixed rate; moisturing control was improved; the composition and moisture of briquettes became stable; thus the production of homogeneous and strong briquettes was realized.

In MF Process the improvement of the blast air was introduced to raise its treating capacity. The strength of briquettes and the decrease of draft resistance in briquette layers in the furnace made an increase in the amount of blast air possible. By equipping a secondary air fan and using the fan only for the primary air, which was previously used both for the primary and secondary air, the amount of blast air was increased. At present, another primary air fan was equipped and the former blast power of 270Nm3/hr was raised to 320Nm3/hr. In addition, by reinforcing the preheater, the blast temperature was raised from 280 °C to 320 °C. By these improvements, the treating capacity of MF Process increased markedly, and its treating amount reached 90,000t/y. Figure 11 shows the increase of the treating amount of MF Process.

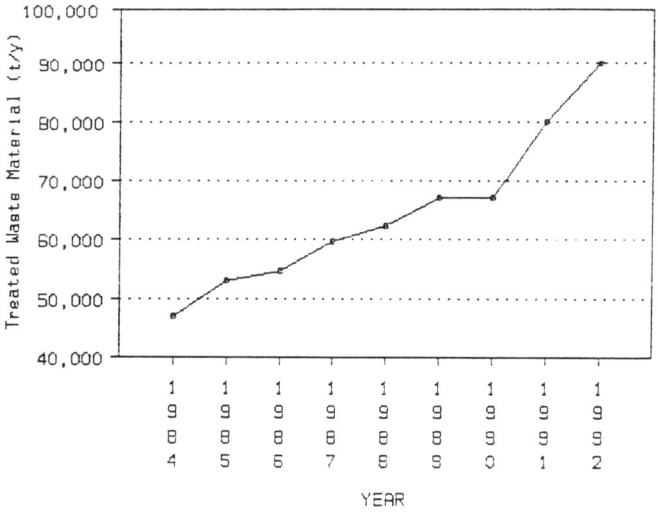

Figure 11 Increase of the Treating Amount of MF Process

Two Types of Operation Used to Produce Two Kinds of Crude Zinc Oxide

Crude zinc oxide produced by MF Process is used as an oxide material for ISF at Hachinohe Smelting Co., while at hot briquette plants the material lower in S is desired. Accordingly, MF has started two types of operation since 1990; one is the operation using for a limited period only zinc leaching residue for materials to produce high S(4-6%) oxide suitable for S-DL; the other is the operation using mostly EAF dust for the remaining periods to produce low S(less than 1.5%) oxide suitable for hot briquette plants. Table VII shows the comparison of each crude zinc oxide.

Table VII Assay of MF Crude Zinc Oxide

Raw Materials	Zn %	Pb %	Ag g/t	S %	Cl %	F %
EAF Dust & Others	56	11	120	1.2	0.5	0.5
Leaching R. only	46	20	250	5.8	<0.1	<0.1

Summary

As environmental problems continue to grow more serious, MF Process, by its original technology, will contribute greatly to the recycling treatment of waste materials, especially those containing nonferrous metals. Our company earnestly hopes to develop MF Process into a more efficient and more economical process.

A WASTE TREATMENT FACILITY IN CORPUS CHRISTI, TEXAS - A CASE STUDY

Dan Mancuso

Encycle/Texas, Inc.
5500 Up River Road
Corpus Christi, Texas 78407-1322

Abstract

This case study is an example of how Encyle/Texas, Inc. (a subsidiary of ASARCO) responded to various socio-economic forces by converting an idle zinc hydrometallurgical refinery into a metals recycling facility. The facility accepts metal-laden process waste streams and produces various "synthetic ores" which are sold back to industry. Acceptance and reactions by industrial, political, economical and regulatory interests towards the project are also discussed.

Residues and Effluents - Processing and Environmental Considerations
Edited by Ramana G. Reddy, William P. Imrie and Paul B. Queneau
The Minerals, Metals & Materials Society, 1991

IRON AND ZINC CYCLES

IN STEELMAKING INDUSTRIES

J.DEFAYS
COCKERILL SAMBRE S.A.
BELGIUM

Introduction

COCKERILL SAMBRE is a steel company producing 4.4 millions tons yearly. This company has a leading role in the production of zinc coated sheets since a long time and more than ever nowadays : our yearly consumption of zinc amounts to 50 thousand tons. This is why we are concerned and deeply involved in the topic of recycling zinc reverts.

Quick evolutions are under way :
. The development of the market of coated steel sheets.
. The growing recognition of environmental concern.

We can illustrate the first point by a picture (1) published by THYSSEN (fig. 1). Though pure zinc coating is not as dominant as in the past, globally zinc and zinc alloys are used more and more for steel coating.

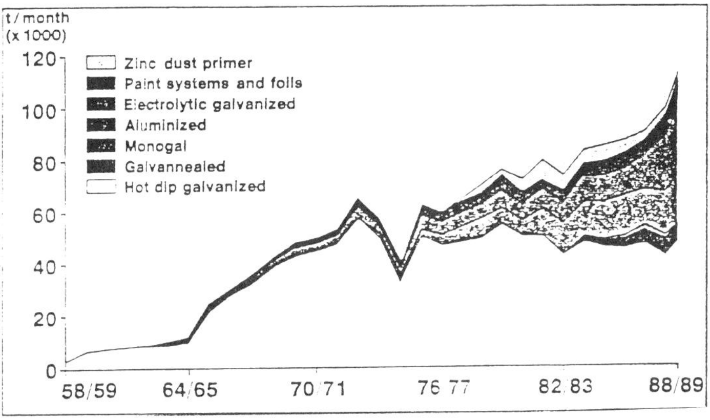

Fif. 1 - Evolution of the production of zinc coated steel sheets by THYSSEN STAHL (Steel Times International - March 1990)

There are many possible illustrations of the new "green mentality", I choose to show you 2) that the steel industry is truly sharing the green ideas through the increased recycling of its products (fig. 2).

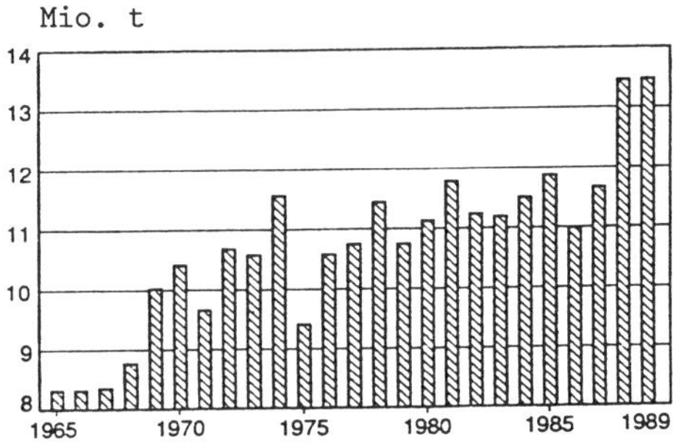

Fig. 2 - Remelting of commercial indigenous steel scraps in Germany

Life of a steel based good

Now, let us examine the life of a steel based consumption good such as a car, a washing machine or a steel shed (fig. 3).

LIFE OF STEEL GOODS

. **ORE MINING**
. **STEEL MAKING**
. **STEEL COATING**
. **STEEL FORMING**)
. **COMPONENTS MAKING**) **MANUFACTURING**
. **ASSEMBLING**)
. **DISTRIBUTING**
. **USING**
. **DISUSING**
. **SCRAPPING**

The story begins with mining of the iron ore which will be processed in the steelmaking industry along hot and cold processes and coating lines to produce the commercial grades of steel : wires, sections, plates and sheets.

The forming industry then creates individual functional steel pieces by forming these wires or sheets. Coating may be delayed until this stage.

In the manufacturing industries, association with a lot of different materials is practiced to a large extent first by the component makers and then on the final assembly lines which deliver goods ready to use.

These will be commercialized by the distributors.

The baby is now 10 to 20, sometimes 24 months old. It will be used for 1 to 30 years and then retired, be left in the corner of the garage for sometime. One day, it will finally be scrapped. To day, we begin to consider scrapping differently : it is not a final death, but the begining of a new life : scrap collection and scrap processing industries are at last recognized as clean and sensible businesses.

Iron and zinc balances

I have tried to measure the present situation of iron and zinc in the steel industries by computing the iron and zinc balances from :

. Well known world production datas
. Generally accepted yields for industrial processes
. Some personal guesses based on litterature datas

This work is not depicting a particular company anywhere in the world.
To produce our reference of 1000 t liquid steel, we need (fig. 4 and 5) 1052 t of iron which we find in 1237 t of ore, 400 t of scrap, 43 t of oxide returns and 7 t of ferro-alloys.

Steelmaking processes share between 70 % Blast Furnace + Basic Oxygen Furnace and 30 % Electric Arc Furnace steelmaking.

In the steelmaking processes, the production of non metallic wastes amounts to 480 t of which 430 are recycled or used in other applications and 45 are dumped.

The production of metallic scraps and chutes amounts to 155 t which are totally recycled. Note that 20 years ago this was already true but with a production above 300 t. Better yields at every stage and the conversion from ingot making to continuous casting enabled this big reduction in the "home scrap" production.

To simplify, I have considered that metal chutes from coating lines, forming and manufacturing industries are not directly recycled in the steelmaking industry but through the scrap processors and the scrap market. This should be kept in mind when speaking about home and purchased scraps.

To evaluate scrap processing, I found some detailed datas for Germany mainly (3). The fig.5 gives the iron balance : each data here is the tonnage of iron contained in the amount of material given on the previous figure.

It is interesting to point out that out of 715 kg of iron in goods, only 110 kg are collected nowadays. Either we stock and increase our "iron capital" or the loss is very important. I believe the first assumption is to day the main one. Anyway, collection will increase in the future due to the environnemental concern.

Now, let us look where iron and zinc are meeting themselves (fig. 6 and 7). Zinc is present at relatively low concentration in iron ore and coke, but we shall see that due to the big tonnage of material, the zinc introduction is important in the steelmaking processes.

Major sources of zinc are the coatings and the zinc alloys used as associated materials. Think about Al-Zn alloys in injected pieces, Cu-Zn in messing ...

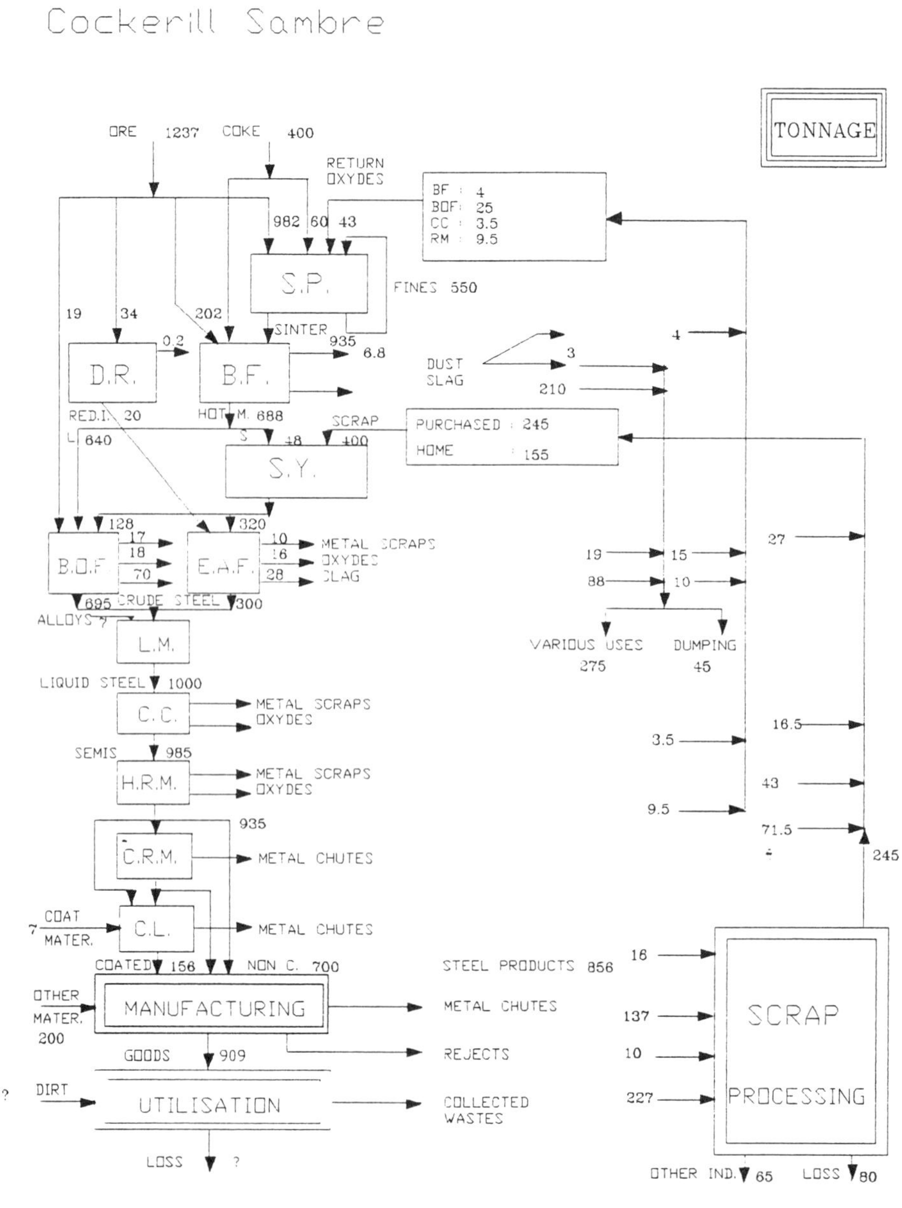

Fig 4 – Main materials flow in the steel industry

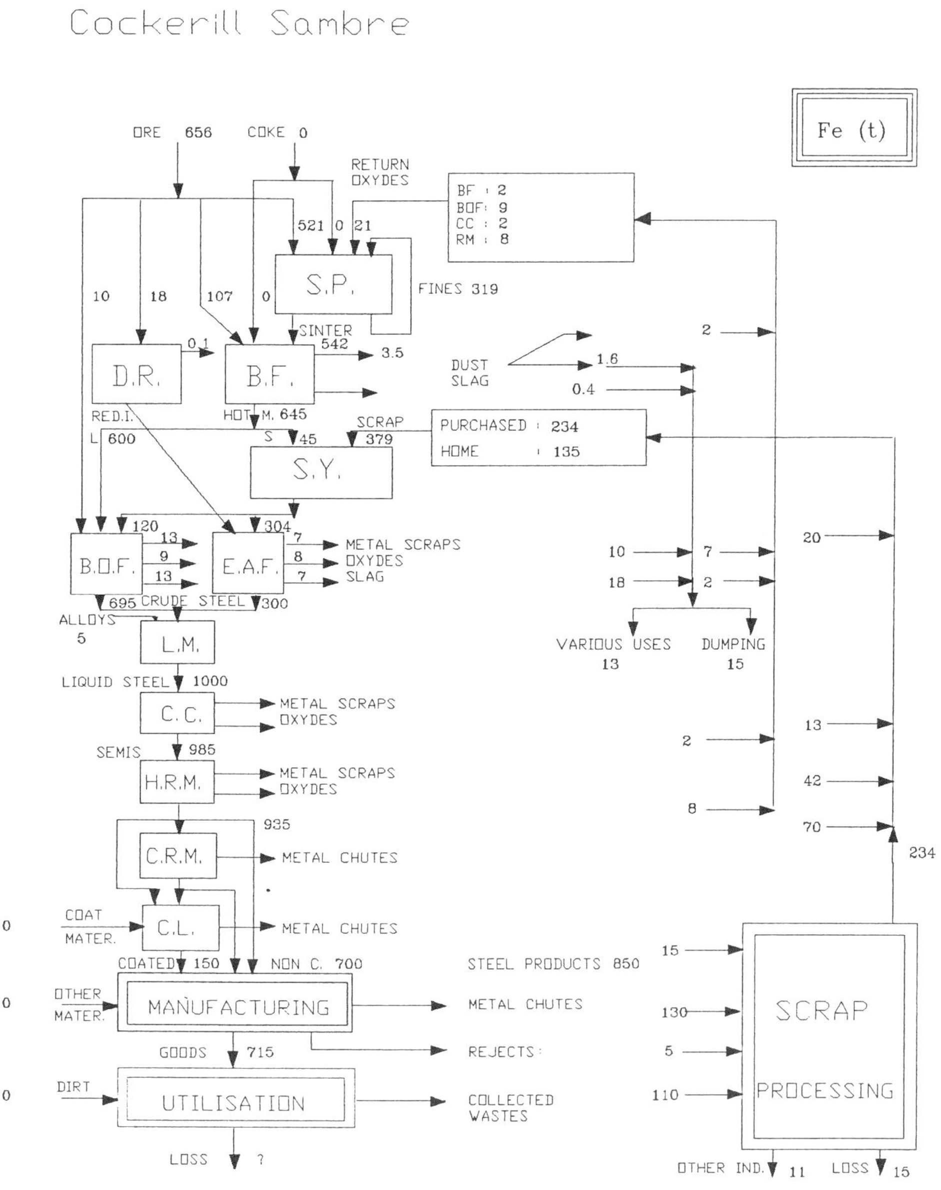

Fig 5 – Iron cycle in the steel industry

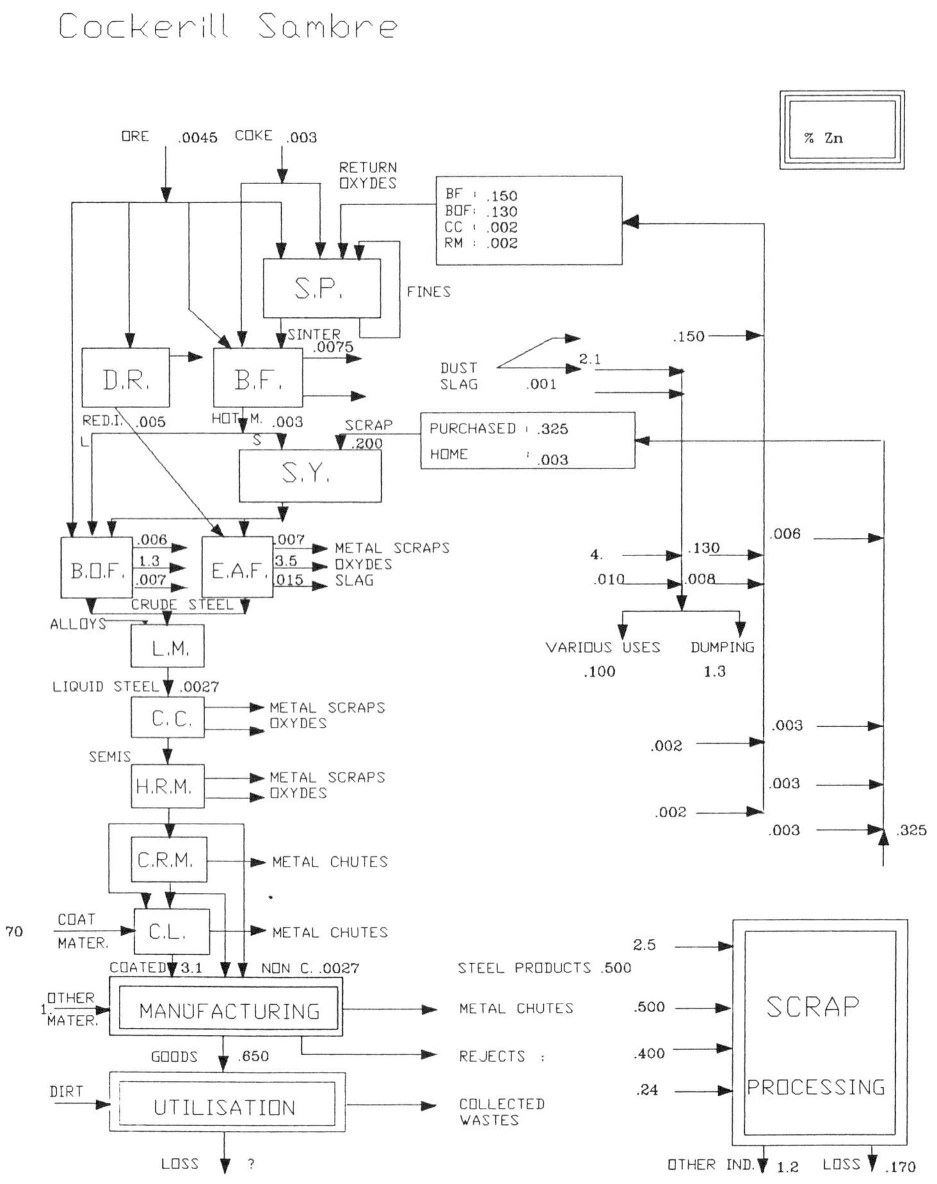

Fig 6 – Zn content of materials in the steel industry

Cockerill Sambre

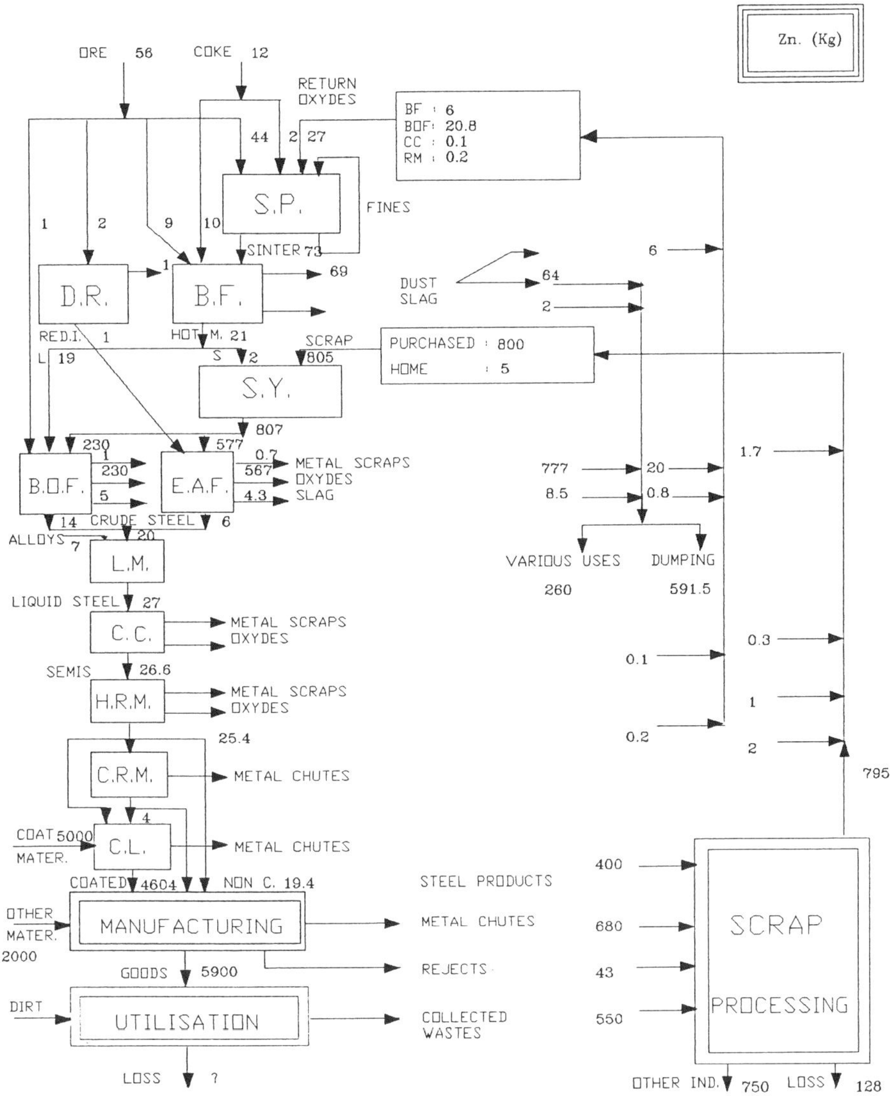

Fig 7 – Zn cycle in the steel industry

Looking at datas from the zinc industry (fig. 8), I believe that almost 75 % of Zn production finally joins steel in the goods we use : 7 t Zn associated with 1000 t steel.

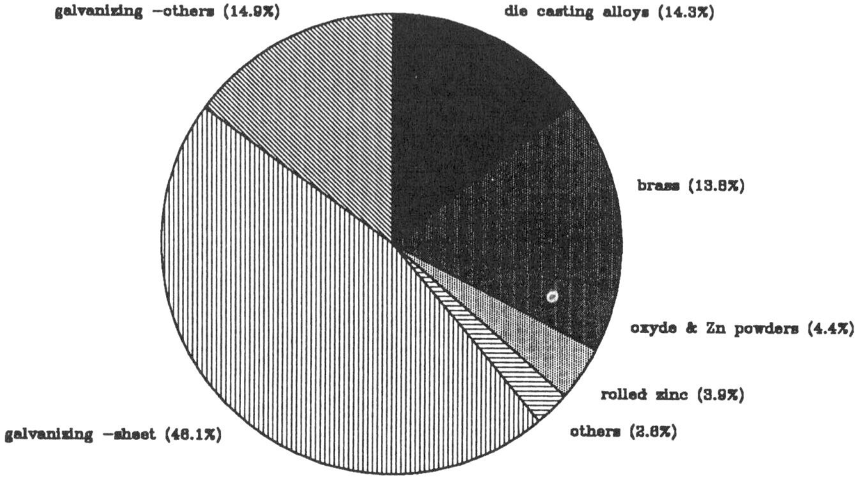

Fig. 8 - Share of Zinc consumption.

As a coating, zinc is a sacrificial element and some is consumed and lost. The remaining zinc will of course follow steel in collection. In scrap processing, the separation is made to a certain degree : to day, separation is based mainly on fragmentation and magnetic selection : this process cannot separate zinc coating from the steel substrate. We assume that to day the % Zn in the collected scrap is only about half of the % Zn in the producted goods.

A survey made in Germany (4) shows that this will rise in the future when Zn coated automobiles will be scrapped (fig. 9).

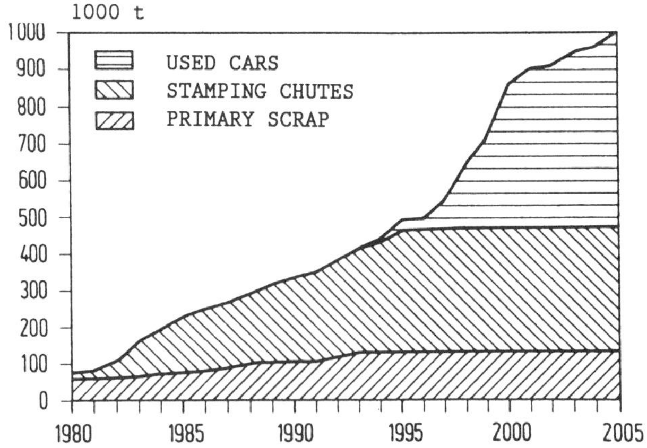

Fig. 9 - Amound of galvanised scrap in the period from 1980 to 2005.

Zinc interference in steelmaking

In the steelmaking processes, major entries for zinc are : iron ore, coke and mainly scrap. The exits are : use of slags and dusts in other industries and landfill dumping.

Due to its low temperature volatilisation, Zn concentrates in the gaseous phase where it is often oxidized as ZnO and collected with other dusts. As these dusts are very rich in iron oxide, steelmakers are willing to recycle them, in the blast furnace mainly via the sinter plant.

And here is where the problem lies : Zn exits from the system are very small. Return oxides are continuously enriched in Zn. Zn condensation occurs in the colder zones of the blast furnace preventing free circulation of the gases. The BF may be blocked. It is well known amongst blast furnace operators that total ZnO charged in BF should be kept below let us say 150 g/t hot metal.

Zinc separation in the steelmaking processes

What can the steelmakers do ?

The relatively small percentage of zinc makes the separation processes very expensive. As zinc concentrates in the dusts mainly, steelmakers have focussed on the different dusts generated by each process.

Ideas to separate zinc oxides from other components in the steelmaking dust are numerous and we shall consider the most promising ones.

Blast furnace dust

In blast furnace dust, zinc and iron oxides remain essentially uncombined. A set of hydrocyclones is the cheapest mean to separate an enriched phase. Experiences are under evaluation at HOOGOVENS (5), BETHLEHEM STEEL, SSAB-OXELÖSUND (6) on blast furnace sludge. A further treatment is still necessary on the Zn rich fraction (a third of the solid, 5 % Zn content).

Basic oxygen furnace dust

In BOF dust, an important part of the zinc is combined in zinc ferrites. This situation entails more difficult separations and it seems unlikely to reach the minimum zinc content compatible with zinc smelting under global economical conditions.

Two methods have been tried in industrial practice to increase the zinc content in the dust :

. Selection of the peak vaporization period of zinc within the blowing phase.
. Recycling of BOF dust in the BOF until the appropriate zinc content is reached.

The first method is usable under particular conditions. With restrictions to low scrap to hot metal ratio, the period of major vaporization of zinc can be reproducible enough to collect a small fraction of the dust containing most of the zinc. The zinc content is not high enough for zinc smelting but there is a real opportunity to use such iron oxide in the cement industry.

There are serious indications by the major cement producer in Belgium that a moderate zinc content in the BOF dust incorporated in the clinker enhance the cement properties. The other big fraction of the dust low in Zn content can be recycled in the blast furnace charge.

The second method is practiced at least in one steel plant in Austria (9) : the fine dust briquetted and recycled collected in the electrostatic precipitator is in the BOF until the zinc oxide content reaches at least 25 %. This occurs in their conditions after four weeks. Then, the dust production of the next week is taken out and sold to a zinc smelter. The coarse dust, poor in zinc, is recycled at the sinter plant.

This steelmaker mentions the advantage of avoiding dumping of any dust in landfill (which is nowadays completely forbidden in Austria) but also the occurence of several problems :

. Clogging of the gaz saturator section of the gas exhauts system
. Steel fuming at tapping, during ladle strirring and also in CC mold
. Coating of the copper mold of the caster by a zinc layer
. Presence of zinc in BOF slag (0.04 %)

Trials made at COCKERILL SAMBRE in CHERTAL WORKS though very limited also entailed deposits in the gaz saturator section. Dust generation increased and could be a problem with the existing dedusting equipment.

Electric arc furnace dust

What is the different in EAF steelmaking versus BOF steelmaking ?

EAF also generates dust but its composition is significantly different of BOF dust. Due to :
. the important vaporization in the high temperature of the arcs
. the high (often 100 %) scrap ratio in the charge
. the remelting of alloyed steels to produce alloyed steels,
the EAF dust is rich in such elements as Cr, Ni, Mo, Mn ... Such dust is therefore more dangerous and should anyway be reprocessed to recover the dangerous but valuable elements.
On counter part, there is less iron and the charging of zinc coated scrap allows to reach relatively easily high zinc content (25 to 40 % Zn). Many of the treatments applied to EAF dust can thus be adapted to recover simultaneously zinc or zinc oxide (10, 11, 12). In this case, the cost of the treatment is balanced by the added values of the different metals recovered which make zinc recovery more interesting from EAF dust compared to BOF dust.

Necessity of dezincing

As the recycling of BOF dust containing zinc have not found a satisfactory solution yet, most BOF steelmakers do not accept steel scrap containing zinc. The severe restrictions to landfilling reinforce this situation.

There is also a limitation in the amount of zinc containing scrap that can be melted in EAF steelmaking. In Germany, THYSSEN Oberhausen have made clear that EAF cannot operate with a 100 % galvanized scrap charge due to liquid steel fuming and the danger of casting problems due to the nucleation of gaz bubbles :

$$P_{CO} + P_{N2} + P\ zH_2 + P_{Zn} > P_{atm} + \sigma/r \qquad (1)$$

To avoid this phenomenon, THYSSEN restricts the amount of coated scrap to 150 kg/t steel. This is why it seems hopeless to find a simple solution as selectively consuming the scraps with a low zinc content in the BOF and the scraps with a high zinc content in the EAF.

Therefore, dezincing of some coated scraps is a necessity. It is my conviction that coated scraps of well known origin and caracteristics, such as industrial and manufactural chutes, should be collected selectively and dezinced before remelting in BOF mainly. Then, the amount of mixed scraps containing dispersed zinc will be reasonnably compatible with EAF remelting capacity.

Trials made in Belgium on an industrial electrolytic detinning line showed that the same line is able to dezinc galvanized scrap with a productivity four times as high as the productivity observed for detinning. However, when tin recovery pays for the operation, the lower value of zinc entails a more critical economical balance. Nevertheless, in Canada and in the US, several dezincing lines are in operation now, if I am well informed.

Other dezincing methods based on selective leaching or vaporization face the same problem. To what extent can the preheating of scraps be effective for at the same time separate Zn from steel and entail electricity savings in EAF remelting ?

Look towards the future

It is certainly to soon to make a conclusion regarding which technology will emerge to enable galvanized steel scrap recycling with minimum material dumping in landfill by the steelmaking industry.

We know that the applicability of many solutions depends on local conditions of existing equipment and more generally on treatment costs : so, several technics will be used on a large scale.

Galvanized steel is a cheap and durable material, its use is generally cost effective. Steel is easily recyclable, the separation of zinc from steel is technically easy if we organize scrap collection properly. This will minimize the cost of the separation and favor coated steel recycling and its futher use.

On a long term basis, selective collection of scraps will increase in importance for all materials by the change in people's mentality. This will contribute to a drastic reduction in material losses and landfill dumping. Larger amounts of scraps and particularly steel scrap will be available. This availability of scrap should entail a modification in the share of production of steel between the blast furnace and basic oxygen steel plants and the electric are steel plants. This is probably the opportunity to create new mini-mills designed for recycling coated steels.

REFERENCES

JOURNAL

1. R. BODE et al., "Thyssen Stahl - a partner for automotive steel sheet". STEEL TIMES INTERNATIONAL, March 1990, 14 sqq.

2. H.W. KREUTZER, "Die Schrottwirtschaft in der Bundesrepublik Deutschland". STAHL und EISEN 110 (1990) Nr 7, 83-85.

3. Eisenhüttentag 1990 "Stahlrecycling". STAHL und EISEN 110 (1990) Nr 12, 44, 52-54, 70.

4. H. GÜNNEWIG, W. ULLRICH, "Recovery of iron - containing materials from scrap automobiles" STAHL und EISEN, special issue "Steel in Motor Vehicle Manufacture" Sept. 1990, 96-99.

SEMINAR

The recycling of zinc containing plant reverts in the metal producing industry. The INSTITUTE OF METALS, London - March 1991.

5. Conditioning of BF dust by hydrocyclone, HOOGOVENS Ijmuiden B.V.

6. Zinc balance at SSAB Oxelosund and SEM investigation of zinc containing dust, SSAB.

7. Problems and results with recycling of zinc containing steelplant dust at Nova Hut Ostrava, NOVA HUT.

8. Zinc containing sludge in Q-BOP steelmaking process in Vitkovice, VITKOVICE State Enterprise.

9. Recycling of dusts in LD steelplant Nr 3 in Linz, VOEST-ALPINE STAHL GmbH.

10. Hi-plas technology for treatment of EAF dusts, DAVY McKee Ltd.

11. Recovery of zinc from EAF flue dust, MULTISERV International Ltd and TETRONICS Research and Development.

12. The recovery of metal from stainless steelmaking dust by plasma smelting technology, BRITISH STEEL Technical.

Criteria for the Selection of a Recycling Process for Low Zinc-Containing Residuals from the Iron / Steel Industry

N. L. Piret, D. Müller

STOLBERG INGENIEURBERATUNG GMBH,
4040 NEUSS 1, FRG

Abstract

Reprocessing of low zinc-bearing residuals with a zinc content of less than 10%, originating from the iron and steel industry, has been given more and more attention in the last few years, due to the potential landfill liability and the increasing cost of disposal by dumping.

Whereas, recycling of hazardous electric arc furnace dust is already being performed commercially on a large scale in the U.S., Japan and Western Europe, almost no reprocessing takes place in the case of these low grade zinc-bearing materials. In blast furnaces and basic oxygen furnaces dusts and / or sludges are generated at a rate of 1 to 2 % of the respective production. The technologies for the retreatment of these types of materials, either commercially available or still being developed are briefly described hereafter. They involve internal recycling, upgrading by physical separation methods, separate hydrometallurgical processing and pyrometallurgical treatment. Their scope and limitations are critically assessed in terms of technological applicability, economic feasibility and environmental compatibility. Since, the possibility of achieving economic profitability is unlikely with this type of material, the main criteria for acceptability of the treatment process are:

- generation of a iron-rich material, suitable for treatment by the iron/steel industry,
- acceptance of the non-ferrous metal-bearing sub-product by zinc smelters,
- generation of non-hazardous waste materials.

Residues and Effluents - Processing and Environmental Considerations
Edited by Ramana G. Reddy, William P. Imrie and Paul B. Queneau
The Minerals, Metals & Materials Society, 1991

Generation of Iron-Bearing Residuals from Iron and Carbon Steel Production

In the production of iron and carbon steel, in an integrated steel mill, the residuals which are generated, amount to 480 kg/t crude steel produced. Out of this figure, slags represent 375 kg/t and are valorised. Iron-bearing residuals are generated at almost every process step , as shown in Figure 1, in the form of dry dust, wet sludges, mill scale and total 60-65 kg/t steel. Most of these iron-bearing residuals (35-40 kg/t steel) are still recycled to the sinter plant, without any further treatment, since, they are low in contaminants which would be harmful to the operation.

The iron-bearing residuals which are contaminated with non-ferrous metals, alkali metals, oil, etc., cannot be recycled to the sinter plant or to any other process step, if they are not priorly submitted to a suitable treatment step, which will substantially lower their level of contaminants. They are also too low in non-ferrous metals to be processable by a non-ferrous metal operation.

Hence, disposal of contaminated iron-bearing residuals in landfills is the only alternative left. Today, this is generally the case, for wet secondary gas cleaning blast furnace sludges and for the dusts collected from gas cleaning of the basic oxygen steelmaking converters. Since, these residuals may also contain contaminants, (such as lead), at a level rendering the material potentially hazardous, special landfilling may be necessary.

Arisings of production residuals in integrated iron and carbon steelmaking plants is presented in Table 1. It is shown that the rate of generation of BF-sludge is 4 to 6 kg/t steel and BOF-dust 15 to 20 kg/t steel (depending on whether a coarse fraction is separated for internal recycling to the sinter plant).

Table I: Arisings of Production Residuals in Integrated Iron and Carbon Steelmaking Plants
(Based on Data from Ref. 1)

	Distri-bution(%)	Specific Amount(kg/tSt)	End-use
1. Slags			
1.1 BF-slags	53	255	-
1.2 BOF-slags	17	82	-
1.3 Other slags	8	38	-
Subtotal	78	375	Valorisation as road-building material, in the cement industry, as fertilizer material
2. Iron-bearing residues			-
2.1 Internally recyclable:			Recycled internally to sinter plant
2.1.1 Sinter plant dust	1.7	8	-
2.1.2 Primary gas cleaning BF gas dust	2.3	11	-
2.1.3 BOF-dust, coarse fraction	0.8	4	-
2.1.4 Mill scale	3.5	16.5	-
2.2 Not internally recyclable:			
2.2.1 Second gas cleaning BF gas dust	1	5	Dumped or separate treatment for upgrading
2.2.2 BOF-dust, fine fraction	3.2	15	Dumped or separate treatment for upgrading
2.2.3 EAF-dust	-	(14-16)[(1)]	Valorisation by non-ferrous metal processors
2.2.4 Mill scale sludge	0.5	2.5	Separate treatment prior to internal recycling
Subtotal	13	62	
3. Other Residues			
3.1 Mill rubble	5	24	-
3.2 Miscellaneous	4	19	-
Subtotal	9	43	Major part to be dumped
4. Total	100	480	-

NOTE (1): Arisings of EAF-dust from scrap treatment is not included in the balance

Fig. 1: Arisings of iron-bearing residuals in the production of iron and carbon steel

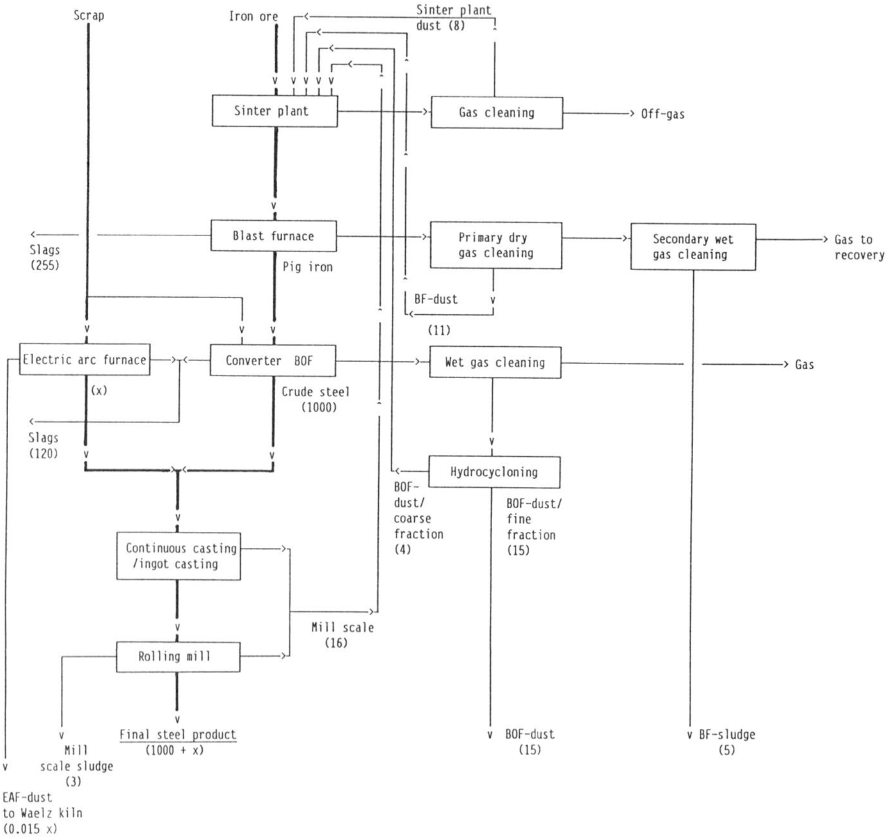

Alternative Schemes for Waste Management of Low Zinc Containing Iron-Bearing Residuals

In Table II, the three possible alternative schemes for the end-use of these type of residuals are summarised:

Table II: Alternative Schemes for waste Management of Low Zinc Containing Iron-Bearing Residuals from Iron and Carbon Steelmaking

Alternative Schemes	Type of Final Product(s)	Factors Affecting Feasibility of Scheme
1. Direct landfill disposal of residual material	- Listed as hazardous waste	* Increasing costs of special landfill capacity * Ever decreasing availability of special landfill capacity * Environmental liability
	- Listed as non-hazardous material	* Increasing cost of landfilling * Environmental liability
2. Decontamination and subsequent landfill disposal of decontaminated iron-bearing residual	- Decontaminated iron-bearing material	* Increasing cost of landfilling * Environmental liability
	- Side-products of decontamination	
	* as by-product(s)	* Treatability of non-ferrous metal-bearing by-products in a non-ferrous smelter
	* as waste product(s)	* Disposal or valorisation possibility depending on characteristics of waste product
3. Conversion to recyclable iron product	- Recyclable iron product	* Quality requirements for iron product
	- Side products of conversion to iron product	
	* as by-product(s)	* Treatability of non-ferrous metal-bearing by-products in a non-ferrous smelter
	* as waste product(s)	* Disposal or valorisation possibility depending on characteristics of waste product

The factors affecting the feasibility of the scheme are:

- straight landfilling
- landfilling after prior decontamination
- retreatment to recover a valorisable iron product

The evolution and forecast of the costs of landfilling, in the case of Germany, is presented in Table III.

Table III: Evolution of Costs of Waste Disposal in Germany (from Ref. 2)

Type of Waste	Non-Hazardous General Industrial/Municipal	Hazardous	
Year (*)	Landfilling/Incineration	Special Landfilling	Incineration
1975	3 - 6	30 - 60	90 - 300
1980	6 - 18	45 - 90	120 - 420
1985	18 - 36	60 - 180	180 - 600
1990	48 - 120	90 - 300	360 - 800
1995	90 - 180	180 - 480	600 - 3000
2000	180 - 300	300 - 720	900 - 4200

* Actual costs converted at exchange rate 1 US $ = 1.70 DM

High actual costs of landfilling and further expected rises, especially in the case of hazardous waste landfilling, as well as the associated liabilities, render the first two alternative schemes more and more unattractive.

The third alternative, the conversion of the residual to a recyclable iron product, can in principle be conducted in two ways:

- internal recycling of the iron-bearing residual subsequent to an on-site treatment being performed to eliminate the contaminants to below the required level
- separate treatment of the iron-bearing residual, generally in an off-site operation.

The criteria for material recyclability of iron-bearing residual are listed in Table IV.

Table IV: Criteria for Internal Recyclability of Iron-Bearing Residuals from Integrated Iron and Carbon Steelmaking Operations

1. CONTENTS OF NON-FERROUS METALS AND OTHER IMPURITIES (ALKALI METALS) IN MATERIAL TO BE RECYCLED INTERNALLY IS COMPATIBLE WITH OPERATING REQUIREMENTS OF PARTICULAR PROCESSING STEP.

2. PHYSICAL CONSISTENCY OF THE MATERIAL TO BE RECYCLED IS IN SUCH A STATE THAT ENVIRONMENTAL REQUIREMENTS CAN BE MAINTAINED UPON RECYCLING OF RESIDUAL.

Thus, in order to fulfill these criteria, there are certain requirements for the process which converts the residual to the recyclable iron product. These requirements are presented in Table V.

Table V: Process Requirements for the Treatment of Iron-bearing Residuals from Integrated Iron and Carbon Steel Making Operations

1. ELIMINATION FROM THE RESIDUALS OF THE NON-FERROUS METALS AND OTHER IMPURITIES (ALKALI METALS) TO SUCH A LOW LEVELS THAT ITS INTERNAL RECYCLING TO A PARTICULAR PROCESS STEP DOES NOT INTERFERE WITH THE PLANT OPERATION.

2. CONVERSION OF THE RESIDUALS TO SUCH A PHYSICAL STATE THAT ITS INTERNAL RECYCLING TO A PARTICULAR PROCESS STEP REMAINS IN COMPLIANCE WITH ENVIRONMENTAL REQUIREMENTS.

3. UPGRADING OF THE NON-FERROUS METAL BY-PRODUCT TO SUCH AN EXTENT THAT IT CAN BE VALORISED BY A NON-FERROUS METAL PROCESSOR, i.e., Zn + Pb CONTENT > 20%.

Due to increasingly more stringent pollution control requirements, permissible emission of solid particles in iron and steel works has been lowered from 150 to 50 mg/m^3. This renders the recirculation, even of non-contaminated residuals to the sinter plant very difficult, despite the improved performance of the existing gas cleaning equipment (electrostatic precipitators), which today are able to achieve a clean gas with a content of 80-100 mg/m^3.

As pointed out in Reference 3, it is doubtful whether the recirculation of non-contaminated residuals in an amount corresponding to 35-40 kg/t steel, as shown in the flowsheet of Figure 1, can be upheld.

The third requirement listed in Table V, namely, the suitability of upgrading the metal by-product to such an extent that it can be valorised by a non-ferrous metal processor, i.e., that the (Zn + Pb) content exceeds 20%, is difficult to achieve by means of internal recycling only.

This is understandable in view of the first requirement. of Table V, namely the removal of the non-ferrous metals to a sufficiently low level permitting internal recycling. As is being discussed later in detail, further upgrading will therefore be necessary.

On the contrary, the separate treatment of the iron-bearing residuals in specially designed equipment is generally able to produce non-ferrous metal by-products with a (Zn + Pb) content in excess of 20%.

Economic Considerations

In the business of recycling zinc-bearing secondary materials, the activity of recycling can be either "economically" induced or "environmentally" forced (Ref. 4), which is also referred to as "conventional" recycling versus "new" recycling (Ref. 2). Obviously, at a given zinc quotation, the border line between "economically" induced and "environmentally" forced recycling depends primarily on the zinc content of the material itself and in the second instance, on the presence of other valorisable or penalty elements.

Depending on the particular conditions of the secondary material prevalent and on the zinc quotation, this border line is situated at a zinc content ranging between 20 and 25%.

For electric arc furnace-(EAF)-dusts, which world-wide are being treated to a long extent by the well-established Waelz kiln process (ref. 2, 5), that has been granted the BDAT (= Best Demonstrated Available Technology) treatment standard, the lower limit is 20-21% zinc (Ref. 6).

The recycling of iron-bearing residuals from iron and carbon steelmaking will, therefore, in any case, fall under the category of "environmentally" forced recycling activities.

If the non-ferrous by-products from the treatment of the iron-bearing residuals can be upgraded to a zinc content of about 20-25%, then similar conditions as for the treatment of EAF-dusts will be applicable.

The economic structure of the "environmentally" forced treatment of iron-bearing residuals will consist of the following terms:

- cost of producing recyclable iron product
- eventual additional cost of upgrading non-ferrous metal by-product to at least 20% Zn
- credit for iron product
- credit due to absence of landfilling costs

There is no credit for the upgraded non-ferrous metal product, for which conditions such as the treatment of EAF-dust are applicable.

In the case of production of a DRI product, it is not expected that this product will command the price of DRI from virgin ore, due to the presence of residual impurities (particularly alkali metals).

The diagram in Figure 2 graphically represents the costs involved in upgrading low grade iron-bearing residuals from iron and carbon steelmaking.

Fig. 2: Simplified Graph Representing the Cost-Revenue Structure for Secondary Zinc Material as a Function of Zinc Content.

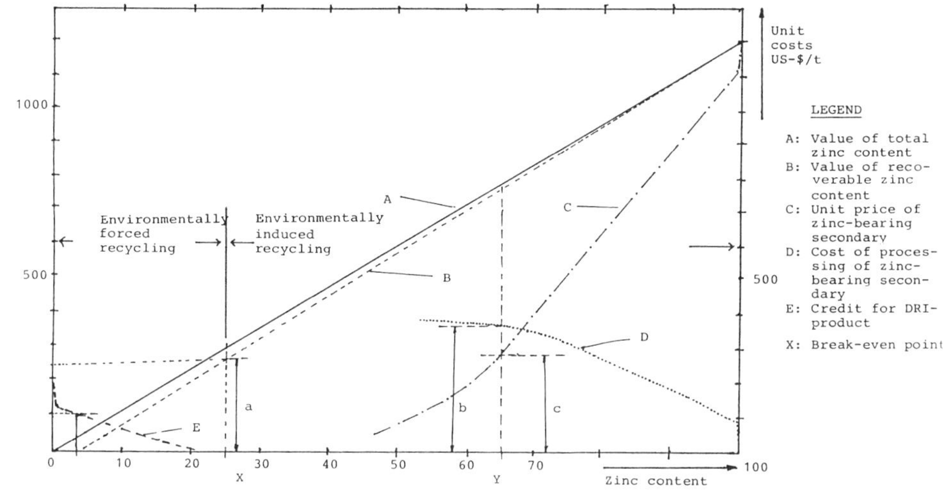

Blast Furnace (BF) Wet Gas Cleaning Sludge

The rate of generation of sludge from wet secondary gas cleaning of BF-gases in an integrated iron and steel plant is normally 4 to 6 kg/t steel produced.

Due to its high impurity content, direct internal recirculation to the sinter plant is not feasible. Its zinc content, usually in the range 3-10%, is, however, too low for valorisation by a non-ferrous metal processor.

1. Characteristics:

1.1 Particle Size Distribution. The particle size distribution varies widely from one blast furnace operation to another, as illustrated in Figure 3. According to investigations by IRSID (Ref. 7), the fineness of the dusts collected in the secondary wet gas cleaning of BF gases depends largely on the type of equipment utilised in the primary dry gas cleaning section. If the primary gas system is also able to collect a larger part of the finer fractions (e.g. by using cyclones), then, obviously, the particle size distribution of the sludge will be finer.

1.2 Chemical Composition. Since, the zinc and lead components of the dust are fumes and not mechanically entrained solid particles, it is obvious that zinc and lead will have the tendency to report to the finer fractions of the sludge. Hence, the finer the particle size distribution of the sludge, the higher the enrichment in zinc and lead can be expected and consequently, the lower its iron content. This is illustrated in Table VI, in which the composition of a range of BF-sludges is shown. The generally high percentage of carbon is an important characteristic of BF-sludge.

Table VI: Composition of Various BF-Secondary Wet Gas Cleaning Dusts

Type	Element %			
	Fe	Zn	Pb	C
A	25-30	2-3	0.5	40-45
B	31-38	2-3	1-1.5	16-20
C	25-30	3-4	2-3	15
D	15-18	4-5	1-2	20-30
E	16-20	5-10	1-2	20-30
F	30	8	3	13
G	11-20	7-10	3	18-30
H	14-18	9-13	4-6	15-20
I	4-6	16-22	3-5	16-19
J	1-7	20-28	8-14	20-37
K	2-5	6-33	2-8	8

(From Ref. 7)

Fig. 3: Particle Size Distribution of Various BF-Secondary Wet Gas Cleaning Dusts (Rosin-Rammler Graph) (Ref.7)

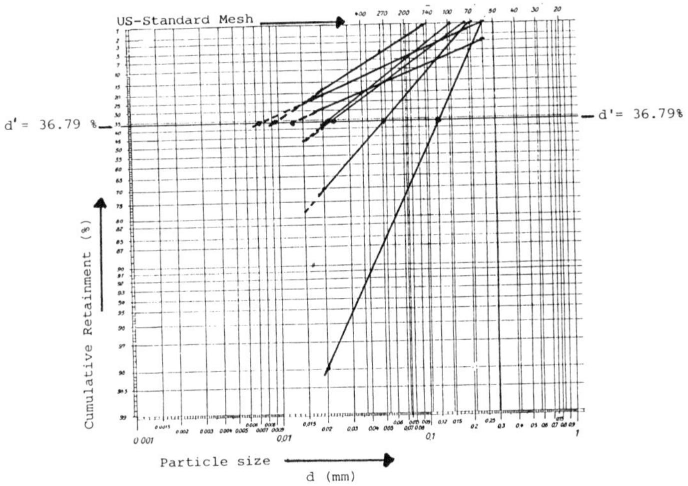

In Figure 4, it is shown how the zinc content of the sludge varies according to the fineness of the sludge which is characterised by its "d'" value:

- sludges with a relatively coarse particle size distribution, such as the ones obtained in systems without cyclone in the primary gas cleaning section, have a low zinc content and vice versa.
- the zinc content of the sludges depends significantly on the zinc content of the BF-feed. However, with the usual concentration of the zinc encountered in integrated iron and steel plants (< 100 g/t feed), the zinc content of the sludge is not expected to exceed 10%, even in the most favourable case.

1.3 Mineralogical Constituents. The mineralogical composition, particularly the form in which zinc and lead are present, is important with regard to the possibility of upgrading the material. There appears to be general agreement that a large part of the zinc and lead (i.e. 50-90%) is present as the simple oxides ZnO and PbO. (Ref. 8, 9, 10, 11), the remaining part being present as sulphide, silicate, zinc ferrite or metal.

2. Processing Methods for BF-Sludges or Dusts

The impossibility of recycling BF-sludges or dusts directly to the sinter plant or to have them treated by non-ferrous metal processors, necessitates the implementation of new treatment technologies to achieve the requirements as specified in Table V.

2.1 Physical Methods of Separation. Flotation, hydrocycloning and air classification are three methods which have been considered. (Ref. 6).

- Flotation has received attention, but, due to the fine particle size, selectivity suffers and reagent composition is high (Ref. 1, p.159; Ref. 12, p.14).
- Hydrocycloning is still receiving widespread attention for upgrading BF-sludge,
  * as described by IRSID (Ref. 7, p.88-89)
  * as industrially applied in Japan (Ref. 7)
  * as commercially implemented at Hoogovens, IJmuiden (Ref. 10).
  * as being planned for installation at Thyssen, Duisburg (Ref. 1, p.160).

  Two or three stage cycloning is performed. The zinc-depleted coarse fraction usually contains less than 1% zinc. The zinc enriched fine fraction, containing up to 80 to 90% of the zinc, will generally contain too low a zinc content for direct further processing by a non-ferrous metal operation.
- An air classification process was developed and installed at the Kokura Steel Works of Sumitomo Metal Industries, (Ref. 13), after implementation of the BF-gas dry gas cleaning system (BDC). The system, which is presented in Figure 5, is claimed to give better results than hydrocycloning

The applicability of physical methods of separation of BF-sludge, particularly of hydrocycloning, for the purpose of upgrading the non-ferrous metal by-product and for recycling of the iron-bearing residual, will depend on the characteristics of the BF-sludge and therefore, vary from BF-operation to BF-operation.

Fig. 4: Relationship between Zinc Content and Fineness of Secondary Wet Gas Cleaning Dust at Two Zinc Input Levels (Ref. 7).

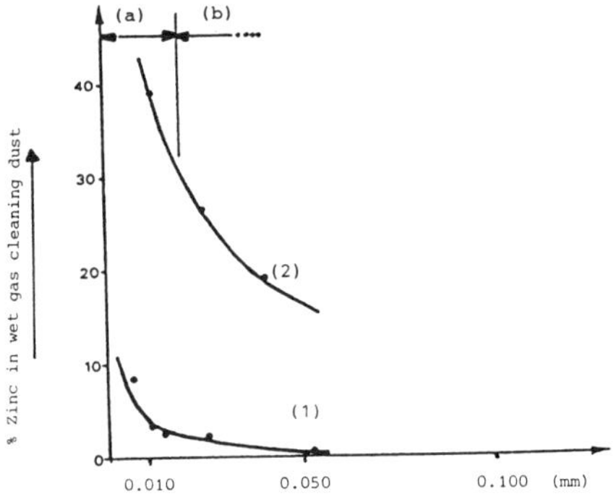

Legend:

- Range of wet gas cleaning dusts after primary gas cleaning:
  a) with hot cyclone
  b) without hot cyclone

- Zinc content of feed to blast furnace:
  1) 100 - 200 g/t Fe
  2) 1,100 g/t Fe

d': Size of sieve, retaining 36.79 % of material

Fig. 5 : Removal of zinc from dry BF-dust by air classification (Ref. 13)

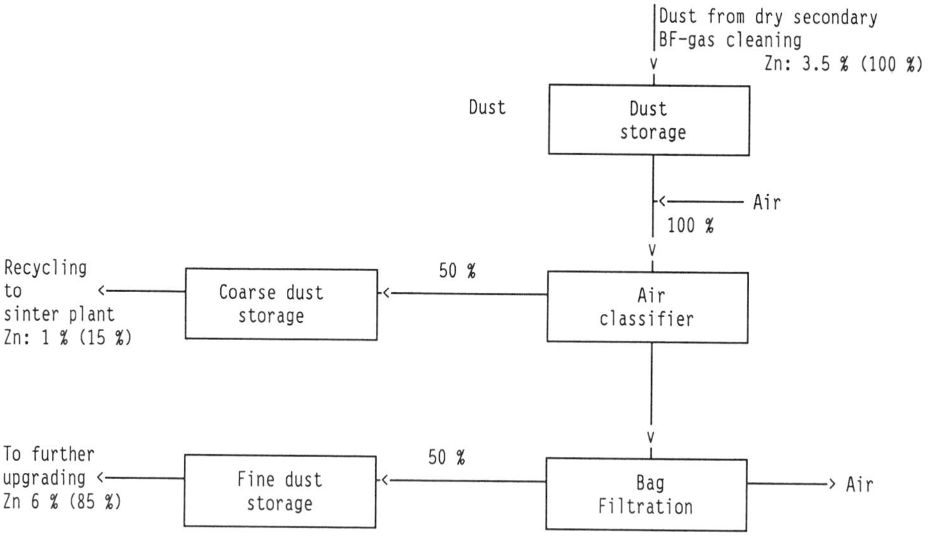

Further upgrading of the non-ferrous metal by-product will be required prior to its processability by non-ferrous metal smelter in any case.

2.2 Hydrometallurgical Methods of Separation. The response of BF-sludge with regard to separation performance by hydrometallurgical methods will largely depend on the mineralogical constituents of the sludge.

Selective leach extraction of zinc is considered to be directly proportional to the fraction of the zinc present as the simple zinc oxide compound. It is a well-known fact from zinc sulphide roast-leach-electrowinning practice that zinc ferrite, if present, requires strong acid leach or reducing conditions, under which conditions the separation of iron and zinc necessitates additional complex process steps. Other zinc compounds such as silicate and sulphide also require leaching conditions under which zinc/iron selectivity cannot be maintained.

- In one process (Ref. 9), selective sulphuric acid leaching of zinc is claimed to have been conducted on a BF-sludge, containing the zinc as zinc oxide. The $PbSO_4$-containing iron-bearing residual is subsequently submitted to hydrocycloning to separate the lead. (Figure 6).
- For the fine fraction enriched in zinc, obtained by hydrocycloning of the BF-sludge produced at Hoogovens, IJmuiden, a process was proposed (Ref. 10, 11) in which high temperature acid leaching in the chloride medium, using spent pickling liquor, is performed under oxidising conditions, to precipitate the iron as hematite (Figure 7). This complex and undoubtedly expensive process was selected because of the presence of 40% of the zinc as difficult-to-leach constituents.
- Caustic leaching of the zinc oxide present in BF-dust as proposed by the "Cebedeau" process (Ref. 14) as well as, by the "Cardiff" process (Ref. 15) (Figure 8). However, the presence of zinc ferrites and other zinc compounds would require additional treatment steps of the residue, such as reducing roasting (Ref. 15), in order to comply with the requirements of recyclability laid down in Table IV.

It is therefore, concluded that only in particular cases where virtually all of the zinc is present as zinc oxide, could leaching be an attractive proposition for separation treatment of BF-sludge.

Fig. 6: Sulphuric Acid Leaching of BF-Sludge (Ref. 9)

Fig. 7; Two-Stage Pressure Leaching of Zinc-Bearing BF-Sludge (Ref. 10. 11)

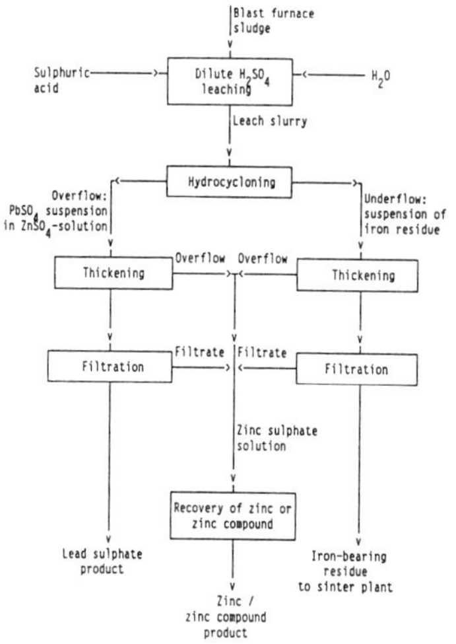

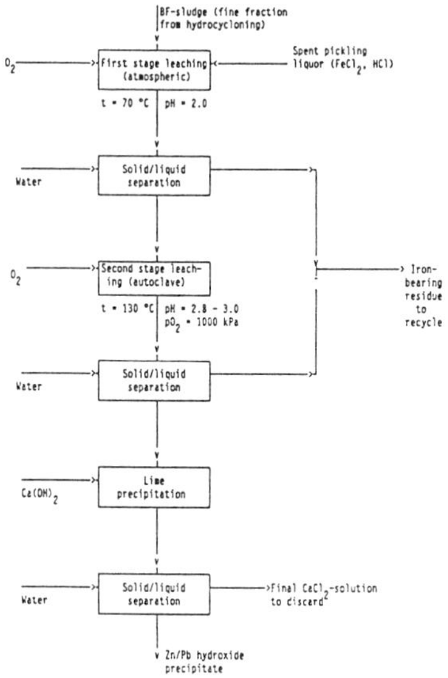

Fig. 8: Caustic Leaching of BF-Sludge (Ref. 14, 15)

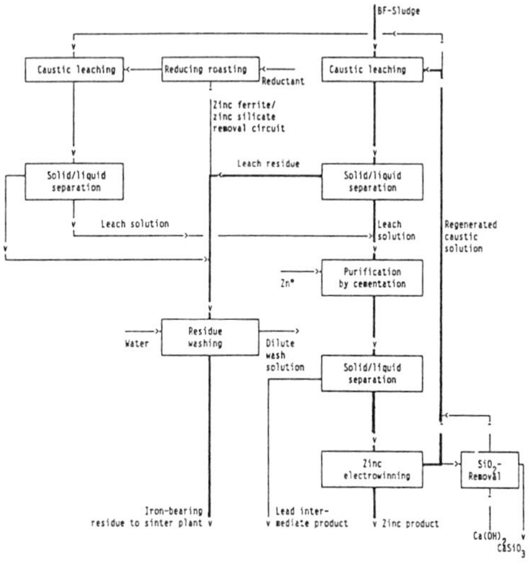

Basic Oxygen Furnace (BOF) Gas Cleaning Dust/Sludge

The rate of generation of this material from cleaning of BOF-gas is 15 to 20 kg/t steel produced. The amount of zinc is related to the content of zinc in the charge. By utilisation of inside steelmaking scrap the zinc content of the dust is obviously low (0.05 to 0.1%). If purchased scrap is processed, it can rise to 5%.

Generally, the content of non-ferrous metals in BOF-dust is too high to be recyclable to the sinter plant. Other treatment methods are necessary, if landfilling is to be avoided.

The dusts are mainly collected by a wet gas cleaning system resulting in wet dusts, but in some operations wet gas cleaning has been replaced by dry gas cleaning systems.

1. Characteristics

1.1 Particle Size Distribution. The particle size distribution of the dusts collected by a wet gas cleaning system is very fine, the "d" value, i.e., the size of the sieve retaining 36.79% of the material is 0.005 to 0.01 mm. However, the particle size of dry gas cleaned dust is much coarser due to particle agglomeration, which will be discussed later.

1.2 Chemical Composition. The chemical composition differs significantly from that of BF-sludge, due to:

- the much higher total content of iron
- the very low carbon content
- the high degree of metallisation especially of the coarse fraction.

The chemical compositions of the fine and coarse fractions of the BOF-dust are also quite different, as will be illustrated later.

1.3 Mineralogical Constituents. The iron is present as fine metal particles and as iron oxide. The degree of metallisation appears to depend on the system of gas cleaning, being very much higher in the case of dry gas cleaning than with wet gas cleaning. The coarse fraction is also more metallised than the fine fraction. Non-ferrous metals are concentrated in the fine fraction. However, unlike in BF-sludges an appreciable part of the zinc is present in zinc ferrite ($ZnFe_2O_4$) which renders physical methods of non-ferrous metal operation almost impossible and hydrometallurgical methods very difficult. Other constituents are CaO, MgO.

2. Processing Methods by Means of Internal Recycling

2.1 Dezincing on Sinterband. This process was developed in the 1980's (Ref. 16, 17). The proposed process is based on the sectional and selective reduction of the zinc oxide on the sinterband and the separate withdrawal and cleaning of the zinc/lead bearing off-gases. The sectional and selective reduction is achieved by separate preparation of the iron and carbon steel making residual with excess reductant and structural depositing of the agglomerated material as a bottom layer on the band. Sintering of the residual occurs only at the end-section of the band, installed as a separate unit, in which reducing conditions are maintained by adjustment of the air/reductant ratio. Figure 9 shows a simplified scheme of the proposed process, which, however, appears complicated in terms of equipment.

Fig. 9: Simplified Scheme of Proposed Sinterband Dezincing Process (Ref. 16, 17)

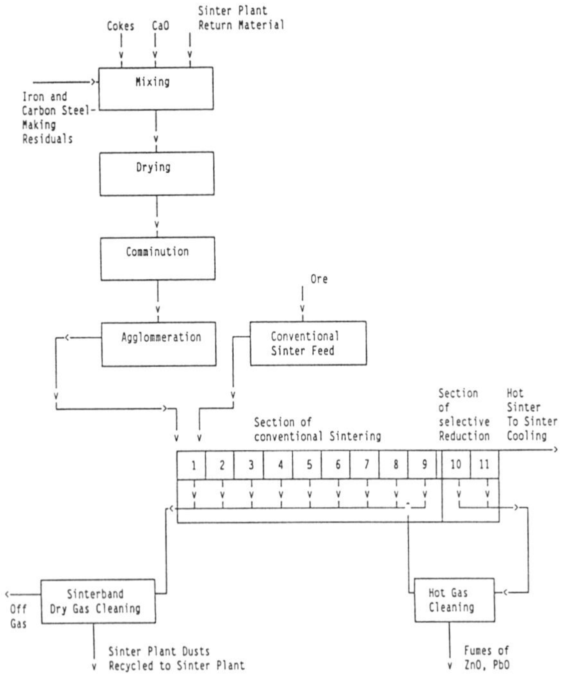

2.2 Elimination of Zinc by Fractional Collection of BOF-Dust. Fractional collection of BOF-dust is based on the finding (IRSID, Ref. 7) that zinc volatilisation from the converter charge takes place during the first stage of the converting cycle. Thus, by dividing the solid-liquid separation circuit of the wet gas cleaning system into two sections and collecting the dust from the first stage in one of the sections and the dust from the second stage in the other section, a wet dust enriched in zinc and one depleted of zinc could be produced. The ratio of zinc contents of these two products varied between 10 to 1 and 5 to 1, depending on the zinc content of the charge. This finding, however, could not be duplicated at another steel works (Ref. 18), where the zinc volatilisation rate was found to occur in parallel with the dust removal rate.

2.3 Internal Recycling of BOF-Dusts from Wet Gas Cleaning. The zinc content of dusts from the basic oxygen furnace, is related to the zinc content of the converter charge (mainly from the input of foreign scrap). It is too low for direct input into a non-ferrous metal operation. The quantities of BOF-dusts generated (15-20 kg/t Steel) and their zinc content (1-3% Zn) preclude their recycling, so that without any special process step this material has to be dumped.

In order to lower the amount of material being dumped, BOF-dusts from wet gas cleaning are frequently submitted to a physical separation step, by means of hydrcycloning to obtain a coarse fraction low in zinc (< 1% Zn) which can be returned to the sinter plant and a fine fraction enriched with zinc which only has to be dumped.

In one particular case, (Ref. 19) hydrocycloning of the BOF-sludge at 60 μm resulted in two fractions with the composition as given in Table VII.

Table VII: Hydrocycloning of BOF-sludge (Ref. 19)

Fraction: Size (μm)	Coarse > 60	Fine < 60
Quantity (kg/t Steel)	3 - 4	12 - 16
Fe-Total %	46.0	56.0
Fe-Metal	24.0	3.5
Fe 2+	12.1	47.6
Fe 3+	6.1	4.5
Zn	< 1.0	4.5
Pb	0.3	1.3
CaO	26.3	5.5
SiO_2	1.6	1.4
Al_2O_3	0.3	0.2
MgO	0.3	0.4

A possible method for internal recycling of the BOF-sludge is based on the sintering of the BOF-sludge after a laborious operation of dewatering and/or drying of the fractions. (Ref. 20). The sinter is used either as cold dope in the BOF (replacing ore or scrap) or in open hearth furnaces (OH), as was the case at Hoesch Hüttenwerke. (Ref. 18). In that particular case, the availability of still operating OH-furnaces proved to be beneficial on account of the high degree of enrichment of OH-furnace dust in zinc/lead, which rendered it suitable as a raw material to a non-ferrous smelter. The Hoesch flowsheet is illustrated in Figure 10.

Fig. 10: Internal Recycling of Wet Gas Cleaned BOF-Dust (Ref. 18, 20)

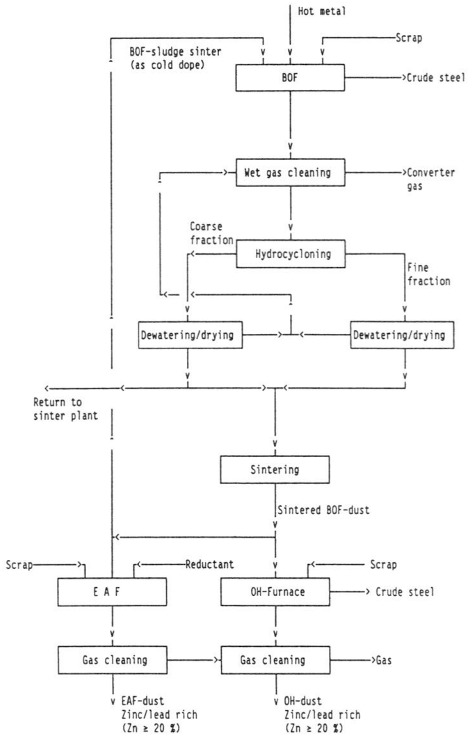

Fig. 11: Hot Briquetting of BOF-Dusts from Converter Dry Gas Cleaning (Ref. 23)

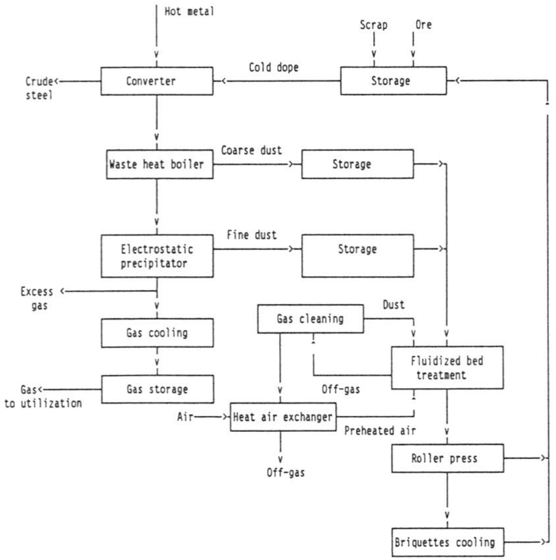

The electric arc furnace (EAF), as a scrap processor, is also suitable for the treatment of BOF-sludge sinter provided an adequate amount of reductant is added.

The EAF dust thereby obtained should also be suitable material for input into a non-ferrous smelter.

2.4 Hot Briquetting of Steelmaking (BOF) - Dusts. In the case of dry gas cleaning of BOF-gases being performed for the purpose of CO-recovery from converter gas, dust is generated in the form of a coarse dust from the boiler section and also in the form of a fine dust from the dry electrostatic precipitator.

For one particular industrial case, the characteristics of these dusts are shown in Table VIII.

Table VIII: Characteristics of dusts from dry gas cleaning of BOF-dusts. (Ref. 21)

	Coarse Dust	Fine Dust
Specific Gravity (g/cm^3)	6.12	5.14
Bulk Density (t/m^3)	2.5	1.0
Specific Surface Area (m^2/kg)	203	2607
Ratio of Generation	1	1.6
Chemical Composition (%)		
Fe - Total	85.4	70.7
Fe - Metal	71.7	19.7
Fe 2+	9.0	21.5
Fe 3+	4.7	29.5
Zn	0.36	2.0
Pb	0.04	0.2
CaO	6.5	3.0
S	0.02	0.1
C Total	1.4	0.7
Degree of Metallisation (%)	84.2	27.8
Size Distribution (%)		
> 2mm	2.4	6.6
1.0 - 2.0	8.4	6.5
0.63 - 1.0	10.5	5.8
0.315 - 0.63	27.1	14.6
0.1 - 0.315	39.2	32.5
0.063 - 0.1	7.3	24.2
< 0.063	5.8	9.5

In this case the total generation of BOF dust amounts to 15 - 18 kg/t steel of which 35 - 40% is coarse dust and 60 - 65% is fine dust.

Table VIII shows that the coarse dust is highly metallised due to the presence of mechanically entrained iron particles and the non-ferrous metals are concentrated in the fine dust as expected. The particle size distribution of the fine dust, collected from the ESP is relatively coarse due to auto-agglomeration of the finer, pyrophoric metallised particles.

The high degree of metallisation of the dust renders it suitable to binder-free hot briquetting. This property is taken advantage of at Thyssen Steel AG. (Ref. 8, 21, 22, 23), where a hot briquetting plant with a throughput of 12 mt/h was industrially implemented in 1987, yearly capacity about 70,000 mt of briquettes. A schematic flow diagram is presented in Figure 11.

The dust, after admixing a suitable amount of coarse dust to the fine dust, is partially oxidised in a fluidised bed at 750°C by means of a controlled supply of air and formed into briquettes in a continuous roller press. After cooling to 70°C, the briquettes are recycled to the steelmaking process as cold dope, replacing ore or scrap.

It is obvious that continuous recirculation of the BOF dust will result in an enrichment, particularly of the fine fractions in zinc and lead. After 5 recirculations the zinc content of the fine dust could be 7 to 10%. How many dust recirculations are possible in practice, without bleed-off of some of the fine dust, is not documented.

It is, however, unlikely that by multiple recirculation of briquettes to the BOF the degree of upgrading of nonferrous metals will be sufficient for direct further processing by a non-ferrous metal processor.

On the contrary, input of the briquettes into an electric arc furnace (EAF), as a processor of scrap, would most likely result in EAF-dust suitable for direct treatment in a non-ferrous smelter.

Effect and Behaviour of other Impurities

Besides the non-ferrous metals (Zn, Pb), other impurities such as alkali metals (Na, K) and halogenides (Cl, F) are also found enriched in BF- and BOF- dusts/sludges.

Processes successfully developed and implemented for the elimination of the non-ferrous metals, do not always succeed in removing to a sufficient extent, other impurities such as alkali metals, of which the behaviour is frequently not well documented.

Upon internal recycling of the iron-bearing residual, subsequent to the treatment for non-ferrous metals removal, further enrichment of these particular impurities would take place. On the other hand, high levels of chlorides and/or alkali metals in the non-ferrous metal by-product to be processed by non-ferrous metal operators are undesirable or penalised. According to one source (Ref. 20), the total amount of alkali metals should not exceed 1%, whereas the total sum of sulphur and halogenides must be lower than 2%.

In order to eliminate alkali metals and halogenides, a process has been industrially applied (Ref 18, 20), consisting of a dilute caustic leach at a pH of 9 to 9.5, resulting in a substantial lowering of the level of undesired impurities in the filtered dust.

Obviously, the suitability of such a process will depend on various factors, such as:

- ease of leachability and achievable extraction of the impurities to be removed
- desired state (wet/dry) of the material, in view of its further processing.

Therefore the adequacy of the implementation of such a process must be investigated from case to case.

Separate/Off-Site Treatment of Iron and Carbon Steelmaking Residuals

Processes for the separate treatment of iron and carbon steel making residuals are in operation and are still being developed with the aim of producing a valorisable iron product in the form of:

- iron oxide
- directly reduced iron (DRI)
- pig iron as well as environmentally stable slags.

These processes are characterised by the following points:

- high capacity operations
- insensitivity to type, origin, consistency, composition of the residuals, at least within certain limits
- off-site operations, due to their high capacity, necessitating transportation of the residuals from the site of origin.

These processes are mainly pyrometallurgical processes, although the hydrometallurgical processses, which are finding consideration, or which have been proposed for the treatment of EAF-dusts, could obviously be applied to the treatment of residuals arising out of the production of iron and carbon steel.

1. Pyrometallurgical Processes for Separate Off-Site Treatment. The three different process schemes, producing directly reduced iron, iron oxide and pig iron will be dealt with hereafter.

1.1 Conversion of Carbon Steel Production Residuals to DRI

1.1.1 Inmetco Process. The Inmetco process (Ref. 24) operated by International Metals Reclamation Co. Inc. (INMETCO) of Ellwood City (Pa.), since 1978, was developed by its parent company, INCO Ltd., in the mid 1970's. The proprietary technology initially aimed specifically at reclaiming iron, chromium and nickel from speciality steel making waste materials but in the mid 1980's, it was extended to include the treatment of low grade zinc bearing secondaries from integrated steel plants.

In this process, the carbon steel production residuals are mixed with coal, ground to less than 0.25 mm and pelletized. The green pellets are introduced into a rotary hearth furnace (RHF), (Figure 12) in which the oxides are reduced. Metallisation of iron reaches 92 %. The non-ferrous metals zinc, lead and cadmium are volatilised and re-oxidized in the furnace atmosphere (Figure 13) and the non-ferrous metal oxides are recovered from the process waste gases. Halogenides and some of the alkali metals are also eliminated.

Fig. 12: Conversion of Carbon Steel Production Residuals to DRI by INMETCO Process (Ref. 27)

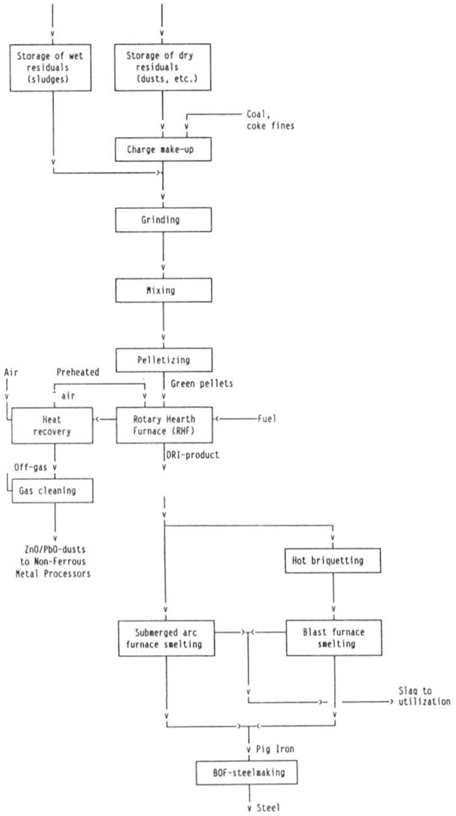

Fig. 13A: Schemeatic Top View of INMETCO Rotary Hearth Process (Ref. 27)

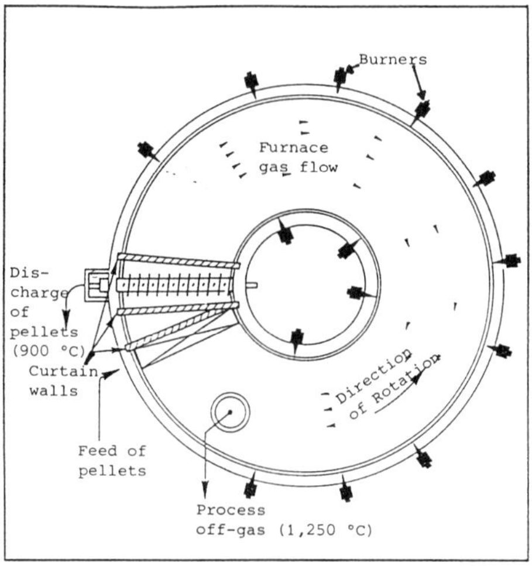

Fig. 13B: Mechanism of Dry Reduction in the INMETCO Rotary Hearth Process (Ref. 27)

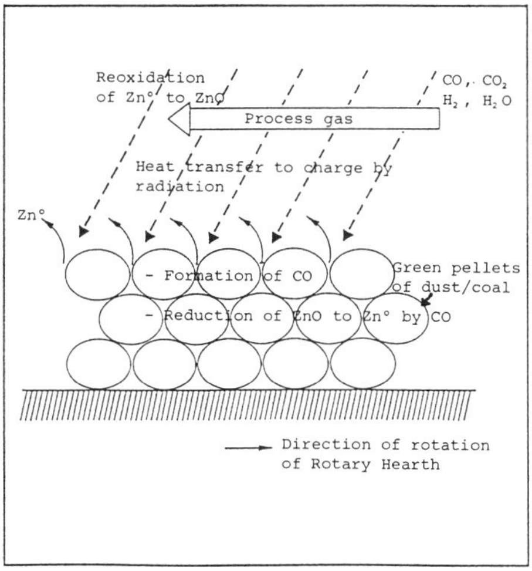

The direct reduced iron sponge containing some residual carbon can be smelted directly in a submerged arc furnace or can be hot briquetted and cooled for later use or for blast furnace smelting.

Typical process parameters are as follows:

Pellets size	6 - 10 mm
Green strength	1 - 2 kg/cm^2
Bed depth	20 - 25 mm
Reducing temperature	1225 - 1250 °C
Hearth pellet loading	30 kg/m^2
Retention time	15 min
Hearth DRI - production rate	0.06 t/m^2 h
DRI / Pellet Ratio	0.60 - 0.65
Process gas / t pellet	5000 Nm^3/t

The capacity of the only existing facility at Ellwood City today is 47,000 mt/a carbon steel production residuals out of which 34,000 mt/a of DRI is produced. The actual capacity of the RHF (170 m^2 hearth area), however, is twice as large. (Ref. 25).

In 1985, a worldwide licensing agreement with Mannesmann Demag of Duisburg was concluded (Ref. 26) and a study was made to set up a 330,000 mt/a treatment facility for carbon steel production residuals generated from integral steel mills in the Saar region (Ref. 24), which however, has not yet been realised.

1.1.2 Lurgi - CFB - Process. The principle of the Circulating Fluidised Bed (CFB) was developed by Lurgi in the 1960's, initially for application to endothermic processes, this was extended to exothermic processes in the '70's. (Ref. 28).

The treatment of carbon steel production residuals in integrated steel mills in a CFB is a follow-up to the technology of the direct CFB reduction of iron ore.

Proven advantages of this technology are:

- excellent mass transfer and heat transfer conditions resulting in high reaction rates.
- good control possibility of process parameters such as retention time, reduction potential and temperature.

Initial testing of the CFB-process to eliminate zinc and lead from carbon steel waste material and to generate a DRI material suitable for unproblematic recycling to the integrated steel mill resulted in a 94% iron metallisation. (Ref. 29).

However, due to the rather limited low operating temperature (830°C) of the particular CFB reactor, zinc elimination was not complete. At the optimum operating temperature (1000°C) residual zinc and lead are expected to be as low as 0.2 and 0.1% (Ref. 5).

A demonstration plant of 20,000 mt/a is being set-up at Thyssen AG, Duisburg, in a joint effort by B.U.S. and Thyssen (Ref. 1, 2).

On account of the flexibility of the CFB to control the reduction potential in the reactor, the process could be operated in two modes:

- complete reduction to DRI
- selective reduction to volatilize Zn/Pb with little iron metallisation.

The latter mode would have the advantage of considerably lowering the process operating costs.

A simplified flowsheet and expected material balance is presented in Figure 14. It can be seen that the enrichment of the non-ferrous metals in the dust is sufficient for further processing by a non-ferrous metal processor.

1.2 <u>Selective Reduction of Low Zinc Containing Carbon Steel Production Residuals</u>. A process has been proposed by IRSID in which selective reduction of zinc and lead oxides takes place without simultaneous metallisation of the iron content. (Ref. 30)

The objective of this development is to replace the DRI-process with a process which has lower operating costs due to the decreased energy requirement. The poor quality of the DRI-product (residual non-ferrous metals, presence of alkali metals), resulting in lower pricing of the product as compared with DRI obtained from primary material and the requirements for handling this type of material prior to its further processing, are additional factors which could favour the selective reduction process.

The conditions under which selective reduction can take place are shown in Figure 15, in case of the reduction in the system CO/CO_2 and in Figure 16 in case of the reduction in the system of H_2/H_2O.

Fig. 14: Circulating Fluid Bed (CFB) Process for the Treatment of Iron/Carbon Steel Production Residuals (Lurgi-Berzelius Umwelt-Service) (Ref. 5)

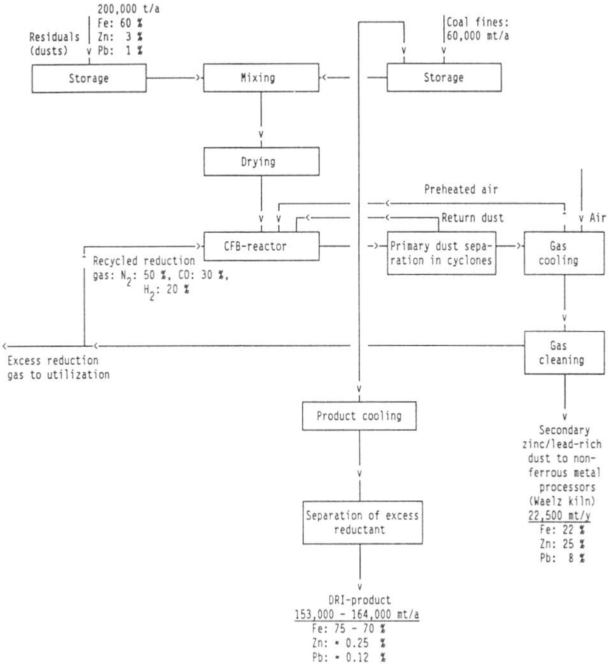

Fig. 15: Vapour Pressure of Zinc (p in atm) as a Function of CO/CO_2 at Various Temperatures

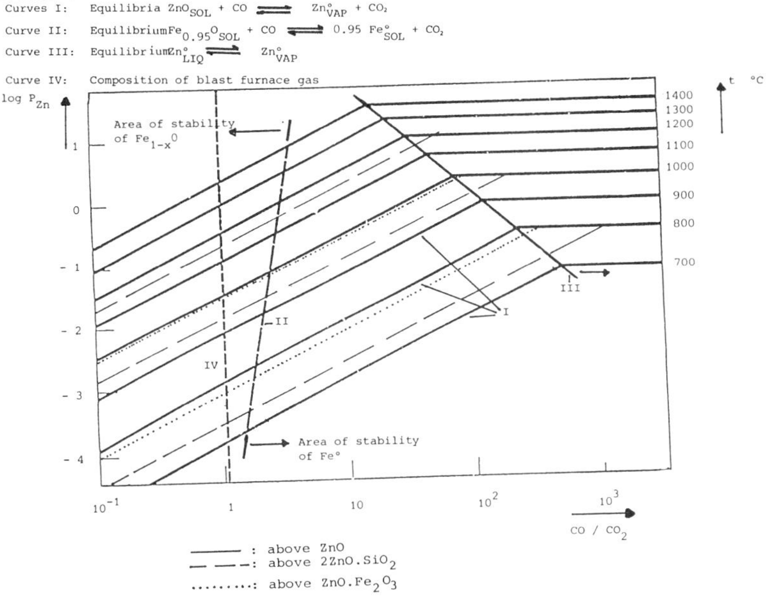

Fig. 16: Vapour Pressure of Zinc (p in atm) as a function of H_2/H_2O at Various Temperatures

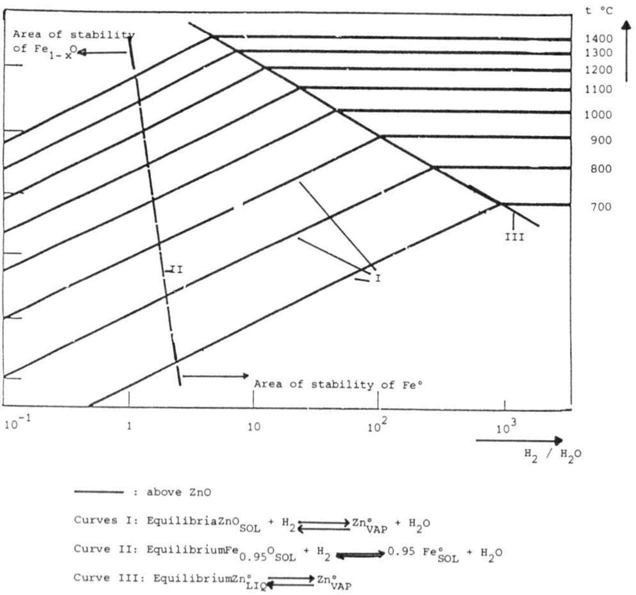

At the ratio CO/CO_2 from 0.5 to 1.5 and in the temperature range 1000 - 1100°C, it is claimed that in excess of 99.0% of zinc and lead are eliminated with less than 2% metallisation of iron (Ref. 7). The higher temperature range is beneficial.

Blast furnace gas is suitable to carry out the selective reduction. The presence of carbon in the material to be treated will render selectivity difficult or impossible. Thus BF-dusts, which are usually high in carbon, are not suitable unless pretreated for C - removal.

The flowsheet proposed by IRSID is shown in Figure 17.

Although there has been no disclosure on the preferred equipment to perform the selective reduction, it is obvious that a CFB reactor would be suitable for this purpose. (Ref. 2).

1.3 Direct Production of Pig Iron. In the production of iron and steel from primary materials, the zinc content of the feed to the sinter plant in large integrated iron and steel mills may not exceed 100 g/t. In smaller blast furnace units, a maximum of about 200 g/t could be tolerated. (Ref 4)

At Duisburger Kupferhütte, Duisburg, FRG, iron-bearing secondary materials from iron and steel mills are directly converted to special grade pig iron by the conventional smelting and blast furnace process. (Ref 31)

Fig. 17: Proposed Flowsheet for Selective Reduction (IRSID, Ref. 30)

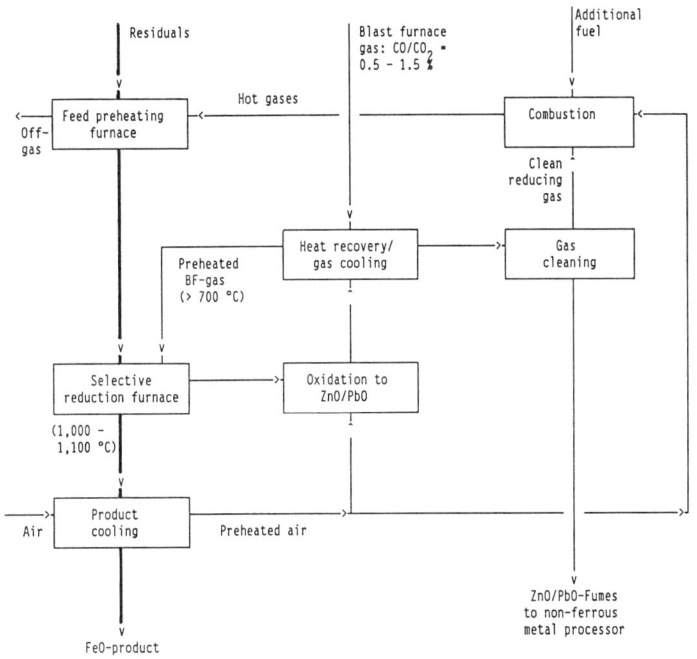

Table IX: Composition of Various Types of Secondary Iron-Bearing Feeds Suitable for Pig Iron production according to the DK Process

ORIGIN OF FEED	% H_2O	WEIGHT % (DRY BASIS)											
		Fe	Zn	Pb	C	S	SiO_2	CaO	Cu	Mn	As	Cr	P
BLAST FURNACE DUSTS/SLUDGES	15-30	36	1.0	0.4	<30	0.4	5	6.0	0.01	0.3	<0.01	0.05	0.01
BOF - DUSTS	10-15	67	2.0	0.2	<1	0.07	1	6.5	0.02	0.6	<0.01	0.05	0.06
BOF - SLUDGES	15-40	55	2.0	0.3	<1	0.2	1	11.0	0.02	0.5	<0.01	0.05	0.06
FOUNDRY DUSTS	1-10	55	<0.5	<0.1	<5	0.02	5	2.0	0.01	1.0	<0.01	0.05	0.02
RESIDUES EX CHEMICAL INDUSTRY AND OTHERS:													
a:	25	67	0.2	0.02	<0.1	0.4	3	0.3	0.01	0.4	0.02	0.12	0.1
b:	18	57	1.2	0.6	<0.1	1.0	12	1.5	0.02	0.1	0.09	0.02	0.02
c:	15-25	59	0.05	0.05	<5	0.3	2	5.0	0.02	0.3	<0.01	0.1	0.2

(From Ref. 31)

Using two small blast furnaces, 450,000 mt/y of secondaries containing up to 2 % zinc and some lead are treated (Table IX), producing 230,000 mt/y of special grade pig iron. Careful preparation of the secondaries is essential so that the feed to the sinter plant does not exceed 1.5 % zinc and that other elements critical to the quality of pig iron such as Ti, Cr, As, P, Cu are below the specified limits. The flowsheet is presented in Figure 18.

Fig. 18: Conversion of Low-Grade Iron and Carbon Steelmaing Residuals to Pig Iron at Duisburger Kupferhütte (DK) (Ref. 31)

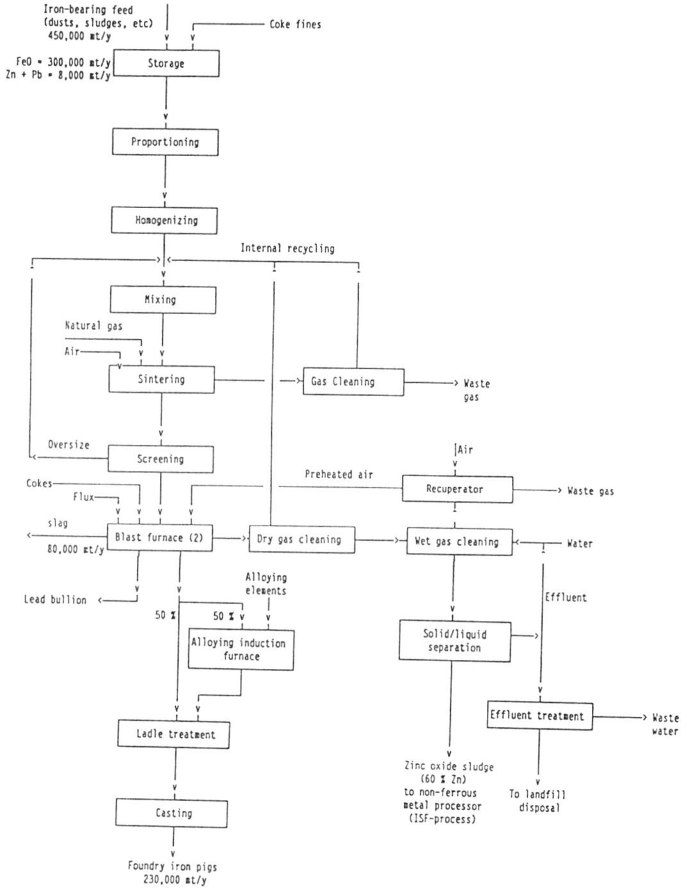

As by-products, lead is recovered in bullion from the bottom of the blast furnace and zinc in a 60 % Zn containing zinc hydroxide sludge from the second stage wet gas cleaning of BF - gases. Both metals are processed further by non-ferrous metal smelters. BF-slags are used for road construction.

Due to the small capacity of the operation and the poor quality (presence of Zn, Pb, alkali metals) of the feed material, operating costs are high on account of:

- the low smelting rate encountered
- the frequent shutdowns for removal of accretions
- the increased coke and energy consumption.

The economic feasibility of the process is based on the following:

- the existence of a largely depreciated facility after discontinuation of the pyrite cinder treatment at DK. (Ref. 31)
- the presence of a well-established market for special grade pig iron for foundry applications
- the plentiful availability of low zinc-containing iron bearing secondaries which are delivered free DK and which command a treatment charge, that, however, is lower than the locally prevailing costs of landfilling.

The restriction of this process is the limited market for special grade pig iron for foundry applications.

2. Hydrometallurgical Processes for Separate/Off-Site Treatment

Due to the various type of residuals and the significant differences in mineralogical composition of the non-ferrous metal compounds, the proposed hydrometallurgical schemes are complex.

In general, hydrometallurgical processes proposed for the treatment of EAF-dust will also be applicable to this type of low grade material. However, the requirement to produce a residue which would still be acceptable as feed to the sinter plant, is expected to be difficult to fulfill.

Table X summarises the different types of proposed processes.

Table X: Proposed Hydrometallurgical Processes for Treatment of EAF Dusts (Zn > 20%) also Potentially Applicable to Lower Grade Iron and Carbon Steel Making Residuals

Medium	Process Steps	Ref./Fig.
1. Two stage leaching in sulphate system	- Neutral leach (pH 3.0) - Oxidative acid leaching at elevated temp. (200°C) in autoclave with simultaneous iron precipitation as hematite	32
2. Two stage leaching in chloride system	- Leaching with HCl (98% extraction) - Pressure oxidation and hematite precipitation in autoclave. - Zn recovery by SX & EW or by precipitation.	10, 11/ Fig. 7
3. Ammonium carbonate leaching	- Selective prereduction of Fe_2O_3 compounds to Fe_3O_4 - Ammonium carbonate leaching - Purification by Zn° cementation - ZnO production by calcination - Zn recovery only 70-75%	12
4. Sodium hydroxide leaching (Cardiff-process)	- NaOH leaching - Purification by Zn° cementation - Electrolytic zinc recovery from zincate solutions as powder - Selective reduction of residue followed by dilute NaOH-leaching - Zn extraction: > 95%	15/ Fig. 8

Whereas the acid-leaching based processes suffer from the lack of selectivity between iron and zinc during leaching, the alkaline based processes require an additional selective reduction step to achieve acceptable zinc extraction.

Conclusions

The steadily increasing costs of landfilling, the decreasing availability of landfill capacity and the environmental liability associated with this practice renders the disposal of the iron-bearing residuals from the iron and carbon steelmaking mills by means of recycling unavoidable. Due to the low value of these residuals and the high cost of retreatment, this activity belongs to the category "environmentally" forced recycling.

One generally-practised method consists of internally recycling of the residual to a suitable step of the process. By recycling to the sinter plant a large part of the residuals, eventually after prior on-site treatment for removal of the contaminants (zinc, lead, alkali metals) which are deleterious to the blast furnace operation, are reprocessed. Although, this route is for obvious reasons the most economic one, more stringent environmental control regulations urge the reduction of the burden of finely-sized material fed to the sinter plant.

A further disadvantage of internal recycling to the sinter plant is the fact that the upgrading of the non-ferrous metal by-products by is normally insufficient for direct input into a non-ferrous metal operation. However, the approach of recycling of suitably agglomerated residuals to the electric arc furnace (e.g. as cold dope) is likely to circumvent this drawback.

Processes for separate treatment of residuals, because of their conceptual design, provide sufficiently upgraded non-ferrous by-products. However, the larger investment costs for their installation will require high capacity units and for this reason, off-site residual treatment in most of the cases.

Although for quite some time, two types of processes for separate treatment of iron-bearing residuals have been in operation, one providing DRI and one producing foundry pig iron, it is considered that the introduction of the CFB may render separate treatment of these types of materials economically more attractive, due to its flexibility to control operating conditions, expected lower energy consumption and higher specific throughput.

References

1. J. A. Philipp, J. Möller and B. Jansen, "Entfall und Verbleib Fe-Haltiger Restoffe in der Eisen - und Stahlindustrie dargestellt am Beispiel der Thyssen Stahl AG", (Paper presented at 1. NRW - Recycling Congress / 4. Duisburger Recycling Tage, April 1990, LWA - Materialien, 2 (1990)) 152-175.

2. H. Maczek, "New Fields of Metal Recycling with particular reference to Zinc", (paper presented at 1990 Annual Commodity Meeting on Zinc, IMM, London, December 1990, Minerals Industry International, July 1991) 23-26.

3. J. A. Philipp and R. Endell, "How German Steel Industry is Managing Waste Disposal", Steel Technology International, (1990-1991) 275-279.

4. N. L. Piret and J. F. Castle, "Review of Secondary Zinc Treatment Process Options" (Paper presented at Symposium on "Recycling of Metalliferous Materials, IMM, Birmingham, April, 1990), 181 - 197.

5. R. Kola, "The Processing of Steelworks Waste", (Paper presented at the International "Lead-Zinc '90" Conference, ed. T.S. Mackey and R. D. Prengaman, The Minerals, Metals and Materials Society, 1990) 453-464.

6. O. Burghardt and E. Mertins, "Zinkentfernung aus Gichtgas-schlämmen", in "Aufbereitungstechnik in Hüttenwerken" H. 54, 21. Metallurgisches Seminar GDMB, (September 1989) 79-93.

7. R. Pazdej and J. M. Steiler, "New Treatment possibilities of BF/BOF Zinc (and Lead) Bearing Dusts", Proc. of First Process Technology Conference, Washington, March 1980, 85-100.

8. H. D. Pflipsen, W. Florin and R. Höffken, "Einsatz aufbereiteter Prozeßstäube in der TBM-Konvertern der Thyssen Stahl AG", Stahl u. Eisen 109 (7) (1989) 350-354.

9. Eur Patent No. 0021922 (1980), R. Pazdej, (IRSID), "Procédé et Dispositif de Traitement Hydro-métallurgique des Poussières Métallurgiques et Application de ce Procédé ou de ce Dispositif au Traitement des Poussières de Hauts Fourneaux"

10. R. Geutskens, "Pressure Leaching of Zinc-bearing Blast Furnace Dust", (Paper presented at the International "Lead-Zinc '90" Conference". ed. T. S. Mackey and R. D. Prenganan, The Minerals, Metals and Materials Society, 1990), 529-545.

11. C. L. Van Deelen, "Recovery of Zinc from Hazardous Waste, Current Status in the Netherlands", (Paper presented at the ILZ - Study Group Meeting "Recycling Lead and Zinc, The Challenge of the 1990's", Sub-Committee on Recycling, Rome, 11-13 June, 1991,) 411-420.

12. E. Peek, "Dissolution of Reduced Zinc Ferrite Components in Ammoniacal-Ammonium Carbonate Solutions", (Thesis, Delft University of Technology, June 1991).

13. Anon, "Dry Removal Method of Zinc from Blast Furnace Dust", Trans. of Iron and Steel Institute of Japan, (26) (1986) (2), 167.

14. J. Frenay, J. Hissel and S. Ferlay, "Recovery of Lead and Zinc from Electric Steel Making Dusts by the Cebedeau Process", in "Recycle and Secondary Recovery of Metals", ed. P.R. Taylor, H. Y. Sohn, N. Jarret, TMS, (Dec. 1985), 195-201.

15. B. I. Wheatley and F. D. Pooley, "Production of Zinc Powder from Arc and Smelter Flue Dusts", in "Recycling of Metalliferous Materials", IMM, Birmingham, (April 1990), 291-299.

16. German Patent Application, DE 3221283 (1984), H. Hilbrans and C. Rao, "Verfahren zum Abtrennen von Nicht-Eisenmetallen aus Eisenhaltigen Sekundärstoffen".

17. P. Hammerschmid et al., "Entzinken und Entbleien von Hüttenwerkstäuben und - schlämmen auf das Sinterband", Stahl und Eisen, 107, (1987), 1, 33-37.

18. M. Kühn, "Actual Process Technology for Re-use of BOF-Sludges at Hoesch Hüttenwerke, AG", Proc. of First Process Technology Conference, Washington, March 1980, 101-103.

19. W. Pluschkell, "Verhalten von Zink bei der Stahlerzeugung im LD-Konverter", 3. Duisburger Recycling Tage, ed. J. Agst, 1988, Informationschrift 39, 233-256

20. M. Kühn and M. Haucke, "Erfahrungen bei der Behandlung und Verwertung von Stahlwerkstäuben und Schlämmen", Stahl und Eisen 101 (1981), 701-705.

21. E. Höffken et al., "Die Entwicklung des Thyssen-Heissbrickettier-verfahrens und seine betriebliche Anwendung", Stahl u. Eisen 108 (14, 15) (1988) 705-354.

22. R. Auth et al. "Heissbrikettierung von Filterstäuben aus der Oxygenstahlerzeugung", Stahl u. Eisen 107 (2) (1987) 80-84.

23. R. Auth et al. "Die Entwicklung des Thyssen-Heissbrickettier-verfahrens und die betriebliche Anwendung", 3. Duisburger Recyclingtage, 1988, ed. J. Agst 93-120.

24. M. Hennig and H. J. Lehmkuehler, "Thermische Aufbereitung Metallhaltiger Hüttenrestoffe mit Hilfe des Inmetcodirekt-reduktionsverfahrens" (Paper presented at the 3. Duisburger Recycling-Tage, 1988, ed. J. Agst), 139-156.

25. Anon, "A Waste Winner", Mining Journal, London, 316 (8127) (June 21st, 1991), 476-477.

26. Anon, "Wertstoffe aus Hütten-Reststoffen: Inmetco Direktreduktion", DST Demag Steel Plant Technology.

27. J. K. Pargeter and H. J. Lehmkuehler, "Recycling of Waste and Flue Dust from the Steel Industry into Hot Metal, using the Inmetco Process", (Paper presented at the Proc. of the 44th Electric Furnace Conference, Vol. 44, 1986), 403-407.

28. Lurgi Annual Report 1991, 33-42.

29. M. Hirsch, A. Kaune and H. Maczek, "Recovery of Zinc and Lead from Steel-Making Dusts in particular by the Circulating Fluid Bed", (Paper presented at the 28th Annual Conference of Metallurgists, August 20th - 24th, 1989, Halifax, Nova Scotia).

30. French Patent No. 7637568 (1976) L. Ivanier and J. M. Steiler, "Procédé d'extraction du Zinc continu dans les Déchets Sidérirguques".

31. H. Planken, "Von Stäuben und Schlämmen zum Roheisen" (Preprint of paper presented at the 5. Duisburger Recycling-Tage, 20th - 21st February, 1991).

32. M. C. Jha et al., "A Two-Stage Leaching Process for Selective Recovery of Zinc from Steel Plant Dusts", Recycle and Secondary recovery of Metals, TMS-Aime, (1985), 143-157.

SENSIBLE CLASSIFICATION OF WASTE CONTAINING

TRACE LEVELS OF RADIONUCLIDES

Christopher R. Casias

Science Applications International Corporation
14062 Denver West Parkway, Suite 110
Golden, CO 80401

Abstract

A basic methodology is presented (employing both standard and theoretical assessment, analysis, and data interpretation techniques) for classifying waste containing trace levels of radionuclides. Despite increased regulatory requirements for handling, transporting, and disposing of such waste: (1) no formal characterization methodology has yet been implemented, and (2) no limits have been set to define the point at which a waste material is considered to be low-level and below which it is "below regulatory concern" (BRC) or can be considered to contain no more than a diminimus level of radioactivity (i.e., is nonradioactive). The concept of diminimus limits specific to waste streams and waste forms is introduced, and a mechanism for evaluating waste in relation to BRC and diminimus limits is outlined. This methodology can be used to perform a complete and adequate characterization of waste for classification according to its level of radioactivity.

Residues and Effluents - Processing and Environmental Considerations
Edited by Ramana G. Reddy, William P. Imrie and Paul B. Queneau
The Minerals, Metals & Materials Society, 1991

Introduction

This paper presents an overall methodology that can be used to classify a waste containing trace levels of radionuclides (1). The presence and concentration of specific constituents (chemicals and radionuclides) in a waste are determined, in many cases, so that the waste can be classified according to its level of hazard and handled and disposed of accordingly. Needs for such a methodology are also driven in part by regulatory requirements for handling, transporting, and disposing of radioactive waste. This paper addresses the issue of waste classification and the approach, for the purpose of evaluation, to categorize wastes by physical form, waste stream, and method of generation. This study was originally conducted for facilities that handle large quantities of radioactive materials, but the methodology is applicable for many other facilities.

The classification of a waste into a specific category is based on its physical, chemical, and/or radiological characteristics. The process by which these characteristics are determined is defined as the waste characterization process. The process by which radioactive waste is characterized involves the determination of the level of radioactivity and may include identification of specific radionuclides and determination of the concentrations of these radionuclides. Methods used to analyze waste, to determine the level of radioactivity and determine radionuclides and radionuclide concentrations, range from nondestructive methods such as process knowledge or analysis of a representative sample of the waste to destructive methods such as incineration of the waste material and identification of radionuclides and their concentration from the remaining residue.

Of particular interest are (1) the level at which a waste contains a detectable level of radioactivity and is defined to not be radioactive (diminimus), and (2) the level at which a waste is defined to be radioactive but below regulatory concern (BRC). This paper introduces the concept of diminimus limits specific to waste streams and waste forms, although, it does not attempt to establish specific BRC or diminimus limits. However, it is postulated that once BRC limits are established by the appropriate regulating agencies, the characterization methodology presented here can be used in conjunction with these limits to properly classify waste. Additionally, this methodology can be used for determining that a waste is not radioactive by demonstrating that it contains no more than a diminimus level of radioactivity.

Background

Currently at many facilities, conventional sampling and analysis techniques (standardized methods for which procedures have been developed and published by the Environmental Protection Agency (EPA) or other recognized agency and which are commonly used by regulators and the industry) cannot be used to characterize radioactive waste due to the heterogeneous nature of the waste material; hence conservative classification of such waste material as radioactive, based on process knowledge, is common. Waste materials for which conventional sampling and analysis techniques are used are mostly homogeneous liquid and solid matrices such as oils, sludges, and soils. Waste that is not amenable to representative and conventional sampling and analysis may be amenable to the use of less conventional techniques such as radiation assaying. Radiation assaying has been used almost exclusively for the assessment of waste containing high levels of radionuclides; the use of radiation assaying to determine radionuclides present and corresponding concentrations in waste containing trace levels of radionuclides is considered to be unconventional.

Characterization Techniques

Process knowledge, as a method of waste characterization, relies on an understanding of the operation(s) producing the waste. The ability to perform a material balance evaluation is often the determining factor in the ability to use process knowledge. Sufficient operational controls must be in place for a high confidence to exist in an assessment based on the material balance. Documentation that substantially supports the methodology used for determining these values is required. In addition to process knowledge, other nondestructive techniques include the passive measurement or assay (using radiation detection instrumentation) of radiation types and energies emitted from the waste package.

Radiation detectors used to measure the radioactive decay of alpha and beta particles and gamma and X-rays can produce accurate results, but the effort and time required to measure these radiations and identify the source radionuclides increase significantly as the radioactivity level of the waste decreases and as the complexity of the waste stream and the potential for radionuclides to be present increases. Active assay methods include interrogation using neutrons (i.e., a waste containing fissionable materials is subjected to a neutron flux) that results in the occurrence of fissions and the release of prompt and characteristic delayed neutrons. The characteristics of the delayed neutrons observed can then be used to accurately and precisely determine fissionable radionuclides and concentrations.

Representative sampling can be employed using either nondestructive or destructive techniques. The ability to obtain a representative sample of a waste depends on the physical waste form composition and waste generation process. For homogeneous waste forms, nondestructive evaluation methods such as representative sampling and radiation assaying of that sample can often be conducted. Knowledge of the waste generation process is critical in determining whether representative sampling can be conducted because often a generation process can produce a highly variant waste for which representative sampling can be difficult. For heterogeneous waste forms, destructive evaluation of the waste package and subsequent analysis is often required.

Selection Factors

The use of any of these techniques can require a significant level of effort, time, and money. In theory, these techniques should produce the information necessary to accurately and precisely determine the radionuclides and their concentrations. In practice, both the application of these techniques and nature of the waste under evaluation must be well understood if the waste is to be adequately characterized. The factors affecting the selection of any particular technique for determining the radionuclides and concentrations of these radionuclides include but are not limited to the physical nature of the waste form, the ability to sample, special hazards associated with the waste, and the regulatory requirements. Regulatory requirements designed to protect workers, the public, and the environment from any significant threat resulting from the generation, treatment, storage, or disposal of radioactive waste may also affect or direct the use of a particular technique.

Waste Characterization Methodology

The characterization of radioactive waste involves: (1) an understanding of the processes, operations, and general physical, chemical, and radiological properties associated with the waste generated; (2) the use of a method to analyze, directly or indirectly, for the presence and concentration of radionuclides; and (3) the placement of such waste into a group or class. This study developed a methodology that implements such an approach for characterizing the radioactive constituents of radioactive waste containing low concentrations of radionuclides.

The characterization methodology presented herein employs conventional techniques for sampling, laboratory analysis, and interpretation of results, whenever possible. This methodology centers on radioactive waste characterization for waste containing low concentrations of radionuclides, and the emphasis is on the types of waste streams that require characterization using less conventional approaches as well as waste streams that have yet to be accurately and precisely characterized. As will be seen, several waste streams can be characterized using basic methods of sampling, analysis, and data interpretation. It is essential that the methodology also encompass this waste because the characterization of more complex waste streams is sometimes associated with assumptions about simpler waste streams. This is especially true for waste streams that are of the same form or have been generated from within the same facility.

General Waste Characterization Considerations

The initial classification of a waste stream as a waste type (i.e., nonradioactive, potentially radioactive, radioactive) is usually based on explicit assumptions about the process generating the waste. The assumptions regarding the waste-type classification are based on knowledge about the input streams for a process; if these input streams are known to contain specified concentrations of radionuclides, the waste derived from this process is known to contain radionuclide concentrations no greater than those of the original input streams.

Waste that contains trace levels of radionuclides is particularly difficult to characterize due to several factors associated with its generation. In many cases, this waste has been generated secondarily to waste containing higher levels of radioactivity as a result of incidental contact. Waste that contains low concentrations of radionuclides might also be generated from general support activities. In other cases, this waste is generated from non-routine operations or as a result of spills and spill clean-up and, consequently, tends to be highly variable; the radionuclides suspected to be present are known with less certainty.

Waste Characterization Principles

The fundamental principles of waste characterization are:

- The development of a waste characterization methodology requires a thorough understanding of the processes generating the waste, without which, the ability to accurately and precisely characterize a waste is impacted significantly;

- It is important to recognize the difference between waste that is potentially contaminated due to indirect causes and waste that is derived from processes which use input streams that contain radioactive materials; and

- The approach used for waste characterization must consider inhibiting factors associated with waste containing low concentrations of radionuclides, such as the ability to measure low levels of radiation and the knowledge of specific radionuclides present.

These factors are not always significant when characterizing non-varying waste or waste containing high levels of radionuclides generated from routine operations, but should nonetheless be considered.

Insights to Previous Waste Characterization Efforts

Information presented in this section is from previous activities that have been conducted in an attempt to characterize waste. These waste streams have been, are currently being, or may be generated at the Rocky Flats Plant located in Golden, Colorado, and the Lawrence Livermore National Laboratory, located in Livermore, California. Several approaches to characterizing waste have been and/or continue to be conducted at these Department of Energy (DOE) facilities. Of these approaches evaluated, the most extensive effort has been conducted at the Rocky Flats Plant.

From 1985 to 1987, nearly all operations at the Rocky Flats Plant were evaluated and waste streams associated with these operations were characterized (2). The strategy involved a combined use of process knowledge to initially classify each waste stream and, if determined to be necessary, subsequent sampling and analysis of specific waste streams to quantify select parameters. This information proved to be useful for general waste classification, but was by no means appropriate to adequately classify waste containing trace levels of radionuclides. In addition, the information remained static while the processes changed, and the information soon became outdated. Then, from late 1989 through 1990, a second attempt to characterize waste streams at the Rocky Flats Plant was conducted (3). The strategy resulted in waste characterization by the development of detailed process descriptions and process flow diagrams based mostly on knowledge of the processes. These efforts are not completely adequate for use in characterizing waste containing trace levels of radionuclides that have been packaged in containers since this waste cannot always be linked back to, and the data cannot be directly correlated to, a specific waste stream. Often this is due to poor quality control or inadequate documentation. In addition, complete material balances were not performed in these efforts, and it can be difficult to apply process knowledge in order to determine precise quantities of constituents without knowing the specific quantities of material originally input in the process.

Characterization efforts at the Lawrence Livermore National Laboratory have also been undertaken at various levels. Waste stream characterization is primarily the responsibility of the person generating the waste, and there has not been an attempt to perform a site-wide characterization effort such as those conducted at the Rocky Flats Plant. Although this approach is appropriate since much of the waste generated is highly variant and irregular, it does cause several problems. Among these is the fact that the level of characterization often varies from area to area, depending on the training and knowledge of the personnel conducting the characterization. Additionally, waste from several generators may be consolidated, and the ability to evaluate the resultant product based on the compilation of several sets of waste characterization data is difficult, particularly for evaluating the waste for trace levels of radionuclides.

The point in presenting this information is that the adequacy of the waste characterization is based on establishing complete objectives. Objectives for obtaining adequate information for waste characterization can often be incorporated into the study of a particular waste, including sampling strategies and analytical parameters, as well as being incorporated into operations and procedures. In both cases, the facilities have had difficulty in receiving authorization for the disposal of waste containing trace level of radionuclides as well as for waste with higher levels of radionuclides.

Those waste streams that contain low levels of radionuclides as identified by waste characterization efforts at both the Rocky Flats Plant and the Lawrence Livermore National Laboratory have been placed into waste form categories based on their physical form and/or nature of generation and are presented in Table I. The purpose of categorizing the waste streams by waste form is to streamline the amount of information that would be required to present a methodology for each waste stream. Rather, the methodology has been developed based on the waste categories and subcategories identified; factors such as homogeneity and the ability to sample are also significant to this categorization.

Development of Waste Classification Methodology

Although it is not the intention of this paper to develop a diminimus or BRC limit, a regulatory threshold (RT)[1] parameter for use with radioactivity and radionuclides is introduced. This parameter serves as the baseline justification for the assessment of concentrations of radionuclides, and it could be applicable towards evaluation of a diminimus or BRC limit. Regardless of this capability, the methodology presented is functional for radioactive waste characterization independent of any specified threshold limits.

On the other hand, if it is assumed that threshold limits exist for radioactive waste classification, analysis using essentially the same approach as that used by the EPA to determine whether or not a waste is hazardous, based on the concentration of the hazardous constituents, can be applied (4). For such an assumption and based on a normal distribution or large sample size (e.g., $n > 30$), the following confidence interval (CI) can be established:

$$CI = x_A \pm t_\alpha s_x, \qquad (1)$$

1 "Regulatory threshold" as it pertains to waste characterization is defined as a level established by a regulating agency that cannot be equaled or exceeded for a waste to be classified within a designated category, where the upper limit of this category is equal to the RT.

Table I - Waste Form Categories

Waste Form Category		Waste Stream Type[1]
Liquids		Acid, Cyanide, Oils, Paints, Decontamination and Decommissioning (D&D) Liquids
Sludges		Bypass Sludge, Cutoff Sludge, Electrochemical Milling Sludge, Sewer Sludge
Solids	Homogeneous Composite	Contaminated Dirt/Soil, Composite Chips, Incinerator Ash, Fluorescent Lights, Metal Chips, Ground Glass, Shot Table Debris[2], Roaster (Uranium) Oxide
	Heterogeneous Composite	Organic Combustibles/Compactible Waste, Oil Dry, D&D Solids Inorganic Contaminated Dirt/Soil, Construction Debris, Insulation, Labware, Metals (Heavy and Light), D&D Solids
	Surface Contaminated	Lead, Equipment, Sheet metal

1. Typical waste streams containing low concentrations of radionuclides at the Rocky Flats Plant, Golden, Colorado. Several of these waste stream types are also generated at the Lawrence Livermore National Laboratory, Livermore, California.
2. Shot Table Debris generated only at Lawrence Livermore National Laboratory.

where x_A is the mean sample activity concentration measurement, t_α is the Student's t statistic at a level of significance, α, and s_x is the standard error of the mean. Assuming that this analysis is used for comparison to an RT limit for gross alpha, gross beta, X and gamma radiation, or a specific radionuclide, a waste would be in excess of the RT limit when the upper bound of the confidence interval is greater than this limit such that:

$$CI_{Upper\ Limit} = x_A + t_{\alpha/2}s_x > RT. \qquad (2)$$

The RTs established would likely include limits for maximum gross alpha, gross beta, X and gamma radiation, and in addition, each radionuclide could conceivably have an RT concentration limit to account for the specific energies of radionuclide decay products. RT limits could also be specific for particular waste forms or processes associated with the generation of the specific radioactive waste. However, since RT limits for waste containing trace levels of radionuclides have not been established, the actual applicability is somewhat subjective at this point in time.

Analysis of radionuclides can be conducted conventionally by chemical and isotopic separation, counting of decay products, radionuclide spectrometry[2] of decay products, or by a combination of these techniques. Alpha, X, and gamma radiation activity concentrations can be determined effectively using spectrometry. Alpha particles have a definite range dependent on the absorbing material and initial energy, and the specific ionization is characteristic along a Bragg[3] curve. X and gamma radiation emitters release a range of photons with characteristic energies and decay probabilities. Beta particles interact erratically, making it difficult to differentiate between specific radionuclides. Because of this, radionuclide concentration determinations for beta emitters normally require chemical and/or isotopic separation prior to counting when several beta emitters are involved.

Radiation counting and its effectiveness relies the use of calibration standards with energies equivalent to those to be measured. Factors affecting the counting efficiency for alpha and beta particles are: geometry, backscatter, and self-absorption. The amount of self-absorption is dependent on the sample and must be established. This step requires the generation of calibration curves using samples of varied weight with known activity concentrations (5). Geometry and backscatter are properties of the detector configuration. For all cases, a lower minimum detectable concentration can be achieved for systems with high counting efficiencies, large sample sizes, and a low standard deviation for the background radiation. Thus, if these factors can be adjusted or modified so as to produce maximum benefit, the lower concentrations of radionuclides can be measured with greater accuracy and precision.

Container counting relies on the same principles as does radiation counting of small discrete samples. For X and gamma radiation, geometry is the most significant factor affecting the counting efficiency. For larger geometries, such as 55-gallon drums, as the sample material increases in density and as the atomic numbers of the elements composing the materials increase, the detection efficiency decreases. Calibration of gamma detectors to be used for waste container configurations can be conducted by counting a standard with known radionuclide concentrations for a given geometry. Specifically, for X and gamma radiation detection, a lower minimum detectable concentration can be achieved for detection of radionuclides with X and gamma ray decays of high energies, for high levels of radioactivity, and for a long length of count time. The minimum detectable concentration is impacted by increased branching ratios of decaying X and gamma rays, number of radionuclides, and overlapping of gamma energies from different radionuclides.

Liquids

Liquid waste can normally be sampled representatively using conventional means such as a Coliwasa or sampling tube. The major consideration associated with sampling liquid waste is the potential for stratification. Provided that a representative unit sample can be obtained, laboratory analysis can also be conducted using conventional means. In general, there are few problems associated with conducting representative sampling and analysis of liquid waste. The problems that arise tend to be with the interpretation of the results.

Waste oils at the Rocky Flats Plant, generated from the replacement of oil in vehicle engines and machinery crank cases, were sampled and analyzed in 1989 as part of a routine procedure to determine the level of radioactivity prior to off-site shipment. Although, there is no reason to believe that these oils were subject to contamination, the results indicated the presence of alpha activity. From a liability standpoint, DOE would not allow the oils to be shipped off site, except as radioactive waste. In final resolution, samples were taken from a large number of drums of new oil coming on plant-site, and these were analyzed to determine the total alpha activity concentration. From these results, a 95% confidence interval was established and the total alpha activity concentration mean value was calculated to be 0.08

---

2 Spectrometry of radionuclides is conducted by measuring the energies of decay products (either particles or photons) and determining the originating radionuclide based on the often unique or characteristic properties of the decay products.

3 The Bragg Curve represents a measurement of the specific ionization of an alpha particle along a distance traveled. The energy loss increases continuously to a maximum and then rapidly drops to zero.

pCi/g (6). Waste oils drums that are determined to have an alpha activity concentration within two standard deviations of the mean value are allowed to be shipped off site to a nonradioactive treatment facility. All other waste oils are currently handled as low-level radioactive waste. This resolution has essentially established the background level for vehicle engine and machinery crankcase waste oil generated to be 0.08 pCi/g. This method can be a valuable approach for other waste that is of the same general form as the original product, and for waste that is not suspected to be contaminated.

In another example of a waste stream at the Rocky Flats Plant, for oil waste destined for incineration in the fluidized bed incinerator, radiochemistry analysis was conducted by ashing 10 g samples to determine gross alpha, gross beta, and uranium and plutonium isotopic concentrations. The minimum detectable concentrations for alpha and beta were determined to be 0.4 pCi/g and 0.9 pCi/g, respectively. For most organic-based liquid samples that can be ashed, aliquots can be prepared and analyzed based on the establishment of a single minimum detectable concentration. This is possible since the minimum detectable concentration is based on physical size of the sample and standard deviation of the background, and these factors can be assumed to be independent for any given number of aliquots taken from the same sample. The counting efficiency for the measurement of the same nuclides as were measured for this study, assuming the sample matrix to be similar (e.g., ash), would be essentially the same since the backscatter factor would be constant and the factors that affect the counting efficiency are independent of the sample.

The Department of Transportation (DOT) defines radioactivity as any material having a specific activity greater than 0.002 µCi/g (2000 pCi/g)[4]. The mean values for oil samples were determined as follows: 23 pCi/g for gross alpha; 10.65 pCi/g for gross beta; 9.3 pCi/g for Pu-239,240; 4.3 pCi/g for U-233,234; 0.21 pCi/g for U-235; and 5.7 pCi/g for U-238. The specific activity can be compared by summation of the mean values for gross alpha and gross beta or by summation of the mean values of individual radionuclides. In doing so, the sum of the gross alpha and gross beta equals 33.65 pCi/g and the sum of the individual radionuclides equals 19.51 pCi/g. The difference can be accounted for based on the better counting efficiency for the gross determinations and due to the omittance of other potentially present radionuclides unaccounted for when summing individual radionuclides. The more conservative determination of 33.65 pCi/g will be used for the remainder of this discussion. For this particular case, it was initially assumed that the waste was radioactive. If a 95% CI is established, then the upper limit of the CI (Equation 2) is 23.75 pCi/g for alpha and 12.67 pCi/g for beta, and the sum is equal to 36.4 pCi/g. Using the DOT definition as the RT limit, the waste would not be classified as radioactive.

Aqueous waste samples can be treated in much the same way and with similar expectations for the minimal detectable concentration limit. The limit, again, will be influenced most by the self-absorption factor of the prepared sample residue. The counting efficiency will be correspondingly affected since the more self-absorption that occurs, the lower the efficiency will be and with greater bias.

Sludges

As with liquids, sludge waste can normally be sampled representatively by conventional means. In many cases, it may be practical to sample the sludge as a fluid prior to solidification. The major consideration associated with sludge waste is the moisture content of the sludge. This factor will have the largest influence on the unit mass of the waste and hence also on the specific activity. As a good practice, the activity concentrations should be analyzed and reported based on the dry sludge weight.

As an example, for a sludge waste generated at the Rocky Flats Plant and classified as radioactive, the mean alpha activity concentration was determined to be 59.85 nCi/g. This level or radioactivity leaves little doubt as to whether the waste is radioactive. In addition, based on process knowledge, it is known that radionuclides are discharged as effluent into the streams that make up the sludge sediment. This was not the case for the oils destined for incineration, which are also generated from processes with

---

[4] The specific activity is defined as the total activity per gram of material (7).

known use of radioactive material but resulted in a low level of radioactivity in the pCi/g range. The basis for this difference is the fact that the radioactive particulate material is concentrated (by design) through the process to produce sludge, whereas the oils are not subjected to this type of treatment. For both cases, the use of the CI (Equation 1) was of benefit in classifying the waste; one waste stream classified as nonradioactive and the other as radioactive. Without this classification ability, both waste streams would typically be handled and disposed as radioactive waste.

A classic example of this situation is the sewer sludge waste generated at the Rocky Flats Plant. All currently identified waste streams that feed to the Sewer Treatment Facility are nonradioactive, and hence the sewer sludge produced should also be nonradioactive. Sewer sludge waste samples were taken in 1986, 1989, and 1990. The results indicate mean levels of radioactivity ranging from 39.78 pCi/g to 75.27 pCi/g for gross alpha and 36.67 pCi/g to 83.7 pCi/g for gross beta. Although these values are low (pCi/g range), the conservative practices followed at the Rocky Flats Plant dictate that since radioactive material is handled at the facility and there is a potential that the measured levels are a result of contamination from radioactive material, there is no current alternative other than to handle the sewer sludge as low-level radioactive waste. Based strictly on a 95% CI and using the DOT limit of 2000 pCi/g as the RT, the waste would not be radioactive; the corresponding upper limits of the CI ranged from 50.66 pCi/g to 92.32 pCi/g for gross alpha and 45.07 pCi/g to 110.4 pCi/g for gross beta. Additionally, a study of sludges from typical wastewater treatment facilities (those that do not receive effluent from facilities that handle radioactive materials) could produce valuable information regarding the levels of radioactivity seen in a typical sewer sludge waste. It is suspected that, due to the presence of radionuclides in the environment and based on the sludge formation process (e.g., particulate material concentration), similar radioactivity levels will be observed.

Solids

The solid waste form category consists of three subcategories: homogeneous composite, heterogeneous composite, and surface contaminated. Each subcategory requires a different approach to assess the radionuclide content. In general, homogeneous waste forms can be assessed using conventional techniques (as already discussed for liquids and sludges), whereas heterogeneous waste form assessments can become exceedingly complex.

Homogeneous Composite. Homogeneous composite waste can often be sampled using either a trier, thief, or auger. The largest volumes and most common form of homogenous composite solid waste tends to be contaminated dirt and soil. The factors affecting characterization of soils are weighted heavily on a determination of background radiation. At the Rocky Flats Plant, an extensive soil sampling program to establish background levels has been ongoing for over fifteen years. Typical background levels of plutonium concentrations in areas west of the Rocky Flats Plant and not suspected to have been contaminated as a result of any activity range from 0.02 to 0.16 pCi/g (8). A detailed discussion of the complex regulatory and legal aspects surrounding this program is purposely avoided here, since it would contribute minimally to the technical components of the characterization methodology. Instead, the discussion will continue to focus on example waste streams.

At the Lawrence Livermore National Laboratory, shot table debris is generated as a result of the detonation of test assemblies using high explosives. Generally the test materials are nonradioactive, although depleted uranium is used in a small portion of the total number of experiments conducted annually. The shot table debris was sampled using a simple scoop and the mean concentration was determined to be 0.34 nCi/g for gross alpha and 0.58 nCi/g for gross beta. These values are between the values measured for the two sludges discussed above. Based on the process, the alpha and beta activity concentrations were expected to have been lower (closer to those for sewer sludge) than the results indicate. These values are also higher than those of the oil waste stream that is known to have been in direct contact with radioactive material. This discrepancy may be due to incorrect assumptions about the amount of radioactive material used in the process or due to the wide dispersion of radionuclides, as a result of detonation, which may increase the probability of finding radioactive material in a randomly selected sample.

Heterogeneous Composite. The most significant consideration in any approach to the characterization of heterogeneous wastes is the variability of the waste stream. For heterogeneous waste forms with low variability, stratified sampling techniques that account for each matrix may be implemented. As an example, a liquid waste containing both an oil-based and aqueous-based component will eventually separate, assuming it is undisturbed, and the layers would be identifiable as separate matrices. Once each matrix has been physically identified, sampling of the matrix is conducted in the same manner as for homogeneous composite waste. The analysis techniques are also essentially the same. Data interpretation should be reported for each matrix separately and as weighted average values based on the mass ratio.

Of the various waste streams that make up the category of heterogeneous composite waste, combustible waste streams, by far, have been the most difficult to characterize. The problem with combustible waste is related to the number of different methods by which a 55-gallon drum or similar container of combustible waste may be generated. Combustible material may consist of anything from wipes and rags to disposable clothing and, with poor controls, administratively and operatively, other waste such as debris and metal pieces may become part of this waste stream. Combustible waste may be generated from direct interaction with radioactive material, but a large portion is only suspected to contain radioactivity due to the indirect contact with possibly contaminated materials or equipment.

EPA has an established method for spatially sampling using a simple or stratified random sampling approach as presented in Chapter 9 of SW-846 (4). This approach incorporates the division of the container into a three-dimensional imaginary grid and random selection of the grid sectors for sampling. Unfortunately, this method cannot account for a high potential for variation within the waste drum, much less for the compounded variation created by the generation of several drums from within the same operation or facility. Adequate representation by sampling requires a large number of samples, and no formal attempts have been made to characterize this specific waste based on these types of sampling approaches. Additionally, due to the problem with obtaining a representative sample from within a single waste drum of heterogeneous waste, characterization techniques that consider the entire contents of the drum are recommended. Subsequently, without supplementary information regarding the activities associated with the generation of a container of waste, such as log books, or without extremely tight operational controls, adequate characterization may require that all containers be sampled in order to obtain statistical validation to any significant level of confidence.

The evaluation of a waste can be conducted using several approaches. The two most likely methods to be employed are (1) nondestructive radiation assaying and (2) destructive evaluation by incineration, to obtain an ash for sampling. If the second approach is used, representative sampling of the ash can normally be conducted. The inherent problems associated with this approach involve, on a drum-by-drum basis, the initial ability to incinerate the combustible waste (and achieve complete destruction) and to obtain a homogeneous ash sample. If the number of drums required to be sampled[5] is exceedingly large, based on the desired level of confidence, it may be pointless to develop an initial characterization strategy for a waste form that will be subjected to additional treatment and chemical transformation. Although, if the initial sampling could be used to determine whether the waste was nonradioactive, then the subsequent treatment could be done on a nonradioactive basis. In the case where the number of drums required to be sampled is exceeding large, the proper initiative would be to develop a two-phase characterization plan: pre-treatment characterization to assure the waste may be safely incinerated, and post-treatment characterization of the ash that could include representative sampling, laboratory analysis, and data interpretation.

A nondestructive approach by radiation assaying may prove to be beneficial for this waste category, because of the effort required to conduct a destructive evaluation. Although no technique for radiation assaying at low levels is currently established, several factors make this approach appear feasible. The first factor is the fact that X and gamma radiation can be detected at low energies in the keV range, the

---

[5] Refer to Chapter 9 of SW-846 (4) for a detailed discussion on how to determine the appropriate number of samples to be collect… Additionally, statistical methods and practical sampling strategies can be found in other sources (e.g., Refs. 9 and 10).

second is the fact that combustible waste is composed of low density materials consisting of low atomic numbers. This is desirable since the attenuation (self-absorption) of X and gamma radiation is directly proportional to the atomic number of a material. In the same way that the counting efficiency for aliquot samples is affected by self-absorption, the counting efficiency for a drum of waste increases for a waste material in which the X and gamma radiation attenuation decreases. Factors affecting the minimum detectable concentration are the sample size, standard deviation for the background radiation, energies of the X and gamma rays, branching ratios of decaying X and gamma rays, number of radionuclides, overlapping of gamma energies from different radionuclides, and length of count time. Of these factors, for a given container of waste, control over the standard deviation of the background and the counting time are obtainable. All other factors are characteristics of the waste or of the radionuclides under consideration.

For a given waste container, the geometry is preset, and for this study the waste container will be assumed to be a 55-gallon drum. The ideal situation would be to use a detector that fully encloses the drum for counting. The practical approach is to place several detectors about the drum. Engineered controls can be built to minimize and control background radiation such that the standard deviation of the background is reduced to a significantly low value. Lastly, the length of time measurement can be increased within the practical limitations of operational feasibility.

A theoretical study was conducted using Microshield®[6] (11) to evaluate the interaction of X and gamma radiation emitted from radionuclides suspected to be present in combustible waste generated at the Rocky Flats Plant. Microshield is used primarily to analyze the shielding of gamma radiation. It requires the development by the user of geometric dimensions, material and shielding properties, source strength, buildup factors and integration parameters. The interaction of photons emitted from the source are analyzed for the specified parameters to determine the resultant flux at a given point. The source strengths by energy group, based on the radionuclides selected are determined by Microshield. The resultant source strength is the sum of the individual gammas from all radionuclides selected, which are sorted into energy groups and are in units of photons emitted per second. The model assumes steady-state conditions (e.g., in this case it assumed that the change in the decay rate of radionuclides in the specified source is negligible). This assumption is based on the long half lives of radionuclides Pu-239, Am-241, and U-235 used in this study. The half lives are 24,360 years, 433 years, and 7.1 x 10^8 years, respectively, and thus this assumption is justified. For the cylindrical geometry, the gamma exposure was calculated by point-kernel numerical integration.

The first step of the assessment consists of the development of a combustible waste with a hypothetical concentration of radionuclides present. Radionuclide concentration ratios for Pu-(239,240), Am-241, and U-235 were based on a sludge waste, assumed to be representative of general plant-wide operations. The average ratio for Pu-(239,240) to Am-241 was calculated to be 0.676, and the average ratio of Pu-(239,240) to U-235 was calculated to be 0.0194. Using these ratios, a hypothetical source for the combustible waste was built for analysis. Further studies to determine more precise ratios of radionuclides in combustible waste may be warranted.

A rather simplistic physical waste form was created to evaluate the attenuation and resulting dose rate at the outer surface of the waste drum. This waste form, in its initial form, possesses the chemical make-up of cellulose and the physical density of paper and is expected to exhibit average properties of a very general category of dry combustible waste containing a high concentration of paper products. For a waste drum of dry combustibles, the two parameters that are most likely to vary are (1) the density of the waste package, due to packing efficiency and void volume, and (2) the actual concentration of the radionuclides present. Parametric analysis by variation of these two parameters was performed to determine the theoretical dose point flux on the surface of the drum at the axial center. The dose point flux variation was nonlinear for a change in density. Based on this observation, it can be said that, for the hypothetical waste form, the dose point flux varies exponentially and it is due to

[6] Microshield is a microcomputer adaptation of the main frame code ISOSHLD and is used to analyze gamma radiation shielding. The solution algorithms and physical data are the same as for ISOSHLD^^ '

gamma attenuation by the waste material. The dose point flux varied linearly for the change in concentration (constant density), as would be expected. This is supported by review of the fundamental understanding of gamma radiation interaction with material represented in the following equation:

$$I = I_o e^{-\mu x} \tag{3}$$

where I is the resultant gamma radiation intensity, I_o is the original gamma radiation intensity, μ is the gamma attenuation coefficient for a given material, and x is the thickness of the material (12). Additionally, $\mu = \mu_m \rho$, where μ_m is the mass attenuation coefficient[7] for a given material and, ρ, is the density of the material. Thus a change in the density ρ would cause the resultant intensity to vary exponentially from the original intensity. Likewise, since the original intensity is a direct result of the concentration, a change in concentration would result in a linear change in the intensity, provided that the density remained constant.

A hypothetical waste form containing a void volume of 50% was used to calculate the dose point flux and dose rate on the outside surface of the axial center of a 55-gallon stainless steel drum containing the source waste. A radioactive source consisting of a concentration of 100 pCi/g of Pu-239 and appropriately proportioned concentrations of Am-241 and U-235 was used. The mass of the waste in the drum, based on a 50% void volume, was calculated to be 75.10 kg (165.57 lbs). The total activity for each radionuclide was 4.2131×10^{-6} Ci, 7.5099×10^{-6} Ci, and 1.6897×10^{-7} Ci, for Am-241, U-235, and Pu-239, respectively. For practical applications, an adaptation of Microshield could be developed to back-calculate the concentration of specified radionuclides based on the observed dose point flux and dose rate. Specifically, the adaptation would allow for the user input of detector efficiencies, geometries, material properties, and specific radionuclides present with corresponding ratios. With this information, the concentrations of each radionuclide could conceivably be determined. Of course, it is suggested that additional analysis of the physical form of each waste (as well as radionuclide concentration ratios) be conducted to determine the feasibility of such implementation. Factors that require assessment include the determination of the detection capabilities of instrumentation for the photon energies of radionuclides at low concentration levels for a given geometric configuration and the ability to establish values for these counting efficiencies. Also, accurate determination of attenuation factors for each material is necessary. The ability to determine the ratio of one radionuclide to another may not be as crucial since often the characteristic photons emitted are distinguishable. Other considerations would include the feasibility of implementing engineered controls to achieve a minimum standard deviation for background radiation and the length of count time requirements.

Without this type of approach, the precision of the characterization of heterogeneous waste is limited. In addition, this approach could be practical for other waste streams that are composed of materials with low densities and low atomic numbers. Other types of waste, such as construction debris and glass waste, are potential candidates.

Surface Contaminated. Contaminated equipment, lead bricks, and sheet metal are the principal types of surface-contaminated waste. The level of surface contamination for lead bricks and sheet metal can be determined by performing smears using absorbent wipes (for a representative surface area) and measuring the removable contamination. In addition, a direct radiation assay (for a representative surface area) can be conducted to measure the fixed or present level of radiation. For this waste, the techniques are well established and the levels of contamination can be evaluated directly against established limits. The level of contamination for equipment such as pumps, motors, and machinery is extremely difficult to determine (due often to the lack of a representative surface area), and there is often no alternative other than to conservatively classify potentially radioactive waste as radioactive. In general, there are no known feasible methods available for quantitatively assessing these types of contaminated equipment.

---

[7] Mass attenuation coefficients can be found in a number of reference sources such as the Radiological Health Handbook (13).

Waste Classification Methodology

The methods for radioactive waste classification are laid out schematically in Figures 1 through 3 (these figures are discussed in the subsections below). As a general rule, the methodology errs on the conservative side by forcing a waste that cannot be positively classified at a given level to be evaluated at the next higher level of radioactivity classification.

Liquids, Homogeneous Composite Solids

The starting point for this characterization method (See Figure 1) is the assumption that two conditions are possible: (1) the waste is only potentially contaminated or, (2) the waste is known to be radioactive based on known contamination or direct contact with radioactive material.

For waste that is only potentially contaminated, it is necessary to first determine whether a direct comparison to the original product or input stream (assuming the input stream is not radioactive) can be conducted. If a direct comparison to a waste stream can be conducted, a waste stream-specific diminimus limit[8] can be established based on the development of a 95% CI for the original product or input stream. The waste stream-specific diminimus limit is the upper bound of the CI. Waste that is below the waste stream-specific diminimus limit is not considered to be radioactive. In addition, if there are waste sample points above the waste stream-specific diminimus limit, the 95% CI for the waste samples must be determined and the upper bound of the CI must be below the waste stream-specific diminimus limit. It is important to ensure that the samples taken for the original product or input stream are representative of the volume of waste to be sampled. The number of samples of original product must be equivalent to or greater than the number of waste samples. If the number of samples is equivalent, a Student's t-paired test is also recommended to evaluate the means. If a direct comparison cannot be made, representative sampling of the waste shall be conducted and a 95% CI developed.

Likewise, if a direct comparison to a waste form can be conducted, a waste form-specific diminimus limit[9] can be established based on the development of a 95% CI for the original product(s) or input stream(s). The probability of having a waste form for which a direct comparison to the original product or input streams can be made is less likely than that for a waste stream. However, the potential exists, so this scenario has been considered. If a general diminimus limit[10] can be established, and if the upper bound of the 95% CI is less than the general diminimus limit, the waste is not considered to be radioactive. For a waste that is classified as radioactive based on known contamination or from direct contact with radioactive material, representative sampling of the waste shall be conducted and a 95% CI developed. If a general diminimus limit can be established, and if the upper bound of the 95% CI is less than the general diminimus limit, the waste is not considered to be radioactive. If a general diminimus limit does not exist or the waste is determined to be above the general diminimus limit, and if a BRC limit can be established and the upper bound of the 95% CI for the waste is below the limit, the waste is a BRC radioactive waste. Lastly, if a BRC limit does not exist, or the waste is determined to be above the BRC limit, the waste must be classified as a low-level radioactive waste.

Sludges

The method for characterizing sludge waste (See Figure 2) differs from that for liquids and homogeneous solids in that sludge waste is not originally introduced in this form as an input stream.

---

[8] A waste stream-specific diminimus limit is specific to the waste stream under evaluation and must be based on establishment of a 95% confidence interval for the original product or input stream.

[9] A waste form-specific diminimus limit is specific to the physical waste form under evaluation and must be based on establishment of a 95% confidence interval for the original product(s) or input stream(s).

[10] A general diminimus limit is general for a waste stream or waste category and may be developed based on similar waste streams, similar waste categories, similar physical waste forms, background, or other criteria.

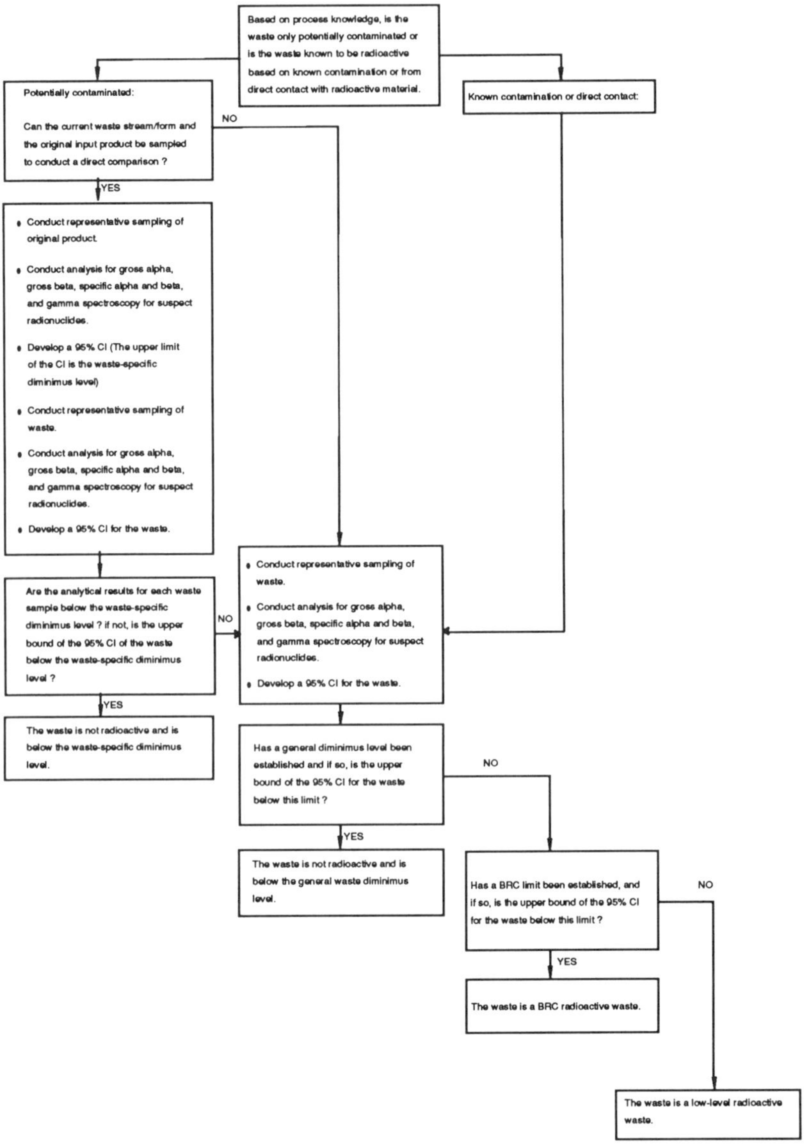

Figure 1. Classification Methodology for Liquid and Homogeneous Composite Solid Waste

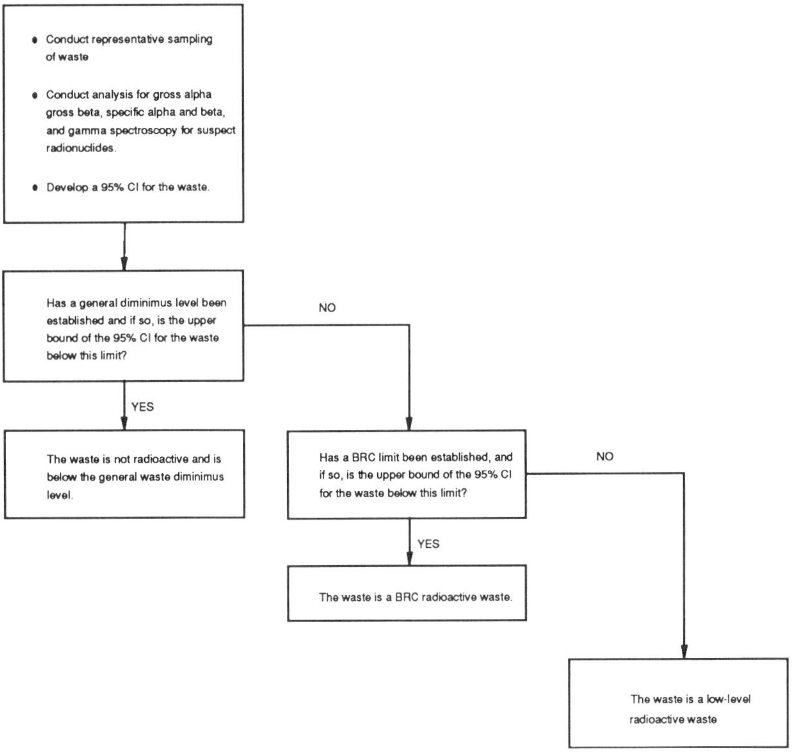

Figure 2. Classification Methodology for Sludge Waste

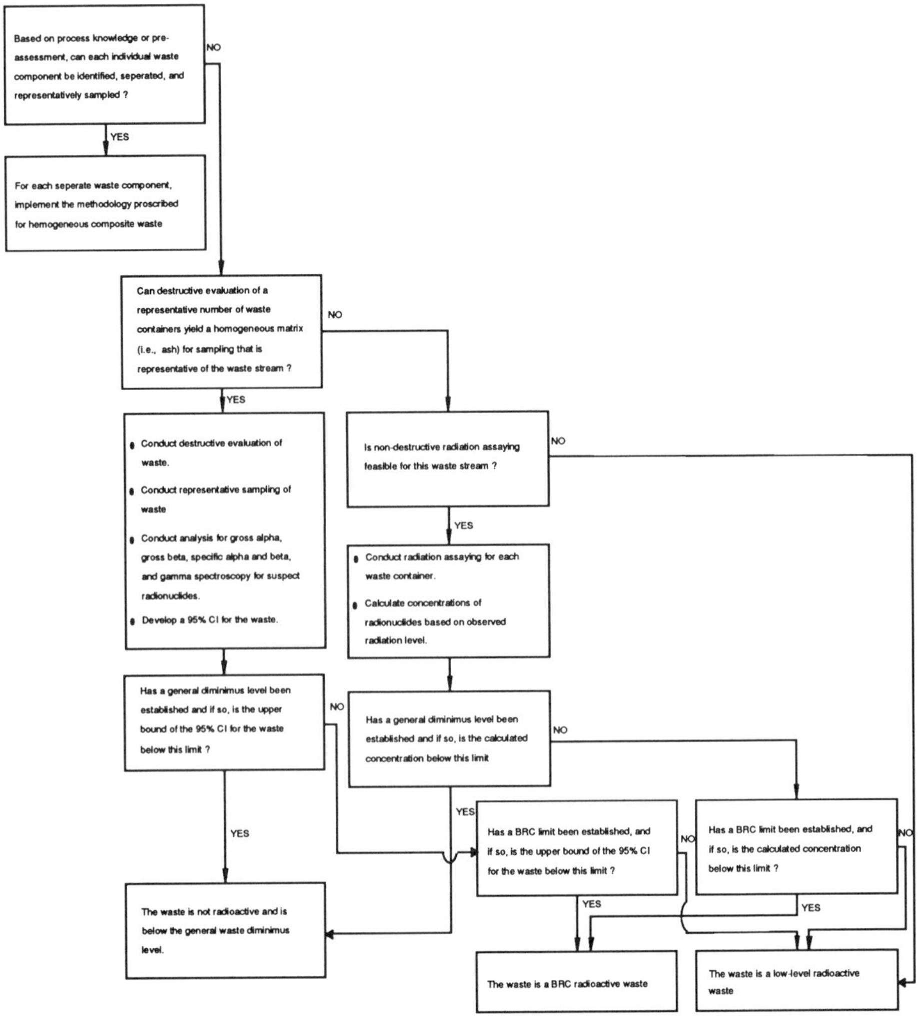

Figure 3. Classification Methodology for Hetergeneous Solid Waste

Thus, a direct comparison to the original product or input stream cannot be made. As previously mentioned, it is recommended that programs be conducted to evaluate sludge waste from facilities that do not handle radioactive waste. Waste processes that concentrate particulate material (including the unintentional concentration of radioactive particulates) can create unwarranted problems, particularly for facilities that handle radioactive waste, because there is always the possibility that contamination has occurred and any indications of elevated levels causes such concern. Although this possibility still exists at facilities that have no association with radioactive materials, the probability is low, and radioactivity levels observed could be used to establish a general diminimus limit.

Heterogeneous Composite Solids

The characterization method (See Figure 3) proposed for heterogeneous waste streams is, by far, the most complex. The initial activity is the assessment of whether, for each individual waste component that makes up the waste, identification, separation, and representative sampling can be conducted. If so, the methodology for homogeneous composite solid waste must be applied for each individual waste component. If not, then a determination must be made as to whether destructive evaluation of a representative number of waste containers will (1) yield a homogeneous matrix that is (2) representative of the waste streams under evaluation. For a specific small inventory of combustible waste, complete incineration could be used to fulfill both criteria. Once these criteria have been met, sampling, analysis and data interpretation would be the same as for homogeneous composite solid waste. Note that, to be effective, a homogeneous matrix must be created and must be representative of the waste stream. This is often impossible. As an example, the incineration of a container of combustible waste will yield an ash that can be thoroughly mixed and will qualify as homogeneous, but the single container is not representative of other combustible wastes, in general, because of the high variability of this waste stream. For an accurate representation of Rocky Flats Plant combustible waste, based on the known high variability, it is assumed that nearly all waste containers would require incineration.

If destructive evaluation is not selected, the feasibility of nondestructive radiation assaying must be assessed. This assessment must include the evaluation of detection capabilities for the suspect radionuclides, radiological properties of the suspect radionuclides, and physical properties of the waste form including the density and bulk gamma attenuation coefficient. The assessment may include actual parametric analysis using those parameters presented above. If nondestructive radiation assaying is determined to be feasible, each container should be assayed (after all appropriate calibrations have been conducted) and concentrations calculated. Based on these concentrations, a 95% CI shall be developed, and if a general diminimus limit can be established, and if the upper bound of the 95% CI is less than the general diminimus limit, the waste is not radioactive. If a general diminimus limit does not exist or the waste is determined to be above the general diminimus limit, and if a BRC limit can be established and the upper bound of the 95% CI for the waste is below the limit, the waste is a BRC radioactive waste.

If nondestructive radiation assaying is determined to be infeasible, process knowledge shall be used to conservatively classify the waste as low-level radioactive.

Surface Contaminated

The amount of surface contamination for solid objects (bricks, sheet metal) can be determined by performing smears using absorbent wipes (for a representative surface area) and measuring the removable contamination. In addition, a direct radiation assay (for a representative surface area) can be conducted to measure the fixed or present level of radiation. For this waste, the levels of contamination can be evaluated against established limits. For surface-contaminated equipment which cannot be representatively sampled and analyzed, the classification can be based only on process knowledge; thus, the waste shall be classified as low-level radioactive.

Results and Conclusions

The information in this study illustrates that there are many adverse factors affecting the overall characterization. Based on the examples presented, it is not likely that a few absolute limits can be established for use in waste classification. It is apparent that the limits must rely on a relative comparison to background levels of radiation, other similar waste forms and waste sreams, and that the methodology must employ sound scientific and engineering principles.

Engineered and operational controls can be implemented to minimize problems that are essentially technical issues. Other factors affecting the overall characterization include regulatory issues. Even though a waste containing a low concentration of radionuclides might be fully or adequately characterized, its classification into a category other than low-level can be inhibited by the lack of established lower limits (diminimus or BRC). As was illustrated, the methodology presented here relies on the ability to establish limits for classification. The initial step is to place the waste into one of two categories: wastes that are either potentially contaminated or waste that are definitely contaminated. Then, conditions for characterization can be placed on each category.

For liquids, sludges, and homogeneous composite waste forms, the techniques for sampling and analysis are well established and require minor adaptation for incorporation into this methodology. These techniques were termed "conventional sampling and analysis" and are commonly used in the industry. Heterogeneous composite waste stream characterization methods (destructive and nondestructive evaluation), although not as well established and by no means simple, could be implemented for a large number of waste streams, as this paper proposes.

Destructive evaluation techniques are recommended for low variant heterogeneous waste streams for which representative samples can be taken. This technique could be applicable to a wide variety of waste streams. Also, destructive evaluation techniques can be used for waste forms that are only occasionally generated or are limited in inventory, and for which radiation assaying is infeasible due to the complexity of calibrating for the waste form and suspect source radionuclides.

Radiation assaying may prove to be a viable method of characterization for many highly variant waste streams. For waste streams that are composed of high density material consisting of high atomic numbers, the ability to detect the source radionuclides becomes increasing difficult. If other factors can be enhanced that allow for a higher detection efficiency, such as reduced standard deviation of background radiation and greater length of counting time, then it may be possible to apply this technique to high density waste streams. Experimentation, using varied radioactive sources and waste forms, and comparison to theoretical expectations is the next obvious step for implementing a sound characterization program consisting of low-level waste radiation assaying.

For potentially contaminated waste, waste stream-specific and waste form-specific diminimus limits were introduced as a means of determining whether a waste is radioactive or not. For waste that can be compared directly to the original product, the basis for establishing a waste stream-specific or waste form-specific diminimus level is sound because the limit can be based on a measurable and discrete background level of radiation. General diminimus limits, by their nature, are expected to be more stringent than waste-specific diminimus limits, since the basis for determination of the limit may be a collective assessment of background radiation levels. This type of approach often requires the inclusion of a significant level of safety. BRC limits, by their definition, can be used in conjunction with this methodology once they are established by the regulators. The high variation in levels of radioactivity observed in this study would imply that BRC limits would need to be specific to the waste stream, waste form, or waste process, since nonspecific limits may not provide any greater subclassification than currently exists. In any case, if the methodology is followed through and it cannot be demonstrated that the waste contains no more than a diminimus level of radioactivity, nor have BRC limits been established, the waste can continue to be conservatively classified as low-level waste; the level at which this waste is currently classified.

Future research is required in several areas if this proposed methodology is to be implemented in full. First and foremost, the establishment of diminimus and BRC limits will be required and will demand a significant level of effort to be placed on determining appropriate and acceptable limits. This effort will undoubtedly challenge both the regulatory and technical community. Additionally, for gamma radiation assaying to become a method that can be commonly implemented for a wide variety of situations, the following steps must be taken:

- A systematic method must be developed for assessing the composition of waste streams, forms, and categories in order to establish consistent gamma attenuation coefficients for the waste in conjunction with calibration methods.

- A method must be established for qualifying potential radionuclides present in a given waste and for determining their relative ratios.

- Microshield (or similar program) must be adapted so that the software can calculate the radionuclide concentrations based on empirical dose point fluxes.

- An assessment must be made of the feasibility of implementing engineered controls to achieve a minimum standard deviation for background radiation such that the lowest possible concentration of radionuclides can be measured.

- An assessment must be made of the detection capabilities for photon energies of radionuclides at low concentration levels for given geometric configurations.

References Cited

1. Christopher R. Casias, "A Waste Characterization Methodology for Determining the Concentration of Radionuclides in Radioactive and Mixed Waste" (T-4020, Master's Thesis, Colorado School of Mines, 1991), T-4020.

2. "Waste Stream Identification and Characterization" (Rocky Flats Plant, Golden, CO: US Department of Energy, 1987).

3. "Waste Stream and Residue Identification and Characterization" (Rocky Flats Plant, Golden, CO: EG&G Rocky Flats Inc., 1990).

4. "Test Methods for Evaluating Solid Waste," SW-846, 3rd edition (PB88-239223, Washington, DC: US Environmental Protection Agency, 1986).

5. "Radiochemical Analytical Procedures for Analysis of Environmental Samples" (EMSL-LV-0539-17, Las Vegas, NV: Environmental Monitoring and Support Laboratory, US Environmental Protection Agency, 1979).

6. Robert Baker, private communications, EG&G Rocky Flats Inc., 1991.

7. "Transportation" Code of Federal Regulations, (49 CFR 173, Washington, DC: US Department of Transportation, 1988).

8. "Rocky Flats Plant Site Environmental Report for 1988" (RFP-ENV-88, Rocky Flats Plant, Golden, CO: Rockwell International Corporation, Aerospace Operations, 1988).

9. Richard O. Gilbert, Statistical Methods for Environmental Pollution Monitoring (New York, NY: Van Nostrand Reinhold, 1987).

10. L. H. Keith, Environmental Sampling and Analysis: A Practical Guide, (Michigan: Lewis Publishers, Inc., 1991).

11. "Microshield," Version 3 (Rockville, MD: Grove Engineering, Inc., 1987).

12. Arthur R. Foster and Robert L. Wright, Jr., Basic Nuclear Engineering, 4th edition, (Boston, MA: Allyn and Bacon, Inc., 1983).

13. Bureau of Radiological Health, ed, Radiological Health Handbook, (Rockville, MD: US Department of Health, Education, and Welfare, 1970).

WASTE MINIMIZATION ACTIVITIES IN THE MATERIALS FABRICATION DIVISION AT LAWRENCE LIVERMORE NATIONAL LABORATORY

J. W. Dini

For: TMS International Symposium on Processing of Residues and Effluents
March 1-5, 1992, San Diego, CA

Introduction

The mission of the Materials Fabrication Division (MFD) is to provide fabrication services and technology in support of all programs at Lawrence Livermore National Laboratory (LLNL). MFD involvement is called for when fabrication activity requires levels of expertise, technology, equipment, process development, hazardous processes, security, or scheduling that is typically not commercially available. Customers are encouraged to utilize private industry for fabrication activity requiring routine processing or for production applications.

MFD is a large diversified service organization including approximately 400 full time employees. The group provides a wide variety of processes including all common machining and fabrications operations, laser and water jet cutting, electrical discharge machining, lapping, electron beam and laser welding, vacuum brazing, heat treating, forming, electroplating and metal finishing, optical polishing and grinding, vacuum coating, vacuum engineering and a variety of plastics operations. These operations occupy more than 1,250,000 square feet of floor space. For more detail on the many fabrication related services available from MFD, see Reference 1.

Waste Minimization Program

Our waste minimization (WM) program has been directed at source reduction and recycling in concert with the working definition of waste minimization used by EPA and presented in Figure 1 (2). The principal focus of WM activities has been on hazardous wastes as defined by RCRA, however, all pollutant emissions into air, water and land are being considered as part of the program. The incentives include: 1-economics, 2-regulatory conformance, 3-public image and 4-environmental concern.

An effective waste minimization program consists of five items: 1-top management commitment, 2-goals, 3-employee training/awareness, 4-continuing attention and 5-materials management (3). All of these elements are in place in our program.

1. Top Management Commitment - Our Director, J. H. Nuckolls, issued a statement to all employees on March 8, 1989, that the Laboratory would

Residues and Effluents - Processing and Environmental Considerations
Edited by Ramana G. Reddy, William P. Imrie and Paul B. Queneau
The Minerals, Metals & Materials Society, 1991

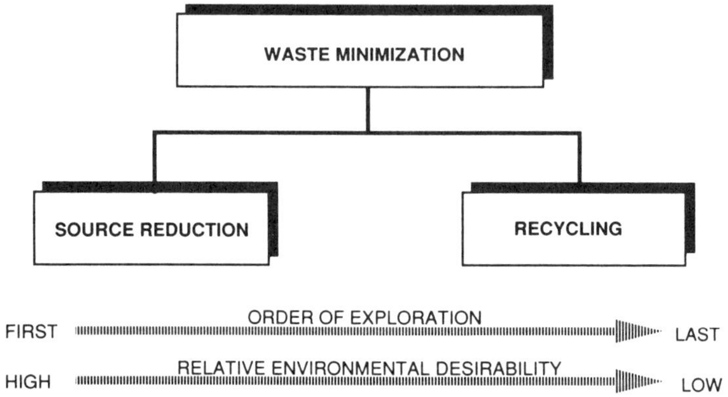

WASTE MINIMIZATION

The reduction, to the extent feasible, of hazardous waste that is generated or subsequently treated, stored or disposed of. It includes any source reduction or recycling activity undertaken by a generator that results in either (1) the reduction of total volume or quantity of hazardous waste or (2) the reduction of toxicity of the hazardous waste, or both, so long as such reduction is consistent with the goal of minimizing present and future threats to human health and the environment (EPA's Report to Congress, 1986, EPA/530-SW-86-033).

SOURCE REDUCTION

Any activity that reduces or eliminates the generation of hazardous waste at the source, usually within a process (op. cit.).

RECYCLING

A material is "recycled" if it is used, reused, or reclaimed (40 CFR 261.1 (c) (7)). A material is "used or reused" if it is either (1) employed as an ingredient (including its use as an intermediate) to make a product; however a material will not satisfy this condition if distinct components of the material are recovered as separate end products (as when metals are recovered from metal containing secondary materials) or (2) employed in a particular function as an effective substitute for a commercial product (40 CFR 261.1 (c) (5)). A material is "reclaimed" if it is processed to recover a useful product or if it is regenerated. Examples include the recovery of lead values from spent batteries and the regeneration of spent solvents (40 CFR 261.1 (c) (4)).

Figure 1 - Waste Minimization Definitions per EPA (from Ref. 2) make a strong commitment to waste minimization (4), and this was followed by a similar policy statement by MFD management.

2. Goals - The Goal was to reduce the Laboratory's hazardous/radioactive waste streams by a major factor in the next three to five years.

3. Employee training/awareness - All employees were made aware of the program and goals and also were given a training class on Hazardous Waste Handling Practices. A member of the senior management team in MFD (J. W. Dini) was appointed waste minimization officer with the responsibility of serving as a focal point for all WM activities and also keeping all personnel informed on WM.

4. Continuing attention - This is done by publishing a bimonthly bulletin entitled *Waste Notes* (Figure 2). These address hazardous waste management, recycling, and waste minimization. Each issue spotlights personnel and operations that have successfully implemented waste minimization efforts. Waste Notes also provides information on regulatory requirements and division efforts regarding waste minimization. In addition, various shops and operations also hold briefings and issue memoranda on waste minimization to inform employees and increase their awareness for waste minimization opportunities. The waste minimization officer also provides informal talks at group meetings on the waste minimization program and its progress.

5. Materials management - This is on-going and is essentially some of the information that will be discussed in this paper.

Waste Minimization Initiatives

Within the past two years waste minimization initiatives implemented in MFD have resulted in cost savings of over $500,000/year. In addition, air quality has improved and water usage has decreased significantly. Table 1 provides a listing of the key initiatives and the savings they provided. These will be discussed in some detail in the following sections.

A - Recycling Machining Coolant

The Main Bay Machining Facility houses 33 major machine tools using 1,100 gallons of coolant that require recycling. In the past, spent coolant was stored in 55 gallon drums, sampled at MFD, and processed through HWM (Hazardous Waste Management). Since January, 1989, a XYBEX 1000 system and its partner the "Yellow Bellied Sump Sucker" have reduced coolant hazardous waste generation by 85%. The system consists of a skid mounted unit containing dirty and clean coolant reservoirs, an automatic coolant proportioning pump, an oil belt skimmer, a high speed centrifuge, electrical controls, and PVC plumbing.

Figure 2 - *Waste Notes,* a bimonthly publication directed at waste minimization topics

Table 1

Waste Minimization Initiatives Undertaken Within the past 1-2 Years in MFD

Initiative	Work Center	Savings ($/year)
Recycling machining coolant	Main Bay Machining	37,000
Relocating cyanide solutions	Plating	320,000
Eliminating chromium plating*	Plating	25,000
Changing rinsing practice	Plating	7,000
Eliminating vapor degreaser	Plating	2,300
Water Jet Cutter	Assembly	140,000
Eliminating radioactive waste stations	NC Machining	11,000
Oil/water separation	LODTM	7,000

*This was a one-time cost saving

The heart of the XYBEX 1000 is a Wesfalia high speed centrifuge operating at 9000 RPM. It removes machine tool lubricant to less than 0.5% concentration and solid matter to 5 microns or less. The centrifuge processes up to 90 gallons of coolant per hour and requires one dedicated cleaning for every 8 hours of use. MFD also purchased a sump cleaning vacuum with a brightly painted underside nicknamed "The Yellow Bellied Sump Sucker". It is a propane operated, portable, twin compartment unit holding 100 gallons of fresh and 100 gallons of used coolant. It is completely compatible with the XYBEX.

Regularly scheduled recycling began in March, 1989. The recycling requires 20 hours per month to complete. Besides regular maintenance, a sample of fresh coolant is analyzed each month for concentration, pH, bacteria count, fungi count, and machine tool lubricant oil percentage. Bacteria count is an important consideration in the system since a high bacterial population can ruin the coolant stored in the machines. A bactericide is occasionally added as a control measure.

The payback period on this system was 1.2 years. A number of other shops are now included in the program and recently a second unit was purchased.

B - Metal Finishing Initiatives

1 - Relocating Cyanide Solutions

All plating solutions containing cyanide were relocated to a separate room and all plating and rinsing associated with these solutions is also confined to this room. By doing this, we avoid having to categorize all wastes from this area as cyanide containing and this noticeably reduces costs for disposal. For example, disposal costs for wastes containing cyanides is approximately $12/gallon compared to $4/gallon for non-cyanide containing wastes (5). Since we typically generate 40,000 gallons of liquid waste per year, a savings of $320,000 is realized by separating the cyanide solutions.

2 - Elimination of Hexavalent Chromium Plating

January 1, 1989 was the date for compliance with Regulation 11, Rule 8 of the Bay Area Air Quality Management District directive which specified control requirements for plating operations emitting hexavalent chromium (6). This regulation is directed at controlling emissions of hexavalent chromium to the atmosphere from hard chromium plating, decorative chromium plating, and chromic acid anodizing. We estimated that it was going to cost us about $25,000 to demonstrate that we were in compliance with this regulation and that continued funding would be required in future years to provide the analytical results required to show continued compliance. Since our chromium plating customers were very few, we decided it was time to quit offering this service in our facility. Therefore, two chromic acid plating solutions (about 300 gallons each) were decommissioned effective January 1, 1989. Customers with requirements for chromium plating have since been referred to shops in the Bay Area.

3 - Changing Rinsing Practices

Prior to 1989, rinsing practices in plating consisted of first rinsing parts over the floor grating (this solution was then sent for disposal as waste), followed by rinsing in a running water rinse tank and finally rinsing in a running DI water tank. This practice generated an average of 40,000 gallons/year of non-sewerable waste. In addition, the water rinse tanks were in continuous operation for an average of 10 hours/day. This water was passed through a system that monitored for pH and total dissolved solids (TDS).

In 1989 rinsing methods were modified to reduce both the volume of the collected waste water and the quantity of hazardous materials contained within it. The wet floor rinse to the central sump was discontinued. This "first" rinse was now performed over the plating tank. In June, 1990 we converted the continuous flow rinse tanks into spray rinse stations. These stations consisted of an empty tank and a hand held, trigger operated spray rinse nozzle, located directly adjacent to each chemical/plating tank. This provided both a first rinse over the tank (greatly reducing dragout) and a second rinse in the empty tank. Several large DI still rinse tanks were retained to prevent any drag-in to the plating processes. At this stage we no longer discharged to the sanitary sewer. Figure 3 shows that as a result of these changes in rinsing practices, we are saving over 2.5 million gallons of water per year.

4 - Elimination of Vapor Degreaser

In May, 1990 we eliminated a large vapor degreaser from our metal finishing facility. This unit utilized about 700 gallons of perchlorethylene per year and also created waste that had to be disposed of via our hazardous waste management group. We've replaced degreasing with soak and electrolytic cleaners and also now use a water pressure cleaning system that recycles the water used for cleaning. On rare occasions, some parts are given a scrub with rags soaked in acetone.

C - Water Jet Cutter

We recently installed a water jet cutter to provide an added dimension in our ability to cut materials. This machine operates by using water under high pressure, and a considerable amount of water is used during the cutting process. For example, assuming 3 hours per day of operation, 700 gallons of water is created in a five day operating period. Since the water becomes contaminated with particulate matter, it has to be handled as hazardous waste. By simply putting a filter (1 μm) on the water system to trap the particulate matter the water is no longer contaminated and can be disposed through sewer system. The approximate yearly savings by using this filter on the water jet cutter is approximately $140,000.

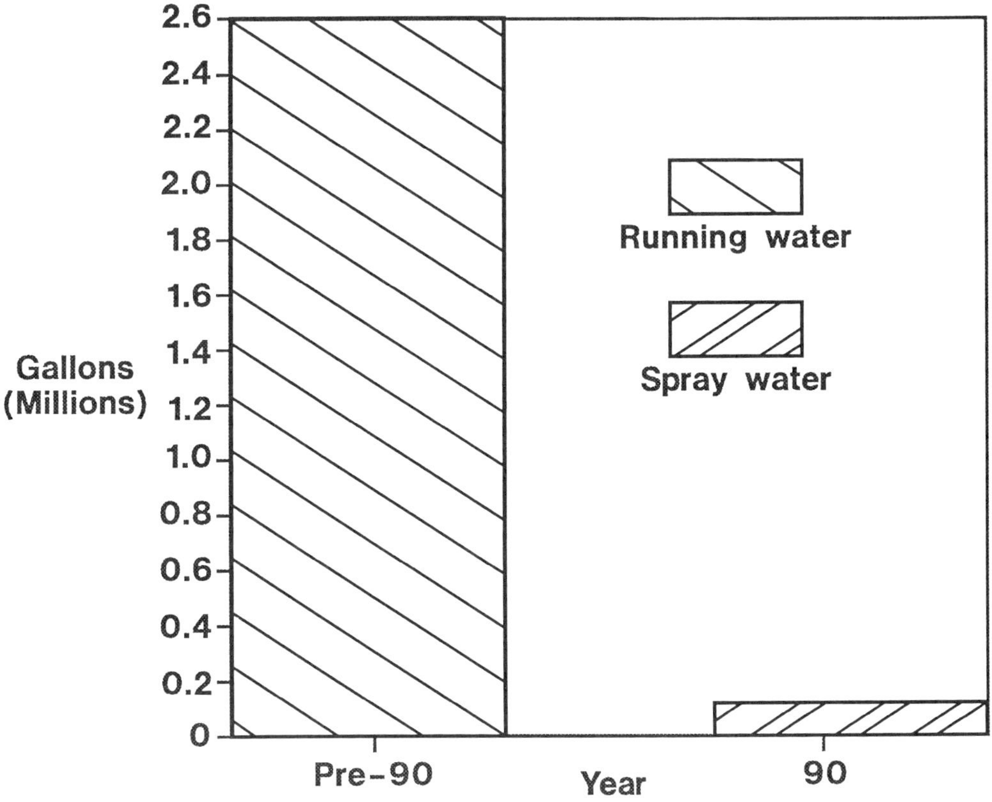

Figure 3 - Reduction of Water Usage in Plating Shop

D - Eliminating Radioactive Waste Stations

The area where radioactive material is machined used to have a container for waste at every machine. These containers were being used for all waste (not just radioactive waste) and had to be treated as mixed waste because of the materials that were handled in the area. This meant that we were creating a large amount of radioactive waste that often had little or no radioactive material. Twenty of the containers were eliminated and the remaining containers were strategically placed throughout the shop so that only radioactive waste was placed in containers designated for this type of material. Because of this change, 100 less drums of mixed waste are now created per year in this facility.

E - Oil/Water Separation

Our Large Optic Diamond Turning Machine (LODTM), one of the most accurate high-precision machines in the world, requires 150 psi of instrument quality dry air supplied by a compressed air system. A 40-HP, 180 psi "Sullair" screw type air compressor removes moisture from the intake air using twin tower, alternatively operating, air dryers. This machine operates 24 hours a day, 365 days a year.

The compressor creates 50 gallons of condensate per month consisting of 49.5 gallons of water and 0.5 gallons of "Sullube 32" (the air compressor lubricant). Despite the fact that the condensate has a very small percentage of oil (1%), we are not allowed to discharge it to the city sewer system. Our HWM disposal cost for the condensate is $600 per drum, amounting to an annual cost of over $7000.

Our Environmental Protection Department advised us that if we purchased and installed an oil/water separator, the waste water could then be discharged to the city sewer system. This would cut the waste accumulation from approximately 600 gallons to 6 gallons per year. Two types of commercial oil/water separators appeared viable, however, the operating costs for the machine supplies and the limited availability of space in the area led to a different solution.

Following the above investigation, the idea of evaporating the condensate occurred to Ray Ewing, facility coordinator for the area. He had noticed that the air compressor had an air cooler that generated a blast of hot air. With the approval of an environmental analyst, a tray was built and installed directly into the hot air exhaust stream (Figure 4). An automatic sump pump was also installed in the condensate catch basin. When the condensate level reaches a pre-established depth, the sump pump starts automatically and transfers an amount of condensate equal to the capacity of the evaporation tray. It then shuts down automatically. (In the unlikely event of system failure, the evaporation tray has a safety overflow system that will discharge the condensate into a 55 gallon drum.) Sufficient heat is generated from the exhaust stream to exceed the compressor moisture output. When the

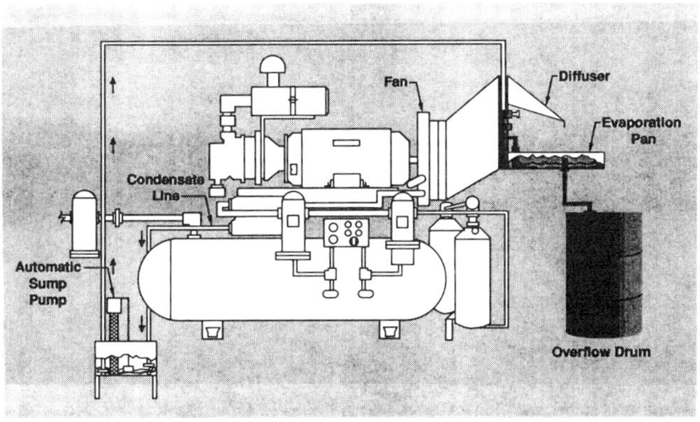

Figure 4 - Evaporation System for LODTM Compressor

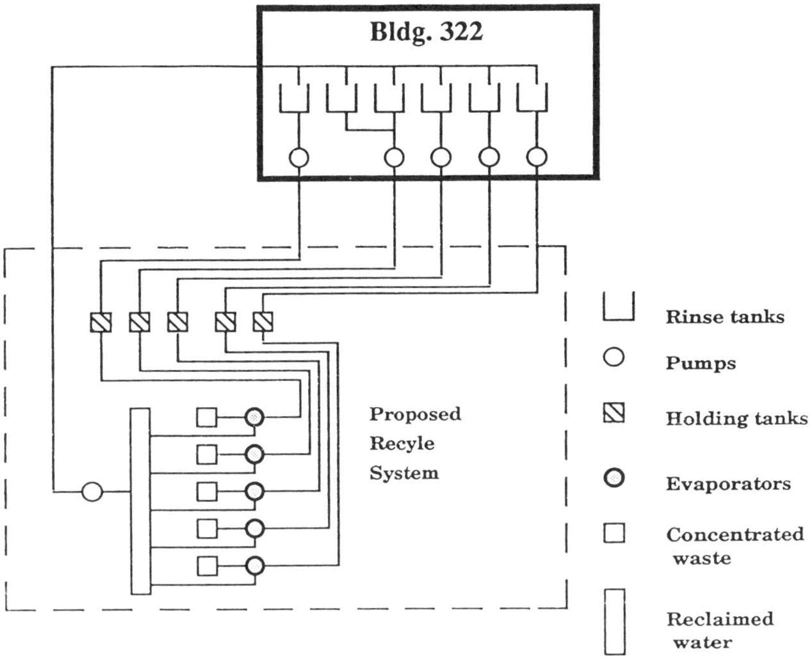

Figure 5 - Proposed system for recycling waste water (under construction)

evaporation tray is dry, the oil residue is removed with a red shop towel. This process only required a $1000 investment (primarily for labor) to build.

F - Other Initiatives

A variety of other steps have been taken which also contribute to minimization of waste:

- Rapid tap was eliminated form the machining areas and was replaced with safety tap which is detergent based and biodegradable.
- Methyl ethyl ketone was eliminated from the shop areas and was replaced with acetone.
- Bottled water has been replaced with water that is treated via reverse osmosis in a number of units strategically located in our areas. Payback on this project is less than one year.
- Although not in the category of waste minimization but rather recycling, aluminum cans are collected in MFD for recycling and LLNL has a Lab-wide program for recycling paper.

Present Initiatives

A - Water Recycling

MFD's Metal Finishing Shop intends to recycle some of the non-sewerable waste that is now disposed of through Hazardous Waste Management. The primary method will be by vapor re-compression evaporation. This will allow the water which is the bulk of the waste to be captured and returned to the shop for reuse. The system will comprise drain lines serving five rows of rinse lines within the building, pumping stations, holding tanks and an evaporator. Presently, these five lines (comprised of eight rinse tanks each) feed into three drain lines going to one common accumulation tank. When the plating shop recycle system is operable, this water will be recycled back into the shop for re-use in rinsing, etc. The proposed design will also segregate waste streams by providing separate drain lines, holding tanks, dirty water tanks, and evaporator chambers for the five rinse lines designated for cleaning (Figure 5). The system has been purchased and installation is in progress.

B - Elimination of Copper Cyanide Plating

Our principal use of copper cyanide plating is as an intermediate step in plating on aluminum. We've shown that excellent adhesion is obtained if aluminum is given a copper strike after zincating; in fact, the adhesion is considerably superior to that obtained when the copper strike is omitted (Table 2). Since the copper cyanide solution is now located in our cyanide room, the processing of aluminum parts has gotten complex since all the other solutions used in the cycle are in the main part of the shop. If a non-cyanide copper strike were available that could provide the results we typically obtain with cyanide copper, it could be located by the aluminum processing line, thus making plating of aluminum parts an easier venture.

A copper pyrophosphate strike has been shown by others to be an acceptable substitute for copper cyanide (7). For example, Table 3 compares ring shear data for various aluminum alloys with a cyanide strike and with a pyrophosphate strike. The results obtained with the pyrophosphate strike are quite comparable. We are in the process of evaluating this type of solution for our operation.

Comments on Waste Minimization Personnel

Another item worthy of some discussion is the issue of finding personnel to champion a waste minimization program. An electroplating facility or printed wiring board manufacturing facility are good places to start a search. The reason for this is that electroplating and printed wiring board operations have several key objectives in the area of environmentalism: avoiding liability by remaining in compliance with regulations; safely handling large quantities of hazardous waste material; conserving water; waste minimization and recycling; and finding suppliers whose products and processes eliminate waste or include methods to handle wastes (8).

Within the DOE complex a number of personnel who were once involved with electroplating or printed wiring board facilities are now heavily involved in waste minimization activities in areas covering a much broader range of technologies than just coatings. The reason for this is related to the fact that the electroplating industry was one of the early and most heavily regulated activities, and thus has served as a training ground for working issues involving regulations, disposal, waste minimization, etc.

Observations and Conclusions

To date, our program of stressing common sense, employee awareness and existing technology has worked quite well in our waste minimization program. Employees are our best source of suggestions for changes and examples of this were discussed in this paper. To make sure to keep interest active in the program, we schedule occasional updates on progress and success stories so everyone can maintain an awareness of waste minimization and its importance. With this program in place, we are saving over $500,000 yearly.

Table 2

Ring Shear Adhesion of Electroless Nickel Deposited on Various Aluminum Alloys (A)

	Shear Strength (B) (MPa)	
	Zincate + Copper Strike + Electroless Nickel (C)	Zincate + Electroless Nickel
Alloy	As-Deposited	As-Deposited
1100	99	68
2024-T351	221	251
5083-0	173	--
6061-T6	184	31
7075-T651	250	107

(A) The pre-plating cycle consisted of vapor degreasing, alkaline cleaning, rinsing, nitric acid pickling, rinsing, zincating for 30 sec., rinsing, nitric acid pickling, rinsing, zincating for 30 sec., rinsing, then either copper striking in cyanide solution or electroless nickel plating. The zincate solution contained 525 g/l sodium hydroxide and 97.5 g/l zinc oxide.

(B) Each reported value is the average of five tests from one rod.

(C) All samples prepared with one proprietary electroless nickel solution, except for the 5083 alloy, which was plated using a different proprietary.

Table 3

Comparison of Ring Shear Data for Cyanide and Copper Pyrophosphate Strikes for Preparing Aluminum Alloys for Plating

	Shear Strength (A,B) (MPa)	
Alloy	Cyanide Copper	Pyrophosphate Copper
1100-F	79	77
2024-T4	288	273
6061-T6	217	222
7075-T6	285	374

A. These data are from reference 7.

B. The zincate solution was Alumon D and the time was 30 seconds. The copper strikes were overplated with nickel in a nickel sulfamate solution. All reported values are the average of 5 samples.

References

1. G. E. Cruz, "Materials Fabrication Division Services Guide", Lawrence Livermore National Laboratory, Report No. 91-038-03 (March 1991)

2. G. Lorton et al., "The EPA Manual for Waste Minimization Opportunity Assessments", PB88-213004, EPA/600/2-88/025 (April 1988), U. S. Dept. of Commerce, National Technical Information Service, Springfield, VA 22161

3. D. Shoemaker, Jacobs Engineering Group, private communication, March 1991

4. J. H. Nuckolls, "Director's Statement on Waste Minimization", Lawrence Livermore National Laboratory Administrative Memo, Vol 19, No 6 (March 8, 1989)

5. M. N. Hayes, Lawrence Livermore National Laboratory, private communication, February 1991

6. Bay Area Air Quality Management District, "Regulation 11 Hazardous Pollutants, Rule 8 Hexavalent Chromium", January 1, 1989

7. H. R. Johnson, W. D. Bonivert and J. Hachman, "Alternatives for Cyanide Copper Plating", Spring Meeting, Materials Research Society, Symposium W, Environmentally Conscious Materials Processing, May 1991.

8. D. E. Johnston, "An Industry-Wide Environmental Audit", PC Fab, 14, 20 (May 1991)

Acknowledgements

Although a multitude of MFD personnel have provided input and assistance in the waste minimization program, I would particularly like to acknowledge: Chris Steffani, Plating; Dick Fasholz, Main Bay Machining; Rey Valentine, NC Machining; Mike Prokosch, Assembly; Ray Ewing, LODTM; Brad Thomson of EPD and Margaret Sands, editor of WASTE NOTES for their help and continued interest. This work was performed under the auspices of the U. S. Dept. of Energy by LLNL under contract No. W-7405-Eng-48.

PROCESS EVALUATIONS FOR URANIUM RECOVERY FROM SCRAP MATERIAL

B.R. Westphal and R.W. Benedict

Argonne National Laboratory
P.O. Box 2528
Idaho Fall, ID 83403-2528

Abstract

The Integral Fast Reactor (IFR) concept being developed by Argonne National Laboratory is based on the pyrometallurgical processing of spent nuclear metallic fuel with subsequent fabrication into new reactor fuel by an injection casting process. During fabrication, a dilute scrap stream containing uranium alloy fines and broken quartz (Vycor) molds is produced. Waste characterization of this stream, developed by using present operating data and chemical analysis, was used to evaluate different uranium recovery methods and possible process variations for the return of the recovered metal. Two methods, comminution with size separation and electrostatic separation, have been tested and can successfully recover over 94% of the metal. Recycling the recovered metal to the injection casting process was evaluated for the different economic and process impacts. The physical waste parameters and the important separation process variables are discussed relative to their effects on the viability of recycling the material. A discussion of the criteria that is being used to establish the acceptable operating limits is also presented.

Residues and Effluents - Processing and Environmental Considerations
Edited by Ramana G. Reddy, William P. Imrie and Paul B. Queneau
The Minerals, Metals & Materials Society, 1991

<u>Introduction</u>

The Integral Fast Reactor (IFR), a sodium cooled, metal-fueled reactor is being developed by the Department of Energy as an advanced power reactor concept [1]. An important feature of the IFR program is the fuel cycle technology which pyroprocesses the spent fuel into new fuel. The fuel cycle, which is shown in Figure 1, consists of three pyrometallurgical processes: electrorefining, cathode processing, and injection casting. The electrorefining process separates the actinides from the fission products by electrochemical techniques. The cathode processor separates, by high temperature distillation, the electrorefiner process fluids from the actinide metals collected on the electrorefiner cathodes. The consolidated actinide ingots are then mixed with zirconium and injection cast at 1500C to form new metallic fuel.

The injection casting process uses an inductively heated furnace to heat, melt, and stir the different fuel constituents into a molten homogeneous metal alloy. Quartz mold tubes are lowered into the melt and the furnace is pressurized with argon. Following the pressurization, the molds are quickly cooled to form new fuel pins. Before the remaining molten alloy pool, called a casting heel, solidifies, the filled quartz molds are raised out of the pool.

After cooling, the filled molds are transferred from the furnace to pin processing where the fuel is separated from the molds. The demolding process produces two streams: unfinished pins and mold scraps. The unfinished pins are sheared to length and qualified to predetermined specifications. The scrap metal produced after shearing is combined with the casting heel and returned to the casting furnace as remelt. The accepted pins are inserted into new fuel element jackets which provide a containment for the fuel metal alloy. This jacket is welded closed forming a new fuel element which then undergoes additional quality checks and is assembled into new fuel assemblies for return to the reactor. The mold scraps are predominantly broken quartz molds although a significant amount of metal alloy remains attached to the molds due to

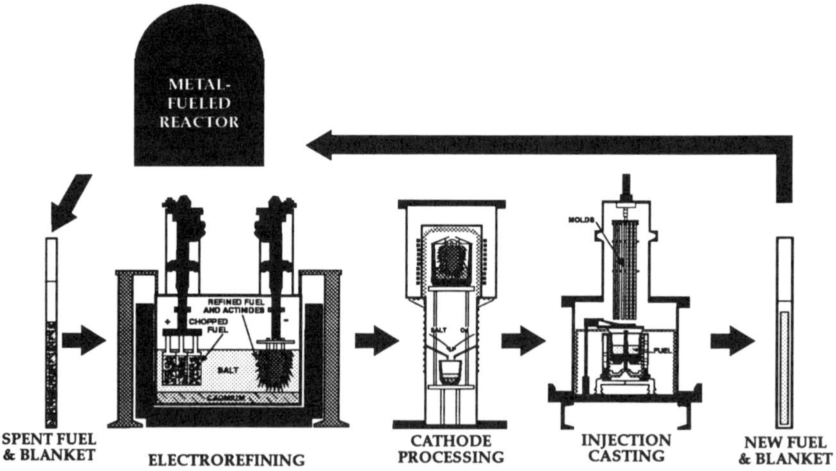

Figure 1. IFR Pyroprocessing Diagram

solidification and thermal effects at the quartz-metal interface. Figure 2 shows the flow paths of the different process streams during casting and pin processing operations.

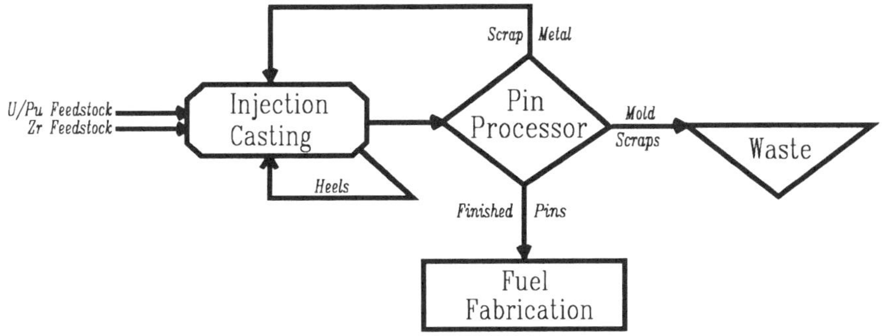

Figure 2. Casting and Slug Processing Diagram

Since the early 1960's, this basic fuel fabrication process has been used to supply fuel for the Experimental Breeder Reactor II (EBR/II) at the Argonne National Laboratory in Idaho. Currently, this fuel is being produced at the Fuel Manufacturing Facility (FMF) using pure uranium feedstocks from the Department of Energy. As part of the IFR demonstration program, the entire fuel cycle process, including all the pyrometallurgical processes, is to be demonstrated in the Fuel Cycle Facility (FCF) [2,3]. This demonstration will differ from the present operation because all work must be performed remotely in shielded hot cells. Also, the commercial aspects of the program require that minimization of material losses, especially actinides, and remote equipment automation be demonstrated.

Although the quartz molds have produced adequate fuel pins for many years, some particular problems are caused by their use. Of particular concern is the effect of silicon oxide on the different pyrometallurgical processes. Thermodynamically, the silicon oxide is less stable than many of the actinide oxides, actinide silicides, and zirconium silicides at the elevated temperatures of the various processes. Consequently, these by-products are formed if the quartz is not isolated from the process. Silicides, oxides, and carbides can potentially remove zirconium from the fuel matrix and lower the fuel alloy melting point. In addition, these compounds would result in the actinides being lost to the standard IFR fuel cycle process and would lower the overall efficiency.

These problems have been minimized by using zirconia mold washes which reduce the interaction of the quartz with the molten alloy in the casting process. This mold wash has been very effective and allows the casting heel and pin scraps to be directly recycled to the casting furnace. However, since the mold scrap stream, a mixture of metal alloy fines, broken molds, and mold washes, contains approximately 5.7% of the finished product, a program was initiated to investigate possible methods for recovery of the metal alloy.

Experimental

Preliminary separation tests have been conducted on the Fuel Manufacturing Facility's mold scraps to model recovery techniques for the IFR program. The first series of tests involved the comminution and sieving of the mold scraps while a second series of experiments utilized an electrostatic separation device. These two methods will be discussed and evaluated for the separation of metal alloy from the mold scraps based on current operational parameters.

Comminution/Sieving

A combination of size reduction followed by sieving was the basis for the first series of separation tests. The differing fracture mechanics of the quartz and metal are such that the quartz is selectively reduced in size while the metal alloy remains relatively intact. Thus, separation of the two constituents by sieving should recover the metal fraction.

A laboratory-size ball mill was used for the size reduction of the mold scraps. Two different types of grinding balls were tested: 0.8125 in. diameter corundum ceramic and 1.0 in. diameter 304 stainless steel. The mold scraps were ground in the ball mill for 30 minutes for both types of grinding media. Following the grinding, the mixture was separated by the sieves into four different size fractions. After separation, the material less than the #20 mesh was set aside. The procedure was repeated two additional times for a total grinding time of 90 minutes.

Table I gives the cumulative uranium recoveries and cumulative quartz removals of the individual sieves for both the ceramic and stainless steel grinding balls. Quartz removals reflect the amount of quartz not recovered with the metal. As would be expected, the quartz removal decreases with increasing uranium recovery. Although the quartz removals for the ceramic test appear to be higher than for the stainless steel test, the higher values are attributed to sampling difficulties encountered during testing. Neglecting sampling problems, the values should be comparable to the stainless steel results. It should also be noted that the quartz removal for the #20 sieve of the stainless steel test is estimated.

Table I. Comminution/Sieving Test Results

Sieve Size	Ceramic: Cumulative Uranium Recovery (%)	Ceramic: Cumulative Quartz Removal (%)	304 SS: Cumulative Uranium Recovery (%)	304 SS: Cumulative Quartz Removal (%)
1/4"	70.39	99.95	---	---
#8	90.87	99.90	84.15*	99.75*
#14	92.59	99.87	88.76	99.72
#20	94.54	99.84	93.65	~98
PAN	100	0	100	0

* Includes both the 1/4" and #8 sieves

Electrostatic

A basic electrostatic separator was used to investigate the recovery of uranium from the injection casting mold scraps. The separation of particles is achieved by exploiting the differences in the electrical conductivities of the quartz and metal alloy. The conducting or metal alloy particles tend to be attracted towards an electrode source and, thus, are readily separated from the non-conducting quartz.

The separator consists of a hopper connected to a vibratory feeder, a rotating drum, a bank of ionizing electrodes, a splitter, two collecting trays (conducting and non-conducting), and a wiping brush to assist in the removal of residual particles. The power supply is 110V AC producing approximately 25kV of stable corona discharge across the electrodes.

Two phases of experimentation have been conducted using the electrostatic separator. The initial scoping phase was performed to determine the practicality and efficiency of the electrostatic unit while the second phase attempted to enhance those recovery techniques established during Phase I. Uranium recovery was improved through reduction of the drum speed, positioning of the splitter, and modified sampling techniques.

A multiple stage cleaning operation was employed for both phases of testing. Following the initial pass of the mold scraps through the separator, the conducting fraction was cycled through the device two additional times producing a final cleaned product. The cleaned product was then analyzed to determine the effects of cycling on uranium recovery.

The cumulative uranium recovery rates per cleaning stage for both series of tests are shown in Table II. Although the uranium recovery values decrease with each stage, the additional stages were required to remove sufficient quartz from the mold scraps. Cumulative quartz removals are not yet available pending analytical results.

Table II. Cumulative Uranium Recovery for Electrostatic Tests

Stage	Phase I	Phase II
1	93.88	96.20
2	85.74	94.43
3	85.67	87.56

Comparison of Separation Techniques

Prior to the selection of a method for the beneficiation of uranium from the mold scraps, a few process criteria should be established. The first, uranium loss, should be maintained at less than 1% of the finished product to ensure a sufficient uranium recovery for the operation. A second, recovered stream quality, should be maximized such that subsequent process parameters are not exceeded, particularly since the recovered material is to be recycled back into the system. The stream quality or stream grade takes into account the amount of impurities, primarily quartz, carried along with the uranium.

For the communition/sieving tests, material greater than the #20 sieve yields about a 94% recovery for both types of grinding media. However, if the stream grade is considered, the stainless steel results appear to be more reliable and, thus, were chosen over the ceramic using the #20 sieve.

For the electrostatic experiments, the results obtained from Phase II, the second stage, appear to be superior based on recovery efficiency and quartz removal. The stream qualities for Phase II appear to be higher than for Phase I although results are not yet available. When compared with the stainless steel/#20 material, the results look quite similar in terms of recovery and stream quality. Moreover, both methods will achieve a uranium loss of less than 1% of the finished product.

A few other issues should also be addressed before the selection of a separation technique. The grinding operation produces a very fine granular product which, in terms of hot-cell contamination, is very undesirable. The powdery characteristic of the ground quartz is the main concern considering the stringent controls kept on the hot-cell environment. Containment would be necessary in order to minimize the spread of glass particulates during any grinding operation. In addition, a comminution/sieving operation would require a significant amount of material handling compared to the electrostatic method. Coupled with the fact that hot-cell operations are performed remotely, the additional manipulation seems unwarranted.

Another concern is that the IFR reactor concept uses two different fuel types: uranium-zirconium alloy for blanket assemblies and uranium-plutonium-zirconium for driver assemblies. Separation testing to date has only been performed on the U-Zr alloy which is less brittle than the ternary fuel. The fracture characteristics of the ternary alloy are very similar to quartz and could cause the comminution process to produce an unacceptable amount of material in the waste stream.

Based on the recovery/grade considerations and the secondary items mentioned above, the electrostatic method shows the greatest promise for the separation of metal alloy from the mold scraps.

Recovered Metal Process Development

The recycle of recovered metal alloy back to the injection casting furnace will be investigated. A conceptual flowsheet for the injection casting process including quartz-metal separation is given in Figure 3. A steady state mathematical model was developed to simulate the modified casting process in terms of casting integrity. The process flow diagram determines the impurity content of a casting based on expected feed streams and current FMF production values. The model was also used to project variations in product quality resulting from fluctuations in operating practices.

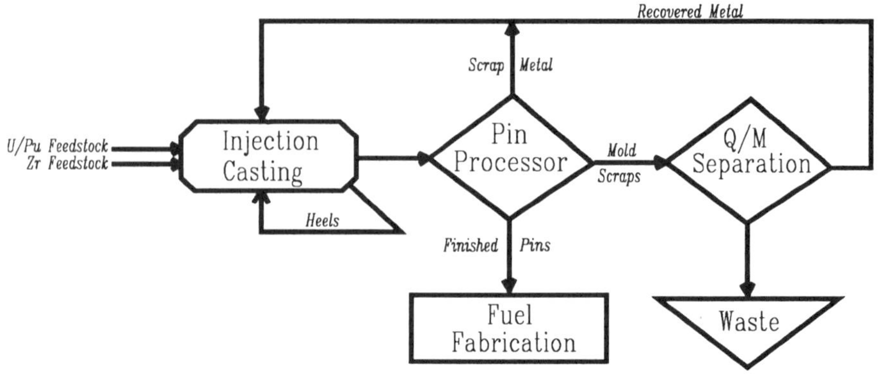

Figure 3. Casting and Slug Processing Diagram with Mold Scrap Separation

Anticipated sources of impurity pickup to the casting stage include: the uranium/plutonium feed, the zirconium feed, the quartz-metal separation product, and the remelt. The remelt and recovered metal stream will be recirculated into the casting to reduce the quantity of feedstock material. A uranium/plutonium ingot, purified by the cathode processor, is combined with the zirconium feed, supplied by Teledyne Wah Chang Albany, and recycled streams to achieve a metallic fuel composition of 70%U/20%Pu/10%Zr.

Another probable, although minor, source of casting contaminants is the impurities picked up through reactions during casting. These impurities are either derived from the various washes used to prevent adhesion of the molten metal or from other materials used in the operation that come in contact with the molten metal. Silicon, carbon, and yttrium are typically the impurities found to react during the process.

Since IFR production rates do not yet exist, the FMF production levels were used as an estimate. Current FMF activities demonstrate a finished pin yield rate of 46.7% with a remelt of 50.6% of the casting. The mold scraps and samples make up 2.7% and less than 0.1%, respectively, of a typical casting.

The process model develops a final impurity value for the casting based on the aforementioned inputs. Applying the FMF chemical specifications for impurities in cast metal fuel, a limit of 2000 ppm has been established per casting. Although conservative, this constraint was found to impede the removal of zirconium from the fuel matrix by casting impurities. Zirconium loss from the matrix results in diminished fuel performance. Thus, the 2000 ppm impurity restriction will be used as an upper bound to assure casting quality.

As an initial test of the model, a final impurity content of 675 ppm was calculated for the current FMF operations. This compares favorably with the results found by analytical methods of 571 ppm considering simulation disparities and analytical errors. For a uranium recovery of 94.5% and a glass removal efficiency of 99.5%, a value of 1903 ppm was calculated by the model when the recovered metal stream was recycled into the casting. The uranium recovery was based on Phase II, stage 2 of the electrostatic tests while the quartz removal value was considered feasible in terms of current test results. The model provides a credible method to evaluate the effectiveness of the quartz-metal separation process on batch operations when analytical data is not available.

In addition, the model has the ability to reliably estimate results based on processs knowledge of the casting operation. The estimation of product quality has been investigated for various anticipated fluctuations in the casting furnace operating parameters. Variables to be considered are the feedstock impurities, separation efficiencies, product yield, and reaction by-products.

The impurities associated with the uranium and plutonium feedstocks can be varied in the model to determine limits for the cathode processor. A uranium impurity value of 470 ppm was found to exceed the 2000 ppm casting limit. For the plutonium feed, 1140 ppm impurities are allowed before the casting limit is met. These levels should be considered for the refining operations at the cathode processor. The zirconium feed impurities were not varied because they are reliably fixed by procurement specifications.

Table III gives an indication of the parameters associated with various quartz-metal separation efficiencies. By varying both the separation efficiency and quartz removal in the model, it can be seen that the quartz removal values are the controlling factors during separation. A quartz removal of greater than 99.46% is necessary in order to meet the 2000 ppm casting specification regardless of separation efficiency.

Table III. Casting Integrity (ppm) Based on Separation Results

Separation	Quartz Removal (%)				
Efficiency (%)	100.0	99.8	99.6	99.5	99.4
100	654	1161	1659	1904	2147
90	656	1163	1659	1903	2145
80	658	1164	1658	1901	2141
40	667	1167	1644	1875	2101
0	675	675	675	675	675

Although the production rates used in the model have been based on FMF data, they can now be varied to simulate anticipated variances during FCF start-up. Prior determination of quartz removal as the controlling element in quartz-metal separation has led to it being varied along with the production yields, see Table IV. Notice that the sum of the pin and remelt streams is the same as current FMF productions assuming mold scrap rates will remain constant. At a constant quartz removal, decreasing the pin yield, or increasing the remelt, results in increasing the impurities to the casting. Also notice, contrary to previous observations, a quartz removal efficiency of less than 99.46% is possible if the pin yield is sufficiently high. Finally, for a given quartz removal, a required pin yield can be found in order to meet the casting specifications. The converse is also true. For a given pin yield, a required quartz removal rate can be found.

Table IV. Casting Integrity (ppm) Based on Production Yields

Quartz	% Pin/% Remelt				
Removal (%)	70/27.24	60/37.24	50/47.24	40/57.24	30/67.24
99.75	1001	1098	1230	1428	1744
99.50	1419	1582	1808	2139	2672
99.25	1826	2054	2370	2832	3576

Conclusions/Summary

Mold scrap recovery studies predict that reduction of the uranium loss from 5.7% to less than 1% of the finished product is possible.

The electrostatic method was found to be the most effective for separation of the mold scraps in terms of recovery, stream quality, hot-cell compatibility, and process impacts. Thus, future research should focus on this technique.

The recovery of metal alloy, followed by the recycle into future castings, is essential for the demonstration of a viable closed fuel cycle.

The reduction of radioactive waste and operating costs is of the upmost concern considering the immediate challenges facing the nuclear industry.

Using the existing model, operational parameters can be varied to see how well the model correlates casting results. Preliminary tests are being conducted to better understand and estimate process operations.

The model has been a useful tool for evaluating possible glass-metal separation techniques and identifying the important parameters which can improve the integrated processes.

Acknowledgments

Support of this research was by the U.S. Department of Energy under Contract W-31-109-Eng-38. The authors would like to acknowledge the assistance of M.F. Adam, R.J. Briggs, G.C. Knighton, J.R. Krsul, and S.A. Parra.

References

1. Y.I. Chang, "The Integral Fast Reactor," Nuclear Technology, 88 (1989), 129-138.

2. M.J. Lineberry and R.D. Phipps, "Preparations for the IFR Fuel Cycle Demonstration," Trans. Am. Nuc. Sci., 60, (San Francisco, CA, November 26-30, 1989), 170.

3. M.J. Lineberry, et al, "IFR Fuel Cycle Demonstration in the EBR-II Fuel Cycle Facility," (Paper to be presented at FR'91 International Conference on Fast Reactor and its Fuel Cycles, Kyoto, Japan, October 28-November 1, 1991).

THE BEHAVIOUR OF POLONIUM-210 IN THE PROCESSING OF TIN-LEAD RESIDUES

A.M. Strachan, J.A. Litten

Capper Pass & Son Limited
P O Box 6
North Ferriby
North Humberside HU14 3HD
England

Abstract

Capper Pass & Son Limited was, until its recent closure, the world's largest producer of tin from secondary materials including solder drosses, non-ferrous smelter flue dusts, tin based metal alloys, residues and slags. Polonium-210, an alpha emitting radionuclide, was present in varying concentrations in some of the numerous feed materials processed by the smelter and refiner. The polonium-210 followed a complex path through the process flowsheet. This paper discusses the behaviour of polonium, both from its arisings in other smelters' process streams and through the entire Capper Pass smelting and refining circuit. The consequential personnel safety and environmental precautions necessarily taken in some areas of the process operation are outlined.

Residues and Effluents - Processing and Environmental Considerations
Edited by Ramana G. Reddy, William P. Imrie and Paul B. Queneau
The Minerals, Metals & Materials Society, 1991

Introduction

Capper Pass & Son Limited was, until its closure in late 1991, the world's largest producer of tin from secondary materials including solder drosses, non-ferrous smelter slags, flue dusts, tin based alloys and residues.

The withdrawal of financial support by the International Tin Council buffer stock manager resulted in the tin price collapse in October 1985. This factor and subsequent over-production around the world, largely as a consequence of rapid expansion of tin production in Brasil, have combined (with the exception of a brief period in 1989) to yield a very low tin price over the last six years. Such an extremely low tin price had two major effects:

(1) it encouraged the miner to upgrade his concentrate at the expense of recovery thereby reducing freight costs and commanding lower treatment charges. Thus smaller tonnages of tin concentrates were available in a very competitive market environment.

(2) because of the inherently low contained metal values, the movement of low grade materials, including residues, around the world for processing and metal recovery was discouraged. This particular effect was exacerbated by the relatively low prices of other base and precious metals prevailing over the same period of time.

In addition to these factors the strengthening of the pound against the US dollar made Capper Pass's treatment charges less favourable than dollar based smelters, a factor which was compounded by entry of the United Kingdom into the European Monetary System.

Capper Pass's major sphere of activity and expertise was in the processing of complex lower grade concentrates and residues through a complex series of metallurgical processes to recover the contained metal values. In recent years the proportion of secondary materials processed at Capper Pass had increased to over 70% of the total with a corresponding reduction in low grade tin concentrates. Unfortunately the diversity of the circuit contributed substantially to the fact that it was expensive when compared with smelters of higher grade and cleaner feed materials, thereby necessitating higher treatment charges. Due to the factors highlighted, the smelting and refining operations were ceased because the forward raw material purchasing program and the resultant marginal income was inadequate to sustain the projected plant operating costs.

Dependent upon their composition and physical form the various raw materials processed at Capper Pass could enter the process by one of nine possible entry points. During the early 1980's the presence of polonium-210, a naturally occurring radionuclide, was identified in some of the materials processed and produced at the smelter. This paper reviews the complex flowsheet which operated at Capper Pass and discusses in some detail the origin of polonium in the feed materials and the pathway taken by this particular impurity through the circuit.

Smelter Raw Materials Profile

Table I reproduces a breakdown of the quantities and types of materials purchased and subsequently treated through Capper Pass during the last two full years of operation (1989/90) and during the 1991 controlled closure program.

Table 1: Raw Materials (tonnes) Processed at Capper Pass 1989-91

Material Description	1989	1990	1991 26 weeks
Primary Tin Concentrates	11,098	22,438	5,672
Flue Dusts and Stannates	4,473	1,537	1,715
Tin Drosses and Secondary Residues	12,258	10,367	3,101
Tin Based Alloys	2,966	1,900	837
Low Grade Fuming Furnace Feed	13,713	19,117	11,459[+]
Lead Drosses and Residues	1,046	1,177	178
Total	45,554	56,536	22,962

[+] Fuming Furnace operated for 40 weeks to process contaminated refractories, residual materials, etc.

Between 61% and 75% of the total smelter feedstock was secondary material and in excess of 30% was first processed through the large slag fuming furnace to pyrometallurgically upgrade the low metal content (typically 3-10% Sn) materials[1].

The various feed materials purchased from numerous countries around the world for processing and metal recovery at the plant contained, in addition to tin, many impurities including lead, antimony, bismuth, cadmium, copper, nickel, indium, arsenic and precious metals. The association of such impurities with tin is not particularly unusual and in fact the smelter process flowsheet was capable of separating and recovering all of the impurities listed above. A somewhat unusual impurity present in some of the feed materials purchased (but not paid for!) was polonium-210.

Polonium-210

Polonium-210 is a naturally occurring alpha emitting radionuclide with a half-life of 138 days. This isotope of polonium is a daughter product of the uranium/radium decay series. The alpha particle emitted in the decay process is of a relatively high energy and caution must be exercised to prevent inhalation or ingestion by plant personnel of any material containing the element.

Polonium in Smelter Feed Materials

Following the detection of the presence of polonium-210 in the Capper Pass circuit it was necessary to construct a special laboratory facility and to develop and improve existing analytical techniques for the various materials examined. These techniques and analytical procedures were subsequently validated by independent third parties.

Polonium-210 was present in many of the raw materials processed but at minute concentration levels. Radioactivity is measured in disintegrations per unit of time and the unit employed is the Becquerel or one disintegration per second. With some few exceptions only polonium-210, and smaller non-equilibrium proportions of lead-210 and bismuth-210 were present in the materials processed rather than many of the other possible radionuclides in the uranium/radium or uranium/thorium series. This was due, particularly in the case of the residues, to the fact that they would have been subjected to previous metallurgical processing probably thermal in nature. Polonium having a tendency to form volatile species, during processing, would volatilise from its parents in the decay series many of which would report in the slag phase of the primary smelting process.

However, despite being at minute concentration levels in some of the feed materials, the polonium did concentrate in certain areas of the smelter circuit and it also progressed through a significant proportion of the possible pathways. It was in those areas where polonium concentrated in the circuit and activity levels rose to more significant levels that precautions had to be taken and these are discussed later in the text.

Table II reproduces some typical polonium concentration levels in the materials processed at Capper Pass.

Table II: Typical Polonium-210 Concentrations in Materials Processed

Materials Description	Origin	Typical Polonium-210 Assay - Bq/g
Primary Tin Concentrates	U.K.	1.7
	Australia	2.1
	Bolivia	0.5
	Peru	0.8
	Canada	4.1
Flue Dusts	U.S.A.	0.4
	Bolivia	2.3
	Europe	66.7
Stannates	Europe	<1
Tin Drosses and Residues		
Slags/Hardheads	World wide	0.5
Anode Slimes	U.S.A.	0.8
	Bolivia	2.0
	S. Korea	4.7
	Texas	15.6
Residues	World wide	1.3
	Europe	16.1
Tin Based Alloys		
Metal	World wide	7.5
Condensate	Australia	33.3
Crystalliser 'Bottoms'	Far East	200-300
Fuming Furnace Feed	World wide	1.0
Lead Drosses & Residues	Europe	4.0

Clearly from the above data there was widespread low level activity in the feed materials processed. However, presented in Table II are some materials which have an activity level significantly greater than the threshold value for a radioactive substance (15 Bq/g) as defined by the United Kingdom Radioactive Substances Act 1960[2] and related Rare Earths and Phosphate Substances Exemption Order[10]. Such materials include:

Condensate - produced from the application of vacuum refining techniques to impure tin to reduce a particular volatile impurity concentration. Being volatile any polonium present will report in the condensate.

Flue Dust; Europe - produced during the recovery of refractory metals from concentrates and primary smelter slags. These flue dusts often contained radionuclides in the thorium decay series.

Crystalliser Bottoms - the Chinese crystalliser is a very effective process for tin refining, exploiting the particular alloy phase relationships. Impurities such as bismuth and lead report in the fluid phase at the bottom end of the crystalliser, as does any polonium present.

As mentioned above, under the Radioactive Substances Act 1960[2] and related Exemption Orders[10] for the specific purpose of registration and waste disposal, any material with a specific activity greater than 15 Bq/g is defined as radioactive. However, a second major piece of legislation relates to the protection of personnel using or processing materials containing radionuclides. Under this legislation, the Ionising Radiation Regulations 1985[3], the definition of a radioactive substance is more complicated and differs significantly from that of the Radioactive Substance Act. Under the Ionising Radiation Regulations, account is taken of possible specific radionuclide volume concentrations which may be created during a process if it involves a dusty operation.

In the circumstances of processing a metal alloy, e.g. the vacuum condensate or crystalliser bottom alloy, provided the treatment creates no dust or fume the particular radionuclide volume concentration in the surrounding air is unlikely to be exceeded and therefore no hazard will be created. This is particularly the case in the examples of alloy given since the polonium present was largely unsupported apart from some lead-210 and did not contain more hazardous or more restricted (as defined by the Ionising Radiation Regulations) higher chain radionuclides.

However, in the case of the European flue dust, in which the presence of more hazardous radionuclides in the thorium decay series was common, it was necessary to immerse this material in water to thoroughly wet it. This action minimised any risk of an elevated highly restricted radionuclide volume concentration occurring in the surrounding air during processing.

Caution had to be exercised even towards substances with considerably less than 15 Bq/g polonium concentration, for example the Canadian tin concentrates at 4.1 Bq/g of polonium. Generally in the case of tin concentrates, other radionuclides in the particular decay series were present and in equilibrium with polonium. Although the Canadian tin concentrates were damp they could not be allowed to dry out as subsequent handling could create an elevated volume concentration of the most hazardous element under the Ionising Radiation Regulations (in this case thorium-232). The problems associated with, and encountered by smelting these particular concentrates are discussed more thoroughly elsewhere[4].

Capper Pass Metal Recovery Circuit

Figure 1 reproduces a synopsis of the smelter flowsheet. The core of the process was the solder circuit. All tin bearing concentrates, fine flue dusts and stannates entered this circuit via the sinter plant where the material was agglomerated on a downdraught two-metre wide sinter strand. The prepared sinter was smelted along with the high tin bearing drosses and residues in a blast furnace producing a very impure solder metal typically assaying 45-65% Sn, 2-8% Sb, 1-3% As, 0-2% Bi, 0-1% Fe, 1-4% Cu, 0.5% Ni, 0.2-0.5% Ag, 0.15% In, balance Pb.

Slag from the blast furnace was treated in a heavy fuel oil fired submerged combustion slag fuming furnace to recover the contained tin, antimony and lead[1]. In addition to slag, this furnace also processed several thousand tonnes per year of the low grade materials recorded in Table I. Tin fume from this process was cycled to the sinter plant for incorporation into the sinter.

Formation of a speiss at the blast furnace was also encouraged to complex the arsenic. At arsenic saturation (~20 wt% As) the speiss was ground and roasted to remove the arsenic from the circuit and the iron-tin calcine was recycled to the sinter plant[5].

Metal from the blast furnace was transferred fluid to the dry refinery where iron was removed by liquation, copper and some of the nickel was removed by liquation then sulphur drossing followed by arsenic and nickel removal again by liquation. The iron and arsenic drosses were cycled back to the blast furnace for speiss formation and the copper dross was processed in the copper circuit.

Following drossing the metal was granulated and electrolytically refined in a unique and complex process employing a sodium hydroxide electrolyte. This tin refinery capacity was 16,000 tpa. The tin level in these granules (or cockles) typically assayed 54-62% tin with the balance being lead, antimony, bismuth and precious metals. The only significant impurity in the cathode tin produced was indium at 0.03-0.05% which was removed by chlorination and the slag processed and electrorefined to produce high purity indium. The product tin had a minimum assay of 99.93% and typically was 99.96% tin.

The tin in the granule was reduced to approximately 2.5 wt% in the refinery slime. This slime became the major feed component for the lead circuit as illustrated in Figure 1. Slimes and purchased residues were smelted in a short rotary furnace, with an oxygen - medium fuel oil burner, producing an impure lead. This impure lead was drossed with sulphur to remove copper and the dross routed to the copper circuit. The detinning dross was cycled back to the solder blast furnace and the lead-antimony oxide dross from softening was smelted in a Lurgi short rotary furnace producing a lead antimony alloy sold for use in battery manufacture. Tellurium was removed as a dross, stored and accumulated for sale.

The dry refined lead was cast into anodes and electrorefined to chemical grade 4 x 9's lead in a standard Betts electrolysis process. Lead electrorefining capacity was 8,500 tpa.

Anode slimes from the lead electrorefining, containing the bismuth and precious metals, were smelted and liquated to yield bismuth and silver rich alloy fractions. Both such alloys were vacuum refined to marketable products.

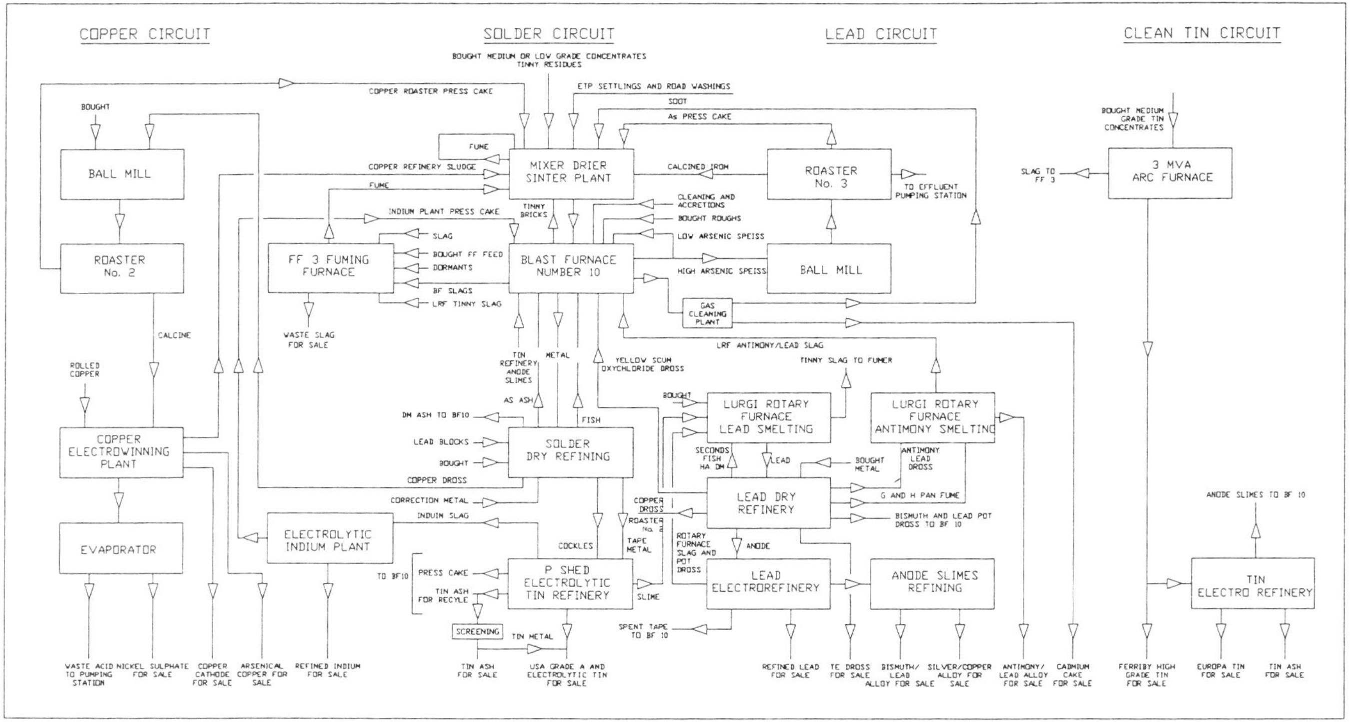

Figure 1 – Metal Recovery Circuit (1991)

As referred to earlier in the text, the copper drosses from the dry refining operations were routed to the copper recovery plant. These drosses were ground and roasted in a fluidised bed roaster to render the copper and nickel soluble and the tin, lead and antimony insoluble in sulphuric acid. The copper-nickel calcine was leached, pressure leached, filtered, the copper electro-won and nickel sulphate recovered by evaporation. The leach residue filter cake containing the tin, lead and antimony was cycled back to the solder circuit via the sinter plant.

In parallel to the main processing circuit the smelter flowsheet included a fully integrated high grade tin circuit capable of smelting clean medium and high grade tin concentrates in a 3 MVA electric arc furnace and fire refining the tin[4]. Higher impurity bearing tin concentrates were also smelted where the tin produced, not being amenable to conventional dry refining processes, was electrorefined in a 3,000 tpa acid electrolysis facility.

Polonium in the Capper Pass Recovery Circuit

Figure 2 reproduces a further basic smelter flowsheet with typical polonium-210 analytical data superimposed on the salient inputs and intermediates produced. In most cases the data is the yearly average or typical monthly average.

Complete interpretation of the data presented in Figure 2 is limited without full information on the tonnages of cycling loads. However, sufficient data is presented to allow the route taken by the polonium to be identified clearly and this path is highlighted in Figure 2. Most of the polonium entering the sinter plant reports in the sinter product and following the blast furnace smelt reports in the metal phase. No significant levels of polonium reported in the slag. A cycling load built up in the sinter plant electrostatic precipitator and relatively high activity levels were reached requiring stringent precautions when entry to the precipitator by personnel became necessary. These precautions are discussed more thoroughly later in the text.

Some polonium recycle inevitably occurred with the dry refinery drosses but the majority continued through to the anode metal for tin electrolysis. An almost complete shunt of polonium occurred through the electrolysis process with negligible amounts reporting in the tin cathodes and all reporting with the lead, bismuth and antimony in the anode slime.

During the subsequent lead circuit smelt and dry refine, the predominant path for polonium was again with the metal phase and negligible quantities reporting in the slag. During the Betts electrolysis in terms of total polonium content there is an approximate 50% partition. Typical lead and slimes product assays were 0.96 Bq/g and 49.7 Bq/g respectively.

The slimes were smelted in a small rotary furnace, liquated to yield bismuth rich and silver rich fractions both of which were vacuum refined. Figure 3 reproduces a more detailed breakdown of this section of the recovery circuit with polonium data again superimposed and the predominant pathway highlighted.

It will be observed that the polonium on this occasion largely followed the bismuth although clearly the divergence was far from total. A cycling load of fume from this smelting process again led to elevated polonium levels in the gas cleaning facility, on this occasion a bag filter plant. As previously, stringent precautions were necessary when personnel were required to effect maintenance, etc. inside the baghouse.

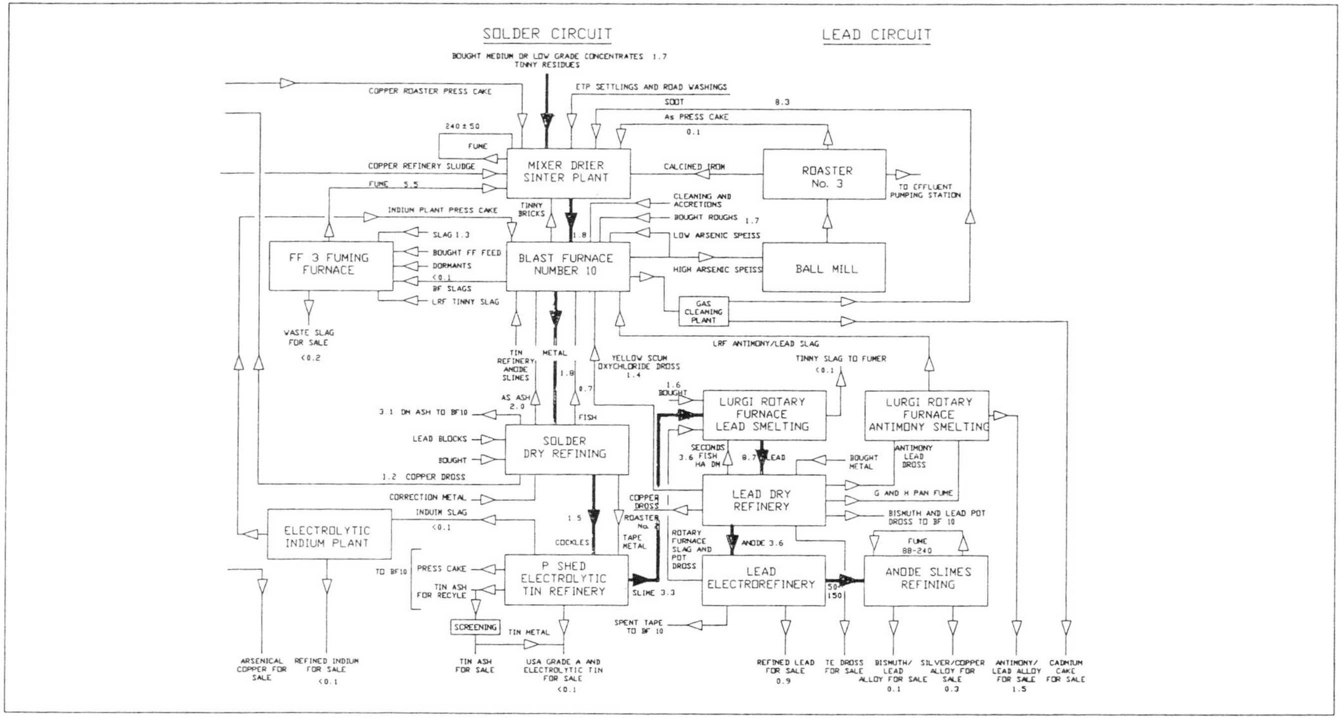

Figure 2 – Metal Recovery Circuit, Showing Route Taken By Polonium (Data in Bq/g)

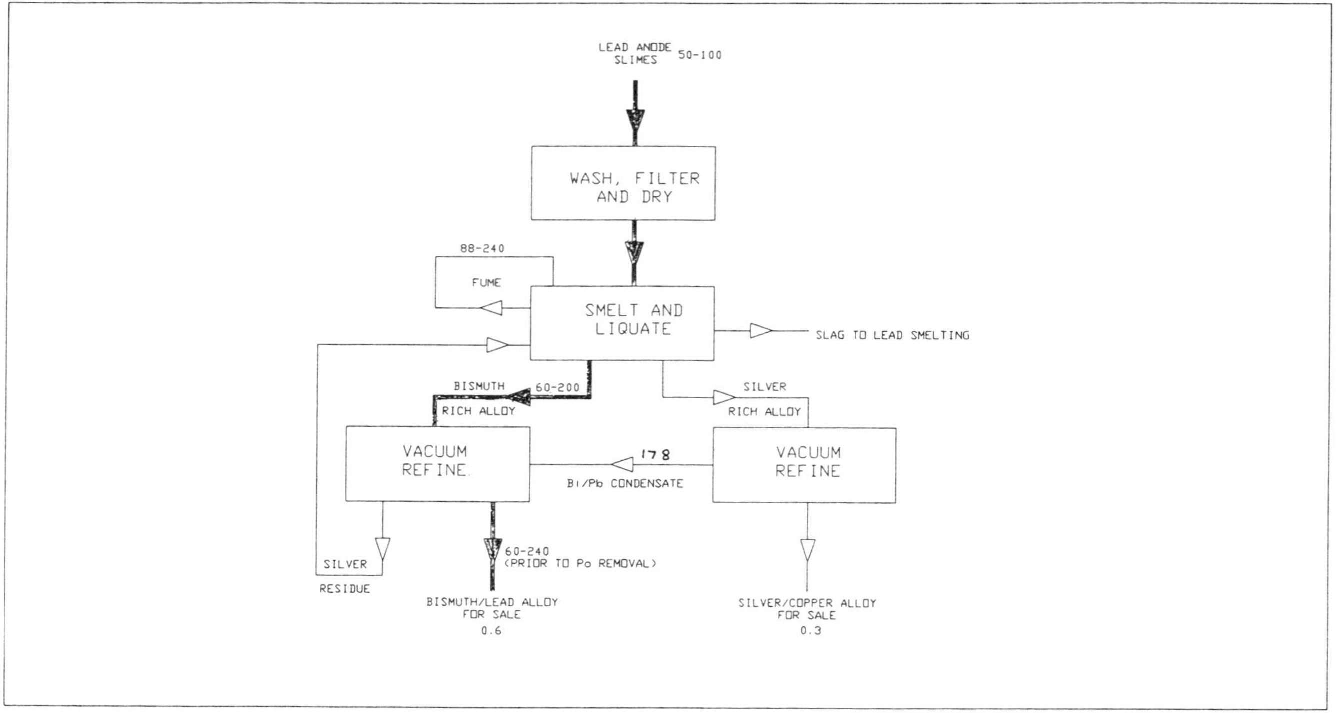

Figure 3 – Lead Anode Slimes Processing, Showing Route Taken By Polonium (Data in Bq/g)

Being relatively volatile any polonium which was present in the silver fraction readily reports in the condensate during the silver distillation. This condensate is then re-distilled with the bismuth fraction and ultimately all the polonium in the lead slimes reported in the final bismuth product.

Clearly the prevailing level of polonium in the final bismuth product was unacceptably high and investigations were carried out at Capper Pass to develop a technique to remove it from the bismuth. A process was developed during the mid-1980's and subsequently applied to all bismuth metal product prior to its sale[6]. The polonium level in the final bismuth product sold during 1989-90 was typically 0.6 Bq/g.

In order to reduce the polonium level in the lead slimes and therefore the elevated level in the cycling load of fume on the slimes smelt, a similar process for polonium removal was developed and introduced prior to lead anode casting[6]. The polonium in anode metal was reduced from typically 8.7 to 3.6 Bq/g. This resulted in some lowering of the polonium level in the slimes but the prevailing level in the slimes smelting fume still remained relatively high.

The removal of polonium from both the impure lead and the vacuum refined bismuth resulted in dross formation with high activity levels. The lead dross was typically 2-2500 Bq/g and that from the refining of bismuth was substantially higher at around 30,000 Bq/g. These drosses were stored in a restricted authorised area to allow the unsupported polonium to decay naturally to below 15 Bq/g. The lead dross was then sold for its tellurium content and the bismuth dross was reprocessed through the lead circuit.

Environmental, Health and Safety Controls

In the United Kingdom the processing of materials containing radioactivity are subject to the legal provisions contained in the Radioactive Substance Act of 1960[2] with related Exemption Orders. Under this Act and the 1963 Exemption Order[10], material containing less than 15 Bq/g of activity is not deemed radioactive. Therefore the vast majority of materials processed at Capper Pass were not radioactive as defined by this Act. However it was necessary for the company to make application to Her Majesty's Inspectorate of Pollution (HMIP) to register to be permitted to store and use substances greater than 15 Bq/g on the site and also for authorisation to cover the 600' stack discharge. Applications for registration and discharge authorisation were accompanied by detailed environmental impact assessments and implementation procedures. To assist in the preparation of these assessments the company employed the expertise of the independent National Radiological Protection Board (NRPB). Continuous monitoring programmes for polonium in the process plant and main stack emissions were carried out and also at nine ground air monitoring stations in the surrounding locality. In practise the normal discharge levels of polonium via the 600' stack rarely exceeded 10% of the authorised level of 400 megabecquerels of polonium-210 per week. No levels of polonium in the air monitoring stations were ever found to be in excess of background levels.

The other most relevant and applicable piece of legislation is the Ionising Radiation Regulations (1985)[3] which deal specifically with the protection of personnel actually working with radioactive materials. In fact Capper Pass began using these regulations prior to 1985 whilst still in draft legislative form. The major impact of these regulations was in areas of the operation where elevated levels of polonium could be encountered. Two such areas already referred to were the sinter plant precipitator and the lead anode slimes smelting gas cleaning facility. The only other such area at the Capper Pass operation was in the process used for the polonium removal[6]. When personnel were required to enter these locations in the plant, 'controlled areas' were instigated specifically to prevent contamination of the operator and prevent any inhalation or ingestion of polonium-bearing dust. Those requiring to enter the area wore respiratory protective equipment and disposable protective clothing and the time spent within the area was recorded. Written systems of work were strictly followed and concomitant environmental air monitoring of the area concerned was also carried out during the work period. The possible dose rate exposure for each individual was computed from subsequent analytical data of the air quality and the duration of entry. Records of all personnel entries into controlled areas and possible exposure data are kept in perpetuity.

The presence of polonium in the Capper Pass circuit had been well known and understood by all of the employees and by all of the relevant government and local regulatory authorities. All legal requirements and procedures were strictly followed. As stated, the company had enlisted the expertise of the NRPB as its radiological advisers and all recommendations made by the NRPB were enacted.

Irrespective of the fact that the presence of polonium was well controlled and fully understood and all regulations strictly adhered to, in the late 1980's the subject was reported by the media and near hysterical, if well intentioned, efforts were made by local pressure groups to link the minute polonium emissions with purported clusters of childhood leukaemia in the populated area to the north east (downwind) of the smelter site. The resultant media hype led to three substantial independent investigations[7,8,9] into the alleged link all of which concluded unambiguously that no such evidence existed. The local health authority in its 1990 Annual Report rejected any linkage and reported that cancer rates in the area were in fact substantially below the national and county levels[11].

Finally, in order to gain some perspective as to the quantity of polonium under discussion, the <u>weight</u> of polonium contained in approximately 50,000 tonnes of raw materials treated annually at Capper Pass was 0.4mg. I leave the readers to draw their own conclusions.

References

[1] A.M. Strachan, I.A. Linton, Improved Efficiency in the Recovery of Tin from Recycled Low Grade Residues in the Capper Pass Sulphide Fuming Furnace. Recycling of Metals and Engineered Materials, eds. J.H.L. van Linden, D.L. Stewart, Y. Sahai. 1990 TMS Fall Extracted Meeting Williamsburg Va. pp 371-81.

[2] Radioactive Substances Act 1960, H.M. Stationery Office, London.

[3] The Ionising Radiation Regulations 1985, H.M. Stationery Office, London.

[4] A.M. Strachan, I.A. Linton, Electric Smelting of Tin Concentrates at Capper Pass Limited, Lead-Zinc-Tin '90, eds. T.S. Mackey, R.D. Prengaman, 1990 TMS Annual Meeting Anaheim Ca. pp 841-56.

[5] J.A. Litten, P. Halsall, J. Hood, Removal of Arsenic from Speisses formed during Smelting of Tin Concentrates and Residues by use of a Fluid-Bed Roaster. Pyrometallurgy '87 IMM London, pp 743-66.

[6] A.M. Strachan, J.A. Litten, The Removal of Polonium-210 from Lead, Solder and Bismuth. Lead and Tin Refining 1992 TMS Annual Meeting, San Diego, Ca.

[7] R.A. Cartwright, F.E. Alexander, P.A. McKinney, Childhood Cancer Aggregations in the Beverley and Kingston-upon-Hull Administrative Districts. Report by Leukaemia Research Fund Centre for Clinical Epidemiology to the East Yorkshire Health Authority 1988.

[8] Her Majesty's Inspectorate of Pollution Investigations into the Disposal of Wastes from the Capper Pass Factory on North Humberside. March 1989, London.

[9] M.S. Baxter, B.W. East, A.B. Mackenzie, E.M. Scott, A Review of Radioactivity in and around the Capper Pass Smelter, Melton Works, North Humberside. Report by Scottish Universities Research and Reactor Centre for the East Yorkshire Health Authority February 1990.

[10] Exemption Order No.2648. The Rare Earths, Phosphate Substances, etc. Exemption Order (1963).

[11] East Yorkshire Health Authority, Annual Report of the Director of Public Health, December 1990, p 17.

TREATMENT AND RECOVERY OF ZIRCONIUM AND HAFNIUM FROM THE EFFLUENTS AT NUCLEAR FUEL COMPLEX

SM RAO, B BALKISHAN, YS SUBRAHMANYAM & N.SWAMINATHAN
DEPARTMENT OF ATOMIC ENERGY
NUCLEAR FUEL COMPLEX
HYDERABAD - INDIA

A B S T R A C T

Nuclear Fuel Complex(NFC), Hyderabad (INDIA), is engaged in the manufacture of nuclear fuel assemblies and zircaloy structural components required for the Indian nuclear power reactors. The production of zirconium oxide, sponge and zircaloy materials through chemical and hydro-metallurgical processes generates many effluents containing valuable zirconium and hafnium. This paper deals with the various processes such as multiple crystalization, phosphate precipitation, ion-exchange, amine solvent extraction etc., for the treatment of these effluents and subsequent recovery of valuable zirconium and hafnium.

INTRODUCTION:

Nuclear Fuel Complex (NFC) Hyderabad is an integrated facility created by Government of India (under the control of Department of Atomic Energy) for the manufacture of nuclear fuel for Indian Reactor Programme. It comprises of chemical and metallurgical plants which

Residues and Effluents - Processing and Environmental Considerations
Edited by Ramana G. Reddy, William P. Imrie and Paul B. Queneau
The Minerals, Metals & Materials Society, 1991

produce fuel assemblies of Uranium di-oxide cladded in zircaloy tubes starting from uranium concentrate and zircon sand. The complex also produces zircaloy hard-ware components required for the power reactors. During hydro-metallurgical processes involved in NFC, a variety of effluents are generated and some of them contain valuable zirconium and hafnium. Table-1 gives details of such effluents. NFC developed processes for the recovery of zirconium and hafnium from these effluents before they are disposed off safely.

TABLE-1: EFFLUENTS THAT ARE BEARING ZIRCONIUM AND HAFNIUM

S.No.	EFFLUENT	SOURCE	Quantity per month
1.	Raffinate (2-3 g/1 of zr containing 30% Hf)	Solvent extraction (Zirconium Oxide Plant)	500 KL
2.	Zirconium Chloride Wastes (Hf free zr=95%)	Chlorination & Reduction (Zirconium Sponge Plant)	2.0 MT
3.	Spent pickling Solution (Hf free zr 60-80 g/1)	Pickling (Zircaloy Fabrication Plant)	12 KL

1. HAFNIUM RECOVERY FROM RAFFINATE:

Zirconium is always associated with about 2% of hafnium in naturally occuring ores. Separation of hafnium from zirconium is a vital step in the manufacture of nuclear pure zirconium because hafnium is a nuclear impurity. Because their chemical properties are very similar separation of zirconium and hafnium is very difficult by normal chemical methods. NFC adopted sol-vent extraction process using TBP as solvent for their separation.

Mineral zircon (Zirconium Silicate) is fused with caustic soda and most of the silicate removed by leaching with water and filtration.

..3

The cake is dried and dissolved in nitric acid to get feed material containing 80 g/l of ZrO_2 and 2 g/l HfO_2 in nitrate form for solvent extraction. Tri-butyl phosphate with 50% kerosene extracts preferentially zirconium and leaves hafnium in the raffinate along with all other impurities. Hafnium gets concentrated to the extent of 30% with respect to zirconium in the raffinate. Raffinate is generally considered as effluent because it contains all the impurities of the feed. Presently this acidic raffinate is being used to neutralize other alkaline effluents. NFC developed a process for recovery of pure hafnium from this raffinate.

During initial trials raffinate was concentrated by evaporation and total acidity of the resultant concentrated solution was adjusted to 8.0 N with additional nitric acid. Both hafnium and zirconium from this were extracted into solvent (50% TBP and 50% Kerosene) and separated by selective stripping using dilute nitric acid and D.M.Water. The flow sheet established is shown in Fig.1.

This process was found to be associated with inherent problems like fouling of heat-exchanger tubes with dissolved silica in the raffinate, very high consumption of nitric acid and poor stability of solvent (TBP) due to high normality of nitric acid in the feed. Because of these problems, after initial trials this process was not adopted on plant scale.

Later another process (1,2,3) using amine as solvent in sulphate media was considered for the enrichment and purification of hafnium from the raffinate. This route was chosen because silica content in the sulphate media becomes negligible and all the earlier problems due to silica can be avoided. However, the presence of nitrate-ion in the feed supresses the extraction by amine. Hence, a suitable nitrate free feed containing all the hafnium and associated zirconium values of the -

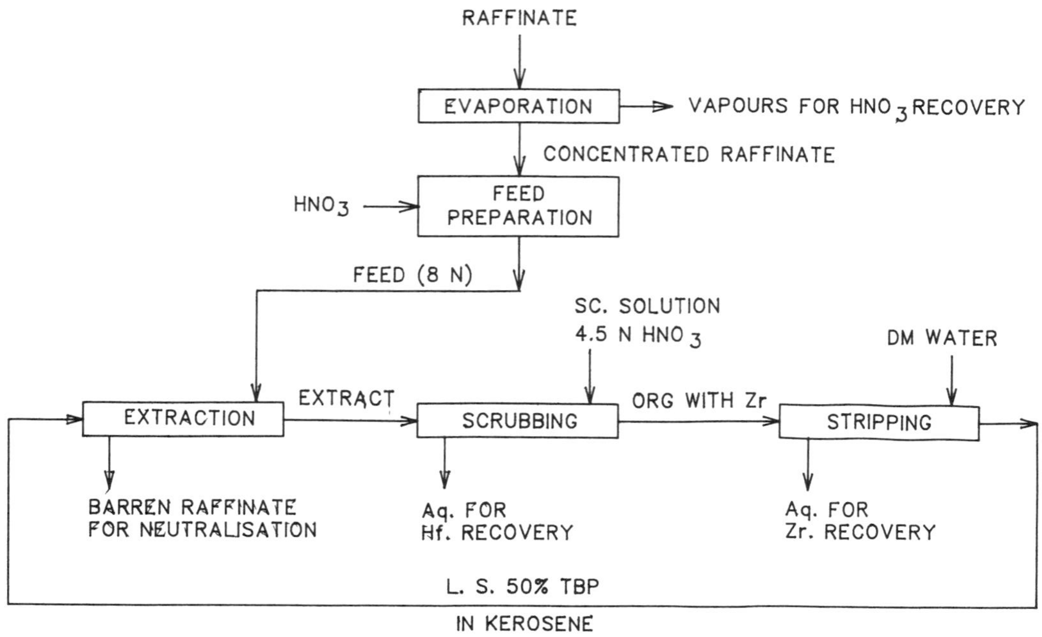

FIG. 1 HAFNIUM RECOVERY FROM RAFFINATE

..5

raffinate is to be made. Alkali precipitation follwed by multiple repulpings and filtrations of the raffinate and dissolution of the resultant wet cake in sulphuric acid gives acceptable quality feed for solvent extracction. But large volume of nitrate bearing effluent is generated which cannot be disposed-off readily. Hence, Ion-exchange method is adopted for denitration and feed preparation after standardization of operating parameters.

Strong acid cationic exchange resins equivalent to DOWEX-50 is used for the purpose. Raffinate is passed through the resin column for loading. Hafnium and zirconium values of raffinate get exchanged and are retained on the resin and almost all the nitric acid of the raffinate with major portion of impurities is retained in the barren liquor. This acidic effluent is used for the neutralisation of alkaline effluent generated in the plant.

The resin column is washed with water to remove loosely bound traces of raffinate and is then eluated with 2.5 N sulphuric acid. The eluate containing 20-25 g/l of (Hf+Zr)O2 with 2.5 N total acidity, is the feed material for the solvent extraction unit for hafnium extraction. Table-2 gives the performance of the ion-exchange process.

TABLE-2: PERFORMANCE OF ION-EXCHANGE PROCESS

	Raffinate	Eluate (Zr-Sulphate feed)
(Zr+Hf)O2	4.26 g/l	25.28 g/l
Hf(Hf+Zr)	32%	33.06%
Total nitrates	191.62 g/l	0.39 g/l
Fe	470 mg/l	138 mg/l
Ti	0.22 g/l	0.14 g/l
Si	780 mg/l	56 mg/l

..6

Hafnium and zirconium from sulphuric acid eluate solution are extracted by amine solvent, leaving most of the impurities. Loaded extract is scrubbed with 2.5 N sulphuric acid to remove titanium impurity and is then stripped out of he solvent to recover hafnium rich (75%) aqueous solution. This is given a reverse scrub with lean solvent to get pure hafnium solution which is precipitated, filtered and calcined to get pure hafnium oxide. Fig.2 gives details of the process for recovery of pure hafnium from raffinate waste.

2. RECOVERY OF PURE ZIRCONIUM OXIDE FROM ZIRCONIUM CHLORIDE WASTES: Zirconium chloride wastes are generated during zirconium sponge production where, pure zirconium oxide together with carbon is chlorinated and the resultant zirconium tetrachloride is reduced with magnesium to get reactor grade zirconium sponge. During this process a considerable quantity of hafnium free zirconium escapes as chloride in the off gases of chlorinator condensers. It is contaminated with impurities like Fe, Si, Al, Sn etc., As major cost of zirconium production is in the separation process of zirconium and hafnium, recovery of zirconium from this hafnium free chloride waste is still economically attractive.

Purification processes such as basic-sulphate precipitation method and phthalate precipitation method were considered initially for recovering zirconium from chloride wastes. Though the product obtained by these methods was pure, they are associated with problems due to the large volumes of effluents generated and the requirement of heating and cooling of chloride bearing acidic liquids. To overcome these difficulties multiple crystalization using common ion effect(4) was used as the waste chloride recovered from chlorinator is readily soluble in water. It is relatively simple in operation and gives minimum amount of effluent. Hydrochloric acid is added to zirconium oxy-chloride

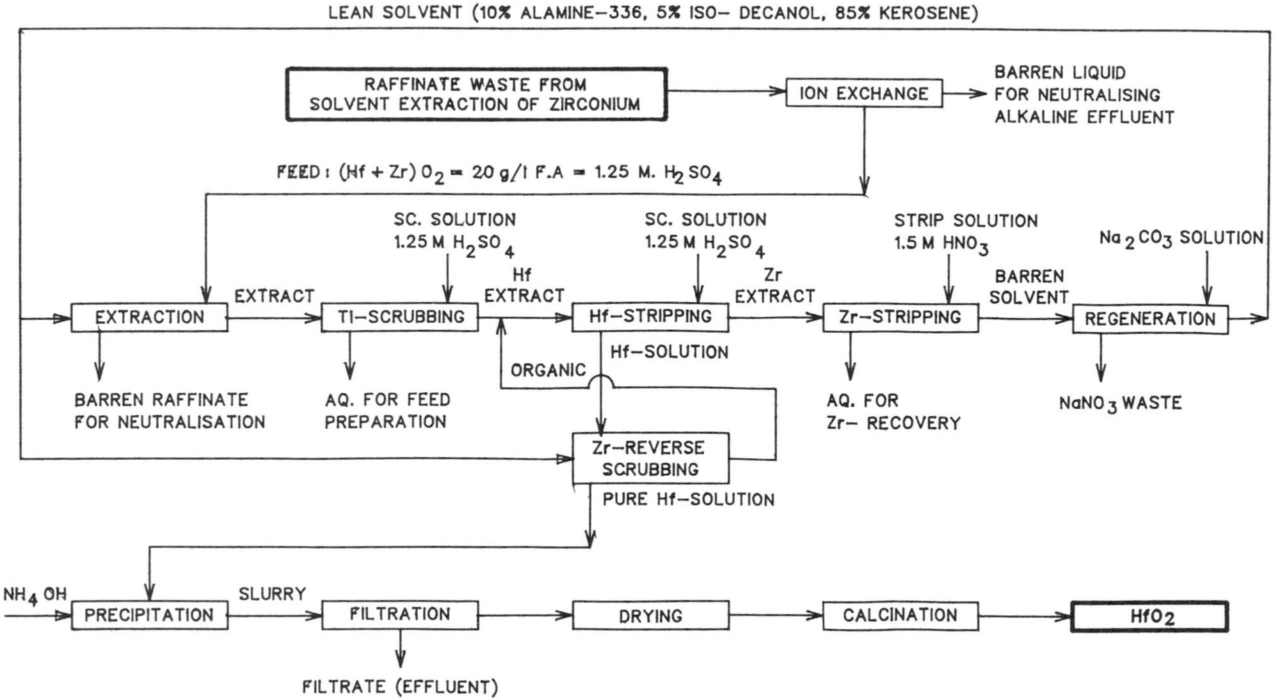

FIG. 2 EXTRACTION OF HAFNIUM FROM RAFFINATE

...8

solution for crystallization by common ion effect and only room temperature operations are involved. Fig.3 indicates various steps involved in the process.

Pure zirconium oxy-chloride crystals obtained after double crystallisation are redissolved in water and zirconium hydroxide is obtained by precipitating with ammonia. This hydroxide is further filtered, dried and calcined to get pure zirconium oxide.

3. RECOVERY OF ZIRCONIUM FROM SPENT PICKLING SOLUTION:

During conversion of ingots of zirconium alloys into various structural components required for nuclear industry, oxidation takes place on the surface at various stages. These layers are removed completely by acid pickling.

The pickling bath for this process consists of 2% HF, 25% HNO_3 and balance water. The pickling rate decreases as the zirconium concentration in the pickling solution increases and after certain use pickling becomes very slow and non-uniform and the bath requires replacement.

Usually spent pickling solution has the following composition: HF=1% W/W, Total acidity = 4-5 N and zirconium = 60-70 g/l expressed as ZrO_2. The disposal of this spent liquor poses pollution problems due to the presence of toxic fluoride and nitrate ions and also it leads to the loss of valuablce hafnium free zirconium. About 12,000 ltrs per month of spent pickling solution is generated in NFC which contains about 750 kgs of zirconium. NFC developed a process to recover this zirconium and also regenerate the bath so as to avoid pollution to the environment.

Initially experiments were carried out to precipitate zirconium after fixing the fluorides with lime, calcium nitrate and aluminium

...9

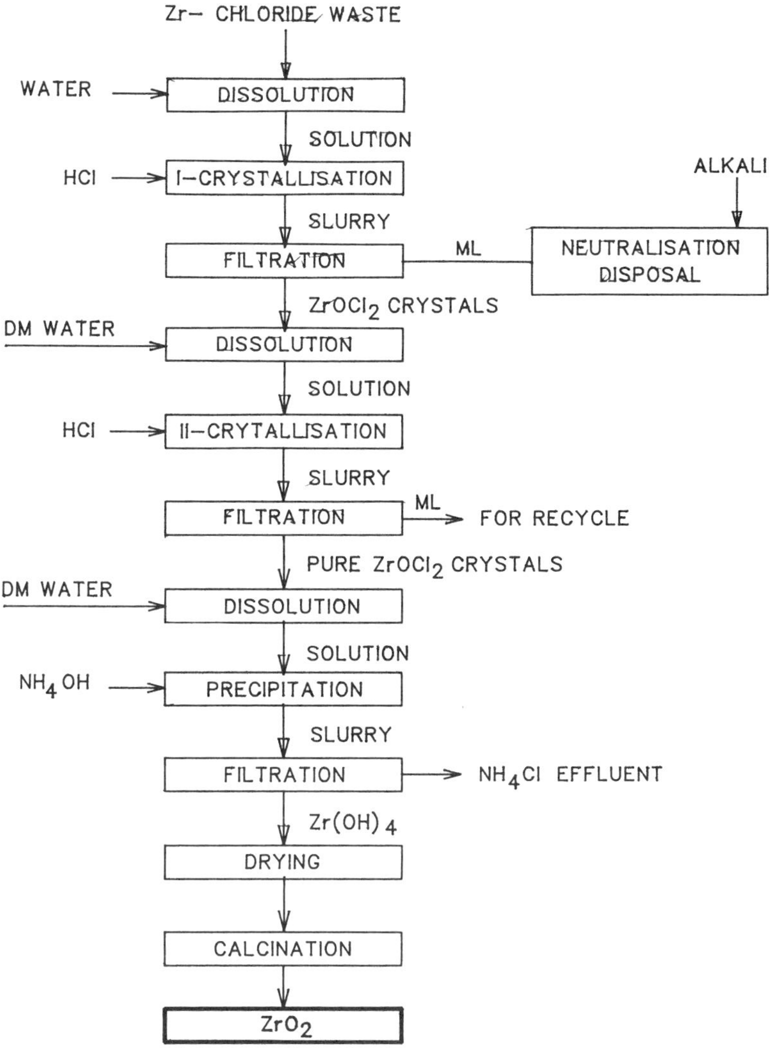

FIG. 3 ZIRCONIUM RECOVERY FROM CHLORIDE WASTES

...10

nitrate. In all the three cases loss of zirconium was very high and the recovered zirconium was found to be heavily contaminated with Ca and Al.

Then ion-exchange technique (5) was tried to recover free acids from the bath by exchanging the zirconium values with anion exchange resins equivalent to DOWEX-1. This also was found unattractive due to very low loading capacity of the anion resins and also due to the generation of large quantities of liquid effluents during washing and later processes.

Finally zirconium values were successfully recovered from the spent liquor by precipitating with phosphoric acid (6). As zirconium phosphate has least solubility even in acidic solutions, zirconium readily gets precipitated with phosphoric acid releasing all the free acids (HF and HNO_3) for recycle to the pickling operations. In this method precipitation of alloying elements such as Sn, Cr, Fe, etc., is absent because of the high acidity of the bath. Zirconium phosphate thus obtained is digested with sodium hydroxide solution to get zirconium hydroxide and tri-sodium phosphate. Zirconium hydroxide is converted to ZrO_2 by calcination. The process is simple, fast and economical. Fig.4 gives the flow-sheet developed by NFC, which is going to be implemented on plant scale shortly.

ACKNOWLEDGEMENTS:

The authors express their sincere thanks to Shri K.Balaramamoorthy, Chief Executive, Nuclear Fuel Complex, Hyderabad for the keen interest and constant encouragement.

...11

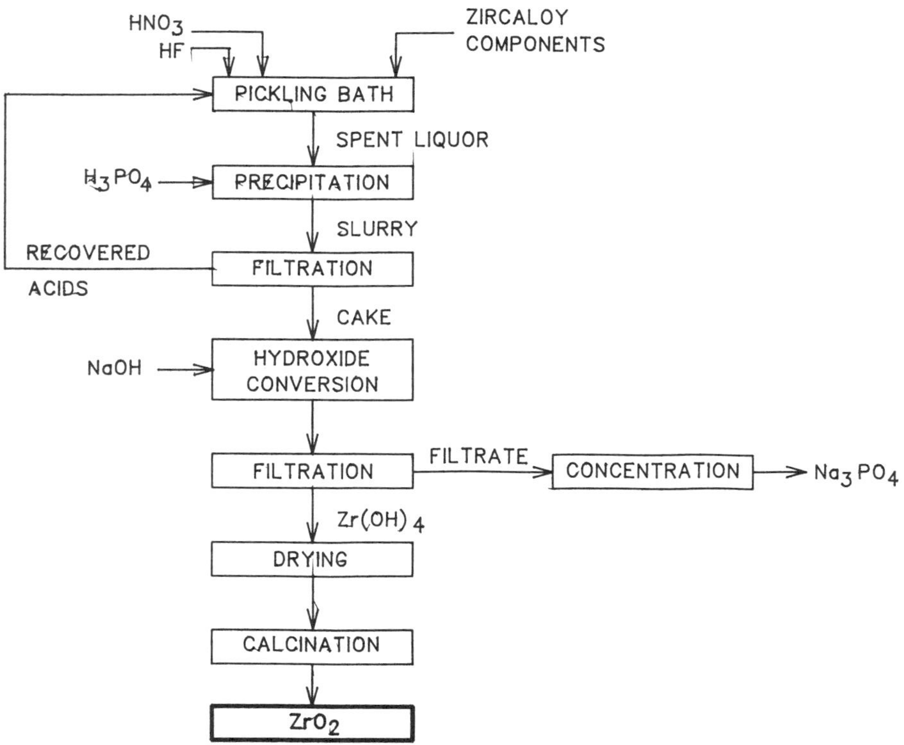

FIG. 4 ZIRCONIUM RECOVERY FROM SPENT PICKLING SOLUTION

...12

REFERENCES:

1. "Amine extraction process for separation of zirconium and hafnium". (Report No:UED/82/5). R.G.Swamy, S.M.Deshpande, R.B Gajankush, N.S.Iyer, K S.Koppiker and T.K S.Murthy, Uranium Extraction Division, Bhabha Atomic Research Centre, Bombay, India.
2. "Recovery of hafnium from ZOP raffinate solution At NFC, Hyderabad". (Report No:UED/87/19). S.M.Deshpande, R.G.Swamy, R.B.Gajankush, N.S.Iyer and K.S.Koppiker, Uranium Extraction Division, Bhabha Atomic Research Centre, Bombay, India.
3. "New Solvent Extraction Process for Zirconium and Hafnium," Zirconium in the Nuclear Industry: Sixth International Symposium, ASTM STP 824. D.G.Franklin and R.B Adamson, Eds., American Society for Testing and Materials, 1984, pp. 45-56,: Takahashi, Miyazaki, H., and Katch, Y.,
4. Project Report -", Recovery of zirconium values from Reduction residues And Duct chlorides from Zirconium Sponge Plant". (April-1985). S.M.Rao, Nuclear Fuel Complex, Hyderabad, India
5. "Process for Recovering acids and zirconium contained in pickling solutions" Indian Patent No:-862 DEL 78, 29 NOV 1978. of UGINE ACIERS a French company of 10, rue du General Fers, 75008 Paris, France.
6. Analytical chemistry of zirconium and hafnium". (Pages 8 and 33), a translation of Analyticheskaya Khimiya Tsirkoniya I Gafniya, Izdatel'stvo "Nauka" Moskua 1965.:S.V. Elinson & K.I.Petrov, Academy of Sciences of the USSR -

NEODYMIUM AND IRON RECOVERY FROM NdFeB PERMANENT MAGNET SCRAP

J. W. Lyman[1] and G. R. Palmer[2]

U.S. Department of the Interior
Bureau of Mines
Salt Lake City Research Center
729 Arapeen Drive
Salt Lake City, Utah

Abstract

The U.S. Department of the Interior, Bureau of Mines, investigated methods to recover valuable rare-earth materials from NdFeB magnet scrap. Although several processing techniques were evaluated, the best separation of rare earths from the NdFeB magnet scrap was obtained by sulfuric acid dissolution followed by precipitation of recyclable rare-earth salts. Both sodium and ammonium intermediate double-salt precipitates that can be converted to a variety of useful neodymium products were investigated. The sodium system showed a neodymium recovery of 98 pct, while the ammonium system yielded 70 pct during double-salt precipitation. The iron-rich effluent following the rare-earth recovery was treated to produce sodium and ammonium iron jarosites that can be disposed of or converted to hematite. By following this new procedure, many materials handling and economic disadvantages found with fluoride or oxalate direct precipitation were avoided.

[1]Chemical engineer.
[2]Supervisory metallurgist.
Salt Lake City Research Center, Bureau of Mines, U.S. Department of the Interior, Salt Lake City, Utah.

Residues and Effluents - Processing and Environmental Considerations
Edited by Ramana G. Reddy, William P. Imrie and Paul B. Queneau
The Minerals, Metals & Materials Society, 1991

Introduction

The U.S. Bureau of Mines is engaged in developing technology to treat a variety of wastes that contain valuable and strategic metal values. Economic recovery from such wastes ensures cost savings for industry and reduces the degree of U.S. dependence on foreign imports. As part of this effort, the separation of neodymium from iron in NdFeB magnet scrap was studied. The objective of this study was the economic separation and recovery of neodymium from this scrap. Since this scrap typically contains nearly 30-wt pct Nd for which the demand is constantly increasing (*1-3*), economic methods for scrap treatment will significantly impact industrial expenditure and materials supply in the expanding area of magnet manufacture.

Permanent magnets are indispensable components of electromechanical and electronic devices such as computers, printers, motors, microwaves, and audio visual components among others. An average home in the United States contains about 40 magnets, and a fully equipped car contains more than 30 (*2,4,5*). In 1979, small but powerful samarium-cobalt magnets began to replace ordinary ferrite magnets in such applications as the Sony "Walkman" series of miniaturized portable stereo headphones and tape players. By 1982, consumption of samarium oxide in the United States was 165 st despite the fact that this market was totally nonexistent in 1978 (*6*). As the demand for permanent rare-earth magnets increased, and due to unrest in Africa which threatened to affect the supply of cobalt, research was undertaken to develop other rare-earth, transition-metal alloys which could combine improved magnetic capabilities with lower raw materials costs (*4*). The fortuitous discovery of a stable $Nd_2Fe_{14}B$ phase met the demand for a lower cost, more readily-available alloy with improved magnetic properties. NdFeB magnet material has the highest energy product of all permanent magnets and exhibits excellent resistance to demagnetization at normal operating temperatures (*7-9*).

The increased power of NdFeB alloy has made possible the manufacture of smaller and more efficient magnets that are now used in a variety of applications including automotive cranking motors, servo and other motors, meters, computers, audio-visual components, hand-held portable tools, and other devices requiring reduced size and weight (*2,4,10,11*). NdFeB magnets are also used in magnetic separators, military and aerospace systems, and in magnetic levitation railways systems which will allow cruising speeds of over 450 km/hr (*12-14*).

Currently, large amounts of scrap from NdFeB magnet manufacture are being stockpiled due to the lack of a cost-effective processing method. Because of the growing new market for individual rare-earth metals such as neodymium, which was valued at $280/kg at year-end 1988 (*2*), and due to the likely continuing increase in demand for permanent magnets, scrap will undoubtedly become an important alternative materials source. It is clear that development of a viable method to process NdFeB scrap will aid domestic manufacturers economically and reduce the necessity for import of rare-earth materials.

The two major methods currently being used to manufacture NdFeB magnets are rapid solidification and powder metallurgy techniques. In rapid solidification, a melted alloy is supercooled by generating melt-spun ribbon which is then crushed and treated in further processing steps such as polymer bonding or hot pressing. By this process, material with extremely fine-grain size and uniformity of microstructure is obtained which has excellent magnetic properties. Typical scrap types generated by this process are slag and spillage materials from alloy melting, discarded ribbon material, and discarded finished magnets in a variety of forms. Powder metallurgy processing involves the sintering of a powder which is aligned in a magnetic field. This process generates grinding swarf in addition to discarded sintered and unsintered (green compact) magnet pieces (*1,4*). Some sintered magnet scrap bears a protective plastic outer coating as well.

The composition of scrap from powder metallurgy processing was found to be similar to that from rapid solidification. All types have the following average composition, in weight percent: 67.3 Fe, 26.3 Nd, 1.85 Pr, and 0.84 B. Because scrap compositions were quite uniform, economical treatment methods were sought which could efficiently separate rare-earth metals from iron in all types of scrap despite variations in scrap geometry and surface oxidation. Initial work by the same authors at the Bureau focused on selective oxidation of neodymium while maintaining a reducing atmosphere for iron (*15*). Although selective oxidation was found to be successful, separation of oxidized neodymium from metallic iron could not be easily accomplished without extensive further processing. For this reason, the focus of the work was shifted to acid dissolution and precipitation of rare-earth salts.

Acid Dissolution

Although sulfuric and hydrochloric acids were found to be equally effective for scrap dissolution, sulfuric acid was used in the bulk of tests due to its lower cost and because it is widely employed in industry for rare-earth concentrate leaching. Precipitation of rare-earth salts and double salts from sulfate systems has been extensively studied and forms the basis for many industrial practices (*16,17*). All types of scrap dissolved readily without heating or agitation over a 24-hr period. Beneficiation of large, uncoated scrap pieces to -1/4" in a jaw crusher reduced the time needed for complete dissolution but was not required. It was found that coated scrap needed to be crushed prior to leaching to render it amenable to attack by the acid. Tests were conducted to determine the typical acid requirements for scrap dissolution and to examine the leaching behavior of different types of scrap.

Tests showed that a critical factor for scrap dissolution was the weight ratio of sulfuric acid to weight of scrap to be dissolved. For all scrap types tested, a weight ratio of acid to scrap of 2:1 or higher was sufficient both to totally dissolve the scrap and to keep the solution pH low enough to prevent the oxidation of ferrous ion to ferric. It is known that acid solutions of ferrous ion oxidize less rapidly with increasing acidity (*18*). Ferric ion is undesirable because it forms precipitates in solution which can contaminate a process stream. For 1*M* acid solutions and an acid-to-scrap weight ratio of 2:1, the pH of the final, clear blue solution was approximately 1.0, and no iron precipitates formed. When the acid-to-scrap ratio was 1:1, the final pH was 3, dissolution was incomplete, and iron precipitated out of the greenish filtrate over time.

Dissolution behavior for two types of scrap can be seen in Figures 1 and 2. These figures show that despite differences in leach behavior, complete dissolution occurs when the acid-to-scrap weight ratio reaches 2.

In addition to determining the amount of acid required for effective dissolution, it was necessary to study the effect of acid concentration while maintaining the acid-to-scrap weight ratio at 2. One-hundred grams of ribbon material was completely dissolved in 1 L of 2*M* (200 g) H_2SO_4 yielding a solution which analyzed 33 g/L Nd and 78 g/L Fe. (Analyses appear higher than those in the original scrap due to loss of water volume from exothermic heating and hydrogen evolution during dissolution.) One-hundred-fifty grams of ribbon material was dissolved in 1 L of 3*M* (300 g) H_2SO_4, but large amounts

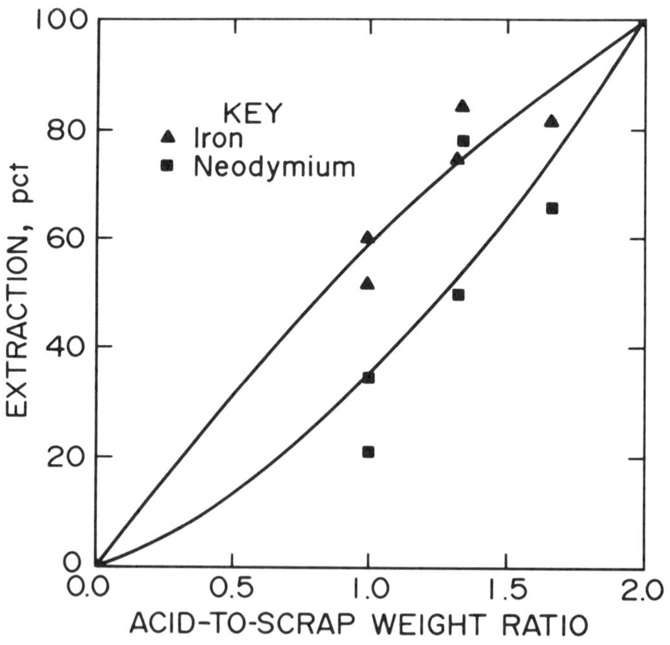

Figure 1 - Leaching of ribbon scrap.

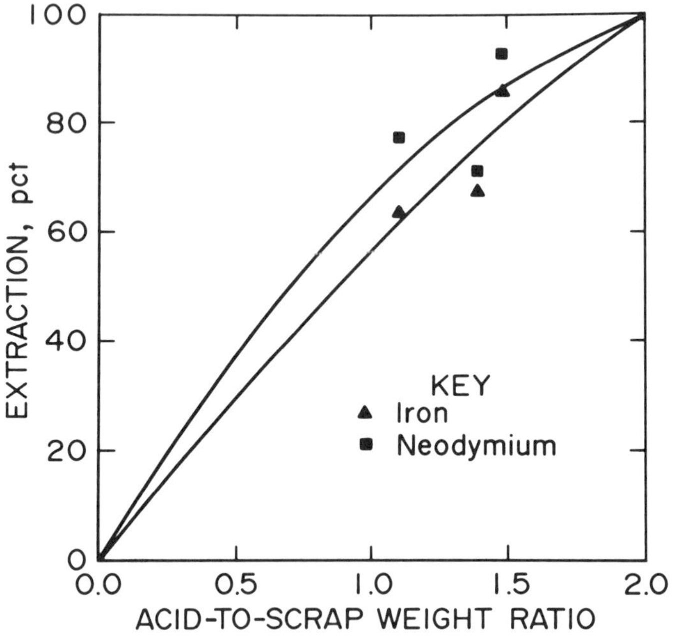

Figure 2 - Leaching of hot-pressed ribbon scrap

of iron and rare-earth sulfate salts precipitated out of solution. Although the original scrap contained approximately 40 g of neodymium and 100 g of iron, the filtrate contained only 23 g/L Nd and 77 g/L Fe. It was evident that sulfate solubilities were exceeded for both iron and neodymium when the sulfuric acid molarity was 3 due to the decreased amount of water used. Operating the system at a molarity of 2 was adequate to effect total sample dissolution while minimizing the water requirement. Other types of scrap such as slag, swarf, and bulk-sintered magnets exhibited similar leaching behavior.

Rare-Earth Precipitation

Once scrap was dissolved in acid solution, separation of rare earths from iron was achieved by precipitation of a rare-earth salt. For acid dissolution to be a viable method for treating NdFeB scrap, a rare-earth precipitation product must be generated that can be recycled at some point into already-existing processes which use neodymium salts as feed materials. Pure neodymium fluorides and oxides are used in laser, glass, and ceramics manufacture, but by far, the largest demand for neodymium salts is in metallothermic or electrolytic reduction to produce neodymium metal and/or alloys for magnet manufacture.

Both NdF_3 and Nd_2O_3 would be suitable precipitation products for recycle to neodymium metal and alloy-producing operations. Fluorides can be precipitated directly from aqueous solutions, and, in fact, this is a recommended practice when large amounts of iron are also present in solution (*17*). Rare-earth oxides are generally obtained by roasting rare-earth oxalates, and although the presence of large amounts of iron is known to inhibit their selectivity, rare-earth oxalates are some of the most important rare-earth salts (*17,19*).

Fluoride and oxalate precipitation, however, have attendant problems. Fluoride precipitates tend to be gelatinous and difficult to filter, and oxalates are costly to produce and must be roasted to yield an oxide product suitable for recycle. A possible solution to the problems attendant with the direct precipitation of fluorides and oxalates is to generate an intermediate precipitate which could then be converted to either a fluoride or oxalate product depending on industrial demand. One example of such an intermediate precipitate is the double sulfate salt of neodymium and sodium which is used industrially for the precipitation and storage of rare-earth materials (*17*).

Precipitation of Neodymium-Sodium Sulfate Double Salt

The formation of a neodymium-sodium sulfate double salt was investigated as an intermediate precipitate which would serve to economically separate rare earths from iron without a filtration problem. Sodium hydroxide was added to acid leach solutions of magnet scrap to raise the pH to 1.5. A light-purple, easily-filtered precipitate formed which is probably a neodymium-sodium sulfate double salt. Such double salts are very common although the composition is variable, and few have been adequately characterized. Double sodium sulfates are listed as having formulas such as $Nd_2(SO_4)_3 \cdot Na_2SO_4 \cdot 6H_2O$ or $NaNd(SO_4)_2 \cdot nH_2O$ where n is 0 or 1 (*17,20*). The precipitate was analyzed to contain, in percent, 35 Nd, 2 Pr, 6.5 Na, 50 SO_4, and only 0.2 Fe. It is interesting to note that the molar amounts of neodymium and sodium are roughly equivalent, corresponding to the formulas above. Figure 3 shows the recovery of neodymium and iron into the double sulfate salt as a function of pH. It can be seen that the recovery of neodymium exceeds 95 pct and reaches its maximum between pH values of 1.5 and 2, while iron recoveries remain low. At pH values above 5, significant amounts of iron begin to precipitate out of solution.

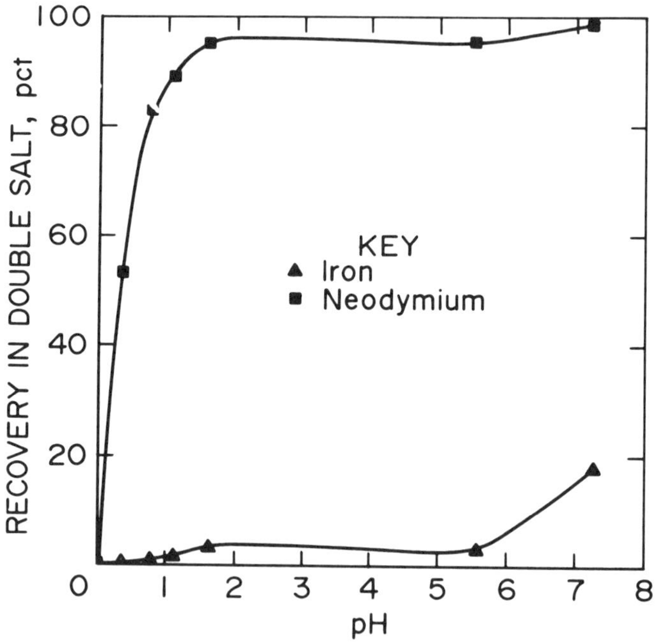

Figure 3 - Recovery of neodymium and iron in the sodium sulfate double salt.

Precipitation of Neodymium-Ammonium Sulfate Double Salt

Instead of NaOH, NH_4OH was added to metal-rich, sulfuric-acid leach solutions to raise the pH to 1.5. A light-purple precipitate formed which again filtered very easily. This precipitate was characterized using X-ray diffraction and identified as $Nd_2(NH_4)_2(SO_4)_4 \cdot 8H_2O$. The precipitate was analyzed to contain, in percent, 34 Nd, 2 Pr, 5 NH_4, 45 SO_4, and 0.2 Fe. Again, molar amounts of neodymium and NH_4 are equivalent. Figure 4 shows the recoveries of neodymium and iron in the ammonium double salt as a function of pH. The maximum neodymium recovery in this salt is approximately 70 pct due to its greater solubility in aqueous solution. As in the sodium system, significant amounts of iron do not precipitate until the pH is well above 2.

Precipitation of Neodymium-Potassium Sulfate Double Salt

A similar, easily filtered purple salt was also formed by the addition of KOH to pregnant leach solutions of dissolved magnet scrap. Ninety-three pct of the neodymium was recovered in this salt which analyzed, in percent, 30 Nd, 2 Pr, 9 K, 50 SO_4, and 0.2 Fe. The molar potassium content is equivalent to that of sodium and NH_4 in other double-salt species previously discussed.

Double-Salt Conversion to Fluoride

Tests were performed to determine whether the neodymium-sodium and ammonium double-salt precipitates could be converted to pure fluoride products by leaching in hydrofluoric acid (HF) solution. It was found that both sodium and ammonium double salts could readily be converted to NdF_3 with no loss of neodymium. A fifteen-minute leach time was found to be adequate for complete

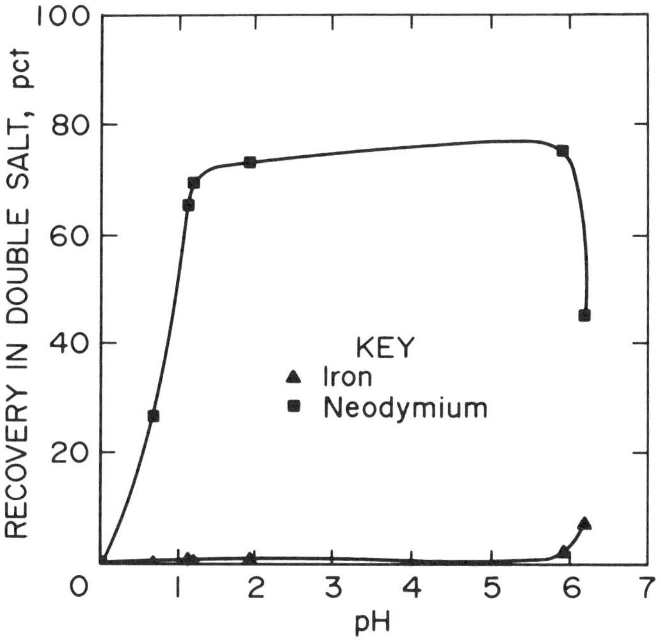

Figure 4 - Recovery of neodymium and iron in the ammonium sulfate double salt.

conversion. Due to their similarity to sodium double salts, potassium double salts were not included in this phase of the research.

Residual amounts of sulfate and alkali-ammonium ions were found in the fluoride products and varied depending on the amounts of HF and water used in the leach. Tables I and II show neodymium concentrations and impurity levels present in NdF_3 products from the conversion of sodium and ammonium double salts, respectively, for a variety of leach conditions. It can be seen from the tables that sulfate content in the converted fluoride varies inversely with the HF-double-salt ratio. When the HF-double-salt ratio is 0.71, sulfate content is independent of the water-HF ratio, although for higher HF additions, the sulfate content does appear to increase when water additions are also low possibly due to ionic strength and/or saturation effects.

Table I - Fluoride Conversion of Sodium Double Salts

HF-double salt Ratio	Water-HF ratio	Fluoride product composition, pct			
		Nd	Fe	SO_4	Na
0.14	100.0	48.8	0.12	27.6	3.7
.29	50	65.0	.13	5.4	2.3
.71	20	67.3	.20	2.3	2.1
.71	10	57.1	.16	2.2	4.1
.71	5	60.2	.26	2.4	2.9
1.43	10	66.6	.12	1.1	1.8
1.43	2.5	46.6	.44	1.3	7.6

Table II - Fluoride Conversion of Ammonium Double Salts

HF-double salt ratio	Water-HF ratio	Fluoride product composition, pct Nd	Fe	SO_4	NH_4
0.14	100.0	66.6	0.12	7.5	<0.1
.29	50	65.7	.11	6.8	.1
.71	20	69.2	.10	2.6	.1
.71	10	67.2	.08	2.8	<.1
.71	5	67.6	.11	2.6	.2
1.43	10	69.3	.14	.6	<.1
1.43	2.5	70.8	.20	1.2	.2
2.86	4.5	68.1	.13	.5	<.1

Ammonium ion concentrations were low in all converted fluoride products, but sodium ion concentrations tended to be much higher, particularly for a high HF addition coupled with less water. It is of interest to note that whenever the sodium content was high, the neodymium recovery was low. This can be explained by the formation of insoluble Na_2SiF_6 which consumes needed fluoride ion. The presence of this compound was confirmed by X-ray diffraction (XRD), and it was found that a second HF leach was sufficient to remove it, lowering the sodium content to 0.9. Formation of Na_2SiF_6 was prevented entirely by performing leaching, precipitation, and particularly conversion stages in nalgene reaction vessels which do not contain silicon.

In general, the purity of NdF_3 obtained from the sodium double salt is not as high as in ammonium double-salt conversions. For both salts, the purest products are obtained at an HF-double-salt ratio of 1.43 or higher and at a water-HF ratio of 10. Residual iron content was found to be independent of the conversion conditions, depending instead on the amount of iron originally present in the double salt. Fluorides obtained from conversion of the double salts are easily filtered, although for the ammonium double-salt conversion, HF must be added to the leach solution prior to introduction of the double salt to prevent the formation of a gelatinous, difficult-to-filter fluoride precipitate. The ease of filtration of fluorides obtained from double salts is one major advantage to this method of processing NdFeB scrap.

Double-Salt Conversion to Oxalate

Oxalate products were also obtained by reacting sodium and ammonium double salts with 0.2*M* aqueous solutions of oxalic acid. The double salt-to-oxalic acid weight ratio was 0.72. Fifteen minutes was found to be adequate for complete conversion. Losses of neodymium were negligible in the conversion process, and the weights of neodymium and praseodymium in the final products indicated that nearly pure oxalate precipitates were obtained which could be roasted to yield pure oxides if desired. The purity of oxalate products was confirmed by XRD. It was concluded that neodymium double salts may be treated with either HF or oxalic acid to generate fluoride or oxide products, respectively.

Iron Jarosite Precipitation

Following rare-earth precipitation, the spent leach solution, which still contains as much as 77 g/L iron, must be treated by an iron removal process prior to disposal or recycle. Because ordinary neutralization leads to the formation of iron hydroxides which cannot be easily filtered, it was decided to investigate precipitation of a jarosite to facilitate filtration. Jarosites are compounds of the type $MFe_3(SO_4)_2(OH)_6$ where M is K^+, Na^+, NH_4^+ or a metal ion such as Ag^+ or 1/2 Pb^{2+}. Iron is present in the ferric oxidation state. Jarosite precipitation chemistry is well-understood and widely used in the zinc industry for removal of iron in the most economical manner possible (*21-23*). Jarosites are easily filtered, exhibit good stability, and can be formed directly from spent leach solutions following double-salt precipitation. In addition, jarosites can be converted to hematite for use in cement production, pigmentation, and iron manufacture (*24*).

Spent solutions were placed in a reaction kettle through which oxygen was bubbled and allowed to react for 6 hr at 90° C to form a yellow iron jarosite which was then removed by filtration. The presence of oxygen ensures oxidation of ferrous to ferric ion. The amount of jarosite which formed depended on the presence or absence of an oxidant, reagent addition, pH of spent solution, and hydroxide species used.

<u>Effect of Oxidant Addition</u>

Tests showed that addition of hydrogen peroxide positively affected jarosite formation in both the sodium and ammonium systems probably due to improved kinetics for the oxidation of ferrous to ferric ion. Figure 5 shows a plot of iron extraction as a function of oxidant addition to spent leach solutions at a pH of 1.5. It was possible to form jarosites from both sodium and ammonium-spent leach solutions without an oxidant, but the amount of jarosite formed was much greater when the oxidant was present. Iron recovery in the ammonium jarosite was slightly higher than in sodium jarosite. Recovery of iron did not increase dramatically for either system beyond a hydrogen peroxide addition of 30 g (50 pct).

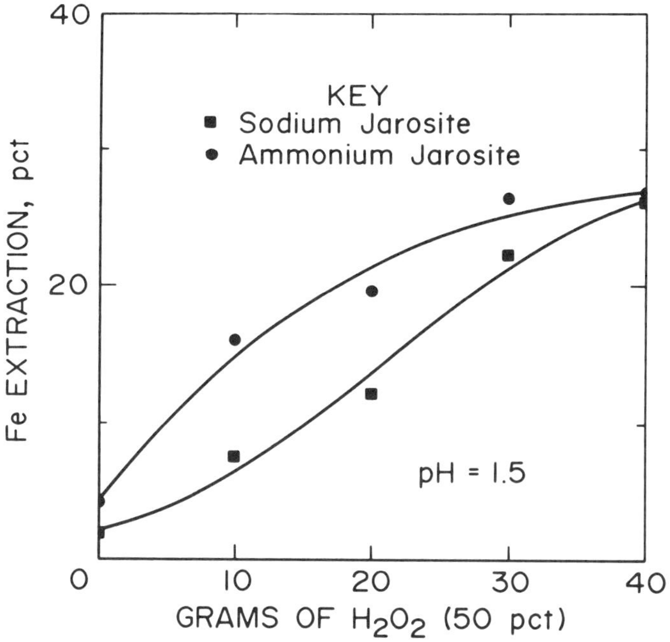

Figure 5 - Iron recovery as jarosite.

Effect of pH and Sodium Sulfate Addition

Varying amounts of sodium sulfate were added to oxidant-containing spent leach solutions at a pH of 1.5 to investigate the effect of increasing the available sodium ion. A positive effect on jarosite formation was observed up to a Na_2SO_4 addition of 30 g as shown in Figure 6. Adding more than 30 g decreased the iron recovery. The most influential parameter was found to be pH. When 1 L of spent leach solution at pH 1.5 was placed in the jarosite reactor with 30 g of 50 pct H_2O_2 and 30 g of Na_2SO_4, the maximum iron recovery was on the order of 20 pct. When the pH was raised to 2 with more hydroxide, however, 70 pct of the iron precipitated as jarosite (other reagent additions were not varied). Clearly, the effect of pH is dramatic, and it is likely that ongoing work will show that the iron recovery is enhanced even further at pH values above 2.

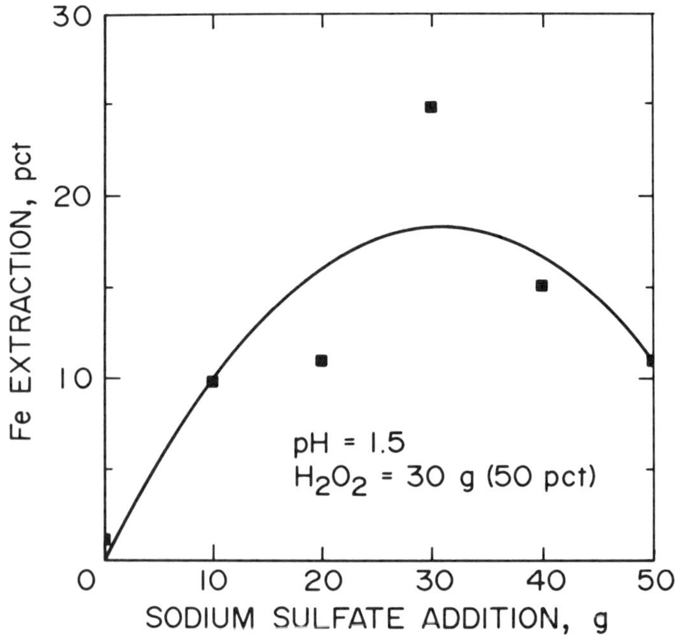

Figure 6 - Effect of sodium sulfate addition on jarosite formation.

Effect of Time on Jarosite Formation

A test series was performed in which small samples of the jarosite slurry were removed at 1-hr intervals and submitted for analysis. The pH of the initial solution was 1.83, and 30 g of 50 pct H_2O_2 were used. Results of this series are shown in Figure 7. Very little jarosite formed after 2 hr of reaction time. This time interval corresponded to an abrupt drop in pH caused by the consumption of OH^- ions as jarosite formed. In order to form more jarosite, a pH adjustment after 2 hr of reaction time would be required. The apparent increase of dissolved species concentrations at 5-6 hr could have been caused by water evaporation from the reactor.

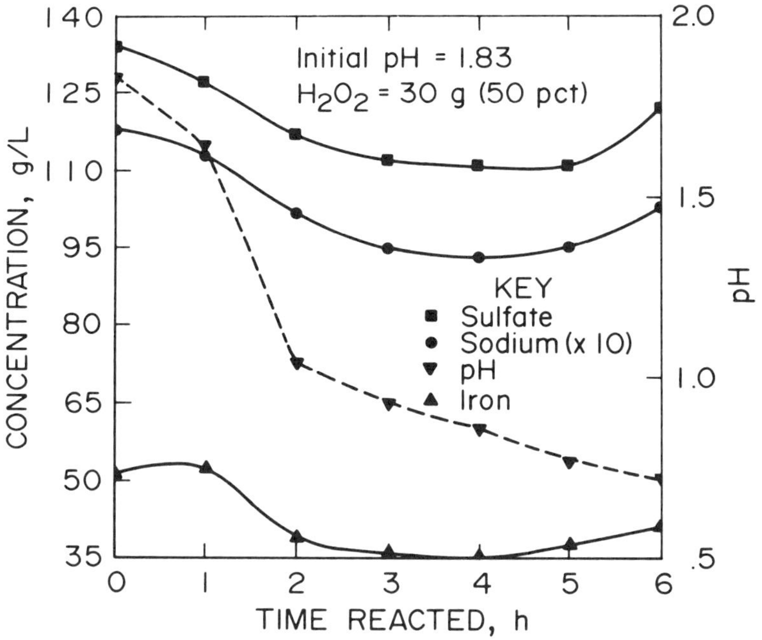

Figure 7 - Variation in jarosite solution over time.

Behavior of Residual Neodymium During Jarosite Formation

In the sodium system, residual neodymium precipitates with the jarosite and is lost; whereas, in the ammonium system, neodymium is retained in the spent solution from jarosite processing and could potentially be recovered in a later step. As was shown previously, neodymium recovery in the ammonium double salt is only 70 pct. The rest remains in solution. It is advantageous that the neodymium remaining in solution does not precipitate in the ammonium jarosite. Spent neodymium-containing solution following iron removal could be recycled to the original leach.

Fluoride Ion Inhibition of Jarosite Formation

Some jarosite precipitation tests were carried out on solutions from which neodymium had been removed by direct precipitation with fluoride ion, and it was found that jarosite failed to form in every test where fluoride ion was present in solution. Instead, a white precipitate formed which analyzed high in fluoride content. It is evident that the presence of fluoride ion inhibits the jarosite process. Direct precipitation of the fluoride from iron-bearing leach solutions is therefore undesirable, not only due to filtration difficulty, but also due to the inhibiting effect of the fluoride ion on subsequent jarosite precipitation.

Behavior of Boron

Boron does not precipitate either with the rare-earth double salt or during subsequent jarosite formation, and its concentration in solution never exceeds approximately 1 g/L. Preliminary tests in which zinc was added to spent solutions from jarosite processing showed that it is possible to precipitate a form of zinc borate hydrate by raising the pH. Zinc borate hydrates are used commercially as flame retardants (*25*). Other salts, such as zinc fluoborate, could also be precipitated.

An overall processing flowsheet is shown in Figure 8 delineating the stages required to separate and recover rare earths, iron, and boron.

Conclusions

Conditions for leaching magnet scrap were effectively determined. Sulfuric acid dissolution was easily accomplished, requiring neither heating nor agitation. An acid weight two times that of the scrap weight was needed for complete dissolution at a molarity of 2. Neodymium double salts of sodium, ammonium, and potassium were produced which show promise as intermediate products from which pure oxalates and fluorides can be obtained by leaching with appropriate reagents. Precipitation of the sodium or ammonium rare-earth salt has no attendant filtration problem and effectively separates neodymium from iron. Fluorides obtained by conversion of the double salts can also be easily filtered, and the contaminating fluoride ion is not introduced into the iron-rich stream resulting from neodymium precipitation. Iron removal from spent leach solutions can be accomplished through precipitation of a potentially useful jarosite product. Finally, boron may be recovered separately in a usable form such as zinc borate hydrate.

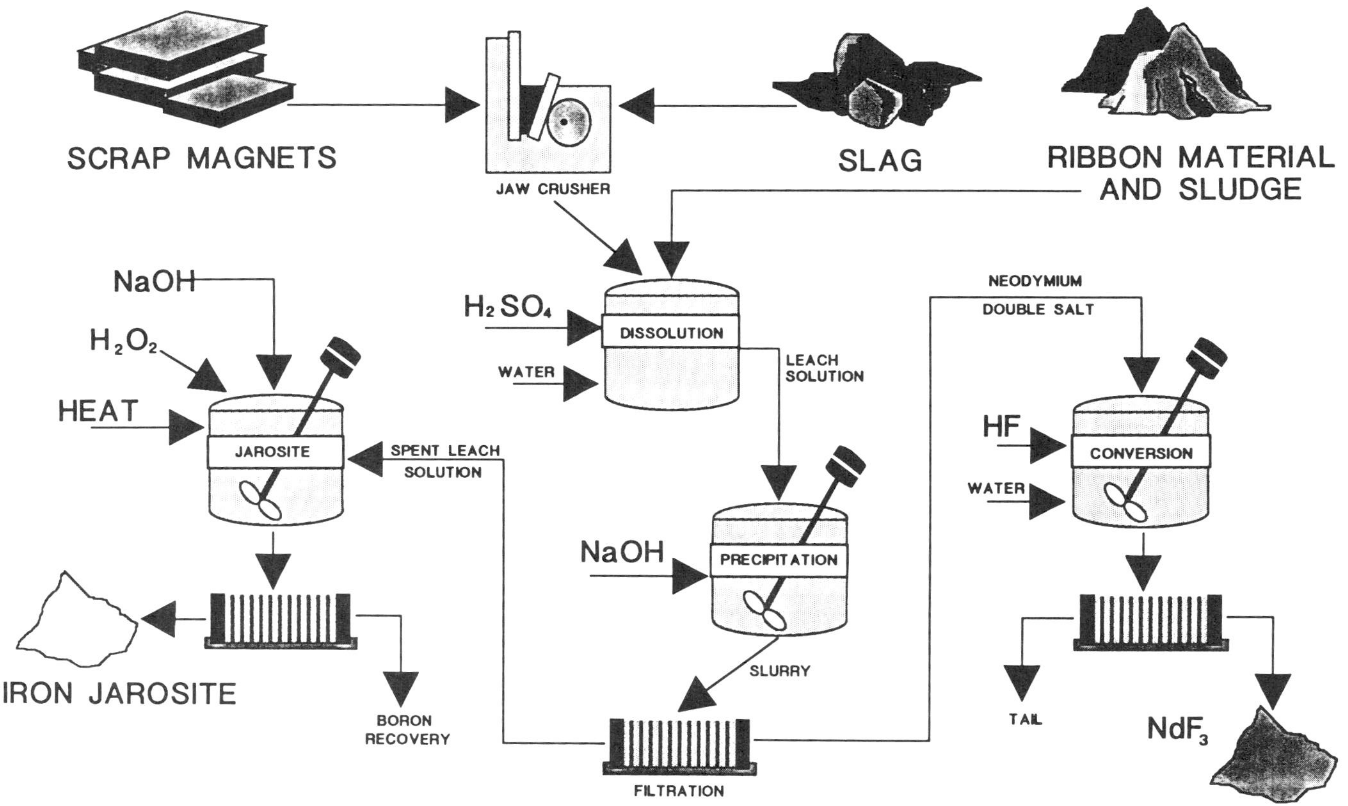

Figure 8 - NdFeB magnet scrap recycle process

REFERENCES

1. Jacqui Robbins, "Magnetic Materials Update," Materials Edge, September/October (1989), pp. 54-60.

2. James B. Hedrick and David A. Templeton, "Rare-Earth Minerals and Metals," BuMines Minerals Yearbook 1988, v. 1, 11 pp. (preprint).

3. James B. Hedrick, "Rare-Earth Metals," Sec. in BuMines Mineral Commodity Summaries 1990, p. 134.

4. J. D. Livingston, "The History of Permanent-Magnet Materials," Journal of Metals, February (1990), pp. 30-34.

5. I. V. Mitchell, ed., Nd-Fe Permanent Magnets: Their Present and Future Applications (New York, NY: Elsevier Applied Science Publishers, 1985).

6. Joseph G. Cannon, "Rare Earths," Eng. and Min. Journal, March (1983), pp. 136-137.

7. Jacqui Robbins, "Magnetic Materials Update," Materials Edge, September/October (1988), pp. 51-52.

8. H. R. Kirchmayr, "Physical Properties of $R_2Fe_{14}B$-Based Alloys," Handbook on the Physics and Chemistry of Rare Earths, ed. K. A. Gschneidner, Jr. and L. Eyring (Amsterdam: North Holland, 1978), pp. 71-132.

9. "EEC Attraction for Rare-Earth Magnet Technology," Materials Edge, May/June (1988), p. 7.

10. Isidor S. Hirschhorn,"Lanthanides," Engineering and Mining Journal, March (1985), pp. 127-128.

11. "Potent Portable," Popular Science, June (1988).

12. "New Rare-Earth Magnets Offer High Strength and Economy," Engineering and Mining Journal, March (1989), p. 77.

13. Michael Sponseller, "Magnetic Train," Popular Science, December (1988), pp. 97-98.

14. Matt Bacon, "Railways - Trains of Thought," Materials Edge, June (1989), pp. 48-54.

15. J. W. Morrison and G. R. Palmer, "Recovery of Metal Values from NdFeB Magnet Scrap," Second International Symposium -Recycling of Metals and Engineered Materials, ed. J. H. van Linden, D. L. Stewart, and Y. Sahai, pp. 593-609, TMS, Warrendale, PA (1990).

16. S. Vijayan et al., "Rare Earths: Their Mining, Processing, and Growing Industrial Usage," Mining Engineering, January (1989), pp. 13-18.

17. I. M. Kolthoff and P. J. Elving, eds., Treatise on Analytical Chemistry, Vol. 8 (New York, NY: Interscience Publishers, 1963), pp. 1-146.

18. F. A. Cotton and G. Wilkinson, Advanced Inorganic Chemistry, 3rd Ed. (New York, NY: Interscience Publishers, 1972), p. 890.

19. "The Lanthanide Series," Comprehensive Inorganic Chemistry, Vol. IV, ed. Sneed and Brasted (Princeton, NJ: D. VanNostrand Co., Inc. 1955), pp. 151-187.

20. L. Niinistö and M. Leskëla, "Inorganic Complex Compounds II," Handbook on the Physics and Chemistry of Rare Earths, ed. K. A. Gschneidner, Jr. and L. Eyring (Amsterdam: North Holland, 1978), pp. 91-320.

21. V. Arregi, A. R. Gordon, and G. Steintveit, "The Jarosite Process - Past, Present, and Future," Lead-Zinc-Tin '80, ed. J. M. Cigan, T. S. Mackey, and T. J. O'Keefe, (AIME, New York 1979), pp. 97-123.

22. J. E. Dutrizac, "The Physical Chemistry of Iron Precipitation in the Zinc Industry," Lead-Zinc-Tin '80, ed. J. M. Cigan, T. S. Mackey, and T. J. O'Keefe, (AIME, New York 1979), pp. 532-564.

23. M. G. Kershaw and R. W. Pickering, "The Jarosite Process - Phase Equilibria," Lead-Zinc-Tin '80, ed. J. M. Cigan, T. S. Mackey, and T. J. O'Keefe, (AIME, New York 1979), pp. 565-580.

24. J. E. Dutrizac, "Converting Jarosite Residues into Compact Hematite Products," Journal of Metals, January (1990), pp. 36-39.

25. Kirk-Othmer, The Encyclopedia of Chemical Technology, Vol. 4, 3rd Ed., (Wiley-Interscience, NY, 1980) p. 106.

3.5
HEAVY METALS, SOLUTIONS AND ACIDS

RECOVERY OF METALS FROM DILUTE EFFLUENT

STREAMS BY BIOSORPTION METHODS

A. E. Torma and W. A. Apel

The Center for Biological Processing Technology,
The Idaho National Engineering Laboratory, EG&G Idaho, Inc.,
P. O. Box 1625, Idaho Falls, Idaho 83415-2203

Biotechnological processes are considered to be cost effective especially for situations where the recovery of metals from very large volumes containing trace metal concentrations (less than 10 ppm) is concerned. Microorganisms, alive or dead, or their products can be highly efficient and selective biosorbents for metal concentration. It is important that the metal bound to the biosorbent can be stripped by suitable agent(s) and regenerated for repeated use. The stripped solution should be concentrated enough to be introduced into the common metal recovery circuit. The industrial acceptance of biosorption processes is dictated by increased competitive advantage over other processes. This paper reviews the pertinent literature, regroups the information, and points out future possibilities for the mineral and metal processing industries.

Residues and Effluents - Processing and Environmental Considerations
Edited by Ramana G. Reddy, William P. Imrie and Paul B. Queneau
The Minerals, Metals & Materials Society, 1991

Introduction

Recovery of metals from industrial effluents is traditionally done by precipitation, reverse osmosis, liquid or solid ion exchange, cementation, and electrodeposition. However, as the metal ion concentration decreases below 10 ppm, the efficiency of the above methods becomes marginal and economically unattractive. Therefore, alternative solutions are sought for the removal of trace concentrations of environmentally toxic metals and radionuclides from the effluent streams. Biological processes offer an alternate approach, which is the subject of the present article.

Metal binding or uptake can occur on the surfaces of viable or non-viable biomass or by intracellular accumulation (1). The surfaces of the microorganisms are not inert and movement of materials into or out of the cell is controlled by a number of active (i.e. enzyme mediated) or passive (i.e. porin mediated, protein mediated) systems, based essentially on the chemistry and structure of the surface (2). Diverse biopolymers (polysaccharides) extracted from seaweeds (3) can be used as immobilization agents for microorganisms or as bioextractant bead material (4). Biosorption methods are especially attractive for industry because they can also selectively remove trace metals from large volumes of effluent solutions (5). For a biosorption process to be economically successful it must (a) have high specific metal uptake capacity, (b) be readily available, (c) be easily producible from inexpensive resources, (d) be stable over a significant range of operating conditions (e.g. pH, temperature, etc.), and (e) be regenerable for continued reuse.

A conceptual approach to biosorption (using diverse biomaterials) for the removal of environmentally toxic heavy metals, and/or high value precious metals from industrial waste streams, is outlined in Figure 1 (6):

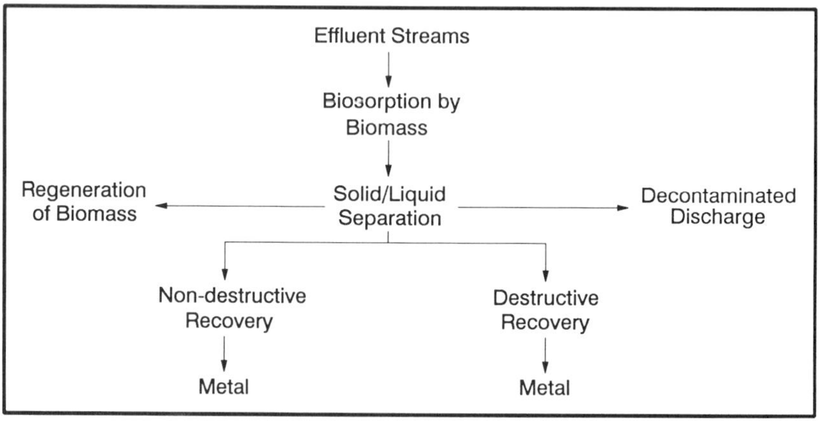

Figure 1 - Basic principles involved in the removal of metals from effluent solutions by biosorption techniques

The biomass in the above process can be living, dead, free, immobilized, unicellular or filamentous cells, films, aggregates or pellets, as well as excreted products or cell extracts. The metal extraction from the loaded biomass (complexes and precipitates) can be achieved by non-destructive (stripping) processes, or by destructive (burning) processes. If living cells are used in the metal removal system, toxicity may lead to poisoning and inactivation (6). Although, resistant microbial strains can be isolated, they may exhibit decreased accumulation or impermeability. The use of dead cells or derived products eliminates the problem of toxicity, not only from dissolved metals, but also from adverse operating conditions (high temperature, low pH) and improves the economics since supplying nutrients to the system is unnecessary.

Metal Uptake Processes

The biosorption processes that can be used in the removal of metals from different effluent streams can be grouped together in the following ways:

Activated Sludge System

Activated sludge systems are used conventionally for the removal of organic matter from wastewater (7), but it is also recognized that in this process heavy metals are efficiently eliminated by the mixed culture of microorganisms indigenous to the sludge. A typical activated sludge treatment process is shown in Figure 2. The biosorbent in this process may consist of an unidentified mixed culture of bacteria, algae, fungi,

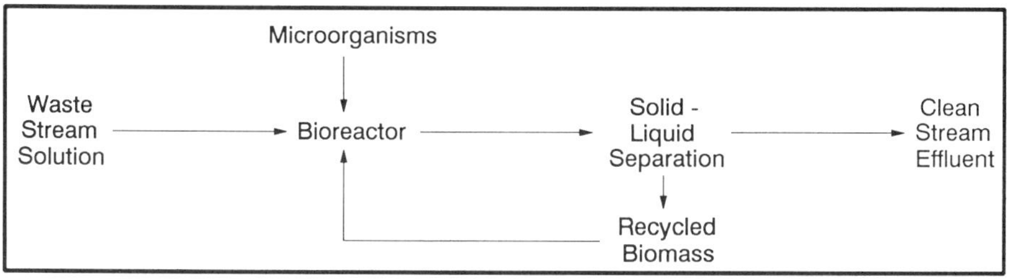

Figure 2 - Generalized flow sheet of an activated sludge biosorbent system

protozoa, rotifers and nematodes (8). In typical municipal sewage treatment operations the metallic constituents, such as cadmium, copper, chromium, manganese and lead are removed with about 92 to 93% efficiency from the waste liquor. However, the opinions on biosorption efficiency of the systems differ from author to author. For example, it was suggested (9) that 5 to 20% of the heavy metal contents are eliminated during the final filtration step in the activated sludge process, rather than from biosorption activities during the reactor treatment.

Several microorganisms have been isolated from activated sludge systems. *Zooglea ramigera* biomass was found to be efficient in sorbing copper, cobalt and iron (10). A mixed culture of bacteria grown in an activated sludge reactor, was reported to sorb large amounts of copper (11) or cadmium (12). Factors affecting the removal of heavy metals in the activated sludge process were reviewed (13). The age of microorganisms was found not to be a factor in the sorption of lead, chromium (3+) and copper (14). Biomass in the activated sludge was shown to absorb radium-226 preferentially (15). A slime producing microorganism, *Sphaerotilus natans*, was found to absorb iron, cadmium, copper and cobalt in the mucilaginous layer on the cell wall (16). Cells of *Pseudomonas aeruginosa* were shown to sorb 10 to 15% dry weight of uranium (17).

Recovery of Heavy Metals by Algae, Yeast and Fungi

Algae are photosynthetic microorganisms comprised of thousands of procaryotic and eucaryotic species. The current system of algal classification recognizes between four and nine divisions of the algal kingdom, which have been organized according to cell wall composition, differences in pigmentation, starch and lipid composition, or presence of flagella and other morphological characteristics (22).

Fungi consist of both filamentous (termed molds) and single cell forms (termed yeast). Their cells are eucaryotic and have a cell wall consisting of hexoses and hexosamines. In many cases, chitin is the main macromolecular unit of the fungal cell wall and is composed of N-acetyl glucosamine residues linked by β-1,4-glycosidic bonds (44). Depending on species and growth stage, fungi are capable of both sexual and asexual reproduction.

Many potential binding sites occur in cell walls and extracellular matrices of algae, yeast and filamentous fungi. Their structural components include polysaccharides, cellulose, uronic acid, and proteins (6). These microorganisms concentrate metals from aqueous solutions in an impressive way (6,17). Reference 6 summarizes the metal sorption data individually for algae, fungi and yeast as given in Table 1. The original references regarding the information in Table 1, are listed in reference 6. The sorption capacity of fungi and yeast may be higher as dead biomass than as living cells.

Compounds (mannans, glucans, melanins, chitin and chitosan) derived from fungal biomass may act as efficient biosorption agents (19). Chitin phosphate and chitosan phosphate were found to sorb greater amounts of uranium than copper, cadmium, manganese, zinc, cobalt, nickel, magnesium, and calcium (20). Many fungal products (citric and oxalic acids) can chelate heavy metals, and hydrogen sulfide produced by yeasts, can result in extensive precipitation of heavy metals in and around cell walls (21). Metal complexation by active or inactive algal cells may follow different mechanisms (23).

Biosorption of Radionuclides

Conventional uranium processing technology results in low-level radioactive wastes. For example, after uranium is extracted from the ore, approximately 95% of the long half life radionuclides Th-230, Ra-226 and Pb-210 remain in the tailings (24, 25). Currently, tailings management consists of

Table 1 - Bioaccumulation of Metals

Algae

Species	Metal	Species	Metal
Scenedesmus obliquus	Cd	*Chlamydomonas sp*	U
Chlorella vulgaris	Cu	*Chlorella sp*	U
Chlorella sp	Hg	*Dunaliella sp*	U
Chlorella regularis	Mn	*Chlamydomonas sp*	U
Chlamydomonas sp	Mo	*Chlorella regularis*	U
Scenedesmus sp	Mo	*Chlorella vulgaris*	Au
Chlorella regularis	Mo	*Euglena sp*	Fe, Al, Ba, Zn, Mn, Cu, Ni, Pb, U, Th
Ankistrodesmus sp	U, Ba		
Selenastrum sp	U, Ba		
Chlorella regularis	U		
Scenedesmus sp	U		

Filamentous Fungi

Species	Metal	Species	Metal
Unidentified Fungi	Cu, Ag	*Aureobasidium pollulans*	Cu
Phoma sp	Ag	*Penicillium sp*	U
Aspergillus niger	Cd	*Neurospora sitophila*	U
Mucor racemosus	Cd	*Penicillium chrysogenum*	U, Th
Penicillium chrysogenum	Cd	*Penicillium lilacinum*	U
Trichoderma reesei	Cd	*Aspergillus niger*	U, Th
Rhisopus arrhisus	Cu, Th	*Rhizopus oryzae*	U
Cladosporium resinae	Cu	*Trichoderma reesei*	U
Penicillium italicum	Cu	*Rhizopus arrhizus*	Cr, Mn, Cu, Zn, Cd, Hg, Pb, U, Ag
Penicillium sp	Pb		
Penicillium spinulosum	Cd, Cu, Zn		

Yeast

Species	Metal	Species	Metal
Unidentified Yeast	Cu, Ag	*Rhodotorula mucilaginosa*	Cd
Saccharomyces cerevisiae	Cd, Zn, Pb, Hg, U, Th	*Candida albicans*	U
Candida utilis	Cd, Cu	*Hansenula anomala*	U
		Rhodotorula glutinis	U

The original references are listed in Reference 6

controlled containment, in which a synthetic liner is placed under the tailings to prevent migration of radionuclides, heavy metals and hazardous chemicals into underlying soils and groundwater. However, early operations in the 1940's through the 1960's were not regulated, and as a result, soils and aquifers were often contaminated (26,27). In addition to conventional chemical decontamination methods (28), the biotechnological approach (24) is becoming an important complimentary activity.

A process has been reported for the removal of uranium, molybdenum and selenium from industrial effluents by algae (29). The process is claimed to be especially applicable to mining operations. The mine drainage water can first be passed through a settling pond, then a solid ion exchange system, where the major portion of heavy metals and radionuclides are eliminated. The remaining trace concentrations of the above metals can be removed in an algal pond, predominantly by *Spirogyra*, *Chara*, and *Oscillatoria* species.

The outline of the process is presented in Figure 3. The reduction of metal

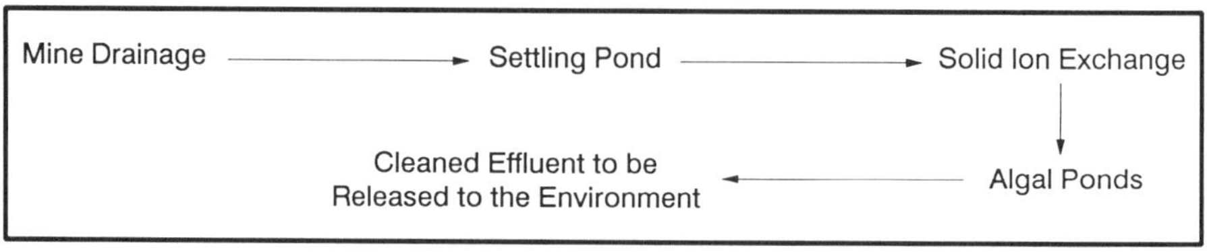

Figure 3. An integrated chemical-biological effluent stream remediation process

contents were given as follows: 5.8 ppm U, 2.3 ppm Mo and 0.3 ppm Se were decreased to only 0.8 ppm U, 0.8 ppm Mo and 0.01 ppm Se.

The removal of uranium from industrial streams has been the subject of many investigations using a number of algal, fungi and yeast species in pure and mixed cultures (6,18). A conceptual flow sheet has been proposed for the recovery of uranium from industrial solutions as shown in Figure 4 (29-31). The best sorption results for uranium uptake have been obtained with biomass from *Penicillium* species and *Rhizopus arrhisus*. The pilot plant test was operated successfully for 100 days.

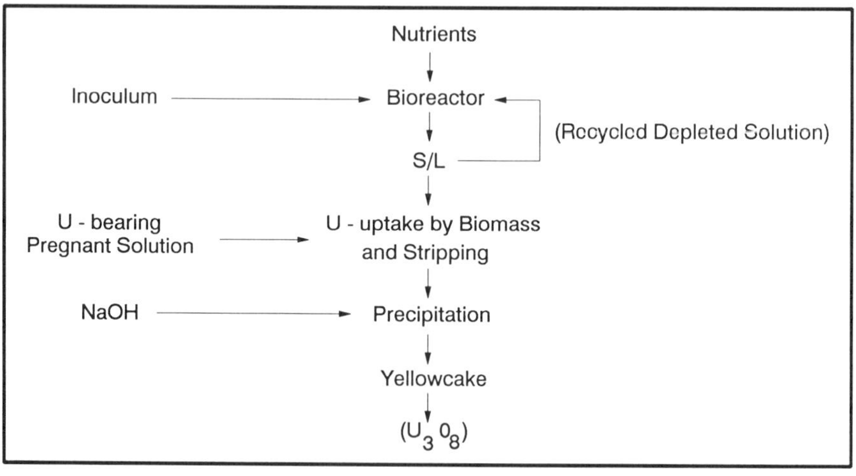

Figure 4. Schematic outline of a continuous biosorption process for the recovery of uranium from mine drainage waters

Removal of heavy metals and radionuclides from contaminated uranium mill tailing sites by chemical and biological techniques has been proposed (4, 32). The process consists of bioassisted *in situ* leaching of contaminated soils. The pregnant solution is pumped to the ground where it is treated with lime and barium chloride to precipitate heavy metals and radionuclides. The remaining trace metallic contamination will be removed from the solution by alginate based biosorbent beads (33).

Biological Treatment of Metal Contaminated Effluents

Acid mine drainage is a common water pollution problem at active and abandoned coal and heavy metal sulfide mines (34). Acid is formed when pyrite and other metal sulfides on the surface or in the ground are allowed to contact oxygen and water. In the absence of neutralizing agents, the acid drainage will dissolve varying amounts of heavy metals, such as iron, copper, zinc, manganese, etc. The dissolution process may involve a variety of indigenous iron and sulfur oxidizing bacteria (*Thiobacillus ferrooxidans, Leptospirillium ferrooxidans, Thiobacillus thiooxidans, Sulfobacillus thermosulfidooxidans, Sulfolobus brierley*) which catalyze the oxidative dissolution process (35). The removal of heavy metal contaminants from aqueous waste streams is one of the most important environmental issues facing many mining areas today. A biosorbent bead material called BIO-FIX beads was developed for extracting toxic metals from mining effluents (36). These biosorbent beads were fabricated from high-density polysulfone dissolved in dimethylformamide and mixed with dried, thermally killed biomass produced by algae, yeast, bacteria, and aquatic flora. The spherical beads were formed by injecting the mixture into water. The BIO-FIX beads were found to be efficient in the removal of arsenic, cadmium, copper, mercury, lead, manganese and zinc. The beads retained their sorption capacity after repeated loading and elution cycles.

Mine drainage is typically acidic (pH = 2-4), with high sulfate, iron, copper and aluminum concentrations. An effective treatment must generate water of neutral pH containing no heavy metals in order to meet regulatory standards for mine effluents. Sulfate reducing bacteria can be used to achieve this effect (37,38). A successful pilot plant demonstration (38) suggested that biological processes based on using sulfate reducing bacteria can be viable alternatives to the conventional lime treatment of acid mine effluents.

Recovery of Precious Metals by Biosorption Processes

Solubilization of gold by the action of heterotrophic bacteria or bacterial cell components (proteins) from lateritic ores and sands was studied intensively during the 1960's by French scientists (39,40). Effects of heterotrophic bacteria and fungi on the solubilization of native gold has been reported (41). These microorganisms release metabolic amino acids, which were found to be effective in solubilizing gold. The dissolution process requires the addition of an oxidizing agent such as sodium peroxide or potassium permanganate.

Microbial biomass (AMT-BIOCLAIM™) was found to be effective in the removal of precious metals (gold, silver, platinum, and palladium) from process wastewater (42). Gold recoveries as high as 99% have been achieved from cyanide solutions. Silver was accumulated from photographic laboratory wastewater, platinum and palladium from synthetic solutions. The biosorption of gold and platinum group metals was found to involve reductive reactions, whereby the metal ions were reduced to elemental forms. The AMT-BIOCLAIM™

was found to be compatible with other wastewater treatment systems (43). The loaded AMT-BIOCLAIM™ can be stripped of its metal contents and reused. As such this system is not burdened with the liability of toxic sludge disposal.

Biological Methods for Chromium Removal

Biosorptive techniques have been used for the removal of Cr(6+), a widespread environmental contaminant of ever increasing importance. Studies with *Zoogloea ramigera* have shown that reasonable amounts of Cr(6+) can be removed from solution under optimum conditions which were defined as pH = 2.0, temperature = 25°C and Cr(6+) concentrations of 75 ppm (45). From the data presented, however, it was difficult to separate the effects of the low optimum pH, which may have lead to the chemical reduction of Cr(6+) to forms of insoluble Cr(3+), from biosorption by the bacterium.

Direct metabolic reduction of Cr(6+) to Cr(3+) which possesses very low water solubility, and as such, lends itself to physical separation techniques, holds considerable potential as a biological process for the treatment of Cr(6+) contaminated water. In the 1970's, Russian scientists reported that bacteria are capable of direct Cr(6+) reduction (46). The earliest of these reports dealt with a *Pseudomonas* sp. which appeared to use Cr(6+) as a catabolic terminal electron acceptor when grown under anaerobic conditions. Later reports showed that a wide variety of bacteria including not only *Pseudomonas* but *Achromobacter, Aeromonas, Arthrobacter, Bacillus, Escherichia, Enterobacter, Flavobacterium, Micrococcus* and *Sarcina* species are capable of Cr(6+) reduction (47,48,49,50,51). This reduction occurs under both anaerobic and aerobic conditions depending on the bacterial species in question and the growth conditions (48). A number of compounds have been shown to act as electron donors for bacterial chromate reduction depending on the species of bacteria used. These include acetate, butyrate, glucose, fructose, glycerol, succinate and ethylene glycol (47, 52). Most of this reduction occurs during the exponential growth phases of the chromate reducing bacteria, and as such, commercialization of this process would be expected to occur using a continuous flow system so that the organisms can be held in a continuous exponential growth phase (53). In addition, strains of chromate reducing bacteria have been shown to reduce both chromate and dichromate nonpreferentially (54).

Conclusions

It was shown that microorganisms have the ability to accumulate metals and radionuclides from aqueous industrial streams. Dead and living cells as well as cell products were found to be effective in metal sorption processes. The literature indicates that the various biosorbent systems reported are versatile enough to be able to deal with complex and mixed metal containing effluents. The biosorbents can also be effective for the removal of low level heavy metal and radionuclide concentrations from high-volume waste solutions, where generally the conventional methods are time intensive and too expensive. The biosorption technologies are relatively new developments, but they have the potential to solve many of the nation's outstanding problems in remediation of contaminated sites. It is expected that in the near future, the mineral and metal processing industries will take advantage of these possibilities and employ biosorption technologies in an integrated fashion to treat their waste-streams.

Acknowledgements

This work was supported under contract No. DE-AC07-76ID001570 from the U.S. Department of Energy to the Idaho National Engineering Laboratory/EG&G Idaho, Inc.

References

1. D. W. Darnall et al., "Recovery of Heavy Metals by Immobilized Algae", Trace Metal Removal From Aqueous Solutions, ed. R. Thompson (London, The Royal Society of Chemistry, 1986), 1-24.

2. R. J. C. McLean and T. J. Beveridge, "Metal-Binding Capacity of Bacterial Surfaces and Their Ability to Form Mineralized Aggregates," Microbial Mineral Recovery, eds. H. L. Ehrlich and C. L. Brierley (New York, McGraw-Hill Publishing Company, 1990), 185-222.

3. K. B. Guiseley, "Chemical and Physical Properties of Algal Polysaccharides Used for Cell Immobilization," Enzyme Microbiol. Technol., 11 (1989), 706-715.

4. J. M. Barnes, et al., "Biosorption of Heavy Metals From Uranium Mill Tailing Effluents," Proceedings Reclamation 2000: Technologies For Success, eds. W. Oaks and J. Bowden (Princeton, VW, American Society of Surface Mining and Reclamation, 1991), 141-150.

5. B. Volesky, "Biosorbent Materials," Biotechnology and Bioengineering Symposium, 16 (1986) 121-126.

6. G. M. Gadd, "Accumulation of Metals by Microorganisms," Biotechnology, ed. H. J. Rehm (Weinheim, Germany, VCH Verlagsgesellschaft, 1988) 401-433.

7. R. S. Ramalho, Introduction to Wastewater Treatment Processes (New York, NY: Academic Press, 1977) 121-342.

8. G. Strachan and R. E. Wong, "Bioadsorption of Heavy Metals From Mineral Processing Streams-Phase I,"(Final Report, DSS File No.:26SQ.23440-3-9139, Allelix Inc. for CANMET, 1984).

9. P. O. Nelson, A. K. Chung, and M. C. Hudson, "Factors Affecting the Fate of Heavy Metals in the Activated Sludge Process," J. Water Pollution Control Fed., 53 (1981) 1323-1333.

10. P. R. Dugan and H. M. Pickrum, "Removal of Mineral Ions from Water by Microbially Produced Polymers," Proceedings of the 27th Annual Purdue Industrial Waste Conference (West Lafayette, Indiana, Purdue University Press, 1973), 1019-1038.

11. G. M. Dunn and A. T. Bull, "Bioaccumulation of Copper by a Defined Community of Activated Sludge Bacteria," European J. Applied Microbiol. Biotechnol., 17 (1983) 30-34.

12. J. Remacle, "Cadmium Uptake by Freshwater Bacterial Communities," Water Res., 15 (1981) 67-71.

13. S. Stoveland and J. N. Lester, "A Study of the Factors which Influence Metal Removal in the Activated Sludge Process," Sci. Total Environment., 16 (1980) 37-54.

14. R. M. Sterritt, M. J. Brown, and J. N. Lester, "Metal Removal by Adsorption and Precipitation in the Activated Sludge Process," Environment. Pollut., 24 (1981) 313-323.

15. M. Tsezos and D. M. Keller, "Adsorption of Radium-226 by Biological Origin Absorbents," Biotechnol. Bioeng., 25 (1983) 201-215.

16. R. T. Hatch and A. Menawat, "Biological removal and Recovery of Trace Heavy Metals," Biotechnol. Bioeng. Symp., 8 (1978) 191-203.

17. G. W. Strandberg, S. E. Shumate, II., and J. R. Parrott, Jr., "Microbial Cells as Biosorbent for Heavy Metals: Accumulation of Uranium by S. cerevisiae and P. aeruginosa," Applied Environ. Microbiol., 41 (1981) 237-245.

18. J. R. Benemann and E. D. Wilde, "Literature Review on the Use of Bioaccumulation for Heavy Metal Removal and Recovery,"(Technical Report WSRC-TR-90-175, Vol. II, Westinghouse Savannah River Company,Aiken, SC, 1991).

19. G. M. Gadd, "Fungi and Yeasts for Metal Accumulation," Microbial Mineral Recovery, eds. H. L. Ehrlich and C. L. Brierley (New York, NY, McGraw-Hill Publishing Company, 1990) 249-275.

20. T. Sakaguchi and A. Nakajima, "Recovery of Uranium by Chitin Phosphate and Chitosan Phosphate," Chitin and Chitosan, eds. S. Mirano and S. Tokura (Tottori, Japan, The Japanese Society of Chitin and Chitosan, 1982) 177-182.

21. J. Ashida, N. Higashi, and T. Kikuchi, "An Electron Microscopic Study on Copper Precipitation by Copper-Resistant Yeast Cells," Protoplasma, 57 (1963) 27-32.

22. B. Greene and D. W. Darnall, "Microbial Oxygenic Photoautotrophs (Cyanobacteria and Algae) for Metal-Ion Binding," Microbial Mineral Recovery, eds. H. L. Ehrlich and C. L. Brierley (New York, NY, McGraw-Hill Publishing Company, 1990)277-302.

23. S. Hunt, "Diversity of Biopolymer Structure and Its Potential for Ion-Binding Applications," Immobilization of Ions by Bio-Sorption, eds. H. Eccles and S. Hunt (Chichester, UK, Ellis Horwood, 1986) 15-46.

24. L. E. Macaskie, "The Application of Biotechnology to the Treatment of Wastes Produced from the Nuclear Fuel Cycle: Biodegradation and Bioaccumulation as a Means of Treating Radionuclide-Containing Streams," Critical Reviews in Biotechnology, 11 (1991) 41-112.

25. A. E. Torma, Extraction of Radionuclides from Low-Grade Ores and Mill Tailings (Santa Fe, NM: New Mexico Research and Development Program, 1981) 1-31.

26. Anonymous, "Annual Status Report on the Uranium Mill Tailings Remedial Action Program," (Report DOE/EM-0001, Dist.Category UC-70, US Department of Energy, 1989)

27. Anonymous, "EPA Workshop on Radioactively Contaminated Sites," (EPA 520/1-90-009, U. S. Environmental Protection Agency, Office of Radiation Programs, 1990) 1-171.

28. I. J. Itzkovich and G. M. Ritcey, "Removal of Radionuclides from Process Streams-A Review" (CANMET Report 79-21, Energy Mines and Resources, Ottawa, Canada, 1979) 1-171.

29. M. Tsezos, "The Performance of a New Biological Adsorbent for Metal Recovery. Modeling and Experimental Results," Biohydrometallurgy, eds. P. R. Norris and D. P. Kelly (Kew Surrey, U. K., Science and Technology Letters, 1987) 465-475.

30. M. Tsezos, S. H. Noh, and M. H. I. Baird, "A Batch Reactor Mass-Transfer Kinetic Model for Immobilized Biomass Biosorption," Biotechnol. Bioeng., 32 (1988) 545-553.

31. M. Tsezos et al., "The Use of Immobilized Biomass for the Continuous Recovery of Uranium from Elliot Lake Ore. First Pilot Plant Results," Biohydrometallurgy, eds. J. Salley, R. G. L. McCready and P. L. Wichlacz (Ottawa, Ontario, Canada, CANMET SP89-10, 1989) 279-287.

32. M. L. Apel, J. M. Barnes and A. E. Torma, "Biosorption Kinetics of Metal Removal from Uranium Mill Tailing Effluents," (Submitted for Publication, 1991).

33. A. E. Torma et al., "Sorption of Radium-226 and Heavy Metals by Biopolymer Gels," (Submitted for Publication, 1991).

34. H. Edenborn, "Biological Treatment of Metal-Contaminated Water," Biological Remediation of Contaminated Sediments, with Special Emphasis on the Great Lakes, eds. C. T. Jafvert and J. E. Rogers (Athens, Georgia, U. S. Environmental Protection Agency, 1991) 145-147.

35. A. E. Torma and P. A. Pryfogle, "Mechanisms of Bacterial Metals Removal from Solids," Biological Remediation of Contaminated Sediments, with Special Emphasis on the Great Lakes, eds. C. T. Jafvert and J. E. Rogers (Athens, Georgia, U. S. Environmental Protection Agency, 1991) 159-165.

36. T. H. Jeffers, C. R. Ferguson and D. C. Seidel, "Biosorption of Metal Contaminants Using Immobilized Biomass," Biohydrometallurgy, eds. J.Salley, R. G. L. McCready and P. L. Wichlacz (Ottawa, Ontario, Canada, CANMET SP89-10, 1989) 317-327.

37. A. L. Mills, P. E. Bell and A. T. Herlihy, "Microbes, Sediments and Acidified water: The Importance of Biological Baffering," Acid Stress and Aquatic Microbial Interactions, ed. S. S. Roa (New York, NY, CRC Press, 1989) 1-19.

38. G. Bechard, S. Rajan and R. G. L. McCready, "Microbiological Process for the Treatment of Acidic Drainage at the Halifax International Airport," (Report MSL 91-54 (OP&J), CANMET, Ottawa, Ontario, Canada, 1991).

39. Y. Pares, "Action de quelques becteries heterotrophes banales dans le cycle del'or," Annales de l'Institut Pasteur, 107 (1964) 573-575.

40. Y. Pares et al., "Influence de la concentration du milieux de culture sur la mise en solution de l'or par voi bacterienne, " Anales de l'Institut Pasteur, 108 (1965) 815-819.

41. S. N. Groudev and V. I. Groudeva, "Microbiological Solubilization of Gold," Precious and Rare Metal Technologies, eds. A. E. Torma and I. H. Gundiler (Amsterdam, Netherlands, Elsevier, 1989) 255-262.

42. J. A. Brierley and D. B. Vance, "Recovery of Precious metals by Microbial Biomass," Biohydrometallurgy, eds. P. R. Norris and D. P. Kelly (Kew Surrey, UK, Science and Technology Letters, 1988) 477-485.

43. J. A. Brierley, C. L. Brierley, and G. M. Goyak, "AMT-BIOCLAIM™: A New Wastewater Treatment and Metal Recovery Technology," Fundamental and Applied Biohydrometallurgy, eds. R. W. Lawrence, R. M. R. Branion, and H. G. Ebner (Amsterdam, Netherlands, Elsevier, 1986) 291-304.

44. B. D. Davis et al., Microbiology (Philadelphia, PA: Harper and Row, Publishers, 1980), 818-827.

45. Y. Sag and T. Kutsal, "The Use of *Zoogloea ramigera* in Wastewater Treatment of Cr(VI) and Cd(II) Ions", Biotechnology Letters, 11 (1989), 145-148.

46. V. I. Romanenko and V. N. Koren'kov, "A Pure Culture of Bacteria Utilizing Chromates and Bichromates as Hydrogen Acceptors in Growth Under Anerobic Conditions", Mikrobiologiya, 46 (1977), 414-417.

47. E. V. Lebedeva and N. N. Lyalikova, "Reduction of Crocoite by *Pseudomonas chromatophila* nov. sp.", Mikrobiologiya, 48 (1979), 517-522.

48. L. H. Bopp, and H. L. Ehrlich, "Chromate Resistance and Reduction in *Pseudomonas fluorescens* strain LB300", Archives of Microbiology, 150 (1988), 426-431.

49. E. I. Kvasnikov et al., "New Gram - Variable Bacterium Which Reduces Chromium and Has a Mixed Type of Flagellation", Mikrobiologiya, 54 (1985), 83-88.

50. P. Wang et al., "Isolation and Characterization of an *Enterobacter cloacae* Strain that Reduces Hexavalent Chromium under Anaerobic Conditions", Applied and Environmental Microbiology, 55 (1989), 1665-1669.

51. E. I. Kvasnikov et al., " Bacteria Reducing Chromium in Nature and In Industrial Sewage", Mikrobiologiya 57 (1988), 680-685.

52. H. Ohtake, E. Fujii, and K. Toda, "A Survey of Effective Electron Donors for Reduction of Toxic Hexavalent Chromium by *Enterobacter cloacae* (Strain HO1)", Journal of general and Applied Microbiology, 36 (1990), 203-208.

53. W. A. Apel and C. E. Turick, "Bioremediation of Hexavalent Chromium by Bacterial Reduction" (Submitted for Publication, 1991).

54. W. A. Apel and C. E. Turick, "Reduction of Hexavalent Chromium in Aqueous Solutions Using Chromate Reducing Bacteria", Bioprocess Engineering Symposium - 1991, (New York, NY: The American Society of Mechanical Engineers, 1991), (In press).

BACTERIAL FLOCCULATION OF PHOSPHATE WASTES USING A HYDROPHOBIC BACTERIUM

R. W. Smith and M. Misra
Department of Chemical and Metallurgical Engineering,
Mackay School of Mines, University of Nevada, Reno, Nevada, USA

J. Dubel
Department of Chemical Engineering,
Technical University of Karlsruhe, Karlsruhe, GERMANY

Abstract

A microorganism will function as a mineral flocculant if it is to some degree hydrophobic and can attach itself in some manner, such as through electrostatic interaction, to the fine mineral particles to be flocculated. The present paper concerns the use of a hydrophobic bacterium, *Mycobacterium phlei*, as flocculating agent for phosphate slimes.

Experimental work on the flocculation of phosphate slimes and of dolomite and clay mineral suspensions was performed. It was found that *M. phlei* functions as an excellent flocculant for a dolomitic slime but does not function as well for mineral suspensions where large quantities of the solids are finer in size than the approximately 1 μm size of the bacterium.

Residues and Effluents - Processing and Environmental Considerations
Edited by Ramana G. Reddy, William P. Imrie and Paul B. Queneau
The Minerals, Metals & Materials Society, 1991

Introduction

It has been shown (1-5) that *Mycobacterium phlei* is both highly hydrophobic (contact angle 65-70°) and highly negatively charged with an isoelectric point (iep) at about pH 2.0 - 2.5. It should, thus, readily adsorb onto hydrophilic mineral surfaces if the mineral is of low negative, neutral or positive charge. The adhesion could also be promoted by a hydrophobic mineral surface (1,2). Where both the organism and the mineral are negatively charged the adhesion can be explained based on DLVO theory (1-5).

Experimental work on the flocculation of hematite using *M. phlei* as a flocculant has been reported on in a previous publication (5) and it was found that the organism functions as an excellent flocculant for the mineral.

Since the bacterium has an rod/ellipsoid/coccal shape [the organism is pleomorphic and can assume different forms depending on growth conditions (6,7)] rather than a long chain structure such as possessed by synthetic polymeric flocculants, certain of its flocculation characteristics are markedly different from those of the polymeric flocculants. While both the bacterium and long chain polymers in part flocculate mineral particles by bridging between mineral particles, the entwining characteristics of the long chain molecules are absent in the case of *M. phlei*. Thus, hydrophobic interactions between adhering microorganisms on adjacent solid particles are of greater importance in the formation of mineral aggregations. One consequence of this mechanism of aggregation is that the aggregations in the case of hematite are very "tight" and dense and contain little water. It is also likely that a reduction of positive electrostatic charge on the mineral by interaction with the negatively charged microorganism cells contributes to the aggregation of the hematite particles.

M. phlei should also accumulate on the surfaces of minerals found in a phosphate processing waste at least under certain pH conditions. The present paper reports on experimental work performed on the flocculation of fine phosphate processing wastes and on the flocculation of suspensions of some of the individual minerals found in these wastes. Electrophoretic mobility measurements were also performed in an attempt to understand mechanisms involved in the flocculation of the suspensions.

Experimental

The *Mycobacterium phlei* used in the experimentation was grown from cultures obtained from Carolina Biological Supply Company. The bacterium is ubiquitous and found in soils and on the leaves of plants, particularly grasses and is sometimes known as hay or timothy bacillus (8,9). It is non - pathogenic in man and in all animals thus far tested (8,10). Several different media were used initially to culture the bacterium. The medium selected for experimentation was one suggested by Pratt (11) and Guirard and Snell (12) because of its simplicity and the rapid growth of the organism. The composition of the medium was (in gm/dm^3 water): glucose 10, beef extract 1, yeast extract 1 and enzyme-hydrolyzed casein 1. Culturing was done in 250 ml flasks continuously shaken in a water shaker bath held at 35 °C.

HCl and NaOH used for pH adjustment in various studies was of reagent grade.

Three different Florida phosphate wastes were studied. They were "Four Corners", a dolomitic material, (phosphate waste #1), "Agrico" (phosphate waste #2) and IMC (phosphate waste #3). The pure minerals studied were purchased from Ward's National Science Establishment, Inc.

The electrokinetic measurements were made using a Komline-Sanderson zeta reader in a conventional manner. A 2 ml sample of the bacteria or hematite solution was diluted with distilled water just prior to measurement and the pH adjusted with either HCl or NaOH.

Sedimentation measurements on phosphate wastes in the presence and absence of *M. phlei* were made by first mixing 1000 ml of the suspensions to be studied in controlled stirring vessels (HS-4 stirrer, Phipps and Bird). To provide good mixing, bacteria were added to the phosphate slime at an impeller speed of 250 rpm. For comparison purposes the slimes not containing bacteria were similarly handled. After 2 min stirring the stirrer speed was reduced to 30 rpm for 5 min for floc growth. The suspension was transferred to a modified Andreasen pipette. Samples were then taken at the start of sedimentation and at various subsequent times at a 4 cm height from the bottom of the pipette. The solid concentration was determined gravimetrically. From the data obtained it was then possible to calculate the relative stabilities of the suspensions as a function of time by dividing the weight % solids in the sample taken at time (t) by the weight % solids in the sample at the start (W_t/W_o). For convenience the values were converted to percentages by multiplying them by 100 (W_t/W_o X 100 values).

Additionally, for two of the phosphate slimes and for a clay slurry handled in a similar manner the weight % solids were determined for the "consolidated" suspensions at the bottom of the pipette as functions of time standing.

Experimental Results

Figure 1 illustrates the zeta potential of *M. phlei*, quartz, kaolinite, apatite and dolomite as a function of pH. It is to be noted that the zeta potential of *M. phlei* is more negative at all pH values than that of either apatite or dolomite and more positive than that of quartz. The organism and kaolinite are similarly charged except at pH values greater than about pH 7 where the organism is slightly less negatively charged.

Figure 2 shows the % stability (W_t/W_o X 100) of phosphate waste # 1 as a function of time. It is to be noted that in the absence of the bacteria the suspension settles little and remains fairly stable (~40 %) after 40 min. In the presence of 10.2 g *M. phlei*/kg solids the stability is less than 3 % after about 15 min. Thus, the settling is much more rapid in the presence of the organism. The photomicrograph of figure 3 confirms the flocculation of the material by the microorganisms.

Figure 4 shows similar data for the other two phosphate slimes. It is to be noted that the presence of *M. phlei* has little effect on the settling (stability) of the suspensions. For phosphate waste #2 the stability in the presence or absence of the bacteria was still 30+ % after 200 hr and for phosphate waste #3 it was still 10+ % after 200 hr.

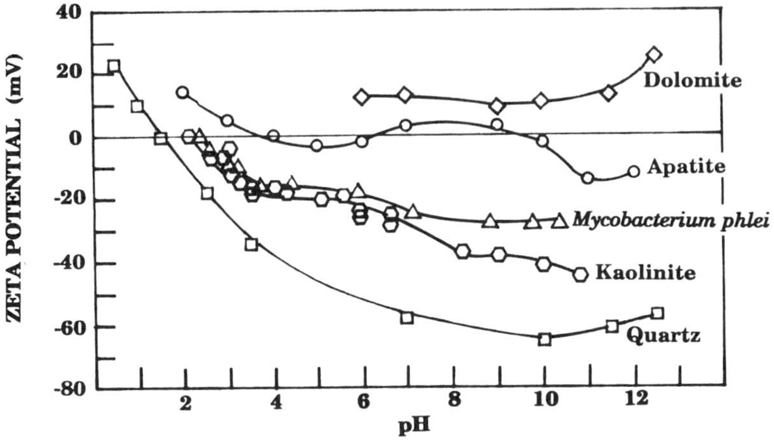

Figure 1 Zeta potentials of *Mycobacterium phlei*, dolomite, apatite, kaolinite and quartz as functions of pH.

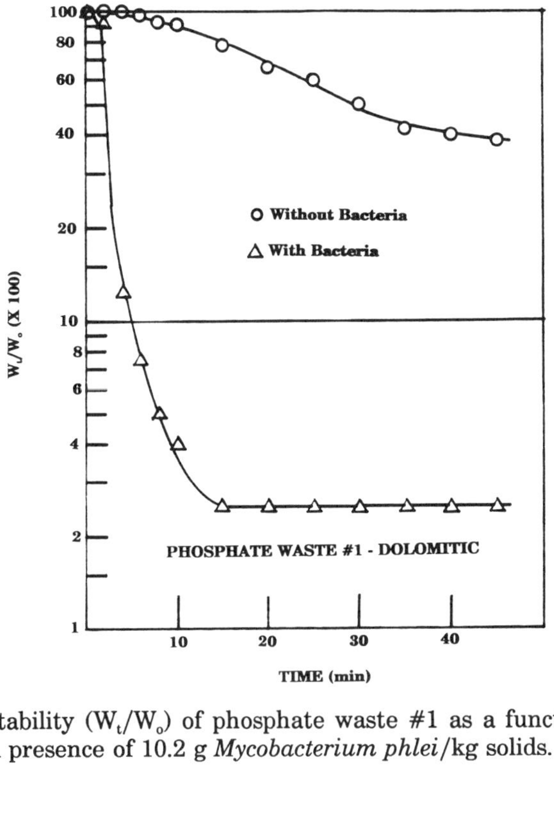

Figure 2 Stability (W_t/W_o) of phosphate waste #1 as a function of time in the absence and presence of 10.2 g *Mycobacterium phlei*/kg solids.

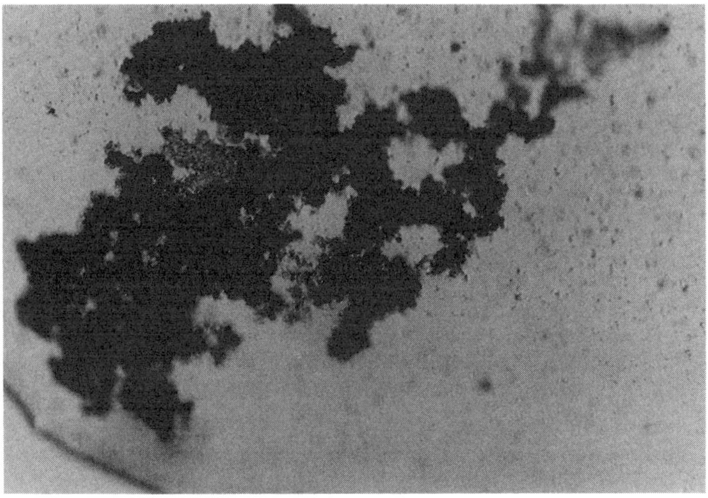

Figure 3 Photo micrograph of phosphate waste #1 in the presence of *Mycobacterium phlei*.

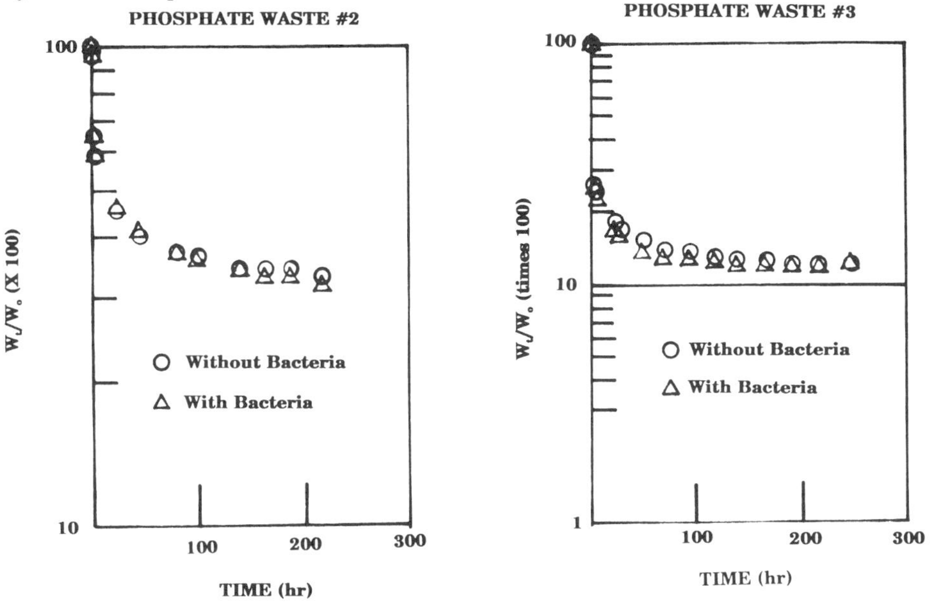

Figure 4 Stabilities (W_t/W_o) of phosphate wastes #2 and #3 as functions of time in the absence and presence of 9.0 and 12.8 g *Mycobacterium phlei*/kg solids.

Figure 5 shows the solids concentration in the "consolidated" suspensions for phosphate wastes # 2 and 3 as a function of time standing both in the presence and absence of *M. phlei*. It is of interest that for waste # 2 the solids concentration is only about 4 % after 225 hr standing and about 22 % for waste # 3.

Similar data are to be found in figure 6 for a montmorillonite clay system.

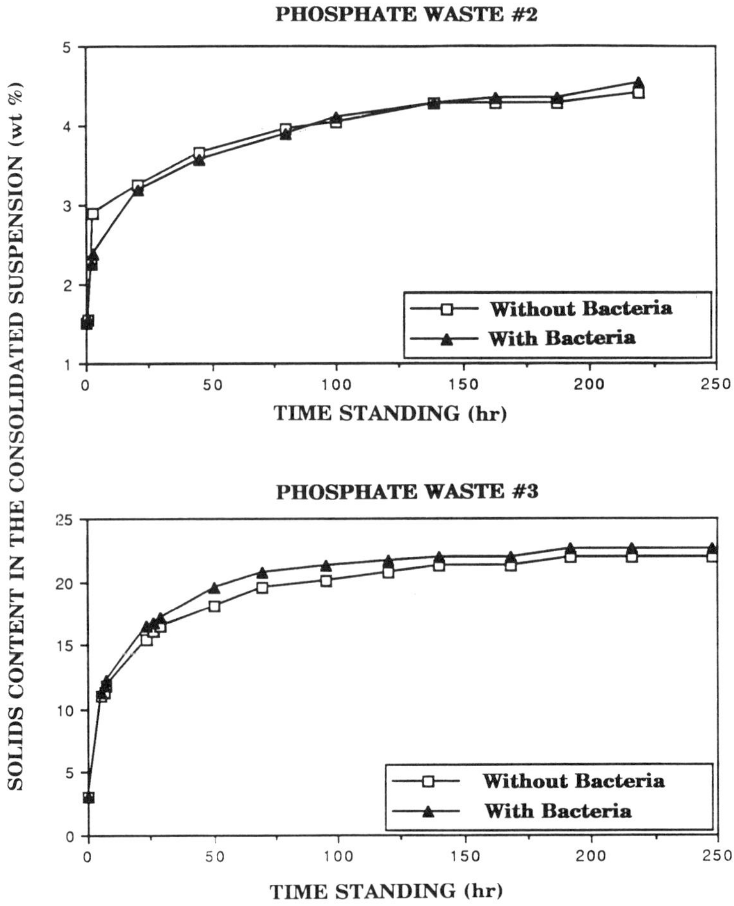

Figure 5 Solids content in the consolidated suspensions of phosphate wastes #2 and #3 as functions of time standing (9.0 and 12.8 g *M. phlei*/kg solids, respecively).

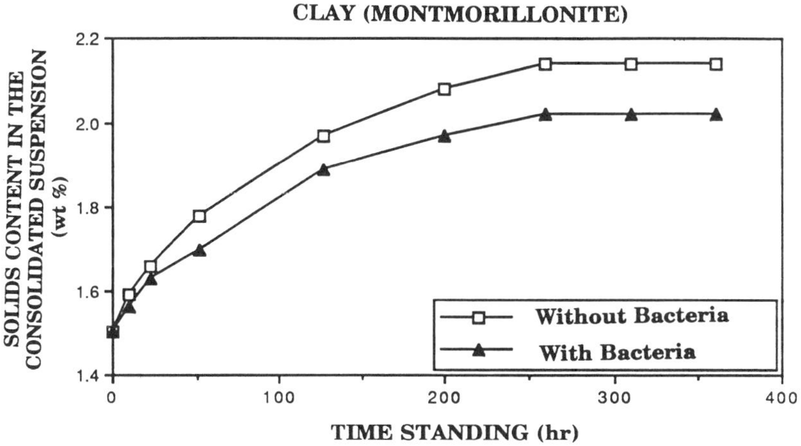

Figure 6 Solids content in the consolidated suspension of a clay (montmorillonite) suspension as a function of time standing (10.4 mg *M. phlei*/kg solids).

In this case the weight % solids in the "consolidated" suspension is, after 350 hr standing, about 2.1 % in the absence of the bacteria and about 2.0 % in the presence of the bacteria.

Discussion

The data of figure 1 indicate that there should be a strong coulombic interaction between *M. phlei* and apatite and dolomite leading to adhesion of the organism to these minerals. Also, there should be only small barriers to adhesion of the organism to kaolinite and other clay minerals [certainly there should be an interaction between the long known positive as well as negative sites on clay minerals (13,14) and the negatively charged bacterium]. In the absence of foreign ions changing the surface characteristics of quartz there should be little coulombic interaction between *M. phlei* and quartz except at quite low pH values. Thus, it might be anticipated that the microorganism would be a flocculating agent for slurries containing apatite, dolomite and probably clay minerals.

The data of figure 2 and the micrograph of figure 3 clearly show that the dolomitic phosphate waste #1 is flocculated by *M. phlei* resulting in greatly enhanced settling of the suspension. Observation of the photomicrograph of figure 3 indicates that there appears to be bridging of the mineral particles by the bacterial cells. This material contains relatively coarse dolomite particles and figure 7 shows a scanning electron micrograph *M. phlei* cells adhering to a dolomite particle. The aggregation of the particles after accumulation of the bacterium must be through the hydrophobic interaction and bridging between the now hydrophobic bacterial surfaces of the mineral particles plus reduction in the magnitude of charge on the minerals.

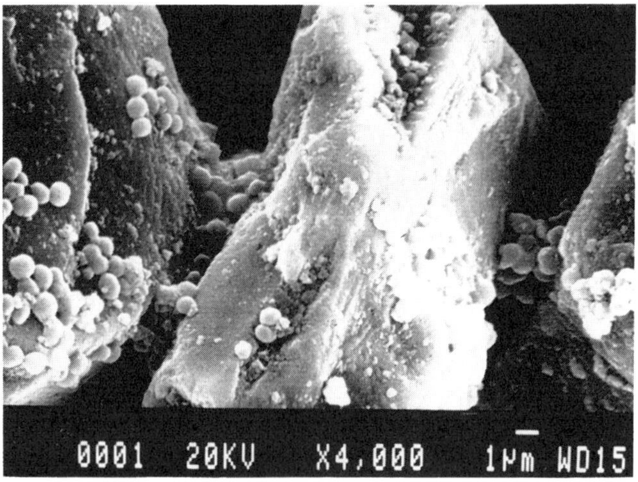

Figure 7 Scanning electron micrograph of *Mycobacterium phlei* attached to a dolomite particle.

The curves of figure 4 show, however, that *M. phlei* does not function as a good flocculating agent for non - carbonate gangue mineral - high clay mineral phosphate slimes. Further, figure 5 indicates that, particularly for phosphate waste # 2, the % solids left in "consolidated" slimes of this type, after settling, is very low regardless of whether the bacteria are present or not. When montmorillonite is studied (figure 6) in an identical manner to the studies represented by figure 5, it is found that the % solids in the "consolidated" material is even less. In addition, the presence of the bacteria actually decreases the amount of solids in the "consolidated" material. An explanation for this phenomenon and the failure of the microorganism to function as a good flocculant for non - carbonate gangue, high clay slimes can be found in the work of Daniels (15).

This author reports that the manner of adhesion of a microorganism to a solid particle depends on the relative sizes of the organism and the particle. If the particles are larger than the organisms the organism cells adhere to various sites on a particle surface, sometimes completely covering the particle. If, however, the microorganism cells are larger than the particles the particles adhere to sites on a cell surface. Thus, we believe that if a slurry is composed of particles larger than the hydrophobic microorganisms the microorganism creates a hydrophobic surface on the mineral particles allowing for bridging of particles. If the reverse is true for relative mineral particle and microorganism sizes the hydrophobic bacteria surfaces become hydrophilic because of their now hydrophilic mineral surfaces. In fact this arrangement may even give rise to a more dispersed suspension. Figure 8 illustrates this effect of relative mineral particle and bacterial cell size on the manner of adhesion mineral particle - bacterial cell and the effect of these sizes on flocculation and dispersion of mineral suspensions.

Conclusions

From the experimental work reported on here the following conclusions can be made:

1. The bacterium *Mycobacterium phlei*, a hydrophobic, highly negatively charged microorganism, functions as an excellent flocculating agent for the dolomitic "Four Corners" phosphate slime (phosphate waste #1).
2. *M.phlei*, however, does not function as a flocculating agent for non - dolomitic phosphate slimes that contain large quantities of very fine clay minerals.
3. If microorganisms and finely divided minerals present in a suspension mutually interact and adhere to one another and the microorganisms are smaller than the mineral particles the organisms will more or less cover the solid particles. If the microorganism cells are larger than the mineral particles the mineral particles will more or less cover the microorganism cells.
4. Thus, a microorganism will function as a flocculating agent for a mineral slurry only if a large portion of the mineral particles are larger in size than the organism.

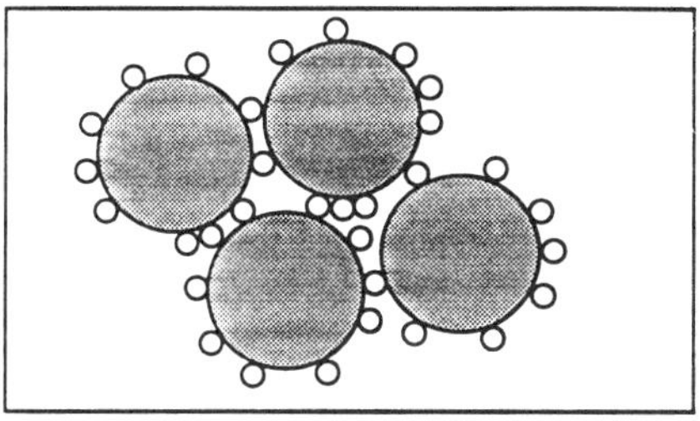

8a
SIZE PARTICLE > SIZE *M. phlei*

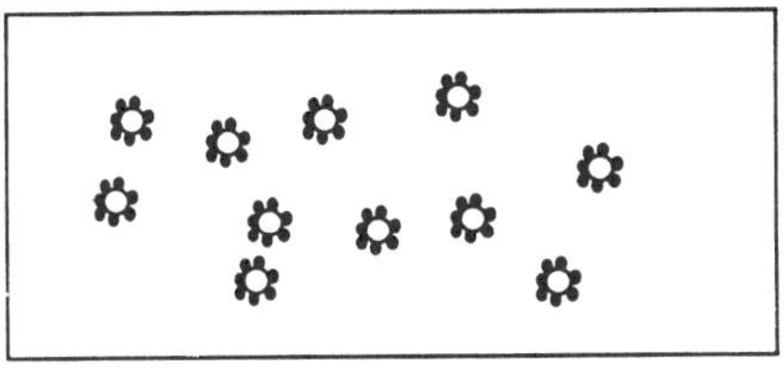

8b
SIZE PARTICLE < SIZE *M. phlei*

Figure 8 Manner of attachment of *Mycobacterium phlei* to mineral particles; 8a: the bacterium cells are smaller than the mineral particles and the mineral particles form aggregations; 8b: the bacterium cells are larger than the mineral particles and the mineral particles remain dispersed; the darkened circles represent mineral particles and the open circles represent bacterium cells.

References

1. van Loosdrecht, M. C. M., Bacterial Adhesion, Doctoral Dissertation, The Agricultural University, Wagenhingen, The Netherlands, (1988).
2. van Loosdrecht, M. C. M., Lyklema, J., Norde, W., Schraa, G. and Zehnder, A. J. B., "The Role of Bacterial Cell Wall Hydrophobicity in Adhesion", Appl. Environ. Microbiol., 53, (1987) 1893-1897.
3. van Loosdrecht, M. C. M., Lyklema, J., Norde, W., Schraa, G. and Zehnder, A. J. B., "Electrophoretic Mobility and Hydrophobicity as a Measure to Predict the Initial Steps of Bacterial Adhesion", Appl. Environ. Microbiol., 53, (1987) 1898-1901.
4. Smith, R. W., Misra, M. and Dubel, J., "Mineral Bioprocessing and the Future", Minerals Engineering, 4, (1991) 1127-1141.
5. Smith, R. W. and Misra, M., "Mineral Bioprocessing, an Overview", to be published in: Mineral Bioprocessing, eds. R. W. Smith and M. Misra, The Minerals, Metals and Materials Society, paper presented at the Engineering Foundation Conference: Mineral Bioprocessing, June 16-21, 1991.
6. Stewart-Tull, D. E. S., "Occurrence of Dimorphic Forms of *Mycobacterium phlei*", Nature, 208, (1965) 603-605.
7. Barksdale, L. and Kim, K-S, "Mycobacterium", Bacteriol. Rev., 41, (1977) 217-372.
8. Buchanan, R. E. and Gibbons, N. E., Bergey's Manual of Determinative Bacteriology, Eighth Edition The Williams and Wilkins Company, Baltimore, (1974) 693-694.
9. Laskin, A. I. and Lechevalier, H. A., Handbook of Microbiology, Second Edition, Volume 1, Bacteria, CRC Press, (1977) 288.
10. Singleton, P. and Sainsbury, D., Dictionary of Microbiology and Molecular Biology, Second Edition, John Wiley and Sons, (1987) 576.
11. Pratt, D., "Nutrition of *Mycobacterium phlei*, I. Requirements for Rapid Growth", Journ. Bacteriol., 64, (1952) 651-657.
12. Guirard, B. M. and Snell, E. E., "Biochemical Factors in Growth", Manual of Methods for General Bacteriology, editor in chief P. Gerhardt, American Society for Microbiology, (1981) 79-111.
13. Thiessen, P. A., "Wechselseitige Adsorption von Kolloiden", Ztschr. Elektochem., 48, (1942) 675-680.
14. Thiessen, P. A., "Kennzeichnung Submikroskopischer Grenxflachenbereiche Verschiedenartiger Wirksamkeit", Z. Anorg. Chemie., (1947) 161-169.
15. Daniels, S.L., "Mechanisms Involved in Sorption of Microorganisms to solid Surfaces", Adsorption of Microorganisms to Surfaces, ed's. Britton, G. and Marshall, K.C., John Wiley and Sons, (1980) 7-58.

SELENIUM REMOVAL FROM SOLUTIONS USING Iota Chips

S. Wang, M. Misra, R. G. Reddy and J. C. Milbourne*
Department of Chemical and Metallurgical Engineering
University of Nevada-Reno
RENO, NV 89557

*Brown & Root Braun, P.O.Box 3
Houston, TX 77001

ABSTRACT

A study was undertaken to remove the Se from waste solutions using a raw form of the biomass called Iota Chips derived from the seaweed, *Eucheuma spinosum*. Removal of Se from solutions was investigated as a function of time, pH, conc. of Se in solution (0.1 to 100 ppm) and Iota Chips to liquid ratios of 10^{-2} to 10^{-5}. For solutions containing initial conc. of 0.1 ppm Se, with the addition of Iota Chips at a solid to liquid ratio of 10^{-3}, the Se conc. decreased to 0.03 ppm in about 3 hours. Iota Chips absorption capacity to the Se and its application to the treatment of the metallurgical processing effluents are discussed.

Residues and Effluents - Processing and Environmental Considerations
Edited by Ramana G. Reddy, William P. Imrie and Paul B. Queneau
The Minerals, Metals & Materials Society, 1991

INTRODUCTION:

Selenium is a geochemically mobile element and is highly toxic to humans, fish and wildlife. At low levels, it is an essential nutrient but at slightly higher concentrations it becomes quite toxic[1,2]. The dietary requirement for animals is estimated to be 40- 100 μg Se/Kg of food (40-100ppb)[3]. No such requirement for humans has been determined. Smith and Westfall[4] reported symptom of chronic toxicity at 10 to 100 μg Se/Kg/day. For a 60 Kg person it would amount to a daily intake of 600 to 6000 μg Se/day. The average daily dietary intake by the American population is reported to be 200 μg Se/day[5]. The United States Environmental Protection Agency[6] set the public drinking waters maximum contaminant limit at 10 μg of Se/liter, based on the assumption that two litres of water per day would be consumed, and the resulted selenium consumption from the water would not be more than the 10% increase in the average daily intake. The Environmental Protection Agency[7] also identified the concentration of selenium in waste that will qualify the waste as "hazardous" under the Resource Conservation and Recovery Act (RCRA) of 1976.

The United States Fish and Wildlife Service (U.S.F. and W.S.) reported that due to selenium toxicity in the Kesterson Teservoir Area, California, high rates of mutation occurred in hatchlings of wild water fowl in the Kesterson National Wildlife Refuge[8]. The dispose of subsurface irrigation drain waters flow from the San Luis Service area (San Joaquin - Sacramento Delta), are stored and regulated from the Kesterson Reservior. Andreson et al.[9] predicted that in agriculture soils, in the regions of low rain fall and alkaline soils, selenium occurs as Selenite (SeO_4^{-2}) and in the regions of acid environment, selenium occurs in terrace hydroxide precipitates as basic ferric selenite [Fe_2 $(OH)_4SeO_3$]. The selenite is more mobile and readily transported in ground waters than the ferric selenite. Total selenium concentration in the Kesterson Teservoir area drain waters were determined to be in the range of 140 to 1400 μg/l[8].

Selenium occurs in conjection with sulfide mineral phases (Zn, Cu-Ni, and uranium) and is eliminated as selenium dioxide during conventional high temperature roasting and smelting operations. The flue gases are cleaned by water scrubbing prior to converting sulfur dioxide to sulfuric acid, and selenium dissolves in the scrubber water. The scrubber water, after aeration to remove much of the absorbed sulfur dioxide, typically contains 5 to 30 ppm of Se, depending on the Se contained in the original sulfide mineral phases. The amount of scrubber waste waters alone are reported to be discharged at a rate of about 250 gallons per minute.

Reported selenium in the discarded waste solutions from mine, mineral, gas scrubber (5 to 30 ppm of Se)[10] and agricultural waste waters (140 to 1400 μg Se/l)[8] are well above the maximum contaminant limit (10 μg Se/l) for the public drinking water set by the EPA[6].

Use of microorganisms in processing of several mineral systems has been reported in the recently held symposium on mineral bio-processing[11]. Reddy[12] discussed the energetics treatment to explain the differences in adhesion of bacteria on mineral surfaces and interfacial interactions of metal ions with bacterium solutions.

Seaweeds belong to a rather not well defined assemblage of plants known as the algae found in the groups of chlorophyceae (green algae), rhodoplyceae (red algae) and phaeophyceae (brown algae)[13,14]. Several investigators[15,16,17] reported the analysis of the seaweeds, surface properties and their application in treatment of the waste waters containing heavy metal ions, but very little is known about properties of seaweeds and its absorption mechanism of toxic ions in the solution at present. Veroy[18] has developed an idealized structure of a typicalseaweed, E. spinosum as shown in Figure 1. The active functional groups are sulfated polysaccharides carrageenan. Toxic nature of these waste waters containing Se and the lack of helpful information concerning treatment methods are important considerations. Fundamental information to develop technology that can economically process these toxic waste waters to eliminate or reduce to an acceptable selenium concentration level for the public use is not available at present.

In this paper, the absorption behavior of Iota Chips (seaweeds) and its capacity to remove the selenium from the waste solutions is reported.

EXPERIMENTAL:

Materials:

Elemental selenium of >99.84% purity was obtained from Fisher Chemicals and SeO_2 of commercial purity was obtained from K & K

Lab.Inc. The average size of the raw biomass Iota Chips (Eucheuma spinosum) used in this study is about 1.1 mm and surface area is about 1.8 cm^2/mg. The stock solutions of 71∓1 wt. pct HNO_3 certified by ACS specifications are used in the present study.

Procedure:

Stock solutions of known concentration of Se in the range of 0.1 to 200 ppm were prepared using the selenium powder. The selenium powder was dissolved using 0.8 ml HNO_3 acid. The SeO_2 powder dissolved readily in distilled water and predissolution of it in the acid as in the case of elemental selenium, was not required. All experiments were done at 298K using glass flasks. The ratio of Iota Chips (seaweeds) to solution was varied from 10^{-2} to 10^{-5}. After predetermined time, the solution samples were taken and then analysed using AA and IC. The pH and the interfacial tension of the solutions containing Iota Chips was measured by using Fisher model 925 pH meter and Fisher model 215 Autotensiomat respectively. The Iota Chips before and after Se loading were analyzed using SEM and X-ray diffraction.

RESULTS AND DISCUSSION:

Amount of selenium removal from solution was experimentally determined as a function of time, Se initial concentration in solution and the ratio of seaweed to solution at 298 K. The experimental results are discussed below.

Figure 2 shows the removal of Se as a function of time at on seaweed to solution ratio of 10^{-3} and at different initial concentrations of selenium in the solution. These solutions were prepared by dissolving the elemental Se in 0.8 ml HNO_3 acid. As seen from the figure, the removal of Se from solutions by Iota Chips (seaweed) increased with increase in time. For example, at a fixed initial concentration of Se (0.1 ppm), the concentration of Se in solution decreased from 0.08 to 0.03 ppm as the time increased from 15 to 240 minutes respectively. About the same percent of Se (about 70 pct) was removed even when the initial concentration of Se was raised from 0.1 to 10 ppm. But at a higher initial concentration of Se (about 100 ppm), only 30 pct of Se was removed from the solution and also the rate of Se removal was also slower. This indicates that available surface area of the Iota Chips play a significant role in removal of Se from the solution.

Effect of ratio of seaweed to solution on removal of Se is shown in figure 3. These experiments were carried out at an initial concentration of 200 ppm Se in solutions. These solutions were prepared by dissolving SeO_2 powder in distilled water. As seen from the figure, at a fixed time, the removal of Se from solution increases with increase in the ratio of seaweed to solution. For example, at a fixed time of 135 minutes and initial concentration of 200 ppm Se, the final concentration of selenium decreased from 109 to 63.4 ppm as the ratio increase from 10^{-4} to 10^{-2}. From this it can be said that the higher the available surface area of the Iota Chips in the solution the larger the amount of Se removal.

It is difficult to calculate or measure the absolute value of absorption of selenium because both the selenium and the seaweed would absorb each other in solution. But, we may able to calculate the relative value of absorption of selenium by knowing the initial and the finial concentration of selenium in the solution[19]. The relative absorption of Se (Γ) in pct can be expressed as:

$$\Gamma = [\ C_o - C\ /\ C_o\ ] \times 100$$

where C_o is the initial concentration of Se and C is the finial concentration of Se at the time t. The relative adsorption of Se on to seaweed in the solution as a function of time and ratio of seaweed to solution is shown in Figure 4. As expected, the amount of selenium absorbed increases with increasing equilibrium concentration of seaweed at each measured time. On the other hand, the results show that the absorption capacity of seaweed increased continuously as the time increases. A maximum of 70 pct relative adsorption of Se was obtained. It can be suggested that this absorption is a chemical absorption and chemical interaction forces play an important role in this process of absorption of Se on to the Iota Chip surface. Similar observations were made by Gebhart and Fuerstenan[20] in a related mineral system.

Zeta potentials:

The zeta potential of the solutions containing Iota Chips is remeined constant value of -40 in the pH range of 3 to 7. But when the pH of the solution decreased from 3 to 1.8, the zeta potentials increased from -40 to 0.

Interfacial tension:

The interfacial tension of the solutions containing Iota chips were measured at 298K. The interfacial tension of the solution increased with increase in ratio of seaweed to solution. For example, the interfacial tension increased from 57 to 69 dynes/cm as the ratio of seaweed to solution increased from 10^{-5} to 10^{-2} respectively.

Figure 5 shows the pH change as a function of seaweed to solution ratio with and without Se in the solution. The pH of the distilled water used is 6.51. As seen from the figure, in the case of without Se addition (curve 1), the pH of the solution increased with increase in addition of seaweed to the solutions. This indicates that seaweed surface is composed of alkaline ionic species and with the result alkalinity of the solution increases as the increase in addition of Iota chips are added. A maximum pH of 9.7 was observed at a ratio of 10^{-2}. The pH of the distilled water decreases significantly with the addition of Se. As seen from the curve 2 (figure 5), the pH of the solution containing 200 ppm Se is about 3.48 (extrapolated). The pH of the Se containing solutions increased with increase in addition of Iota Chips, and reaches a maximum pH of 8.14 at the Iota Chip to solution ratio of 10^{-2}. Figure 5 (curve 3) shows the relationship between pH and seaweed to solution ratio for the solution containing 200 ppm Se. This solution was prepared by dissolving elemental Se in 0.8 ml HNO_3. As seen from the curve 3 (figure 5), the solutions are very acidic and pH is about 1.43. Even with the increase in seaweed to solution ratio of 10^{-2}, the pH remained constant. It is interesting to observe that in both the cases, Se in solution obtained by dissolving SeO_2 (figure 5, curve 2), and by dissolving elemental Se (figure 5, curve 3), though the pH values were significantly different largely acidic to alkaline in nature, the absorption of Se on to Iota Chips from both types of solutions (figure 2 and 3) obtained was only 70 pct maximum. No clear explanation can be given regarding this observation at present.

Selenium (Se) is known to exist in both organic (Se^{2-}) and inorganic forms of Se (+4 and +6) in the solutions. The question of reduction of Se level in the waste solutions is complicated by the chemistry of Se itself in the oxidation states of +6 to -2. The form of Se in the solutions (Se^{+4} or Se^{+6}) or whether it is in the organic (Se^{2-}) form is not yet fully known. In mine runoff waste waters from molybdenum operations, Se is known to be entirely in the form of Se^{+6}. But in the alkaline and acidic solutions, oxidation states of Se are not clear. The oxidation state of Se is also not clear when the Se concentration in the solution is very low (i.e. in the range of ppb to ppm)[9]. The stability of various Se complexes in solution and effect of different parameters i.e., anions, cations in solution, temperature and pH of the solution is not known at present. Further work is in progress on determination of changes in surface and chemical properties of Iota Chips in solutions containing Se.

CONCLUSIONS:

The removal of selenium from solution using Iota Chips (seaweed) was investigated. The effect of time, selenium initial concentration and seaweed to solution ratio on selenium removal at 298K were determined. From the present study, the following conclusion can be made:

1. The removal of selenium from solution increased with increase in time and also increase in seaweed to solution ratio.

2. For the initial concentration of 0.1 ppm Se containing solutions, the selenium in the solution was removed to a lowest level of 0.03 ppm Se using seaweed to solution ratio of 10^{-3} in about 2 hours time.

3. The addition of Iota Chips increases the alkalinity of the solution. The pH of the solution increased to about 9 as the seaweed to solution ratio increases to 10^{-2}.

4. Zeta potentials of the solution containing Iota Chips remained constant about -40 in the pH range of 3 to 7. At a lower pH value of about 1.8, the zeta potential of the solution is determined to be zero.

5. The interfacial tension of the solutions increased with increase in seaweed to solution ratio. The interfacial tension values of solutions increased from 57 to 69 dynes/cm as the seaweed to solution ratio increases from 10^{-5} to 10^{-2} respectively.

ACKNOWLEDGEMENTS:

This work was supported by Pyrometallurgical Processing and

Extraction of Metals (a consortium of Metal Industries) program, Grant # 5302-804. One of the author (S. Wang) expresses his appreciation to the program for the financial support. Special thanks are extended to Mr. P. Lechler and Mr. M. Desilets, Nevada Bureau of Mines and Geology for their help in the chemical analysis.

REFERENCES:

1. I. Rosenfeld and O. Beath; Selenium: Geobotany, Biochemistry, Toxicity and Nutrition, Academic Press, London, pp. 114, 1964.
2. H. Vokal-Borek ; Selenium, USIP Rep. 79-16, University of Stockholm, 1979.
3. K. Schwarz, Nutrition Reviews, Vol. 18, pp. 193-197, 1960.
4. M. Smith and B. B. Westfal, Public Health Reports, Vol. 51, pp. 1496-1505, 1937.
5. V. C. Morris and O. A. Levander, J. of Nutrition, Vol. 100, pp. 1383-1388, 1970.
6. U. S. Environmental Protection Agency, National Interim Primary Drinking Water Regulations, Office of Water Supply, EPA-570/9 - 76-003, pp.159,1977.
7. U. S. Evironmental Protection Agency, Federal Register, Vol. 45, pp. 33063-33285, 1980.
8. T. S. Presser and I. Barnes, U. S. Geological Survey, RI 84-4122, pp.26.
9. M. S. Anderson, H. W. Lakin, K. C. Beeson, F. F. Smith and E. ^Thacker, Agriculture Handbook No. 200, U. S. Agricultural Research, pp. 159, 1977.
10. W. N. Marchant, R. O. Dannenberg, and P. T. Brooks, U. S. Bureau,

Water Regulations: Office of Water Supply, EPA-570/9 - 76-003, pp. 65, 1961.

11.R. W. Smith and M. Misra editors, proceedings of Engineering Foundation Conference: Mineral Bioprocessing, held on June 16-21, 1991.

12. R. G. Reddy, "Thermodynamics of Bacterial-Mineral Reactions", to be published in " Mineral Bioprocessing ", eds. R. W. Smith and M. Misra, TMS, 1991.

13. V. J. Chapman and D. J. Chapman, Seaweeds and their uses, 3rd ed.,London, New York, Chapnan and Hall, pp. 18, 1980

14. I. A. Abbott and E. Yale Dawson, How to know the seaweeds, 2nd ed., Dubuque, Iowa: H. C. Brown Co., pp. 5, 1978

15. D. M. Chizhikov and V. P. Shchastlivyi, Selenium and selenides, Collet's (Publishers), LTD., London and Wellingborough, pp.39,1968.

16. J. O. Nriagn and H. K. Wang, Nature, Vol. 301 (6), Jan. 1983

17. T. Nakahara, S. Kobayaashi, T. Wakisara and S. Musha, Appl. Spectroscopy, vol. 34, Nov. 1980.

18. R. L. Veroy, et al, "Studies on the Binding of Heavy Metals to Algal Polysaccharides from Phillippine Seaweeds, 1. Carrageenan and the Binding of Lead and Cadmium" Botanica Marina, Vol. xxIII, p. 59, 1980

19. X. C. Fu and R. H. Chen, Phy. Chem., pp. 301, 1979

20. J. E. Gebhart and D. W. Fuerstenau, Interfacial phenomena in Mineral Processing, United Engineering Trnstees Inc., pp. 175-188, 1982.

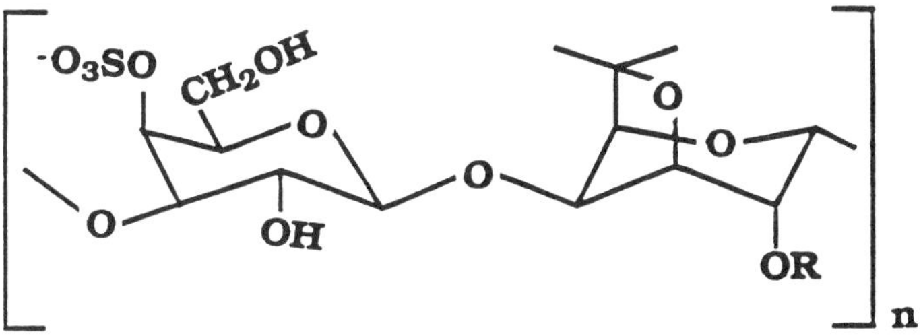

Figure 1: Idealized structure of carrageenan. K-carrageenan when R is H and t-carrageenan when R is SO_3^-

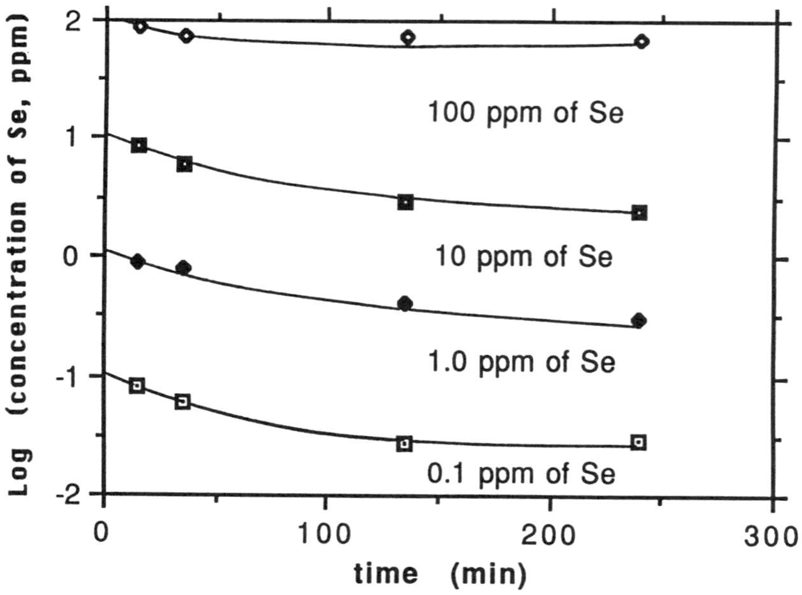

Figure 2: Concentrations of Se as a function of time at a ratio of seaweed/solution = 10^{-3}

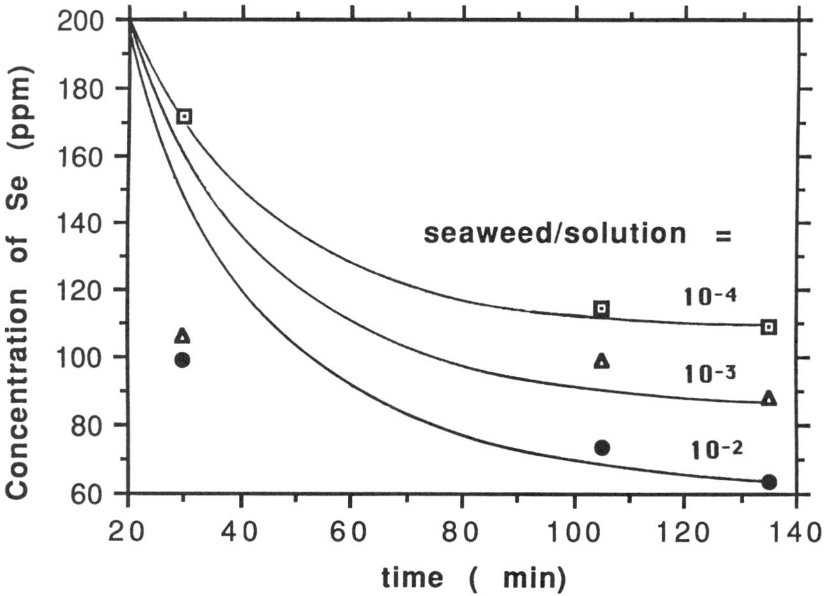

Figure 3: Concentrations of Se as a function of seaweed to solution ratio at an initial concentration of Se 200 ppm

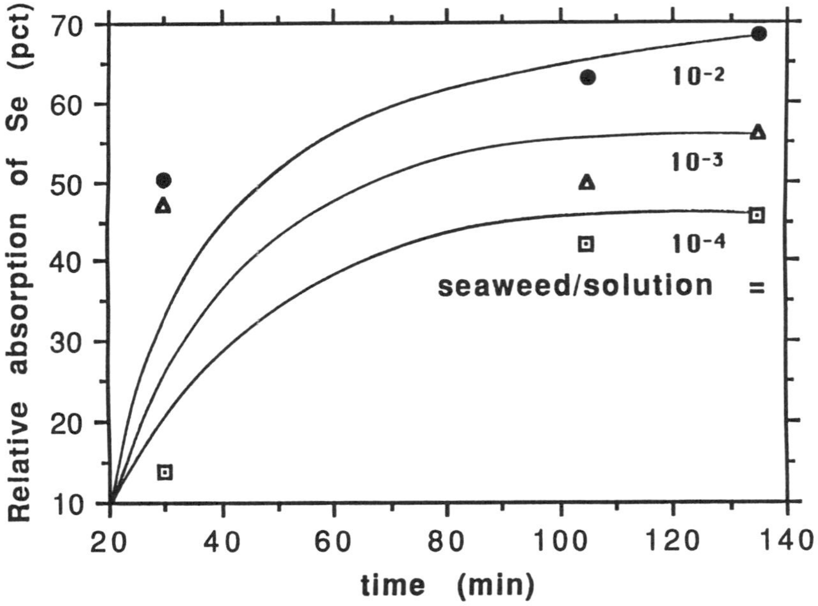

Figure 4: Relative absorption of Se as a function of time and intial concentration of Se=200 ppm

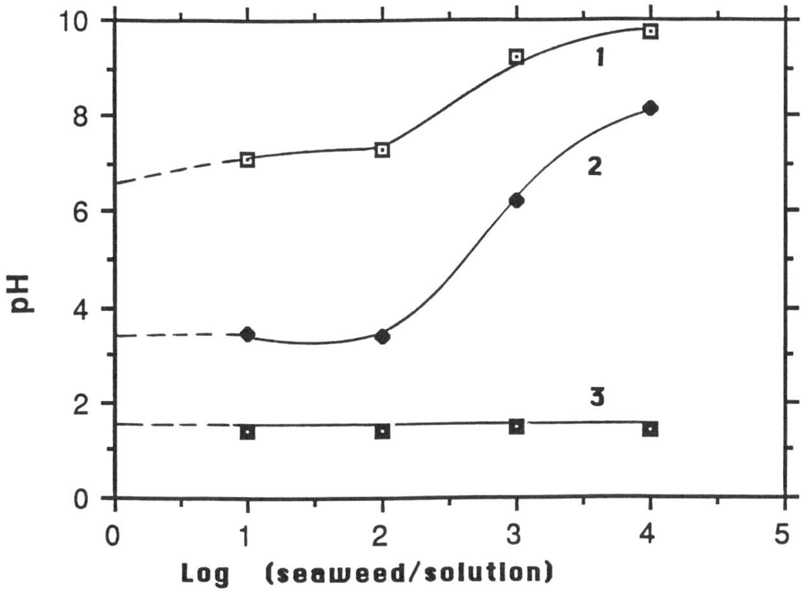

Figure 5: pH as a function of seaweed to solution ratio. (1) without Se in solution (2) with 200 ppm of Se (as SeO_2) (3) with 200 ppm of Se (as dissolved in HNO_3)

ACID LEACHING MECHANISM AND KINETICS OF SELECTED SECONDARY MATERIALS FOR ALUMINA RECOVER

S. Vukcocic, M. Stamatovic & R. Vukcevic

Institute for Technology of Nuclear and other Mineral Raw Materials
86 Franshe d'Eperey St.,
Yu-11000, Belgrade
Yugoslavia

R. Vracar

Faculty of Technology and Metallurgy
4 Karnegijeva St.
Yu-11001, Belgrade
Yugoslavia

Abstract

A long term research project is in progress to establish the parameters of an acid leach process to recover alumina from high-iron and/or low-iron white bauxites, kaolinite, coal ash and copper mine tailing. In this paper the results of acid leaching of copper flotation tailing for the removal of alumina are presented. The tailing is very complex material which contains 25% Al_2O_3 in the -10 μm fraction. X-ray diffraction indicates presence of quartz, kaolinite and minor alunite. After classification and separation of the -10 μm fraction, the raw material is fired to 550°C and leached at 90°C. Filtration is applied for the separation of white silicate residue form leach liquid. After the purification step, aluminum salt is crystallized from the solution, and calcined to alumina. Kinetic correlation and regime of the processes are defined and presented in the paper.

Residues and Effluents - Processing and Environmental Considerations
Edited by Ramana G. Reddy, William P. Imrie and Paul B. Queneau
The Minerals, Metals & Materials Society, 1991

ELECTRORECOVERY OF METALS FROM DILUTE PROCESS SOLUTIONS IN ELECTROPLATING

Roland Kammel

Institut für Metallurgie - Metallhüttenkunde -
Technical University Berlin

Abstract

In combination with various physical and chemical methods also electrochemical techniques are most frequently used to improve the recovery of metals from aqueous dilute process streams or effluents e.g. in electroplating, metal finishing, mining and metallurgical plants.

To insure cathodic metal recoveries with favorable current efficiency and high throughputs per unit of cell volume several types of electrolysis cells have been developed comprising extended electrode surface areas and/or intensive electrolyte agitation. The practical experiences with these units indicate especially for the electrolytic recovery of metals whose values are high, that the treatment cost permits profitable operations with relativly short payback periods for the units. Anodic reactions can be utilized for the removal of undesired anion contents or buildups of impurities.

Appropriate combinations of these recovery methods are used to improve the versatility and the overall economy. Industrial applications reveal their potentional as addidtive techniques for existing processes to reduce the metal content of effluents to values that meet the guide lines of environmental legislation whereas integrated in-operation recovery methods are nowadays increasingly introduced to approach nearly close-loop operations.

Residues and Effluents - Processing and Environmental Considerations
Edited by Ramana G. Reddy, William P. Imrie and Paul B. Queneau
The Minerals, Metals & Materials Society, 1991

Introduction

Under increasing pressure of enviromental legislation with stringent limitation for metal emissions the metal finishing and electroplating industry has to reduce or avoid metal losses in the course of the processing steps by new technologies which may be based e.g. on an optimization of existing processes to low-emission operation; on an in-operation recycling resp. on an nearly closed-loop procedure; on a novel process technology.

As summarized in figure 1 many methods for the recovery of metals from dilute solutions in metal finishing and electroplating industry have been developed, which involve the separation or concentration of the metal contents as well as the winning of markable compounds or metal deposits 1,2).

The most commonly used technologies nowadays are:

Physical methods:	(with heat supply) evaporation, volatilization; (with heat removal) crystallizaon; (with membranes) reverse osmosis, ectrodialysis.
Chemical methods:	precipitation, cementation, ion-exchange, retardation, liquid-liquid extraction.
Electrolytic methods:	with/without electrode compartments; fixed/not fixed contacted electrodes.

Each method may have its special application. Usual combinations are used to achieve efficient and favourable recovery performances e.g. ion exchange, retardation or liquid-liquid extraction can be used to concentrate the metal content from very dilute solution and enable then the electrochemical metal recovery from the concentrated eluates rsp. regenerants. In combination with various recovery methods, several aqueous electrochemical process techniques have shown their potential for saving materials and water, for preventing pollution, and for avoiding or minimizing harmful wastes.

Different from conventional electrolytic techniques utilizing parallel flat-plate electrodes e.g. in electrowinning and electrorefining, where metals are cathodically deposited from concentrated aqueous electrolytes with high metal contents, the electrochemical recovery methods have to be adjusted for the treatment of diluted solutions having only metal contents of few g/l or few ppm.

Sources of these dilute metal-containing solutions can originate most commonly from following metal industries:

electroplating:	rinse water, spills, drag out, spent bath;
metal finishing:	cleaning-, etching- and electrochemical mashining solutions;
mining/metallurgy:	leach liquors, mine waters, spent electrolytes, process solution drag out.

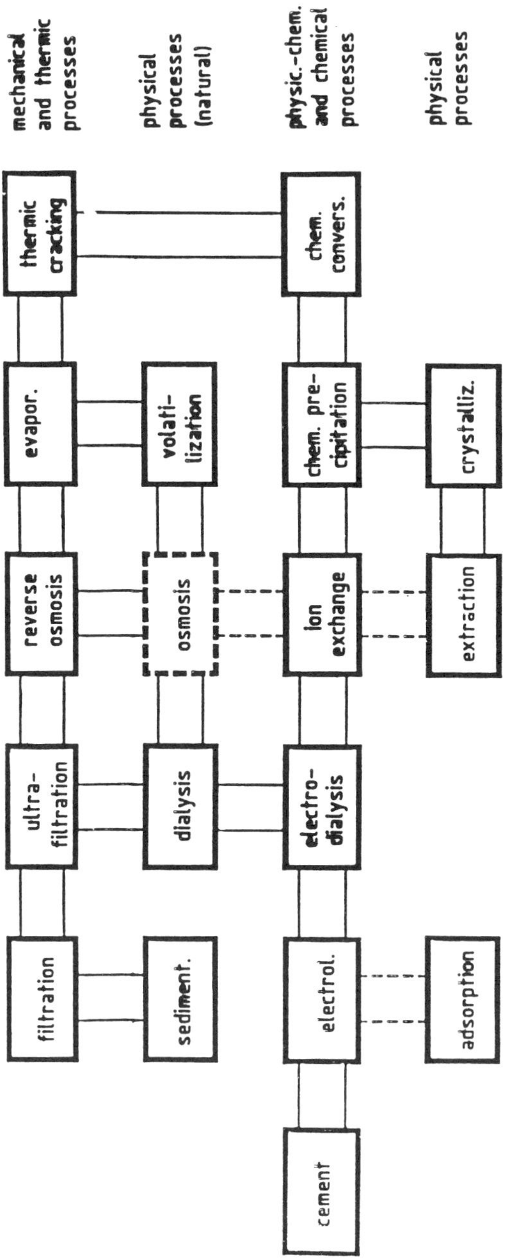

Figure 1: Processes to recover metals from dilute aqueous process streams

Electrorecovery Cell Systems

As indicated in figure 2 for the electrorecovery of metals from dilute solutions two process routes have to be considered namely: the recovery of metals down to predetermined values, after which the solution can be recirculated into the process e.g. for closed loop reduction and adjustment of the metal contents in rinsing operations;

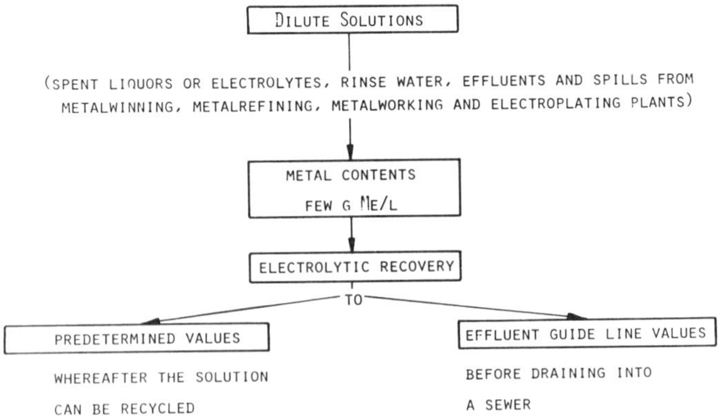

Figure 2: Treatment of dilute solutions by electrorecovery

the removal of metals before draining into a sewer to residual contents that meet the guide lines of environmental legislation, which means to metal values of about 1 ppm or less.

Conventional electrowinning cells equipped with vertical electrode plates are unpractical and uneconomical for electrorecovery operations. The fast occuring depletion of the metal ions in the solution adjacent to the cathode can form polarized layers and leads to poor quality deposits with codeposition of hydrogen or to frequent shortcuts by dendrites and spongy deposits. From the equation in figure 3 it can be elucidated that high space-time yields during the electrorecovery of metals from dilute solutions can be achieved either by increasing the cathode surface area and/or by rigorous electrolyte agitation, which minimizes the cathodic boundry layer-thickness and depletion of metal ions by improved mass-transfer conditions 3).

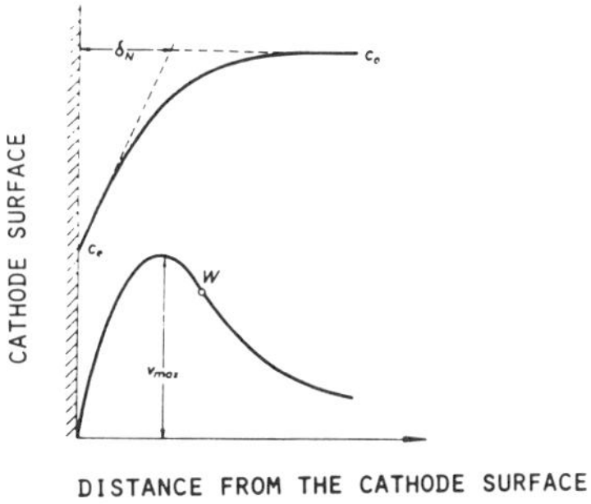

Figure 3:

Schematic representation of the concentration and velocity profile at the electrode/electrolytic interface 3)

$$I = K \cdot Q \cdot \frac{C_O - C_E}{\delta_N}$$

I = CURRENT

K = FACTOR DEPENDENT ON PROCESS PARAMETERS AND CURRENT EFFICIENCY

Q = EFFECTIVE CATHODIC SURFACE

C_O = METAL CONTENT OF THE SOLUTION

C_E = METAL CONTENT OF ELECTROLYTE AT THE CATHODE SURFACE

δ_N = THICKNESS OF BOUNDARY LAYER AT THE CATHODE SURFACE

As reviewed and discussed in a previous paper (4) general following two types of recovery cellsystems can be distinguished: Electrorecovery concentrator cells, which produce in a primary concentration step from dilute solutions metal deposits from which the metal is recovered by chemical dissolution to yield a concentrated solution or the metal deposits are anodically stripped and deposited on cathodes as in conventional electrorefining processes. The deposits can be also recovered as molten metals by direct pyrometalurgical treatment which allows to remove the cathode base materials e.g. graphite, steel wool or polymeres. Most of these cells are equipped with stationary electrodes and extended cathode surface areas of different shapes and dimensions. Typical cell devices have large three-dimensional flow-through type cathode surface areas of the fixed-bed version packed e.g. with carbon fibers, steel wool, carbon particulate beds and metallic coated polymere foams or balls.

Direct-electrorecovery cells which allow to recover the metal contents on the cathode either in form of non-adherent or coherent deposits, or as powder or particulates. Improved electrorecovery conditions with stationary plate or mesh cathodes have been achieved with intensive electrolyte agitation e.g. by air sparging in the cathode area, with a non-conducting bed of glass beads as turbulence promoter or by impellers located at each end of the cathode compartment.

Other recovery cells combine electrolyte agitation with cathode motion e.g. a fluidized bed of conducting cathode particles; a rapid rotating cylinder system on which above limiting current density metal powders are deposited; a mechanical rotated tubular reactor with a particulate bed as cathode.

Recovery results

As an example of the numerous industrial applied cell types in figure 4 the s.c. impact rod reactor is shown 4).

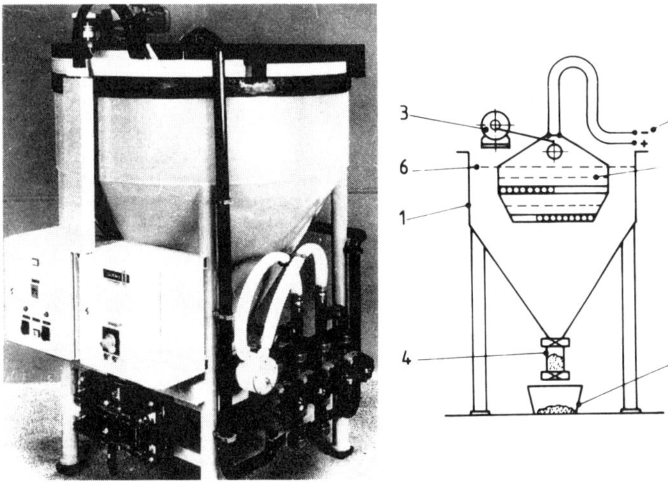

Figure 4: Impact rod electrorecovery cell (Goecomet)
1 cell tank, 2 impact rod system,
3 mechanical agitation,
4 and 5 collecting reservoir with sluice chamber,
6 liquid level, 7 source of electricity

In this recovery cell the mechanical wiping agitation of the electrode rack with the rolling cathode rods and the electrolyte flow improves the mass transfer conditions. Owing to the impacting rolling cathode rods, the metal depositis peel off after a certain thickness and are collected at the bottom of the cell in a reservoir.

From many metal recovery investigations in this cell system in figure 5 the cadmium removal from acid as well as cyanide electroplating rinse waters are shown.

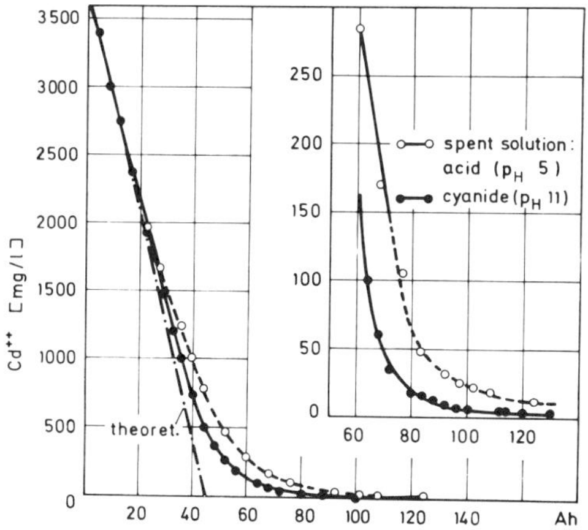

Figure 5: Electrolytic recovery of cadmium from acid and cyanide spent solutions
Acid: 2.5 g Cd/l, 4 g/l H_2SO_4
Cyanid: 3.75 g Cd/l, 12 g/l NaCN, 5 g/l NaOH

The cadmium recoveries from these spent solutions reveal that the sytem can be operated economically to low residual metal values of few ppm. Calculations show that significant savings can be achieved compared with commonly used precipitation techniques for the treatment of effluents and the electrolytic recovery of cadmium is repaying fast the investment costs.

Figure 6:

Cost for recovery of cadmium by electrolytic and ion exchange treatment as a function of the cadmium content in waste water

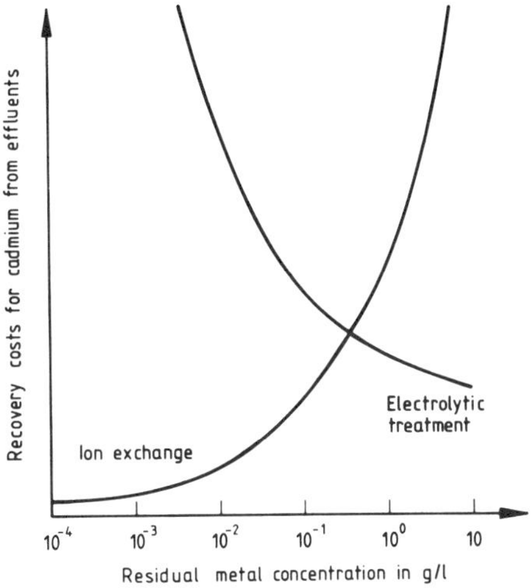

Despite developments of new cell design, several research activities have been lately focussed on improvements of the electrode materials as well as anode and cathode processes. With the aim to save energy new anode materials e.g. dimensionally stable anodes of low potential type with metallic bonded coatings of precious metals and their oxides as well as high conducting polymer electrodes are nowadays tested. Furthermore, research activites with hydrogen type anodes, which could allow to avoid undesired anodic reactions and distinct energy savings have shown promising results. Substantial savings by anodic reactions are also achieved by decomposition of undesired organic complexes in the bath or by detoxification of cyanides. Cathodic investigations are focused to avoid rapidly decreasing current efficiency at very low rediual concentration levels e.g. with a controllable power supply whose control signal is obtained via a reference electrode in the cathode area.

The advantages of a combination of the electrorecovery of cadmium with ion exchange technique are demonstrated in figure 6. From the dependency of recovery cost from residual metal concentration it can be seen that the electrorecovery of cadmium is favourable with increasing cadmium contents, where as the ion exchange unit can be operate more economically at low concentration levels.

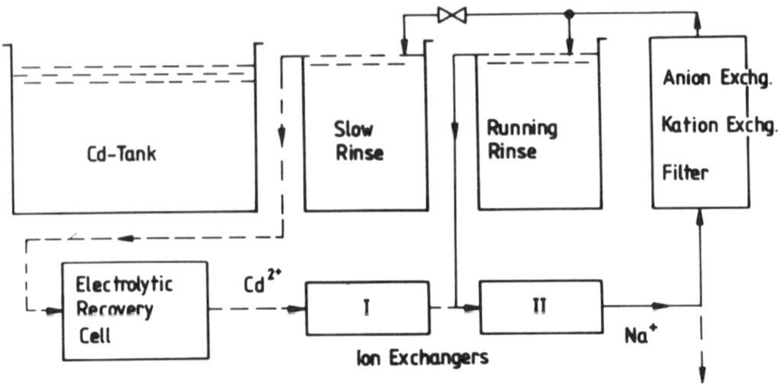

Figure 7: Recovery of cadmium from a reclaim tank by electrolytic treatment and ion exchange

A typical industrial applied combination of these two processes is schematically shown in figure 7. The overflow from the slow rinse tank with higher cadmium contents is directly treated in an electrorecovery cell. After the electrorecovery treatment the residual cadmium content is removed by ion exchange down to values of 0.05 ppm. The concentrated eluate of the ion exchange can be returned and treated in the electrorecovery cell.

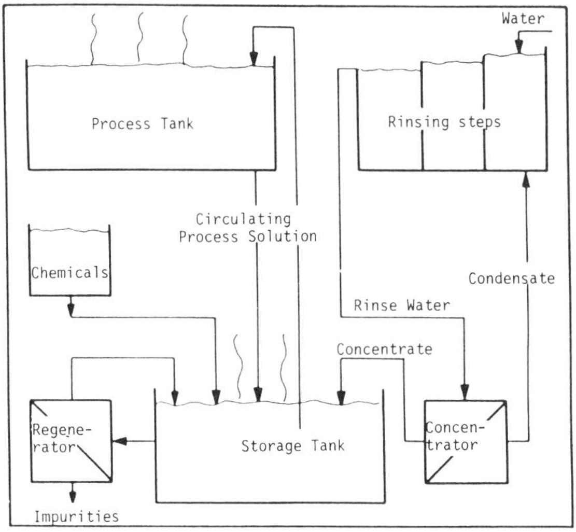

Figure 8: Electroplating process with integrated regeneration and water recirculation to avoid effluents

Since lately discussed environmental guide lines with extrem limitations for metal and other constituents in effluents as well as the previous anouncements e.g. "Zero Discharge" by EPA, USA, for future legislation plans can't be met even by best available additive 'end pipe' technologies, the metal industry is forced to develop integrated nearly closed-loop procedures to avoid closures or shut downs.

With combination of a concentration unit for the water balance and a regeneration unit for cleaning the circulating electrolyte, as schematically shown in figure 8, it is intended to perform the electroplating process without effluents and with only small amounts of imputities which have to be discarded.

The first large scale zinc plating plant to be operated without any waste water and sludge is just beeing assembled in our country. For the introduction of s.c. "clean" production lines the electrorecovery of metals to predetermined values is of great importance. This unit also allows simultaneously the removal of undesired impurities without chemical additions. However, still many problems have to be overcome regarding the proper selection or the treatment of bath additions and their build-up, e.g. especially the complexing or levelling agents in the electrolyte to maintain the necessary electrolyte quality for close-loop operations.

Conclusion

The emerging and already industrial applied electrorecovery of metals which is favourable performed in combination with various physical and chemical processes, has successfully demonstrated a new perspective to prevent pollution, and to avoid or minimize harmful wastes.

Electrochemical units designed with extended electrode surface and/or agitation of the electrolyte have contributed in the metal industry to recover metals from diluted process solutions and effluents. As advantages of these electrorecovery methods may be summarized: no additions of chemicals are required; no sludge is generated; the consumption of water is reduced by closed loop process approach; the investment and operating costs are low. These benefits lead also to attractive payback periods even for the recovery of less expensive metals.

Literature

1) Kammel, R., and Lieber, H.-W.: Möglichkeiten zur Behandlung galvanischer Abwässer unter Vermeidung von Sonderabfällen; Galvanotechnik, vol. 68 (1977) pp. 57-63; pp. 123-130; pp. 413-418; pp. 710-715; pp. 789-794; pp. 883-886; vol. 69 (1978) pp. 317-324; pp. 624-630; pp. 687-696.

2) Hartinger, L.: Abwasser reinigen und vermindern; Entsorgungspraxis 7-8 (1987) p. 362-369.

3) Lieber, H.-W.: Beitrag zur Kenntnis der kathodischen Abscheidungsbedingungen bei der Raffinationselektrolyse von Kupfer; Dissertation, TU Berlin (1966).

4) Kammel, R.: Recent Developments in the Design of Electrochemical Cells for the Recovery of Metals; Proceedings, International Conference on Mineral Science and Technology -MINTEK 50- Sandton, SA (1985) vol. 2, pp. 443-452.

5) Lieber, H.-W.: Private Information

KINETIC MODELLING OF THE OXIDATION OF CYANIDE IN AQUEOUS SOLUTIONS WITH HYDROGEN PEROXIDE

Alexandre Bonan, Liuz A. Teixeira, Helio M. Kohler

Catholic University of Rio de Janeiro
Dept. of Materials Science and Metallurgy
CP. 38008 - Rio de Janeiro, RJ-CEP. 22453

Abstract

The paper reports on the results of a kinetic study on the oxidation reaction of aqueous free cyanide ions with H2O2, in the presence of Cu(2+) catalyst. The variables and respective ranges studied were the following:

$(CN)_o$ = 10 - 100 mg/l
pH = 10 - 11
(H2O2/CN) mol ratio = 1 - 3
(Cu/CN) mol ratio = 0 - 0.25

The results lead to the development of a semi-empirical model which incorporates both the useful oxidation reaction and the parallel alkaline decomposition of H2O2.

(*) In cooperation with Peroxidos do Brasil Itda, a company member of the INTEROX Group

Residues and Effluents - Processing and Environmental Considerations
Edited by Ramana G. Reddy, William P. Imrie and Paul B. Queneau
The Minerals, Metals & Materials Society, 1991

A COST-EFFECTIVE, "ZERO" DISCHARGE HEAVY METAL REMOVAL SYSTEM:

Lary L. Speakman

Harrison Western Environmental Services, Inc.
1208 Quail street
Lakewood, Colorado 80215

Abstract

Typical heavy metal removal system generate a great deal of waste that must then be discharged somewhere. Using Bio-Fix Beads developed by the Bureau of Mines in combination with other waste stream minimization processes, Harrison Western Environmental Services has developed a "zero" discharge operation. Data from current on-site operations will demonstrate the effectiveness of the Harrison Western approach. For 1992 Annual TMS meeting: "Future Trends and Innovations" Material category Key words: Residues and Effluents Technical category Key words: Extraction and Processing.

Residues and Effluents - Processing and Environmental Considerations
Edited by Ramana G. Reddy, William P. Imrie and Paul B. Queneau
The Minerals, Metals & Materials Society, 1991

ION AND PRECIPITATE FLOTATION OF HEAVY METAL IONS FROM SOLUTION:

J.W. Perez

Research and Engineering Division
Freeport McMoran
Box 910, Belle Chasse
LA 70037

F.F. Aplan

Mineral Processing Section
Penn State University
University Park
PA 16802

Abstract

In the treatment of process effluents containing metal ions, it is sometimes desired to remove all of the heave metal ions while at other time it is desiredto remove one heavy metal ion from another, selectively. Ion and precipitate flotation processes were evaluated to perform these functions using the copper-iron aqueous system as a model. Because preliminary foam fractionation experiments proved to be relatively inefficient, most of this work has focused on precipitate flotation using both hydroxide and sulfide precipitation. Test results indicate that by proper control of reagents and operating conditions, it is possible to substantially reduce the froth volume over that achieved with foam fractionation and to separate copper from iron selectively. Mechanisms for collector attachment to precipitates were studied.

Residues and Effluents - Processing and Environmental Considerations
Edited by Ramana G. Reddy, William P. Imrie and Paul B. Queneau
The Minerals, Metals & Materials Society, 1991

IMPROVED TECHNOLOGY FOR PRECIPITATING METAL HYDROXIDES

Roger N. Kust

TETRA Technologies, Inc.
9391 Grogans Mill Road, Suite A-6
The Woodlands, TX 77380

ABSTRACT

Removal of heavy metals from industrial process streams is becoming increasingly important for the recovery and reuse of values in process tails. Technology has been available for several years which removes heavy metals from dilute (0.1-3.0 g/L) streams, as hydroxide or hydrated oxide sludges, allowing discharge of washed cakes with up to 60% solids. This paper reports on improvements to this technology which now allows treatment of streams with metal concentrations in excess of 60 g/L while eliminating the need for thickeners. Streams containing copper, iron, aluminum, chromium, nickel, zinc, cobalt and lead have been successfully treated.

Residues and Effluents - Processing and Environmental Considerations
Edited by Ramana G. Reddy, William P. Imrie and Paul B. Queneau
The Minerals, Metals & Materials Society, 1991

I. INTRODUCTION

For several years TETRA Technologies, Inc. has offered a proprietary metal hydroxide densification technology under the name HDS for treating metal bearing waste streams. With the addition of lime, caustic or other source of hydroxide, many of the metals which precipitate as hydroxides can be removed to levels commensurate with the solubility limit of the metal hydroxides. This technology was originally developed by Bethlehem Steel to treat acid mine drainage. TETRA now owns the rights to this technology which is covered by U.S. and foreign patents. TETRA designs, constructs, and operates plants using this technology and has granted licenses internationally.

II. DESCRIPTION OF HDS TECHNOLOGY

The original HDS technology which has been described in several previous publications, was essentially a solids recycle process which recycled a controlled mass of thickened sludge back into the metal precipitation process allowing particle growth. Although many of the metal hydroxides or hydrated metal oxides are considered amorphous, the HDS process is similar to crystallization processes wherein crystalline products are recycled to improve particle size and settling and filtering properties. But whereas crystallization processes rely upon the control of a margin of supersaturation, the HDS process relys upon the controlled addition of precipitant at a location removed from the site of the precipitation process thus allowing surface effects to be fully exploited. Commercial treament facilities have been constructed and are currently operating for removing iron(II), iron(III), chromium, nickel, zinc and cadmium. Additionally pilot and/or mini-pilot operations have been successfully completed for aluminum, cobalt, copper, magnesium, titanium and lead.

II.A Process Flow Diagram

A process flow diagram for the original HDS process is shown in Figure 1. The generally acidic metal bearing stream enters the precipitation reactor where the metal is precipitated by the alkalinity entering the reactor with the recycled sludge. The slurry is discharged to a thickener. A carefully regulated amount of the underflow from the thickener is fed into the alkalization reactor where hydroxide ions from caustic, lime or other source of alkalinity are absorbed on the surfaces of the particles; the remaining underflow is discharged, usually to some type of filter. The alkalized sludge then is fed to the precipitation reactor where the metal ions are encouraged to precipitate on the surface of the recycled particles. Overflow from the thickener is discharged either directly to sewers, to storage for use as plant water or to a polishing filter.

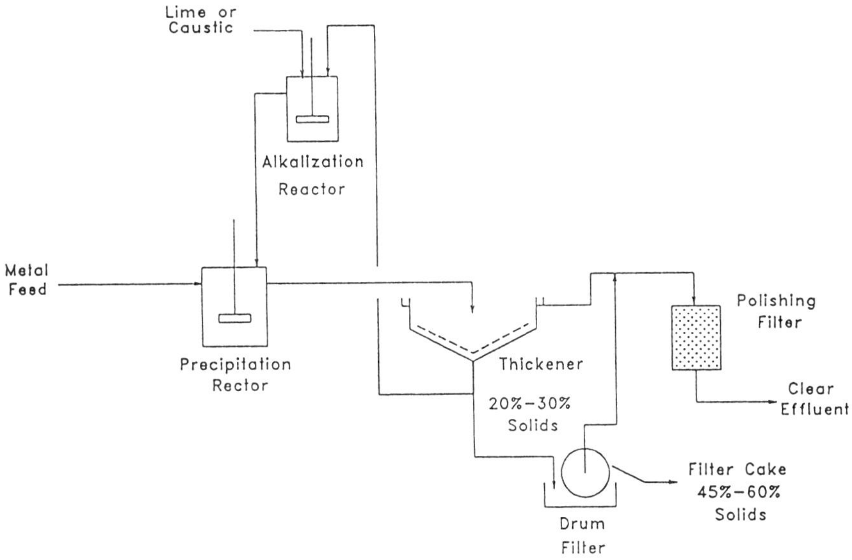

Figure 1. Original HDS Process

II.B Advantages and Disadvantages

The advantages of using the HDS process for removing heavy metals from process or waste streams lie in the improved characteristics of the precipitated solids. Table 1 compares the typical properties of conventionally precipitated hydroxides to those obtained from the HDS process.

TABLE 1. Properties of Conventional and HDS Hydroxide Precipitates.

Property	Conventional Precipitates	HDS Precipitates
% Solids in thickener underflow	2-5	20-35+
Thickener unit area (no floc./floc.), m2/MTPD	?	2.5-3.8 / 1.0-2.0
Sludge Compressibility	Compressible	Slightly compressible
Particle size	.01-0.1 microns	10-20+ microns
Filtration Rate on rotary vacuum filters, 1 cm cake	Not applicable-pressure filtration mostly used	150-500+ kg/m2-hr
% Solids in Filter Cake Discharge	15-25	45-60

Another apparent advantage of the HDS process is the significant decrease in gypsum scale formed when lime is used with sulfate-based systems. In several HDS installations much of the gypsum formed is encapsulated in the precipitated metal hydroxide sludge.

The major disadvantage of the HDS process as described is that the recycled solids all pass through the thickener. The amount of solids recycled will vary with the particular application but will be between 10 and 30 times the amount of metal hydroxides precipitated from a unit volume of feed. Because thickeners are generally sized on the mass of solids fed, thickener diameters can become very large as the metal concentration of the feed stream is increased. Figure 2 portrays the general relationship between the concentration of metals in the feed stream and the diameter of thickener required for the original HDS process. Clearly the type of metal, the alkalinity source, the amount of recycled solids, and the use or non-use of a flocculant all affect the thickener specifications. The curve in Figure 2 attempts to show in general how the thickener size will increase as metals concentration increases. The basis for the curves is: flow rate = 1.0 m3/min, thickener underflow = 25% solids, a recycle ratio of 20, a 10% lime slurry as precipitant and a sulfate anion (leading to the formation of gypsum.) The curve labeled unflocculated has been calculated with a unit area requirement of 3.4 m2/MTPD; the curve labeled flocculated has been calculated with a unit area requirement of 1.54 m2/MTPD.

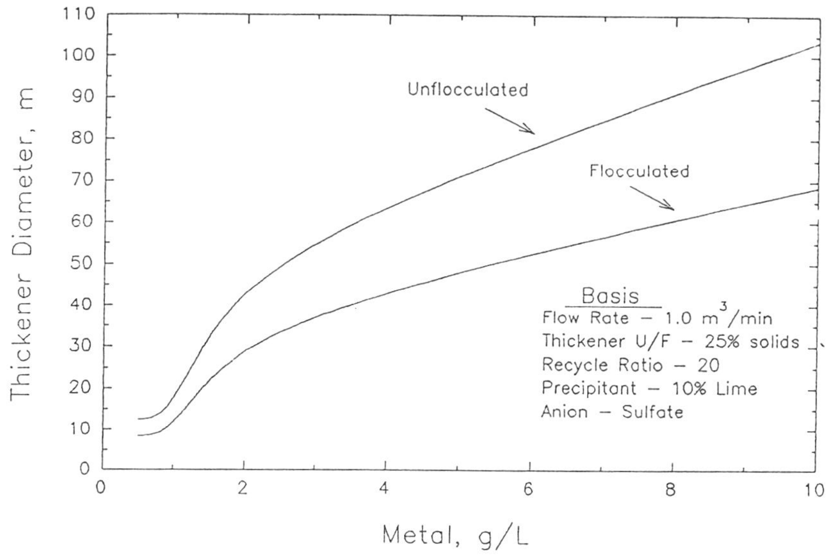

Figure 2. Approximate Thickener Diameters Required for Feed Streams with Different Metal Loadings in the Original HDS Process.

III. DIRECT RECYCLE HDS

The Direct Recycle Option to the HDS process was developed specifically in response to the problem of thickener size. This option, which has recently been granted a patent by the U.S.Patent Office (Aug.13, 1991, #5,039,428), decreases the size of thickener needed dramatically and in some cases even eliminates it altogether.

III.A Process Flow Diagram

The Direct Recycle HDS Process is represented in the flow diagram in Figure 3. That fraction of the solids leaving the precipitation reactor which is required for the recycle is taken off before entering the thickener. Thus the only solids entering the thickener are those equivalent to the amount formed from the metals in the feed stream. The amount of solids processed by the thickener and hence the size is decreased.

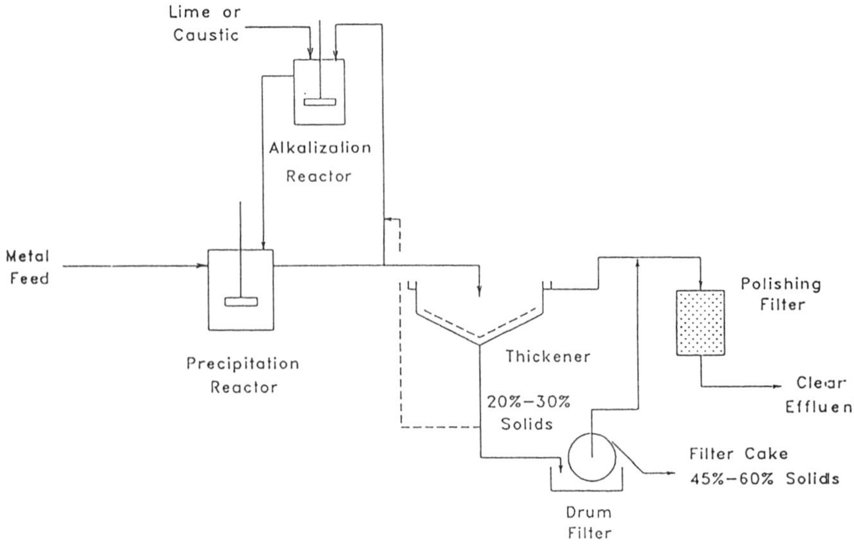

Figure 3. Direct Recycle HDS Process

Figure 4 illustrates the relationship between metals concentration in the feed stream and the thickener diameter. The same basis as used in Figure 2 has been used in Figure 4. For a feed stream containing 10 g/L metals without the use of a flocculant a thickener of about 18 meters in diameter is needed with the Direct Recycle HDS Process as compared to the 104 meter diameter thickener needed in the original HDS Process.

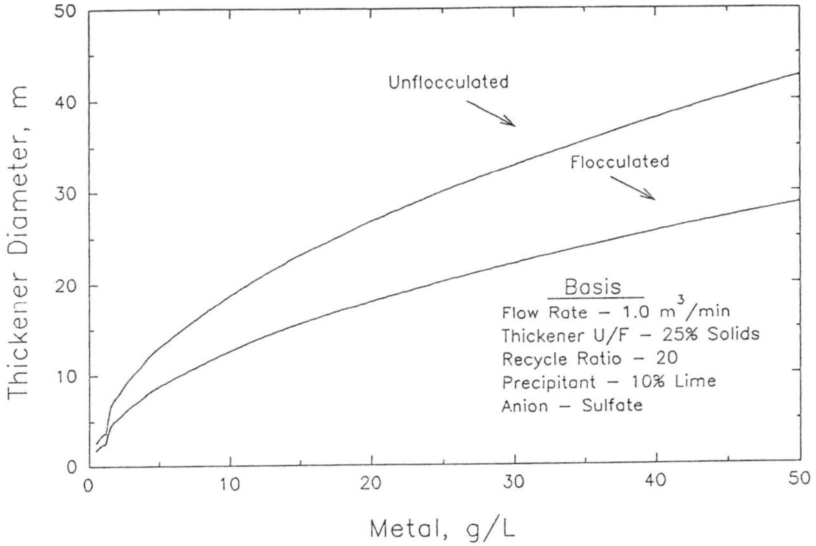

Figure 4. Approximate Thickener Diameter Required for Streams with Different Metal Loadings in Direct Recycle HDS.

When the concentration of metals in the feed stream becomes sufficiently high, the solids content of the effluent slurry becomes large enough to allow direct feeding to a rotary vacuum filter, thus eliminating the thickener altogether. Exactly where this point is depends to a large degree on the specific system. TETRA is currently operating a Direct Recycle Process with a feed stream in the range of 60-70 g/L metals which feeds directly to a vacuum filter. Filter discharge is generally +50% solids.

The Direct Recycle HDS Process also can use hydrocyclones to classify solids in the effluent from the precipitation reactor. Currently pilot tests are being carried out to determine what size cyclones and what configuration are needed. These cyclones would lessen the need for thickeners even in the more dilute feed stream regions.

III.B Disadvantages of Direct Recycle

By recycling directly from the precipitation reactor into the alkalinity reactor, the thickener size required is decreased but the size of the precipitation and alkalization tanks is increased. A detailed engineering analysis must be made to determine the relative costs for reactor tanks and thickeners for a specific system. Generally, however, estimates indicate that the total cost of large reactors required with 100% direct recycle is less than the increased cost of thickeners with no direct recycle.

Fractional Direct Recycle HDS is possible where intermediate amounts of the required recycle solids are obtained from thickener underflow. This allows the determination of an optimum configuration based on cost.

IV. CONCLUSION

The High Density, process in both its original form and in its Direct Recycle modification, provides a method for recovering a variety of metals from industrial streams. The metals are precipitated in a form which is easily filtered and in most instances can be washed free of soluble materials, producing a filter cake with a high percent of clean solids. Disposal of the cake becomes less burdensome. More importantly the washed materials might be acceptable as recycled feed to the plant which produced them in the first place or be sold as feed stock for a different process altogether. The net result will be a major decrease in the amounts of materials disposed of in land-fills or deep wells.

SOLUBILITY OF Ni IN SPENT PICKLING SOLUTIONS

R. G. Reddy, B. Chen and S. Wang
Department of Chemical and
Metallurgical Engineering
University of Nevada-Reno
RENO, NV 89557

G.L. Horter and J.B. Stephenson
U.S. Bureau of Mines,
Rolla Research Center
P.O.Box 280
ROLLA, MO 65401

ABSTRACT

Nickel solubilities in spent pickling solutions (i.e. HF-HNO_3-H_2O solutions) were measured as a function of HF and HNO_3 acid concentrations. The chemical analysis of the industrial pickling solutions and sludges were made. The stable phases observed in the sludges are $FeF_3 \cdot 3H_2O$ and (FeCr) $F_3 \cdot 3H_2O$. Pure Ni, and stainless steel type 316 were used in this study. The results showed that the solubility of Ni increased with increase in concentrations of HF in pickling solution at a fixed concentration of HNO_3. An excellent agreement between the experimental solubility data and the theoretical data was obtained. X-ray diffraction analysis showed that the solid solubility product formed is mainly $NiF_2 \cdot 4H_2O$ compound.

Residues and Effluents - Processing and Environmental Considerations
Edited by Ramana G. Reddy, William P. Imrie and Paul B. Queneau
The Minerals, Metals & Materials Society, 1991

1. INTRODUCTION:

Production of stainless steels requires pickling in aqueous solutions to obtain the desired surface properties. Mixed-acid pickling solutions of two or more acids are used for cleaning the oxide covered stainless steels, which are formed during the annealing process. The most common pickling solutions contain nitric and hydrofluoric acids. In practice, the stainless steels may be left in the pickling solution longer than necessary, causing excessive dissolution of the bulk steel, resulting in losses of several thousand tons of nickel annually(1).The dissolved metals also increase the use of acids in the pickling bath by complexing of precipitating acid salts(2-5). The high cost of waste acid disposal and loss of the valuable metals they contain are important considerations. A study, therefore, was undertaken to investigate the solubility of nickel in HF-HNO_3-H_2O solutions.

2. EXPERIMENTAL

2.1 Composition of industrial pickle solutions

In stainless steel industrial pickling processes, composition and temperatures of pickling solutions, varies significantly from one industry to another. Chemical analysis of 10 samples from the representative companies are shown in Table 1. As can be seen from the table, the composition range for the samples analyzed are within the reported range of industrial pickling solutions, compositions 10-40 weight percent HNO_3 and 2 to 16 weight percent HF(6). Temperatures of pickling solutions are reported to vary from 298-350K. As expected, the concentration of chromium and nickel in solution, as shown in Table 1, increases with an increase in total concentration of acids in the pickling solutions. Figure 1 shows the SEM-EDAX micrograph of the crystalline phases in pickling sludges. The chemical analysis of the phases identified are shown in Table 2. The phases observed are FeF_3 $3H_2O$ and $(FeCr)F_3 \cdot 3H_2O$. these observations are in agreement with the analysis of fluorides based on standard Gibbs free energy change. FeF_3 and CrF_3 are more stable (larger negative free energy) compared to NiF_2, FeF_2 and CrF_2.

Table 1. Chemical analysis of industrial pickling solutions

Industry Sample #	Chemical composition wt. pct.				
	Fe	Cr	Ni	NO_3^-	F^-
1	1.353	0.070	0.300	11.740	1.127
2	8.907	0.627	0.620	10.510	3.119
3	2.667	0.680	0.560	13.690	2.949
4	2.880	0.340	0.780	11.460	3.102
5	4.960	0.786	0.620	13.500	5.080
6	4.710	0.850	0.500	16.000	5.800
7	1.440	0.140	0.260	16.300	1.120
8	3.130	0.440	0.770	12.900	3.440
9	2.710	0.522	0.411	8.250	5.910
10	2.430	0.496	0.291	14.110	4.670
Range	1.35-8.9	0.07-0.85	0.26-0.78	8.25-16.30	1.12-5.91

Table 2. Chemical analysis of industrial pickling sludges

Industry Sample#	Chemical Composition wt. pct.			Comments
	Fe	Cr	Ni	
1	31.0	0.78	0.038	X-ray diffraction and SEM-EDAX analysis. Phases: $FeF_3 \cdot 3H_2O$ and $(FeCr)F_3 \cdot 3H_2O$
2	30.5	0.73	0.041	
3	28.0	1.50	0.057	

2.2 Materials

Stock solutions of 71∓1 wt. pct. HNO_3 and 49∓1 wt. pct. HF certified by ACS specifications are used with no further additions. The alloy (stainless steel type 316) and pure Ni metal are used in the present study.

2.3 PROCEDURE:

Solubility of nickel in HNO_3-HF-H_2O solutions was determined at a temperature of 298K. All experiments were carried out with a minimum equilibration period of 7 days. The experimental method used was described in detail elsewhere (7) and a brief description is given here. A known amount of metal was added to the 500 ml teflon bottles containing a known concentration of HNO_3 and HF acids. The concentration of HNO_3 and HF used were in the range of 1 to 8 molality. After equilibration, the solution samples were analyzed for Ni^{++}, NO_3^-, F^- using AA and IC. The crystalline phases were analyzed using SEM and X-ray diffraction.

3. RESULTS AND DISCUSSION:

3.1 Pure Nickel:

The solubility of nickel in solutions was determined as a function of concentration of HF and HNO_3. The results are shown in Figure 2. As seen from the figure, at a fixed concentration of HF, the solubility of nickel increased with increase in concentration of HNO_3 in solution. For example, at a fixed concentration of HF (2 molality), the solubility of nickel increased from 3.7 to 16.5 weight percent as the concentration of HNO_3 in solution increased from 1 to 8 molality respectively. For a given fixed HNO_3 concentration, no significant increase in solubility of nickel in solution was observed with increase in HF concentration from 1 to 8 molality.

3.2 Alloy

The solubility of nickel from 316 stainless steel in HNO_3-HF-H_2O solutions is shown in Figure 3. The chemical composition of 316 SS in wt. pct is: 16-18 Cr, 10-14 Ni, 2.0 Mn, 1.0 Si, 0.08 C, 0.045 P, 0.03 S and the balance is Fe. As seen from the figure, the solubility of Ni in solution increased with increase in concentration of HF and also increased with increase in HNO_3. For a fixed concentration of 4 molality of HNO_3, the solubility of nickel increased from 0.18 to 0.55 weight percent as the concentration of HF in solution increased from 1 to 8 molality. This is in very good agreement with the solubility of Ni observed in the industrial pickling solutions. As expected, the solubility of pure nickel in solutions as compared to 316 stainless steel, is about an order of magnitude higher. It is clear that the solubility of nickel is dependant on the elements such as Fe, Cr which are present in the alloy. The solubility data are compared in Figure 4. The data indicates that the presence of elements Fe, Cr and other cations reduce the solubility of nickel.

3.3 Theoretical consideration

X-ray diffraction analysis of the solid substance formed at higher acid concentrations in solutions are determined to be metallic fluorides. These metallic fluorides are mainly hydrated nickel fluoride crystals of NiF_2 $4H_2O$. The solubility of nickel fluoride at a specific temperature and pH can be calculated using solubility product or standard Gibbs free energies of formation. The solubility of solid in equilibrium with solution containing appropriate ions can be written in the general form as:

$$AB_{z\ (solid)} + zH^{+}_{(aq)} = A^{Z+}_{(aq)} + zHB_{(aq)} \qquad (1)$$

The standard Gibbs free energy of reaction (1) can be written in terms of equilibrium constant as:

$$\Delta G_f^o = -RT \ln K \qquad (2)$$

and equilibrium constant (K) is given as:

$$K = [HB]^z [A^{Z+}] / [AB_z] [H^+]^Z = \exp (-\Delta G_f^o / RT) \qquad (3)$$

When the system is in equilibrium (i.e. several solids and ions coexist), and if the standard Gibbs free energies of all the species are known, the solubility relationship can be deduced from the mass balance of the system containing nickel ions, the probable fluoride complexes reaction sequence can be written as follows:

$$HF = H^+ + F^- \quad (4)$$
$$HF + F^- = HF_2^- \quad (5)$$
$$Ni^{2+} + 2F^- = 2NiF^+ \quad (6)$$
$$NiF^+ + F^- = NiF_2 \quad (7)$$

and the probable nitrate complexes are

$$HNO_3 = H^+ + NO_3^- \quad (8)$$
$$Ni^{2+} + NO_3^- = NiNO_3^+ \quad (9)$$

The thermodynamics data for these species are presented in Table 4. The species considered are mainly aqueous fluoride complexes. The solubility of nickel in solutions was estimated using Pitzer's (8,9) equation along with the interaction parameters for complex species of fluoride reported by Fernando (5) and Wilson and Taube (10). The material and mass balance equations were established and solved by an iterative method to yield the equilibrium concentration of the species. The calculated concentrations of nickel ions for pure nickel and 316 SS alloys along with the experimental data at a fixed 4 molality of HNO_3 are presented in Table 4 and also in Figure 4. As can be seen, an excellent agreement between the calculated and the experimentally

Table 3. Thermodynamic data at 298K

Species	ΔG_f^0 (K cal/mol)	Reference
F^-	-66.64	(11)
HF	-70.95	(11)
HF_2^-	-138.18	(11)
Ni^{+2}	-10.9	(11)
NiF_2	-142.9	(11)
NO_3^-	-26.61	(11)
HNO_3	-26.61	(12)

Table 4. Calculated and Measured Solubility of Nickel at 298K

Species	pH	Ni^{2+}	NiF^-	NiF_2	Ni (wt%) cal.	Ni (wt%) Expt.
316	1.90	0.2400	0.0000	$1.7x10^{-8}$	0.24	0.25
316	1.60	0.3400	0.0000	$2.8x10^{-8}$	0.34	0.33
316	1.49	0.5000	0.0000	$5.7x10^{-8}$	0.50	0.51
316	1.45	0.5500	0.0000	$7.6x10^{-8}$	0.55	0.55
Nickel	5.40	10.0000	0.0001	0.0005	10.00	10.00
Nickel	5.35	11.0000	0.0001	0.0006	11.00	11.10
Nickel	3.80	10.8000	0.0001	0.0007	10.80	11.00
Nickel	3.65	10.4000	0.0001	0.0008	10.40	10.60

measured solubility data is obtained. The solubility of nickel, as seen from the figure increased with an increase in concentration of HF. The calculated results confirm that at higher concentrations of HF, nickel fluoride crystals are precipitated. The measured experimental data, however , showed that the precipitates are formed at 4 molality of HF for 316 SS and at 6 molality for pure nickel in pickling solutions.

4. CONCLUSIONS:

The solubilities of nickel in pickling solutions (i.e. HNO_3-HF-H_2O solutions) were measured as a function of concentration of HNO_3 and HF. The solubility experiments were carried out at a concentration of HNO_3 and HF in the range of 1 to 8 molality at 298K. From the present study the following conclusions can be made:

1. The solubility of nickel increased with an increase in concentration of HNO_3 and also with an increase in concentration of HF in the concentration range of 1 to 8 molality of each HNO_3 or HF acids.

2. SEM and X-ray diffraction analysis showed that the solid solubility product formed is mainly NiF_2 $4H_2O$ compound in the case of nickel. The stable crystalline phases observed in the industrial stainless steel sludges are $FeF_3 \cdot 3H_2O$ and $(FeCr)F_3 \cdot 3H_2O$.

3. An excellent agreement between the experimentally measured solubility of nickel data and the calculated data is obtained.

5. ACKNOWLEDGMENTS

The authors wish to acknowledge the financial support of the United States Bureau of Mines under the Mining and Minerals Research Institute Generic Center program (grant number G1125132-Mineral Industry Waste Treatment and Recovery Generic Center). Special thanks are extended to Mr. Paul Lechler and Mr. M. Desilets, Nevada Bureau of Mines and Geology for their help in the chemical analysis.

6. REFERENCES

1. J. B. Stephenson, J. C. Hogan and R. S Kaplan, Environ. Prog. , 1984, vol. 3 (1), pp. 50

2. D. P. Whittle, and G. C. Wood, J. Electrochem. Soc., 1967, vol. 114, pp. 986-993

3. B. Vicentini and G. Bombara, Electrochem. Metall., 1968, vol. 3, pp. 313

4. B. S. Covino, Jr., J. V. Scalera, T. J. Driscoll and J. P. Carter, Metall. Trans. A, 1986, Vol. 17, pp. 5-9

5. L. A. Fernando, Metall. Trans. B, 1990, Vol. 21B, pp 5-9

6. H. S. Cole, Allied chemical corp., 1974, ICP-1036, pp.42

7. S. Wang, M.S. thesis, University of Nevada, Reno

8. K. S. Pitzer, Activity coefficients of electrolyte, ed. by R. M. Pytkowicz, CRC press, Boca Raton, FL, 1979, p. 157

9. M. A. Hughes and H. Sunghou, J. Chem. Eng. Data, 1986, vol. 31, pp. 4

10. A. S. Wilson and H. Taube, J. Amer. Chem. Soc., 1952, vol. 74, pp 3509-12

11. F. T. Wall: Chmical Thermodynamics, 3rd ed., by W. H. Freeman and Company, 1974, pp. 463-478

12. J. F. Zemaitis, Jr., D. M. Clark, M. Rafal and N. C. Serivner: Handbook of Aqueous Electrolyte Thermodynamics, AIChE, New York, NY, 1986, p. 726

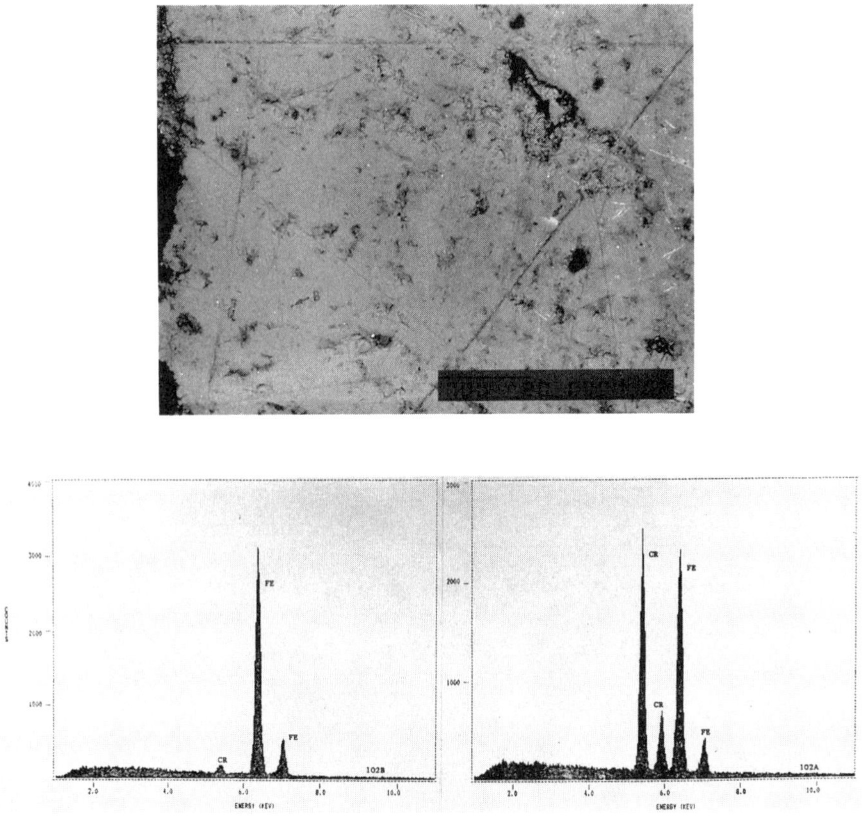

Figure 1. SEM-EDAX diffraction patterns and micrograph of industrial pickling sludges.

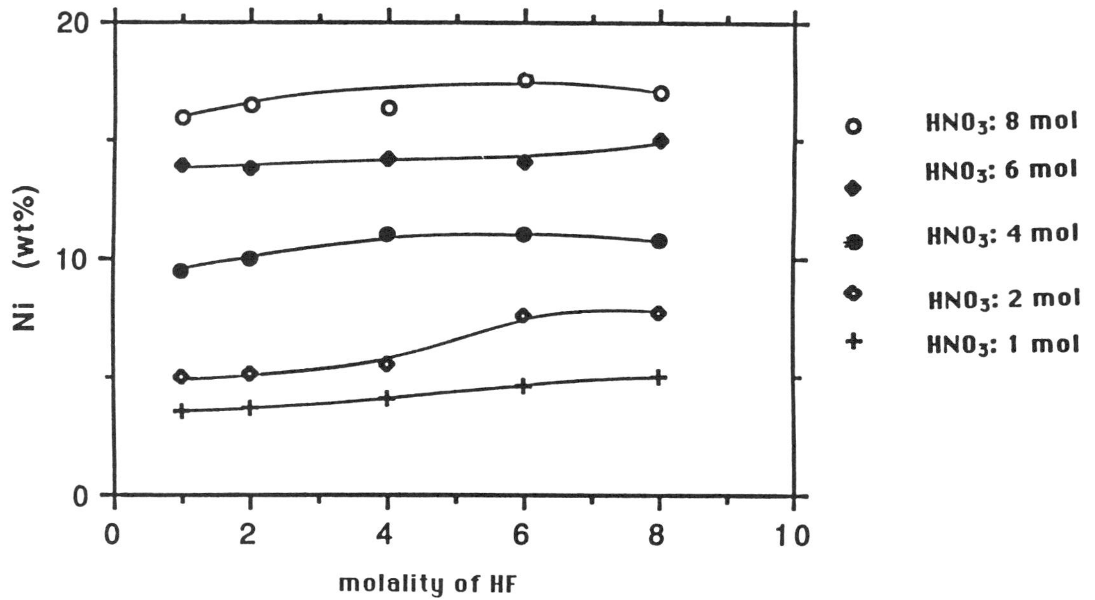

Figure 2: Solubility of pure Ni in HNO_3-HF-H_2O solutions at 298K

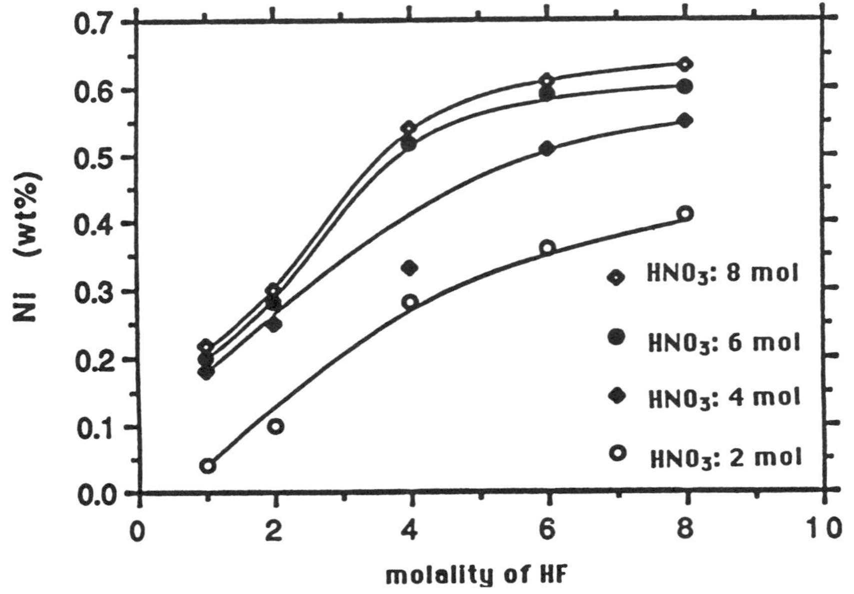

Figure 3: Solubility of 316 ss in HNO_3-HF-H_2O solutions at 298K

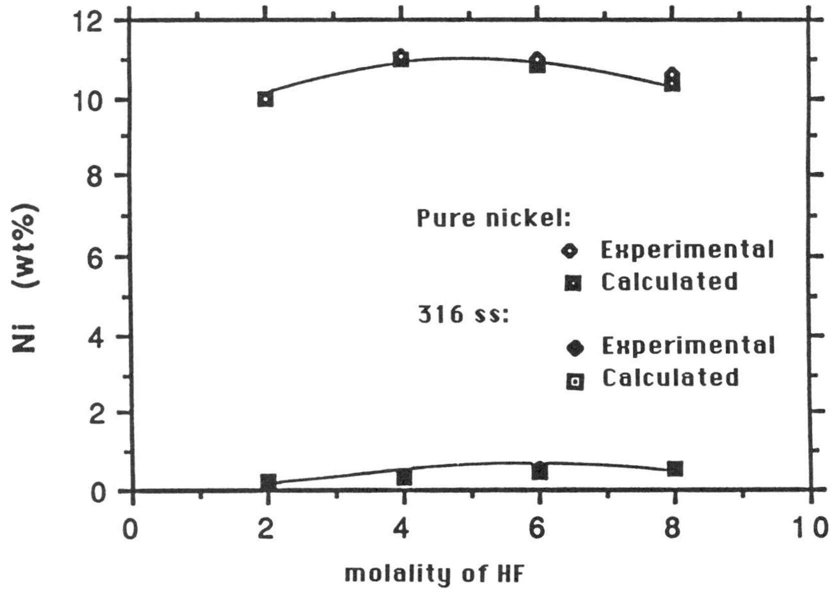

Figure 4: Calculated and measured solubility of Nickel in HNO_3-HF-H_2O solutions at 298K and molality of HNO_3=4 mol.

RECOVERY OF ACID VALUES FROM METALLURGICAL ACID PLANT BLOWDOWN

A.A. Trickett, C.M. Evans, S. Kozary

Chemetics International Company Ltd.
200 Town Centre Boulevard
Markham, Ontario L3R 8H8, Canada

ABSTRACT

In the treatment of gases from pyrometallurgical processes prior to acid manufacture, volatile components, sulphur trioxide and particulates are removed in wet gas cleaning systems. In the process a weak acid stream is generated, typically less than 10% H_2SO_4, contaminated with volatiles such as arsenic, fluoride and chloride, plus metallic particulates.

This effluent stream has, in the past, been disposed of in a variety of different ways. As a result of changing environmental legislation, smelters are having to review alternative solutions for treating this effluent stream. This paper will describe a process developed by Chemetics which treats this effluent stream, generating a small dry metal sulphate stream plus a clean acid stream which is suitable for recycling to the smelter acid plant.

Residues and Effluents - Processing and Environmental Considerations
Edited by Ramana G. Reddy, William P. Imrie and Paul B. Queneau
The Minerals, Metals & Materials Society, 1991

INTRODUCTION

The performance of a metallurgical acid plant is dependent on the performance of the gas cleaning section. If the gas cleaning system is effective in removing impurities from the incoming gas, then the performance of the contact section will be enhanced. Although the design of wet gas cleaning systems may differ widely, depending on the level and nature of contaminants in the incoming gas streams, they all have the following requirements and features in common:

- Solid contaminants must be removed, giving a final gas quality of 1-2 mg/Nm^3.
- Sulphur trioxide content of the gas must be reduced to the level of 15-25 mg/Nm^3.
- Fluorides and chlorides must be removed to prevent damage to downstream tower brickwork and converter catalyst.
- The gas must be cooled to a temperature of approximately 35-40°C to satisfy the acid plant water balance.
- As a result of sulphur trioxide in the incoming gas stream being absorbed in the gas scrubbing process, the liquor in the scrubber loop will become acidic, typically 1-10% H_2SO_4.
- The gas stream will initially be adiabatically saturated, then cooled, and will exit the gas cleaning section in a saturated state.

The scrubbing liquor is recirculated resulting in an accumulation of the separated impurities. As a result of this, it is necessary to purge scrubbing liquor from the system in order to maintain performance. This liquor is normally referred to as 'blowdown acid' and is a weak acid stream contaminated with halides and heavy metals. In the past, this stream has been treated and disposed of in a variety of ways, none of which have involved complete recycling. Smelters are increasingly being forced to reduce or eliminate such effluents by more stringent environmental legislation. The purpose of this paper is to present a process developed by Chemetics for the treatment of blowdown acid based on the vapour phase technology originally developed for the treatment of titanium dioxide waste acid.

Process Alternatives

Blowdown acids may vary in acid strength and level and type of contaminants but all such acids contain halides and heavy metals. Analyses of two acids are shown in Table I, one from a copper smelter and the other a combined stream from a lead/zinc complex.

In the past, blowdown acid streams have typically been neutralized but it is becoming apparent that neutralization followed by storage of the resultant by-products is not inexpensive and also has the potential to create a significant future liability exposure. Frequently the neutralization approach has involved some prior attempt at volume reduction by increasing the strength of the acid effluent. This can be achieved either by modifications to the gas cleaning system or by vacuum concentration of the weak blowdown acid to the 40-50% range.

TABLE I

TYPICAL BLOWDOWN ACID ANALYSES

Copper Smelter Blowdown Acid	Component	Lead/Zinc Smelter Blowdown Acid
20.00%	H_2SO_4	13.90%
77.92%	H_2O	78.94%
1200 ppm	Fe	--
40 ppm	Al	--
180 ppm	Cr	--
250 ppm	Ni	--
2100 ppm	Cu	--
1500 ppm	Zn	4130 ppm
140 ppm	As	5160 ppm
400 ppm	Cd	--
70 ppm	Bi	--
--	Hg	4130 ppm
--	Se	80 ppm
--	Pb	520 ppm
--	Te	20 ppm
1800 ppm	F^-	5680 ppm
410 ppm	Cl^-	5.16%
2000 ppm	SO_2	--
1000 ppm	Other	--

A refinement of the vacuum concentration approach is to carry out several consecutive stages of vacuum concentration with intermediate precipitation/filtration steps. Coupled with chemical addition/precipitation as appropriate to reduce arsenic and mercury levels, this can be an effective technique for treating blowdown acid and producing a relatively strong, clean product. This product will still contain impurities at the solubility level but, depending on local circumstances, may be suitable for internal use or external sale. This technology is offered by Chemetics.

In 1990 Chemetics were approached by a client in the U.S.A. who wished to treat blowdown acid from a copper smelter to produce a product which was equivalent to commercial grade acid or at least suitable for blending with the main acid plant production. Upon discussions between the client and Chemetics, it was recognized that the 'vapour phase' acid recovery process originally developed by Chemetics for the recovery of commercial grade sulphuric acid from titanium dioxide waste acid could potentially be applied to blowdown effluent acid. Therefore, an extensive bench and pilot scale development program was performed in order to confirm that the base process developed for the treatment of titanium dioxide waste acid could be modified to process blowdown acid.

Process Background

In 1982/83 Chemetics was contracted by Gulf and Western to critique a study on waste acid recovery processes for titanium dioxide waste acid. As a result of the study, Chemetics were awarded a process patent[1] and a paper on the conceptual process was presented at Sulphur '85[2]. The proposed recovery scheme involved total evaporation of the waste acid with separation of the solid metal sulphates from the vapour phase and subsequent recovery of 93-96% product acid by a novel partial condensation. The heat released by the partial condensation is used to preconcentrate the weak feed acid. A generalized flowsheet of the process is shown in Figure 1.

In 1987 this vapour phase technology was demonstrated on an industrial scale when a 110 MTPD demonstration plant was constructed at Sorel, Quebec by Chemetics on behalf of a consortium comprising Tioxide Canada Inc., NL Chem Canada Inc. (now Kronos Canada Inc.), Q.I.T. Fer et Titane Inc. and Chemetics. The plant, as originally erected, is shown in Figure 2.

Demonstration plant operation commenced in the fall of 1987 and continued until early 1990. The construction and operation of the plant was a major venture of several years' duration and with a total expenditure by the consortium members in excess of $15 million. The operation of the plant confirmed the validity of the original process concept but, as would be expected, was a major learning experience which resulted in a significant amount of process optimization, particularly with regard to materials of construction and certain specific items of equipment.

This experience is covered in detail in a paper by G. Lebel (Tioxide Canada Inc.)[3] which was presented at Sulphur '90 and reviews the entire operation of the demonstration plant, including the major technical challenges which were presented and the means by which they were overcome.

The major problems encountered and their solutions are fully discussed in the paper. These were acid mist emissions, gas-solids separation, condensation train materials and spray dryer wall wetting. It is finally concluded that all of the original objectives of the pilot program had been achieved, namely:

- To confirm the production of greater than 93% acid suitable for recycle to the pigment process.
- Confirmation of the overall process energy requirements.
- Generation of solids samples for:
    - Solids Handling Trials
    - Roasting Trials
    - Disposal Tests on Roasted Solids
- Confirmation of the quality of gaseous effluents from the process.
- Acquisition of chemical engineering data for the design and construction of a full scale plant.
- Acquisition of data to confirm the operability of a full scale plant.

The Chemetics process, as finally developed, for the recovery of titanium dioxide waste acid is as depicted in the flowsheet shown in Figure 1. Weak acid effluent from the pigment process can first be preheated by indirect heat exchange with the product acid and is then introduced into the economizer tower, where the acid is preconcentrated. The preconcentration (water removal) is achieved by indirect heat exchange, recovering the heat released during the

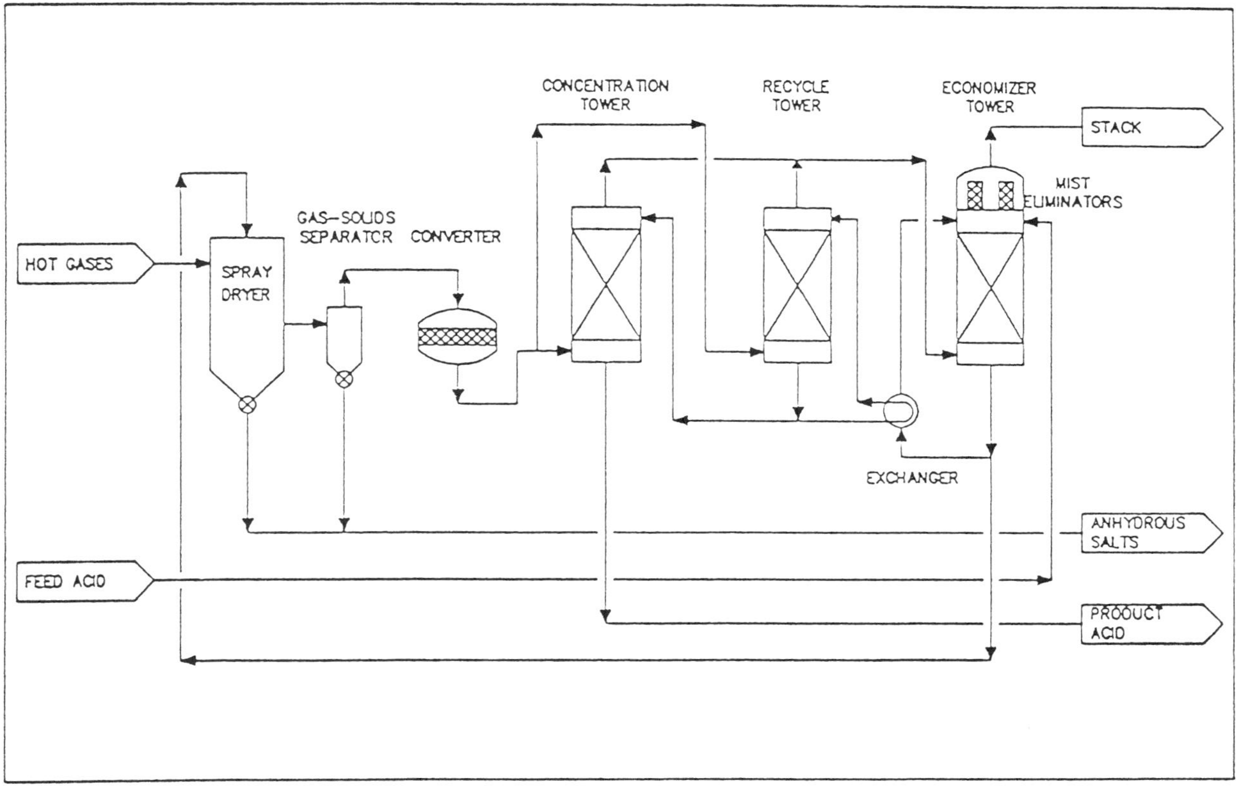

FIG. 1: GENERALIZED FLOWSHEET OF CHEMETICS TiO_2 WASTE ACID PROCESS

FIG. 2: TiO_2 WASTE ACID RECOVERY PILOT PLANT

partial condensation of the product acid, as well as by direct heat exchange, through recovery of sensible heat from the gases passing through the spray dryer and partial condensation systems. The preconcentrated acid is then pumped to a spray dryer where the liquid is atomized and contacted with high temperature gas at 1000°C. In the spray dryer the water and free sulphuric acid are vaporized and the solid metal sulphates are then separated from the vapours in a sintered metal filter.

The essentially solid-free gases exiting the sintered metal filter then enter a converter containing vanadium pentoxide catalyst to convert any sulphur dioxide, formed in the spray dryer by ferrous sulphate oxidation, to sulphur trioxide. From the converter the gases are split, part going to the concentration tower and the remainder to the recycle tower. Acid of greater than 60% strength is recovered from the recycle tower and fed to the concentration tower where it further contacts gases from the spray dryer. As a result, acid is condensed from the gas and water evaporated from the liquid producing acid of 96% strength. The off-gases from both towers are fed to the economizer tower for further heat recovery and then pass to a stack via mist eliminators.

The successful development of the process as described was a major exercise involving significant effort and expenditure over a period of years. The process concept was proven in every respect and all engineering and materials issues studied in depth. A complete front-end design package for a commercial scale plant for Tioxide was completed. This project was eventually cancelled.

The challenge for Chemetics was then to improve and adapt the technology, proven for the treatment of titanium dioxide waste acid, to process blowdown acid.

Application of Vapour Phase Technology to Blowdown Acid Recovery

As previously mentioned, the acid strength of the acid plant blowdown stream is typically less than 10% H_2SO_4. For the vapour phase recovery process described in the previous section, the feed strength needs to be in the region of 20% H_2SO_4 for optimum energy recovery. In some cases, it is possible to achieve a blowdown acid strength of 20% by modifying the operating conditions in the gas cleaning system. In other cases, water removal may be necessary prior to feeding the acid to the vapour phase plant. This can be achieved using standard vacuum evaporation technology to preconcentrate the acid. For the particular case of blowdown acids, which normally contain fluorides, this can be achieved in a single evaporation stage using plastic as the prime material of construction, rather than glass-lined steel which is the norm for non-fluoride containing acids. A flow diagram of a typical Chemetics evaporation system is shown in Figure 3. Chemetics has completed pilot plant testing programs on smelter blowdown acid on the liquid phase vacuum concentration step in addition to that performed on the vapour phase process.

In order to gain a full appreciation of the similarities between the spent acid from a pigment plant and the blowdown stream from the gas cleaning section of a metallurgical acid plant, a comparison of the two acid streams is presented in Table II.

In general the streams are remarkably similar, the major differences being:

- There are significantly less inorganic sulphates present in the blowdown acid than in the TiO_2 waste acid.

- There are small amounts of potentially volatile metals (As/Cd/Bi) in the blowdown acid which are not present in the TiO_2 waste acid.

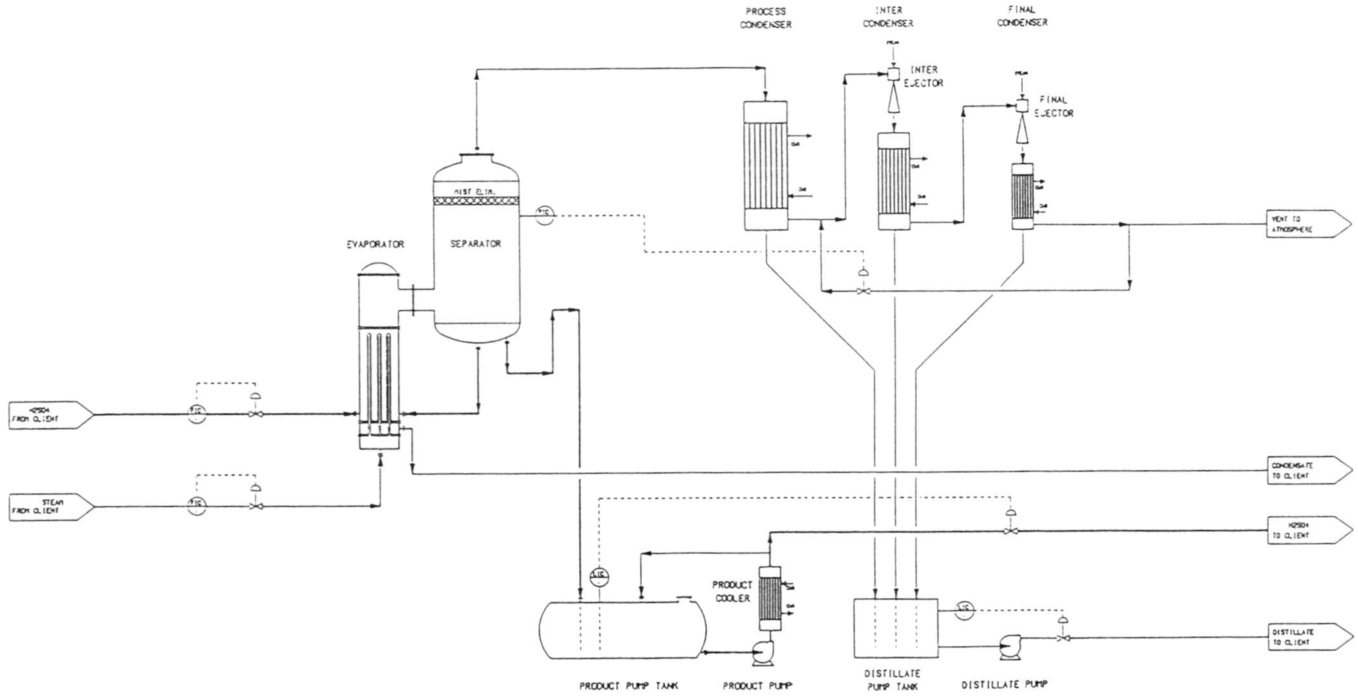

FIG. 3: TYPICAL SINGLE STAGE SULPHURIC ACID VACUUM CONCENTRATOR

- There are fluorides present in the blowdown acid but not in the TiO_2 acid. However, there are significantly less chlorides.

- There is SO_2 present in the blowdown acid but not in the TiO_2 acid.

TABLE II

ACID QUALITY COMPARISON

TiO_2 Waste Acid	Component	Blowdown Acid
21.44%	H_2SO_4	20.00%
65.52%	H_2O	77.92%
14,400 ppm	Fe	1,200 ppm
6,550 ppm	Mg	--
2,630 ppm	Al	40 ppm
590 ppm	V	--
210 ppm	Cr	180 ppm
250 ppm	Mn	--
2,500 ppm	Ti	--
80 ppm	Ni	2,100 ppm
--	Cu	2,100 ppm
--	Zn	1,500 ppm
--	As	140 ppm
--	Cd	400 ppm
-	Bi	70 ppm
--	SO_2	2,000 ppm
3,600 ppm	Cl^-	410 ppm
--	F^-	1,800 ppm
2,000 ppm	Other	1,000 ppm

The SO_2 can be readily removed by stripping prior to feeding the acid to the vapour phase plant. However, the presence of volatile metals and fluorides are potential problems for the following reasons:

- Volatile metals may volatilize in the spray dryer, pass through the sintered metal filter and appear in the product acid.

- Unless fluorides can be readily removed from the feed acid then there are two possible concerns:

  - Fluorides may report to the product acid jeopardizing product quality.

- The acid towers in the TiO_2 waste acid process were of brick-lined construction with ceramic packing. Both of these materials would be attacked by fluorides.

As a result of these concerns, an extensive laboratory and pilot scale testing program was performed, in co-operation with the client, with the objectives of addressing the fate of volatile elements and developing a means of removing fluorides from the acid.

Pilot Plant Development of Vapour Phase Process for Blowdown Acid Recovery

Chemetics maintains laboratory and pilot plant facilities in Mississauga, Ontario and McMasterville, Quebec. Recognizing the potential importance of the vapour phase technology, a decision had been made in 1989 to construct a small pilot vapour phase plant at the McMasterville facility (Figure 4). The pilot plant includes a spray dryer, sintered metal filter (Figure 5) and acid condensation unit and was extensively used during the blowdown acid testing.

The major objectives of the development program were as follows:

- To develop a process by which fluorides could be removed from the acid.
- To confirm data from titanium dioxide waste acid processing on the removal of chlorides from the acid.
- To confirm that the blowdown acid could be spray dried to give a dry, free-flowing solids stream exit the sintered metal filter.
- To obtain solids samples from the sintered metal filter for evaluation by the client.
- To confirm the distribution of potentially volatile metal components in the spray dryer/sintered metal filter/acid absorber system.

Removal of Halides

It was known from the operation of the titanium dioxide waste acid demonstration plant that chlorides were readily stripped from the feed acid during the preconcentration step in the economizer tower. Laboratory and pilot tests confirmed that this was also the case for the blowdown acid. However, only partial stripping of the fluorides was obtained under the conditions prevailing in the economizer tower.

After bench scale reaction tests, it was discovered that pretreatment of the feed acid with silica prior to feeding to the economizer tower resulted in formation of the volatile species SiF_4, the majority of which can be stripped in the economizer tower. This was confirmed in subsequent pilot plant testing.

Based on this work, the full scale design concept for a vapour phase plant for recovering blowdown acid includes a silica reaction stage prior to the economizer tower. To guard against upset/breakthrough conditions, the materials of the relatively low temperature economizer tower have been changed to appropriate plastics rather than the brick and ceramic used in the titanium dioxide waste acid case. In addition, the materials of construction of the recycle tower have been upgraded to include a carbon brick lining.

Gas-Solids Filtration

Experience with titanium dioxide waste acid has proven that >99.99% removal efficiency of all metal sulphate particles was achieved in the sintered metal

FIG. 4: CHEMETICS' VAPOUR PHASE PILOT PLANT

FIG. 5: PILOT PLANT SINTERED METAL FILTER

filter. However, the blowdown acid contained the potentially volatile elements arsenic, cadmium and bismuth which were not present in the pigment waste acid (see Table II). There was a concern that these metals may be sufficiently volatile to pass through the sintered metal filter and report to the product acid thus jeopardizing the acid quality. Therefore, several pilot plant tests were carried out in which blowdown acid was spray dried and a detailed mass balance performed. The results from these trials were as follows:

- The blowdown acid can be successfully spray dried with the production of a dry, free-flowing solid.
- The heavy metal contaminants in the feed acid report almost exclusively to the sintered metal filter solids.
- Of the potentially volatile components, essentially all of the cadmium and bismuth reported to the sintered metal filter solids and none to the acid condensation system. On average, 90% of the arsenic reported to the solids, the remainder passing through to the acid product. However, there were strong indications that close control of spray dryer conditions could result in an increase in arsenic capture up to 99%.

Implications for Full Scale Design

The client for this particular acid recovery case wished to obtain a product acid of around 70% strength for use as 'dilution water' in the main acid plant rather than direct production of 96% acid from the recovery plant. In order not to jeopardize the quality of the acid plant product, the analytical criteria established for the 70% recovered acid are as shown in Table III below:

TABLE III

CLIENT QUALITY REQUIREMENTS FOR RECOVERED ACID

Strength	=	70% H_2SO_4
Cl^-	=	< 10 ppm
F^-	=	< 120 ppm
Me^{++}	=	< 35 ppm
As	=	< 20 ppm

The results of the pilot plant tests had indicated that this quality could readily be achieved with the possible exception of arsenic content. However, additional laboratory testing had shown that, if necessary, further reduction in arsenic content can be achieved by chemical addition techniques.

The flowsheet developed as a result of the pilot testing is shown in Figure 6. It is essentially the same as that described previously for the pigment waste acid case and shown in Figure 1 with the exception of the elimination of the concentration tower and the inclusion of silica addition facilities. The flowsheet could readily be modified to include a concentration tower, thus giving the ability to produce 96% product acid of essentially commercial grade.

Effluents

Two effluents are produced from the process; one a dry, free-flowing stream of

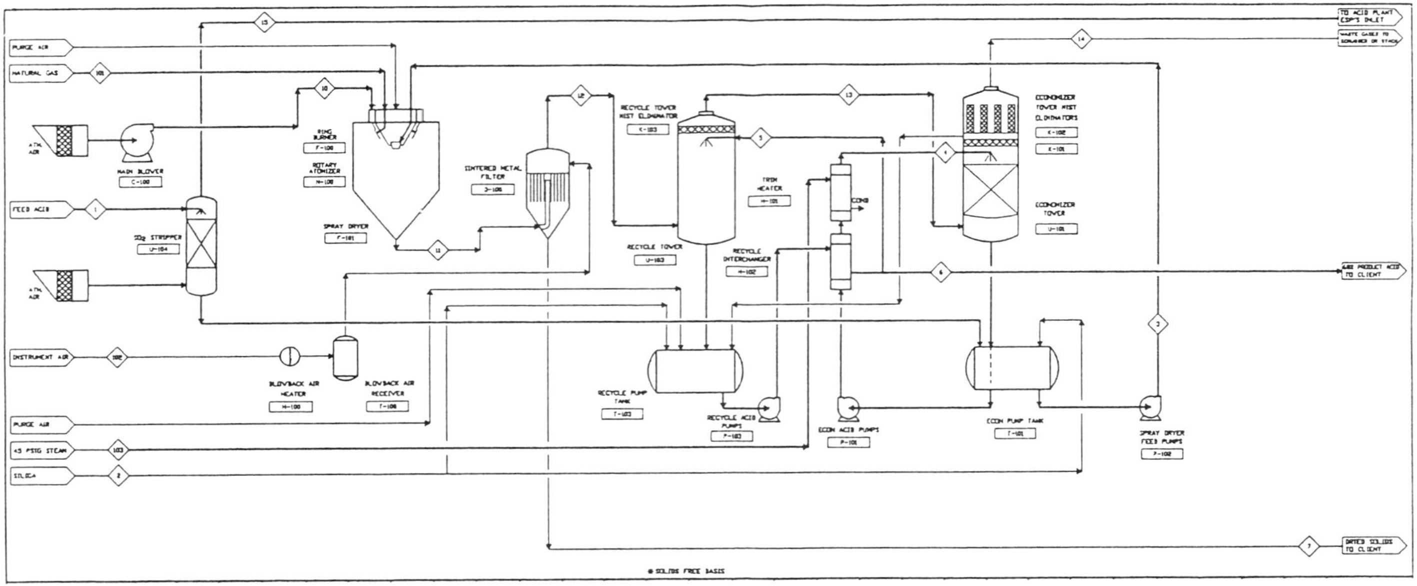

FIG. 6: VAPOUR PHASE BLOWDOWN ACID RECOVERY PROCESS

metal sulphates, and the other a gaseous stream containing fluorides and chlorides. Treatment of these streams is very site specific and it is not appropriate to discuss them except in general terms.

The solids stream is a small, easily handled stream with the potential for further recovery of valuable metals. The gaseous stream may be suitable for discharge to the atmosphere depending on the local regulations. However, it is more likely that scrubbing would be required which, depending on the exact system used, would most likely produce a solid fluoride by-product for possible return to the smelter and a chloride containing aqueous effluent stream.

Economics

Some simple economics of a generic blowdown acid recovery process based on vacuum preconcentration and vapour phase technology, are presented in Table IV. This example is based on a smelter with an acid plant of 1500 STPD of 100% H_2SO_4 and an inlet SO_3 concentration of 0.3%. The gas cleaning circuit is operating at a strength of 10% and the product acid from the vapour phase unit is 68% strength, and is recycled to the acid absorber pump tank as dilution water. Vacuum concentration is used to preconcentrate the feed acid to 20% prior to feeding to the vapour phase plant.

TABLE IV

BLOWDOWN TREATMENT PLANT ECONOMICS

DESIGN BASIS:

Vacuum Concentrator:

Feed Strength	:	10% H_2SO_4
Feed Rate	:	11,900 lbs/hr.
Product Strength	:	20% H_2SO_4
Steam Requirement	:	7,700 lbs/hr @ 150 psig

Vapour Phase Unit:

Feed Rate	:	5,950 lbs/hr.
Product Strength	:	68% H_2SO_4
Natural Gas Usage	:	6.7 GJ/hr.

OPERATING COSTS:

		Cdn $/Yr. VPOD + SAC
Labour	:	$310,000
Steam	:	$305,000
Electricity	:	$100,000
Natural Gas	:	$280,000

ACID CREDIT:

100% H_2SO_4 Basis	:	5000 STPY

CAPITAL COSTS:

Concentrator:	Single Stage SAC:	Cdn. < $1.0 MM
Vapour Phase Unit:		Cdn. $4.5 MM

Conclusion

As a result of increasing environmental legislation, smelters are having to review their approach to treating weak acid effluent streams from wet gas cleaning systems and move towards a zero discharge approach. A number of alternative process options exist for treating such streams.

The preferred process choice for each smelter will be site specific depending on local regulations, the quality of the weak acid effluent stream and the options available for sale or internal use of the recovered acid stream.

Chemetics have further developed the vapour phase technology originally proven for the recovery of titanium dioxide waste acid to process gas cleaning blowdown effluent to give a product suitable either for direct sale or for use as dilution water in the smelter acid plant. This technology can be used on a stand alone basis or in conjunction with vacuum preconcentration as required.

REFERENCES

1. U.S. Patent 4,547,353

2. I.M. Smith, H.C. Peterson and G.M. Cameron Sulphur Value Recovery from TiO_2 Waste Acid (Sulphur '85).

3. G. Lebel, Pilot Plant Development of the Chemetics TiO_2 Waste Acid Recovery Process (Sulphur '90).

4. A.A. Trickett, Acid Plant Design and Operations (Canadian Institute of Mining and Metallurgy, August 1991).

REMOVAL OF ARSENIC FROM WASHING ACID

BY THE SACHTLEBEN-LURGI PROCESS

Gunther Klamp, Dieter Wanner

Lurgi Energie und Umwelttechnik GmbH
Lurgi Allee 5
D-6000 Frankfurt am Main/Germany

Abstract

Disposal of arsenic compounds from washing acid in metallurgical plants is becoming a severe problem when using lime neutralization. A further problem is the disposal of relatively large quantities of gypsum which is produced during this neutralization step and which is also contaminated with other toxic metal compounds. The Sachtleben-Lurgi process removes arsenic and other metals from washing acid in the form of insoluble sulphides. The purified acid contains less than 0.5 ppm of arsenic and can be used for standard sulphuric acid applications. The process is demonstrated commercially at Sachtleben's acid plant since more than five years.

Residues and Effluents - Processing and Environmental Considerations
Edited by Ramana G. Reddy, William P. Imrie and Paul B. Queneau
The Minerals, Metals & Materials Society, 1991

Description of the arsenic problem

Copper, zinc, and lead ores contain arsenic in appreciable quantities. During smelting operation, highly volatile arsenic components arrive at the SO_2 gas treatment section and contaminate the washing acid as arsenic oxide (As_2O_3).

Neutralization of washing acid with lime and disposal of the relatively large quantities of gypsum containing calcium arsenate is now widely accepted as unfeasible because of conversion of calcium arsenate to calcium carbonate under the influence of atmospheric carbon dioxide, with subsequent redissolution and mobilization of arsenate.[1]

Increasing sensitivity of the public with regard to industrial activities cause environmental pressure on legislation in most industrialized parts of the world. Actual emissions in Germany for the non-ferrous industries are limited to 1 mg/Nm^3 for gaseous components and to 0.1 mg/l for water when fed to public sewers.

Sachtleben-Lurgi-Process

The Sachtleben-Lurgi-Process removes arsenic and other metals from washing acid in the form of insoluble sulphides and allows the purified acid to be used for standard sulphuric acid application. The process is patented in major countries and proven for more than five years in a commercial size plant.

Chemistry

The extremely low solubility of metal sulphides is utilized to precipitate metals from aqueous solutions. The low pH-value of washing acid allows for the precipitation of arsenic (As_2S_3), antimony (Sb_2S_3), tin (SnS), mercury (HgS), lead (PbS), bismuth (Bi_2S_3), copper (CuS), and cadmium (CdS) only, whereas nickel, cobalt, iron, manganese and zinc cannot be removed by the process.

Basically hydrogen sulphide is used as precipitation agent, but to achieve a very low arsenic content of less than 0.5 ppm under large scale operating conditions a small amount of hydrogen sulphide ions are added to the reactor as well.

The process principle can therefore be demonstrated by the following overall equations:

$$As_2O_3 + 3\,H_2S \dashrightarrow As_2S_3 + 3\,H_2O \qquad (1)$$

$$As_2O_3 + 3\,HS^- \dashrightarrow As_2S_3 + 3\,OH^- \qquad (2)$$

---

1 P. Comba, D.R. Dahnke, and L.G. Twidwell, Removal of arsenic from process and wastewater solutions. Department of Metallurgical and Mineral Processing Engineering, Montana College of Mineral Science and Technology, Butta, Montana, Arsenic Metallurgy Fundamentals and Application p.305-319

Process Description

The reactions (1) and (2) take place in a stirred tank reactor. The block diagram in figure 2 shows the steps of which the process consists in more details:

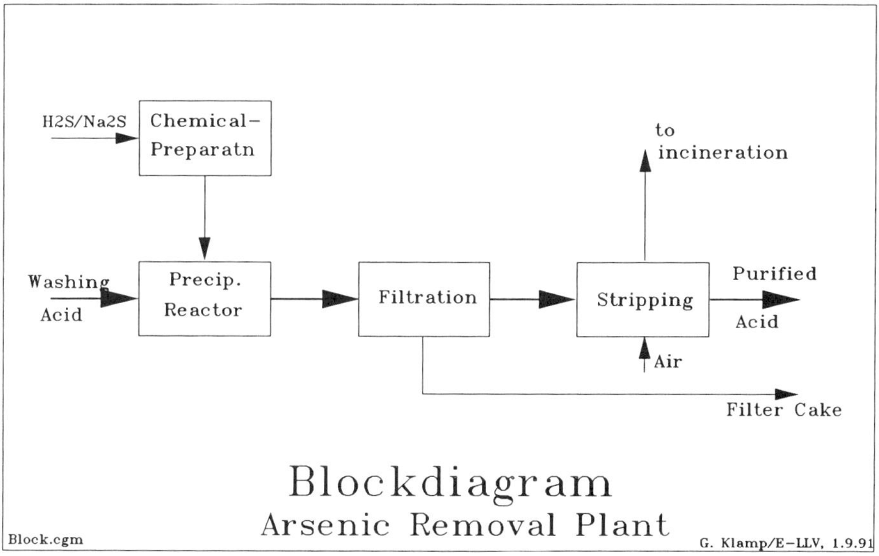

Figure 1 Block diagram of arsenic removal plant

- preparation of precipitation chemicals
- reaction and precipitation of arsenic sulphide
- removal of arsenic sulphide by filtration
- air stripping of purified washing acid

Characteristics of purified washing acid

The quality of the washing acid depends on the inlet composition of the washing acid. Typically a sulphuric acid concentration of 5 to 30% can be treated with arsenic contents up to the solubility of arsenic oxide (As_2O_3) in sulphuric acid at given operating temperature.

The treated sulphuric acid contains less than 0.5 ppm (guarantee) of arsenic and is clear.

Composition of removed arsenic sulphide

The filter cake contains the precipitated arsenic sulphide, other suspended solids from the treated acid and filter aids. The filter cake has a high solids content of more than 50% and can easily be handled.

In Germany this filter cake is considered as special waste which has to be stored in an underground safety disposal. Therefore the filter cake is filled into steel drums which are lined with a plastic foil.

Treatment of stripp-air

The treated washing acid contains dissolved hydrogen sulphide which has to be removed by stripping with air. The off-air containing 1 to 1.5% by volume of hydrogen sulphide is burnt replacing combustion air in power generation or other process sections.

Operational and safety aspects

The Sachtleben plant in Duisburg/Germany is in operation for more than five years. The plant is incorporated into the sulphuric acid production facility. The arsenic removal plant is **operated by one operator**. The plant has been built according to high safety standard to prevent explosion, fire, corrosion and the formation and escape of toxic material.

Cost estimate

Estimated operating cost are given in table 1 for a plant having a **throughput of 30 t/h** of 25 to 30% H_2SO_4 washing acid removing 700 TPY of arsenic.

The total investment cost based on actual German conditions are in the range of 4 million US$.

Utility Cost	Consumption/t	Unit Cost	$/t of acid
Chemicals	5 - 10 kg	30 ¢/kg	2.25
Process water	0.1 m^3	20 ¢/m^3	0.02
Electricity	7.5 kWh	4 ¢/kWh	0.30
Disposal	8 l	12 ¢/l	1.00
TOTAL			approx. 3.60

Table 1 Utility cost for arsenic removal plant

Maintenance cost for material and labour are about 1.5 - 2 %/Yr of TIC.

This type of emission control plant does not contribute to the economics of a smelter, on the contrary it adds on investment and operating costs, but **it could be a key to earn money in the business of non-ferrous metallurgy in industrialized countries under tightening environmental conditions.**

4.
WASTE UTILIZATION

METAL RECOVERY FROM SPENT ACID SOLUTIONS

AND BAGHOUSE BAGS USING THE INMETCO PROCESS

R. H. Hanewald, W. A. Munson, Jr. and D. L. Schweyer

Abstract

The INMETCO process has been modified so as to recover nickel, chromium and iron from a wide variety of metal containing waste streams. This now includes the recovery of metals from spent acid solutions such as chromic acid etching liquids, nickel and chromium plating solutions, stainless and superalloy pickling solutions, etc. In addition, hazardous electric furnace stainless steel and superalloy baghouse bags can be processed so as to recover the valuable metallics. A detailed review of the above new recovery techniques will be discussed as well as the current operating methods used at INMETCO.

Residues and Effluents - Processing and Environmental Considerations
Edited by Ramana G. Reddy, William P. Imrie and Paul B. Queneau
The Minerals, Metals & Materials Society, 1991

The INMETCO process has been modified so as to recover nickel, chromium and iron from a wide variety of metal containing waste streams. This now includes the recovery of metals from spent acid solutions such as chromic acid etching liquids, nickel and chromium plating solutions, stainless and superalloy pickling solutions, etc. In addition, hazardous electric furnace stainless steel and superalloy baghouse bags can be processed so as to recover the valuable metallics. A detailed review of the above new recovery techniques will be discussed as well as the current operating methods used at INMETCO.

THE INMETCO PROCESS

The INMETCO Process consists of three basic steps: 1) feed preparation, blending and pelletizing; 2) reduction; and 3) smelting and casting. (See attached Figure 1 flow sheet.)

The equipment used is relatively well known in the primary metal producing industry. The feed preparation equipment is specifically suited to the feed material, utilizing a rod mill for mill scale and a table feeder and rotary breaker for the swarf. For blending, mass flow bins with electronic speed controls are employed to move the finely prepared material, via a screw feeder, to the covered conveyor and into screw conveyors. The screw conveyor discharge is carried by conveyor to a 4.3 m diameter pelletizing disc where water is added and the pellets formed (approximately 1.2 cm in diameter). The first reduction step is carried out in a 16.7 m diameter Rotary Hearth Furnace. Smelting and final reduction is accomplished in a 6.3 MVA, 6.4 m diameter submerged Arc Electric Furnace. Casting is performed by pouring the hot metal from a 12 metric ton ladle into a twin strand pig caster. Pig weight is about 14 kg.

The first step in the process involves receiving the various wastes and pretreating them, where necessary, to insure a uniform size material. These pretreated wastes are stored in separate silos from which they are fed at a carefully controlled rate. The wastes are blended with coke or coal and water in a screw conveyor. The mixture proceeds to a disc pelletizer which produces green waste pellets strong enough to resist disintegration in the subsequent thermal operations. Since the pellets are not going to be tumbled in the reduction process, the strength requirement is minimal.

The second major step in the process is the partial reduction of the metal oxides in a Rotary Hearth Furnace. In this furnace, some of the carbon in the pellets react with oxygen in the waste to produce reduced metal. A portion of the Zinc, Lead and Halogens contained in the flue dust are exhausted into the off-gas treatment system. Hot, metallic, sintered pellets are transferred in sealed containers to the Electric Arc Smelting Furnace where the third major operation is performed. In this stage, the pellet is melted and chromium oxides are reduced by the residual carbon in the pellet. Lime, silica, alumina and magnesia separate to form a liquid slag which assists in cleansing the metal bath. Metal and slag are tapped periodically from the furnace. The metal is cast from a refractory lined ladle into pigs which are sold to the steel mills. The slag is treated in a special area in order to obtain a sized material that can be used locally as fill or ballast. The process water is treated and yields a filter cake that is recycled in a separate processing plant to recover the Zinc values.

The INMETCO plant at Ellwood City is capable of treating 50,000 metric tons of raw material per year. At this feed rate, the plant can produce 21,000 metric tons of pig metal with a composition similar to that shown in Figure 2.

FIGURE 2: TYPICAL METAL ANALYSIS

Ni	-	8.0	Co	-	0.5
Cr	-	13.9	S	-	0.2
Fe	-	70.1	C	-	3.8
Mn	-	1.8	Si	-	0.1
Mo	-	0.9	P	-	0.05
Cu	-	0.7			

Spent baghouse bags from the electric furnace arc melting of steels (carbon and stainless) are classified hazardous by the EPA (K061) since they contain the actual dust generated during melting. INMETCO has installed the appropriate equipment to size these bags by a shredding technique so that the sized bags can be fed into our furnaces where the metals can be recovered. As can be seen in Table 1, significant quantities of metals are contained in these bags which can be recovered by our recycling.

Typically, hazardous landfill operators would have difficulty in treating baghouse bags so that they would not leach chromium, lead, cadmium, nickel, etc. prior to landfilling. The INMETCO process recovers all of these metals, plus other metals, in its process and yet recycles another hazardous waste avoiding landfill. This aids steel mills in achieving <u>Waste Minimization Goals</u> while avoiding landfill liability.

NICKEL AND CHROMIUM LIQUID SOLUTION RECYCLING

Nickel, chromium and iron can be recovered in the INMETCO process from virtually any type waste such as dusts, sludges, solids, cakes and liquids. Currently, we can accept full tank truckloads of 5,000 gallons or 55 gallon drum shipments of spent acid solutions containing nickel and/or chromium. Nickel solutions such as Ni Cl, Ni NO_3, Ni SO_4 and nickel stripper solutions have been recycled with nickel contents up to 327,000 PPM (See Table 2). In addition, INMETCO restricts the copper content to about 1800 PPM and phosphorous to 860 PPM due to the specification requirements of its stainless steel remelt alloy. The pH range virtually covers the full range of 1 to 13.2 as can be seen also in Table 2.

A large amount of chromic acid solutions used in etching, pickling or plating can be recycled. Chromium contents up to 161,000 PPM have been successfully recycled (See Table 3). Again, note that the copper and phosphorous levels are below our internal specifications.

INMETCO recycles these liquids by adding them into our pelletizing process. Typically, our waste feeds, particularly EAF baghouse dust and pickling cakes, contain large quantities of lime. When the spent acid is added, some neutralization occurs on the pelletizing disk during pellet making. However, we have resurfaced the pelletizing deck with grade 316 stainless steel to resist corrosion. The chromium and nickel liquids replace water previously used in the pelletizing process.

When the pellets containing the liquid nickel and chromium compounds are processed through the Rotary Hearth Furnace (RHF), the acid radicals are fumed off and captured in the wet scrubber sump area where they are neutralized with lime prior to our in-plant waste water treatment plant.

PICKLING WASTE PROCESSING

INMETCO has modified its processing system whereby pickling wastes from the specialty steel industry can be consumed while the metals are being recovered. We can receive either the liquid acid solutions of nitric, hydrofluoric-nitric, or HCl. These spent acids can be received in bulk tanker or in drums.

The metal containing spent acids will be pumped at a controlled rate into our pelletizing disk and used in the production of waste pellets as a replacement for water. In order to avoid corrosion of our pelletizing disk, we have added a 316 stainless steel deck. Approximately 11% of the pellet weight is made up of spent acid containing nickel, chromium and iron. INMETCO can consume 70,000 gallons per month of this material.

Most specialty steel mills, as well as superalloy producers, now lime neutralize their spent acid solutions and precipitate the metals as hydroxides or carbonates. Figure 3 shows the chemical analysis from five different specialty mills of typical pickling wastes. Generally, there are sufficient quantities of nickel, chromium and iron to warrant recovery.

Figure 4 shows the chemical analysis from two superalloy mills from their spent acid and lime neutralized wastes. Both the superalloy and stainless pickling cakes are processed in a similar fashion at INMETCO. If the material is dry, INMETCO can receive it pneumatically similar to flue dust. If it is delivered as a "wet" filter cake with chemically combined water, we receive it in plastic lined roll off boxes or dump trailers. This material is fed at a controlled rate after adding about 10% carbon, generally as coke fines, via a variable speed screw feeder to INMETCO's Rotary Hearth Furnace (RHF).

INMETCO has designed a 10' (3.3 m) table feeder to feed several 28' (8.5m) long x 24" (0.62m) diameter ribbon flight screw conveyors that mix the carbon fines and filter cake prior to delivery to the RHF. The carbon will react with the nickel and iron hydroxides for purposes of reduction prior to transfer to the Electric Arc Furnace. Basically, the reactions are as follows:

$$NiO \bullet H_2O + Fe_2O_3 \bullet H_2O + C \ \Delta\Rightarrow \ Ni + Fe + CO + CO_2 + H_2O \quad (1)$$

Note, about half of the lead and zinc will fume off, along with a portion of the halogens which are captured by the wet scrubber and removed in our waste water treatment plant.

Another pickling waste is the metal scale that drops to the bottom of the tanks during pickling. Figure 5 shows the analysis of that scale. This material can be blended with regular mill scale at INMETCO and processed through our system for full metals recovery.

One final waste product from pickling operations involves the carbon brick linings that are used in many of the pickling baths. Figure 6 shows the non-metallic analysis of what some of these bricks look like. In addition, these bricks will be contaminated with hexavalent chromium from pickling operations. INMETCO can receive these carbon bricks and process them back into the INMETCO system. We require no pretreatment or rinsing by the specialty or superalloy mills prior to consumption at our Ellwood City, Pennsylvania facility.

INMETCO will crush these carbon bricks to -2" and feed them as a reductant into our electric furnace. The hazardous hexavalent chromium will be converted to the tri-valent state and recovered in our stainless steel remelt alloy.

ADDITIONAL WASTE FEED MATERIALS

Flue Dust - Furnace baghouse dust or flue dust is generated during the melting and refining steps in the manufacture of stainless steel. Flue dust is the most metallurgically complex and contaminated waste. These dusts are usually collected in baghouses and consequently are dry. The principal constituents are the major components of steel and slag. Flue dust particles are extremely fine, typically one micron particle diameter, and flow readily. Stainless steel flue dusts are considered hazardous by the Environmental Protection Agency because they are strongly alkaline and contain high levels of Cr^{+6} and lead. While flue dust is the most metallurgically complex and contaminated waste, it is one of the easiest materials for our system to handle. Since it is dry, it can be collected in pneumatic truck tankers and transported to the INMETCO facility.

When the tanker arrives at INMETCO, the dust is blown into storage silos from which it is fed by a mass flow system into the pelletizing line. No additional processing is required. Due to the fact that the dust is collected, stored and transported under carefully controlled conditions, the dust is uniform and free from foreign objects. The flue dust handling system is entirely enclosed to contain the dry dust and to prevent it from escaping and becoming an environmental problem.

Mill Scale - Mill scale results from oxidation of the stainless steel surface during operations such as soaking or preheating the steel, forging and hot rolling. This waste has a flaky appearance and is quite variable in particle size.

Collection of mill scale in the steel plants is not as well controlled as the collection of flue dust. Many tramp elements such as crushed barrels, scrap iron, rags and sundry other items are contained in the mill scale collection areas. At INMETCO, the larger tramp elements are removed by hand and smaller objects are removed by means of a grizzly screen in the mill scale processing circuit.

Mill scale is quite variable in size and, thus, must be ground to 0.03 mm before it is fed into the pelletizing circuit. Since the mill scale is very hard and abrasive, all wear parts in the grinding circuit must be made of Ni-hard cast iron.

Grinding Swarf - Swarf, or oily grindings as identified here, is a product of the belt grinding of stainless steel sheet and strip. This material presents special handling problems.

Grinding swarf is metallic and not oxidized, thus no reduction or oxygen removal is required. The swarf contains oil which is a source of fuel for the process. There are some foreign objects in the swarf that must be screened out. The oily, fine strands that make up the swarf tend to form into wads or clumps. These wads are too large for the pellet feed circuit and must be broken up into smaller pieces. INMETCO has developed a "swarf breaker" which promotes breakdown of the swarf by a tumbling action similar to autogenous grinding of mineral ores. The result is a 0.25 mm material that can be stored and transferred to pellet making without excessive reagglomeration.

Spent Dolomitic Brick from A.O.D. vessels can be used as a slag additive to the INMETCO process. This material replaces an operating supply and again recycles another waste, avoiding landfilling.

Spent Chromium Refractories, such as chrome-magnesite bricks, can be recycled so as to recover the chromium back into the stainless steel remelt alloy and use the magnesia as a slag additive. Most spent chromium refractories are hazardous due to the leaching of hexavalent chromium. The INMETCO process recovers the chromium, consumes a hazardous waste and avoids landfilling of yet another non-renewable valuable resource.

Baghouse Bags from Electric Furnace Dust Collection Systems containing hazardous metals can be recycled through the INMETCO process to recover the valuable metals while avoiding landfilling with this hazardous waste.

Super Alloy Wastes from either standard melting or rolling mill operations, along with electro chemical machine (ECM) operations, can be recovered/recycled back into the INMETCO process. Once again, the Nickel, Chromium and Iron will be recovered while avoiding landfill.

Hexavalent Chromium Containing Rinse Waters - Industry generates rinse waters of various types that contain some hexavalent Chromium which needs to be treated before disposal. At INMETCO, we have routinely taken this water with Cr^{+6} and have converted it to the environmentally safe Cr^{+3} and in doing so have also reconverted this Chrome in our product so that it is recycled back to industry.

Plating Wastes Liquids and Cakes - In the plating industry, one of the difficulties faced is in the disposal of spent electrolyte. On a regular basis, the nickel platers need to dump their plating tanks. If they have in-house treatment facilities, they have filter cakes containing such things as Nickel, Chrome and Iron, as well as some Sulfates or Nitrates. INMETCO can receive and treat either the neutralized Nickel solutions or Nickel Nitrate solutions while recovering the metallics. The filter cakes normally are hydroxides of Nickel, Iron and Chrome. These cakes appear dry to the eye but may contain up to 70% chemically attached water. These cakes can be processed with our normal feed stocks so that the water is driven off in our Rotary Hearth Furnace and the metallics are again recovered in our final smelting step.

Spent Catalysts - Many nickel and chrome containing catalysts are used in the chemical, petroleum refining and food processing industries. Once the useful life of these catalysts is gone, they need to be disposed. INMETCO can consume these spent catalysts by a variety of methods. Dry rings and beads of nickel catalysts can be directly fed to our Electric Furnace where the nickel, chromium and other valuable metallics will be recovered. Liquid containing catalysts, such as those containing activated Nickel, will be consumed in our blending circuit and roasted in the Rotary Hearth Furnace prior to smelting. Again, the valuable metallics are recovered and recycled back to industry. Figure 7 shows many but not all of the waste streams that can be processed by INMETCO.

Nickel-Cadmium Batteries - In 1990, the EPA stated in a new ruling that High Temperature Metal Recovery was BDAT (Best Demonstrated Available Technology) for the treatment of nickel-cadmium batteries. INMETCO is the only North American operator of such a process. The reclaimed nickel is used back into the stainless steel remelt alloy, while the cadmium is fumed off with zinc and lead and sent to another firm for recovery.

RECENT OPERATING EXPERIENCE

The first commercial heat was produced at INMETCO's Ellwood City facility in December 1978. During 1979, the plant went through a process development and equipment shakedown period. In 1980, regular production commenced and plans were made to modify certain equipment and processes to achieve and exceed design capacity. These changes were completed in late 1980 and early 1981. Since then, INMETCO has demonstrated that it can run at 100% design capacity on a routine basis.

A major portion of the design, construction and operational start-up costs of the INMETCO plant were assigned to determine the best methods, both environmentally and physically, of handling the waste feed stocks.

RECOVERIES

Through the reduction and reclaiming steps followed at INMETCO, we have become highly efficient in the recovery of metallics. Some actual recovery values are indicated below.

Ni	Mo	Co	Cr	Fe	Mn
98%	92%	97%	86%	96%	60%

While achieving the above metal recoveries, we also have removed the Zn, Pb and most of the Sulfur.

COMPOSITIONAL CONTROL

INMETCO's customers, the specialty steel producers that provide the larger percentage of the raw material used at INMETCO, desire a quality remelt alloy that can be used to produce material similar to type 304 stainless steel (18% Cr - 8% Ni - balance Fe). When Molybdenum is present, the pig could be used for the manufacture of the higher valued type 316 stainless.

INMETCO's quality control group has developed the necessary procedures to sample the various nickel-containing waste liquids and solids that are processed here. This insures complete and accurate chemistry control. This group specifies what material can be used in our process, as well as the processing circuit to be employed and the consumption rate.

The INMETCO procedure is to analyze each lot or truckload as it arrives at the INMETCO plant. The material is then assigned to a particular storage bunkered area based on its Nickel and Chromium content. The contents of the various bunkers are carefully blended to produce a mix with satisfactory processing characteristics and uniform alloy content. Any shortfall in Nickel content from the target level is made up by adding nickel alloy wastes. The INMETCO Electric Arc Furnace contains a heel of about 120 metric tons of hot metal which helps smooth out product compositional fluctuations.

RESOURCE RECOVERY PLANT WASTES

A resource recovery plant is, of course, subject to the same environmental concerns as the industry it serves. INMETCO does not landfill any residues from its process. Our non-hazardous slag co-product passes the EPA TCLP test, does not leach, and is sold in the same way as other slags have been since the turn of the century.

INMETCO management observes the stringent air quality standards of the state of Pennsylvania. The gaseous effluent from the Rotary Hearth Furnace contains significant quantities of Zinc, Lead and Halogens. A wet gas scrubbing system was specified to treat this effluent. The scrubber water is treated to meet all the applicable quality standards. After treatment, it is discharged to the Ellwood City municipal sewage treatment system. The treatment system produces about 2,000 metric tons per year of an alkaline filter cake containing significant quantities of Zinc and Lead (see Figure 8). INMETCO recycles this cake through a secondary zinc producer. The Electric Furnace is attached to a standard baghouse for collecting the remainder of the zinc and lead. (See Figure 8 for compositional information.)

FIGURE 8: INMETCO BY-PRODUCT COMPOSITION

Constituent	Composition Weight %	
	Scrubber Filter Cake	EF Baghouse Dust
$SiO_2 + Al_2O_3$	7%	3%
CaO + MgO	13%	5%
F	3%	1%
Cl	0.5%	9%
Pb	6%	4%
Zn	22%	28%
Fe	9%	2%
Cr	3%	0.5%

The slag that is generated is poured from the slag pot, while still molten, down a ramp into a thin layer. This is cooled and broken up with a front end loader. The broken slag is screened and sized. This sized material is chemically inert and is sold for roadway beds and construction material. Figure 9 gives the analysis of this slag.

FIGURE 9: 1990 CHEMICAL ANALYSIS OF INMETCO'S SLAG CO-PRODUCT

SiO_2	25%	CaO	31%
MgO	18%	Al_2O_3	10%
Cr_2O_3	5%	Fe_2O_3	3%

COMMENTS ON THE INMETCO OPERATION

This paper is presented to demonstrate how the INMETCO process works with pickling wastes, baghouse bags, and solutions containing nickel and chromium, as well as other metal bearing materials. Over 1800 tons of nickel, 3000 tons of chromium and 16,000 tons of iron are annually recovered and recycled. These non-renewable resources are kept from landfills. The environment is improved by processing hazardous wastes into non-hazardous valuable, recyclable remelt alloy.

INMETCO started out merely as a reclaimer of Nickel, Chromium and Iron, but now serves a very important role in the environment. We are an environmental service company first, who also recovers valuable metallics in the process. We help other nations by helping companies avoid hazardous waste landfilling and producing valuable and useful products.

INMETCO is committed to the concept of sustainable development, which requires balancing the need for economic growth with good stewardship in the protection of human health and the natural environment. INMETCO will strive to minimize any potentially adverse impacts of its operations and products on its employees, customers, the general public and the natural environment, and will seek to not only meet but, if possible, surpass the standards set by relevant legislation, by diligent application of technically proven and economically feasible environmental protection measures throughout all phases of operations.

INMETCO

Detailed Manufacturing Process

Figure 1

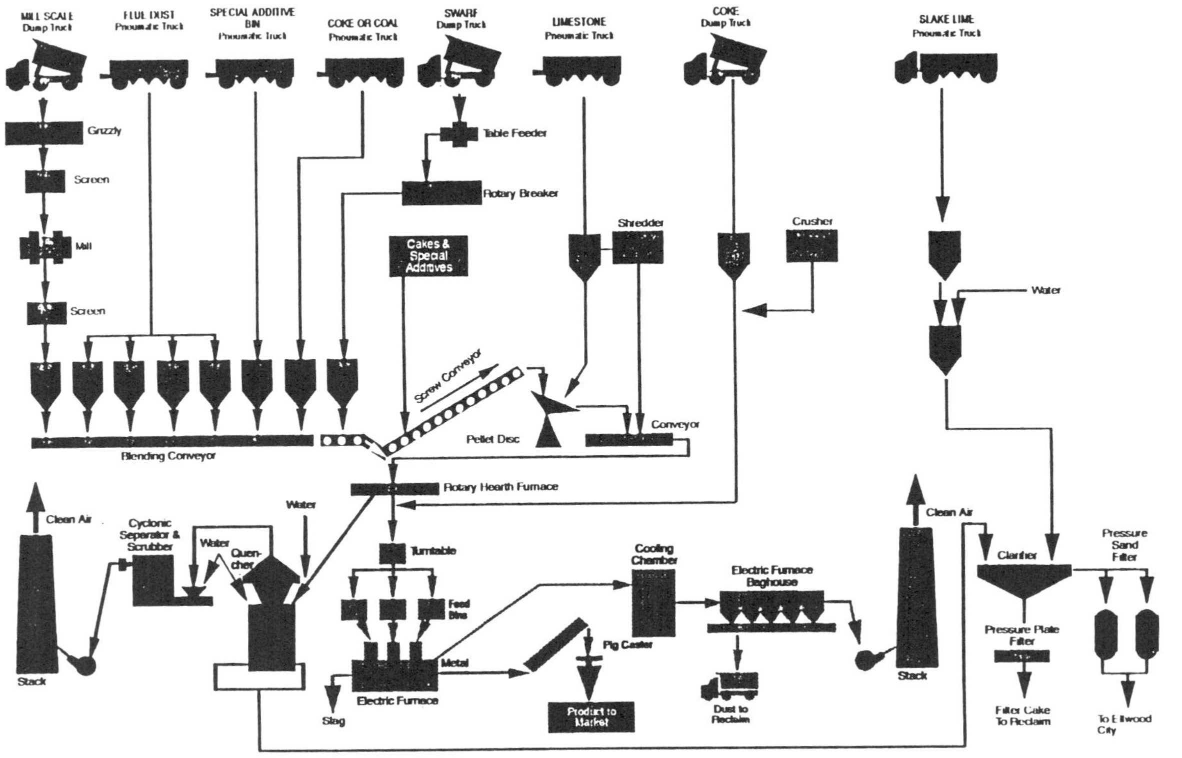

FIGURE 3

CHEMICAL ANALYSIS OF

STAINLESS STEEL PICKLING CAKES

Client Mill #	Cd	%Mo	%Pb	%Zn	%Cu	%MgO	%Ni	%Co	%Fe	%Cr	%Mn	%CaO	%SiO2	%P	%Al2O3	LoI
1.	0.03	0.12	0.04	0.02	0.11	0.50	2.29	0.08	20.04	4.22	0.16	24.61	1.6	0.026	0.5	63.3
2.	0.07	0.22	0.1	0.16	0.33	7.12	4.66	0.16	34.59	11.28	1.04	0.8	0.96	0.023	0.21	56.8
3.	0.07	0.35	0.19	0.5	0.44	2.02	3.74	0.37	34.08	8.42	0.92	13.04	3.37	0.036	0.68	0.2
4.	0.06	0.17	0.1	0.11	0.17	0.32	5.3	0.2	38.38	8.91	0.67	0.39	4.79	0.039	0.36	60.8
5.	0.03	0.19	0.06	0.02	0.12	0.37	1.46	0.08	9.1	1.61	0.08	16.93	45.8	0.015	1.89	63.4

March 1992

FIGURE 4

CHEMICAL ANALYSIS OF

SUPER ALLOY PICKLING CAKES

Client Mill #	Cd	%Mo	%Pb	%Zn	%Cu	%MgO	%Ni	%Co	%Fe	%Cr	%Mn	%CaO	%SiO2	%P	%Al2O3	LoI
1.	0.08	0.42	0.09	0.12	0.38	0.64	5.22	1.04	35.37	8.15	0.40	13.54	1.36	0.028	0.51	57.4
2	0.07	0.58	0.12	0.29	0.14	3.25	6.83	0.68	3.58	3.09	0.18	36.65	5.52	0.197	1.86	67.7

March 1992

FIGURE 5

CHEMICAL ANALYSIS OF

PICKLING TANK BOTTOM SCALE

Client Mill #	Cd	%Mo	%Pb	%Zn	%Cu	%MgO	%Ni	%Co	%Fe	%Cr	%Mn	%CaO	%SiO2	%P	%Al2O3	LoI
1.	0.08	0.6	0.13	0.08	0.1	0.7	1.10	0.1	25.40	24.1	1.4	3.6	12.6	0.022	1.4	24.2
2	0.05	0.28	0.22	0.04	0.08	0.22	1.00	0.08	27.88	19.66	1.83	0.78	2.49	0.038	0.30	49.00

March 1992

FIGURE 6

CHEMICAL ANALYSIS OF

CARBON BRICK FROM STAINLESS STEEL PICKLING TANKS

Client	Fixed Carbon	Sulfur	H_2O	% Volatile	* % Ash
Mill #1	97.7	0.83	4.2	0.96	1.3
Mill #2	88.3	1.1	12.5	6.5	5.2

* Is hazardous for hexavalent chromium.

FIGURE 7

EPA WASTE DESIGNATIONS ACCEPTED AT INMETCO

D001	D006	F006
D002	D007	K061
D003	D008	K062

Ni & Cr Solutions
Ni-Cr Plating Wastes & Bags
Nickel Stripper Solutions
Chromic Acids
Ni & Cr Sludges & Pickling Cakes
Ni-Cr Dusts & Grindings
Ni Cd Batteries
Ni Fe Batteries
ECM-EDM Ni Cr Fe Solids
Ni & Cr Catalysts
Raney Nickel Catalysts
Chrome-Magnesite Refractories
Dolomitic Refractories
Carbon Brick from Acid Tanks
CaO and MgO By-Products
Carbon & Coke Fines & By-Products
Magnesium Powders & Machinings
K061 Baghouse Bags
EAF Baghouse Dusts

3/92

TABLE 1

CHEMICAL ANALYSIS OF

BAGHOUSE BAGS

Client Mill #	%Cd	%Mo	%Pb	%Zn	%Cu	%MgO	%Ni	%Co	%Fe	%Cr	%Mn	%CaO	%SiO2	%P	%Al2O3	LoI
1		0.72	0.03	0.05	9.83	3.20	21.6	0.1	8.7	12	2	4.5	2.1	0.008	4.4	24
2	0.04	0.69	0.84	2.16	0.56	5.96	2.41	0.11	23.89	9.42	4.32	7.06	4.56	0.021	0.47	16.35
3	0.03	0.26	0.67	8.9	0.46	3.4	2.18	0.09	24.28	10.40	4.00	4.3	6.85	0.022	0.50	15.8
4		9.07	0.13	0.38	0.26	20	9.51	1.32	2.59	7.25	1.39	9.9	2.4	0.018	2	21.3
5	0.06	0.11	0.44	8.39	0.32	0.73	0.3	0.1	33.3	1.17	5.03	2.13	21.32	0.04	1.39	75

Mar-92

TABLE 2

CHEMICAL ANALYSIS OF

NICKEL SOLUTIONS

Client Mill #	pH	Fe (ppm)	Ni (ppm)	Cr (ppm)	Cu (ppm)	Co (ppm)	P (ppm)
1	1.20	3160	33000	63	17	19	6
2	13.22	9	23200	32	4	460	40
3	1.38	65	57600	131	174	33	53
4	1.90	784	73560	212	410	107	435
5	1.00	632	83330	435	927	92	160
6	12.68	13	327169	906	110	350	15

Mar-92

TABLE 3

CHEMICAL ANALYSIS OF CHROMIC ACID

Client Mill #	pH	Fe (ppm)	Ni (ppm)	Cr (ppm)	Cu (ppm)	Co (ppm)	P (ppm)
1	1.00	4870	215	98400	162	65	50
2	1.00	8760	64	161000	224	12	14
3	1.00	14000	32	132600	1210	5	16
4	1.00	29262	153551	101496	1320	391	359
5	1.00	131	320	142400	1060	6	13
6	1.28	3629	39	40820	130	19	262
7	1.00	1491	63	140490	1427	54	39

Mar-92

DEVELOPMENT OF COAL FLY ASH AS A HYDRAULIC AND CHEMICAL BARRIER

Cam T. Nhan and Donald W. Kirk

University of Toronto
Dept. of Chemical Engr. & Applied Chemistry
200 College Street
Toronto, Ontario M5S 1A4
Canada

Abstract

The potential use of coal fly ash as a construction material has been recognized for many years. In this work, an admixture of the ash residue is investigated for use as a landfill liner. In particular, its performance as a chemical barrier to the leachate from the municipal solid waste (MSW) is determined. The test specimens are a water compacted admixture of class F fly ash, low in calcium content, which is blended with lime dust and calcium bentonite. MSW is passed through the compacted fly ash specimens under a gravitational head. The diffusion of elements in the leachate through this material is monitored as a function of time. The distribution profiles of the retained elements in the compacted admixture provide prediction of the long term performance of the compacted admixture directly affect the leachate diffusion rate, hence, the mobilities of the retained elements.

Residues and Effluents - Processing and Environmental Considerations
Edited by Ramana G. Reddy, William P. Imrie and Paul B. Queneau
The Minerals, Metals & Materials Society, 1991

USE OF INDUSTRIAL BY-PRODUCTS CONTAINING HEAVY METAL CONTAMINANTS IN AGRICULTURE

John J. Mortvedt

National Fertilizer and Enviromental Research Center
Tennessee Valley Authority
Muscle Shoals, Alabama 35660-1010

Abstract

Application of some by-products of the metals industry which contain plant nutrients to agricultural lands could constitute resource recovery instead of costly disposal. Some industrial by-products contain appreciable concentrations of heavy metal contaminants, such as Cd, Cr, Ni, and Pb. After land application, these metals may be absorbed by plants and enter the human food chain or be leached to groundwaters. Plant uptake of these heavy metals varies with plant species, soil pH, soil texture, and the particular metal. Certain leafy vegetables can absorb significant amounts of heavy metals, while absorbed metals are usually poorly translocated to cereal grains. Allowable heavy metal concentrations in fertilizers now are not regulated, but EPA is considering such regulations. Heavy metal impurities in animal feed supplements are regulated by the states, so some by-products may not be permitted for this use.

Residues and Effluents - Processing and Environmental Considerations
Edited by Ramana G. Reddy, William P. Imrie and Paul B. Queneau
The Minerals, Metals & Materials Society, 1991

Introduction

Accumulation of waste products is inherent in any manufacturing or production process. While attempts are made to minimize waste streams, disposition of wastes is an important part of any process. Some of these wastes contain metals or other elements which have value as the intended product, but processing them into usable form may not be cost effective.

In the past, many of these low-value wastes were stored in piles or landfills near their site of production. However, shortages of satisfactory landfill space as well as increased public concern about the long-term effects of disposing some industrial wastes in landfills have led to reviews of disposal methods. Increased recycling activities also reduce the amounts and types of wastes.

Because many industrial by-products of the metals industry contain elements which are nutrients for plants or animals, it may be possible to use some of these by-products in agriculture. Such use would constitute resource recovery instead of costly disposal. However, some by-products which contain the nutrient elements needed by plants or animals may also contain heavy metal contaminants such as Cd, Cr, Ni, and Pb. Use of such by-products containing unacceptable levels of these contaminants could cause plant or animal toxicities or otherwise result in entry of these metals into the human food chain.

The purpose of this paper is to discuss the types of waste products which might be used in agriculture, either as fertilizers for crops or in trace mineral supplements for animal feeds. Some of the environmental risks as well as economic benefits of using industrial by-products which contain certain heavy metal contaminants in agriculture also will be discussed.

Wastes as Fertilizers

Plant nutrients

Optimum plant growth results from adequate light, moisture, temperature, and soil conditions, and a supply of essential nutrients. There are 16 elements which are essential for plant growth. Of these, C, H, and O are obtained from air and water. The other elements are divided into three nutrient groups as follows: primary (N, P, and K), secondary (Ca, Mg, and S), and micronutrients (B, Cl, Cu, Fe, Mn, Mo, and Zn). The elements in the above three groups are so classified by their relative concentrations in plants and fertilizer application rates, as shown in Table I.

These plant nutrients are obtained from the soil or from fertilizers. The plant nutrients in crop residues are recycled to the soil and eventually become available to succeeding crops through decay and mineralization. Supplemental nutrients (fertilizers) may be applied as organic or inorganic sources. However, plant roots absorb only inorganic forms of nutrients. Organic sources of plant nutrients are first mineralized in the soil to produce inorganic ions.

Plant uptake of nutrients varies considerably with crop, yield, and nutrient element. Because much of the above-ground portions of each crop is not removed from the field at harvest, it is important to recognize that crop residues provide considerable amounts of plant

Table I Ranges in Tissue Concentrations and Recommended Plant Nutrient Applications for Crops

Plant Nutrient	Avg. conc. in plant tissue	Recommended rates kg/ha[1]
	-----%-----	
N	1.0 - 4.0	50 - 250
P	0.2 - 0.5	15 - 50
K	1.0 - 3.5	30 - 200
Ca	0.4 - 1.5	[2]
Mg	0.2 - 0.8	10 - 40
S	0.1 - 0.6	10 - 40
	----ppm----	
B	10 - 200	0.5 - 2.0
Cl	100 - 500	[3]
Cu	5 - 30	1.0 - 5.0
Fe	100 - 500	[4]
Mn	20 - 30	2.0 - 10.0
Mo	0.1 - 2.0	0.1 - 0.3
Zn	20 - 150	1.0 - 10.0

[1] Rates vary with crop, soil, and climate (1.12 kg/ha = 1 lb/Ac).
[2] Usually applied as lime ($CaCO_3$) to increase soil pH.
[3] Applications generally not recommended.
[4] Soil application not recommended for most conditions.

nutrients. Amounts of each of the micronutrients contained in above-ground portions of field crops usually are <1 kg/ha at normal yield levels. Because most of each micronutrient is contained in the plant stems and leaves, removal of only the grain portion of corn, soybeans, and wheat from the field recycles most of these nutrients back to the soil.

Fertilizer rates and use

The recommended fertilizer rates for each plant nutrient are related to the intended crop yield and expected soil supply, and vary considerably with nutrient (Table I). The highest recommended rates are for the primary nutrients (N, P, and K), and these are the main nutrients listed on fertilizer labels. Calcium and Mg usually are provided in lime, which increases soil pH to optimum levels. Sulfur is supplied in many forms, but mainly as sulfates at rates of 10 to 40 kg/ha. Rates of the micronutrients generally range from 1 to 10 kg/ha.

Fertilizers usually are manufactured specifically for agricultural use, although significant tonnages also are used for turf, lawn, gardens, and other horticultural uses. Most of these products are inorganic compounds, but some are organic fertilizers either naturally occurring or industrially processed.

The Crop Reporting Board of the Statistical Reporting Service at the U.S. Department of Agriculture began publishing data of annual micronutrient use in the U.S. in 1968. Data in Table II were obtained from known producers of Cu, Fe, Mn, Mo, and Zn fertilizers. Boron use data were not included because there was so few producers of B fertilizers. However, annual use of B for agriculture is estimated at 4,300 metric tons. Amounts of Mn and Zn fertilizers are highest because these micronutrients are recommended for major crops such as corn, wheat, and soybeans.

Table II Amounts of Micronutrients Sold as Fertilizers in the U.S. (1)

Year	Cu	Fe	Mn	Mo	Zn
	----------------1000 metric tons[1/]--------------				
1967-1968	2.2	3.1	10.5	0.07	13.1
1971-1972	0.6	1.2	11.2	0.09	14.4
1975-1976	0.5	2.3	8.1	0.10	14.7
1979-1980	1.5	5.1	13.2	0.13	40.8
1983-1984	1.1	5.9	15.1	0.01	37.3

[1/]Amounts expressed on elemental basis.

Significant increases in Mn, and especially Zn, usage since 1980 are mainly related to inclusion of data from companies which sell industrial by-products as micronutrient fertilizers. While collection of micronutrient use data was discontinued by USDA in 1985, it is known that industrial by-products have become the major fertilizer source of the metallic micronutrients (Cu, Fe, Mn, and Zn) in the U.S.

Wastes as sources of fertilizers

Waste products have been used as fertilizers for centuries. Crop residues were considered as wastes and were allowed to rot in the field and be incorporated into the soil. Manures from human and animal populations also are recycled to the soil to provide plant nutrients. The first commercial fertilizers, as they are called today, were ground bones which consist of calcium phosphates. Their value as fertilizers was widely recognized around 1800. Soon after, guano (bird excrement) was found to provide N and P for crops. The early history of fertilizers as well as an interesting and complete review of the U.S. fertilizer industry to 1980 was recently published (2).

Many industrial wastes contain plant nutrients which could be potentially used as fertilizers. For example, H_2SO_4 and H_3PO_4 are used to clean metals prior to processing. Eventually, these acids are "spent" and must be replaced in the cleaning process. Significant quantities of spent H_2SO_4 are produced by the U.S. galvanizing industry. This waste product contains Zn, Fe, and S as plant nutrients, but it may also contain heavy metal contaminants, notably Cd. Some processes have been developed to separate heavy metals from Zn and Fe sulfates by differential precipitation. Disposal of this waste as a fertilizer on agricultural land would result in at least part of the contained Zn, Fe, and S being recovered by crops. These plant nutrients have value in proportion to their cost as manufactured fertilizers. In 1985 for example, it was estimated that spent acid from the galvanizing industry in Nebraska contained about 700 metric tons of Zn, 1,400 metric tons of Fe, and 1,600 metric tons of S with an estimated value of $6.5 million. Other sources of Zn fertilizers are baghouse dusts from Zn smelters and the steel industry and spent acids from other industries.

Because many of these by-products contain heavy metal contaminants, it is important to know their availability to plants. Such availability is related to soil and plant factors. Availability of heavy metals, such as Cd, Ni, and Pb, is higher in coarse-textured (sandy) soils than in fine-textured (clay) soils and increases with increasing soil acidity. Some plant species, mainly leafy vegetables, tend to absorb larger amounts of heavy metals than other species, such as cereal crops. For example, six Zn fertilizers made from industrial by-products were compared with reagent grade $ZnSO_4$ applied to an acid (pH 5.8) soil in a greenhouse pot experiment (3). Uptake of Cd, Ni, and Pb each was higher in Swiss chard leaves than in corn forage. Uptake of Cd by each crop was highest from those fertilizers containing the highest Cd concentrations. Like the micronutrients discussed earlier, heavy metals do not translocate significantly from forage to grain tissues in most plants.

Application rates of heavy metals in industrial by-products to soil are determined by multiplying the heavy metal concentration in the by-product times the fertilizer application rate. Because most of the heavy metal concentrations in these by-products are in the parts per million range, their application rate to soil would be low. This especially is true for micronutrient fertilizers, such as Mn and Zn, which are applied to soil at low rates (Table I). For example, application at a rate of 5 kg of Zn/ha as $ZnSO_4$ (36% Zn) containing 100 ppm of Cd would result in a Cd application rate of 0.0014 kg/ha. Total Cd in the surface 15-cm layer (6 inches) soil generally is 1-2 kg/ha. Therefore, it would take about 1,000 years of annual applications of the above Zn fertilizer to double the Cd concentration in the surface soil.

The above comparison assumes no Cd losses by plant uptake or leaching in the soil and gives no information on relative bioavailability of the applied Cd versus the native soil Cd. Yet, the calculation shows that even though a given product contains an apparently significant concentration (100 ppm) of a heavy metal contaminant in a micronutrient fertilizer, the agronomic consequences may be minimal, depending on the crop and soil situation. However, the long-term consequences of possible heavy metal buildup of using such by-products in agriculture are unknown. Mobility of most heavy metals in soil is low, so these metals will mainly remain in the surface soil layer.

Some knowledge on the long-term effects of applying Cd to soil as a contaminant in fertilizers can be obtained by a study of nine long-term (>50 years) soil fertility experiments in the U.S. Samples of plant tissues and soils were analyzed for Cd and some plant nutrients. Samples of plots treated with P fertilizers were compared with those from unfertilized soil. Results showed that plant uptake of Cd contaminants (which came from the phosphate rock) in P fertilizer containing <10 ppm Cd was negligible even after 50 years' annual applications (4). The estimated annual Cd application rates ranged from 0.3 to 1.2 g/ha or a total of 17 to 60 g/ha (0.017 to 0.06 kg/ha) for the 50-year period.

Wastes as Trace Mineral Supplements for Animal Feeds

The animal feed industry markets significant tonnages of metals and other nutrients to feedlots and other animal feeding operations. These elements generally are known as minerals, trace minerals, or mineral supplements. The products are mixed with other feed ingredients such as corn, cottonseed meal, sorghum, and soybean meal to provide balanced rations for animals. The main minerals are $CaCO_3$, $CaHPO_4$, and NaCl, but other sources of Ca, Mg, and P also are used. The trace minerals are various sources of Co, Cu, Fe, I, Mn, Se, and Zn (5). Some sources of these trace minerals were listed by Nelson (6) at a previous TMS symposium on recycling of metals and engineered materials. Trace elements that are essential for animal nutrition are usually found in concentrations <100 ppm on a dry matter basis.

Most of the minerals used in the feed industry contain varying levels of other elements or compounds as contaminants. While many of these contaminants are innocuous, others may cause deleterious effects to animals if they are present in significant concentrations. Even if they do not adversely affect the health of animals, certain metals could enter the human food chain if they are deposited in muscle, organs, or other tissues consumed by humans. Concentrations of heavy metals in muscle tissues (meat) vary but generally are lower than those in organs such as liver and kidney tissues.

Few reports have been published on the effectiveness of industrial by-products relative to that of commonly used sources in trace mineral supplements in animals. Miller et al. (7) compared a metallic Zn dust (average particle size was 6 µm, -325 mesh) from the metal smelting and refining industry with ZnO, which is the usual Zn source in trace mineral supplements for swine. While average daily gains in weight of young pigs were increased by supplemental Zn, growth was similar with either source and level of Zn in the feed ration. However, blood serum Zn was higher in swine fed the metallic Zn than in those fed ZnO. These data suggest that when ZnO is in short supply, metallic Zn by-products could be acceptable Zn sources for the feed industry.

The metallic Zn product contained 99.3% Zn and 0.02% (200 ppm) each of Cd and Pb. Concentrations of both heavy metals were well below the suggested upper limits for these heavy metal contaminants. The ZnO powder contained 0.002% (20 ppm) Pb. No data were given for the levels of Cd or Pb in blood serum of swine treated with each Zn source, so there is no indication of the bioavailability of these heavy metals.

Guidelines for contaminant levels in mineral supplements

Regulation of ingredients for animal feeds is a function of state government, and each state has a designated Feed Control Official responsible for such regulation. The Association of Animal Feed Control Officials (AAFCO) is a national group composed of the state control officials. One of its functions is to promote uniform regulations for the various states, so that interstate commerce of feed products is simplified.

A mineral feed committee of the AAPFCO recently has developed guidelines for contaminant levels permitted in mineral feed ingredients (8). These guidelines were based on a report on mineral tolerance of domestic animals by the National Academy of Sciences which included toxicity data of various elements to animals. Because nutrient requirements and toxicities of biological species vary considerably, precise contaminant limits are not set at the brink of toxicity, but rather at lower levels. The guidelines developed by this committee and accepted by the AAFCO must be incorporated into each state's regulations before they become official.

The guidelines have classed all known contaminants (non-nutrient elements) into four categories: highly toxic, toxic, moderately toxic, and slightly toxic. Three ranges of each element within each group were set as follows: (i) contaminant levels below which labeling was not required, (ii) a permitted range with required labeling, and (iii) level above which the product is prohibited for use in animal feeds. The list of contaminant elements and concentration ranges for each by groups is shown in Table III.

Table III Guidelines Suggested for Contaminants in Individual Mineral Feed Ingredients (Adapted from Reference 8)

Category	Maximum Tolerance Conc. in Complete Feed[1/]	Conc. not Required on Label	Conc. Required on Label	Conc. Prohibited for use in Feeds
	ppm			
1. HIGHLY TOXIC	1-9	<5	5-500	>500
Cadmium	0.5			
Mercury	2.0			
Selenium	2.0			
2. TOXIC	10-40	<100	100-1000	>1000
Cobalt	10			
Molybdenum	10			
Barium	20			
Tungsten	20			
Vanadium	10			
Copper	25			
Lead	30			
3. MODERATELY TOXIC	41-100	<500	500-2000	>2000
Arsenic	50			
Nickel	50			
Iodine	50			
Antimony	70			
4. SLIGHTLY TOXIC	101-1000	<2000	>2000	--
Boron	150			
Aluminum	200			
Bromine	200			
Zinc	300			
Bismuth	400			
Manganese	400			
Chromium	1000			

[1/]Dietary level that, for a limited period, will not impair animal performance and should not produce unsafe residues in human food derived from that animal.

Also included in Table III is the maximum tolerance level (concentration) in a complete feed ration. The maximum level permitted (MLP) in feed ingredients without labeling is calculated as the product of the continuous feeding level times the dilution factor, divided by a safety factor. For example, in the highly toxic category, the maximum tolerance level for Cd in the complete feed ration is 0.5 ppm. If the dilution factor (DF) for a feed ingredient was 25 and the safety factor (SF) was 2.5, the MLP for Cd in the feed ingredient would be 5 ppm, as shown in Table III. This procedure provides a systematic approach to establishing contaminant limits in animal feed ingredients. The equation used to set these limits was designed to handle the worst case situations. It was based on the most toxic element for each group and assumes the lowest dilution factor of the ingredient (DF=25). While the safety factors are subject to some judgment, they provide an additional margin of safety.

Transferring Industrial By-products for Use in Agriculture

How can those in the metal industries involved in processing and disposal of industrial by-products effectively transfer the proper by-products to agriculture? Because agriculture is in another sector of the economy, many may not have the proper contacts to determine which products could be safely used in agriculture and what are the desired physical and chemical forms. The following groups should be considered when planning alternate means of disposal of residues and effluents. Effective use of safe by-products in agriculture will constitute resource recovery versus costly disposal.

Waste exchanges

Waste exchanges have developed in the U.S. in recent years because of the increased recycling of industrial by-products. These exchanges serve as brokers for trading of wastes between producers and potential users. Periodic newsletters or catalogs are published which include lists of wastes with addresses and telephone numbers of the producers and distributors. Waste recyclers also may be listed along with the types of wastes which they handle. These newsletters or catalogs usually are available free of charge. Costs are supported by fees to list wastes or to advertise waste recyclers. There are ten waste exchanges which cover the U.S. and five waste exchanges in Canada. Waste exchange information is available from most state Departments of Environmental Management.

Fertilizer and feed industries

Many fertilizer and feed producers use by-products wherever possible because they usually are less costly. Addresses of many basic fertilizer producers can be obtained in the *Farm Chemicals Handbook* (9). Lists of brokers or traders who are involved in handling industrial by-products and other wastes which may be used by the fertilizer industry also are compiled annually (10, 11). The *Directory of Fertilizer Plants in the U.S.* (12) contains lists of all retail fertilizer plants in the U.S., along with State Fertilizer Control Officials and national and state fertilizer industry associations. A new edition of this directory will be available in 1992.

While there are fewer companies involved in producing trace mineral supplements for animal feeds, these companies could be contacted for possible sales of suitable by-products. Most brokers would be aware of the needs of these companies. The Animal Feed Control Official in each state also would be an excellent contact.

Regulations for By-product Use in Agriculture

Any paper concerning use of by-products in agriculture would not be complete without discussing the status of regulations on such use. At this time, there are few regulations except those defining and controlling hazardous wastes. Certainly, everyone is interested in managing the environment to prevent the entry of undesirable levels of certain metals into the human food chain. The Food and Drug Administration (FDA) has conducted an extensive study to determine normal "background" levels of Cd, Pb, and other elements in the major field crops and vegetables grown in the U.S. (13). Periodic analyses will be continued to determine if such levels are affected by changes in agricultural practices.

Much research has been done since 1970 to assess the benefits and risks of using municipal sewage sludge on cropland. Results of this research now are being used as the basis for city, state, and federal regulations concerning safe use of such wastes. Regulations governing use of industrial by-products in agriculture may be based on the same research data and concepts which govern the safe use of municipal sewage sludge.

One of the main conclusions of the sewage sludge research is that a "no observed adverse effect level" quality sludge has been defined (14). While sewage sludge can be beneficially used for cropland, certain sludges may contain unacceptable levels of heavy metals. Hazardous wastes are defined in CFR 40. Allowable cumulative metals and organic pollutants in soils are governed by the Clean Water Act. New research data are being used to formulate changes in amendments to this Act. Twelve pathways for risk assessment (i.e., sludge→soil→plant→human) were considered, along with the types of individuals most exposed to possible effects of contaminants in sludge. Maximum tolerable levels of elements in domestic livestock diets also were included in this assessment. All of these data were used to formulate sludge management regulations. It is entirely probable that similar regulations governing use of inorganic industrial by-products as fertilizers will be developed in the future. Such regulations have been or are being developed for animal feed ingredients, as was discussed earlier.

References

1. United States Dept. of Agriculture, Crop Reporting Board, Statistical Reporting Services, Commercial Fertilizers, Consumption for Years Ending 1968, 1972, 1976, 1980, and 1984, (Washington, D.C.: U.S. Government Printing Office).

2. L. B. Nelson, History of the U.S. Fertilizer Industry, (Muscle Shoals, AL: Tennessee Valley Authority) 1990.

3. J. J. Mortvedt, "Plant Uptake of Heavy Metals in Zinc Fertilizers made from Industrial By-product," J. Environmental Quality 14 (1985), 424-427.

4. J. J. Mortvedt, "Cadmium Levels in Soils and Plants from Some Long-term Soil Fertility Experiments in the United States of America," J. Environmental Quality 16 (1987), 137-142.

5. E. R. Miller, X. Lei, and D. E. Ullry, "Trace Elements in Animal Nutrition," In J. J. Mortvedt et al. (ed.) Micronutrients in Agriculture 2nd Ed. (Madison, WI: Soil Science Society of America), 1991, 593-662.

6. J. Nelson, "The Animal Feed Industry as a Possible Market for "Clean" Metal Salts and Compounds from Reclamation Operations," In J.H.L. van Linden, D. L. Stewart, Jr., and Y. Sakai (ed.) Proc. Second Intl. Symposium Recycling of Metals and Engineered Materials. (Warrendale, PA: The Minerals, Metals, and Materials Soc.) (1990) 477-484.

7. E. R. Miller, P. K. Ku, J. P. Hitchcock, and W. T. Magee, "Availability of Zinc from Metallic Zinc Dust for Young Swine," J. Animal Sci. 52 (1981), 312-315.

8. A. Hanks, et al., "Guidelines for Contaminant Levels Permitted in Mineral Feed Ingredients," Official Publ. Assn. of Animal Feed Control Officials (1991) 170-175.

9. C. Sine, (ed.) Farm Chemicals Handbook, (Willoughby, OH: Meister Publishing Company) 1991, B70-B94.

10. K. Baker, (ed.) Who's Who in Fertilizer and Ag-Chemical Supply, Tenth Ed., (Houston, TX: Stalsby/Wilson Press) 1990.

11. Anon., "Fertilizer Traders Directory 1990," Fertilizer Focus 7 (6) (1990), 16-34.

12. J. T. Berry, et al., Directory of Fertilizer Plants in the United States - 1988, Association of American Plant Food Control Officials (available from Joel S. Padmore, North Carolina Dept. of Agriculture, 4000 Reedy Creek Rd., Raleigh, NC 27607).

13. K. A. Wolnik et al., "Elements in Raw Agricultural Crops in the United States. 3. Cadmium, Lead, and Eleven Other Elements in Carrots, Field Corn, Onions, Rice, Spinach, and Tomatoes," J. Agricultural & Food Chem. 33 (1986), 807-811.

14. R. L. Chaney, "Twenty Years of Land Application Research," Biocycle 31 (9), (1990) 55-59.

COAL DUST WASTE REDUCTION

Larry B. Fiske

GEORGIA-PACIFIC CORPORATION
Chemical Division
1754 Thorne Road
Tacoma, Washington 98421

Elimination of dust and processing of fines for both thermal and metallurgical coals is addressed. The efficiency of treatment with lignosulfonates compares favorably to oil in thermal coal. Effect of treatment with various lignosulfonates upon cokability of metallurgical coal was determined by Free Swelling Index, Gieseler Plasticity and Arnu Dilatation.

Contributors in this work were Dr. Hope Huntington of the UEC COAL AND COKE LABORATORIES in Monroeville, Pennsylvania, and Dr. Richard Buchholz and Ms. Sara Madsen of GEORGIA-PACIFIC CORPORATION, Chemical Division.

Residues and Effluents - Processing and Environmental Considerations
Edited by Ramana G. Reddy, William P. Imrie and Paul B. Queneau
The Minerals, Metals & Materials Society, 1991

Introduction

The abundance and distribution of coal makes it a natural choice for use in electrical generation, gasification and liquefaction to make alternate fuels. Since it is a polluter, providing ways to economically eliminate the problems of gaseous and particulate emissions are vitally important in its future utilization. The particulate emissions, while localized, have been found difficult to eliminate due to economic, physical and use constraints (1). Current allowable treatment costs are one-to-two dollars a ton. Moisture, which varies uncontrollably, affects the efficiency of the dedusting treatment. The treatment should not greatly affect the moisture, sulfur content, chloride content, ash and BTU value, as well as specialized physical properties such as cokability for metallurgical coal.

The use of lignosulfonate solutions obtained from acid sulfite pulping of wood as binders has been well documented (2,3,4,5). Their prices are likewise attractive, being only a few cents per pound. Their application to controlling dust normally lost to wind or in-process waste streams is the scope of this research. Two case histories will be discussed. The first involves coking or metallurgical coal and the second deals with baghouse fines from a fluidized bed coal dryer.

Metallurgical Coal

A great deal of coal fines was lost from the surface of a large coal stockpile being constantly fed by rail and shipped out by water. The coal was quite hydrophilic, thus easily treated with water, but the surface dried out rapidly to resume losing fines to the atmosphere. The elimination of coal dust losses during transfer and storage was attempted by the uniform application of fermented or unfermented calcium lignosulfonate liquor. Metallurgical coal from Western Canada was found to require about 1.5% of lignosulfonate solids to remain dust free following air-drying. The moisture content of these coals was 7-9%, except on the surface, where it rapidly decreased to 1-2%. Chloride content, which may average around 0.01%, would not be affected by the addition of lignosulfonates containing 100 ppm or less chloride. The increase in sulfur content, even if all the coal was treated, would amount to 6% of 1.5%, or 0.09%. Since most Western Canadian coals have about 0.3-0.6% sulfur, this may be detectable. However, since surface treatments alone would be effective, a much lower total treatment level would be used, making any increases in sulfur below the level of detection. A decrease in BTU value for the completely treated coal would amount to about 80 BTU/lb due to the lignosulfonate, which has a fuel value of 8800 BTU/lb. Metallurgical coals average about 14,000 BTU/lb. Other analyses would not be affected. The residual dust was determined by three methods: sieve analysis down to 400 mesh, fugitive dust captured and weighed on a millipore filter above tumbling coal, and visual observation of the treated coal dropped near an electric fan. All three methods had good correlation, showing a decrease from 5-10 percent dust to less than 0.5% on the treated coal.

The coal treated with the most effective lignosulfonate products was also tested for cokability by means of Free Swelling Index, Gieseler Plasticity and Arnu Dilatation. Unfermented calcium lignosulfonate gave the most encouraging results.

Table I Cokability Tests on Coal

Sample	Untreated	0.86% U-CaLS	1.68% U-CaLS
FSI	7.0	8.0	8.0
Gieseler Plasticity			
Maximum Fluidity	4.6	20.4	11.4
Solid. Temp.	449	488	492
Range, °C	52	58	58
Arnu Dilatation			
Total Dilution, %	2	--	8

Fermented calcium lignosulfonate at 1.0% of the coal (solids on solids) gave less encouraging results, decreasing the FSI to 6.5, the Gieseler Plasticity maximum fluidity to 1.0 and plastic temperature range to 32°C. Free Swelling Index is the amount that a compacted coal pellet expands upon controlled heating to about 500°C. Gieseler Plasticity gives the rheological characteristics of a compacted coal while being heated through its plastic range under constant torque, and Arnu Dilatation shows the linear expansion or contraction of a compacted pencil of coal during similarly controlled heating. A high FSI is desirable, a high fluidity and large plastic range important in the Gieseler Plasticity, and no change or an increase in dilatation is wanted in the Arnu Dilatation. Although these tests do not substitute for a full scale or laboratory coking test, such as set forth by ASTM D2014-85, it is considered indicative of changes in cokability if the appropriate standard is used.

Since coal properties may be greatly dissimilar, it is important to deal with each separately, but the results obtained with unfermented calcium lignosulfonate are encouraging. The major factors are still economic and environmental, with the environment winning slowly.

Thermal Coal

Alleviating dust in dried thermal coal was found to be much more difficult, even at much higher moisture levels, perhaps due to the electrostatic charge on the extremely fine dust generated. These fines, which amount to about three percent of the total coal, are swept from the product by the heated air. They are collected in a series of baghouses and are recombined with the coal prior to treatment with an oil dedusting agent. However, the product is not satisfactory to the electrical generating plants when these fines are recombined. This dust contained about 7% moisture and was currently a very difficult disposal problem. It was largely composed of fines passing a 500 mesh screen. The current treatment for the dried coal was Bunker C oil, which appeared to eliminate the dust immediately following treatment but could not control dust following the coal's transfer by railcar to the electrical generating plant. The dust problems thus were twofold, involving both the disposal or recycling of the fines and satisfaction of the customer's dust problem in the dried product.

This very fine dust, with a moisture content of 7%, was extremely difficult to treat. Wetting with water was difficult and liberated a large amount of dust during mixing. Treating the dust with dilute calcium lignosulfonate solutions did not eliminate the loss of a large amount of dust during treatment. Combining the fines with coarser coal made a slight improvement in treatability. Treatment of the coarser coal prior to adding the fines avoided the large losses of dust. An equal weight of the coarse coal was found to be minimum.

At this point, it became necessary to more accurately determine the quantitative amount of dust in a laboratory sample in order to more easily assess the effectiveness of different products. An optical dust analyzer was built using a 30-inch free-fall tower which utilized a helium-neon laser and a photo detector to determine when the extremely fine dust had passed the detection point. The chart print-out showed a large, slowly decreasing peak for extremely dusty materials and a small, sharp peak for non-dusty ones. This instrument could give information about the size distribution of the dust particles through the slope of the curve as it returned to the base line. The rate of settling could also be quantified by determining the time necessary for the curve to return to the base line. Sensitivity was much greater than earlier methods, and testing time was also reduced. A complete test could be made in less than five minutes.

Figure 1

OPTICAL DUST ANALYZER

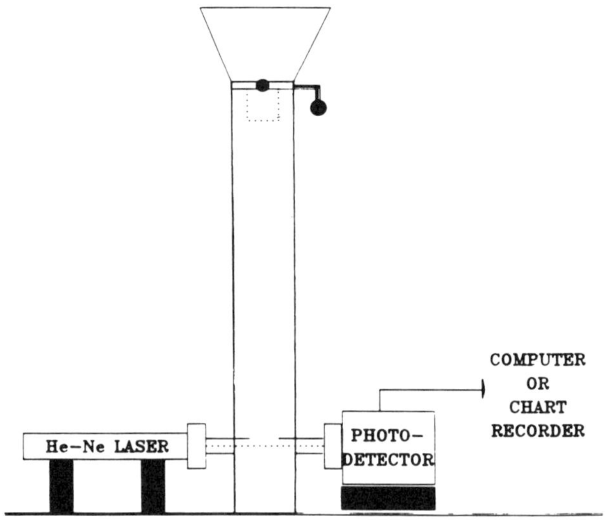

To simulate the tumbling and separation of fines found during transporting in a railcar, an attrition test was devised which helped predict the durability of the bond formed with the dust. This used a baffled container which was rotated at a set speed for one hour. Work on the finished coal was continued using our latest techniques. The dust in this product was most effectively reduced by the addition of a modified unfermented lignosulfonate containing a tackifier. The product use level to achieve the stability towards attrition was quite high, 10 gallons per ton of dried coal. Additionally, the moisture level could not be much below 12% for the product to perform adequately following the attrition test.

Table II Analysis of Treated Coal Samples

Sample	% Water	Transmission, % 15 Sec.	Transmission, % 60 Sec.	Transmission, % After Attrition 15 Sec.	Transmission, % After Attrition 60 Sec.	Product
1	8.6	74	82	24	40	3.6% LS
2	9.0	60	73	23	79	3.2% LS
3	12.0	79	91	54	79	Oil
4	12.0	98	99.5	90	94	3.6% LS
5	12.0	98	99.5	50	70	3.2% LS

It is extremely difficult to remove all dust from the dried coal, requiring 3.6% of a modified unfermented lignosulfonate to eliminate all but a trace of dust following the attrition test. At this use level, the treatment is probably economically unjustified, but it is the only material that has thus far succeeded in eliminating the dust problem.

REFERENCES

1. Joseph Haggin, "Coal," Chemical and Engineering News, 69 (24) (1991), 32-35.

2. R. S. Shallenberger, and G. G. Birch, Sugar Chemistry (Westport, CT: Avi, 1975), 158-159.

3. M. Grayson, ed., Kirk-Othmer: Encyclopedia of Chemical Technology, 1 (3rd Ed) (New York, NY: J. Wiley, 1979), 488-510.

4. W. A. Knepper, ed., Agglomeration (New York, NY: J. Wiley, 1962), 379-414.

5. M. Grayson, ed., Kirk-Othmer: Encyclopedia of Chemical Technology, 21 (3rd Ed) (New York, NY: J. Wiley, 1979), 77-104.

BENEFICIAL REUSES FOR

SPENT BRIDGE PAINTING BLAST MATERIAL

Thomas E. Weyand

Pittsburgh Mineral & Environmental Technology, Inc.
700 Fifth Avenue
New Brighton, PA 15066

Abstract

Prior to 1975, lead-based paints were commonly used in protecting steel structures. Because of the toxic effects of lead on human health and the environment, their use has been significantly reduced. However, problems with these paints persist when they are removed during normal maintenance by pneumatic blasting with mineral abrasives.

Technologies are developed to contain the spent abrasive and paint during blasting operations. This paper reports on a study to identify beneficial reuses for the spent blasting material. Nine samples were gathered from different projects within Pennsylvania. The samples were thoroughly characterized with regard to composition, morphology, and EP and TCLP leach test performance. Potential reuse applications will be reviewed with regard to technical adequacy, economics, and environmental constraints.

This work was supported by a grant from the Pennsylvania Department of Transportation and the Federal Highway Administration Research Project 89-02, Mr. James Merrill, Contract Manager.

Residues and Effluents - Processing and Environmental Considerations
Edited by Ramana G. Reddy, William P. Imrie and Paul B. Queneau
The Minerals, Metals & Materials Society, 1991

Introduction

PennDOT (Pennsylvania Department of Transportation) maintains 15,000 bridges in its system, 5,500 of which are steel structures. This number does not include the bridges owned and maintained by counties, municipalities, and other agencies in the State. In the mid 1980's, PennDOT recognized the seriousness of the lead paint removal problem and instituted an aggressive program to control air and water pollution resulting from blasting operations. Compliance with existing rules have increased painting costs significantly and consumed excessive amounts of staff time in assuring compliance.

Proper disposal of the spent blasting material remains a costly, time-consuming problem. As of this date, a permitted hazardous waste disposal site does not exist in Pennsylvania. In addition, disposal of nonhazardous waste containing lead remains quite difficult, and proposed residual waste regulations for the State will require disposal of nonhazardous lead-bearing residual waste in double-lined landfills similar in construction to those required for hazardous waste disposal.

Prior to this project, various potential reuse applications were proposed as an alternative to land disposal(1)(2)(3)&(4). In fact, a PennDOT contractor proposed the use of spent blasting abrasive as flowable backfill, researchers in North Carolina studied uses in asphalt and cement concrete paving formulations, and others have suggested a variety of methods to minimize the volume of waste generated.

The goal of this research project was to identify a direct, beneficial reuse that would minimize or eliminate spent abrasive disposal. Additionally, the reuse should comply not only with present but envisioned regulations as well as provide an economical alternative to land disposal alternatives.

Regulatory Summary

From a federal point of view, the land disposal ban(5) and the change from the EP Toxicity to the TCLP test(6) are the primary motivating forces toward identifying alternatives to land disposal.

However, within the State, the PADER has published proposed(7) regulations regarding the disposal of residual wastes. These proposed regulations may well indicate why Module 1 permits for disposal of lead-bearing nonhazardous wastes in Pennsylvania sanitary landfills have been so difficult to obtain. The new regulations will be far more stringent than previous with regard to wastes containing lead and arsenic. Such material must be disposed in Class 1 sites which will have operating requirements quite similar to those for hazardous waste disposal sites.

The Pennsylvania DER requires notification of intention to recycle hazardous material. The recycling process must be outlined in the notification, and the product chemical and physical characteristics must meet those of similar products used for the same purpose.

Sample Characterization

Nine samples from three regions of the State were provided by PennDOT. Each of the samples were screened at 1/4 inch to remove various articles of trash. Samples were then thoroughly blended and riffle split into samples for the following analyses:

Bulk Chemical Analyses
Screen Size Analyses
Microscopic Characterization/Modal Analysis
X-Ray Diffraction Analyses
Scanning Electron-EDX Analysis
Environmental Leach Testing (EP, TCLP)

The purpose of this detailed sample characterization was to identify any significant anomalies in the abrasive or paint compositions in each of the sample groups. Furthermore, this characterization provided important background information necessary to support any reuse applications and the estimation of the amount of upgrading required to meet reuse product specifications.

Bulk Chemical Analyses

In general, the samples effectively represent a very broad span of physical and chemical properties--as expected from this type material. Based on the bulk sample chemistries presented in Table I, Samples 104-1, 2, 3, 5, 6, & 8 are coal slag abrasives. The elevated copper and zinc levels indicate that Samples 104-4, 7 and 9 are copper slag abrasives. Regardless of the paint content of the samples, the copper slag samples contained the greatest amount of total lead in the group.

Table I - Quantitative Bulk Chemical Analysis (Wt. %)

SAMPLE #	Pb	Cu	Zn	Ti	Al	Si	Ca	Fe
104-1	0.15	0.00	0.02	0.6	10.8	22.5	2.9	16.7
104-2	0.26	0.00	0.03	0.6	10.7	21.7	2.8	18.4
104-3	0.47	0.00	0.04	0.6	10.8	22.2	2.9	16.5
104-4	0.84	0.94	3.75	0.4	6.0	14.1	6.6	23.6
104-5	0.20	0.04	0.27	0.6	11.4	21.6	2.9	15.5
104-6	0.28	0.07	0.42	0.6	11.6	21.8	2.8	15.1
104-7	0.52	0.69	3.25	0.4	7.4	15.5	5.4	23.1
104-8	0.50	0.01	0.16	0.7	11.2	20.2	1.2	20.1
104-9	0.92	0.56	3.15	0.4	7.1	16.1	4.4	23.4

No other classes of mineral abrasives, for example silica sand, garnet, or nickel slags, were represented in this group of samples.

Screen Size Analysis

A sieve analysis was performed for each sample. In addition to the weight fractions for each screen interval, analytical and mineralogical samples were prepared to determine the lead distribution and mineralogical composition by screen fraction.

Figure 1 represents the results of one of the samples. Although the screen size distributions varied considerably from sample to sample, no particular bias was found in the distribution of lead in the various screen fractions. These data clearly indicate why size classification methods are not effective in reducing the amount of lead present in spent blasting abrasives.

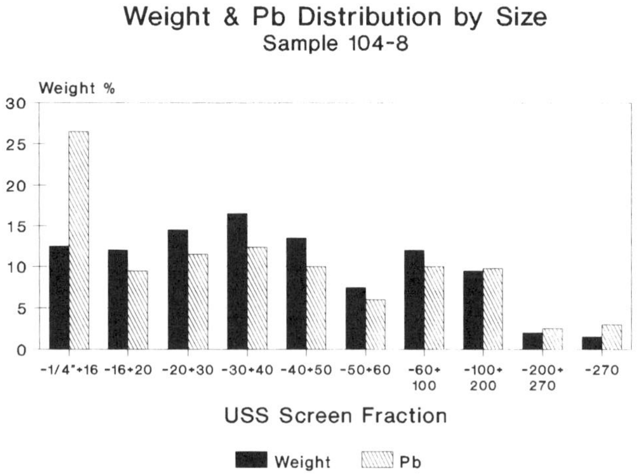

Figure 1

Microscopic/Modal Analysis

A modal analysis of each screen fraction was determined for each sample. Figure 2 presents the results of one sample. Note that in this sample, the coarse screen fractions contain a majority of abrasive material, while the finer screen fractions consist of a majority of metal, rust, and soil contaminants.

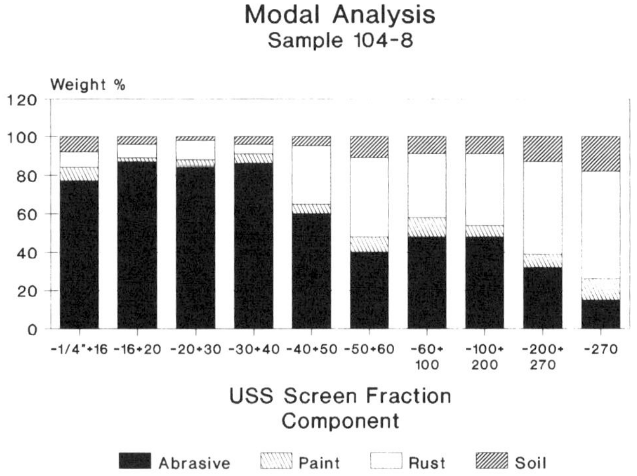

Figure 2

X-Ray Diffraction Analyses

Samples of abrasive particles were ground and analyzed by X-ray diffraction methods. This method identifies the crystalline compounds present. The results of these tests indicated that the abrasives are mostly amorphous (glassy) phases, as expected in these slag systems. All samples did contain trace to minor amounts of crystalline SiO_2 (quartz). In addition, the three copper slag samples, 104-4, 7 and 9, contained trace amounts of the following crystalline phases:

FeO	(wüstite)
$FeAl_2O_4$	(hercynite)
and $Mg(Al,Fe)_2O_4$	(spinel)

SEM/EDX Analysis of Paint Chips

Paint chips from each of the samples were mounted such that a cross section of the chip could be analyzed. The chips generally exhibited several layers of coatings, with the "bottom" surface identified easily by the orange/red color of the primer while the "top" surface exhibited the pigment of the top coat layer. Sections of the top and bottom portions of the chips were analyzed using an EDX spectrometer. The results of these analyses are presented in Figure 3. The paint chips generally contained much higher lead content on the bottom sample than the top coating. No materials other than normally expected paint filler and pigment compounds were identified in this analysis.

SEM/EDX Analyses of Paint Chips

Element	"Top"	"Bottom"
Pb	7-40%	21-90%
Zn	0-23%	0-3%
Fe	0-3%	0-40%
Cr	0-4%	0-2%
Si	4-11%	0-14%
Al	3-37%	0-10%
Ti	0-29%	-

Figure 3

Environmental Leach Testing

Environmental leach testing was performed by a commercial lab with US EPA CPL credentials. The results are shown in Table II. Based on the EP leach results, three samples (104-1, 2, 5) passed the leach test, three marginally failed the leach test (104-3, 6, 7), and three failed by a large margin (104-4, 8, 9). In both the EP and TCLP tests, **the only hazardous metal extracted in significant amounts was lead.** The EP leach data correlated loosely with the total lead in the samples, providing an r squared correlation of 0.79 for the equation:

$$\text{EP lead (ppm)} = 16.17 * \text{Total Pb (wt \%)}$$

Table II - Summary of EP/TCLP Data

Coal Slag	EP Pb ppm	TCLP Pb ppm	Ratio TCLP: EP	EP Xtrn %	TCLP Xtrn %	Total Pb %	Total Fe %	Fe Metal %
104-1	1.8	6.9	3.8	2.4	9.2	0.15	16.7	0.08
104-2	3.1	22.0	7.1	2.4	16.9	0.26	18.4	0.23
104-3	5.8	25.0	4.3	4.5	10.6	0.47	16.5	0.30
104-5	4.0	11.0	2.8	4.0	11.0	0.20	15.5	0.21
104-6	5.4	13.0	2.4	3.9	9.3	0.28	15.1	0.25
104-8	9.7	23.0	2.4	3.9	9.2	0.50	20.1	0.15
Cu Slag								
104-4	19.0	25.0	1.3	4.5	6.0	0.84	23.6	0.71
104-7	5.2	0.7	0.1	2.0	0.3	0.52	23.1	0.67
104-9	13.0	1.5	0.1	2.8	0.3	0.92	23.4	0.84

All TCLP results were in excess of 5 ppm lead, ranging from a low of 6.9 to a high of 25 ppm, except for Samples 104-7 (0.73) and 104-9 (1.5). Curiously, these samples were both in excess of 5 ppm on the EP leach test and contained rather high levels of total lead, 0.52 and 0.92%.

The leach test procedures used were carefully rechecked, and no obvious errors in technique or calculations were found. Nor can this anomaly be ascribed to the fact that 104-7 & 9 were copper slags since Sample 104-4 was also a copper slag with similar chemical composition. The EP and TCLP data for lead were recalculated to present the % extraction. The term "% extraction" refers to the amount of lead analyzed in the leach solutions versus the amount (total) available in the solid samples. The first observation possible from this data is that in the EP Leach test, the lead responded in the same relative fashion for both types of slag. Therefore, if lead is present in the copper slag abrasive, it is responding with similar rates of extraction as the coal slags where lead was only present in the paint particles. This similar response permitted the loose correlation between the lead content of the sample and the EP lead values shown above.

However, the performances of the two slags are considerably different in the TCLP test. The TCLP % extractions were clearly greater in the case of the coal slags versus the copper slags. Since the performance was similar in the EP Leach test, the data in the TCLP does not seem related to lead morphology, for example lead encapsulated in the glassy slag matrix. Therefore some type of interference is affecting the leaching process in the TCLP test.

Iron has long been shown to interact with heavy and precious metals(8). The mechanism is commonly encountered in the primary extraction of precious metals where tramp iron generated during ore milling scavenges the heavy and precious metals from the process solutions. The use of iron filings was reported in the past as one method to render brass foundry sands contaminated with lead as "nonhazardous."

Total iron and metallic iron analyses for the nine samples are presented in Table II. As a group, the copper slags have higher total iron and metallic iron contents. Therefore the iron values present in the copper slags may have some effect on the leaching response. However, satisfactory correlations against the various iron analyses or combinations thereof failed to demonstrate a good fit against lead extractions.

Based on these data, the two abrasive slag types responded differently to the TCLP test. The lower lead leachability seems to be related to the presence of iron or iron compounds present in the spent abrasives. The fact that EP leach test results provided a more uniform set of lead extraction data suggests that the anomalies found in the TCLP data relate to solution chemistry effects rather than differences in lead morphology in the abrasives. Based on these preliminary data and prior literature citations, iron is reducing the dissolved lead from solution by a cementation reaction. However, insufficient samples and test data were available from this project to conclusively correlate the exact mechanism of the interference noted here.

It is concluded that iron filings present in the spent abrasive samples are cementing dissolved lead from the leach solutions. Such fixation is only temporary in nature. Once the iron is oxidized due to environmental stress, the lead sponge will readily redissolve(9).

Beneficial Reuse Options

Eight different reuse applications were evaluated in this study. The feasibility of each application was determined based on the technical performance of the spent abrasive in the reuse, anticipated regulatory acceptance of the reuse, and the potential market volume available for the application.

The results of this study are summarized in the data in Table III.

Table III - Beneficial Reuse Summary

REUSES	TECHNICAL PERFORMANCE	REGULATORY CONSIDERATIONS	MARKET
Portland Cement	Marginal-Poor	Stab. Concern	Large
Asphalt Concrete	Marginal	Stab. Concern	Large
Cement Kiln Feed	Acceptable	Part B Needs	Large
Blasting Abr. Reuse	Marginal to Acceptable	Acceptable at Site	Driven by Eco.
Polishing Abrasives	Poor	No Problems	Small
Airport Antiskid	--	Not Permitted	--
Pb Smelter Feed	Acceptable	Slag Former	Small
Structural Clay Products	Acceptable-Superior	Should be Acceptable	Large

The paving applications yielded marginal to poor performance in the testwork. It is believed that improved performance can be achieved with modifications to the mix formulas. Of greater concern however, is that the lead-bearing constituents in the spent abrasive are mechanically encapsulated in the paving mixes. If the pavement deteriorates, then leaching of the lead compounds can take place. In Pennsylvania, the environmental regulators will have difficulty in accepting this application.

Testwork indicates that spent abrasive can be used as a feed material in cement manufacturing kilns up to possibly 10% of the kiln feed. Clinker made with spent abrasive demonstrated equivalent properties to that prepared from virgin materials. The cement industry has ample capacity to utilize the spent abrasive generated annually, however, there is some reluctance to date for the environmental regulators to permit cement kilns to accept spent abrasives testing hazardous.

Eight of the nine samples in this testwork yielded coarse abrasives of sizes suitable for reuse in the 40-50% range. Therefore, reuse of spent abrasives after a size classification step is a technically viable approach. (Debris removal may be required prior to reuse to ensure that cleaned surfaces are not contaminated.) However, this reuse only has economic impact on the blasting operation since ultimately the fine spent abrasive generated must be disposed or reused in another application.

The use of spent abrasive as lead smelter feed probably will be judged "sham" recycling since the lead content of this material is so modest. However, smelters utilizing blast furnaces purchase slag-forming ingredients similar to those found in the spent abrasive. With flux additions, spent abrasives are an appropriate blast furnace feed for their slag requirement. Unfortunately the market for this application is very modest with each blast furnace requiring no more than 2000 tons per year.

Reuse in polishing abrasives was not technically successful. Direct reuse in airport antiskid application was considered a direct land application which would not be permitted.

Clearly the most promising reuse application identified for spent abrasive is in structural clay applications.

The spent abrasive was ground to pass a 20-mesh screen, no upgrading of the material was performed beyond grinding. Ground spent abrasive was mixed with standard brick mix in loadings of 20 and 40 weight %. Table IV contains the specifics of three brick mixes produced as part of this evaluation. Lower moisture was required as the % of spent abrasive increased in the test mixes. This resulted in slightly reduced shrinkage of the bricks on drying. After firing at 1200°C, the test bricks demonstrated lower shrinkage against the standard and significantly higher strength.

Table IV - Bar Properties
Spent Abrasive (104-8) & Brick Mix

Mixture	Test #1	Test #2	Test #3
Brick Mix (%)	100	80	60
Ground Abrasive (%)	0	20	40
Moisture (%)	19.6	18.5	15.8
Dry Properties			
Shrinkage (%)	4.67	4.67	3.89
Density (g/cc)	1.97	1.93	1.99
Fired (1200°C) Properties			
Shrinkage (%)	3.40	2.39	3.11
Density (g/cc)	2.08	2.04	2.09
Strength (MPa)	11.50	15.70	16.09

The fired test bricks demonstrated a uniform red color (Figure 4) and no indication of bloating or glassy exudations. A microstructure of the fired material reveals a uniform structure, there is no evidence of the paint chips in the final structure, lead was detected in the cordierite phases using SEM/WDX methods. This lead is very resistant to subsequent environmental leaching. A brick sample from composition #3 was pulverized to pass a 100-mesh screen and subsequently tested on the TCLP protocol. Lead in the TCLP filtrate was below detection (<0.02ppm).

Figure 4 - General View of Bricks

In summary, spent abrasive can be incorporated into brick mixtures in significant volumes, yielding bricks with improved mechanical properties and similar appearance to those with virgin constituents.

The lead compounds in the spent abrasives are chemically altered during the brick firing process, yielding potentially very stable silicate compounds. The tri-state area of Pennsylvania, New Jersey, and New York produces approximately 1.2 million tons of bricks annually. At least on a theoretical basis, this market alone could absorb one half million tons of spent mineral abrasives annually.

In our opinion, of the reuses evaluated, the structural clay product reuse application is the best fit for Pennsylvania. Spent abrasives added to the virgin brick mix yield improved technical performance at very significant loadings and the hazardous components in the spent abrasive are physically and chemically altered into more stable compounds that apparently yield improved strength at potentially lowered processing costs. Lastly, the market is large and presumably can easily accommodate the production of at least the region's spent abrasive production.

References

1. Paul Khosla and M. Leming, "Recycling of Lead Contaminated Blasting Sand in Construction Materials," in B. Appelman, L. Smith, J. Zappa (eds.), Lead Paint Removal From Steel Structures, Steel Structures Painting Council, Pittsburgh, 1988.

2. William Medford, "Containment and Beneficial Reuse of Blasting Sand in Asphalt Concrete," North Carolina Dept. of Transportation, September, 1989.

3. William Medford, "Containment and Beneficial Reuse of Blasting Sand in Asphalt Concrete: A Case History," Journal of Protective Coatings & Linings, p. 36, Vol. 7, No. 1, January 1990.

4. Letter from L. Lunsk, PADER, Norristown, PA to William Ward, IA/Buckley Company, November 25, 1988.

5. "Land Disposal Restrictions for Third Scheduled Wastes, Proposed Rule," USEPA, 40 CFR Part 148 et al., Federal Register, November 22, 1989.

6. "Hazardous Waste Management System; Identification and Listing of Hazardous Waste; Toxicity Characteristics Revisions," USEPA, 40 CFR Parts 261, 265, 268, 271, and 302, Federal Register, March 29, 1990.

7. "Proposed Rulemaking--Residual Waste Management," PADER, Pennsylvania Bulletin, Vol. 20, No. 8, February 24, 1990.

8. T. R. Ostrom & P. K. Trojan, "Lead Compound Formation in Alloy CDA 84400 and the Effect of Iron on Leachable Lead," AFS Transactions, 86-133.

9. Wolfgang Baum, et al., "The Use of Process Mineralogy for Plant Optimization at the El Indio Gold-Silver-Copper Operation (Chile)," ERZMETALL, 42, (1989) No. 9, P. 373.

Studies on Commercial Utilization of Industrial Wastes -

Magnesite Dust and Fines

S. Prakash and K.N. Gupta

National Metallurgical Laboratory

Burmamines Jamshedpur
Bihar, India 831007

Abstract

The paper concerns an unconventional, yet promising, use of waste magnesite dust and fines. In India a significant amount of this dust is generated from some of the integrated iron and steel industries, refractory and cement plants, during calcination of raw magnesite. However, it is possible to utilize this dust, which is normally a waste , conveniently and economically for the production of magnesium chemicals.

A bench scale study was carried out to determine the rate of formation of sulfated compounds of Mg,Ca,and Fe, inter alia, present in the dust system containing MeO constituents (in which Me=Mg,Ca,Fe,Si,Al etc.) . Effects of various process parameters, which affect the reaction rate as also the purity of the product ,were also studied . Some work carried out to develop the process flow charts involving the salient steps to produce magnesium chemicals commercially is reported in the paper .

Residues and Effluents - Processing and Environmental Considerations
Edited by Ramana G. Reddy, William P. Imrie and Paul B. Queneau
The Minerals, Metals & Materials Society, 1991

Introduction

In recent years utilization of process wastes for the production and / or recovery of valuable product has attracted considerable industrial interest ,particularly ,in the developing countries.The interest in secondry sources is ,primarily ,due to use of parent resources leading to relatively high energy consumption, high unit cost and subsequent increase in the cost of the production.

In India the total consumption (1) of calcined magnesite is around 1.5 MTPA and 20-30% of this is being presently imported, particularly, sea-water magnesite (2) for its purity. In the process of magnesite calcination in rotary/shaft kiln, the off-gases are passed through the cyclone, for dust separation. As high as 10-20% of the calcined magnesite goes directly as trapped dust in the cyclone.

As most of these wastes contain appreciable amount of magnesium oxide (MgO=50-60%) along with calcium oxide (CaO=5-6%); ferrous oxide (FeO = 10-12%) and silica (SiO_2 = 10 - 15%), it is certainly worthwhile exploring ways to recover magnesium chemicals from these wastes. Integrated iron and steel industries, refractories and cement plants are some of the major industries generating this type of dust in India.

Applications

Of the several magnesium chemicals, the most widely used chemical is magnesium - sulfate-heptahydrate ($MgSO_4$, $7H_2O$). Its applications include manufacture of quality magnesite refractory bricks, micronutrients to tea and cotton growing areas, fire-proofing compositions and as tanning agents etc. The use of magnesium sulfate, as the common denominator, ranging from agriculture to other industries are illustrated in figure 1.

Commercial Production of Magnesium Salts

Commercially, magnesium salts are mostly produced from the magnesite bearing minerals such as Natural-Keiserite, Langbeinite, Brucite, Sea-water concentrates, Natural-Brines, Magnesite and Dolomite.

In the normal commercial process of producing magnesium sulfate, magnesium minerals are decomposed by heating at 850-900 deg C to magnesium oxide. Sulfur di oxide (SO_2) is then absorbed in the slurry of magnesium oxide to form magnesium-bisulfate solution, which is followed by aerial oxidation at 50-60 deg C in the presence of metallic catalyst. The conventional process is evidently quite expensive. Besides it requires extensive purification to recover magneium sulfate in pure form, the energy requirement for decomposition of mineral is in astronomical proportions.

Alternatively a method of double decomposition of gypsum and magnesium carbonate can be used to produce magnesium - sulfate -heptahydrate. The reaction being reversible, it becomes difficult to carry out the process to completion , unless magnesium sulfate is constantly removed from the system.

In the above processes, the end product is magnesium sulfate. This can be used as starting material for other magnesium chemicals. Magnesium carbonate, amongst others, got its commercial boost, owing to its outlets for the development of more important uses : starting material for the preparation of magnesium nitrate solution; as an antacid, in pharmaceutical industries; in paint and printing ink industries; in heat insulation compositions etc.

However, for industrial use magnesium carbonate is manufactured from two primary sources: natural sources, magnesite and dolomite ,and slurries of magnesium hydroxide or hydrated magnesium oxide .The latter may be obtained from sea-water.

The process for magnesium carbonate production, utilizing sea water as the raw material, comprises absorption of carbon dioxide (CO_2) in magnesium hydroxide slurry and subsequent boiling of the resulting magnesium carbonate trihydrate ($MgCO_3$, $3H_2O$).

Another commercial method of preparing magnesium carbonate, from dolomite, is by calcination of dolomite in rotary or stack kilns. The calcined product (CaO, MgO) is slurried in water and carbonated with the calcination stack gases, which are scrubbed and collected for this use. The resulting slurry contains magnesium-bicarbonate in solution , with

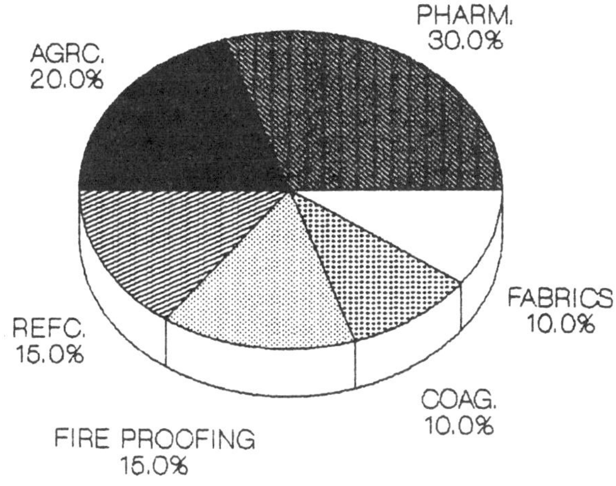

Fig.1 USES of MAG-CHEMICALS MAG-SULF AS A COMMON DENOMINATOR

calcium carbonate and other impurities in the suspended solids which are separated by sedimentation. The liquor is then heated for precipitation of voluminous basic carbonate. The magnesium hydroxide slurry obtained from sea-water is relatively in the purer form.

NML Proposed Process

In this process, developed at NML, the waste dust, containing 50-60% MgO; 5-6% CaO; 10-12% Fe_2O_3; 10-15% SiO_2, in the size range of 60-80 mesh, is treated with sulfuric acid in a stirred tank reactor under controlled conditions. This is followed by vaccum crystallization under meticulously maintained conditions, due to numerous metastable phases formation and their slow equilibrium attainments, Fig.2(3), during crystallization of Mg-salts, to obtain acceptable $MgSO_4$, $7H_2O$.Successively, the crystallized material is fed into the centrifuge for separation of crystals from mother liquor. The process flow chart involving the salient steps to produce magnesium-sulfate-heptahydrate and magnesium carbonate are presented in Figs. 3 & 4 respectively.

By operating at controlled conditions, approaching the optimum experimental levels, the dissolved impurities would limit their reaction, provided a large excess of residual acid is not present (4).

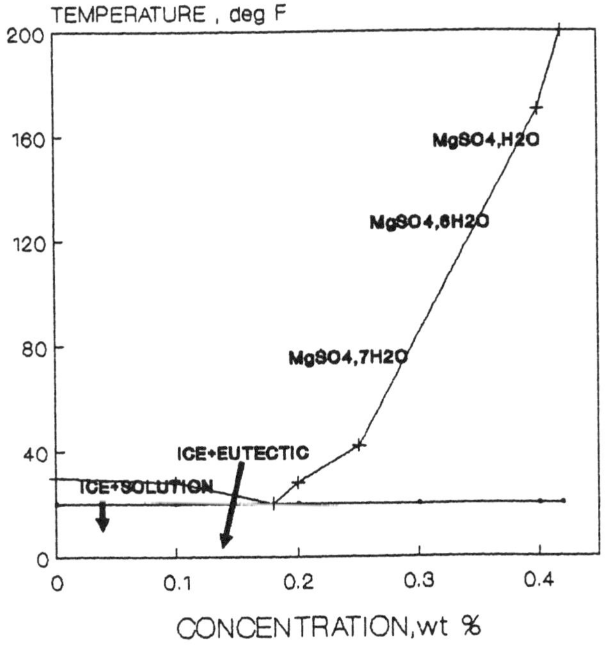

Fig.2 PHASE DIAGRAM :MgSO4,H2O SYSTEM

The mother liquor from the centrifuge after separating crystals of $MgSO_4$-$7H_2O$, can be utilized for producing magnesium carbonate. The liquor is reacted with sodium bicarbonate. This will give sodium sulfate as by product.

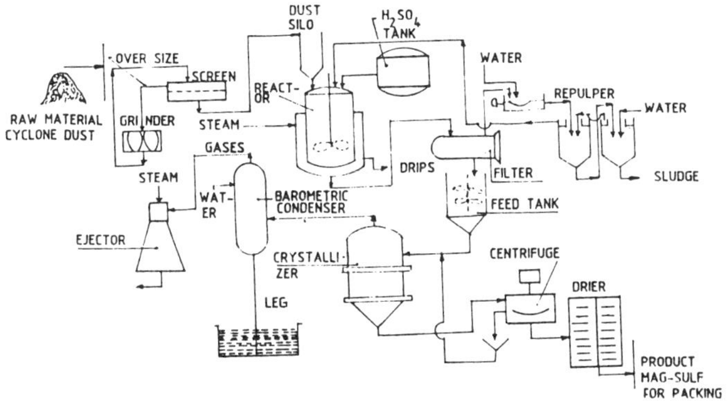

FIG. 3 PROCESS FLOW DIAGRAM FOR PRODUCTION OF MAGNESIUM SULFATE HEPTA HYDRATE

Bench scale experiments were conducted to study the reaction rate and mechanism. Accordingly, the optimum process conditions were ascertained. Based on these results several operative practices were tested at commercial scale with an aim to achieve reproducibility of the results .

Experimental

Magnesite dust/fines from Indian Integrated Iron and Steel industries' magnesite kiln was used in the present investigation . It was treated with various concentrations

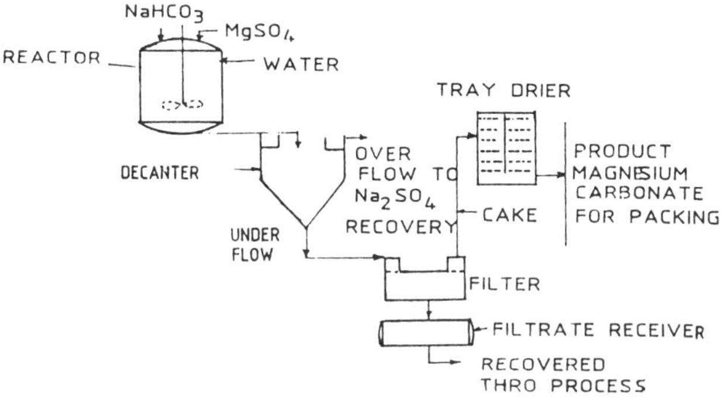

FIG. 4 PPROCESS FLOW DIAGRAM FOR MAGNESIUM CARBONATE PRODUCTION FROM MAG-SULF.

of sulfuric acid in a stirred tank type reactor at about 60-70 deg. C. Experiments were conducted employing a standardized method : using, a fixed particle size; volume of acid; reactor geometry and standardized stirring and sampling methods. Mg^{++}, Ca^{++} and Fe^{++}, constituents of the dust system, were analysed using Atomic Absorption Spectrophotometer (Pye Unichem, Model SP-2900) and SO_4^{--} and CO_3^{--} analysed by conventional chemical methods. The leached residue, the leached solutions at different intervals and, in some cases, the unleached residue were analysed for proper understanding of reaction rates and mechanism.

Results and Discussion

Leaching studies have shown the rates at which the different constituents of the dust/fines react with sulfuric acid of different concentration.

The rates of formation of sulfated components are plotted in the figures 5,6 and 7. These compounds are expressed in terms of cation (Me:Mg^{++}, Ca^{++}, Fe^{++}) concentrations, present in the leached solution of the dust systems, containing MeO constituents.

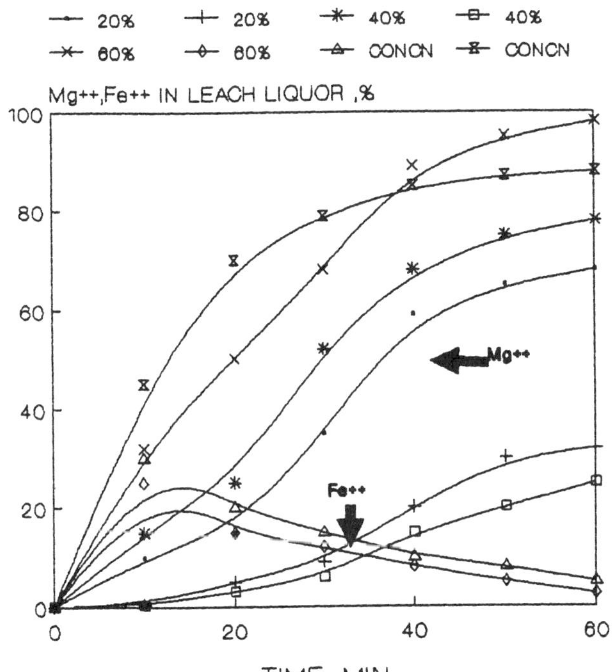

Fig.5 DEPENDENCE OF ACID CONC. ON REACTION RATE

These figures show the effect of sulfuric acid concentration on the time variation of Mg^{++}, Fe^{++} and Ca^{++} in the leach liquor. The reaction rate of formation of $MgSO_4$ (Fig. 5) is initially low. It increases gradually and attains a maximum value in almost 35-60 minutes from the start of the reaction. The rate, at which the maximum yield of $MgSO_4$ can be attained, depends immensely on the acid concentration, inter alia, parameters - temperature, particle size, stirring speed etc.

It is also seen in these figures 5 & 6 that while the yield of $MgSO_4$ increases, progressively, until it attains a maximum value, this is not so with other sulfated compounds.

Figs. 5 and 6 collate the reaction rates of MgO with FeO and CaO, respectively. These figures, however, show that the rates of reaction of FeO and CaO increase rather rapidly in the beginning of the reaction, it decreases after attaining a maximum value, unlike in the case of MgO. Some of these features that appear for the time variation of Fe^{++} in the leach liquor are identical with those of Ca^{++}. It ensues that the acid concentration influences, significantly, the rate of reaction. The reaction rates of FeO and CaO decrease rapidly beyond 40% of acid concentration, while the reaction rate of MgO increases progressively until maximum conversion is achieved.

The reasons for such phenomena is not very clear. However, it implies that, there exists an optimum condition at which the magnesium - sulfate yield is maximum with minimal impurities. A typical optimum acid concentration, at which the maximum Mg^{++} can be recovered, is featured in Fig.7 . It shows that the maximum yield of $MgSO_4$ can be

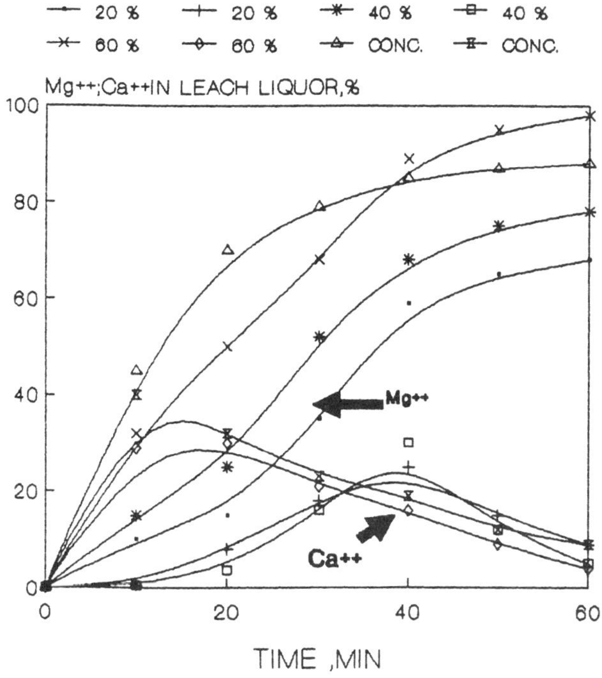

Fig.6 DEPENDENCE OF ACID CONC ON REACTION RATE

achieved by treating the dust with appropriate acid concentration under controlled conditions. The reaction rates of impurities such as FeO and CaO can be reduced provided a large excess of residual acid is not present.

Conclusions

It may be concluded that the magnesium chemicals can be produced utilizing the waste dust and fines. The reaction rates of the impurities can be controlled through control of acid concentration, temperature, stirring rate etc. so as to make the product suitable for specific applications.

The laboratory scale experimentation has indicated that the process not only gives better recovery and high purity, but it is also techno-economically viable for up-scaling on a commercial scale (5000 TPA Mag-sulf) plant. Preliminary estimates have shown that, for the unit at 100% capacity utilization, the percentage return on capital investment and working capital is approximately 60%. The break-even-point, however, works out to be at 45%.

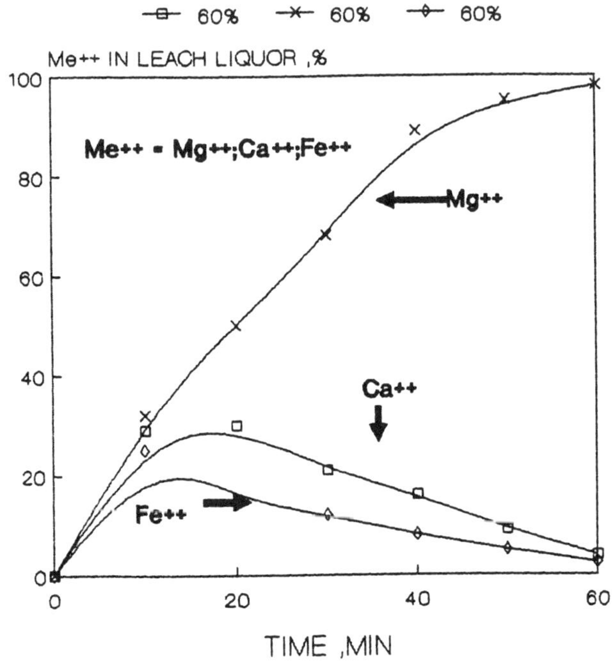

Fig.7 TYPICAL OPTIMUM ACID CONCENTRATION (60 %)FOR MAXIMUM RECOVERY OF Mg^{++}

The process has been, commercially, tried successfully. It is seemingly a plausible solution to the utilization of the waste magnesite dust and fines, having converted into valuable chemicals.

Acknowledgement

The authors wish to thank Prof. S. Banerjee, Director, National Metallurgical Laboratory, Jamshedpur, for providing facilities to conduct the experiments and according permission to publish the paper.

References

1. G Banerjee and KK Bhattacharjee, Indian Ceramic Industry and Overview, 50th Annual Session of Indian Ceramic Society,Calcutta, Feb. 7-9 1986, 20-41

2. Directorate General of Technical Development, Panel on Refractories, 1990, Govt. of India, New Delhi, India.

3. Kirk Othmer, Magnesium Compounds, Encyclopedia of Chemical Technology, 2nd Ed., Interscience Publishers, John Wiley & Sons Inc., Vol. 12, 1967, 711-715.

4. S. Prakash and K. N. Gupta, Commercial Production of Magnesium Chemicals from Industrial Waste - Dust and Fines, Chemical Engineering World (Project Report), Vol. 22(1), 1987, 39-41.

Session Chairpersons

Yrjo Anjala, President
Outokumpu, Chile SA,
San Sebastian 2839
Casilla 53030-Santiago 1
Chile

M.V. Chaubal
Manager, Reynolds Metal Company
P.O. Box 9911,
Corpus Christi, TX 78469

J.P. Hager
Colorado School of Mines
Dept. of Metallurgical Engg.
Golden, CO 80401

D.D. Harbuck
USBM
Salt Lake City Research Center
Salt Lake City, UT 84108

W.P. Imrie
Bechtel
50 Beatle St.
San Francisco, CA 94105

A.P. Lingras
Manager
Metals and Industrial Processes
Gas Research Institute
8600 W. Bryn Mawr Ave.
Chicago, IL 60631

S. Malhotra
Arthur D. Little Inc.
35 Acorn Park
Cambridge, MA 02140

Cheryl Mardock
Albany Research Center
U.S. Bureau of Mines
1450 Queen Avenue SW
Albany, Oregon 97321-2198

M. Misra
Mackay School of Mines
University of Nevada, Reno
Reno, NV 89557

P.B. Queneau
Hazen Research Inc.
4601 Indiana St.
Golden, CO 80403

V. Ramachandran
ASARCO Inc.
3422 South 700 West
Salt Lake City, UT 84112

J.M. Rapkoch
Texas Copper Corporation
1201 Logan Street
P.O. Drawer 3868
Texas City, TX 77590

R.G. Reddy
Mackay School of Mines
University of Nevada, Reno
Reno, NV 89557

D. Robertson
Dept. of Metallurgical Engg.
University of Missouri, Rolla
Rolla, MO 65401

D.R. Sadoway
Department of Materials
Science & Engg.
MIT, B-109
Cambridge, MA 02139

R.E. Siemens
Director, Reno Research Center
USBM,
Reno, NV 89557

R.W. Smith
Mackay School of Mines
University of Nevada, Reno
Reno, NV 89557

S.B. Smith
Vice President
Hudson Bay Mining &
Smelting Company
P.O. Box 1500, Flin Flon,
Manitoba R8A lN9,
Canada

H.Y. Sohn
Department of Metallurgy
University of Utah
412 Browning Blvd.
Salt Lake City, UT 84112

M.D. Strachan
Techpro Mining & Metallurgy
International House
Dover Place, Ashford, Kent
TN23 lEX, UK

P. Taylor
Department of Metallurgical Engg.
University of Idaho
Moscow, ID 83843

G.P. Tyroler
Inco Ltd.
CCNR Complex
Copper Cliff, Ontario,
Pomino, Canada

M. Van Camp
Metallurgy Hoboken Overpelt
Research, Development &
Technology
A. Greinerstraat 14
Hoboken, B2660, Belgium

Author Index